Theoretical and Practical Solutions of Mineral Resources Mining

NEW DEVELOPMENTS IN MINING ENGINEERING 2015

Theoretical and Practical Solutions of Mineral Resources Mining

Editors

Genadiy Pivnyak
Rector of National Mining University, Ukraine

Volodymyr Bondarenko
Department of Underground Mining, National Mining University, Ukraine

Iryna Kovalevska
Department of Underground Mining, National Mining University, Ukraine

CRC Press
Taylor & Francis Group
Boca Raton London New York

CRC Press is an imprint of the
Taylor & Francis Group, an **informa** business

A BALKEMA BOOK

CRC Press/Balkema
P.O. Box 447, 2300 AK Leiden, The Netherlands
e-mail: Pub.NL@taylorandfrancis.com
www.crcpress.com – www.taylorandfrancis.com

First issued in paperback 2020

ISBN 13: 978-0-367-57541-0 (pbk)
ISBN 13: 978-1-138-02883-8 (hbk)

Although all care is taken to ensure integrity and the quality of this publication and the information herein, no responsibility is assumed by the publishers nor the author for any damage to the property or persons as a result of operation or use of this publication and/or the information contained herein.

Typeset by Vasyl Lozynskyi, Kateryna Sai & Kostiantyn Ganushevych, Department of Underground Mining, National Mining University, Dnipropetrovsk, Ukraine

Visit the Taylor & Francis Web site at
http://www.taylorandfrancis.com

and the CRC Press Web site at
http://www.crcpress.com

Publisher's Note
The publisher has gone to great lengths to ensure the quality of this book but points
out that some imperfections from the original may be apparent.

Theoretical and Practical Solutions of Mineral Resources Mining – Pivnyak, Bondarenko & Kovalevska (eds)
© 2015 Taylor & Francis Group, London, ISBN: 978-1-138-02883-8

Table of contents

Theoretical and Practical Solutions of Mineral Resources Mining – Pivnyak, Bondarenko & Kovalevska (eds)
© 2015 Taylor & Francis Group, London, ISBN: 978-1-138-02883-8

Efficiency increase of heat pump technology for waste heat recovery in coal mines

G. Pivnyak, V. Samusia, Y. Oksen & M. Radiuk
National Mining University, Dnipropetrovsk, Ukraine

ABSTRACT: The nature and extent of the influence of hydraulic losses in evaporator and condenser on the energy efficiency of heat pumps have been investigated. Based on the analysis of the thermodynamic cycle of the heat pump with minimum of external irreversibility it has been shown that the increase of refrigerant pressure losses up to a certain limit in the sections of boiling in the evaporator, condensation and subcooling of the condensate in the condenser does not affect the performance and energy efficiency of the heat pump. The method of determination of the limiting values of the pressure losses has been determined.

1 INTRODUCTION

Heat pump technology is becoming more popular in Ukraine (Samusya et al. 2013). To ensure its high efficiency, more attention is paid to the development of theoretical foundations and principles of heat pump systems design (Pivniak at al. 2014, Oksen & Samusia 2014, Samusia 2015).

The energy efficiency of thermal systems is determined by irreversible losses occurring during thermodynamic processes. There are external and internal irreversibility. In the single stage heat pumps the external irreversibility is determined by the heat transfer at the finite temperature difference between the heat pump's refrigerant and the low-grade as well as high-grade heat. Internal irreversibility is caused by hydraulic losses in the basic elements of the heat pump (compressor, evaporator, condenser, throttle) and the connecting pipes. Until recently, as the evaporators and condensers of heat pumps with heating rate more than 100 kW mainly shell and tube heat exchangers in which the refrigerant flows in the annulus have been used. Losses of refrigerant pressure in such devices are small, so their calculations in heat pumps have been usually negligible. In modern plate evaporators and condensers recently used, the pressure losses on the refrigerant side can reach more than 300 kPa (Cennik 2015). It leads to a significant difference between real and isobaric thermodynamic processes occurring in them.

The aim of this study is to investigate the nature and extent the influence of hydraulic losses in evaporators and condensers on the energy efficiency of heat pumps. Let us consider a single-stage heat pump (Figure 1) that heats the water for a hot water supply system.

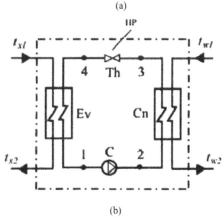

(a)

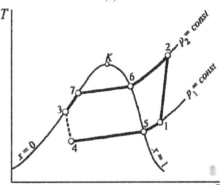

(b)

Figure 1. Diagram (a) and thermodynamic cycle (b) of heat pump: *HP* – heat pump; *Ev* – evaporator; *Cn* – condenser; *C* – compressor; *Th* – throttle.

In this case, there is a possibility of significant subcooling of refrigerant condensate at the expanse of heat exchange with cold water being heated. A diagram and thermodynamic cycle of the heat pump with condensate subcooling and vapor superheating without losses of pressure in the evaporator and condenser (processes $4-5-1$ and $2-6-7-3$ – isobaric) for zeotropic working fluid in the T, s – coordinates (T – temperature, s – entropy) are shown in Figure 1.

Figure 2 shows a diagram of changes in temperature of refrigerant and heat-transfer agent in the evaporator and condenser of the heat pump.

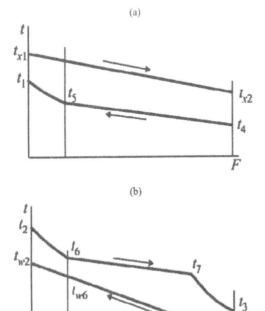

Figure 2. Diagram of changes in temperature of refrigerant and heat-transfer agent in the evaporator (a) and condenser (b) of the heat pump.

2 HEAT PUMP MODELING

Given that " the reporting cycle is a cycle with a minimum of external irreversibility. In this case, at the distinguishing sections of the condenser (sections that begin where the vapor condenses) as well as at the hot end of the evaporator, the temperature differences are equal to given ones. At the cold side of the evaporator temperature difference may be equal to a given one, or greater than it:

$$t_6 - t_{w6} = \Delta t_{6\,min}, \tag{1}$$

$$t_3 - t_{w1} = \Delta t_{3\,min}, \tag{2}$$

$$t_{x1} - t_1 = \Delta t_{1\,min}, \tag{3}$$

$$t_{x2} - t_4 \geq \Delta t_{4\,min}, \tag{4}$$

$$t_{w6} = t_{w2} - (t_{w2} - t_{w1})\frac{i_2 - i_6}{i_2 - i_3}, \tag{5}$$

where t_1, t_3, t_4 and t_6 – refrigerant temperature at 1, 3, 4 and 6 points of the cycle; t_{x1} and t_{x2} – temperature of low-grade heat refrigerant at hot and cold end of the evaporator; t_{w1}, t_{w2} and t_{w6} – temperature of high-grade heat refrigerant at the cold end of the condenser and at the section where vapour condensation begins; $\Delta t_{1\,min}$, $\Delta t_{3\,min}$, $\Delta t_{4\,min}$ and $\Delta t_{6\,min}$ – given temperature differences at the sections where refrigerant temperature is equal to t_1, t_3, t_4 and t_6, i_2, i_3, i_6 – enthalpy of refrigerant at the corresponding points of the cycle.

Since in a given cycle external and internal irreversibility are minimal, it will be the most efficient cycle at the given temperature conditions (at the temperatures of high-grade and low-grade heat transfer agent and temperature differences in the specific sections of the heat exchangers). In the p, i – coordinates (p – pressure, i – enthalpy) cycle is shown in Figure 3.

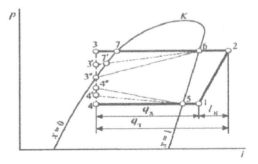

Figure 3. Heat pump thermodynamic cycle.

Shown in Figure 3 the corresponding segment specific cooling capacity q_x, heating capacity q_h and heat pump compressor work l_c are determined by:

$$q_x = i_1 - i_4, \tag{6}$$

$$q_h = i_2 - i_3, \tag{7}$$

$$l_c = i_2 - i_1. \tag{8}$$

Efficiency of the heat pump is determined by the coefficient of performance:

$$COP = \frac{q_h}{l_c}. \tag{9}$$

Let us consider how q_x, q_h and l_c change when there are losses of pressure in certain areas of the evaporator and condenser.

Let us assume that the pressure losses occur only in sections of boiling, condensation and subcooling of the refrigerant, and in sections of vapour superheating at the evaporator and its cooling to the saturated state in the condenser are equal to zero.

Condition of the refrigerant at the outlet of the condenser depends on the temperature t_3 that is determined from the equation (2), and pressure p_3, so as the pressure decreases due to hydraulic losses at the condenser point 3 of the cycle will move downward along the isotherm t_3. In the area of liquid isotherms on p, i – diagram are almost vertical. In order to simplify the analysis, we assume that they are strictly vertical and coincides with izoenthalpy. Then, when the pressure of the refrigerant leaving the condenser at $p_{3'}$, point 3 takes the position of point 3' on izoenthalpy $i_3 = i_4 = const$, point 7 takes the position of 7' on the saturated liquid line and the process of changing the state of the refrigerant in the condenser will be represented by $2 - 6 - 7' - 3'$ (Figure 3). The figure shows that q_x, q_h, l_c of the cycle $1 - 2 - 6 - 7' - 3' - 4 - 5 - 1$ are the same as for the loop $1 - 2 - 6 - 7 - 3 - 4 - 5 - 1$ for which the hydraulic losses in the evaporator and condenser are assumed to be equal to zero. Thus, the analysis shows that within certain limits an increase in pressure losses in the areas of condensation and supercooling of the refrigerant in the condenser does not reduce the performance and energy efficiency of the heat pump. This is due to the fact that the increase of pressure losses in the condenser of the heat pump is compensated by their decrease in the throttle.

Critical decrease of the final refrigerant pressure in the condenser at which the maximum efficiency of the cycle remains maximum is determined by the pressure of saturated liquid at the temperature t_3, i.e. by pressure at point 3":

$$p_{3'} = p_{sat}(t_3). \tag{10}$$

Accordingly, critical pressure losses in the condenser at the condensation sections is:

$$\Delta p_{6-3\,max} = p_6 - p_{3'}. \tag{11}$$

Losses of pressure in the boiling sections of the evaporator $4 - 5$ leads to the increase of pressure at the point 4 and the displacement of the point on izoenthalpy $i_4 = i_3 = const$ to point 4'.

Figure 3 shows that the cycle $1 - 2 - 6 - 7 - 3 - 4' - 5 - 1$ has the same energy indicators as cycle $1 - 2 - 6 - 7 - 3 - 4 - 5 - 1$ for which the hydrau-

lic losses in the evaporator and condenser are assumed to be equal to zero. Thus, the increase of the hydraulic losses up to some value at the boiling sections of the refrigerant in the evaporator does not affect the efficiency of the heat pump, because this increase is compensated by a corresponding losses reduction in throttle.

The maximum possible displacement of the point 4 due to the pressure losses in the evaporator is limited by point 4", in which the temperature:

$$t_{4'} = t_{x2} - \Delta t_{4\,min}, \text{ if} \tag{12}$$

$$t_{x2} - \Delta t_{4\,min} < t_{3'}, \ t_{4'} = t_{x2} - \Delta t_{4\,min}, \tag{13}$$

or by point 3", when:

$$t_{x2} - \Delta t_{4\,min} \geq t_{3'}, \tag{14}$$

$$t_{4'} = t_{3'}. \tag{15}$$

The maximum pressure losses in the evaporator for evaporating area, at which the energy efficiency of the heat pump is maximum, will be equal to:

$$\Delta p_{4-5\,max} = p_{4'} - p_5, \tag{16}$$

$$p_{4'} = p(t_{4'}, i_{4'}). \tag{17}$$

Pressure losses in the evaporator in the refrigerant vapour superheat part (in the process $5 - 1$) leads to the pressure decrease at the end of the process, i.e. at the inlet of the compressor. Pressure losses in the condenser in the part of cooling refrigerant vapour up to saturation (in process $2 - 6$) leads to an increase in pressure at the beginning of the process, i.e. at the outlet of the compressor. Pressure reduction in the inlet and its increase at the output of compressor leads to the increase of the compressor work and, accordingly, to the energy efficiency reduction of the heat pump.

For the estimation of the influence of the pressure losses Δp_{5-1} and Δp_{2-6} at the heat transfer sections $5 - 1$ and $2 - 6$, the calculation of the heat pump cycle has been made. As initial data given: the initial temperature of low-grade heat source $t_{x1} = 15\,°C$, the initial and final temperature of the high-grade heat source are $t_{w1} = 12\,°C$ and $t_{w2} = 49\,°C$, the minimum temperature difference in the evaporator is $\Delta t_{1\,min} = \Delta t_{4\,min} = 3\,°C$, the minimum temperature differences in the condenser are $\Delta t_{3\,min} = 4\,°C$ and $\Delta t_{6\,min} = 7\,°C$. The temperature difference for the low-grade refrigerant in the evaporator is $\Delta t_x = 4\,°C$. Working fluid is freon R407C, isentropic efficiency of the compressor – 0.66. The main results of the calculations are shown in the Table 1.

3

Table 1. Results of the heat pump cycle calculation.

Δp_{5-1}	Δp_{2-6}	COP	Δp_{5-1}	Δp_{2-6}	COP
kPa	kPa	—	kPa	kPa	—
0	0	5.088	0	0	5.088
1	0	5.082	0	1	5.086
2	0	5.076	0	2	5.085
5	0	5.059	0	5	5.081
8	0	5.042	0	8	5.076
16	0	4.996	0	16	5.064

The Table 1 shows that the pressure losses in the sections under investigation affect the COP slightly. Pressure losses increase in the area of the superheated vapour of the evaporator Δp_{5-1} up to 16 kPa at $\Delta p_{2-6} = 0 = 0$ results in the decrease of COP by only 0.092. The same pressure losses increase in the cooling refrigerant vapour of the condenser Δp_{2-6} at Δp_{5-1} results in an even smaller COP decrease is only up to 0.024. Thus, the maximum pressure losses in refrigerant boiling part of the evaporator and the condensate supercooling in the condenser, that do not lead to a change in COP, are $\Delta p_{4-5max} = 158$ kPa and $\Delta p_{6-7-3max} = 1057$ kPa.

To estimate the possible losses at the heat transfer areas of the condenser and evaporator for the given temperature conditions the heat pump with screw compressor CSH 7571 90 (Y) by Bitzer has been designed. Heat power output of the heat pump is 275 kW. When calculating the plate evaporator, condenser and subcooler condensate, the plates of type AC 500, AC 230 and CB 30 by Alfa Laval have been taken. Estimated number of plates in these devices has been respectively 132, 130 and 122. The calculated values of the pressure losses in the areas of heat transfer are as follows: $\Delta p_{4-5} = 85.7$ kPa, $\Delta p_{5-1} = 0.7$ kPa, $\Delta p_{2-6} = 0.9$ kPa, $\Delta p_{6-7-3} = 4.0$ kPa. In this energy conversion efficiency of the heat pump was $COP = 5.083$. The most significant pressure losses are in the boiling part of the evaporator $\Delta p_{4-5} = 85.7$ kPa. But they do not decrease COP of the heat pump since they are less than the maximum $\Delta p_{4-5max} = 158$ kPa. Those losses which could decrease COP are so insignificant ($\Delta p_{5-1} = 0.7$ kPa, $\Delta p_{2-6} = 0.9$ kPa) that lead to the decrease of COP compared to the cycle with no losses only by 0.005.

Thus, the analysis performed shows that the losses of hydraulic pressure to overcome the resistance of the evaporator and the condenser at the different portions of these devices have different influence on the energy efficiency of the heat pump. At sections of condensation, supercooling and refrigerant evap-oration the pressure losses increase to some extent does not affect the energy efficiency of the heat pump, and superheating sections and cooling to a state of saturation causes a reduction of efficiency.

The weak dependence of the energy efficiency of heat pumps and the pressure losses in the evaporator and condenser causes relatively independent systems of low- and high-grade heat pump plants and allows for autonomous optimization of their parameters.

It will also be appreciated that the pressure losses of the refrigerant at the evaporating and condensing sections significantly affect their temperature and the temperature difference in the evaporator and condenser, which determines in turn the heat transfer area and the cost of these devices. Therefore, despite the small effect of pressure losses in the evaporator and condenser on the energy performance and efficiency of heat pumps, their inclusion in the calculation of heat pump systems is very important.

3 CONCLUSIONS

Based on the analysis of the thermodynamic cycle of the heat pump with minimum of external irreversibility it has been shown that the increase of refrigerant pressure losses up to a certain limit in the sections of boiling of the evaporator, condensation and subcooling of the condensate in the condenser does not affect the performance and energy efficiency of the heat pump. The method of determination of the limiting values of the pressure losses has been determined.

REFERENCES

Cennik. 2015. [Electronic resource]: access mode: http://www.beijer.pl/pdf/cennik/Cennik_ 2013_druk.xls. Title from the screen.
Oksen, Y. & Samusia O. 2014. Economic efficiency of heat pump technology for geothermal heat recovery from mine water. Progressive Technologies of Coal, Coalbed Methane, and Ores Mining. The Netherland: CRC Press/Balkema: 191 – 194.
Pivniak, G., Samusia, V., Oksen, Y. & Radiuk, M. 2014. Parameters optimization of heat pumps units in mining enterprises. Progressive Technologies of Coal, Coalbed Methane, and Ores Mining. The Netherland: CRC Press/Balkema: 19 – 24.
Samusya, V. 2015. Determination of rational parameters of heat pump plant for mine water waste heat recovery. Metallurgical and mining industry, No 1: 126 – 131.
Samusya, V., Oksen, Y. & Radiuk, M. 2013. Heat pumps for mine water waste heat recovery. Mining of mineral deposits. The Netherland: CRC Press/Balkema: 153 – 157.

Theoretical and Practical Solutions of Mineral Resources Mining – Pivnyak, Bondarenko & Kovalevska (eds)
© 2015 Taylor & Francis Group, London, ISBN: 978-1-138-02883-8

Anchor's strengthening of rock walls of extraction mine workings

I. Kovalevska & O. Malykhin
National Mining University, Dnipropetrovsk, Ukraine

M. Barabash & O. Gusiev
PJSC "DTEK Pavlohradvuhillia", Pavlohrad, Ukraine

ABSTRACT: The determined regularities of influence the geomechanical parameters of extraction mine working support on condition of the main elements of its support system allowed to find out the connections of the stress-strain stress (SSS) of support system elements between each other and to develop the recommendations of the rational parameters of wall anchors according to mining and geological conditions of extraction mine working exploitation in period after longwall pass.

1 INTRODUCTION

The results of analysis of a series of computing experiments unambiguously indicate to the essential relation on the SSS of frame support and anchors even without existence of constructive connection between them; the interaction of indicated elements dramatically increase in case of combining wall anchors with frames in one load-carrying construction (Bondarenko et al. 2010, Kovalevska et al. 2010, Sdvyzhkova et al 2010).

First of all, the parameters of wall anchors are determined for which search of their rational value is carried out. In this case the parameters of frame support are established in the form of the most widely applied (on Western Donbas mines) of typical cross-section with TYSLP (Tent-shaped Yielding Support with Lengthen Props). Variable parameters of wall anchors are height and angle of dip, anchor length and bearing capacity.

2 RESEARCH TECHNIQUE

The methodology of search of rational parameters of wall anchors is carried out by task-oriented overrun of variants of geomechanical models of computation experiments with help of finite-element method (FEM). The phrase "task-oriented overrun" means consequent calculation taking into account:

– results of two previous calculations where value of investigated parameter is get in interval complying minimum stress intensity in frame support;

– operational experience in anchor installation schemes in different mining and geological conditions of Western Donbas and analysis of extraction mine working conditions while their reusing;

– results of current investigations of operating regime of frame-anchor support and existing recommendations on anchor installation schemes.

3 RESEARCH RESULTS

Search of the rational parameters of wall anchor installations is carried out starting from lower wall anchors that are placed along height of bottom ripping of extraction mine working. By the results of computation experiments and operational experience, the lower resin-grouted anchor is installed on height $0.4 - 0.6$ m from drift floor with required construction connection with frame leg. Given height of lower wall anchor installation is determined the most significant values of influence of its reaction (through flexible connection) on restriction of material limit equilibrium state field in lower part of frame leg: reducing the height of anchor installation leads to curving size growth in area along coal seam thickness (support construction); increasing the height of anchor installation enhances leg bending deformation in step zone. Anchor is installed horizontally due to primary horizontal shifting of border wall rocks in step zone of frame support. Thus, horizontal arranged anchor will expend maximum strength reaction because it works on stretching without essential bending its "armature". The length l_i of anchor is chosen by condition of key part back from the boundary of wall rock strengthening area. Mathematically this condition can be express by following equation:

$$l_{l,m}^F \geq l_{key} + b_{l,m}^F + l_t, \qquad (1)$$

where l_{key} – length of anchor key; l_t – length of tail part of anchor outgoing in mine working cavity is taken $l_t = 0.1$ m; $b_{l,m}^F$ – width of wall rock weakening zone is determined by formulas (2) or (3) depend on considered side of mine working (Kovalevska et al. 2015):

– from goaf side:

$$b_l^F = \frac{3.6 l_s^{0.73}}{R^{0.1}\left(R_1^F\right)^{0.21}}\left[1 - exp\left(-5.9 \cdot 10^{-3} H\right)\right]; \qquad (2)$$

– from massif side:

$$b_m^F = \frac{9.7}{R^{0.18}\left(R_1^F\right)^{0.45}}\left[1 - exp\left(-3.2 \cdot 10^{-3} H\right)\right], \qquad (3)$$

where H – location depth of mine working; l_s – width of support strip; R – average calculation compression strength of nearing floor rocks; R_1^F – compression strength of immediate bottom rocks.

Calculation (Bondarenko et al. 2011) and tests are showing that load-bearing capacity of resin-grouted roof bolt approximately 200 kH is provided under key length $l_{key} = 0.6 - 0.7$ m. At least greater than or equal to load-bearing capacity of rope bolts reaches under length less than or equal to $1.0 - 1.1$ m. Then using formulas (1), (2) and (3) it is possible to divide rational use field of wall steel and rope anchors for their installation along mine working floor ripping depth. Here, the criterion of division is appearance of technological difficulties of resin-grouted anchors length more than 3.2 m. If in difficult mining and geological conditions the extensive area of weakening rocks in walls of mine working (more than 2.5 m) it is rational to replace the resin-grouted to rope bolt for which there are no such technological restrictions.

For mathematical notation of this condition, the dependence (2) for weakening area width b_l^F of walls of mine workings from goaf side is used by two reasons. First, the size of given parameter, as a rule, exceeds analogical size from massif side that allows with something reserve to design wall rock strengthening in non-operational mine working side. Secondly, in case of mine working reusing the coal massif will be working off from the side of another extraction mine working that inevitable increase strengthening area width. Eventually the expression for determination of boundary value $(R_1^F)_b$ of calculated immediate floor compressive resistance is received.

Under $R_1^F \geq (R_1^F)_b$ it is rational to use resin-grouted anchors; under $R_1^F < (R_1^F)_b$ it is rational to use rope anchors with length not less than value determined by formula (1).

$$\left(R_1^F\right)_b = 5.68 \frac{l_s^{3.48}}{R^{0.48}} \times$$
$$\times \left[1 - exp\left(-5.9 \cdot 10^{-3} H\right)\right]^{4.76}, \text{ MPa.} \qquad (4)$$

The boundary condition (4) for type anchor selection is shown on Figure 1, where area of a combination of geomechanical parameters located above adequate lines determines resin-grouted anchors application and lower area is cable anchors.

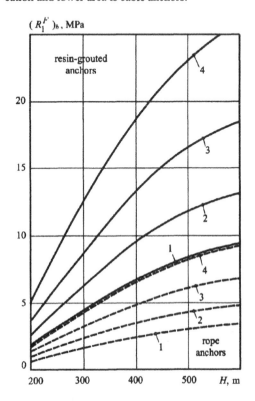

Figure 1. Lines of boundaries of rational resin-grouted or steel anchor application (in depth of bottom ripping) under average calculated compressive resistance: $1 - R = 40$ MPa; $2 - R = 20$ MPa; $3 - R = 10$ MPa; $4 - R = 5$ MPa; —— $l_S = 2.0$ m; – – $l_S = 1.5$ m.

The last four emphasized parameters of lower wall roof bolts are their required resistance reaction equal in a limit of anchor "armature" load-bearing capacity. Here it is lays emphasis on anchor "armature" tearing resistance because its key fastening strength by polymer compounds, as a rule, higher then provided appropriate length of roof bolt key part by introducing in borehole necessary amount and required ampule lengths with polymer compound.

Researches show, that occurring of plastic state on certain section of anchor "armature" length doesn't lead yet to its disruption because unloading of its section takes place due to axial yielding of "armature" on one side and presence of essential reserve of steel resistance growth on stage of strengthening (Baclashov & Timofeev 1979, Skramtaev et al. 1953) up to value of steel temporary tearing resistance σ^t on the other side. In this regard it is rational to use for estimation of roof bolt load-bearing capacity the parameter as steel temporary tearing resistance σ^t and some reserve of roof bolt operation reliability is provided by introduction of reserve coefficient ($k_{res} < 1$). Then it is possible to use the following condition:

$$\sigma = k_{res}\sigma^t, \tag{5}$$

as criterion of achievement of anchor of the maximal compression resistance. Under this criterion, the diameter of roof bolt armature (in certain condition) is elected by following way.

Technique of carrying out of computation experiments was designed in such a way that each calculation executes for specific diameter d of anchor "armature". If under the reduced stress σ in no matter which point exceeded the value in condition (5), the nearest bigger diameter d was accepted and calculation was repeated. Under this overrun of calculation variants, the diameter of roof bolt "armature" suitable to certain condition of operation was determined. Therefore, in contrast to analytical computation methods directly determines connection the diameter anchor "armature" with geomechanical parameters its work under already specified earlier parameters of roof bolt placements in marginal rocks of extraction mine working in intact and lower wall roof bolt particularly.

Fragments of searching results of parameter dependences d^F_l from geomechanical factors of extraction mine working are shown in the diagrams on Figure 2 for resin-grouted and steel anchors.

Roof bolt distribution on two groups is determined by the different strengthening characteristics of steel of its "armature" and different standard diameters using for mine working supporting. Nevertheless, connection regularities of required diameter d^F_l with depth H of location depth are practically same for resin-grouted and cable anchors if it takes into account non absolutely but relative value in plan of intensity growth under increasing H.

In addition, it is taken into account stable growth of anchor loading on the goaf side in comparison with anchors on the massif side. Therefore on diagrams (and following regression equitations) are shown regularities for more loading anchors (on the goaf side), it is correctly supposed that anchors from massif side will be working with some reliability reserve.

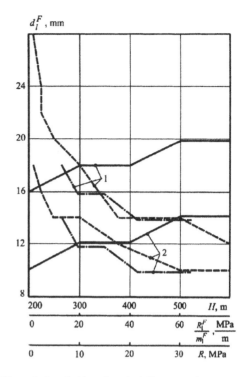

Figure 2. Regularities of required diameter variation "armature" of lower wall anchors d^F_l from location depth H of mine working (——), relations R^F_1/m^F_1 of calculated compressive resistance to the thickness of immediate floor (- -) and average calculated compressive resistance R (- · -) of nearest coal-bearing massif: 1 – resin-grouted anchors; 2 – cable anchors.

The level of influence of average calculated compressive resistance R of nearest massif on the diameter d^F_l of resin-grouted and steel anchors also insignificantly varies from each other on variation relative value. The similar situation is observed for parameter R^F_1/m^F_1, that is entirely logical as, an our opinion, tendencies of geomechanical factor influence slightly depend on material of anchor "armature". In contrast, in absolute sizes of diameter d^F_l the steel of cable anchors have higher tearing resistance than resin-grouted anchors; it follows thence decreasing size of required diameter of cable anchor "armature" (Figure 2).

By the results of revealed regularities the regression equitation for calculation of minimum required anchor diameter is developed:

$$d^F_l = \frac{1.81}{\sigma^t} \cdot 10^3 \left\{ \left(3.8 + 6.2 \cdot 10^{-3} H\right)\left(\frac{17.2}{R^{2.4}} + 0.52\right) \times \right.$$

$$\left. \times \left[0.25 + 0.73\, exp\left(-0.1\frac{R^F_1}{m^F_1}\right)\right]\right\}^{0.5} , \text{mm}, \tag{6}$$

where σ^f – steel temporary tearing resistance of which anchor "armature" is made. In formula (6) initial parameters have following units of measurement: σ^f, R, R^F_1 – MPa; H and m^F_1 – meters.

4 CONCLUSIONS

Following the results the full range of required dependencies and constants for calculation and selection of rational parameters of wall anchors installing in the extraction mine workings in depth of bottom ripping are received.

REFERENCES

Baclashov, I.V. & Timofeev, O.V. 1979. *Constructions and calculation of arches and lining*. Moscow: Nedra: 263.
Bondarenko, V., Kovalevs'ka, I. & Cherednichenko, Yu. 2010. *Substantiation of design and installation technology of tubular rock bolts by explosive method*. New Techniques and Technologies in Mining. The Netherlands: CRC Press/Balkema: 9 – 14.
Bondarenko, V., Kovalevs'ka, I., Svystun, R. & Cherednichenko, Yu. 2013. *Optimal parameters of wall bolts computation in the united bearingsystem of extraction workings frame-bolt support*. Mining of Mineral Deposits 2013. The Netherlands: CRC Press/Balkema: 5 – 9.
Kovalevska, I., Fomychyov, V. & Vyvcharenko, A. 2010. *Calculation substantiation of the yield lock model of the polygonal yieldable support with elongated props by means of experiment*. New Techniques and Technologies in Mining. The Netherlands: CRC Press/Balkema: 83 – 87.
Kovalevska, I., Symanovych, G., Gusiev, O. & Snigur, V. 2015. *Formation of limit equilibrium zone in marginal rocks of excavation mine working*. Materials of the conference "Szkoła Eksploatacji Podziemnej". Krakow, February 23 – 27.
Scramtaev, B.G., Popov, N.A., Gerlivanov, N.A. & Mudrov, G.G. 1953. *Construction materials*: textbook. Moscow: Promstroiizdat: 643.
Sdvyzhkova, O.O., Shashenko, O.O., Kovrov, O.S. 2010. *Modeling of the rock slope stability at the controlled failure*. Proceedings of the Rock Mechanics Symposium (EUROCK). London: Taylor & Francis Group: No 1. 581 – 584.

Theoretical and Practical Solutions of Mineral Resources Mining – Pivnyak, Bondarenko & Kovalevska (eds)
© 2015 Taylor & Francis Group, London, ISBN: 978-1-138-02883-8

Digging plasticity water-saturated soils

C. Drebenstedt
Technical University "Mining Academy Freiberg", Freiberg, Germany

T. Shepel
National Mining University, Dnipropetrovsk, Ukraine

ABSTRACT: In the paper the bucket filling process while digging plasticity water-saturated soil is considered. It is highlighted the necessity of taking into account physical-and-mechanical and rheological properties of excavated soil, as well as cutting parameters to determine the rational geometrical parameters of the bucket, which provide high value of its fill factor. Mathematical models for defining the ultimate length and height of the soil body inside the bucket have been developed. The results of experiments on digging marine sapropel and coccolith sediments in laboratory conditions are given. The mathematical models have been verified on the basis of comparison of the calculation results and experimental data.

1 INTRODUCTION

Earthmoving machines with bucket work tools are used widely to perform different operations in underwater condition. They are used during the construction of hydraulic structures and development of underwater mineral deposits.

In shallow water there might be used hydraulic excavators, draglines, underwater scrapers, bucket dredgers and submersible earthmoving machines (amphibians). Nowadays several types of seabed mineral deposits are of some economical interest. Among them are deposits of polymetallic nodules in the Pacific and Indian Ocean, metal-rich silts in the Red Sea, and organic-mineral sediments in the Black Sea. The depth of their stratification varies from 0.5 to 6 thousand meters. To extract minerals at the great depth such mining equipment with bucket work tools as continuous line bucket system, dragline dredgers, and seabed excavating machines were developed. Although only a few of them were tested in the field operating conditions.

Because of limited speed of the work tool movement in water due to significant hydrodynamic resistance, the primary way of increasing the productivity of excavating machines is to enlarge the bucket capacity. As a rule, it results in rising power consumption of digging process. But achievement of desired productivity is not guaranteed by the reason of the fill factor decrease. This paper is dedicated to a question of interaction between the bucket and plasticity water-saturated soil while digging in order to define the rational geometrical parameters of the bucket, which provide the high value of its fill factor.

2 ANALYSIS OF RECENT RESEARCH

As a rule, when designing buckets for excavation of soil in underwater condition empirical dependences based on experience of onshore earthmoving machines exploitation are used. For instance, to determine geometrical parameters of an underwater scraper bucket it is recommended to use dependences developed by D. Fedorov for the dragline bucket (Gilev & Shejn 2011): $L = 1650\sqrt[3]{V}$; $B = 1050\sqrt[3]{V}$; $H = 750\sqrt[3]{V}$, where L, B, and H – length, width and height of the bucket respectively, mm; V – the bucket capacity, m^3.

For the case of development of water-logged placer deposits, on the basis of analysis of standard scraper buckets empirical dependences of the bucket geometrical parameters on its mass were established (Dobrecov & Opryshko 2006): $L = 306.36M^{0.283}$; $B = 220M^{0.312}$; $V = 0.0021M$, where M – the bucket mass, kg.

The process of mining lake sapropel with torus-shaped scraper bucket is described in the paper (Bulik 2011). There is stressed the necessity of taking into account such properties of excavated soil as humidity and adhesion when designing buckets for development of clay soils. These properties affect sufficiently on the bucket fill factor.

The process of cutting plasticity water-saturated soil was studied in detail at Kyiv National University of Construction and Architecture (Ukraine). It was established that when cutting of the soil with a plane blade the cutting process might be changed into the process of ground spreading without chip separation. It occurs when the following condition is fulfilled (Sukach & Magnushevskij 2005):

$$p \geq 2\tau_0,\qquad(1)$$

where p – is the pressure on the soil surface in front of the blade in ABC zone (Figure 1).

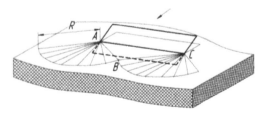

Figure 1. Scheme of cutting plasticity water-saturated soil with the plane blade.

The size of the plasticity zone R is determined as:

$$R = \frac{b}{\sqrt{2}},\qquad(2)$$

where b – the blade width.

Since the bucket might be considered as the combination of a blade and a receptacle, the condition (1) must be fulfilled for the case of digging the plasticity water-saturated soil with a bucket. At some parameters of the dragging prism, which presses on the soil surface in front of the bucket, the filling process must be stopped, so the maximal volume of soil inside the bucket is limited even if the cut chip does not reach the back and top walls of the bucket. To avoid decreasing the fill factor, the bucket filling

process while digging plasticity water-saturated soils should to be studied more in detail.

Aim of the paper is determination of analytical dependences of the rational geometrical parameters of the bucket, under which the high value of the bucket fill factor is provided, on the cutting parameters and properties of the excavated soil.

3 DEVELOPMENT OF THE MATHEMATICAL MODELS

There coined the following terms in the current paper:

1) long bucket – non-inclined to the horizon bucket of limited width with other dimensions of any size;

2) ultimate bucket filling condition (UBF condition) – the condition of the long bucket filling under which the process of filling stops and the ground cutting process changes into the process of ground spreading without chip separation;

3) ultimate filling parameters – parameters of filling the long bucket when the UBF condition.

The process of digging the plasticity water-saturated soil might be considered as follows (Figure 2): stage I – bucket penetration in ground massive 1; stage II – increasing the soil body inside the bucket 2 and the dragging prism 3 in size; stage III – achievement of the maximal size of the soil body inside the bucket and the dragging prism; stage IV – changing the cutting process into the process of ground spreading without chip separation under which the ground piles 4 are formed to the both sides of the trench.

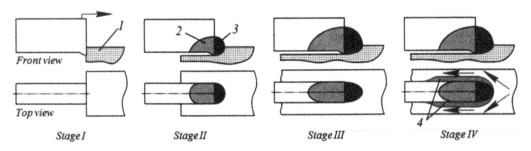

Figure 2. Scheme of the bucket filling while digging the plasticity water-saturated soil: 1 – ground massive; 2 – the soil body inside the bucket; 3 – the dragging prism; 4 – ground piles.

At the stages III and IV the soil body inside the long bucket has its ultimate length and height. Let's establish the dependences of the ultimate length and height of the soil body inside the long bucket on the cutting parameters and physical-and-mechanical properties of the excavated soil.

Statement of the task: the long bucket with rectangular cross-sectional shape moves rectilinearly with the constant speed v_0, cutting layer of the plasticity water-saturated soil. The cutting depth is h_0, bucket width is B. It is necessary to establish the ultimate length and height of the soil body inside the bucket.

The following assumptions have been accepted: soil is the homogeneous viscoplastic medium; the value of critical chip stress is constant at any length of the chip; the form of both the ground body inside the bucket and the dragging prism follows the relevant equations of the ellipse and stays unchanging in the course of time, but their geometrical dimensions change (according to the results of experimental investigation (Drebenstedt et al. 2014)); while the ultimate bucket filling condition ground piles formed on the both sides of the trench have a triangular cross-sectional shape.

The ultimate height of the soil body inside the bucket by the strength properties of the excavated soil h_{ut} may be determined from the calculation scheme in Figure 3 by using the condition (1) and expression (2). The result obtained is:

$$h_{u\tau} = \frac{\sqrt{(\Delta\rho bg\,tg\,\phi)^2 + 16\tau_0^2}}{2\Delta\rho g} + h_0, \tag{3}$$

where $\Delta\rho$ – density of soil in water; g – free fall acceleration; τ_0 – soil cohesion; $tg\varphi$ – quotient of height of the dragging prism by its length.

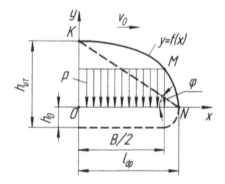

Figure 3. Scheme for calculation the ultimate height of the soil body inside the bucket by the strength properties of the excavated soil (side view of the dragging prism is shown).

From the condition of equilibrium of the cut ground flow rate and the ground flow rate in the piles, which are formed to the both sides of the trench while the ultimate bucket filling condition, the ultimate height of the soil body inside the bucket by the natural friction angle of the cut ground (in loosened condition) $h_{u\gamma}$ was determined (Figure 4). The established expression is:

$$h_{u\gamma} = \sqrt{h_0 B\,tg\,\gamma} + \frac{1}{2}B\,tg\,\gamma + h_0, \tag{4}$$

where γ – is the natural friction angle of the cut ground in loosened condition.

The ultimate height of the soil body inside the bucket h_u equals to the smaller value between $h_{u\tau}$ and $h_{u\gamma}$.

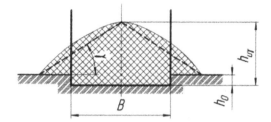

Figure 4. Scheme for calculation the ultimate height of the soil body inside the bucket by the natural friction angle of the ground in loosened condition (front view of the dragging prism is shown).

To define the ultimate length of the soil body inside the bucket it was studied the chip deformation process. The process might be considered as follows. While penetration of the bucket in ground massive it occurs separation of the continuous chip, which has solid and plasticity ranges. Moving into the bucket, the friction between the bucket and the chip rises. It results in changing the chip stress condition. At some value of the chip length the shear stress in the solid range of the chip exceeds the soil yield stress. Because of this the solid range changes into the plastic one. Considering that the value of critical chip stress is constant at any length of the chip, it was established the dependence of minimal thickness of the chip on its length (to define the critical chip stress the differential equations of a soil element static equilibrium and the equations of equilibrium on a solid body surface have been used). From the established dependence, taking into account the ultimate height of the soil body inside the bucket, the expression for determining the ultimate length of the soil body has been found:

$$l_u = \frac{1}{k_p b_{II}}(h_u - h_I)(b_{II} + 2h_{II}), \tag{5}$$

where h_I, h_{II}, and b_{II} – thickness of the plastic range and solid range of the chip and width of the solid range of the chip respectively (while penetration of the bucket into ground massive); k_p – coefficient of proportionality, which takes into account the increase in the friction area of the chip on the side walls of the bucket while the chip deformation. Values of parameters b_{II} and h_{II} are determined by the following dependences:

$$h_{II} = \left(\cfrac{1}{\cfrac{2}{3} + \cfrac{1}{3(Q+1)^2}} - \frac{K}{Q} \right) h_0 , \qquad (6)$$

$$b_{II} = \frac{B}{Q+1} . \qquad (7)$$

Parameter h_I may be calculated as follows (Sukach 2004):

$$h_I = \frac{K}{Q} h_0 . \qquad (8)$$

Similarity parameters K and Q are defined by the expressions (Sukach 2004):

$$K = \frac{\mu_m v_0}{h_0 \tau_0} , \qquad (9)$$

$$Q = \sqrt{K + \frac{K^2}{9} + \frac{K}{3}} , \qquad (10)$$

where μ_m – plastic viscosity of soil.

To provide high value of the fill factor, the bucket geometrical parameters should be determined by using the following dependences:

$$H = k_1 h_u , \qquad (11)$$

$$L = k_2 l_u , \qquad (12)$$

where k_1 and k_2 – coefficients which don't exceed 1 and may be chosen by the way to provide the required capacity of the bucket.

For the case when the desired capacity does not provided, the cutting depth or the bucket width should be increased.

4 VERIFICATION OF THE MATHEMATICAL MODELS

To verify the mathematical models (3), (4) and (5) the laboratory tests were carried out. The tests were done on the basis of the three-dimensional cutting test machine (Figure 5) in TU Bergakademie Freiberg (Freiberg, Germany). Laboratory equipment included a container for soil samples and digging tools (bucket scaled-down models).

When experimental studying the process of digging plasticity water-saturated soils, one of the most important questions is taking into account both the edge effect and scale effect.

In the work (Sukach 2004) there is proven that when digging plasticity water-saturated soil the edge affect may not be taken into account at the ratio $B/h_0 > 4...6$.

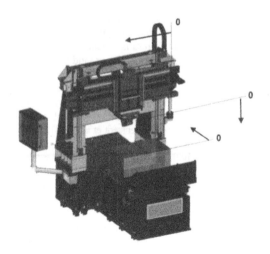

Figure 5. General view of the three-dimensional cutting test machine.

For the chosen minimal cutting depth of 10 mm the minimal bucket width is 40 mm.

To avoid the influence of the scale factor caused by the ratio of linear dimensions of the soil particles to the one of the cutting tool, it was used the following expression for defining the critical cutting depth (Moiseenko 1987):

$$h_c \approx \sqrt{\frac{5 \cdot 10^6 \delta^3}{B}} = \sqrt{\frac{5 \cdot 10^6 0.012^3}{40}} = 0.46 , \qquad (13)$$

where h_c – critical cutting depth, mm; δ – average linear size of soil particles (for the clay soil it might be accepted equal to 0.012 mm).

Accepted minimal cutting depth raises the critical cutting depth in 21.74 times. So, the modeling mistake caused by the scale factor must be insufficient.

Taken values of bucket width were: 40, 60, 80, and 100 mm. The width of the container for samples was calculated by using the dependence (2) for the bucket of the maximal width:

$$B_c = B_{max} + 2R\cos 45° =$$

$$= 0.1 + 2 \cdot 0.07 \cdot \frac{\sqrt{2}}{2} = 0.2 , \qquad (14)$$

where B_c – width of the container for soil samples, m; B_{max} – maximal width of the bucket (0.1 m).

The cutting depth was equal to 10, 20 and 30 mm.

The coccolith and sapropel samples from the Black Sea were used as the plasticity water-saturated soils. Humidity of the coccolit (sapropel) sediments was equal to 188.43% (222.68%), density 1269 kg/m^3 (1219 kg/m^3), cohesion 187 Pa (156 Pa), viscosity 226.8 Pa·s (807.4 Pa·s).

Tests were carried out in atmospheric and underwater conditions (Figure 6). While digging the sediments in atmosphere condition the separation of continuous chip was observed. The chip moved into the bucket and, being deformed, it formed soil body inside the bucket (Figure 7). At some value of the digging way the bucket filling process had been stopped and the cutting process changed into the process of soil spreading without chip separation, under which continuous ground piles were being formed to the both sides of the trench. While digging the sediments in underwater conditions, some cracking of the cut chip was observed. It was caused by the worsening coagulation conditions of soil particles due to less weight of the soil in water. At some value of the digging way the ultimate bucket filling condition was achieved. While this, non-continuous ground piles were being formed to the both sides of the trench.

(a)

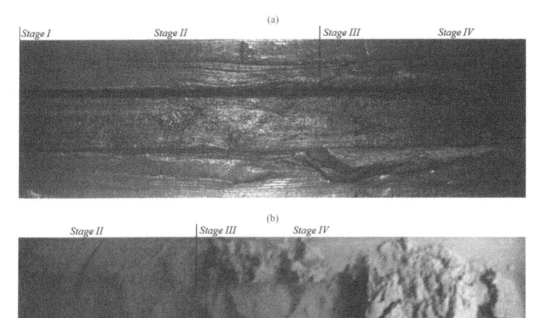

Figure 6. Digging the sediments in atmospheric (a) and underwater (b) conditions.

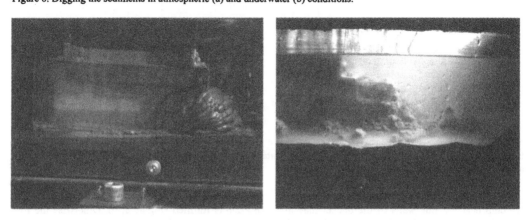

Figure 7. Deformation of the soil chip while digging in atmospheric (a) and underwater (b) conditions.

Comparison of calculation results and experimental data for the case of digging the sapropel sediments in atmospheric conditions is shown in Figure 8. The graphs show that cutting speed in the range from 0.05 to 1.0 m/s influences insufficiently on the ultimate length and height of the soil body inside the bucket (Figure 8, a, b). Rising of the cutting depth and the bucket width results in increase of the mentioned parameters (Figure 8, c – f).

Graphs of functions $l_u(h_0)$ and $h_u(h_0)$ (Figure 8, c, d) don't cross zero of coordinate system. Physically it means that when cutting depth is equal to 0 the soil body inside the bucket and the dragging prism are in equilibrium (their geometrical parameters are unchanged). For instance, it may occur while decreasing of the cutting depth from some value to 0 (while recovery of the bucket out from the work face), or when the bucket, moving without penetrated into the soil cutting edge, meet a soil pile on its way.

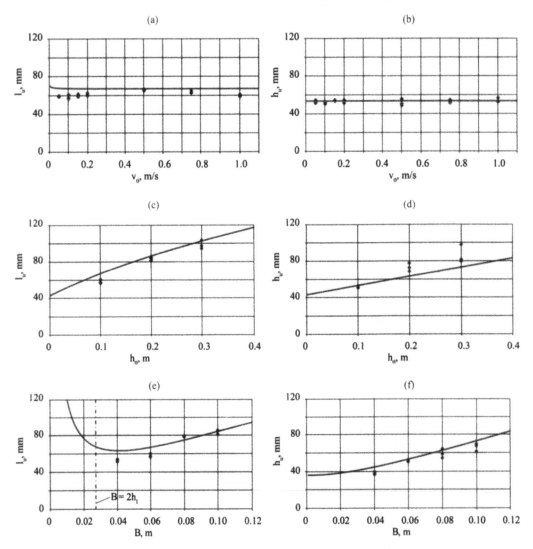

Figure 8. Comparison of calculation results and experimental data for the case of digging the sapropel sediments in atmospheric conditions.

Summarized thickness of the plasticity range of the chip near the side walls of the bucket may not exceed the bucket width. Because of this the dependence (5) must be used only when the condition $B \geq 2h_l$ is fulfilled (Figure 8, e). Otherwise the following approximation should be used:

$$l_u = \frac{l'_u}{B'}B,\qquad (15)$$

where B' – the bucket width under which condition $B' \geq 2h_l$ is fulfilled; l'_u – the ultimate length of the soil body inside the long bucket when value of the bucket width is equal to B'.

On the basis of statistical analysis of almost 100 tests it was established that while digging the marine sapropel and coccolith sediments the average deviation of the calculated results from the experimental data for parameters h_u and l_u does not exceed 12% and 16% respectively. Maximal deviation does not exceed 22% and 27% respectively at the probability belief of 0.95.

5 CONCLUSIONS

1. While digging plasticity water-saturated soils the ultimate bucket filling condition occurs. In this condition the bucket filling process stops and the cutting process changes into the process of ground spreading without chip separation, under which ground piles to the both sides of the trench are formed.

2. To provide high value of the fill factor, the bucket height and length should not exceed the ultimate height and length of the soil body inside the long bucket respectively. For determination of the mentioned ultimate parameters the mathematical models have been developed.

3. For the case of digging the sapropel and coccolith sediments, maximal deviation of the data calculated by using the developed mathematical models from the experimental one does not exceed 27% at the probability belief of 0.95.

REFERENCES

Bulik, J.V. 2011. *The process and machine parameters justification for mining the sapropel.* Abstract of a PhD (Eng) thesis. Luck: State Technical University in Ternopol: 24.

Drebenstedt, C., Franchuk, V. & Shepel, T. 2014. *Experimental investigation of digging the organic-mineral sediments of the Black Sea.* Mining of Mineral Deposits. The Netherland: CRC Press/Balkema: 99 – 108.

Dobrecov, V.B. & Opryshko, D.S. 2006. *Dragline mining of the small continental and shoreline stream gravel.* Mining informational and analytical bulletin (scientific and technical journal), No 16: 229 – 239.

Gilev, A.V. & Shejn, F.J. 2011. *Justification of dragline work tool for water-logged deposit development.* Mining equipment and electromechanics, No 8: 2 – 5.

Moiseenko, V.G. 1987. *Determination of the forces while using earthmoving machines in special operating conditions.* Kyiv: Vishha shkola: 194.

Sukach, M.K. 2004. *Workflows deep machines.* Kyiv: Nauk. Dumka: 364.

Sukach, M.K. & Magnushevskij, V.I. 2005. *The model of cutting the plastic water-saturated soil with the blade of an earthmoving machine.* KhNAHU collection of proceedings, No 29: 74 – 79.

Substantiating parameters of process design of contiguous seam mining in the Western Donbas mines

V. Busylo, T. Savelieva, V. Serdyuk, A. Koshka & T. Morozova
National Mining University, Dnipropetrovsk, Ukraine

ABSTRACT: Distance between stopes where longwall faces of contiguous seams operate safely and effectively at given parting thickness in the Western Donbas mines is determined on the basis of analysis of stress and strain state within the area of stratified inhomogeneous massif near stopes using obtained correlation ratio . Rational location of development working driven along the underworked seam is stated. Rational value of stope advance rate is determined.

1 INTRODUCTION

Development strategy of coal industry in Ukraine as it is pointed out in the National Energy Program is to increase operation efficiency of the mines and to reach the output of coal mining required to satisfy the demands of the national economy.

There is the tendency in Ukraine to reduce the amount of effectively working coal producers that is stipulated by the mines' unprofitability under the current coal mining technology. That is why it is required to increase the output of coal mining by the re-equipment of mines with highly productive technique and to raise the intensity of mining operations.

Re-equipment of the stopes by the complexes of the new technological level must provide sufficient load increase to the longwall face at least up to 1500 tons per day. At such loads the rate of the stopes advance is increased from 2...4 to 8...12 meters per day. It will lead to the certain changes in developing stress and strain state of rock massif within the zones of mining effect. By the current time the results of the study of effecting technogenic factors on the geomechanical processes which take place in advance of the stope as well as around it are practically absent. Technogenic factors are following: dimensional and geometrical parameters of the stope and working area, method and the rate of seam mining, rate of the stope advance and other technological parameters of coal seam mining.

The most popular calculation and abstract models cannot completely take into account massif structure, physical and mechanical coal and rock properties, parameters of stopes and worked-out area as well as dynamics of seam mining while calculating technological parameters of mining operations. The most well-known analytical and numerical solutions are obtained for the dead face. Some scientific theories take into account stope shifting within the stressed medium (real conditions are idealized). Numerous solutions don't have generalized dependences that make difficult their use even in these conditions.

Combined approach developed in the National Mining University (Novikova et al. 2001) allows to take into account the rate of stope shifting, but hypothesis of forming additional load on powered roof support by its increase in the edging part of the coal seam should be confirmed experimentally.

2 DETERMINING THICKNESS AND DISTANCE BETWEEN STOPES

Distance between stopes where longwall faces of contiguous seams operate safely and effectively at given thickness of the partings in the Western Donbas mines is determined in this paper on the basis of analyses of stress and strain state within the area of stratified inhomogeneous massif near stopes using obtained correlation ratio. Rational location of development working driven along the underworked seam is also stated. Safe distance between stopes on contiguous seams and parting power capability are interacted parameters. Area size of support pressure in advance of stopes and the value of maximum stresses within these zones, area sizes of disintegrated rocks behind the stopes within the roof and the floor of each seam as well as maximum roof and floor convergence are the basic ones. These parameters are determined on the basis of analysis of the rock massif stress-and- strain state relating to the condition of overworking (Novikova et al. 2005).

In the process of parameter determination distance L between stopes along contiguous seams and parting stress h are varied.

The values of physical and mechanical coal characteristics are following: elasticity module is $E = 3.0 \cdot 10^3$ MPa, the Poisson ratio is $v = 0.36$, density is $\gamma = 1.4$ t/m^3 compression strength is $\sigma_c = 30$ MPa, tensile strength is $\sigma_t = 3$ MPa.

Parameters of inclosing rock (mudstone) had the values: $E = 2.9 \cdot 10^3$ MPa, $v = 0.3$, $\tau = 3$ t/m^3, $\sigma_c = 30$ MPa, $\sigma_t = 3$ MPa.

Estimation of the stress state is performed by the P.P. Balandin criteria (Gabdrahimov et al. 1966) according to which equivalent stresses are determined by the formula:

$$\sigma_{eq} = \frac{(1-\psi)(\sigma_1 + \sigma_3)}{2} + \frac{\sqrt{(1-\psi)^2(\sigma_1 + \sigma_3)^2 + 4\psi(\sigma_1^2 + \sigma_2^2 - \sigma_1\sigma_3)}}{2}, \quad (1)$$

where σ_1, σ_2, σ_3 – the basic stresses, MPa; $\psi = \sigma_t / \sigma_c$; σ_t, σ_c is strength rock limit as for tension and compression, MPa.

Figure 1 shows the values obtained by the given criteria σ_{eq} within the field which is under consideration. As this figure shows the range of the limiting value area σ_{eq} within the roof of the upper layer according to the coordinate x is about 18 m. The size within the roof of the lower seam is 30 m.

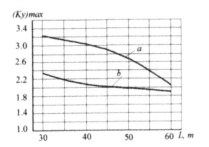

Figure 2. Maximum coefficients of stress concentration σ_{yy} within the roof of the lower layer (a) and the upper layer (b).

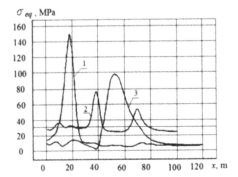

Figure 1. Stresses σ_{eq} according to the P.P. Balandin criteria: 1 – within the roof of the upper layer; 2 – within parting area; 3 – within the floor of the lower layer.

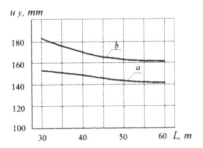

Figure 3. Roof and floor contingence at a 4 meter distance from the stope of the lower layer (a) and the upper layer (b).

Limiting values of stresses take place within the parting area covering its entire thickness and is about 30 m alongside the axis Ox.

Figure 2 shows estimated values of the maximum concentration ratio $(K_y)_{max}$ of the stresses as for the level yH ($H = 300$ m) in the roof of the contiguous seams within the area of support pressure in advance of stopes at various values of a distance L and the fixed thickness $h = 15$ m. As this figure shows the roof of the lower seam is more loaded $((K_y)_{max}$ is 1.36 times more than within the roof of the upper layer).

Figure 3 shows the values of the floor and roof contingence of the lower and upper layers at a distance of 4 m apart from the stope off the worked-out seam at various distances L between the stopes. It can be seen that at the same shape of the curves large shifting takes place within the roof of the seam.

Figure 4, a and b shows epures of equivalent stresses σ_{eq} within the roof and the floor of the seam C_6 if parting thickness is $h = 5$ m. The curves l show that $L = 30$ m, the curves 2 show that $L = 70$ m. Figure 5, a and b shows epures σ_{eq} within the roof and the floor of the lower layer if $h = 15$ m. The curves 1 in this figure are obtained for $L = 40$ m and the curves 2 is for $L = 60$ m.

Figure 4 shows that at a distance between faces $L = 30$ m along the contiguous seams and parting thickness $h = 5$ m the area of potential disintegration within the roof of the lower layer in the stope is 10 m, within the floor is 13 m; if $L = 70$ m equivalent stresses don't exceed the limit of the rock thickness of the roof and floor. The same situation is within the roof of the lower layer if $L = 30$ m and $L = 60$ m and if $h = 15$ m (Figure 5, a, b). Stresses within the floor in this case are not hazardous.

18

(a)

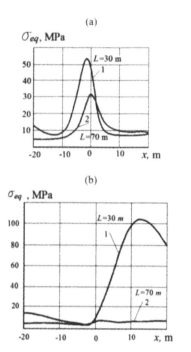

(b)

Figure 4. Stresses σ_{eq} according to Balandin criteria: (a) within the roof of the lower layer; (b) within the floor of the lower layer at parting thickness $h = 5$.

(a)

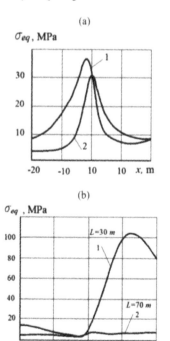

(b)

Figure 5. Stresses σ_{eq} according to the Balandin criteria: (a) within the roof of the lower layer; (b) within the floor of the lower layer if parting thickness is $h = 15$ m.

Therefore, the value of parting thickness $h = 5$ m according to the factor of the rock pressure for the Western Donbas mines can be acceptable when the distance between stopes on the contiguous seams will be not less than 70 m. When parting thickness is $h = 15$ m stopes of the contiguous seams should be at a distance $L \geq 60$ m from each other.

In the first case maximum values of convergence within the stope are determined by the method of the fictitious loads (Crouch & Starfild 1987) are $(\Delta u_{up})_{max} = 131$ mm along the upper layer and $(\Delta u_l)_{max} = 66.7$ mm along the lower one. In the second case the values are $(\Delta u_{up})_{max} = 123.7$ mm and $(\Delta u_l)_{max} = 65.47$ mm. So, shifting in both cases doesn't exceed accepted values which are about $0.3\, m_l = 210$ mm.

3 DETERMINING STRESS CONCENTRATION FACTORS WITHIN THE ROOF OF THE DEVELOPMENT WORKING

Flat calculation model is used to analyze stress and train state of the investigated area of the rock massif performed to find out rational location of the development working in overworking conditions. Furthermore, it is impossible to take into account mining effect. That is why load ratio is introduced into calculated algorithm. The last one is determined according to the parameters of increased rock pressure and is the relationship of stresses σ_{yy} within the sections of this area with the level γH. Load ratio is determined on the basis of the calculation model (Figure 6). Resolving system of equations is formed taking into account boundary conditions for the elements of free boundaries and for the contacting elements by the method of fictitious loads.

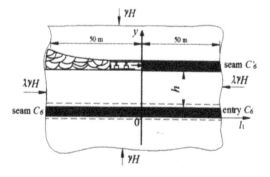

Figure 6. Determining stresses concentration factor σ_{yy} within the roof of the development working of the lower rock layer.

19

Maximum values of load ratio within the roof along the axis of development working located on the lower seam at various distance l_1 from the stope of the upper layer required for the further calculations are determined using this solution (Figure 7).

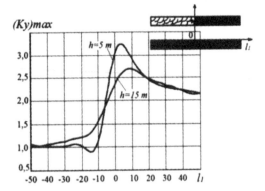

Figure 7. Maximum load ratio within the roof of the development working.

4 DETERMINING RATIONAL LOCATION OF THE DEVELOPMENT WORKING IN OVERWORKING CONDITIONS

One of the principle conditions of contiguous seams effective mining is to provide operating stability of the development workings. Stability of development workings on the contiguous seams first of all depends on the place of their location, regularity of seam mining, parting thickness, and physical and mechanical properties of inclosing rocks.

Special attention is paid to the choice of the extraction drift location along the lower seam mined secondarily as it is subjected to the additional rock pressure due to the mining operations performed along the upper layer. Obviously, workings will be less loaded if they are located within the unloading area formed after mining of overlying seam. However, after extraction of the edging part of the overlying seam these workings can be squeezed due to the sharp increase of the rock pressure. It is known that working drivage within the area of the increased pressure is less negative than pressure rise effecting on the driven working due to the mining operations (Jacobi 1987).

The method of extinguishing workings protection on the contiguous seam is of great importance. Practice and results of scientific investigations show that a pillar leaving to protect workings at any contiguous seams has negative effect on the state of the extraction drift of the next mining seam. Therefore, pillarless protection of the extraction drifts should

be applied on the contiguous seams (Glushko et al. 1975). As for mining regularity, there is the descending order of seam mining in the Western Donbas mines. Seam preparation is individual. Contiguous seams are mined together with the advanced extraction of the overlying ones.

Workings of the lower layer are better to locate below worked-out area. The distance between them is determined by the mining depth and parting thickness. The distance between extraction drifts on the contiguous seams should be 10...15 m if parting thickness is 7...8 m and 400...500 m mining depth.

While driving working along the lower seam next to the stope at the 300 m level it is recommended to keep the distance which is not less than 200 m. But in this case the drift will be within the area of increased rock pressure. Active shifting stage of the overworked rock thickness is ceased at a distance equal to the length of stope mining.

The main task is to find out such place for the development working which could exclude additional mining effect performed along the overlying seam and provide sufficient maneuver to plan mining operations along the overworked lower bed. For this purpose minimally acceptable distance, in terms of development working stability, from the boundaries of the overworking area (the edge of the overlying seam) is determined.

While finding out adequate parameters it is required to remember that working stability should be provided long enough .Therefore, the change of physical and mechanical properties of inclosing rocks in time, rock massif structure and coal seam thickness should be taken into account in the calculated algorithm.

This paper considers calculation model showed in the Figure 8 while determining rational distance value l from the working contour of the lower seam up to the edging part of the upper one.

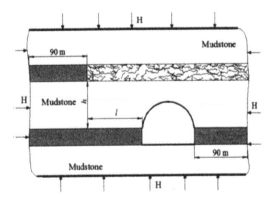

Figure 8. Determining rational location of the development working in overworking conditions.

Calculations were performed for two values of the parting thickness typical for the Western Donbas mines: $h = 5$ m and 15 m. Load ratio marked by the points in Figure 7 was used while defining boundary conditions. They had the values $(K_y)_{max} = 3.2$ if $h = 5$ m and $(K_y)_{max} = 2.7$ and if $h = 15$ m and were the most critical section of the working located at a distance $l_1 = 5$ m and $l_1 = 10$ m from the stope.

The model of linear- inherited medium with the Abel creep kernel was accepted for the massif. According to this model while calculating stresses and shifting rock elasticity modules within the roof and the floor at various times are determined by the formula:

$$E_t = \frac{E}{1 + F_t}, \qquad (2)$$

$$F_t = \frac{\delta}{1 - \delta} t^{1-\alpha}, \qquad (3)$$

where E – is elasticity module; F_t – is creep function; α and δ – are elasticity parameters; t – time, sec.

The results of calculations for $H = 300$ m depth are given below. Figure 9 shows epures of the equivalent stresses within the working roof and the floor if $l = 0$ and parting thickness is $h = 5$ m. Maximum stresses are within the roof section $\varphi \approx 120°$ (250 MPa) that exceeds σ_c of the rock. Epures of the stresses given below are built taking into account the time of extraction pillar mining.

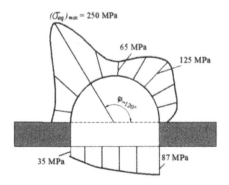

Figure 9. Epures of the equivalent stresses alongside the working contour if $h = 5$ m and $l = 0$.

Figures 10 and 11 show the behavior of the equivalent stresses within the roof of the working alongside of its contour. According to the epures (Figure 10) if parting thickness is $h = 5$ m and $l = 0$ (working is located below the edging part of the upper layer) stresses σ_{eq} within the roof exceed the limit of the rock strength as for compression $\sigma_c = 30$ MPa everywhere and the area $110° \le \varphi \le 140°$ is the most loaded. Such location of

the development working is impossible. If the working will be at a distance $l = 60$ m, operating stresses will not be hazardous.

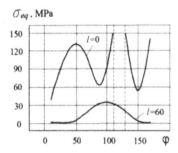

Figure 10. Stresses σ_{eq} within the roof of the working if $h = 5$ m.

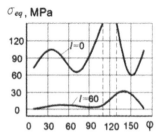

Figure 11. Stresses σ_{eq} within the roof of the working if $h = 15$ m.

Stresses σ_{eq} within the roof also exceed σ_c. The most hazardous area is $110° \le \varphi \le 130°$ but in case if $l = 40$ m. In this case stresses don't exceed the limit of the rock strength.

To generalize the results of theoretical study maximum equivalent stresses are determined within hazardous sections of the working for the parameter l, the value of which were changed from 0 to 80 m with 10 m increment.

Sections where load ratio had maximum values – $(K_y)_{max} = 3.2$ in section $l = 5$ m if $h = 5$ m and $(K_y)_{max} = 2.7$ in section $l = 10$ m if $h = 15$ m were the most hazardous.

All calculations were performed for the depth $H = 300$ m and were adapted to the time $t = 16$ months.

The results were subjected to the statistical analysis. Following correlation dependences of the maximum equivalent stresses $(\sigma_{eq})_{max}$ as for parameter l were obtained:

Within the floor if $h = 5$ m:

$$\left(\sigma_{eq}\right)_{max} = 63.675 - 3.075\sqrt{l} - 0.003l^2, \text{ MPa}, \qquad (4)$$

if $h = 15$ m:

$$\left(\sigma_{eq}\right)_{max} = 25.163 - 2.126\sqrt{l} + 0.001l^2 \text{, MPa,} \quad (5)$$

within the roof if $h = 5$ m:

$$\left(\sigma_{eq}\right)_{max} = 177.974 - 21.662\sqrt{l} + 0.007l^2 \text{, MPa,} \quad (6)$$

if $h = 15$ m:

$$\left(\sigma_{eq}\right)_{max} = 128.010 - 15.966\sqrt{l} + 0.004l^2 \text{, MPa.} \quad (7)$$

Correlation coefficients are given in the Table 1.

Table 1. Figures of high correlation ratio $(\sigma_{eq})_{max}$ and l.

Correlation coefficient	Floor, h		Roof, h	
	5 m	15 m	5 m	15 m
R	0.83	0.97	0.93	0.96

Figures 12 and 13 show calculated values of each variant by the points. Straight lines are the curves built according to the dependences (4) – (7).

Taking into account epures $(\sigma_{eq})_{max}$ within the floor and according to the P.P. Balandin (1) criteria (distance up to the working contour from the edging part of the upper layer should be $l = 70$ m if $h = 5$ m and if $h = 15$ m) stresses at any l are much less than compression strength if the strength margi $K = 1.2$. It means that they are not dominant.

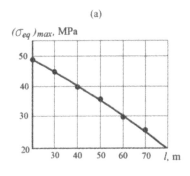

(a)

(σ_{eq})$_{max}$, MPa

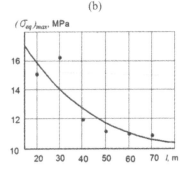

(b)

(σ_{eq})$_{max}$, MPa

Figure 12. Maximum equivalent stresses within the floor of the working if: (a) $h = 5$ m; (b) $h = 15$ m.

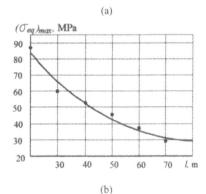

(a)

(σ_{eq})$_{max}$, MPa

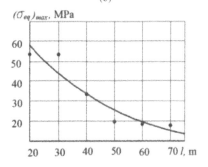

(b)

(σ_{eq})$_{max}$, MPa

Figure 13. Maximum equivalent stresses within the roof of the working if: (a) $h = 5$ m; (b) $h = 15$ m.

Taking into account stresses within the roof if $h = 5$ m l should be not less than 80 m, and if $h = 15$ m permitted value is $l = 50$ m.

Thus, $l > 80$ m if $h = 5$ m and $l > 50$ m if $h = 15$ m are the rational ones for considered mining and geological conditions.

However, it is required to analyze the values of maximum shifting on the contour of the given working. It has been done in the paper (Novikova et al. 2005).

Shifting calculation on the working contour was performed in two stages – at first for the large area which was covered by the calculated model and then small part of the massif around the working where stressed were given as an operating load obtained at the first stage was considered. Sizes of small neighborhood were 45 m across and 7 m down. Working was located symmetrically as for vertical axis. The distance from the floor of the working up to the lower edge of the area was 2 m and up to the upper one is 5 m. Calculation was performed for the values l, which were changed from 0 to 80 m with 10 m step.

Table 2 shows shifting values of the roof and the floor obtained from each calculated mode.

Table 2. Shifting of the roof and the floor of the working.

	$h = 5$ m				$h = 15$ m		
l, m	u_{up}, mm	u_l, mm	Δu, mm	l, m	u_{up}, mm	u_l, mm	Δu, mm
0	174.8	248.5	423.3	0	74.3	105.6	179.9
10	179.7	256.3	436.0	10	88.2	125.4	213.6
20	137.1	194.9	332.0	20	65.3	92.8	158.1
30	122.0	173.7	295.4	30	59.1	84,0	143,1
40	115.5	164.1	279.6	40	55.8	79.3	135.1
50	107.9	153.4	261.3	50	52.2	74.2	126.4
60	94.6	134.5	229.1	60	51.1	69.0	120.1
70	89.2	108.8	198.0	70	50.1	63.0	113.1
80	87.1	105.2	192.3	80	49.8	62.9	112.7

According to the data results and on the basis of statistical analysis correlation dependences connecting convergence of the roof and the floor $(\Delta u_{up})_{max}$ with the distance l were determined for $h = 5$ m:

$$(\Delta u)_{max} = 477.07 + 20.21\sqrt{l} -$$
$$-8.81l + 0/05l^2 , \text{ mm} \qquad (8)$$

for $h = 15$ m:

$$(\Delta u)_{max} = 182.61 + 20.39\sqrt{l} -$$
$$-5.53l + 0.03l^2 , \text{ mm} \qquad (9)$$

Coefficients of correlation of these dependences are 0.91 and 0.96 correspondingly. The curves obtained according to these dependences are shown in Figures 14 and 15. The results of correspondent calculated variants are marked by the points. Table 2 and Figures 14 and 15 show that large shifting takes place if $l = 5$ m. If parting thickness is $h = 5$ m, then convergence will be 440 mm whereas permitted value is 300 mm. If $l = 30$ m and $h = 5$ m Δu_{up} the value of 300 mm is assumed, i.e. in this case l should be not less than 30 m. If the parting thickness is $h = 15$ m, convergence in hazardous section doesn't exceed permitted value. Calculations by the formulas (8) and (9) give the same results.

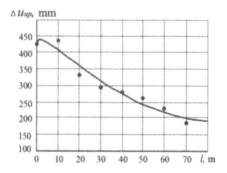

Figure 14. Contingence of the roof and the of the floor of the working: $H = 300$ m; $t = 16$ months; $l_1 = 5$ m; $h = 5$ m.

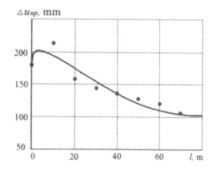

Figure 15. Contingence of the roof and the floor of the working: $H = 300$ m; $t = 16$ months; $l_1 = 10$ m; $h = 15$ m.

Thus, calculation in terms of stresses is the main factor determined rational distance l if $h = 15$ m. Required distance l should be not less than 80 m if $h = 5$ m and not less than 50 m if $h = 15$ m. Values of roof and floor convergence don't exceed permitted ones and are 192 mm and 112 mm.

5 DETERMINING RATIONAL RATE OF STOPE ADVANCE

Calculation of the process time of shifting underworked rock massif within the stope is performed on the basis of the theory of layer by layer blocked shifting described in monograph written by M.S. Chetverik and E.V. Androschuk (Chetverik & Androschuk 2004). Taking into account stope advance stay period of tensile deformations above worked-out area is determined.

As tensioning stimulates rock foliation and foliation areas are working water sources, the height of foliation areas above worked-out space is determined, and the rate of stope advance is calculated. In this case tensile deformations have no time to overspread along the plane of rock disintegration up to the water bearing level.

The rate of developing tensile deformations along the caving plane depends on the mining depth, time of active disintegration and slope angle of caving plane as for the level. In given mining and geological conditions (rocks are of the medium metamorphism) if mining depth is $H = 400$ m, slope angle of the caving plane ω is $63...66°$ and the rate of tensile deformations v_d is $10...15$ meters per day. Therefore, time required to develop deformations along the caving plane from the worked-out area up to the Earth surface is:

$$t = \frac{H}{v_d \sin \omega} \approx 34 , \text{day}. \tag{10}$$

The important stage of deformation development is undermining time of two blocks (time of two caving steps) at which tensile stresses within the roof of the worked-out area are changed by the compression stresses. This time is determined by the formula:

$$T_{ch} = \frac{2L}{v_{fa}}. \tag{11}$$

If caving step is $L = 25$ m and the rate of face advance is $v_{fa} = 3$ meters per day undermining of two blocks is 17 days. If the area of tensile deformations of two blocks at the time of undermining could be widespread along the whole height H the tension area would be above worked-out area for some time and would be stope water source. But in this case:

$$T_{ch} < t \left(\frac{2L}{v_{fa}} < \frac{H}{v_d \sin \omega} \right). \tag{12}$$

Therefore, tensile deformations have time to widespread only up to the stratification area where tension is changed by the compression.

Sizes of the tensile area and zone of rock massif stratification above worked-out area are determined by solution of geomechanical task using the method of boundary elements on the basis of calculated model (Figure 16). The size l is assumed to be equal to 50 m (two steps of the basic roof caving).

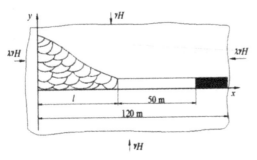

Figure 16. Calculated model of determining rational value of the face advance rate.

Calculations have shown that the extention of the tensile stress area down is 35 m from the floor of the longwall face. At the rate of v_d $10...15$ meters per day it will reach in a time:

$$T = \frac{3H_t}{v_d \sin \omega} = 8...12 , \text{day}. \tag{13}$$

While determining value T it was taking into account that the period of deformations of stresses is one third of the time of rock disintegration. It was determined on the basis of instrumental observation data as for surface shifting (Chetverik & Ozerov 1998).

Rational rate value of the face advance whereby zone of tensile deformations have no time to overspread along the plane of massif disintegration up to the water bearing level. It is determined by the time equation T_{ch} of two blocks underworking and the time T of reaching foliation zone. That is:

$$\frac{2L}{v_{fa}} = \frac{3H_t}{v_d \sin \omega}, \tag{14}$$

$$v_{fa} \geq \frac{2L v_d \sin \omega}{3H_t} = 4.2...6.3 , \text{m per day}. \tag{15}$$

6 CONCLUSIONS

Quantitative estimation of the upper layer longwall face effect on the development working driven along the overworked lower seam was performed by the numerical method. Load coefficients are intended to use as initial data while studying stability of extensive overworked heading.

Correlation ratio between maximum equivalent stresses within the roof and the floor of the overworked heading and the distance from the cross-section of the last one up to the edging part of the upper layer was obtained.

While receiving data of dependences mining (stoping) effect along the upper layer was taking into consideration by the load coefficient.

Correlation ratio between maximum contingence of the drift roof and floor within the overworking area and the distance from its cross-section up to the edging part of the upper layer was determined.

Given study shows that the value of the parting thickness $h = 6$ m for the Western Donbas mines is permitted as for the rock pressure factor if the distance L between stopes of the contiguous seams is not less than 70 m; if the parting thickness h is more than 15 m stopes of contiguous seams should be at a distance of $L \geq 60$ m from each other.

Due to results of performed analysis of stress and strain state of rock massif it was determined that in the Western Donbas mines at the continuous descending order of contiguous seams underworked development working should be located within overworked area at a distance l from the stope of the upper layer which is not less than 80 m if $h = 5$ m and not less than 50 m if $h = 15$ m.

Ration rate value of stope advance at which the area of tensile deformations have no time to overspread along the plane massif disintegration up to the water bearing level was determined.

REFERENCES

Chetverik, M. & Ozerov, I. 1998. *Earth surface displacement and geomechanical processes within underworked massif while developing flat coal seams*. Geotechnical mechanics, No 9: 64 – 70.

Chetverik, M., & Androschuk, E. 2004. *Theory of rock massif displacement and deformation process control while underground coal mining*. Dnipropetrovsk: RIA «Dnepr-VAL»: 148.

Croutch, S. & Starfild, A. 1987. *Methods of boundary elements in mechanics of solids*. M.: Mir: 398.

Gabdraximov, I., Dedyukin, M., & Pozdeyev, A. 1966. *On phenomenological theory of carnolith strength and calculation of durable pillar stability*. Problems of rock mechanics. Alma-Ana: Science: 107 – 112.

Glushko, V., Tsai, T. & Vaganov, I. 1975. *Protection of deep mines workings*. M.: Nedra: 200.

Jacobi, O. 1987. *Practice of rock pressure control*. M.: Nedra: 566.

Novikova, L., Busylo, V. & Nalivaiko, Ya. 2001. *Prognosis of rock pressure manifestation while underworking and overworking coal seam*. Vestnik HSTU. Kherson, Addition. 12-P: 193 – 196.

Novikova, L., Zaslavskaya, L. & Vasiliev, V. 2005. *Determining permitted parting thickness while working-out contiguous seams*. Geotechnical mechanics: No 53: 105 – 110.

Shyrokov, A. 1963. *Regularity of coal accumulation in Western Donbas*. M.: Gosgortehizdat: 452.

Theoretical and Practical Solutions of Mineral Resources Mining – Pivnyak, Bondarenko & Kovalevska (eds)
© 2015 Taylor & Francis Group, London, ISBN: 978-1-138-02883-8

An overview and prospectives of practical application of the biomass gasification technology in Ukraine

V. Bondarenko, V. Lozynskyi, K. Sai & K. Anikushyna
National Mining University, Dnipropetrovsk, Ukraine

ABSTRACT: Mining of coal deposits nowadays in Ukraine has to include not only coal extraction but following processes that accompany coal extraction. In the article the potential of biomass usage for additional heat and power generation during coal seam gasification is proposed. The innovative technologies in mining such borehole underground coal gasification with combination of underground biomass gasification is examined. World share of bioenergy in the world primary energy mix and total primary energy supply by resource are given. The conclusions are drawn at the end of the paper with perspective trend for future studies.

1 INTRODUCTION

Consumption of energy resources in the nearest time will provoke their rapid rise, accompanied by their gradual exhaustion. This situation requires a search of internal reserves based on renewable energy. Energy crisis prompts European countries for searching an alternative source of renewable energy. Important part of those efforts is conducting of integration research and development of road maps for sustainable energy use for whole continents.

Energy production can be provided with two resources (exhaustible and inexhaustible). To exhaustible resources refers: coal, oil, oil sands, natural gas, heavy oil, natural gas, associated petroleum gas etc. To inexhaustible resources refers: wood, straw, energy crops, solid waste, bio-gas and landfill gas. Biomass by definition is an organic matter of vegetable or animal origin that can be used as an alternative energy source.

2 THE MAIN PART OF THE ARTICLE

The supply of sustainable energy is one of the main challenges in modern world. Biomass can make a substantial contribution to supplying future energy demand in a sustainable way.

Biomass fuels can be produced from agricultural, forestry and municipal wastes and residues, as well as from crops such as sugar, grain, and vegetable oil. Crops grown for use as biomass fuel can be grown on degraded, surplus and marginal agricultural land, and algae could, in the future, be exploited as a marine source of biomass fuel. Thus, biomass could provide a significant proportion of the heat, electricity and fuel demand of the future. Bioenergy is the largest global source of renewable energy, and contributes an estimated 10% of global primary energy production in particular as a direct source of industrial and domestic heat (Figure 1) (Gadonneix 2013).

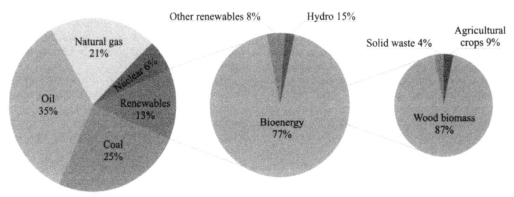

Figure 1. Share of bioenergy in the world primary energy mix.

The share of renewable energy in the global energy production today make up 13% of these 77 accounts for bioenergy main raw material which is wood biomass (87%). The global crisis has stepped up the change of the world energy priorities to nuclear fuel (Khomenko & Rudakov 2010, Vladyko 2012). Share in renewable energy according to the World Energy Recources (Survey of energy resources) for 2013 was 11%, at that according to optimistic indicators for 2020 it should make 22% (Figure 2).

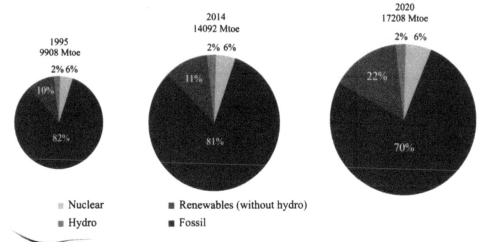

Figure 2. Total primary energy supply by resource.

From Figure 1 the acronyms "Mtoe" means the million tonne of oil equivalent. It is a unit of energy defined as the amount of energy released by burning one tonne of crude oil. It is approximately 42 gigajoules or 11.63 MWh, although as different crude oils have different calorific values, the exact value is defined by convention; several slightly different definitions exist. The toe is sometimes used for large amounts of energy.

The supply and use of energy have powerful economic, social and environmental impacts. Not all energy is supplied on a commercial basis. Fuels, such as fuel wood or traditional biomass are largely non-commercial. Fuel wood is playing a leading role in the developing countries, where it is widely used for heating and cooking. Universal access to commercial energy still remains a target for the future. In many countries, especially in Africa and Asia, the pace of electrification lags far behind the growing demand. It is imperative to address this major challenge without further delays, in particular taking into account the impact access to electricity has on peoples' lives and well-being, economic growth and social development, including the provision of basic social services, such as health and education (Gadonneix 2010).

Annual growth of biomass in the world is estimated at 200 mlrd tons in terms of dry matter, that is equivalent to 80 bln tons of oil. One of the sources of biomass is forests. During added-value wood processing 3 – 4 bln tons make waste, energy equivalent of wood waste make 1.1 – 1.2 bln tons of oil.

Ukraine is one of the most energy intensive countries in the world. In Ukraine the primary energy consumption has been quite the same during the last few years. The primary energy consumption in the year 2014 was 180 Mtsf. The acronyms "Mtsf" in Ukraine means million tonne of standard fuel. It is unit of energy defined as the heating value of 1 kg of coal and is equal to 29.3 MJ or 7000 kcal. About 79 % of the total energy consumption in Ukraine in 2014 was produced by fossil fuels like coal, oil and natural gas.

According to expert estimates of official statistics for 2014, the theoretical potential of biofuels in Ukraine make 50 Mtsf, technical – 36.2 Mtsf and cost-effective – 27.27 Mtsf. That is the basis of the current level of total primary energy consumption in Ukraine (180 million tons of standard fuel standard fuel in 2014) economic potential of biomass can meet about 15% of Ukraine's market needs in the energy sector (Tytko & Kalinichenko 2010). Detailed power potential of biomass in Ukraine is represented in Table 1.

Ukraine has great potential of biomass available for energy use. By bringing this potential to energy production it is possible to satisfy 13 – 15% of the country demand in primary energy in the nearest perspective. Development of bioenergy sector in Ukraine should be conducted consistently and reasonably, taking into account possible impact on national economy and environment. The basic components of biomass potential are agricultural residuals and forest biomass (Geletukha 2010).

Table 1. Detailed power potential of biomass in Ukraine.

Biomass	Power potential, Mtsf		
	Estimated	Technological	Economically viable
Straw of grain-crops	10.39	5.21	1.34
Straw of rape (Brassica napus)	1.07	0.75	0.75
Wastes of corn production	5.7	2.99	2.79
Wastes of sunflower production	4.27	2.86	2.86
Wood biomass	2.13	1.66	1.48
Biodiesel	0.50	0.50	0.25
Bioethanol	2.33	2.33	0.86
Biogas from animal wastes	3.27	2.45	0.76
Biogas from solid waste landfill	0.77	0.46	0.26
Biogas from aeration plant	0.21	0.13	0.09
Energy crop:			
– poplar. acacia. alder, willow	14.58	12.39	12.39
– rape (straw)	1.65	1.15	1.15
– rape (biodosel)	0.78	0.78	0.78
– sunflower (biogas)	1.59	1.11	1.11
Peat	0.77	0.46	0.40
At all	50.01	36.23	27.27

Agricultural biomass is concentrated in the central, southeastern and southern regions, in places with the most fertile soils, while forest biomass may be produced in the northern parts of the country, which is by 25 – 30% covered with pine forests, and in the western part – Ukrainian Carpathians, where the dominant forest species are spruce, beech, fir and oak.

There are a lot of biomass energy sources in Ukraine. The biomass fuels are mainly wood residues from forest and agriculture industry. Bioenergy could supply a substantial part of the energy needs of individual communities in Ukraine. It could improve energy security and, in particular, improve energy supply to areas with poor energy infrastructure. Agriculture, real estate and transport are likely to be the big-gest consumers of energy from renewable sources. Bioenergy can offer new opportunities for job creation, especially in areas of high unemployment. Further, land that, because of contamination, is unsuitable for growing edible plants could be used for energy crops. Development of bioenergy and, in general, the renewable energy sector, could help resolve many environmental problems associated with the energy sector in Ukraine, as well as in other countries.

According the bioenergy potential of biomass Ukraine far ahead of all other European Union countries, including developed countries such as France, Germany, Spain etc. (Tytko & Kalinichenko 2010). The problem is only to use its energy potential, particularly in the older approaches to its use (Figure 3).

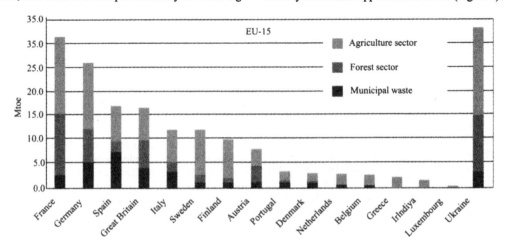

Figure 3. Bioenergy potential of EU member state in comparison with Ukraine.

There are many bioenergy routes which can be used to convert raw biomass feedstock into a final energy product. Several conversion technologies have been developed that are adapted to the different physical nature and chemical composition of the feedstock, and to the energy service required (heat, power, transport fuel). Upgrading technologies for biomass feedstocks (e.g. pelletisation, torrefaction and pyrolysis) are being developed to convert bulky raw biomass into denser and more practical energy carriers for more efficient transport, storage and convenient use in subsequent conversion processes (Gadonneix 2010). The bioenergy routes are shown on Figure 4.

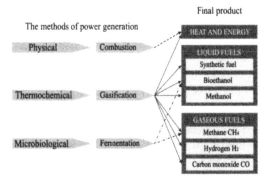

Figure 4. Bioenergy routes of power generation.

The most effective technology of biomass usage in bioenergy last one is the direct combustion, gasification, anaerobic fermentation and all. In general, organic energy is received using physical or thermo-chemical or microbiological methods. Power generation using direct combustion of organic waste is physical methods. The basis of thermo-chemical method is the use of chemical gasification process. The spread of the world is microbiological methods of biogas receive using anaerobic digestion.

As a result of anaerobic fermentation (bioconversions) appear: a biogas, which consists of methane (65 – 80%), carbon dioxide CO_2 (15 – 25%), carbon monoxide CO (2 – 3%), nitrogen (N_2), oxygen (O_2), sulphurhydrogen (H_2S), and high-efficiency environmentally clean disinfected fertilizer. One ton of it is equivalent 3 – 4 tons of nitric-phosphoric fertilizers which are produced industry. In the process of methane fermentation pathogenic microorganisms perish and the unpleasant smell of wastes is neutralized. The method of biogas reception is widely widespread in a number of countries with a warm climate, a high enough temperature, conditioned natural terms, is needed in which for the effective flowing of bioconversion (India, Brazil, Chinese Folk Republic and other). Most active activity of bacteria at temperature 35 – 45°C.

Fluidizers anaerobic fermentation serve as the effective mean of agricultural and stock-raising wastes processing (leaves, stems of plants, weeds, straw, sunflower wastes, corn heads, pus of different farms and dung of poultry factories), and also communal wastes on purification buildings and other. Their application allows deciding three tasks, important from the power, agricultural and ecological points of view, namely: to get a biogas; to convert wastes into high-showy profits (fertilizers); an environment impact (Bondarenko et al. 2010).

Undoubtedly the most promising is thermochemical method of energy production of biomass through gasification process. A variety of opportunities to obtain products using gasification process is much broader than with physical or microbiological method of obtaining energy.

In general gasification is the sequence of exothermic and endothermic chemical reactions between an organic fuel and an oxidant with steam and/or oxygen edition. It is a partial oxidation, in which the raw fuel is converted into a combustible synthesis gas, consisting mainly of CO and H_2. Addition products of gasification include heat, CO_2 and H_2O.

Almost any fossil fuels can be gasified, resulting generator gases are received with a wide range of usage – as fuel for domestic heating and the various processes of the industry. With the wide use of biomass gasification technology can achieve significant savings of fossil fuels.

Underground coal gasification can be examined as one of examples of generator gas reception which carry out new principle – combination of coal extracting with its simultaneous processing in the unique technological process. For today's day in the world there is the new interest to development of new mine and borehole underground coal gasification technology (Bondarenko et al. 2010).

Currently industrial-scale gasification is primarily used to produce electricity from fossil fuels such as coal, where the syngas is burned in a gas turbine. Gasification is also used industrially in the production of electricity, ammonia and liquid fuels (oil) using Integrated Gasification Combined Cycles, with the possibility of producing methane and hydrogen for fuel cells.

There are few examples of commercial gasification plants, and the deployment of this technology is affected by its complexity and cost. In the longer term, if reliable and cost-effective operation can be more widely demonstrated, gasification promises greater efficiency, better economics at both small and large-scale and lower emissions compared with other biomass-based power generation options (Gadonneix 2010).

Despite the positive aspects of biomass gasification today the problem of alternative energy is its high cost. But that's only because the alternative energy is not developed, no mass distribution, which is why their use should invest in most cases greater than in both the conventional. In addition, there are others obstacles that prevent technological development in Ukraine. Basically, among these are:

– low productivity of biogas unit;
– discrepancy between the quality parameters of biomass transport over long distances;
– limitations provided in preferences for reducing CO_2 emissions;
– discrepancy between demand and supply.

On the department of underground mining of national mining university for the last 15 years was conducted a lot of investigation that belong to alternative technologies of energy receive. Such technologies are underground coal gasification, gas-hydrate formation from coalbed methane etc. which, together with other alternative technologies can reduce the consumption of expensive natural gas (Falshtynskyy et al. 2012, Falshtyns'kyy 2013, Ovchynnikov et al. 2013, Koshka et al. 2014).

So, by employees of underground mining department were developed the technical documentation of the demonstrative laboratory unit and calculated heat-thermal balance for different types of coal and blowing mixture (contract with Department of international commerce and industry of Japan and firms JCOAL and Mitsui 2000 – 2001). Also was developed technological schemes of experimental gasifier for the conditions of mine Barbara during implementation the general project – Hydrogen Oriented Underground Coal Gasification for Europe – HUGE (Falshtyns'kyy 2013). A three-year research project financed by the European Commission's Research Fund for Coal and Steel. Project gathers 11 entities from Poland, Belgium, Czech Republic, Netherlands, United Kingdom, and Germany. It was led by Central Mining Institute (Poland).

Combination the technology of borehole underground coal gasification with biomass gasification that is proposed to consider is quite promising direction of alternative energy development. Firstly it will reduce the capacity cost including expensive equipment of surface biomass gasification, as all thermochemical processes take place in underground gasifier at the same time in his place of occurrence. Secondly it allow to resolve the issue of environmental and ecological cleanliness of the process, as underground coal gasification technology developed on a new level in a closed environment-friendly cycle. The latest investigations of UCG in the NMU are described in the works (Lozynskyi et al. 2015, Falshtynskyi et al. 2014).

3 CONCLUSIONS

Analysis of the availability of biomass in Ukraine and in the world shows the potential and feasibility of using organic compounds as alternative energy sources.

In order to carry out of biomass gasification technology combined with coal seam gasification it is only necessary to make some adjustments in the existing technological schemes. Fragments of biomass are injected into the gasifier along a controlled pipeline in the reaction channel, where it will convert to combustible gases. It will allow not only get more energy, but also solve the problem of organic waste utilization in some regions of Ukraine.

Suggested technology can become an alternative to traditional energy, successfully working out technological and social problems, and can become the active ecological hospital attendant of our degrading environment as a result of negative influence of existent power complex.

REFERENCES

Bondarenko, V., Tabachenko, M, Wachowicz, J. 2010. *Possibility of production complex of sufficient gasses in Ukraine.* New Techniques and Technologies in Mining. The Netherlands: CRC Press/Balkema: 113 – 119.

Falshtyns'kyy, V., Dychkovs'kyy, R., Lozyns'kyy, V., Saik, P. 2013. *Justification of the gasification channel length in underground gas generator.* Mining of Mineral Deposits. The Netherlands: CRC Press/Balkema: 125 – 132.

Falshtynskyi, V., Dychkovskyi, R., Lozynskyi, V., Saik, P. 2014. *Some aspects of technological processes control of in-situ gasifier at coal seam gasification.* Progressive Technologies of Coal, Coalbed Methane, and Ores Mining. The Netherlands: CRC Press/Balkema: 109 – 112.

Falshtynskyy, V., Dychkovskyy, R., Lozynskyy, V., Saik, P. 2012. *New method for justification the technological parameters of coal gasification in the test setting.* Geomechanical Processes During Underground Mining. The Netherlands: CRC Press/Balkema: 201 – 208.

Gadonneix, P., Barnés de Castro, F., Franco de Medeiros, N., Drouin, R. et al. 2010. *Survey of energy resources.* London: World energy council: 618.

Gadonneix, P., Nadeau, M., David Kim, Y., Birnbaum, L. et al. 2013. *Survey of energy resources.* London: World energy council: 468.

Geletukha, G., Zhelyezna, T., Lakyda, P., Vasylyshyn, R. et al. 2010. *Potential of biomass for energy in Ukraine.* Kyiv: 25.

Khomenko, O., Rudakov, D. 2010. *The first Ukrainian corporative university.* New Techniques and Technologies in Mining. The Netherlands: CRC Press/Balkema: 203 – 206.

Koshka, O., Yavors'kyy, A., Malashkevych, D. *Surface subsidence during mining thin seams with waste rock storage.* Progressive technologies of coal, coalbed methane, and ores mining. The Netherlands: CRC Press/Balkema: 229 – 234.

Lozynskyi, V.G., Dychkovskyi, R.O., Falshtynskyi, V.S., Saik, P.B. 2015. *Revisiting possibility to cross the disjunctive geological faults by underground gasifier.* Naukovyi Visnyk Natsionalnoho Hirnychoho Universytetu, No 4.

Ovchynnikov, M., Ganushevych, K., Sai, K. 2013. *Methodology of gas hydrates formation from gaseous mixtures of various compositions.* Annual Scientific-Technical Colletion-Mining of Mineral Deposits 2013. The Netherlands: CRC Press/Balkema: 203 – 205.

Saik, P.B. 2014. *On the issue of simultaneous gasification of contiguous low-coal seams.* Naukovyi Visnyk Natsionalnoho Hirnychoho Universytetu, No 6: 33 – 37.

Tytko, R., Kalinichenko, V. 2010. *Renewable energy (Polish experience for Ukraine).* Warsaw: 533.

Vladyko, O., Khomenko, O., Khomenko, O. 2012. *Imitating modeling Stability of mine workings.* Geomechanical Processes During Underground Mining. The Netherlands: CRC Press/Balkema: 147 – 150.

A method for evaluation of tailings hazard

G. Winkelmann-Oei
Federal Environment Agency, Dessau, Germany

D. Rudakov
National Mining University, Dnipropetrovsk, Ukraine

G. Shmatkov
Centre of Environmental audit and Clean Technologies LLC, Dnipropetrovsk, Ukraine

I. Nikolaieva
Innovation Center "Ecosystem", Kyiv, Ukraine

ABSTRACT: The Tailings Hazard Index (THI) has been developed for prompt and preliminary evaluation of the large amount of tailings management facilities (TMFs) on the national level by ranking their hazard. The THI is a dimensionless parameter calculated by summing up the partial indices quantifying the significance of various TMF-related hazards including tailings capacity, toxicity of tailing materials, TMF management quality, natural conditions (geological, seismological, and hydrological conditions) specific to the TMF site, and dam safety. The results of THI method application to ranking of Ukrainian TMFs by hazard are analysed. The THI method can be used for creation of national databases of hazardous sites.

1 INTRODUTION

Last two decades the concern is growing on environmental degradation caused by unintended large-scale movement of hazardous materials as a result of failures of tailings management facilities (TMFs) where large amounts of mining wastes are stored (Classification of mining waste facilities 2007).

The accidents at TMFs may frequently lead to long-term water and soil pollution, damage biota and have negative after-effects to human health. Failures may result in uncontrolled spills and releases of hazardous tailings materials. The negative impacts of such incidents on humans and the environment and severe trans-boundary consequences have been demonstrated by recent accidents in ECE-countries; the most known occurred at tailings in Baia Mare, Romania (2000), aluminium sludge tailings in Kolontar, Hungary (2010), at the Talvivaara Mining Company in Finland (2012).

In 1983 potash fertilizers were released in the Dniester River at Stebnikovskiy plant "Polimineral" in Western Ukraine. In 2008 due to dam failure waste products were again dumped from potash fertilizers tailings at the Kalush chemical plant into Dniester, which caused the concern of Government of the Republic of Moldova. In January 2011 the tails had dried up at the alumina refinery plant near the city of Nikolaiev (Southern Ukraine) and stored wastes were dispersing as dry red dust. As a result, topsoil, atmosphere, ground and surface water, and settlements were affected over the area of tens of square kilometres.

Many efforts have been undertaken recently by the international expert community to improve TMF safety through strengthening safety requirements, for instance, by putting into practice the modern safety standards and the advances in remediation technologies in mining (Safety Guidelines .... 2014, Tailing pits and sludge stores 2012, Reference Document... 2004, Peck et al. 2005). Advances in Earth sciences in the field of geological, seismic, hydrological, and climate risks have been also taken into account for design and operation of TMFs. Nevertheless, many tailings in East Europe and the former USSR still urgently need taking measures for improving their safety.

In 2013, German Federal Environment Agency has initiated the project "Improving the safety of industrial tailings management facilities based on the example of Ukrainian facilities". The project aims to develop an effective methodology for improving TMF safety using both Checklist approach and the method based on Tailings Hazard Index (THI). This should be a useful toolkit for competent authorities, inspecting bodies, and TMF operators in European Community and former Soviet Union countries, responsible for the safety of facilities storing hazardous mining wastes.

The main task of THI method is to develop a simple and easy-to-use procedure to rank a large number of tailings by their hazard using rough criteria for basic and critical parameters of TMFs, primarily, on the national level. Additionally, the THI method can be used for development of national databases on hazardous TMFs, which requires a systematic approach to ranking facilities.

Both international and national requirements have to be accounted for development of the hazard evaluation method as well as recent experience in mining facility operation and closure.

The hazard of TMFs as complex geotechnical systems can be characterized by various indices of different nature. Most of previous assessment approaches were focusing on some individual aspects of tailings features and impacts. The specifics of the proposed THI consist in keeping the balance between growing demand on consistency and trustworthy of environmental assessment on the one hand, and data availability and accessibility to final users, on the other hand.

2 METHOD ESSENCE

In analysis of the TMF impact on the environment and public health and safety we clearly discriminate the terms "risk" related to likelihood of damage and "hazard" related to actual threats to the environment and human beings. Thus, the approach to evaluate TMF impact is based on the assessment of real threats rather than probabilities of damage.

The THI method is intended for the use by state competent authorities as a toolkit to briefly overview the potential threats posed by a large number of TMFs by analysis of a few critical parameters based on the documentation available. Apart from creation and/or update of the country's catalogue of TMFs the THI method is capable to identify the most dangerous TMFs (the TMFs of highest concern) and to optimize the usage of limited financial and institutional resources aimed at improvement of TMF safety.

The THI is a dimensionless parameter calculated by summing up the partial hazard indices quantifying the significance of various impacts on TMF safety. These are:
- capacity of tailings;
- toxicity of substances in tails;
- TMF management quality;
- natural conditions (e.g. geological, seismological, and hydrological conditions) specific to the TMF site;
- and dam safety.

THI is calculated stepwise by the formula:

$$THI = THI_{Cap} + THI_{Tox} + THI_{Manag} + THI_{Site} + THI_{Dam}, \quad (1)$$

where THI_{Cap} – the hazard caused by the amount of tails (TMF capacity), THI_{Tox} – the hazard caused by toxicity of substances contained in tailings, THI_{Manag} – the hazard caused by improper, management of facilities, THI_{Site} – the hazard induced by location of the TMF in the area with specific geological and hydrological conditions, THI_{Dam} – the dam failure hazard (weaknesses in structural and component integrity and functionality).

Some data needed to quantify THI properly may be unavailable in the TMF country's database, e.g. Factor of Safety as the most representative and critical parameter for dam stability. In this case the other (alternative) rougher criteria should be applied that require other parameters, much easier to obtain and commonly available in TMF databases. This way makes TMF Checklist users more flexible in applying the appropriate criteria regarding to data availability.

The calculation procedure includes five steps. In case if some values unavailable or impossible to identify the maximum values for relevant hazards have to be specified. Thus, the hazard is expected to be higher if relevant information is absent.

The other parameters important to environmental safety can be derived on the base of THI such as the overall country THI, the average country THI, and these two indices related to the country area. It should be noted that the overall THI proportional to all TMF capacity will be obviously higher for the countries with extensive mining and large amount of TMFs like USA, Australia, Ukraine, and Canada. In contrast, the average THI might be higher for small countries with a few TMFs storing dangerous materials, like toxic non-ferrous metal ore wastes.

3 HOW TO CALCULATE THI?

1st Step: Capacity. The TMF capacity hazard is assumed to correlate to the growing volume of stored tails by logarithmic relation with the base of 10. Thus, increasing the volume of stored tailing materials by 10 times (one order) will increase this hazard value by 1.0, which allows grading the TMF capacity hazard in easier and understandable way.

The TMF capacity hazard is calculated by the formula:

$$THI_{Cap} = log_{10} [V_t], \quad (2)$$

where V_t – the volume of tails in the TMF (or TMF capacity), Mio m^3.

Examples. For a large TMF with $V_t = 10$ Mio m^3 we obtain $THI_{Cap} = log_{10}[10000000] = 7$. For a small TMF with $V_t = 0,01$ Mio m^3 we obtain $THI_{Cap} = log_{10}[10000] = 4$.

2nd Step: Toxicity. The tailings toxicity hazard is evaluated according to the national classification. The compatibility of two widely used toxicity classifications is shown in Table 1. The Ukrainian classification is applicable also in most of former USSR countries. According to Table 1 the notations "WGK 3" or "CH 1" relate to maximum toxicity of substances, the notations "WGK 0" or "CH 4" relates to minimum substance toxicity.

Table 1. Evaluation of THI_{Tox}.

Classification		Value of
WGK (WHC)[*]	Class of Hazard[**]	THI_{Tox}
"0"	"4"	0
"1"	"3"	1
"2"	"2"	2
"3"	"1"	3

[*] WGK = Wassergefährdungsklasse (WHC = Water Hazard Class), German classification;
[**] CH = Class of Hazard, Ukrainian classification.

3rd Step: Management. The hazard caused by improper management of TMF – assumed to be higher if the facilities are abandoned or orphaned. The value of THI_{Manag} – determined according to Table 2.

Table 2. Evaluation of THI_{Manag}.

TMF is identified as	Value of THI_{Manag}
Active and operated, or Non-active and cared and maintained	0
Abandoned or orphaned	1

4th Step: Site hazards. The site-specific hazard of the TMF includes the contributions of seismic and flood hazard, which are the most critical for TMF safety among natural impacts

$$THI_{Site} = THI_{Seismicity} + THI_{Flood}. \qquad (3)$$

The seismic hazard $THI_{Seismicity}$ – evaluated using geophysical data and maps on quake intensity. Commonly, these data are available in domestic requirements to construction safety. The value of $THI_{Seismicity}$ depends on seismic event magnitude at the probability 10% in scales MSK-64 or EMS-98. The seismic hazard to the TMF is evaluated according to Table 3.

Table 3. Evaluation of $THI_{Seismicity}$.

Seismic risk at the TMF location area	Magnitude of seismic events during last T_{Ret} years	Value of $THI_{Seismicity}$
Low	<6	0
Moderate or high	>6	1

Here T_{Ret} – returning period of earthquakes that can be defined by international or national requirements if the latter are stricter than international ones.

Example. According to international requirements T_{Ret} is established at 475 years (Cruz et al. 2004, EUROCODE 8 2003), the Ukrainian standard (Construction in seismic regions of Ukraine 2006) requires $T_{Ret} = 500$ years for 10% probability of higher quake magnitude; thus, both standards can be applied to Ukrainian tailings facilities.

The flood-induced hazard THI_{Flood} – evaluated using statistical data on frequency of floods and, specifically, the parameter HQ_{100} that quantifies flood event frequency with a one-hundred-year return period (floods with a probability of 1 in 100). The flood-induced hazard at the TMF location area is evaluated according to Table 4. The levels of HQ_{100} have to be updated annually regarding to climate changes.

Table 4. Evaluation of THI_{Flood}.

TMF location	Value of THI_{Flood}
In the area of HQ_{100}	1
Out of the area of HQ_{100}	0

5th Step: Dam. The dam failure hazard THI_{Dam} can be calculated in two ways.

1. *Preferred criterion.* If Factor of Safety (FoS) (Coduto 1999, Fredlund et al. 2012) is available in TMF databases for all facilities THI_{Dam} is calculated using two criteria based on slope stability (FoS) and TMF age by the formula:

$$THI_{Dam} = THI_{FoS} + THI_{Age}, \qquad (4)$$

where THI_{FoS} – the hazard of dam failure due to slope instability evaluated according to Table 5 (FoS has to be calculated already at the TMF design stage); THI_{Age} – the hazard caused by aging the dam.

2. *Alternative criterion.* If Factor of Safety is unavailable in TMF databases THI_{Dam} is calculated using the criteria based on dam material and geometry and TMF age by the formula:

$$THI_{Dam} = THI_{DamMaterial} + THI_{DamWidth} + THI_{Age}, \qquad (5)$$

where $THI_{DamMaterial}$ – dam embankment material; $THI_{DamWidth}$ is dam crest width, m.

Table 5. Evaluation of THI_{FoS} (preferable criterion).

FoS range	Value of THI_{FoS}
FoS > 1.5	0
1.2 < FoS < 1.5	1
FoS < 1.2	2

The dam failure hazard is assumed to increase for aged facilities, as shown in Table 6.

Table 6. Evaluation of THI_{Age}.

TMF age	Value of THI_{Age}
≤ 30 years	0
> 30 years	1

For the alternative criterion (eq. (5)) the hazards caused by improper dam material $THI_{DamMaterial}$ and narrow and weak dam width have to be evaluated by Tables 7 and 8.

The embankment constructed of a hard/blast rock is assumed to be more stable than the embankment of non-hard rocks or soils (earthen dams). In case if this material is unknown it can be identified by tensile strength at uniaxial compression σ_{DC}. For hard rocks $\sigma_{DC} > 5$ MPa, for non-hard rocks and soils $\sigma_{DC} \leq 5$ MPa.

Table 7. Evaluation of $THI_{DamMaterial}$ (alternative criterion).

Embankment material	Value of $THI_{DamMaterial}$
Hard rocks	0
Non-hard rocks and soils	1

The dam is assumed to be more stable if the dam crest width (and obviously, the dam basement) is sufficiently large to retain stored tails in the impoundment.

Table 8. Evaluation of $THI_{DamWidth}$ (alternative criterion).

Dam width	Value of $THI_{DamWidth}$
> 10 m	0
< 10 m	1

4 THI METHOD APPLICATION

For the correctness of THI calculation all TMFs should have the same dataset. In case of absence of some information the missing data are replaced with the values meeting the worst case in terms of TMF safety. For example, if there are no data on tailings materials the user has to specify the highest value for their class of hazard. If no specific information provided for coal ash tailings facilities the user defines the class of waste hazard by accepting the typical value for this material.

The special spreadsheet and template have been designed in MS Excel to calculate the tailings hazard index for TMFs. The file contains the available data on facilities, geology, hydrology, seismicity, site management, and the dam. The spreadsheet includes for each TMF site its name, location specified as latitude and longitude in world coordinates, capacity, information on the material stored, TMF status de-

pending on how the site is managed, maximum seismic activity, flood frequency parameter HQ_{100}, FoS (if available) commissioning year, embankment material and crest width.

Using Excel spreadsheets allows calculating total THI and partial hazard indices as well as plotting the results automatically. Then the user can select the top hazardous TMFs by applying appropriate numerical filters.

The database used for testing the THI method contains information on 153 Ukrainian TMFs (data provided by Kharkiv's Research Institute of micrographs State Archival Service of Ukraine under the contract №1/12-14 dated 21-08-2014), which is likely the most representative dataset collected on the national level (Figure 1). The results of evaluation (Figure 2) allow discriminate several groups of TMFs depending on the THI value.

The majority of tailings are situated in two eastern regions of the country on the Donetsk coal basin area (78.4% of total amount). There is a group of small facilities of less than 1 Mio m³ capacity near mines and chemical plants using coal as raw material. In general, their THI is below 8.

The second group consists of mid-scale TMFs of a few Mio m³ capacity located mostly in the east of the country with THI ranging from 8 to 12 (total 87 TMFs).

The 13 TMFs of highest concern with THI > 12, are situated throughout the country; among them 5 facilities in eastern regions, 5 facilities in the west, 2 facilities in Centre and South, and 1 facility in the North of Ukraine.

The analysis of THI distribution shows that capacity and toxicity of materials contribute 60 – 80% to the total THI for almost all TMFs; these parameters are considered the most critical in terms of environmental impact of tailings.

There are a number of other factors to be taken into account in the future to improve and refine the developed THI method. Drying up tails may threaten the environment and population as a result of blowing out dangerous particulate; this hazard typical for arid zones can be evaluated by adding the criteria based on wind velocity and precipitation-evaporation balance. Close location of TMFs to settlements and residential areas amplifies the tailings hazard, which can be considered by the criteria evaluating distances from tailings to potential targets.

Proper assessment of all TMF-caused hazards requires both development of new robust criteria and accumulation of relevant data.

Figure 1. Locations of Ukrainian TMFs and their hazard evaluation.

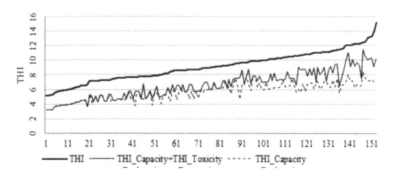

Figure 2. Distribution of Ukrainian TMFs by Tailings Hazard Index. Bottom axis labels show TMF enumeration in the database sorted by THI increase.

5 CONCLUSIONS

The Tailings Hazard Index (THI) has been developed and introduced as the overall parameter capable to rank the large amount of TMFs by their hazard on the national level. The THI integrally considers the contributions of the partial hazard indices quantifying the significance tailings capacity, toxicity of tailing materials, TMF management quality, natural conditions specific to the TMF site. THI method is available in Excel format, which facilitate its practical use.

The analysis of THI method application to ranking the hazards of Ukrainian TMFs enabled discriminating three TMF groups by capacity and total hazard. Besides, the TMFs with highest THI values have been selected as the target group to urgently take appropriate safety measures.

The approval of the THI method on the governmental level will promote to primary check of all TMF, identification of existing hazards and creating the country's catalogue of TMF. In the period ahead the THI method can be implemented as a GIS for making decision in environmental policy and used both on the national level and by international organizations for identifying transboundary threats to population in the areas with hazardous sites.

ACKNOWLEDGEMENT

The authors are greatly acknowledged to Mr. J. Vijgen (International HCH & Pesticide Association), and PhD K. Okhotnyk for ongoing support and kind assistance.

REFERENCES

Classification of mining waste facilities. 2007. Final Report. Prepared by DHI Water Environment Health in cooperation with SGI, Swedish Geotechnical Institute and AGH, University of Science and Technology, Krakow. European Commission, DG Environment: 204.

Coduto, D.P. 1999. *Geotechnical Engineering: Principles and Practices.* Prentice-Hall. New Jersey: 759.

Construction in seismic regions of Ukraine. National Standard of Ukraine. 2006. State Construction Regulations B.1.1-12:2006. Kyiv: 84.

Cruz, A.M, Steinberg, L.J., Vetere Arellano, A.L. et al. 2004. *State of the Art in Natech Risk Management.* Joint Research Center, European Commission: 59.

EUROCODE 8. 2003. *Design of structures for the earthquake resistance. Part 1: General rules, seismic actions and rules for buildings*: 89.

Fredlund, D.G., Rahardjo, H., Fredlund, M.D. 2012. *Unsaturated Soil Mechanics in Engineering Practice.* Wiley-Interscience: 944.

Peck, P.C. et al. 2005. *Mining for Closure: Policies and Guidelines for Sustainable Mining Practice and Closure of Mines.* UNEP, UNDP, OSCE, NATO. Geneva: 97.

Reference Document on Best Available Techniques for Management of Tailings and Waste-Rock in Mining Activities. 2004. UNECE: 517.

Safety Guidelines and Good Practices for Tailing Management Facilities. 2014. UNECE. New York and Geneva: 34.

Tailing pits and sludge stores. 2012. National Standard of Ukraine. State Construction Regulations B.2.4-5:2012. Part I. Planning. Part II. Building. Kyiv: 70.

Features of underlying levels opening at "ArsellorMittal Kryvyi Rih" underground mine

N. Stupnik, V. Kalinichenko, S. Pismennij & E. Kalinichenko
Kryvyi Rih National University, Kryvyi Rih, Ukraine

ABSTRACT: Nowadays the development of oxide ore deposits of Kryvyi Rih iron ore basin is mined by underground methods. All ore deposits have been opened by vertical shafts which are located in the footwall of the deposit. Under this opening scheme increasing of mining depth leads to rising of capital and running costs of opening workings drifting. The authors propose a technique for further mine development by using inclined haulage working. These opening schemes use load-haul-dump equipment. Taking into account that inclined opening working has large rock cross-section and intersects the layers with different physical and mechanical properties it is necessary to ensure its stability for the whole period of operation. Danger zones on the underground working contour depending on the physical and mechanical rock properties are identified by performed studies.

1 INTRODUCTION

There are about 23.2 billion tons of iron ore deposits, which are currently developed both open pit and underground methods in Kryvyi Rih basin. Oxide iron ore deposits (iron content is more than 55%), is developed by underground methods and their re-serves is about 2.3 billion tons up to the depth of 2000 m (Stupnik et al. 2013)

At present mines of Kryvyi Rih iron ore basin operate below the 1135 m level. Balance reserves of iron ore within the existing iron ore enterprises with underground mining is about 1.1 billion tones (Table 1).

Table 1. Iron ore reserves distribution (Stupnik et al. 2014).

Mining enterprise	Mining plant	Depth, m		Balance reserves, mln. t
		mining operations	ore reserve calculation	
PJSC "ArcellorMittal Kryvyi Rih"	Artem mine	1135	1315	120
PJSC "Kryvbaszhelezrudkom"	Mine "Rodina"	1390	1765	160
PJSC "Kryvbaszhelezrydkom"	Mine "Oktyabrskaya"	1265	2015	570
PJSC "Euraz Sukha Balka"	Frunze mine	1135	1580	40
PJSC "Euraz Sukha Balka"	Mine "Yubileynaya"	1260	2060	160
PJSC "Kryvbaszhelezrudkom"	Mine "Gvardeyskaya"	1270	1990	100
PJSC "Krivbaszhelezrudkom"	Lenin mine	1350	1500	80

As Table 1 shows Artem mine, Frunze mine and Lenin mine are approaching to the final counting depth of oxide iron ore balance reserves. At present development scheme the above mentioned mines can operate from 2 to 4 levels.

Mining in the underlying levels consists of following basic operations: shaft sinking, hoists re-placement, construction of pit bottom with crushers and bunkers on the haulage level, crosscut, haulage gate and ort drifting.

According to the condition of iron ore deposits development by underground method it is feasible to perform the construction of main level in conditions when level working lasts for 5 – 10 years, and level reserves are 45 – 60 million tons or more.

Iron ore mined by enterprises of Kryvyi Rih basin is characterized by relatively low price compared with foreign enterprises and high content of harmful impurities, silica in particular.

Iron ore market is characterized by volatility in the demand and pricing, as well as highly sensitivity to general economic cycles. Therefore, further mining of ore deposits at deep level using obsolete opening and development schemes, development systems, as well as technological backwardness and heavy wear of main production capacity, significantly reduce the financial flexibility of mining companies and increase the sensitivity of the company to crisis (Stupnik et al. 2012).

In order to maintain the volume of exports and strengthen its position in the global market of iron-ore raw materials, mining companies need to reduce the production cost; eliminate the previous backlog of new deep levels construction; improve product quality and increase production capacity.

2 STATEMENT OF THE PROBLEM

One of the technical solutions for reduction the cost of ore mining is to change the way of opening new levels that will not only eliminate the backlog of new levels construction, but also successfully increase the production capacity of enterprises.
For this purpose, the authors have analyzed some alternatives of opening new levels, which differ by methods of ore mined transportation along main opening workings, as well as the used type of mining equipment.

Let's examine some variants of opening new levels using inclined opening workings with the use of conveyor transport and powerful self-propelled equipment by an example of Artem mine, PJSC "ArcelorMittal Kryvyi Rih".

3 THE RESULTS OF STUDIES

At present, the iron ore mining at Artem mine is performed at the depth of 1045 m with the stoping face development of 1135 m level. According to the geological conditions, iron ore extends to the depth of 1315 m, the balance reserves in projected 1315 – 1045 m levels is not more than 120 million tons (Table 1).

Deposits of iron ore of Artem mine is opened by the main, servicing and air vertical shafts. At present, the height of the level is 90 m. The estimated reserve distribution on the levels is 1135 – 1045 – 60 million tons; 1225 – 1135 – 35 million tons; 1315 – 1225 – 25 million tons.

The traditional opening scheme of vertical shafts and their further sinking to the 1315 m level will inevitably lead to an increase in production costs, and in some cases reserve development will be impractical.

Let's examined three alternatives of opening of 1225 m mining level using an existing 1045 m main haulage level and the following:
– inclined haulage working with conveyor track from level 1225 m to 1045 m (Figure 1);
– inclined haulage working from level 1225 m to 1045 m, and transportation of ore mined by underground dump with capacity of 40 – 50 t (Figure 1);
– spiral inclined ramp to 1045 m main haulage level from level 1225 m (Figure 2).

It should be noted that above mentioned opening schemes of underlying levels do not involve sinking of the main wind up shaft.

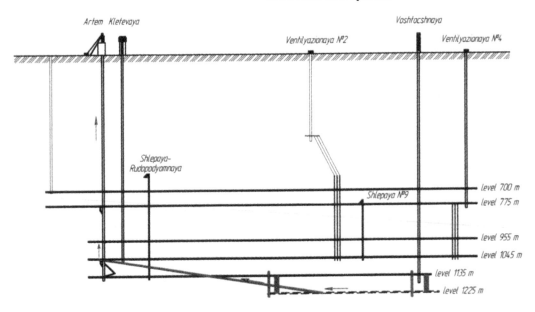

Figure 1. Opening schemes of underlying levels using inclined haulage working with conveyor track or transportation of ore by powerful self-propelled equipment.

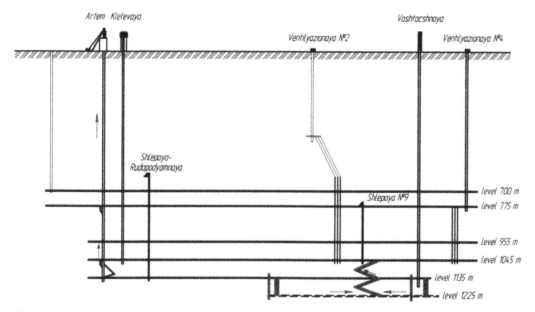

Figure 2. Opening schemes of underlying levels using inclined spiral ramp and transportation of ore by powerful self-propelled equipment.

The essence of the first variant is as follows: there is inclined haulage working from 1045 m level to 1225 m level. The mouth of inclined working is located near the measuring bunker which is below the 1045 m level. To ensure high technical and economic indices of production, as well as ensuring minimum capital and running costs it is necessary to drift an inclined working at an angle of 16 – 18 degrees, with a maximum length of 300 – 500 m.

When using powerful self-propelled equipment it will be necessary to build the spiral inclined ramp for delivering self-propelled equipment from the 1045 m main haulage level to 1225 m mining level.

The advantages of this opening scheme variant are as follows:

– the possibility of increasing the annual capacity of the underground mine up to 3.0 – 5.0 million tons or more;

– fast putting into operation of the mining level;

– reducing the length of mine workings: main crosscut 1.5 – 2.0 times

The essence of the second variant is as follows: opening the level at 1225 m is carried out by straight way inclined working from 1045 m level to 1225 m level. Inclined haulage working is drifted by section of 20 – 24 square meters at an angle of 20 – 24 degree.

Mined rock mass from stoping faces of 1225 m level is transported by underground dump to the 1045 m level to the dumping stations and then is transported to the main shaft by the locomotives.

The length of the inclined haulage working compared with the conveyor track is reduced by 1.2 – 1.5 times.

The advantages of this opening scheme are as follows:

– increasing the annual performance of the underground mine up to 5.0 million tons or more;

– decreasing the date of putting into operation of the mining level by 1.5 – 1.7 times;

– reducing the length of main crosscut by 1.7 – 2.5 times.

The essence of the third variant is as follows: opening the level at 1225 meters is carried out due to driving of inclined spiral ramp from the level of 1045 m on the mining level. Directly from the mouth of the spiral ramp on the 1045 m level dumping station is installed.

Rock mass issue is carried out by underground dump from stoping faces of 1225 m level along the spiral ramp to dumping stations located on the 1045 m level and then loaded onto trucks that deliver ore to the main shaft.

The advantages of this opening scheme variant are as follows:

– the possibility of increasing the annual capacity of the underground mine up to 3.0 – 5.0 million tons or more;

– fast putting into operation of the mining level;

– reducing the length of mine workings: main crosscut 1.5 – 2.0 times and the haulage working between the levels 2.0 – 3.0 times.

Feature of the given schemes of opening is that, compared with the conventional scheme opening (vertical shaft) mine performance with increasing depth of development remains constant or increases.

The results of the feasibility study for the proposed options are listed in Table 2 and shows that the most cost-effective option is opening of 1225 m level using a spiral ramp. Under this option, you must take a minimum number of underground workings at 5023 m; the given costs will also be minimal and up to 26.86 million UAH.

Operating expenses for the traditional pattern of opening is more in 1.7 – 2.0 times. This is connected to the fact that the annual performance of the mine does not exceed 1.5 million tons, and the time on working off reserves increases by 5 – 10 times.

It should be noted that using a scheme of opening by spiral ramp material costs for the purchase of a powerful self-propelled equipment, its repair and depreciation is increased 3 – 7 times compared with traditional schemes of opening by vertical shafts.

Table 2. Feasibility studies for drifting of underground development according to schemes of opening.

Indices	Scheme of opening			
	by vertical shaft	by inclined conveyor shaft	by inclined haulage working	by spiral ramp
The total length of the mine workings, m including:	6791	5857	5942	5023
– vertical, m	1710	1330	1330	1330
– horizontal, m	5081	4107	4232	3443
– inclined, m	—	420	380	250
Capital costs, mln. UAH.	273.8	210.3	202.2	192.0
Operating costs, mln. UAH.	23.0	16.3	12.6	11.5
Given costs, mln. UAH.	44.90	33.12	28.78	26.86
Annual output, mln.t	1.0 – 1.5	3.0 – 15.0	5.0 – 10.0	5.0 – 10.0
Given costs for level opening at 1225 m, UAH/t	96	99	88	86

The proposed schemes of opening have one major drawback is providing the sustainability of main opening inclined working for the entire term of reserves development to the final depth.

The use of powerful self-propelled machinery for underground mining operations will require a significant increase in cross-sectional area of mine workings. Cross-sectional area of mine workings in underground mining for various types of mobile equipment is given in Table 3 (Korolenko et al. 2012).

Table 3. Rock cross section depending on equipment used.

Name of underground mine workings	Rock cross section using, m^2	
	traditional equipment	self-propelled equipment
Capital	12 – 14	18 – 24
Preliminary	8 – 12	16 – 20
In the block:		
– drilling	8 – 10	12 – 18
– cutting	4 – 6	8 – 10
– raising	2.25 – 4	2.5 – 4
– ore delivering	4 – 7	10 – 12
– dumping rooms	—	22 – 25

From the above table 3 it is shown that the cross sectional area of mining using powerful mobile equipment is increased by 1.5 – 2.0 times in comparison with the use of traditional equipment.

The increase in cross-sectional area of mining leads to an increase in capital and operating costs due to increased area of exposure and reduce the stability of the mine workings.

Rock mass of Kryvyi Rih basin is heterogeneous, so on the same depth in the mine workings influence different forces that lead to different strains and as a consequence – to increase the cost of maintaining the mine workings.

In order to reduce operating costs, consider the stress distribution around mine workings with the identification of dangerous areas on the working line.

According to research tension and contraction stresses act around working (Stupnik et al. 2012). If the contraction stresses in the working sides exceed the tensile strength of rocks in compression, it is destroyed, thus there is an increase in the span of exposure and flattening of the line of arch part that in turn leads to the appearance of the roof of the tensile stress with subsequent formation of roof arch.

Carried out analyses yielded the equation of destructive pressure generated by stresses in the mass on the arched working line for homogeneous rocks taking into account current tectonic stresses arising in the rock mass:

$$P_{w.l.} = \pm \frac{r \cdot \tau_0 \cdot \sin \delta}{\sin 2\delta - r^2 \cdot \cos \beta \cdot tg\rho}, \qquad (1)$$

where r – the radius of arched part of mine working, m; τ_o – initial shear resistance, t/m^2; δ – the angle at which the load acting on the working outline, degree; β – angle of rock displacement, degree; ρ – angle of internal friction of rocks, degree.

If the destructive pressure determined by the expression (1) is greater than the normal stress generated by rocks in the mass, the working will be stable when the value of the normal stress in the rock mass will be more than destructive pressure, the working will be subject to deformations on the line under the angle δ. The results of calculations according to expression (1) are shown in Figure 3.

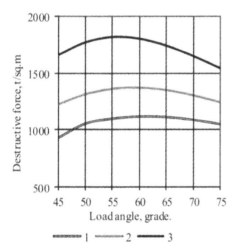

Figure 3. Dependencies of the value of the current stress on the working outline above zone of influence of room works from the corners of the stress and displacement of rocks in a homogeneous rock: 1 – 3 respectively the angle of displacement of rocks 55.60 and 65 degrees.

From the graph shown in Figure 3, it is seen that the limit equilibrium is disturbed by the working line at an angle of 55 to 65 degrees depending on the angle of rock displacement. So at displacement angle of 55 degrees maximum shear force on the working line acts at an angle of 65 degrees, and with increasing angle of displacement up to 65 degrees angle of shear force decreases and acts at an angle up to 55 degrees.

In bedded rocks on the working line inrush formation area is formed, the cause of which is the decrease in strength between beds. The values of rock strength on the contacts between the beds are much smaller, so the destruction of the rock will be at the contact between the rocks. Inrushes border around the perimeter of the arch of mine working is limited by angles θ_{xz1} and θ_{xz2} determined by the formula:

$$\begin{cases} \delta_{1,2} = \dfrac{arcsin\left(\sin \rho + \dfrac{\tau_0}{P} \cos \rho\right) \pm \rho}{2}. \\ 2 \cdot \sin 2\delta \geq 0 \end{cases} \qquad (2)$$

The height of the forming arch corresponds to the new working line, in which acting tensile stress is equal to the rock strength on the uniaxial tensile. Then the height of the arch, after deformation of rocks on the line of mine working is determined by the formula:

$$\Delta h = \frac{a \cdot (1 - \mu)}{\mu} \cdot \left(\frac{1 - 2 \cdot \mu}{1 - \mu} - \frac{\sigma_p}{\sigma_z}\right) - h_v . \qquad (3)$$

where Δh – deformation of the original height of the arch generatric of unstable equilibrium of tensile stress, m; h_v – height of forming arch, m; μ – Poisson ratio; σ_p – tensile strength of rock, t/m^2.

Thus, having determined the zone of maximum acting destructive force on the working line, as well as arching parameters, taking the necessary measures to prevent the destruction of the of mine working. These activities include: installation in a particular area of reinforced fixing, roof bolts installation, the use of injections to prevent inrushes.

4 CONCLUSIONS

The studies found that the use of the offered options of opening in finalizing balance reserves at deeper depths will allow: to reduce capital and operating costs for the maintenance of mine workings; to increase mine productivity with increasing depth of the mine development.

In order to ensure the sustainability of mining in the operation parameters and zone of inrushes on the working line are defined, as well as the location of the most dangerous sections of the possible destruction of the working line.

The necessary measures to prevent the destruction of the mine workings and keeping them in a safe condition are given.

REFERENCE

Korolenko, M., Stupnik, N., Kalinichenko, V., Peregoudov, V., Protasov, V. 2012. *Expanding the resource base underground Kryvbass by involving the extraction of magnetite quartzite.* Dionysus: Kryvyi Rih: 236.

Stupnik, N., Kalinichenko, E., Kalinichenko, V. 2012. *Feasibility study of the usefulness of self-propelled machinery for mining Kryvbass.* Naukovyi Visnyk Natsionalnoho Hirnychoho Universytetu, No 5: 39 – 43.

Stupnik, N., Kalinichenko, V., Kalinichenko, E. 2012. *The economic assessment of the risks of possible violations of geomechanical earth's surface in the fields of mines Kryvbass*. Naukovyi Visnyk Natsionalnoho Hirnychoho Universytetu, No 6: 126 – 130.

Stupnik, N., Kalinichenko, V., Kolosov, V, Pismenniy S., Fedko, M. 2014. *Testing complex-structural magnetite quartzite deposits chamber system design theme*. Metal-lurgical and Mining industry: Scientific and Technical journal, No 2: 89 – 93.

Stupnik, N., Kalinichenko, V., Pismenniy, S. 2013. *Pillars sizing at magnetite quartzites room-work*. Mining of Mineral Deposite. The Netherland: CRC Press/Balkema: 11 – 15.

Increasing of yielding of frame-anchor support steadiness

G. Symanovych & D. Astafiev
National Mining University, Dnipropetrovsk, Ukraine

O. Vivcharenko
Department of Coal Industry of Ukraine, Kyiv, Ukraine

V. Snigur
PJSC "DTEK Pavlohradvuhillia", Pavlohrad, Ukraine

ABSTRACT: Frame-anchor support's construction and mathematics equations for calculation of parameters that conduct high resistance to side loads without increasing of frame's metal-content is given. The conclusions are drawn at the end of the paper.

1 INTRODUCTION

Experience of conducting and supporting of extraction workings with hard mine-geological conditions (Bondarenko 2010) shown that in most cases significant loads on frame support appear from the side of mine workings bottom. In this regard appears plastic deformations of frame props that decreases bearing capacity of support totally.

Analysis of reasons of mine workings destruction and conducted mine research (Kovalevska 2011) allowed to develop rational for hard mine-geological conditions construction of support from special profile with inclined props that forced by anchors with possibility of limited-and-yielding displacements on bed plates.

2 CONSTRUCTION, INSTALLATION AND FUNCTIONING OF SUPPORT

Support includes frame from special profile that consists of props *1*, beam *2*, yielding units *3*, demountable bed plates *4* with anchors *5* and headboards *6* (Figure 1).

Demountable bed plate *4* executed as pressed trapezium-shaped element with bent up side limit platforms *7*, horn *8* and anchor hole *9* (Figure 2).

Frame support is installed in the following way. Bed plates are settled on the working bottom *4*, turned beam up and small foundation to the middle of mine working, where installed props of frame 1 support with further beam fastening *2* by means of yielding unit *3*. The position of prop on bed plate fixes horn *8*.

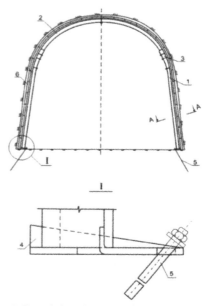

Figure 1. General view of tent-shaped frame-anchor support.

Then through holes *9* anchor bores *5* are drilled, ram to bores binding solution, add rod and connect with bed plates. After that headboard *6* is installed on support contour and backfilled area.

Support works in following way. Props *1* of frame under the influence of moving inside rock walls of mine working load and slide on hard fixed by means of anchor *5* bed plates *4*, contact with internal surfaces of limit platforms and according to growth of side pressure overcome reaction. Then moving to bed plates in hard mode before props blocking between limit platforms and anchor is happened.

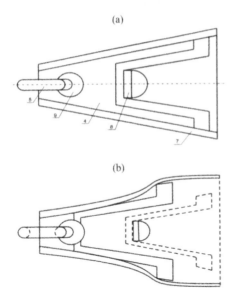

(a)

(b)

Figure 2. General view of bed plate before (a) and after (b) yielding exhaustion with further interaction with support prop in hard mode.

Whereupon props are working in hard mode and fully transmit the load, but this load significantly less than initial because of hard-yielding mode of support work. Additional reaction of prop near foundation increases frame steadiness at all.

3 SUBSTANTIATION OF SUPPORT PARAMETERS

Bearing capacity of unclosed frame support decreases under the influence of bottom rocks heaving that displaced props of support in consequence of rock deformation in mine working's cavity (Bondarenko 2014). At this, props recline against moving rock layer and displaced together with it by means of friction forces. Hence in hinges at the support A and B (Figure 3) appear additional force that acting on frame props and is equal to (in the limit) multiplication of vertical reaction of prop at the side of foot piece on friction coefficient f_{fric} of metal.

According to the stated, computational scheme for uniform distribution of pressure q that differs from well-known schemes for double-hinged frame by means of horizontal reaction in hinges A and B and is equal to $\dfrac{ql}{2} f_{fric}$ is built. At such load distribution (even at insignificant absolute value) in dangerous crosscuts of frame significant moments of resistance which cause plastic deformation with further destruction are acting.

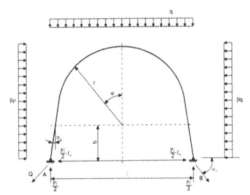

Figure 3. Computational scheme of frame-anchor support.

Frame props in new construction of support are installed incline to vertical on the angle α_1, in places of supporting on rock with angle α_2 for prevention such event. At this, during change of anchor Q reaction value we can lead to minimum acting maximal moments M_{max} of resistance in frame.

Moments of resistance maximum are situated whether in frame arch key ($\varphi = 0$) or in crosscuts with coordinate φ_0 symmetrically in sides of mine working. That is why it is necessary to minimize absolute values M_{max} through the parameter Q and to determine optimal value Q_0 which is corresponded to minimal value of moment of resistance in possible dangerous crosscuts of frame.

Parameter Q_0 is determined at combined equations solution that describe functions $M_{max}(Q)$ for all dangerous crosscuts of frame:

$$Q_0 = q \frac{ll_1}{\cos \alpha_1} \left(C_1 - \sqrt{C_1^2 - C_2} \right), \tag{1}$$

where: $C_1 = \beta h + \dfrac{l}{2} f_{fric} + (1 - \beta)(2h + r),$ (2)

$$C_2 = \left(\beta h + \frac{l}{2} f_{fric} \right)^2 + 2(1 - \beta) \times$$
$$\times \left[\frac{\beta}{2}(h + r)^2 + \frac{Bh^2}{2} - \frac{l^2}{2} + \frac{r^2}{2} + \frac{l^2}{2} f_{fric}(2h + r) \right], (3)$$

where l_1 – distance between frames; l – mine working's width; h – height of frame arch key center; β – ratio of side load on support to vertical one.

Calculations have shown that value of anchor reaction significantly impact on change of maximal moments of resistance along the frame contour. Thus, deviation Q from the value Q_0 on $20-30\%$ increases M_{max} in $2-3$ times and more for which reason it is necessary to use frames with increased bearing capacity (consequently with more metal-content).

Minimal values of maximal moments of resistance $(M_{max})_{min}$ are depend upon set of parameters, including mine working's width l, height of frame arch key center, props angle of inclination to vertical axis of mine working α_2 and coefficient β.

Analysis of deviation dependence $\dfrac{(M_{max})_{min}}{q}$ to support props angle of inclination α_2 shown that with increasing of α_2, ratio $\dfrac{(M_{max})_{min}}{q}$ is decreased (Figure 4). At this, beginning from $\alpha_2 = 10°$ curves are flattened, in consequence of which further increasing α_2 is inappropriate and leads to unjustified increasing of mine working's width. That's why in support TSSS-1 (Tent-shaped Specialized Support) props angle of inclination is equal to $\alpha_2 = 10°$.

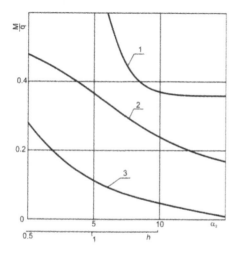

Figure 4. Dependence of ratio $\dfrac{(M_{max})_{min}}{q}$ from props angle of inclination at $1 - \beta = 0.25$; $2 - \beta = 0.5$; $3 - \beta = 0.75$.

Moment of resistance's maximum increases with the growth of parameters l and h (Figure 5). At this, the lowest values of ratio $\dfrac{(M_{max})_{min}}{q}$ is possessed in limits $h = 0.5 - 1$ m. That's why in developed standard sizes of support height of arch key center changes in given interval.

Dependence of moments of resistance form ratio β of sided and vertical load on mine working support is clearly observed. Ratio decreasing $\dfrac{(M_{max})_{min}}{q}$ with the growth of β is caused by means

of anchors in support props are designed to resistance of side loads and bottom heaving.

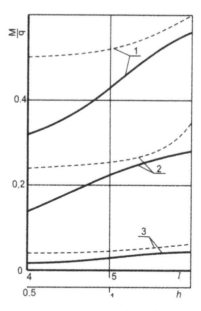

Figure 5. Dependence of ratio $\dfrac{(M_{max})_{min}}{q}$ from mine working width l (- - - -) and height of its arch key h (———) at: $1 - \beta = 0.25$; $2 - \beta = 0.5$; $3 - \beta = 0.75$.

That's why the more such loads in total balance of forces, the better to manage of stress state of support with help of anchor reaction.

Value of anchor reaction Q is caused by frame props deformation. System "frame-anchor" strives to get in balanced condition with minimal internal energy that depends upon stress-strain state. In this case bearing capacity of anchor strives to optimal value Q_0, at which stresses in dangerous crosscuts of arch are minimal. At anchor bearing capacity less than Q_0, optimal deformation mode of system is impossible for realization, because it can break or will slip over the walls of shot and contributes to development of plastic deformations in dangerous crosscut of frame. That's why bearing capacity of anchor should be more or equal to Q_0.

4 CALCULATION OF ANCHOR PARAMETERS

Anchor diameter is determined by the formula:

$$d = 2\sqrt{\dfrac{Q_0}{\pi \sigma_{liq}}}, \tag{4}$$

47

its strengthening length in shot (5):

$$L = L_1 + \frac{d}{4\tau_{fric}} \left(\frac{4Q_0}{\pi d^2} - \frac{\kappa}{F\dfrac{d^2}{4d_s} + cthFL_1 + \dfrac{d^2}{d_s^2} n f_{fric}} \right), \quad (5)$$

where:

$$L_1 = \frac{1}{F} ln \left[\frac{k}{\tau_{fric}} + \sqrt{\left(\frac{k}{\tau_{fric}} \right)^2 - 1} \right], \quad (6)$$

$$F = \frac{2}{d} \sqrt{2E \left[\frac{d_s - d}{d_s G_t} + \frac{d}{d_s G_r} \right]}, \quad (7)$$

$$n = \frac{\mu}{E} \left(\frac{d_s - d}{d_s E_t} + \frac{d}{d_s E_r} + \frac{1}{E} \right)^{-1}, \quad (8)$$

where d_s – shot diameter; σ_{liq} – liquid limit of reinforcing bar; τ_{fric} – tangential stresses of permanent friction; k – adhesion of strengthening layer with rock walls of shot; G_d and G_r – transverse modulus of strengthening layer and rock; E, E_c and E_n – elastic modulus of reinforcing bar material that strengthening layer and rock; μ – coefficient of material transverse deformation of reinforcing bar; f_{fric} – friction coefficient of strengthening layer against rock.

5 CONCLUSIONS

1. Construction of frame-anchor support with increased resistance to side loads is developed for hard mine-geological conditions with predominant side displacements of near-the-contour rocks of mine workings.

2. Decreasing of side load on frame-anchor support is conducted by limit-yielding mode of displacement along the horizontal of frame props that increasing steadiness of mine working at all during minimization of frame metal-content.

REFERENCES

Bondarenko, V., Kovalevs'ka, I., & Fomychov, V. 2012. *Features of carrying out experiment using finite-element method at multivariate calculation of "mine massif – com-bined support" system. Geomechanical processes during underground mining.* The Netherlands: CRC Press/Balkema: 7 – 13.

Bondarenko, V., Kovalevs'ka, I., Svystun, R. & Cherednichenko, Yu. 2013. *Optimal pa-rameters of wall bolts computation in the united bearingsystem of extraction workings frame-bolt sup-port. Mining of Mineral Deposits 2013.* The Netherlands: CRC Press/Balkema: 5 – 9.

Kovalevska, I., Fomichev, V. & Chervatuk, V. 2011. *The problem with increasing metal-content of a development working's combined support.* Technical and Geoinformational Systems in Mining. The Netherland: CRC Press/Balkema: 23 – 29.

Theoretical and Practical Solutions of Mineral Resources Mining – Pivnyak, Bondarenko & Kovalevska (eds)
© 2015 Taylor & Francis Group, London, ISBN: 978-1-138-02883-8

Solution of nonlinear programming problem by Bellman method while optimizing the two-level mining of benches in deep open pits

S. Moldabayev
Kazakh National Research Technical University named after K.I. Satpayev, Almaty, Kazakhstan

B. Rysbaiuly
International Information Technologies University, Almaty, Kazakhstan

ABSTRACT: An algorithm is proposed to optimize the values of current overburden ratio via optimization of overburden and ore volumes by mining stages. Universality of such method provides a solution to a problem of nonlinear optimal control by Bellman dynamic programming method. At each step of constrained optimization the fulfillment of sufficient optimality condition in stationary points of objective function is proved. Thus at each step of iteration the optimal solutions of problem do not leave the range of permissible values. As a result the simplified algorithm of mining operations' schedule optimization has been generated for designing a dynamic pit model.

1 INTRODUCTION

At the current stage of pits development the preference is given to create their dynamic models, which are based on the optimal schedules of mining mode. The best technical and economic parameters are achieved for the operational period of designing while providing their equal annual performance by rock mass at every stage of open pit mining. As a result of executed researches it was found that the dependence between over-burden and mining operations can be reduced through the provision of commensurate development of mining operations between all the benches of working area with minimal current volumes of the first one.

One way to achieve such mining development is to jointly use two new technologies: a two-level mining of benches by crosscut panels with the construction of temporary openings to surface in certain locations of open-pit field (Rakishev & Moldabayev 2012) and the formation of work front of work zones' benches perpendicular to the work front of benches of abrupt loading face (Rakishev & Moldabayev 2014). The predesign studies showed that in this case the conditions for effective use of high-power excavating-automobile complexes to great depths are created while developing the steep-dipping deposits (Moldabayev et. al. 2012).

The leading world open pits achieve the best technical and economic parameters due to high performance of mining-and-transport equipment ($350 - 380$

thous. m^3 per 1 m^3 of excavator dipper). Significant thickness of overburden in Lomonosov iron ore deposit ($100 - 130$ m) made it difficult to choose the most acceptable variant of the primary open pit construction. At priority area of Lomonosov iron ore deposit (Figure 1) the implementation of proposed technologies will reduce the average overburden ratio by $5.5 - 11.1\%$, depending on the variant of tracing.

Due to reduction of open pit walls quarry in the limiting position the ore reserves were increased accordingly by $35.6 - 42$ mln. tons, even with a decrease of overburden stripping volumes in closing contour by $4.14 - 19.52$ mln. m^3.

According to two-level technology of benches mining by crosscut panels, the sinking of subsequent initial cuts on underlying horizons is different. Their location is determined by the width of panels at each stage of mining, as well as on the side of hanging and lying walls of ore deposit (Rakishev & Moldabayev 2014), as initially all top parts of benches, and then all lower parts of benches are mined simultaneously.

Then, the stabilization of production capacity with a decrease of mining is achieved at minimal current overburden mining through uniform distribution of accessed deposits reserves at all pit walls. The re-searches carried out in this direction, showed the necessity to create a universal technique for optimal schedule of mining mode creation. The possibility of its creation can be traced on pit cross-section considering the specified requirements to pits development.

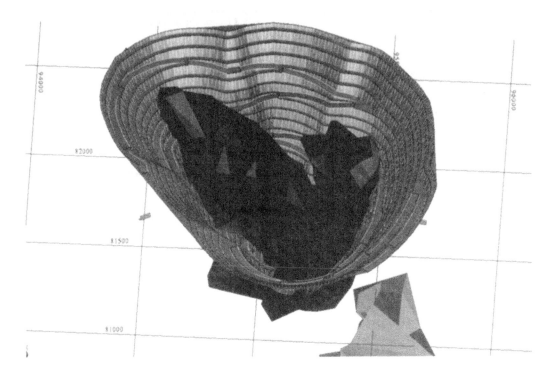

Figure 1. A fragment of top-priority pit dynamic model during development of Central Lomonosov deposit area.

2 OBJECT OF RESEARCH DESCRIPTION

It can be assumed that the cross-section of a pit is a trapezoid ADFE used to simplify the presentation of created technique of mining operations optimization (Figure 2).

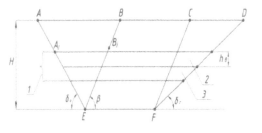

Figure 2. A cross-section of a pit with the contour of steep-dipping deposit: 1 – a horizon under the number t; 2 – first subhorizon under the number $j = 2t - 1$; 3 – second subhorizon under the number $j = 2t$.

Immediately we note that the cross-section type has no effect on the technique. The triangles ABE and CDF indicate the area of rock, and a parallelogram BCFE – an ore deposit.

The volume of mined rocks at t stage of mining we denote as $u(t)$, and the volume of mined ore at t stage of mining we denote as $v(t)$. Taking into account the

introduced notations an overburden ratio is determined by the formula: $k(t) = \dfrac{u(t)}{v(t)}$, $t = 1, 2 \ldots n$.

The main goal of our research is to minimize the stripping ratio values $k(t)$. This study (Moldabayev et. al. 2013) shows the distribution of overburden and mining operations on horizons and subhorizons depending on the stage of mining. For this distribution, taking as a variable the optimization of work area width, a mathematical model of optimization problem was constructed. As a result we obtained the nonlinear programming problem, some numerical results of which are given in the study (Moldabayev et. al. 2014). In this study, we solve a global problem, namely the optimal distribution of rock and ore at each stage of mine development. Two-dimensional optimization problem is solved by dynamic programming method. As a control, we took the volumes of rock and ore at each stage of mining.

3 OPTIMAL CONTROL PROBLEM

It is known that at positive $u(t)$ and $v(t)$ the minimum values of functions $k(t)$ and $k^2(t)$ are achieved at the same points in domain of functions $u(t)$ and $v(t)$. Therefore, the following task is solved:

$$J(u,v) = \sum_{t=1}^{n} \frac{u^2(t)}{v^2(t)} \rightarrow min, \qquad (1)$$

$$(t) = x(t-1) + u(t), \quad y(t) = y(t-1) + v(t),$$

$$t = 1, 2, \dots n, \qquad (2)$$

$$x(n) = S_w, \quad y(n) = S_r, \qquad (3)$$

$$x(t) \in [x(0), S_w] = X, \quad y(t) \in [y(0), S_r] = Y, \qquad (4)$$

$$u(t) \in [0, S_w - x(0)] = U,$$

$$v(t) \in [0, S_r - y(0)] = V, \qquad (5)$$

Here $x(t)$ and $y(x)$ – are respectively the total volumes of extracted rock and ore up to the stage tinclusively; through X, Y and U, V denoted a variety of allowable values respectively to conditions $x(t)$, $y(t)$ and controls $u(t)$, $v(t)$. Through $x(0)$ and $y(0)$ the volumes of extracted rock and ore at zero stage of mining are accordingly denoted.

The optimal solution of the system (1) – (4) is found in such a way where such inclusions will be following:

$$x^*(t) \in X, \quad y^*(t) \in Y, \quad u^*(t) \in U, \quad v^*(t) \in V,$$

$$t = 1,2,3\dots n, \qquad (6)$$

As the function $J(u,v)$ is a separable, then we apply the method of dynamic programming, developed by Bellman for the solution of optimization problem (1) – (6). As a result of these studies we obtained a solution of problem (1) – (6) specified below.

The optimum values of volumes of overburden and ore at zero stage are:

$$x^*(0) = x(0), \quad y^*(0) = y(0), \qquad (7)$$

Sequentially considering that $t = 0, 1, 2, \dots, n-1$, the optimum values of controls and conditions' functions are calculated by the following formulas:

$$u^*(t+1) = \frac{S_w - x^*(t)}{n-t}, \quad v^*(t+1) = \frac{S_r - y^*(t)}{n-t}, \qquad (8)$$

$$x^*(t+1) = x^*(t) + u^*(t+1),$$

$$y^*(n-t+1) = y^*(t) + v^*(t+1), \qquad (9)$$

The optimal value of the current overburden ratio is calculated by the formula:

$$k^*(t) = \frac{x^*(t)}{y^*(t)}, \quad t = 1,2\dots n-1, n, \qquad (10)$$

4 NUMERICAL CALCULATIONS

As an example, we took a cross-section of a pit, shown in Figure 2. The initial data is given in Table 1.

Table 1. Initial data for calculations.

δ_1	δ_r	β	H	M	$X(0)$	$Y(0)$	n
49°	33°	63°	400	600	75	0	10

The calculations show that the average overburden ratio is equal to $k_{sr}(t) = 0.370709$, and the current overburden ratio $k(t)$ in our case is a decreasing variable. This dependence is shown in Figure 3.

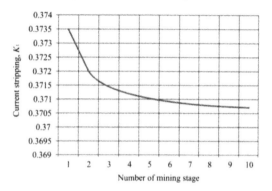

Figure 3. Diagram of current overburden ratio change by stages of mining.

The comparison of numerical data shows that the current overburden ratio is greater than the average overburden ratio. If we increase the volume of overburden at zero stage of mining $x(0)$, the optimal values of current overburden ratio become different, but the monotonous nature of functions $k(t)$ remains the same. At last stage of pit development the optimal value of current overburden ratio is equal to 0.726557. This value is equal to ratio S_w/S_y. This equality is necessary to control the calculation algorithm, because it allows comparing the ratio of total optimal volumes of rock and ore with an average overburden ratio. A comparative dependency graph is shown in Figure 4.

The researches of changes of rock and ore volumes, depending on the stage of mining have been separately conducted. A comparative graph is shown in Figure 5. A linear dependence of optimal value of rock and ore is observed, depending on the stage of mining.

If we increase an inclination angle of pit wall on hanging wall of deposit, than it will directly effect on overburden ratio. For example, if instead of $\delta_1 = 49°$ we will take $\delta_1 = 50°$, the decrease of stripping ratio will be observed. The change of optimal value of the current overburden ratio is shown in Figure 6.

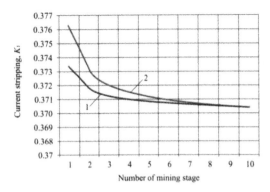

Figure 4. A comparative graph of the current optimal overburden ratio at different values of $x(0)$: $1 - x(0) = 75$; $2 - x(0) = 150$.

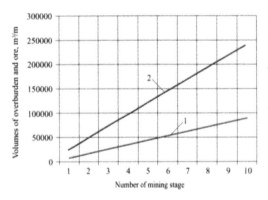

Figure 5. The average overburden ratio at different inclination angles of pit wall on hanging wall of deposit: $1 - y(t)$; $2 - x(t)$.

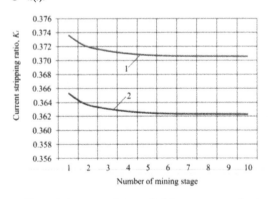

Figure 6. A comparative graph f the current overburden ratio at different inclination angles of pit wall on hanging wall of deposit: $1 - Bl = 45$; $2 - Bl = 49$.

A graph shown in Figure 6 suggests that the graphics of the current overburden ratio at different inclination angles of pit wall on hanging wall of deposit are the equipotential lines. A change of pit depth H also influences on the optimum values of overburden ratio. A comparative graph at values $H = 400$ and $H = 420$ again show s a couple of equipotential lines.

Now we will change the width of ore body M. Suppose that $M = 650$. In this case, the value of average overburden ratio significantly decreases and becomes equal to $k_{sr} = 0.351232$. When $M = 600$, this ratio is equal to $k_{sr} = 0.370709$. The absolute change is $\Delta = 0.019477$, and the relative change is

$$\delta = \frac{0.019477}{0.370709} \cdot 100 = 5.25 \text{ \%}.$$ The relative change of

ore body width is $\delta = \dfrac{50}{600} \cdot 100 = 8.3 \text{ \%}$.

The Tables $2 - 4$ have been filled for studies' presentation.

Table 2. The deviation during determination of $k(t)$ with a change of H.

Deposit depth	Initial value	Changed value	Relative deviation	Deviation $k(t)$
H	400	420	5 %	2.64 %

Table 3. The deviation during determination of $k(t)$ with a change of δ_1.

Inclination angle of hanging wall	Initial value	Changed value	Relative deviation	Deviation $k(t)$
δ_1	49^0	50^0	2 %	2.24 %

Table 4. The deviation during determination of $k(t)$ with a change of V.

Ore body width	Initial value	Changed value	Relative deviation	Deviation $k(t)$
M	600	650	8.3 %	5.25 %

A comparison of Tables $2 - 4$ points out that the change of inclination angle of pit wall on hanging wall of deposit has a strong effect on the overburden ratio $k(t)$ compared with the changes of pit depth and ore body width. An inclination angle of pit wall on lying wall of deposit has a similar effect.

5 CONCLUSIONS

As we know, the current overburden ratio is determined by the formula $k(t) = \dfrac{x(t)}{y(t)}$. Here $u(t)$ – a volume of extracted rock at t stage of mining, and $v(t)$ – a volume of extracted ore at tstage of mining. The main goal of our work was to determine the optimal values of $k^*(t)$ overburden ratio. By using the well-known fact that the optimal values of functions $k(t)$ and $k^2(t)$ are achieved in the same area, we choose the following target function:

$J(u,v) = \sum_{t=1}^{n} \dfrac{u^2(t)}{v^2(t)}$. As a result we were able to

generate a nonlinear optimal control problem (1) – (5). As a control we selected the volume of extracted rock and ore mining at each stage of mining.

On the one hand, the separability of objective function $J(u, v)$ and on the other –' a suitable selection of function type $j(u, v)$ allowed to successfully apply the algorithm of solving the optimal control problem by Bellman dynamic programming method. We proved the fulfillment of sufficient optimality condition in stationary points of objective function at each step of constrained optimization. Another important factor is that we proved that the optimal solution of problem (1) – (6) do not leave the range of permissible values at each iteration step. Finally, we got a solution of the problem (1) – (6) that is convenient for practical calculations and shown of closed form in the formulas (7) – (10).

The numerical results show that the optimal current overburden ratio $k^*(t)$ for investigated case – is a variable decreasing function (Figures 3 and 4). The optimum values of total volume of rock and ore are the linear functions, depending on the stage of mining (Figure 5). When changing the parameters of δ_1, H pit and More deposit, the monotonous nature of optimal value of overburden ratio is preserved, but the comparative graphs become the equipotential lines (Figure 6).

In order to assess the response of optimal overburden ratio to input parameters' change, we drew up the Tables 2 – 4. Their analysis shows a greater degree of the optimal current overburden ratio k_{sr} response to changes of inclination angles of pit walls on hanging and lying walls of deposit, rather than to changes of linear dimensions H and M.

REFERENCES

Moldabayev, S., Rysbaiuly, B. & Sultanbekova, Zh. 2013. *Justification of Countours Belonging to the Stages of Mining Steeply Dipping Deposits Using the Solution of the Problem of Nonlinear programming*. Mine Planning and Equipment Selection. Proceedings of the 22nd MPES Conference. Dresden: Vol. 1: 125 – 132.

Moldabayev, S.K., Sultanbekova, Zh.Zh., Aben, Ye. & Gumenik, I.L. 2014. *Equalization of traffic flows of benches in the working areabenches with excavator-truck complexes*. Progressive Technologies of Coal, Coalbed Methane, and Ores Mining. CRC Press/Balkema: 327 – 332.

Moldabayev, S.K., Sultanbekova, Zh.Zh., Aben, Ye. & Rysbaiuly, B. 2014. *Ways to achieve the optimal schedule of the mining mode of double subbench mining*. Progressive Technologies of Coal, Coalbed Methane, and Ores Mining. CRC Press/Balkema: 35 – 40.

Rakishev, B.R. & Moldabayev, S.K. 2012. *The method of open development of inclined and steep mineral deposits*. The RK innovative patent for invention No 26485 under request No 2012/0049.1 dated 11.01.2012. Publ. on 14.12.2012, Bull. No 12.

Rakishev, B.R. & Moldabayev, S.K. 2014. *The method of open development of steep mineral deposits with the transition to an internal stacking*. The RK innovative patent for invention No 29038 under request No 2013/1337.1 dated 11.10.2013. Publ. on 15.10.2014, Bull. No 10.

Rakishev, B.R. & Moldabayev, S.K. 2014. *The method of open development of steep mineral deposits*. The RK innovative patent for invention No 29037 under request No 2013/1335.1 dated 11.10.2013. Publ. on 15.102014, Bull. No 10.

Theoretical and Practical Solutions of Mineral Resources Mining – Pivnyak, Bondarenko & Kovalevska (eds)
© 2015 Taylor & Francis Group, London, ISBN: 978-1-138-02883-8

Application of the Pert method in parameters matching within the evaluator group module

Zh. Sultanbekova
Kazakh National Research Technical University named after K.I. Satpayev, Almaty, Kazakhstan

ABSTRACT: Possibility to use the three assessment method or so called the Pert method which is based on defining pessimistic and probable assessment of project fulfillment time and applied widely in the project management theory is being observed for the first time. This method in this work is used for adjusting the independent computing blocks parameters. As a result of this sufficiently reliable outcomes were achieved if to compare with the reference values.

1 INTRODUCTION

Two approaches for adjusting parameters were developed to process results of computation made by the evaluators group in order to get the most approximate base value. One of these approaches: the Pert Method (Project Evaluation review technique) was applied which is used in the project management theory for planning on a basis of possible network models. (Evgenev 2006). In some publication this method is known as a three evaluation method (Kuzmin 1991). Parameters matching with the help of three evaluation method is studied in this work and assessment of compu-tation results with application of traditional approach and the approach based on distributed information processing followed by matching the computation results is provided.

2 MATHEMATIC MODEL OF THE MATCHING METHOD IS DESCRIBED ON STEP-BY-STEP BASIS.

Step 1. Define the mean arithmetic value of three shift-based production rate values as per the following formula:

$$P_1 = \frac{Q_{sh1} + Q_{sh2} + Q_{sh3}}{3}. \tag{1}$$

Step 2. Define an absolute error of shift-based production rate calculated by independent valuators against the reference value of the shift-based production rate:

$$\Delta_1 = \frac{|Q_{sh\_et} - Q_{sh1}|}{100}, \quad \Delta_2 = \frac{|Q_{sh\_et} - Q_{sh2}|}{100},$$

$$\Delta_2 = \frac{|Q_{sh\_et} - Q_{sh3}|}{100}, \tag{2}$$

Step 3. Reference mean error against the reference value is to be calculated similarly:

$$\Delta_2 = \frac{|Q_{sh\_et} - P_1|}{100}. \tag{3}$$

Our purpose is to reduce the errors between reference values and estimate indicators received from three independent evaluators.

Step 4. The Pert Method (Project Evaluation review technique), plans evaluation and review method is used for evaluation of project length and has the distribution β. For each work within the network model of the project three evaluations of performance duration should be made: the most probable performance time M; optimistic time evaluation O; pessimistic time evaluation P as shown in Figure 1.

Since the experimental values distribution function in our work has the distribution β the existing Pert Method can be updated and an idea of this method can be used as evaluation of the dump truck shift-based production rate.

For optimistic evaluation the biggest value out of three values for pessimistic – the less value and for the most probable evaluation – the value which is closest to the reference indicator, should be taken.

As a result the following formula is generated:

$$E = \frac{Q_{sh1} + 4 \cdot Q_{sh2} + Q_{sh3}}{6}, \tag{4}$$

where Q_{sh1} – probable value of the parameter under study; Q_{sh2} – optimistic value of the parameter under study; Q_{sh3} – pessimistic value of the parameter under study.

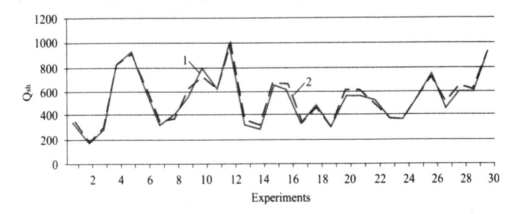

Figure 1. Results of the first matching approach.

Similarly to formulas (1) and (2) define an error in the indicator of adjusted shift-based production rate value Δ_4 received.

Step 5. Define the expected result for each work as follows:

$$R = \frac{(T_p - T_o) \cdot 2}{6},$$ (5)

where T_p – real parameter value; T_0 – expected work result.

This evaluation of dispersion characterizes level of uncertainty associated with the work result evaluation process. The higher a dispersion, the higher the uncertainty.

To define the dispersion apply the formula (2), where the E is resulted from the formula (4) and used in a capacity of T_p and therefore Q_{sh_et} is to be considered as T_o, then we get the following:

$$R = \frac{|E - Q_{sh\_et}| \cdot 2}{6}.$$ (6)

Step 6. The value resulted is to be used to reduce the matched parameter of the shift-based production rate E as per the following formula (3):

$$E_{11} = E - R.$$ (7)

By implementing this algorithm we get a Table with the matched parameters. Relative error of matching results against the reference value can be defined.

Let's test the developed parameter matching methods and show the results in diagrams. One can see a significant difference in results on the diagrams and one can conclude that the second matching algorithm is the most successful. Therefore the second matching method is used as a base for development of algorithm and program module. As a result of parameter matching by two approaches we get two value options relevantly. To see a full picture of matching results let's show them in a form of state series (Figures 1 and 2).

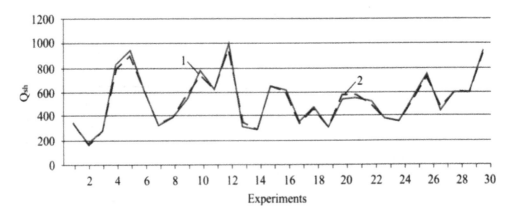

Figure 2. Results of the second matching approach.

On a basis of the first matching option results on 30 experimental tests an absolute error is equal to 12.72 nd for the second option – 5.17. The average value of absolute deviations against the reference value shows that the second matching method results are the most appropriate if to compare with the first matching approach. On a basis of the abovementioned method (second method) a parameter matching algorithm the block diagram of which is provided on the Figure 3 and implemented in MS Visual Basic programming, was developed.

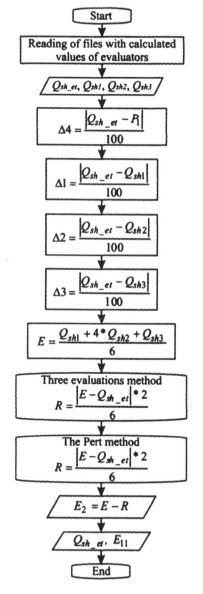

Figure 3. Block diagram of the parameter matching algorithm.

3 CONCLUSIONS

A form view with friendly interface allowing to enter resulted values received from three computing blocks and to get the matched value from different estimate indicators easily was developed to facilitate the coordinator's work in the program.

The parameter matching method may be used in a capacity of the computing interval matching method for the evaluators group model members of in case if the whole process of opencast mining design is executed with application of the distributed information processing method since for parameter matching an idea of plans evaluation and review method (Pert) where works duration has the distribution β, was used.

In addition in calculation of velocity of travel of dump truck along the roads three options defining the desired indicator are used. If to match the resulted values of all three options with the help of the proposed method we get reliable indicators identical to basic indicators and this proves a practical value of the matching method developed. The developed method may be used in the process of solving the other technological parameters, i.e. for matching parameters of the delivered and removed rock mass volumes to sorting depots or shipping sheds but within the frame of this document it is impossible to cover the solving process of such global task.

The method developed for matching results of independent computations with use of mathematic simulation methods and experimental data analysis is implemented in the form of complex and structural programs with database providing with the most reliable results required for the project.

REFERENCES

Evgenev, G.B. 2006. *Systemiology of Engineering Knowledge.* Information Science in Technical Institutes. Moscow: Publishing House of the MHTU after N.E. Bauman.

Kuzmin, I.A., Putilov, V.A. & Filchakov, V.V. 1991. *Distributed Information Processing in Scientific Researches.* Saint Petersburg: Publish House-PGU.

Theoretical and Practical Solutions of Mineral Resources Mining – Pivnyak, Bondarenko & Kovalevska (eds)
© 2015 Taylor & Francis Group, London, ISBN: 978-1-138-02883-8

On the issue of analytical and empirical criteria application for rock failure assessment

O. Shashenko & O. Kovrov
National Mining University, Dnipropetrovsk, Ukraine

B. Rakishev & A. Mashanov
Kazakh National Research Technical University named after K.I. Satpayev, Almaty, Kazakhstan

ABSTRACT: A comparative analysis of analytical and empirical criteria for rock failure is carried out. A new approach to the development of unified failure criterion on the basis of common values that characterizes complex stress-strain state of the rock mass is proposed. New mathematical expressions of empirical characteristics for cohesion and friction angle, as input data of Mohr-Coulomb failure criterion, based on analytical theories of rock strength.

1 INTRODUCTION

Selection and application of appropriate rock failure criteria is a key step when performing the analysis of geotechnical systems stability. Via numerical modeling the researcher can investigate geomechanical processes occurring in soft rocks (or soils) in situ. The better this accepted theory of rock strength reflects the nature and character of rock failure, the closer this theory to real object.

Various failure criteria are commonly used to describe geomechanical processes inside the rock massif. However all of them use different sets of input data and physical and mechanical characteristics of rocks needed to describe the geomechanical processes in the rock massif. For example, Mohr-Coulomb, Drucker-Prager, Bishop failure criteria describe well the behavior of soft rocks or soils, so traditionally applied for geomechanical assessment of slope stability, bulk massifs, open pits, dams, ground geotechnical systems, etc. Some failure criteria, for example Hoek-Brown, Parchevsky-Shashenko are widely used in the practice of numerical simulation to describe the stress-strain state in hard rocks.

The criteria with various physical and mechanical properties as input parameters are used to assess the strength of soft rocks. Thus, the Mohr-Coulomb criterion is based on cohesion and the angle of internal friction as main characteristics, and Hoek-Brown criterion uses dimensionless quantities, that take into consideration genesis, structure and strength of the rock mass. These criteria are essentially empirical and despite the wide range of practical applications they contain a certain degree of subjectivity.

For example, the angle of internal friction in the Mohr-Coulomb criterion is a complex and mostly contradictory characteristics, which doesn't have any conclusive definition so far. Ambiguous interpretation of physical nature of this value leads to the fact that it is very unstable even for a certain lithological rock differences. Also scientists have different opinions concerning the variations of this value depending on rock depth (Shashenko et al. 2008).

The procedure of cohesion value determination from the straight line of the Mohr envelope based on the results of experimental shear tests of soils and soft rocks looks also controversial. Here the accuracy of the calculations depends on the point where the straight tangent line intersects the curved line of the Mohr envelope that gives the main physical and mechanical properties of soft rocks. Such approach also has well-known an element of subjectivity because of uncertainties of this intersection point location.

The Hoek-Brown failure criterion utilizes a range of input parameters such as compressive strength, geological strength index and design coefficients for the intact rock massif, which are determined by approximately tabular data. The objective to take into account as much as possible features of considered rock mass makes empirical relationships much more cumbersome and less accurate, taking into account the spread of values for each of the input parameters (Seituly et al. 2014).

In the absence of a universal theory of strength for solids, the logical questions arise: Which one from the existing theories of strength the most accurately describe rock failure process? Whether these theories are sufficiently suitable for practical application?

The objective of this paper is to justify new mathematical expressions for empirical characteristics of cohesion and angle of internal friction, as main parameters of Mohr-Coulomb failure criterion, on the basis of classical analytical theories of rock strength.

2 ANALYTICAL CRITERION OF ROCK STRENGTH

A range of modern theories of strength were obtained on the basis of the model of solid bodies failure from the general functional dependence (Ponomarev et al. 1956), which unites into one relationship stress intensity and spherical tensor components:

$$\sigma_i^2 + aJ^2 + bJ = c , \tag{1}$$

where a, b, c – some parameters determined from the tests at the elementary states of stress:

$$\sigma_i = \frac{\sqrt{2}}{2}\sqrt{(\sigma_1-\sigma_2)^2+(\sigma_2-\sigma_3)^2+(\sigma_3-\sigma_1)^2} , \tag{2}$$

$$J = \sigma_1 + \sigma_2 + \sigma_3 . \tag{3}$$

Here σ_1, σ_2, σ_3 – the principal stresses.

Following the Mohr hypothesis, let us assume that the strength of the material depends almost only on those terms of expressions (2) and (3) which determine the difference and the sum of the largest and smallest stress components. Then from the equation (1) and by $a = 0$, we obtain:

$$\frac{1}{2}(\sigma_1-\sigma_3)^2 + b(\sigma_1+\sigma_3) = c . \tag{4}$$

Parameters b and c from equation (4) can be obtained via experimental tests of rocks at the simplest state of stress. Thus, by uniaxial compression in the limiting state, we obtain:

$$\frac{1}{2}R_c^2 + bR_c = c , \tag{5}$$

by uniaxial tension:

$$\frac{1}{2}R_p^2 + bR_p = c . \tag{6}$$

Resolving equations (5) and (6) simultaneously, we obtain:

$$b = -\frac{1}{2}R_c^2(1-\psi), \quad c = \frac{1}{2}R_c^2\psi , \tag{7}$$

where $\psi = R_p/R_c$.

Substituting the values of the parameters from (7) in the equation (4), we obtain the following condition for rock strength:

$$(\sigma_1-\sigma_3)^2 - R_c^2\psi - (1-\psi)R_c(\sigma_1+\sigma_3) = 0 . \tag{8}$$

or for general case of stress-strain state:

$$(\sigma_x-\sigma_y)^2 + 4\tau_{xy}^2 - R_c^2\psi - (1-\psi)R_c(\sigma_x+\sigma_y) = 0 . \tag{9}$$

From the equation (8) we obtain:

$$\sigma_1 - \sigma_3 = 2k , \tag{10}$$

where:

$$k = 0.5\sqrt{R_c^2\psi + (1-\psi)R_c(\sigma_1+\sigma_3)} . \tag{11}$$

It is remarkable that under axisymmetric stress distribution $\sigma_1 + \sigma_3 = \sigma_r + \sigma_\theta = $ const.

It follows therefrom that equation (10) is essentially represents Tresca-Saint Venant state of strength. In general case of the stressed state, the right term of (10) depends on the values of the stress components.

For materials that equally resist compression and tension, $\psi = 1$. In this case, from the expression (10), we obtain the Coulomb strength theory.

Let us denote the maximum shear and compressive stresses as $\tau = (\sigma_1 - \sigma_3)/2$ and $\sigma = (\sigma_1 + \sigma_3)/2$, then the expression (8) with consideration of adopted denotations will assume the form:

$$4\tau^2 - 2\sigma(1-\psi)R_c - R_c^2\psi . \tag{12}$$

Equation (12) in the coordinate system «$\tau - \sigma$» represents parabola equation.

Thus, the resulting strength state meets the requirement of Drucker's postulate (Drukker 1957) and corresponds to modern conceptions about the nature of solid bodies the failure (Pisarenko & Lebedev 1969).

Based on the relationship (1) and $a = 0$, the theory of strength of P.P. Balandin (Balandin 1937) can be obtained. Its analytical criterion for the limiting state has the following form:

$$R_c\frac{1-\psi}{2}(\sigma_1+\sigma_2+\sigma_3)+\frac{1}{2}\sqrt{(1-\psi)^2(\sigma_1+\sigma_2+\sigma_3)^2+2\psi[(\sigma_1-\sigma_2)^2+(\sigma_2-\sigma_3)^2+(\sigma_3-\sigma_1)^2]}. \tag{13}$$

Expression (8) can be reduced to the same form:

$$R_c = \frac{1-\psi}{2}(\sigma_1+\sigma_2)+\frac{1}{2}\sqrt{(1-\psi)^2(\sigma_1+\sigma_3)^2+2\psi(\sigma_1-\sigma_2)^2} . \tag{14}$$

From the formula (13) and $\sigma_2 = 0$ we obtain the strength criterion for plane stress state:

$$R_c = \frac{1-\psi}{2}(\sigma_1 + \sigma_3) + \frac{1}{2}\sqrt{(1-\psi)^2(\sigma_1 + \sigma_3)^2 + 2\psi\left[\sigma_1^2 + \sigma_3^2 + (\sigma_3 - \sigma_1)^2\right]}. \tag{15}$$

Comparison of expressions (14) and (15) shows that these are different criteria, and the difference between them increases as far as the plastic properties of the material grow ($\psi \to 1$). Both criteria describe the behavior of brittle materials ($\psi \to 0$) equally and fairly well (Ponomarev et al. 1956). In the same analytical studies of elastoplastic state the expression (8) is more convenient.

The value of ψ for the most of rocks seldom exceeds 0.1…0.2. As the analysis of the formula (8) shows, ψ can be taken equal to zero without violation of sufficient accuracy in the wide range of compressive stresses. This phenomenon is clearly presented in Figure 1, where the limit curves for different values of ψ are shown in a dimensionless coordinate system. Limit line equation for envelope line is obtained from equation (12) by dividing of all its terms to R_c.

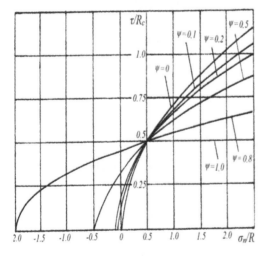

Figure 1. Envelope lines of the limit Mohr's circles for materials with different degree of plasticity.

By $\psi = 0$, we obtain the following particular case of the theory of strength from the expression (8):

$$\sigma_1 - \sigma_3 = \sqrt{R_c(\sigma_1 + \sigma_3)}. \tag{16}$$

Let us consider the behavior of the curves in the field of tension (Figure 1).

If the value of $\psi < 0.5$, the graphs cross the X-axis at a point located from the origin at a distance of less than ψ, i.e. less than R_c. The point of intersection shows equal biaxial (net) tension. Consequently, the strength of materials, which because of structural features approximate to brittle bodies, is de-

termined by the stress of each of the components smaller than the value of uniaxial tensile strength. Thus, such materials resist less to biaxial net tension than uniaxial tension. Such an interesting feature for brittle rocks is described in (Bondarik 1974), but it requires experimental verification.

If the value of $\psi > 0.5$, limit curves intersects the X-axis at a point located at a distance from the origin that is more than ψ, i.e. more than R_p value. Thus, increasing the resistance of plastic properties to the net biaxial tension of the material is more than resistance to uniaxial tension. This phenomenon was confirmed by well-known experiments of G.V. Uzhik with ductile metals (Uzhik 1963).

Similar behavior of the curves in the range of tensile forces, in application to rocks, predicted and substantiated M.M. Protodiakonov (Protodiakonov 1962). From the standpoint of resistance to external forces, water-saturated clays behave as plastic metals: the diagram of deformability explicitly shows the yield plateau. At insignificant value of uniaxial compression strength, clays have almost linear dependence of "$\sigma - \varepsilon$".

Failure of brittle materials is well described by the theory of Griffith, which according to Murrell (Murrell 1963), can be presented in the coordinate system "$\tau - \sigma$" by the following equation:

$$4\tau^2 - 2R_c\sigma - 0.25R_c^2 = 0. \tag{17}$$

Let us assume $\psi = 1:8$ in the formula (12) according to Griffith's theory and compare it with (17). It appears that analytical expressions of two strength theories, derived from different ideas about the nature of physical failure, are almost identical.

Strength condition (8) with $\psi = 1$ has been tested by Baushinger who showed that this formula well describes the process of plastic materials destruction. Previously similar experiments were carried out by Tresca (Frenkel 1954).

Figure 2 shows the theoretical curve obtained from the equation (8) in dimensionless coordinates ($X = \sigma_1/R_c$; $Y = \sigma_3/R_c$) and results of rocks testing obtained by A.N. Stavrogin (Stavrogin & Protosenya 1979).

Despite some scattering experimental points at the graph, that is inevitable when tested such structurally heterogeneous materials as rocks and concretes, it is clear that proposed strength condition (8) describes quite satisfactory the process of the material destruction in the state of three-dimensional compression.

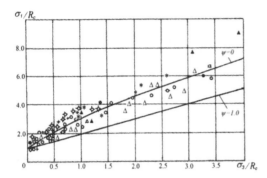

Figure 2. Comparison of the analytical criterion (8) with the results of testing rocks and concretes: ○ – limestone; Δ – argillite (mudstone); ▲ – talc chlorite; -○- – marble 1, -○- – marble 2; + – diabase; * – diorite; ◻ – siltstone D-19; ● – concrete.

Stress components included in the strength condition belong to the limiting curve that separates elastic and plastic range. Therefore, there must be simultaneously performed basic relations and both the theory of elasticity and theory of plasticity. In the case of axisymmetric problem (for example, underground mine working with round cross-section in the equal-component field of external loads) stresses are easier to consider in the polar coordinate system. In this case, the expression (16) assumes the following form:

$$\sigma_\theta - \sigma_r = \sqrt{R_c^2 \psi + (1-\psi)(\sigma_\theta + \sigma_r)R_c} , \qquad (18)$$

where σ_θ and σ_r – tangential and radial stress components respectively. The sum of elastic stresses on the limit curve in this case is an invariant, numerically equal to double external load according to Lame task (Timoshenko & Goodyear 1975):

$$\sigma_\theta + \sigma_r = 2\gamma H , \qquad (19)$$

where γ – volume weight of rocks, H – the depth from the earth's surface.

According to equation (19) the criterion (18) assumes the form:

$$\sigma_\theta - \sigma_r = \sqrt{R_c^2 \psi + 2(1-\psi)\gamma H R_c} = 2k . \qquad (20)$$

Limit depth, at which the failure process of the contour rock mass begins, is determined from (20) at $\sigma_r = 0$:

$$H_{lim} = R_c / 2\gamma , \qquad (21)$$

The formula (20) can be rewritten according to (21) in the following way:

$$\sigma_\theta - \sigma_r = R_c(r)\sqrt{\psi + (1-\psi)\frac{H}{H_{lim}}} . \qquad (22)$$

3 MOHR-COULOMB CRITERION

Mohr-Coulomb failure criterion with a straight envelope line of principal stress circles is the most commonly used for analytic studies of elastic-plastic state (Stavrogin & Protosenya 1979, Shashenko et al. 2001). Its wide application is convenient because it always provides a closed-form solution. Strength characteristics in this criterion, based on the straight envelope line of limiting circles of principal stresses, are the angle of internal friction and cohesion or uniaxial compression strength. In Rock Mechanics, which usually studies hard rocks, angle of internal friction as the strength characteristic was transferred from Soil Mechanics, which utilizes the linear Mohr-Coulomb relationship at assessment of limit states, and have the form:

$$\tau = c + \sigma_n \tan\rho . \qquad (23)$$

Here τ and σ_n are shear stress and normal stress respectively (Shashenko et al. 2008, 2001).

Y.M. Lieberman, studying elastic-plastic state of the rock mass in the vicinity of mine workings, noticed that the straight envelope line must be tangent to the curved line (Lieberman 1969). The accuracy of precise calculation of cohesion and friction angle values depends on determination of the point of tangency. In this interpretation, the angle of internal friction does not represent mechanical constant of the material but geometrical parameter of the straight line in the specified coordinate system.

Let us show how using such approach proposed by Y.M. Lieberman, the transition from curved to straight envelope can be accomplished without altering physical laws of rock failure (Figure 3).

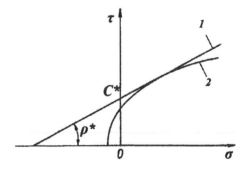

Figure 3. Transition from the curved envelope line of limit Mohr's circles to straight line: 1 – straight line; 2 – curved line.

To do this, the formula (12) should be rewritten as follows:

$$\tau = 0.5\sqrt{R_c^2 \psi + 2R_c(1-\psi)\sigma} . \qquad (24)$$

On the assumption of the fact that the straight envelope line represents a tangent line to real curve of failure, let us differentiate the expression (23) on σ to obtain the formula for determining the angle of internal friction ρ taking into consideration equation (19). This expression serves as the link between linear (23) and nonlinear (24) envelope lines:

$$\rho^* = \frac{1-\psi}{2\sqrt{\psi + 2(1-\psi)\dfrac{\gamma H}{R_c k_c}}} . \qquad (25)$$

Cohesion value c^* is determined by the following equation (Figure 3):

$$c^* = R_c k_c \frac{1 - sin\,\rho^*}{2\,cos\,\rho^*} , \qquad (26)$$

where k_c – the coefficient of structural and mechanical weakening of rock massif.

Figure 4 presents dependencies of the angle of internal friction ρ^* and cohesion c^* on dimensionless parameter $R_c k_c/\gamma H$.

It follows from the Figure 4 that the angle of internal friction decreases with the depth as a result of increasing rock plastic properties (Balandin 1937). Furthermore, its value essentially depends on the rock structural characteristics defined by the parameter $R_c k_c/\gamma H$.

And only in the absence of spherical components of the stress tensor ($H = 0$), the angle of internal friction becomes a constant associated functionally with the main strength characteristics of rocks namely uniaxial compression strength and uniaxial tensile strength:

$$\rho^*_{H\to 0} = arctan\frac{1-\psi}{2\sqrt{\psi}} = arctan\frac{R_c - R_p}{2\sqrt{R_c R_p}} . \qquad (27)$$

Figure 4 shows that obtained experimentally the friction angle value should be in the range of 15...30° (marked as crosshatched region) for the vast majority of rocks. This assertion originates from the assumption that the ratio ψ has usually the values in the range of 0.1...0.2 (Ilnitskaya et al. 1969). Although for water-saturated clays the angle of internal friction may have significantly lower values (Uzhik 1963, Panyukov 1978). A comparison of the $\rho^*_{H\to 0}$ values obtained by calculations with the test data presented in (Melnikov et al. 1975) shows their close match.

From the equation (24) at $H = 0$ follows the relationship between principal mechanical constants:

$$R_r = 0.5\sqrt{R_c R_p} . \qquad (28)$$

(a)

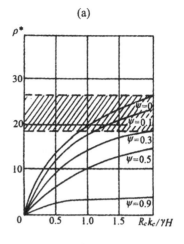

(b)

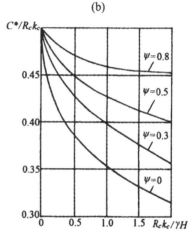

Figure 4. Dependence of the angle of internal friction (a) and cohesion (b) on dimensionless parameter $R_c k_c/\gamma H$.

Thus, equations (25) – (28) obtained above, allow using both linear and non-linear envelope circles of limiting principal stresses as a condition of rock failure depending on the nature and complexity of geomechanical problems. Also, such analytical approach enables deeper understanding the nature of such widespread and commonly used in practical calculations strength characteristics, as cohesion and angle of internal friction (Shashenko et al. 2001).

4 CONCLUSIONS

To assess precisely geotechnical objects stability both analytical and empirical strength criteria are used. Such approach allows simulate the nonlinear process of rock failure with close approximation.

Widely used for numerical simulation of geomechanical processes Mohr-Coulomb failure criterion takes into account the cohesion of rocks and angle of internal friction. In real conditions these parameters vary widely even within the same lithological layers and their value depend on many factors that determine the strength of rock massif (water saturation, joints, layering, etc.), which reduces the calculation accuracy.

Despite the apparent ease of using such criterion in geomechanics, its application invokes some justified objections primarily due to the replacement of the curved envelope line by straight line. As a result, experimental determination of the values of cohesion and friction angle for soft rock or soils looks to a considerable degree subjective.

The presented above analytical dependences allow taking into consideration the features of both linear and non-linear envelope lines of Mohr's circles that improves the accuracy of computations in variable geomechanical conditions.

REFERENCES

Balandin, P.P. 1937. *On the question of the strength hypotheses*. Bulletin of engineers and technicians. No 1: 122 – 127.

Bondarik, G.K. 1974. *Methods of determining strength of clay rocks*. Moscow: Nedra: 216.

Drukker, D. 1957. *On the uniqueness of solutions in the theory of plasticity*. Mechanics. Periodical bulletin of foreign papers, No 4: 72 – 80.

Frenkel, Ya.I. 1954. *Introduction to the theory of metals*. Moscow: Fizmatgiz: 368.

Ilnitskaya, E.I., Teder, R.I., Vatolin, E.S. et al. 1969. *Properties of rocks and methods of their determination*. Moscow: Nedra: 392.

Lieberman, J.M. 1969. *The pressure on the support of capital mine workings*. Moscow: Nauka: 165.

Melnikov, N.V., Rzhevskiy, V.V., Protodjakonov, M.M. et al. 1975. *Reference book (inventory) of rock physical properties*. Moscow: Nedra: 279.

Murrell, S.A. F. 1963. *A criterion for brittle fracture of rocks and concrete under triaxial stress, and the effect of pore pressure on the criterion*. Moscow: Rock Mechanics: 357 – 359.

Panyukov, P.N. 1978. *Engineering geology*. Moscow: Nedra: 296.

Pisarenko, G.S. & Lebedev, A.L. 1969. *Strength of materials to deformation and failure under complex stress state*. Kiev: Naukova dumka: 209.

Ponomarev, S.D., Biederman, V.L., Likharev, K.K. et al. 1956. *Strength computations in mechanical engineering*. Moscow: Mashgiz: Vol. 1: 884.

Protodiakonov, M.M. 1962. *Generalized equation of the envelope line to the limit stress Mohr circles*. Moscow: Studies of physic-mechanical properties of rocks in relation to the issues of control of rock pressure: 27 – 38.

Seituly, K., Shashenko, A.N. & Kovrov, O.S. 2014. *Contemporary approaches to slope stability evaluation while surface mining*. Naukovyi Visnyk Natsionalnoho Hirnychoho Universytetu, No. 3: 103 – 110.

Shashenko, A.N., Sdvizhkova Ye.A. & Gapeyev S.N. 2008. *Deformability and strength of the rock massif*: Monograph. – Dnipropetrovsk: National Mining University: 224.

Shashenko, A.N., Tulub, S.B.& Sdvizhkova, E.A. 2001. *Some problems of statistical geomechanics*. Dnipropetrovsk: National Mining University: 286.

Stavrogin, A.N. & Protosenya, A.G. 1979. *Plasticity of mining rocks*. Moscow: Nedra: 301.

Timoshenko, S.P. & Goodyear, Dj. 1975. *Theory of elasticity*. Moscow: Nedra: 576.

Uzhik, G.V. 1963. *Resistance to tearing-off strength of materials*. Moscow: Mir: 217.

Rock state assessment at initial stage of longwall mining in terms of poor rocks of Western Donbass

O. Sdvyzhkova, D. Babets & K. Kravchenko
National Mining University, Dnipropetrovsk, Ukraine

A. Smirnov
DTEK company, Kyiv, Ukraine

ABSTRACT: The research is focused on the initial phase of longwalling when a "setup room" (assembly chamber) is excavated in the rear of the panel to provide space for assembling the plow equipment. The main goal is to determine the changes of rock stress strain state caused by constructing the assembly chamber and develop a technique for calculating the support bearing capacity considering the rock pressure increasing at the moment when longwall starts moving from the assembly chamber. Stresses and displacements in rocks around assembly chamber and entry face are determined by finite element method based on plastic deformation model. The procedure of strain accumulation is implemented at the sequential change of cavity size. Yielding zones are determined according to the Hoek-Brown strength criterion. Multiple regression and analysis of variance are used to build general regularities.

1 INTRODUCTION

Mining of thin coal seams in Ukraine is carried out under difficult geological conditions. To increase the volume of production the high technologies are involved. One of the challenges is reducing the time of site preparation and equipment dismantling.

The company "DTEK Pavlogradvugillia" introduced the first plow longwall using DBT plow complex under weak rocks conditions of the Western Donbas. Assembling and dismantling of plow longwall are multioperational and complex process, which is carried out under cramped conditions of excavation space. Therefore, the special requirements of stability and technology operations are applied to chambers used for equipment assembly and disassembly. After all, these types of excavations are exposed the extremely load during mining.

This research is focused on the initial phase of longwalling, when a so called "setup room" (an assembly chamber) is excavated in the rear of the panel to provide space for assembling the plow equipment.

The assembly chamber under conditions of thin coal seams of Western Donbass is usually driven in two stages. Firstly the standard arch excavation (assembly chamber itself) is driven and lined with metal arch sets. Then (or in parallel) at one side of this excavation a rectangular face entry (preparatory face) is driven in the coal seam. The width of face entry is equal to the length of the shields installed there (Figure 1).

Figure 1. Face entry supported with timber lining.

Individual timbering is applied to support the roof until the shields of powered support will be installed.

The observation in situ shows that if the shields have not been set just-in-time then ground control problems take place. In particular, the floor heaving occurs under conditions of West Donbass poor and watered rocks. This causes the floor ripping 2 – 3 times prior to powered support installation.

Ensuring the stability of the assembly chamber is a great challenge, however this issue is not clarified enough in Ukrainian regulations. They include only technological scheme of the equipment assembling but does not contain recommendations and guidance concerning the assessment of rock pressure nearby the assembly chamber and face entry.

2 THE STRESS-STRAIN STATE OF ROCKS NEARBY THE ASSEMBLY CHAMBER AND FACE ENTRY

In terms of solid mechanics the construction of assembly chamber can be represented as creating an opening (cavity) in the rock that causes stress redistribution relatively the initial stress-strain state of the rock mass. The face driving results in increasing the cavity size and stress concentration. The combined action of normal and shear stresses results in formatting an area of broken rocks (yielding zone) (Hoek et al. 1998). Within this zone rocks have lost the cohesion between layers and general part of rock mass. The weight of rocks within a yielding area creates the load on the excavation support (Sdvyzhkova et al. 2014). This concept is adopted in Ukrainian standards and regulations. Therefore, the main issue of designing the assembly chamber is to locate the yielding zones by analyzing the components of stress-strain state of rocks nearby the assembly chamber and face entry (Figure 2).

The rock stress-strain state should be determined using one of the numerical methods. We apply the finite element method (FEM) proven in geomechanics problems in combination with the strength theory in the nonlinear formulation (Bondarenko et al. 2012).

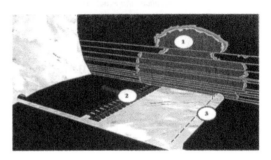

Figure 2. Formation of yielding zone nearby assembly chamber: 1 – yielding zone; 2 – plow longwall; 3 – assembly chamber.

3 NUMERICAL SIMULATION OF CHANGING THE STRESS-STRAIN STATE OF ROCK MASS AT VARIOUS STAGES OF MINING

The software Phase-2 developed by Rocscience laboratory is used to carry out the FEM analysis of the rock stress-strain state in the area including assembly chamber and entry face. The rock mass is modeled as a layered body with plastic properties. Using non-linear model of solid body the different stages of longwall retreat are simulated (Figure 3).

(a)   (b)   (c)

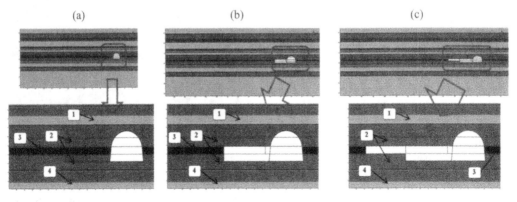

Figure 3. Calculation schemes for determining the stress-strain state of the rock mass: 1 – argillite; 2 – siltstone; 3 – coal; 4 – sandstone.

Initially the intact rock mass loaded with the own weight of overlying layers is modeled. Then the cavities of various sizes are formed in the rock mass by changing the boundary conditions. Stage 1 relates to simulation of assembly camber creating (Figure 3, a), stage 2 shows that additional cavity (face entry) is formed (Figure 3, b), stage 3 supposes that the goaf is simulated as a cavity of 5 m size (Figure 3, c). The strains and displacements calculated at previous stage are taken into consideration at subsequent stage.

The yielding area nearby the excavations is estimated at each simulation stage. Well known empirical Hoek-Brown criterion developed for weak and fractured rocks is used as a condition of rock failure.

The Hoek-Brown failure criterion is used widely in geomechanical practice. Over the years, from 1980, the Hoek-Brown rock mass failure criterion has undergone numerous revisions. The latest version of the criterion is shown below (Hoek et al. 2002):

$$\sigma_1' = \sigma_3' + \sigma_{ci}\left(m_b\frac{\sigma_3'}{\sigma_{ci}} + s\right)^\alpha, \qquad (1)$$

where σ_1' and σ_3' – the major and minor principal effective stresses at peak strength; σ_{ci}' – the uniaxial compressive strength of 50 mm diameter specimen of intact rock with 1:2 diameter: length ratio; m_b – the value of the Hoek-Brown constant m for the rock mass; s and a – constants which depend upon the rock mass characteristics.

For the intact rock pieces that make up the rock mass, this equation simplifies to:

$$\sigma_1' = \sigma_3' + \sigma_{ci}\left(m_i\frac{\sigma_3'}{\sigma_{ci}} + 1\right)^{0.5}. \qquad (2)$$

Comparing the intact rock failure criterion and the rock mass failure criterion, we can see that the first one is modified to predict the strength of rock masses by the introduction of the new parameters, m_b, s and a.

These new parameters are calculated using the value of the Geological Strength Index (GSI) and the disturbance factor $(0 < D < 1)$ by the following expressions:

$$m_b = m_i e^{\frac{GSI-100}{28-14D}}, \qquad (3)$$

$$s = e^{\frac{GSI-100}{9-3D}}, \qquad (4)$$

$$a = \frac{1}{2} + \frac{1}{6}\left(e^{-\frac{GSI}{15}} - e^{-\frac{20}{3}}\right). \qquad (5)$$

Therefore, to use the Hoek-Brown criterion for estimating the strength of jointed rock masses, four parameters of the rock mass have to be estimated. These are: uniaxial compressive strength σ_{ci} of the intact rock specimen; value of the Hoek-Brown constant m_i for these intact rock pieces; value of the Geological Strength Index for the rock mass and disturbance factor D (a factor which depends upon the degree of disturbance due to blast damage and stress relaxation. It varies from 0 for undisturbed rock masses to 1 for very disturbed rock masses).

The parameters of the Hoek-Brown criterion are defined for coal and rocks under conditions of 167-th plow longwall of "Stepna" mine on the basis of the data provided by the mine geological service (Table 1) and visual estimation of rocks during in situ observations. Based on this studying the GSI is taken equal to 50 which corresponds the "very blocky" rock mass structure. The disturb-

ance factor D is equal 0 since at assessing uniaxial compressive strength σ the structural factor k_c is taken into consideration.

The reliability of numerical simulation depends on the reliability of determining the physical and mechanical properties of rocks. As we mentioned above, the Western Donbas rocks can be characterized as poor quality rocks as they are jointed and the compressive strength significantly decreases due to watering. Decrease in the strength of the intact rock should be considered with usage of structural factor k_c defined as:

$$k_c = \frac{\sigma_m}{\sigma_{ci}}, \qquad (6)$$

where σ_m – a compressive strength of real rock mass and $\overline{\sigma_{ci}}$ – an average compressive strength of tested samples.

Structural factor is determined depending on the degree of rock heterogeneity according to statistical strength theory (Shashenko 2008). Its values are shown at Figure 4 depending on a variation of strength value (η) obtained by the rock specimen laboratory testing.

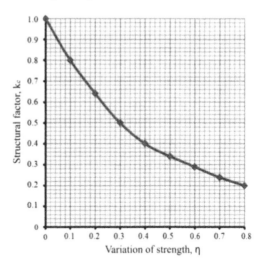

Figure 4. Structural factor depending on variation of strength value.

The Hoek-Brown constants and structural factor values as well as rock physical and mechanical properties are referred to the Table 1.

The simulations in terms of "Stepna" mine are carried out considering the initial stress field equals to 15 MPa. As a result the failure area spreading is obtained depending on increase of excavation space (Figure 5).

Table 1. Physical and mechanical properties of the rocks.

Characteristic	Argillite	Siltstone	Coal	Caved rocks
Young's modulus, MPa	3193.0	2981.7	11755.2	300.0
Poisson' ratio	0.3	0.3	0.3	0.3
Compressive strength of intact rock, MPa	32	43	37.5	7
Structural factor	0.5	0.45	0.6	—
Compressive strength reduced, MPa	16.0	20.0	22.5	7
Hoek-Brown parameter, m_b	1.17	1.13	2.66	0.98
Hoek-Brown parameter, a	0.51	0.51	0.5	0.51
Hoek-Brown parameter, s	0.001	0.001	0.016	0.0007

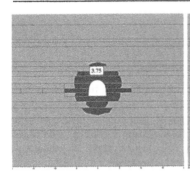  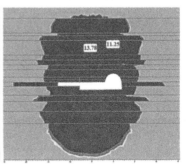

Figure 5. Failure zone spreading.

The yielding rock area (failure zone) around the excavation has an almost parabolic shape that is close to an arch of natural equilibrium by M. Protodiakonov (Shashenko & Pustovoytenko 2004). However, the form of the failure zone is irregular due to the different strength of rocks. Therefore, the main characteristic of failure zone is a height h_p directed normally to the layering.

A plot at the Figure 5 shows that the height of the failure zone above the assembly chamber is 3.4 m at the first stage of the simulation. This value increases up to 8 m as a result of formation of the additional excavation (face entry). We can see it at the second stage of simulation. Further increase of mined out space results in growth of the height h_p up to 11 m above assembly chamber and 12 m above face entry. It means that the load on the excavation support increases as well because it should be obtained as the weight of yielding rocks.

Thus, in terms of engineering calculation the height of failure zone is the main geomechanical parameter. The size of this area is determined by many factors such as a depth of the excavation, dimensions of the cavity and the rock strength properties. These parameters specify various mining and geological conditions of coal mining.

The algorithm described above allows executing multiple calculations for various geological and mining conditions. The multivariate simulation enables to generate simple engineering formulas to determine the height of yielding rock area.

4 MULTIVARIATE SIMULATION. DERIVATION OF THE GENERAL FORMULAS FOR ASSEMBLY CHAMBER DESIGN

Input parameters are:
 − B/h − a ratio of the width to the height of excavation;
 − γh − a normal (vertical) component of the initial stress field;
 − R_c − an average strength of the rock mass;
 − m − a coal seam thickness.

To obtain the failure zone height as a function of the input data the method of nonlinear estimation is used. It generalizes two methods: multiple regression and analysis of variance. Nonlinear estimation involves a pre-selecting type of the target functions of the initial variables. It could be logarithmic, exponential, power, or any composition of elementary functions.

In general, all of the regression model can be written as a formula:

$$y = F(x_1, x_2, \ldots, x_n). \tag{7}$$

The main issue of nonlinear regression analysis is that the relationship between the target function and the initial variables exists, i.e. the dependent variable (function) is connected with the set of independent variables (arguments). Summarizing the results of the numerical simulation the model of exponential growth is used as one of the methods of nonlinear estimation. It looks like:

$$y = a + exp(b_0 + b_1 z_1 + b_2 z_2 + ... + b_m z_m),$$ (8)

where y – a target function; a, b_i – unknown coefficients; z_i – initial parameters.

To assess the adequacy of the model the Pearson's chi-squared test is used. If the value of chi-squared test statistic is significant, we reject the null hypothesis and accept the independent variables that affect significantly the target expression.

As a result of the regression analysis the empirical formulas are derived.

The height of failure zone above assembly chamber:

$$h_p = e^{(0.4\frac{B}{h}+1.27\frac{\gamma H}{R_c}+0.005m+0.24)-0.1}$$ (9)

The height of failure zone above preparatory face:

$$h_p = e^{(0.18\frac{B}{h}+1.67\frac{\gamma H}{R_c}+0.006m+0.75)+3}.$$ (10)

The visualization of formulas (9) and (10) are shown on the Figures 6 and 7 respectively.

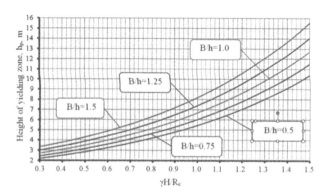

Figure 6. The height of failure zone above assembly chamber.

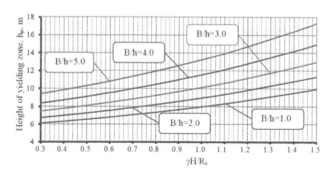

Figure 7. The height of failure zone above face entry.

5 ENGINEERING APPROACH TO THE ASSEMBLY CHAMBER DESIGN

As it has been indicated above, the weight of the rocks within the failure zone creates a load on the support according to the formula:

$$P = \gamma \cdot S_m,$$ (11)

where S_m – the area of failure zone over the assembly chamber or face entry; y is a rock gravity.

In engineering calculations the magnitude S_m should be determined by a simple mathematical formula. Despite the known fact that the area of failure zone has irregular shape, it can be approximately represented as a rectangle with a height h_p, and a base equal to the width of excavation B (Figure 8). Thus, some margin of safety is provided.

69

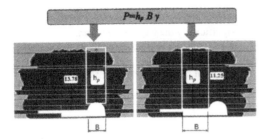

Figure 8. The scheme for determining the load on the support.

Then the load on excavation support generated by failed rocks is approximately equal to:

$$P = \gamma \cdot S = \gamma \cdot B \cdot h_p . \qquad (12)$$

In accordance with the value (12) the assembly chamber support has to be designed to ensure the reactions to this pressure.

In particular, in terms of 167-th plow longwall of "Stepna" mine according to the simulation results the loading on the assembly chamber support is 360 kN/m at the first stage (before the entry face driving). To compensate the rock pressure the arched metal sets are installed with a distance of 0.7 m in the assembly chamber. But at the second stage (when the face entry is excavated) the load on the assembly chamber support increases up to 840 kN/m. The load on the entry face support is 1400 kN/m at this stage. So, the simulation shows that rock pressure increases significantly when the preparatory face is driven. The bearing capacity of the metal arch (280 kN/m) will be run out at this stage. This fact has been proven by in situ observations.

Therefore, the assembly chamber roof is additionally anchored with full bounded rock bolts. The bolt length is 2.5 m. Twin bolts should be installed in a conjugation of the assembly chamber and entry face roof since the rock falls occur often in this place. In addition the plain strand cables of 6 m length are installed as well with distance of 1.4 m.

The pressure of yielding rocks over the entry face should be compensated by installation of the coupled timber bars of 3.8 m length with distance of 0.5 m and vertical timbering props (it should be not less than 11 props per meter). Spreading the failure zone up to the considerable height (about 11.2 m) and a high probability of rock falls demands to anchor the roof of preparatory face with 5 rock bolts of 1.5 m length with distance of 1.0 m.

6 CONCLUSIONS

Visual observations in situ and multivariate numerical simulation show the significant grow of yielding area over the assembly chamber and rock pressure on the support at creating the additional cavity (face entry) and the longwall retreat. This fact is not clarified enough in Ukrainian regulations which include only technological scheme of the equipment assembling but does not contain recommendations and guidance concerning the assessment of rock pressure nearby the assembly chamber and face entry.

Using FEM analysis and nonlinear regression analysis the formulas for an assessment of rock pressure on the support of the assembly chamber and preparatory face have been derived. These formulas should be used at engineering calculations to design the assembly chamber and face entry to ensure the stability of the excavations.

REFERENCES

Bondarenko, V., Kovalevs'ka, I., & Fomychov, V. 2012. *Features of carrying out experiment using finite-element method at multivariate calculation of "mine massif – combined support" system. Geomechanical processes during underground mining.* The Netherlands: CRC Press/Balkema: 7 – 13.

Hoek, E., Marinos, P. & Benissi, M. 1998. *Applicability of the Geological Strength Index (GSI) classification for very weak and sheared rock masses. The case of the Athens schist formation.* Bulletin of Engineering Geologyand the Environment: 57 (2): 151 – 160.

Hoek, E., Carranza-Torres, C. & Corkum, B. 2002. *Hoek-Brown criterion – 2002 edition.* Proc. NARMS-TAC Conference, Toronto: 1: 267 – 273.

Sdvyzhkova, O.O., Babets, D.V., Smirnov, A.V. 2014. *Support loading of assembly chamber in terms of Western Donbas plough longwall.* Scientific bulletin of National Mining University, No 5: 26 – 32.

Shashenko, O.M., Pustovoytenko, V.P. 2004. *Mehanika gornyh porod.* Kyiv: 400

Shashenko, O.M., Sdvyzhkova, O.O. & Gapeev, S.N. 2008. *Deformatsyonnye modeli v geomekhanike.* Dnipropetrovsk: National Mining University: 223.

Theoretical and Practical Solutions of Mineral Resources Mining – Pivnyak, Bondarenko & Kovalevska (eds)
© 2015 Taylor & Francis Group, London, ISBN: 978-1-138-02883-8

Study of dust cloud spraying parameters in terms of its suppression

A. Yurchenko, A. Litvinenko & T. Morozova
National Mining University, Dnipropetrovsk, Ukraine

ABSTRACT. Results of theoretical study of dust cloud spraying parameters after single blast in iron-ore quarries are given. The nature of reducing dust concentration within the cloud in terms of its spraying, level of effect of water drop diameter and its consumption on the dust suppression efficiency is determined. Coefficient of dust particle capture with the help of water drops to spray solutions of the surface-active substances is studied. Optimum values of these parameters to suppress the cloud of iron-ore dust are substantiated.

1 INTRODUCTION

Technological processes of many branches of industry are followed by the considerable dust formation. Dust atmosphere spraying in the place of dust cloud formation is the most popular method of dust suppression. The results of numerous experiments and theoretical studies show that the same efficiency of dust suppression can be reached at various parameters of spraying. Furthermore, in different situations spraying will have various values of such important factors as water consumption, minimum size of captured particles (that determines in great extent social effect of applying dust suppression equipment), and dispersion degree of spraying water, power consumption, reliability and difficulty of maintaining spraying systems.

Moreover, the basic spraying parameter is efficiency of the dust suppression which is in general determined by the ability of dispersed water to capture dust particles (Berlyand 1985). This ability is characterized by the coefficient of dust capture with the help of spherical water drop. It can be possible to increase capture coefficient using water solutions as spraying water with special wetting supplement agents. It enables to increase efficiency of dust suppression by the small volume of spraying water.

2 PROBLEM SETTING

The task of spraying parameter study in terms of suppression the cloud of fine iron-ore dust at single-blast in the quarry as its content within the cloud at the boundary of sanitary protection zone is 90.3% was set up in this paper (Tischuk et al. 2007). It is required to determine capture coefficient of the fine, more hazardous dust for various diameters of the water drops. It will enable to determine optimum size of the drops in terms of capture coefficient will be maximal. It is also very important to determine the effect of the wetting agents on capture coefficient. All these investigations will allow predicting flow intensity of spraying fluid including wetting agent to reach expected efficiency of dust suppression.

3 THE RESEARCH MATERIALS

Spraying is a very complex process of dust suppression where inertial dust capture by the drops of the sprayed water, electrostatic interaction, diffused phoresis, wetting and other chemical and physical phenomena take place. All of them more or less appear when dust particles and water drops are close to each other and can come into contact.

Probability of water drops contact with the dust is proportional to the product of their quantity contained in the volume unit multiplied by effective cross-section $\pi(r + R)^2$ (here r – the radius of dust particle, m; R – the drop radius, m), relative rate of particle and drop movement $\omega = \vartheta - V$ (here ϑ – the rate of dust particle deposition, meters per second; V – the rate of the drop deposition, meters per second), and the time of their interaction. If this product is multiplied by the capture coefficient E, efficiency of dust particles and the drops will be taken into account. As interaction of dust particles with drops depends on their sizes to simplify analysis of dust-removing spraying it is assumed that dust particles and drops are mono-dispersed. Then the change of dust particles concentration dv with the radius r at a time dt while moving with relative rate ω as for the water drops with the radius R can be written as:

$$-dv = vN\pi(r+R)E\omega dt, \qquad (1)$$

where v and N – concentration indexes of dust particles and water drops

As N and ω depend on the time of drop movement and E depends on ω, equation solution (1) if boundary conditions are $v = v_0$ at $t = 0$ will be:

$$v = v_0 e^{-\pi(r+R)^2 \int\limits_0^t E\omega N dt}, \qquad (2)$$

Or in relative form:

$$P = \frac{v}{v_0} = exp\left[-\pi(r+R)^2 \int\limits_0^t E\omega N dt\right]. \qquad (3)$$

In formulas (2) and (3) the value v_0 – initial (before spraying) dust index. Formula (3) shows relative decrease of dust concentration indexes in terms of the dust cloud volume if they accidentally interact with spray. Reducing dust concentration within the cloud will be performed according to the exponential law. The larger degree of dispersed drops, capture coefficient, relative rate of dust particle and drop movement, drop concentration index (water consumption while spraying dust cloud), the more intensive it will be.

Capture coefficient E is a complicated parameter defined by mechanism of drops and dust particles interaction and capture strength. It is determined as the ratio between cross-section of the current tube formed by the extreme (limited) trajectory of particle gravity center and midlength section. It is assumed that air flow is stationary as for drop. Mass concentration of dust particles is small (less than 1 g/m³). It doesn't effect on the movement of the air and drops. The size of the drops is much larger than the size of the particles. Particle shape is spherical. Even under these assumptions determining coefficient E by the equation of particle movement and equation of continuity is a very complex task.

Capture coefficient E depends on the Stokes number (K) and the mode of air flow washed the drop at inertial precipitation of the dust particles on it. Critical value K when dust particles begin to be captured by the drop at potential flow is 1/12 (Belousov 1988), that is $E = 0$, if $K < 1/12$. To calculate capture coefficient while increasing Stokes number Langmuir offered the following formulas:

$$E = 0 \text{ if } K < 0.083, \qquad (4)$$

$$E = K^2 / (K+0.5)^2 \text{ if } K > 0.2, \qquad (5)$$

$$E = \left(\frac{1+0.75 \ln 2K}{K-1.214}\right)^{-2} \text{ if } K > 1.214. \qquad (6)$$

Given formulas allow to calculate capture coefficient of the dust particles with the help the drops while spraying. For this purpose it is necessary to determine the value of Stokes number by the following formula:

$$K = \frac{\rho_p g \omega d_i^2}{18 \mu d_d}, \qquad (7)$$

where ρ_p – density of the dust particle, kg/m³; d_i – diameter of the dust particles of i-fraction, m; $\mu = 181.9 \cdot 10^{-7}$ H·c/m² – dynamic air viscosity; d_d – diameter of the drops; ω – relative rate of the drop and dust particle movement.

Calculations of coefficient of the dust particles capture with the help of the drops while spraying will start from determining the rate of the water drop precipitation for the fixed diameters of the drops: 20; 50; 100; 300; 600; 1000; 1500; 2000; 2500 um. The drop in a stationary air at free fall quickly reaches such constant rate when aerodynamic resistance affected on the drop is equal to its weight. If the air density and effect of air inertia replaced by the dust particle are neglected, the rate of spherical drop can be calculated by Stokes law (Belousov 1988):

$$u = \frac{\rho g d_d^2}{18 \mu}, \qquad (8)$$

where p – drop density, kg/m³; d_d – drop diameter, m.

Table 1 shows results of calculations of drop deposition rate for its selected diameters.

Precipitation rate of the fine fraction of dust cloud particles with 3 um diameter is determined according to this method. Particle precipitation rate of this fraction is $\vartheta = 0.00108$ m/c. Stokes number, while spraying particles of iron-ore dust with 3 um diameter for the accepted water drop diameters is determined from the equation (7). Table 2 shows calculation results.

There are all initial data now to determine capture coefficient E for iron-ore dust particles with 3 um diameter while changing water drop diameter within the range under study. Calculation of capture coefficient is performed by the equation (5). Table 3 shows its results.

As calculation results show, capture coefficient of fine particles of iron-ore dust nonlinearly depends on diameter of water drops. Its maximal value is 0.461 and is reached at 600 um diameter of the water drops.

Low coefficient of dust particle capture with the help of the water drop can be explained by the fact that water has relatively high surface strain $(72.75 \cdot 10^{-3}$ H/m) (Gogo 2008). Limiting wetting angle is quite high and dust particle is not wetted completely. So, dust particles when they contact with water drop are gathered on its surface.

Table 1. Water drop precipitation rate.

Drop diameter, m	Stokes precipitation rate, $u = \dfrac{g\rho d_d^2}{18\mu}$, m/s	Reynolds number $Re = \dfrac{\rho_0 u d_d}{\mu}$	Archimedes' number $Ar = \dfrac{d_d^3 \rho_0^2 (\rho - \rho_0) g}{\mu^2}$	Re number for Stokes particles $Re = \dfrac{Ar}{18 + 0{,}61\sqrt{Ar}}$	U, m/s, for stokes particles $u = \dfrac{Re\,\mu}{\rho_0 d_d}$
$20 \cdot 10^{-6}$	0.012	0.016 < 1	—	—	—
$50 \cdot 10^{-6}$	0.075	0.25 < 1	—	—	—
$100 \cdot 10^{-6}$	0.299	1.07 > 1	42.6	0.968	0.147
$300 \cdot 10^{-6}$	2.69	5.32 > 1	1150.2	29.73	1.50
$600 \cdot 10^{-6}$	10.76	42.6 > 1	9201.6	15.29	4.00
$1000 \cdot 10^{-6}$	29.90	197.0 > 1	42600.0	284.19	4.31
$1500 \cdot 10^{-6}$	67.28	665.0 > 1	143775.0	576.97	6.04
$2000 \cdot 10^{-6}$	119.60	1576.3 > 1	340800.0	926.59	7.03
$2500 \cdot 10^{-6}$	186.87	3078.8 > 1	665625.0	1290.57	7.83

Table 2. Stokes number at water drop diameter under study.

Particle diameter, um		Water drop diameter, um								
		20	50	100	300	600	1000	1500	2000	2500
3	ω	0.011	0.074	0.146	1.50	4.00	4.31	6.04	7.03	7.83
	K	0.593	1.595	1.574	5.390	7.187	4.646	4.341	3.789	3.376

Table 3. Total coefficient of dust particle capture for the considered water drop diameters.

Dust particle diameter, um	Drop diameter, um								
	20	50	100	300	600	1000	1500	2000	2500
3	0.094	0.093	0.087	0.442	0.461	0.452	0.434	0.396	0.361

While water drop moving under the action of gravitation forces disturbance of precipitated dust particle takes place. Moreover, screening effect of dust particles fell down on the drop occurs. To increase efficiency of dust capture by means of the drop solutions of the surface-active substances (SAS) (but not the water) which reduce limiting wetting angle are used. Dust particle is wetted completely and penetrates inside the drop.

Thus, addition of surface-active substances increases dust capture while spraying due to the efficiency upgrading of dust and drop contact and capture strength determined by the capture coefficient.

At present, the following kinds of water solutions are used as wetting agents to increase efficiency of dust capture in the quarries of Kryvyi Rih iron-ore basin (Tyschuk et al. 2009):
– water solution of coal-alkali reagent containing complexes with humolith;
– water solution "ECOM" containing non-ionic surface-active substances – hydroxyethylated alkylphenols which are capable to reduce the strain of the surface solution;
– water solution of ore dust wetting agent containing anionic surface-active substances – hydroxyethylated nonylphenol as well as wetting agent

OD-5, composed of hydroxyethylated spirit of $C_8 - C_{10}$ fraction. Chemical name of ore dust wetting agent is $\textit{н}$ – nonyl spirit.

Value of water-stable dust units obtained in terms of solid and liquid phase interaction is the criteria of solution adhesive ability. It was determined that one-percent water solutions in terms of coal-alkali reagent are the most efficient to capture the dust at a single-blast in quarries. Their application enables to improve the efficiency of dust suppression two times as much compared with water used for spraying.

Thus, conclusions can be drawn about the fact that using one-percent water solutions of coal-alkali reagent as surface-active substances is able to increase capture coefficient of the dust particles with the help of spraying water drops. Table 4 shows dust capture coefficient for the fine dust with the help of the drops of various diameters.

These studies allows make the conclusion that maximum efficiency of dust cloud suppression while its spraying is reached in terms of spraying water dispersion with drop size up to 600 um.

The next stage of the study is determining required quantity of water to reach effective purification of the dust and gas cloud against the dust while its spraying.

Table 4. Total coefficient of dust particle capture for studied diameters of the water drops while using coal-alkali reagent.

Dust particle diameter, um	Drop diameter, um								
	20	50	100	300	600	1000	1500	2000	2500
3	0.188	0.186	0.174	0.884	0.922	0.904	0.868	0.796	0.722

Expression for determining efficiency of dust capture with the help of the water drops is obtained in the paper (Yurchenko 2014) in the form of:

$$\eta = 1 - exp\left(-\frac{3}{2}m\frac{\omega}{V}\frac{H}{d_d}E\right), \qquad (9)$$

where $\eta = \Delta C / C$ – efficiency of the dust capture with the help of the water drops while cloud spraying (here ΔC – the part of dust concentration which is suppressed by means of the cloud spraying, mg/m³; C – initial dust concentration within the cloud, mg/m³); $m = V_f / V_p$ – the ratio of the dust and gas cloud spraying (here V_f – required quantity of water for cloud spraying, m³; V_p – the size of the cloud dust particles, m³); ω – relative rate of drop and dust particle movement, m/sec; H – the height of the dust cloud, m; E – the capture coefficient; V, d_d – the rate the drop precipitation, m/sec, and drop diameter, m, correspondingly.

Ratio of spraying dust and gas cloud is determined by equation:

$$m = ln\frac{1}{1-\eta} \cdot \left(\frac{3}{2}\frac{\omega}{V}\frac{H}{d_d}E\right)^{-1}, \qquad (10)$$

Water quantity to reach required efficiency of dust suppression is determined by the equation:

$$m = \frac{V_f}{V_p}, \text{ then } V_f = ln\frac{1}{1-\eta} \cdot \left(\frac{3}{2}\frac{\omega}{V}\frac{H}{d_d}E\right)^{-1} \cdot V_p.$$

Calculation results of required spraying water specific volume to suppressed dust cloud at a single blast with the yield of 500 t explosive are given in the Table 5. Flow intensity to reach desirable efficiency of dust suppression V_s, kg/kg is the most informative ratio. It can be determined as a quotient of division of required water consumption by the common emission of the fine dust under a single blast.

Table 5. Required quantity of water specific quantity to spray the cloud at assumed efficiency of the dust capture.

Spraying ratio	Specific water consumption at E								
	0.1	0.2	0.3	0.4	0.5	0.6	0.7	0.8	0.9
Specific water consumption, kg/kg	0.130	0.278	0.447	0.637	0.865	1.141	1.497	2.010	2.875
Specific consumption of coal-alkali reagent solution, kg/kg	0.065	0.139	0.223	0.318	0.432	0.570	0.748	1.005	1.438

4 CONCLUSIONS

Reducing dust concentration within the cloud while spraying takes place according to the exportential law. The larger degree of the drop dispersion, capture coefficient, relative rate of dust particle and drop movement, water consumption while spraying dust cloud, the more intensive it.

Capture coefficient of the fine particles of iron-ore dust nonlinearly depends on the water drop diameter. Its maximum value is 0.461 if water drop diameter is 600 um.

Flow intensity for suppression of one kilogram of the fine iron-ore dust is up to three liters in terms of its dispersion to the drop size of 0.6 mm. While using one-percent solution of the surface-active substance of coal-alkali reagent solution specific consumption is 1.5 liters.

REFERENCES

Belousov, V.V. 1988. *Theoretical basics of gas purification processes*. Moscow: Metallurgy: 256.

Berlaynd, M.E. 1985. *Forecast of air contamination control*. Leningrad: Gidrometeoizdat: 272.

Gogo, V.B. 2008. *Hydrodynamic dust suppression in coal mines*. Donetsk: Theory and engineering solutions. GHAE "Don-STD": 240.

Tyschuk, V.Yu., Ermak, L.D., Kovaliov, V.M. & Kotov, Yu.T. *Study of water solutions of surface-active substances effect on efficiency of dust suppression under a single blast in quarries within water stemming boreholes*. Book of scientific papers of the National Mining University, Book 2, No 33: 125 – 133.

Tyschuk, V.Yu., Eudokimenko, M.F., Ermak, L.D., Guba, M.N., Gorobets, M.N. & Kuzmenko, P.K. 2007. *Study of air contamination level on the boundary of quarry sanitary and protective zones after single blast*. Kryvyi Rih: Book of scientific papers, Addition 9: 85 – 98.

Yurchenko, A.A. 2012. *Studying coefficient of aerosol particle capture with the help of water drops while spraying*. Book of scientific papers of the National Mining University, No 39: 187 – 194.

Theoretical and Practical Solutions of Mineral Resources Mining – Pivnyak, Bondarenko & Kovalevska (eds)
© 2015 Taylor & Francis Group, London, ISBN: 978-1-138-02883-8

Technological and environmental aspects of the liquidation of coal mines

V. Buzylo, A. Pavlychenko, O. Borysovs'ka & V. Gruntova
National Mining University, Dnipropetrovsk, Ukraine

ABSTRACT: The features of the restructuring of the coal industry in Ukraine are analyzed. The mechanisms and regularities of changes in the state of the environment in areas of coal mines liquidation are defined. A set of technologies aimed at preventing deformation and subsidence of the earth's surface, areas swamping and destruction of buildings and infrastructure is proposed.

1 INTRODUCTION

Coal has a significant place in the global economy among the natural energy resources and it occupies the third position by volume of use after oil and natural gas. Ukraine has considerable potential for the extraction of coal and ranks the first place in Europe and the eighth place in the world in its reserves, which are estimated at 120 billion tons. Most of the reserves are concentrated in three coal basins: Donetskyj, Dniprovskyj and Lvivsko-Volynskyj.

The coal industry is an important component of the industrial potential of Ukraine and it ensures the operation of the leading sectors of the economy. The further sustainable development and national energy security depends on of the efficiency and stability of coal industry functioning (Pivnyak et al. 2005, Jiang & Jin 2010, Li et al. 2011, Wang et al. 2013, Zhang & An 2014).

As a result of years of coal deposits development the transformation and pollution of the environment takes place (Cao et al. 2012, Tian et al. 2013, Zhang et al. 2013). Living conditions and health status of population in the coal mining regions are deteriorating herewith (Ma et al. 2012, Riznyk et al. 2006, Gorova et al. 2012, Krupskaya et al. 2013, Zvereva et al. 2013, Saksin & Krupskaya 2005). The load on the environment which is created by activities at all stages of mine operation, is one of the most important factors that must be considered when determining the directions of further development of the coal industry (Qi 2010, Solovenko et al. 2014; Guo 2014, Yang & Liu 2012, Zhang 2013, Hu et al. 2013, Han et al. 2013).

Exhaustion of balance reserves, unprofitable of most mines, difficult geological conditions and unstable economic situation have led to massive mine closure. Mining fund becomes outdated in Ukraine:

more than 70 percent of all state-owned mines have seen no reconstruction for over 30 years. The volume of public investment in mining fund restoring is insufficient; coal mines are mostly unprofitable and can not finance investment projects independently. In terms of restructuring of coal mining industry the closure of mining enterprises became widely spread.

Ukraine occupied a leading place in the world for the extraction of coal until mid-70 of twentieth century, but recent years were marked by the decline of the coal industry of the state. In 1995 – 2000 years 143 coal enterprises were transferred for closing as a result of restructuring but liquidation workings fully were performed by only at 50 enterprises.

The liquidation of mines leads to unpredictable economic, environmental, technological and social consequences. Large-scale nature of this problem causes catastrophic changes and degradation of environmental objects. Ignoring requirements of technogenic and ecological safety during mine closure leads to significant changes in the quality of groundwater and surface water, flooding and swamping of areas, salinization and pollution of rivers, subsidence of the earth's surface, increased risks of physical destruction of residential and administrative buildings, infrastructure objects.

All this is due to the fact that a comprehensive scientific and prognostic assessment of environment state was not carried out in these regions at the close of mining companies. In addition, no consequences of further impact of already liquidated mines on environmental components were studied. Insufficiently reasoned closure of mines resulted in disastrous effects for the environment and infrastructure (Arefieva et al. 2014, Šofranko 2014, Krupskaya, et al. 2014, Sliadnev 2001, Hoshovski 2000, Pack 2009).

2 FORMULATING THE PROBLEM

Carrying out of researches is aimed at implementing the main provisions that are formulated in important state documents: Energy Strategy of Ukraine till 2030, National Action Plan on Environmental Protection for the period up to 2015, Basic Principles (Strategy) of the State Environmental Policy for the period 2020, Concept of Development of the Coal Industry, Concept of Sustainable Development of Human Settlements and others.

Because of the slow restructuring of the coal industry a significant number of small and medium unpromising and unprofitable mines are in exploitation. In 1991 there were 278 mines in Ukraine. Mining fund in Ukraine at present is one of the oldest in the world that has not been updated for decades. Almost 96% of mines operate without substantial reconstruction more than 20 years.

The average development depth of coal seams is 720 m, over 20% of mines are working at horizons 1000 – 1400 m. Average thickness of coal seams is around 1 m. Coal mining is accompanied by a number of factors, namely: 90% of mines are dangerous for the emission of methane, 60% – due to explosions of coal dust; 45% – because of sudden emissions; 22% – due to spontaneous combustion. During the closure of coal mines above mentioned processes become unpredictable, and this, in turn, significantly reduces the level of environmental and technogenic safety in coal mining regions.

Restructuring of the coal industry through the elimination of mining companies has led to subsidence of the surface deformation and destruction of buildings and engineering networks at the territories of disposal operations, as well as to formation of explosive methane-air mixtures. This in turn leads to the possible occurrence of natural and man-made emergency situations.

The main environmental impacts of coal mining and removal of mines from exploitation are problems that caused by: water lifting in the mine workings at the flooding process after mine closure; drop of highly mineralized mine water; rock dumps, which often continue to burn and have no vegetation; methane gas emissions; deformation of the earth's surface and so on.

Technogenic and environmental safety of mines liquidation is in direct dependence on the timeliness of development and implementation of environmental measures. Prevention of environmental problems through their timely detection and neutralization, in most cases, is less expensive method compared to fixed costs to overcome their consequences.

The liquidation of mining companies does not guarantee discontinuation of their negative impact on environmental components. In this connection there is a need to develop assessment system for levels of environmental threats and man-made risks, as well as the scientific substantiation of measures to protect the environment in the areas of influence of operating coal mines or mines that are subjects to liquidation.

Therefore, the purpose of this work is to identify technogenic and environmental risks arising at different stages of the coal mines liquidation as well as the development of technologies to minimize them.

3 GENERAL PART

The following aspects were considered to study the negative impact of the closed coal mines on hydrogeological, geomechanical and ecological parameters of the environment:

– the mechanisms of rock mass and the earth surface deformation at different ways of coal mines liquidation (depending on the depth and working conditions of coal reserves);

– regularities of deformation of the rock mass and the earth surface during the flooding of mines in the application of "wet conservation" of coal mines;

– mechanisms of activation of the earth surface subsidence and deformation of industrial facilities and residential buildings, located in the zone of mine fields influence of closed mines;

– regularities of changes in hydrodynamic structure of the rock mass at flooding of coal mines;

– mechanisms for mine gases emission on the earth surface in the territories of the mass liquidation of mines and so on.

Operation and, subsequently, the liquidation of coal companies led to changes of geomechanical processes both in the depths, and at the earth's surface. The shift of rock takes place, the intensity of movement of groundwater and gases raises as well as seismic parameters of massif change. As the water level at the flooding process of mine workings gradually rises up in developed space, wetting of fallen-in rocks occurs as well as their movements down and additional compression. This increases the plastic properties of the rock mass.

The deformation of the rock mass causes damage, collapse and shift of the earth surface. At the liquidation of coal mines with development depth about 150 – 300 m the surface subsidence above workings reaches over 20%, and in case of the depth of over 1000 meters it is 5 – 7%. The depth of the subsidence of the earth surface above mines being liquidated is 0.2 – 1.2 m in average and in some places – more than 5.0 m.

The negative consequences of the mass liquidation of mines are deformation of the rock mass and significant area of subsidence of the earth surface, leading to flooding of territories and physical destruction of communications engineering, buildings and structures. Area of subsidence in coal mining regions is more than 8000 km². Flooding is fixed in Dnipropetrovs'k region in the area of 0.74 km², in the Donetsk region – about 1.66 thousand km². There are 30 cities in the area of flooding with a total area of 230 km². In Chervonograd mining region flooded area is 62 km² and covers all settlements in the region (Informational…2011).

Mass closure of coal mines provokes the rise of underground water and soaking of rocks around workings, causing activation of shifts and failures above them. As a result of flooding of mines there is a change of hydrogeological and geotechnical conditions in large parts of the mine fields. Complete flooding of mines is accompanied by exits of mineral waters to the surface and displacement of acid mine water from old mine workings located at shallow depths is observed. Mine waters, which in the final period of flooding come to the surface, have the highest mineral content and aggression and pose a significant risk to groundwater and surface water and agricultural land.

There is a change of hydrogeological and engineering-geological conditions in large parts of the mine fields as a result of flooding of mines. Changing of engineering, geodynamic and geofiltrational conditions in the mountain massif is accompanied by flooding and swamping of large areas. Subsidence of the earth surface causes flooding of residential areas; flooding of wells, cellars, basements, communications; soaking of foundations of residential and industrial buildings; damping-off of trees; overgrown of large areas with reeds; reducing the flow rate of rivers; swamping of floodplains and etc.

As a result of studies is found that most mines have hydraulic connections between them. Therefore, at the flooding of mines taken to liquidation the water from them can flow through existing hydraulic connections or by filtration through rock mass to existing mines. Flooding of developed space of coal mines with ground waters leads to wetting of massif. In this case the reduction of its strength happens, making the processes of deformation and collapse of developed space renewed. In this regard, the minimum allowable depth of raising the level of groundwater at flooding of mines should not exceed mark of 150 m of a minimum mark of surface.

The consequence of changes in the hydrodynamic conditions is the development of local-regional depression of groundwater surface and deepening of active water exchange zones from 150 – 250 to 450 – 550 m with increased infiltration of precipitation and surface water flows in water horizons from

rivers, reservoirs, etc. Balancing of hydro conditions by mixing of surface and ground water leads to increased salinity due to leaching salts from the rocks.

During the flooding process of mines the chemical composition of mine waters undergoing significant changes because due to the termination of drainage the water exchange rate decreases in mines and the contact area between groundwater and rock mass significantly increases.

It should be noted that the issue of mine water treatment in mining enterprises have been neglected for decades. Also, issues of purification of mine waters that are in storage ponds of active and closed mines remain unresolved (Horova et al. 2012).

Emission of mine gases on the earth surface, to the buildings and structures is one of the dangerous consequences of mines liquidation. Closure of mines greatly intensifies the processes of mine gases migration that can cause technogenic emergency situations and losses of life.

Also, at the liquidation of unprofitable mines issues of management of large-tonnage coal waste remain unresolved; these wastes occupy large areas and they are a source of constant negative effects on components of the environment and public health.

It should be noted that at present time there are no clear mechanisms for mine closure, which would take into account all environmental and technological consequences from the moment of termination of technical equipment operation to the development of strategy for sustainable function of coal-mining regions. In most cases the environmental problems that arise at different stages of liquidation of unprofitable mines significantly affect the state of the environment and further development of adjacent areas (Table 1).

Analysis of the data of Table 1 found that liquidation mines had an adverse impact on virtually all components of the environment and biota. Lithosphere and hydrosphere are most damaged in the process of liquidation and flooding of mines. Therefore, there is a need for the development and implementation of environmentally highly efficient technologies while closing unprofitable mines.

As a result of research in the coal mining regions the following issues developed:

– technologies of management of stress-strain state of the rock mass in "dry" and "wet" method of coal mines liquidation;

– technological schemes of redemption of mine shafts depending on the degree of environmental hazards;

– technologies of liquidation of mine workings in closed mines using plugging solutions;

– technologies of developed space stowing using available materials in the territories of liquidation of mines, including man-made;

– technological schemes of reducing the negative impact of water inflows of closed mines on the work of existing mining enterprises;

– technologies of engineering protection of earth surface from flooding and swamping;

– technologies of disturbed lands reclamation.

The results of years of research are the basis for management system for levels of technogenic and environmental safety in liquidation of the coal mines. The technological solutions obtained are particularly important in terms of large-scale destruction of infrastructure of coal mines at the territory of Donbas.

Table 1. Types and objects of negative processes influence that occur during the liquidation of mines.

Types of impact	Objects of influence				
	atmosphere	hydrosphere	lithosphere	soils	biota
The deformation of the rock mass	—	+	+	+	0
Reducing the sustainability and strength of rocks	—	+	+	+	0
Subsidence of the earth surface	—	+	+	+	+
Water saturation of rock mass	—	+	+	+	0
Violation of hydrological regime	—	+	+	+	0
The exhaustion of aquifers	—	+	+	+	0
Pollution of water bodies	—	+	+	0	+
Mixing mine water with drinking water from underground horizons	—	+	+	0	+
Flooding and swamping of land	—	+	+	+	+
Environmental pollution by combustion products and dusting of dumps	+	0	+	+	+
Deformations and destruction of buildings and infrastructure objects	—	—	+	+	+
Formation of potentially explosive methane-air mixtures	+	—	+	+	+
Emissions of mine gases to the earth surface	+	0	+	+	+
The deterioration of the environment in the territories of liquidation works	+	+	+	+	+

* "—" no effect; "0" the minimum or indirect influence; "+" the continuous direct impact

4 CONCLUSIONS

The mechanism of the rock mass and the earth surface deformation at different ways of coal mines liquidation are defined depending on mining, geological and technological conditions of development of coal deposits.

The regularities of deformation of the rock mass and the earth surface during the flooding of mine workings in the application of "wet" conservation of mines are studied.

The activation mechanisms of surface subsidence, deformation of industrial facilities, residential buildings and infrastructure objects above closed mines are defined.

Forecasting of changes in hydrodynamic and ecological situation on the territories of the coal mining regions with into account the impact of the mass liquidation of unprofitable and unpromising coal mines is completed.

Thus, using the results of years of research will allow solving problems associated with significant changes in state of geological and hydrogeological environment and ensuring the stabilization of the ecological situation in coal regions of Ukraine.

REFERENCES

Arefieva, O.D., Nazarkina, A.V., Gruschakova, N.V. & Sidorova, D.V. 2014. *Impact of Waste Coal on Chemical Composition of Soil Solutions in Industrial Areas of Abandoned Coal Mines* (*Evidence from Avangard Mine, South of the Russian Far East*). Applied Mechanics and Materials, No 448 – 453: 402 – 405.

Cao, D.Y., Huang, C.L., Wu, J., Li, H.T. & Zhang, Y.D. 2012. *Environment Carrying Capacity Evaluation of Coal Mining in Shanxi Province*. Advanced Materials Research, No 518 – 523: 1141 – 1144.

Gorova, A., Pavlychenko, A. & Kulyna, S. 2012. *Ecological problems of post-industrial mining areas*. Geomechanical Processes During Underground Mining. The Netherlands: CRC Press/Balkema: 35 – 40.

Guo, J. 2014. *Research on Dynamic Mechanism of Green Coal Mine Construction, Applied Mechanics and Materials*, No 535: 610 – 613.

Han, W.H., Lv, H. & Yang, X.J.T. 2013. *The Major Environmental Policies' Impact on the Resource-Based Regional Economy: A CGE Modeling Analysis.* Advanced Materials Research, No 869 – 870: 718 – 725.

Horova, A.I, Kolesnik, V.Ye. & Kulikova, D.V. 2012. *Physical modeling of precipitation process of the suspended materials in physical model of sedimentation tank for mine water treatment.* Naukovyi Visnyk Natsionalnoho Hirnychoho Universytetu, No 4: 92 – 98.

Hoshovski, S.V. 2000. *Hydrogeological and Geochemical Problems at Liquidation of the Coal Mines.* The Coal of Ukraine,No 7: 37 – 38.

Hu, C.C., Chen, C., Yu, Y., Zhou, X.B., Lin, S. & Zhou, S.J. 2013. *Study on the Direction of Land Reclamation and Ecological Reconstruction in Jiaozuo Coal Mining Area under the Angle of Central Plains Economic Zone.* Advanced Materials Research, No 726 – 731: 4991 – 4995.

Informational Yearbook regarding Activization of Hazardous Exogenous Geological Processes in Ukraine, according to the monitoring of EGP. 2011. Public Service of Geology and Mineral Resources of Ukraine: 88.

Jiang, W.J. & Jin, Z.G. 2010. *Problems and Countermeasure Research of Modern Coal Enterprises to Implement ERP,* Key Engineering Materials, No 455: 365 – 368.

Krupskaya, L.T., Zvereva, V.P. & Krupskiy, A.V. 2013. *Restoration of disturbed lands as the factor of environment safety and population health in mining towns in the South of the Russian Far East.* Advanced Materials Research, No 813: 242 – 245.

Krupskaya, L.T., Zvereva, V.P., Bubnova, M.B., Chumachenko, E.A. & Golubev, D.A. 2014. *Environmental monitoring of ecosystems under the impact of gold and tin mining wastes in the Far Eastern Federal District.* Pleiades Publishing Ltd.: Russian Journal of General Chemistry, Vol. 84, No 13: 2616 – 2623.

Li, Y.F., Lu, G. & Chen, W. 2011. *Sustainable Development Strategy for Energy in China.* Advanced Materials Research, No 361 – 363: 1315 – 1327.

Ma, W. Zhang, L. & Yang, J.L. 2012. A *Study on the System and Mechanism of Ecological Compensation for Coal Mining Areas: Based on the Theory of Complex System.* Advanced Materials Research, No 524 – 527: 2987 – 2991.

Pack, P. 2009. *Risk Assessment in Donieck Basin: Closure of Mines and Waste Heaps.* UNEP, GRID Arendal: 21 – 22.

Pivnyak, G.G., Pilov, P.I., Bondarenko, V.I., Surgai, N.S. & Tulub, S.B. 2005. *Development of coal industry: The part of the power strategy in the Ukraine.* Mining Journal, No 5: 14 – 18.

Qi, Q.S. 2010. *Cause Analysis and the Counter-Policies for Ecological Environment Problems of Coal Resource-Based Cities.* Applied Mechanics and Materials, No 29 – 32: 2809 – 2814.

Riznyk, T.O., Luta, N.G. & Sanina, I.V. 2006. *Assessment coal enterprises impact to changes of stream flow parameters in Donbas territory.* Bulletin of Kyiv National University named after Taras Shevchenko. Geology, No 38 – 39: 73 – 76.

Saksin, B.G. & Krupskaya, L.T. 2005. *Regional estimation of the effect of mining production on the environment.* Mining Journal, No 2: 82 – 86.

Sliadnev, V.A. 2001. *Factors of Impact of Mass Closure of the Mines on Ecological and Geological State of Donbas.* The Coal of Ukraine, No 7: 18 – 20.

Šofranko M. 2014. *Methodology of Risk Analysis of Endogenous Fire in Coal Mines.* Advanced Materials Research, No 962 – 965: 1153 – 1157.

Solovenko, I.S., Trifonov, V.A. & Nagornov, V.I. 2014. *Russian Coal Industry Amid Global Financial Crisis in 1998 and 2008.* Applied Mechanics and Materials, No 682: 586 – 590.

Tian, J.Y. & Yao, W. 2013. *Research on the Ecological Re-storation of Coal Mining Subsidence Area in Shanxi Province.* Advanced Materials Research, No 827: 384 – 388.

Wang, W., Zhang, C. & Liu, W.Y. 2013. *Impact of Coal-Electricity Integration on Chinas Power Grid Development Strategy.* Advanced Materials Research, No 860 – 863: 2600 – 2605.

Yang, Q. & Liu, W.S. 2012. *The Study on Land Reclamation Technology and Ecological Control for Coal Mining Subsidence Area.* Advanced Materials Research, No 446 – 449: 2973 – 2977.

Zhang, D. & Wang, H. 2013. *The Ecological Risk Assessment of Forest Damage in "A" Coal Mining Area.* Advanced Materials Research, No 726 – 731: 1183 – 1189.

Zhang, L. 2013. *Ecological Risk Assessment of Yulin Coal Mining Area: Based on the PETAR Method.* Advanced Materials Research, No 726 – 731: 1115 – 1120.

Zhang, L.Y. & An, W.D. 2014. *Cointegration Analysis on the Concentration Degree and Energy Efficiency of China's Coal Industry.* Advanced Materials Research, No 962 – 965: 1830 – 1835.

Zvereva, V.P., Lysenko, A.I. & Kostina, A.M. 2013. *Estimation of Effect of Hypergene Processes Proceeding in the Mining-Industrial Technogene System on the Hydrosphere of Dal'negorsk District Using Physicochemical Modeling Method (Primorsky Region).* Advanced Materials Research, No 813: 246 – 249.

Theoretical and Practical Solutions of Mineral Resources Mining – Pivnyak, Bondarenko & Kovalevska (eds)
© 2015 Taylor & Francis Group, London, ISBN: 978-1-138-02883-8

Application of temporary stall timbering works roadway in the process of roadway development while using continuous mining method

S. Felonenko, G. Symanovych & V. Gubkina
National Mining University, Dnipropetrovsk, Ukraine

ABSTRACT: The choice of the main parameters of temporary gripper-fixing devices in the process of opening development by using cutting machine is grounded.

1 INTRODUCTION

It is required to use temporary face support while carrying out roadway development and permanent mine opening in unstable and weak rocks. In that case it would be expedient to apply a design based on standard hydraulic cylinders with one or two hydraulic separations in the form of stall timbering works fixed on the mounting frame of tunneling machine (Malevich 1978, Malevich 1980).

These devices are also appropriate for increasing the stability of tunneling machine in the horizontal plane while face developing and providing its safe maintenance in mining openings (Lusenko & Ivanov 1999).

These carrier devices are fixed by trunnions to the frames of caterpillar trucks. The trunnions with fixed hydraulic cylinders and stretchers with boot drums are screwed to the brackets. Stall timbering works can be installed at an angles of $45 - 90°$ relatively a center line of tunneling machine. Outriggers are screwed to fixed conveyor section.

Working pressure in hydraulic cylinders is defined by the diagram of rock massif strain (it should not increase 70 kg/cm^2 to avoid the possibility of rolling over a tunneling machine).

2 CALCULATION OF SUPPORT PARAMETERS OF SINGLE MINE WORKING

After investigating the conditions of mining openings supports carried out by Donetsk Coal Institute named after Skochynskyi O.O. and All-Union Institute of Mine Surveying (Coal mining of Ukraine... 1998, Mostakov 1979) it was determined that cross-lined openings are more stable than openings along the strike (Baklashov & Kartoziya 1975). Based on data obtained the values of α and β indexes for the

openings of latter type, where shifting motion increases, the greatest ratio are defined. In addition some overestimation of the value of necessary compliance for cross-headings in comparison to roadways takes place.

Rock features of the developments are the most essential for rock pressure manifestation. The same indexes are not important for the rock layers located at the distance of $2 - 2.5$ m from development radius.

Thus, to calculate the roof rocks such values as reliability R and weighted average layer height being 1.5 of the development width are used. To calculate the sides weighted average rock reliability along the development height is used.

Define the development roof shift:

$$u_d = 0.1 \cdot 2 \cdot r_d \cdot \left(e^{\frac{\gamma H}{\sigma}} - 1 \right), \text{ m} \tag{1}$$

where γ – weighted average rock volume, t/m^2; H – depth of the development gobbing, m; σ – rock strength under pressure; where $f \leq 4$ – equal to $\sigma \leq 40$ MPa; r_d – development radius, m.

Define bearing capability of supports with rigid structure:

– for the development roof:

$$q_{dr} = 8 \cdot \gamma \cdot \sqrt[3]{r_d} \cdot u_d^2, \text{ t/m}^2, \tag{2}$$

– for the working walls:

$$q_{ww} = q_{dr} \cdot tg^2 90° - \frac{\varphi}{2}, \text{ t/m}^2, \tag{3}$$

where φ – angle of inner friction.

When flexibility increases, required support bearing capability reduces greatly. However, it does not mean that it could be possible by setting greater flexibility to take in rather small bearing capability as a support must provide sufficient rock support of an inelastic deformation area i.e. it must support the weight of the rocks being located in this area.

Thus, to determine necessary support bearing capability one should know the geometry of the area of inelastic deformation. Empirical dependence (Baklashov & Kartoziya 1975) of roof shifting u_d on the zone b geometry of inelastic deformation area has been found from the results of instrumental observations:

$$\frac{b}{r_d} = \frac{12 \cdot \sqrt[3]{u_d^2}}{r_d^2}, \qquad (4)$$

$$b = \frac{12 \cdot \sqrt[3]{u_d^2}}{r_d^2 \cdot r_d}, \text{ m.} \qquad (5)$$

Support pressure load of the rocks located in the area of inelastic deformation depends on their condition. Making some conscious increase of loading it is assumed that the rocks in the area of inelastic deformation have lost the connection with rock massif and become loose medium. In such a case the support will experience the pressure from the weight of the rocks comprising the area which is confined by a square parabola.

Thus, the loading of the rocks on the roof area per 1 m² is:

$$q(\alpha) = q_{ww} + (q_{dr} + q_{ww}) \cdot sin(\alpha), \qquad (6)$$

where α – angle of the axis tilt of outriggers to be determined per 1 m of the roadway area.

$$Q = q(\alpha) \cdot S = q(\alpha) \cdot 2 \cdot r_d \cdot l = 52.4 \cdot q(\alpha), \qquad (7)$$

$$2 \cdot N_1 \cdot sin(\alpha) = Q_v, \qquad (8)$$

where N_1 – thrust force in the front outriggers.

Hence, $N_1 = Q_v / 2 \cdot sin(\alpha)$. $\qquad (9)$

Upon integrating the formulas we get

$$Q_v' = \int\limits_{90°}^{0°} q(\alpha) \cdot r_d \cdot d \cdot l \cdot sin(\alpha) -$$

$$- r_d \int\limits_{90°}^{0°} (q_{ww} \cdot sin(\alpha) + (q_{dr} - q_{ww}) \cdot sin^2(\alpha)) =$$

$$= r_d \cdot q_{ww} + r_d \cdot (q_{dr} - q_{ww}) \int\limits_{90°}^{0°} sin^2(\alpha) d\alpha. \qquad (10)$$

Then, the total vertical loading is:

$$Q_v' = r_d \cdot (q_{ww} + \pi / 4 (q_{dr} - q_{ww})) \cdot l, \qquad (11)$$

where l – length of unsupported site in the face opening, m.

Thus, rock loading on stall timbering works in terms of the angle of its arrangement is determined in the following way:

$$N_1 = r_d \cdot l \cdot (q_{dr} + \pi / 4 (q_e - q_{ww})) / 4 \cdot sin(\alpha), \qquad (12)$$

Overturning moment calculated from the condition:

$$M_{over} > G \cdot x_1, \qquad (13)$$

where G – weight of extraction combine, H.

Stable moment is calculated from the condition:

$$M_{st} = N_1 \cdot \delta_1, \qquad (14)$$

$$a = \frac{h_1}{tg(\alpha)}, \qquad (15)$$

$$\delta_1 = a \cdot sin(\alpha). \qquad (16)$$

Consider the analytical model to make further calculations (Figure 1).

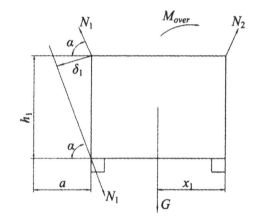

Figure 1. Calculation scheme of loading: δ_1 – force arm of force N_1; a – parameter; h_1 – height of outrigger arrangement; N_1 – thrust force in front outriggers; φ – an angle of internal friction, degree; M_{over} – moment of overturning.

Then $\delta_1 = h_1 \cdot cos(\alpha)$.

The condition for the stable and overturning moment can be calculated from formulas (17) and (18).

$$M_{st} = 4 \cdot N_2 \cdot h_1 \cdot cos(\alpha), \qquad (17)$$

$$M_{over} = G \cdot x_1 + 4 \cdot N_2 \cdot h_1 \cdot cos(\alpha). \qquad (18)$$

Using formula (18) calculate thrust force in backside outriggers:

$$N_2 = (M_{over} - G \cdot x_1) / 4 \cdot h_1 \cdot cos(\alpha), \qquad (19)$$

$$M_{over} = P \cdot l_{on}, \qquad (20)$$

where $l_{on} = h \cdot sin 45° = 5,5 \cdot sin 45° = 5,09$ m,

$M_{over} = 100\,000 \cdot 5.09 = 509\,000$ H·m. (21)

Then $N_2 = (M_{onp} - G \cdot x_1)/4 \cdot h_1 \cdot cos(\alpha)$.

$$N_2(0°) = (509000 - 490000)/4 \cdot 2 \cdot cos(0°) = 2375 \text{ H}, \qquad (22)$$

$$N_2(30°) = (509000 - 490000)/4 \cdot 2 \cdot cos(30°) = 2750 \text{ H}, \qquad (23)$$

$$N_2(45°) = (509000 - 490000)/4 \cdot 2 \cdot cos(45°) = 3360 \text{ H}, \qquad (24)$$

$$N_2(60°) = (509000 - 490000)/4 \cdot 2 \cdot cos(60°) = 4750 \text{ H}. \qquad (25)$$

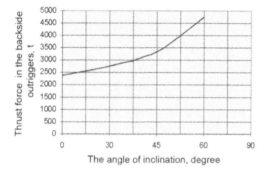

Thrust force in the backside outriggers, t — The angle of inclination, degree

Subsidence, mm — Length of gasifier, m

Using data obtained a diagram is plotted (Figure 2).

gers. The width of face space with temporary support is sufficient for carrying out all activities of roadway development cycle.

Required force in outriggers does not exceed 334 kH which is easily reached using hydraulic advancing cylinder of modern construction.

REFERENCES

Lusenko, A.M. & Ivanov, Y.P. 1999. *Some aspects of developing mining industry of Ukraine and the measures aimed at stabilizing coal extraction output.* Coal of Ukraine, No 4: 63.

Malevich, N.A. 1978. *Creating unified types of heading machines.* Coal, No 2: 9 – 12.

Malevich, N.A. 1980. *Heading machines and complexes*: textbook for higher educational establishments. 2nd edition, revised and enlarged. Moscow: Niedra: 348.

Mostakov, V.A. 1979. *The rigidity of heading machine founded on outriggers.* Mechanization of mining workings. Moscow: Issue 13: 5 – 8.

Baklashov, I.V. & Kartoziya, B.A. 1975. *The mechanics of mining rocks.* Moscow: Niedra: 271.

Coal mining of Ukraine in 1997. 1998. Coal of Ukraine, No 4: 53 – 56.

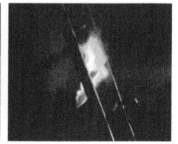

Theoretical and Practical Solutions of Mineral Resources Mining – Pivnyak, Bondarenko & Kovalevska (eds)
© 2015 Taylor & Francis Group, London, ISBN: 978-1-138-02883-8

Assessment of economic efficiency AMS-A (anchor – meshwork – shotcreting) support structure in terms of coal mines

O. Grigoriev, R. Tereschuk & L. Tokar
National Mining University, Dnipropetrovsk, Ukraine

ABSTRACT: Criterion selection to assess stability of underground mine workings has been substantiated. Efficiency of frame and rock bolting has been proved as it is that support system which is the cheapest in the context of its mounting and maintenance if mine workings are sufficiently stable. Observation results as for permanent working state with various support types in Geroiv Kosmosa mine (DTEK Company) have been set out. Comparative analysis of calculated total expenditures connected with sink of the working ant its maintenance for every experimental site has been carried out. Dependence of stability factor on direct expenditures connected with construction of the working and its maintenance for various support types has been developed.

1 INTRODUCTION

As it is known mine working support is mine technical construction built to ensure its stability and technologic safety, and to control mine pressure. Technologic safety is to attain design size of a mine working for the whole period of its operation in terms of safe passages and clearances, ventilation, arrangement of mining and maintenance facilities etc. Depending upon coal deposits deepening and mining and geological conditions complicating, cost of mining and timbering of underground working running meter experiences substantial increase due to increase in cost and wastage of support as well as labour intensity connected with its erection.

Complicated operation of underground mine workings results in increase of operating costs for their maintenance. The matter is that due to rock pressure manifestations, it becomes necessary to perform a number of extra operations (replacement of deformed or ruined support, bottom ripping or its section enlargement etc.) to protect operational factors of a mine working. Combined with expenditures connected with construction of mine workings, today the expenditures are 45% of coal products cost (Solodiankin et al. 2011a).

2 MAIN PART

Generally practical importance of research results depends on its efficiency; in the simplest case it is assessed using calculation of difference between costs of initial solution and proposed one. If certain set of compared results takes place, then one of the approaches to arrive at the most effective (balanced) alternative is an approach of certain function optimum one-dimensional search. It helps determine the most reasonable combination between cost and long-term operational stability of a structure. In our case, the structure is extended permanent mine working. At a first approximation, such a function is:

$$\omega = f(E), \tag{1}$$

where ω – a factor of a mine working stability, and E is total expenditure connected with timbering and following maintenance of the mine working.

Despite apparent simplicity, stability factor ω determined in (Shashenko 2004) as ratio between lengths of mine working sites where supports are maintenance-free and its complete length is quite sufficient to characterize state of mine working as a whole:

$$\omega = \frac{L - L_k}{L}, \tag{2}$$

where L and L_k – complete length of mine working and length of sites with support being out of keeping with safety rules respectively.

In this context if $\omega = 0$ then mine working is completely unstable; if $\omega = 1$ then it is completely stable. Practice of underground mining operations quite accurately confirms that sooner or later those sites within which supports have essential deformations being out of support pattern ($\omega \rightarrow 1$) pass into unstable state. For such unstable mine workings (starting from certain value ω determined individually for specific mining and geological conditions) proper alternations are introduced into calculation procedure for support parameters while developing new mine workings. For mine workings under operation, a value of stability coefficient helps plan expenditures as well as their volumes for future financial periods, budgeting them for mining enterprise.

In formula (1) total expenditures connected with timbering and maintenance of mine working at a first approximation are calculated as follows:

$$E = K + M ,\tag{3}$$

where K and M – expenditures connected with timbering (capital expenditures) and maintenance (operational expenditures) respectively. In any case, expenditures connected with mine working construction are considered as similar ones despite support alternative.

It goes without saying that values of capital costs and operating ones are interdependent. It is quite reasonable that large investment can provide long-term stability of any mine working without substantial operating costs (under otherwise equal conditions). On the contrary, miscounts in the process of support designing or baseless loss in supporting capacity to reduce capital costs result in repeated time-consuming maintenance activities.

As a rule, ω is determined under full-scale mining conditions. To perform the research, measuring sites were determined within main capital mine workings in mines of PJSC "DTEK Pavlogradvugillia". The mine workings are distinguished by the analogy of enclosing coal and rock mass making it possible to characterize them as uniform. Each of the mine workings has TYPS (tent yielding prolonged props support) – based frame support or frame bolting. Their finished cross-sections are 14.4 to 17.7 m².

Surveying of mines performed measurements within following mine workings:

– main western conveyor drift No 2 in Geroiv Kosmosa mine. Pitch of floor-arch support mounting is 0.5 m. A site with circular support (density of support frames mounting is 2 f/m) is also available;

– conveyor crosscut No 3 (425 m level of "Zahidno-Donbaska" mine); frame mounting pitch is 0.5 m;

– 2nd haulage crosscut of 325 m level in "Blagodatna" mine; frame rock bolting with 1.24 f/m and 7 roof bolts within a crosscut roof with 0.8 m mounting pitch is applied;

– 9th western drainage entry of 170 m level in "Samarska" mine; in practice, two sites having different mounting pitches for frame rock bolting elements are available: 1 m + 9 roof bolts and 1.25 m + 11 roof bolts (two roof bolts are mounted into walls of the drift);

– main eastern haulage drift of 490 m level in "Stepna" mine; mounting pitch of frame support is 0.33 m;

– main ventilation drift of 490 m level in "Stepna" mine; five roof bolts were mounted into roof when density of support frame was 1.5 f/m.

The research involved total observation of the state of mine workings; measuring points recorded displacements of the mine working shape according to related procedure (Solodiankin et al. 2011b, Martovitski et al. 2011, Shashenko et al. 2013) as well as support deformations; maintenance registers listed volumes of blasting, retimbering, and inrush liquidation. Observation results helped calculate value ω adequate for each of variations to support mine working (Table 1).

Table 1. Stability parameter of mine workings in mines of "DTEK Pavlogradvugillia" PJSC depending upon support types.

Site No	Mine working. Mine	Pitch of frame mounting, m	The number of roof bolts, pieces	ω
1	Main western conveyor drift Nr. 2 in Geroiv Kosmosa mine	0.5 + Floor arch	—	0.15
2		0.5 (ring)	—	0.05
3	Conveyor crosscut No 3 (425 m level of "Zahidno-Donbaska" mine)	0.5	—	0.7
4	2nd haulage crosscut of 325 m level in "Blagodatna" mine	0.8	7 (roof)	0.5
5	9th western drainage entry of 170 m level in "Samarska" mine	1	9 (roof)	0.4
6		1.25	11 (9 – roof, 2 – floor)	0.3
7	Main eastern haulage drift of 490 m level in "Stepna" mine	0.33	—	0.1
8	Main ventilation drift of 490 m level in "Stepna" mine	0.67	5	0.6

Actual expenses depend on mining and geological conditions of mine working construction and operation. Moreover, due to different approaches as for labor-intensity definition often depending on internal rules of mine resulting cost parameters under the same conditions may vary. Equipment systems are also different for the same techniques of mining, timbering and maintenance, categories of involved specialists, and norms of extra consumption. Thus, averaged cost indicators of resource elemental estimate standards (UNCR D.2.2-35-99 2000) being standard to settle investors documentation for Ukrainian mines were used.

Local cost estimates were formed to calculate capital costs for timbering and support maintenance for each alternative of supports within observational sites (Table 1).They are based upon regulations effective as on 01.01.2014 (DSTU C.SD.1.1-1:2013 2013) involving "Stroitelnye tehnologii – SMETA" software system being licensed and recommended by Ministry of Regional Development, Construction and Municipal Affairs and Housing of Ukraine. With the help of (2) stability factor ω was calculated (Table 1) and function (1) graph was formed (Figure 1).

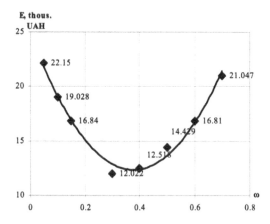

Figure 1. Dependence of changes in stability parameter on expenditures connected with mounting and maintenance of mine workings on (1...8) variations according to Table 1.

Results of the calculations performed to determine minimum expenditures made it possible to develop dependence of a value of stability parameter ω on total capital and operational costs (Figure 1) which analysis helps deduce that:

– maximum of total expenditures connected with mounting and maintenance is when values of stability parameter ω are either maximal or minimal. Sure enough that high stability of mine workings ($\omega{\rightarrow}0$) depends upon significant capital expenditures connected with mine working timbering. Predeterminedly poor supporting capacity in the context of

underfinancing factors into decrease of a mine working stability ($\omega{\rightarrow}1$) and costly measures to make the mine working stable;

– the least total expenditures are typical for frame rock bolting; that once again justifies recent idea for wide use of roof bolts while timbering mine workings. In the context of Western Donbass frame rock bolting is 1.8 times economically sound to compare with frame one;

– curve is of the form of quadratic function of $y = 2.48 - 6.6x + 8.7x^2$ type;

– if $\omega = 0.3...0.45$ then expenses connected with mounting and maintenance of mine workings is minimal.

Thus, roof bolting combined with frame one is the most economically sound providing high stability of mine workings. However, dependence in Figure 1 has been developed on the basis of statistics of state of mine workings under similar but different mining and geological conditions. Moreover, frame roof bolting like any other has reinforcement reserve owing to such known engineering solutions as tamponage, shotcreting protection etc. therefore, to determine the most effective type of support with adequate reinforcement components according to "price-quality" criterion, one should assess a state of one mine working with various support types.

To determine capital expenditures connected with timbering and operational expenditures connected with maintenance, nine observation sites have been established within the second main western haulage gate (2nd MWHG) of 370 m level in Geroiv Kosmosa mine (PJSC "DTEK Pavlogradvugillia"). Within 100 m test site (point 1 in Table 1) support of mine working corresponded to reference support pattern being TYPS-M17, 7 frames made of specific interchangeable shape (SIS) No 33 with 0.5 pitch and reinforced-concrete filler member.

Within other sites (points 2 to 9) both systems of support reinforcement and sections of SIS varied. For example, within site No 2 with total length of 40 m, density of TYPS-M 17,7 (SIS No 33) frame setting decreased to compare with test site (2 f/m down to 1.5 f/m) with extra support reinforcement using tamponage of space behind the support. Within 35 m site No 3, shape No 2 with 0.5 m frame mounting pitch was used; grid system with further shotcreting coating was applied as filler member. Similar support system was also used for 35 m site No 4 but with 0.75 m support pitch.

To reinforce support, roof bolts were applied for sites 5 to 9. Five roof bolts were mounted within interframe area; grid system with further two-layer shotcreting coating was used as filler member. Their total thickness was 250 mm. Within 50 m site No 5 a pitch of SIS No 27 was 1m. Sites No 6 and No 7 were supported using TYPS support made of SIS No 22. Support density was 1.5 f/m and 1.0 f/m respectively. For sites No 8 and No 9 (with 40 m length each) SIS No 19 was applied and support pitch was 0.75 m and 1.0 m respectively.

Surveying of the mine performed observations from 2012 to 2014. Measurement results were entered in specific registers where soil upward/downward gradient, support component damage, tramroad curvature as well as destruction of filler member (for sites No 1 and No 2) and shotcreting layer destruction (for sites No 31 and No 9) were recorded.

As observational results illustrate, heaving became apparent at each site as well as destruction of support components. Table 2 demonstrates total amounts of slashing and retimbering in terms of 100 running meters of a mine working.

With the help of (2) stability factor ω was calculated (Table 2) and function (1) graph was formed (Figure 2).

Table 2. Characteristics of support, maintenance amounts, total expenditures connected with timbering and maintenance, and stability factors for observational sites of 2nd MWHG in Geroiv Kosmosa mine.

Site No	SIS No	Support frame mounting pitch	Type of filler member / Elements to reinforce support	Capital cost of support, K, UAH mln/ 100 r.m.	Capital cost of maintenance, M, UAH mln/ 100 r.m.	Total cost, E, UAH mln/ 100 r.m.	Stability factor ω
1	33	2	reinforced concrete/ no	1.713	0.954	2.667	0.38
2	33	2	reinforced concrete / tamponage	2.278	0.417	2.696	0.56
3	27	2	grid / shotcreting	2.236	0.247	2.584	0.63
4	27	1.5		2.033	0.472	2.505	0.61
5	27	1.0		1.874	0.288	2.162	0.68
6	22	1.5	grid /	2.047	0.394	2.442	0.65
7	22	1.0	five roof bolts,	1.755	0.473	2.228	0.52
8	19	1.5	shotcreting	1.938	0.501	2.438	0.54
9	19	1.0		1.682	0.670	2.352	0.49

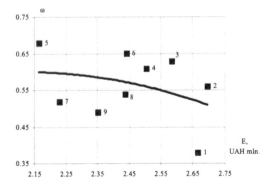

Figure 2. Dependence of stability factor ω of total expenditures connected with timbering and maintenance of observational sites within 2nd MWHG in Geroiv Kosmosa mine: (1 to 9 are alternatives of site timbering (Table 2).

Following quadratic function describes the dependence of total expenditures on stability factor:

$$y = ax^2 + bx + c , \qquad (4)$$

where $a = -2.57$, $b = 1.08$, $c = -5.32$ are approximation coefficients.

3 CONCLUSIONS

Analysis of the dependence and observational results help mention following features:

1. Stability of sites with roof bolting is preferable as compared to those without it. That concerns both cost of timbering and maintenance.

2. Test site with nonreinforced nominal support is the least stable; it is the most expensive together with site No 2.

3. Alternative 5 is optimal: TYPS-M-17.7 support made of shape No 27 with 1.0 m mounting pitch with five roof bolts and shotcreting as filling member. Graphically, this very point is the nearest to mathematical maximum of a function

$$(max = c - \frac{b^2}{4a} = -5.21).$$

4. Within stability factor, approximating curve experiences no more than 10% change. However, in this context cost reduction is UAH 534.000 per 100 r.m. When obtained absolute values of stability factor and expenditures connected with optimal (site No 5) and nominal (site No 1) alternatives of support are compared, it becomes obvious that increase in mine working stability is 1.79 times.

5. If under otherwise equal conditions a pitch of support mounting increases, mine working stability experiences its decrease; however, total expenditures experience their certain decrease too.

Therefore, economic benefit being a result of implementation of TYPS-M 17.7 support with extra reinforcement owing to five roof bolts and shotcreting is UAH 5.34 thousand per r.m. or USD 334 per r.m. at 01.01.2014 values to be very important from the viewpoint of measures taken to reduce net cost of coal products.

REFERENCES

DSTU C.SD.1.1-1:2013. 2013. *Rules to Determine Construction Cost.* RPC "Inproekt" – Efficient from 01.01-14 2013. Kyiv: "Minregion Ukrainy": 88.

Martovitskiy, A.V., Eremin, N.S., Panchenko, V.V., Khalimendik, A.V. & Ivanov, A.S. 2011. *Practice to support mine workings driven by means of counter stopes in the context of "Stepnaia" mine.* Materials of the international conference "Miner's Forum – 2011". Dnipropetrovsk, October: 43 – 49.

Shashenko, O.M., Khalimendik, O.V. & Korol, G.Yu. 2013. *On the problem of determination of criterion to estimate a state of extended mine workings in deep coal mines.* Materials of the international conference "Miner's Forum – 2013". Dnipropetrovsk, October: 126 – 130.

Shashenko, A.N. 1988. *Stability of Underground Mine Workings in Terms of Inhomogeneous Rock Mass.* Dnipropetrovsk: Doctoral Dissertation (Technical Sciences): 507.

Solodyankin, A.V., Gapieiev, S.N. & Raskidkin, V.V. 2011a. *Stabilization of mating of extended mine workings.* Messenger of Kremenchuk Mykhailo Ostrohradskyi National Univercity, No 5: 100 – 105.

Solodyankin, A.V., Khalimendik, A.V. & Kravchenko, K.V. 2011b. *Analysis of extended mine workings state in Western Donbass mines.* Materials of the international conference "Improvement of construction method for mines and underground structures". Donetsk, April: 78 – 80.

UNCR D.2.2-35-99. 2000. *Mining Workings: Construction Standards. Volume 35.* RPC "Inpoekt", Management of Price Determination Reforming, Evaluation Technique and Control of Construction Cost of State Construction of Ukraine. Efficient from 26-05-00. Kyiv: "Gosstroy Ukrainy": 488.

Theoretical and Practical Solutions of Mineral Resources Mining – Pivnyak, Bondarenko & Kovalevska (eds)
© *2015 Taylor & Francis Group, London, ISBN: 978-1-138-02883-8*

Development the concept of borehole underground coal gasification technology in Ukraine

R. Dychkovskyi, V. Falshtynskyi, V. Lozynskyi & P. Saik
National Mining University, Dnipropetrovsk, Ukraine

ABSTRACT: In the article the development concept of technology borehole underground coal gasification based on the modern technological and technical accomplishments of scientists are presented. During last 25 years researches of the Underground Mining Department improve the technology of coal seam gasification. The results of such improvement are a list of performed researches work. Technological schemes of experimental in-situ gasifier opening and preparation for different coal deposits are described. Conclusions are given according to described investigation.

1 INTRODUCTION

Modern development of coal-fired power industry in Ukraine, due to the development of non-conventional energy production technologies based on technology borehole underground coal gasification (BUCG) should be an integral part of energy and chemical complex that ensuring cost-effectiveness and safety during coal development in complex environment with receive power and chemical product.

Nowadays coal is the type of fossil fuel used by all leading countries in developing and reliability of energy, creating fuel from underground coal deposits. According to proven coal reserves in the world Ukraine ranked seventh place – 34.2 bln tons, total reserves are estimated at 117 bln tons, while oil and gas account – 2.4% of the coal reserves. Coal reserves that are fitness to BUCG according to eligibility criteria amount to 40.07 billion tons from total reserves of coal in Ukraine – 117 bln tons. The evaluation of resources that was conducted by institute "Donhiproshaht" make 20.2 bln tons balance and 3.8 bln tons of non commercial coal reserves and 1.1 bln tons commercial and 0.2 bln tons of non commercial lignite reserves, that make 45 – 55% of coal reserves, which primarily can be used for the purposes of underground coal gasification.

It should be noted, that coal are located in coal seams with thickness less 1.5 m make 90 % of all coal reserves. During the reserves depletion this rate will be decrease. Moreover, the depth of development increases and in some cases make up to 1380 m. At the same time Mining industry of Ukraine is one the most polluting industries by level of formation and emission of harmful matters into the atmosphere (Khomenko 2012, Khomenko et al. 2013).

According to "The State Service of Mining Supervision and Industrial Safety" during 2013 in the coal industry was injured 3147 miners, of which 100 – deadly. In general, the ratio of fatal accidents per 1 million tons of coal produced in 2013 was 1.2. By the number of injuries and loss of life in mines Ukraine make the top three.

Ukraine is the largest consumer of natural gas consumption and takes the sixth place. By imported gas supply Ukraine takes the third place in the world after countries such as Germany and the United States. Having an imbalance in energy resources, Ukraine only 28 – 32% of energy produced from coal. In world such rank of energy production from coal make 40%. This is due to low quality indicators of thermal coal, pricing, decreasing of lignite reserves mining, etc.

Therefore, the important factor is not only to increase the coal production but improving the technology of its processing and use. For these essential conditions are to use the technologies without the presence of people in the working mines. Usages of advanced technology of plow and augering coal extraction only for a time solve the problem. It is necessary to come on alternative clean coal technology and complex processing of solid fuels, recycling and disposal of greenhouse gases. Borehole underground coal gasification is one of the areas of such technology (Bondarenko et al. 2010, Falshtyns'kyy et al. 2013).

Development and implementation of these technologies for the extraction and complex processing of coal seams in-situ to obtain heat, electricity and chemical products, are a technological breakthrough, which will cause a qualitative change in the development of coal, power and chemical industry.

Conceptual development of borehole underground coal gasification associated with a number of technological solutions provides advanced technology applications, safety and efficacy of technological schemes in the underground part of the gasifier and in the surface coal gas purification system.

Ensure environmental cleanliness of BUCG process due to its controllability, tightness of underground gasifier and complex use of cogeneration technologies in a closed purification cycle of generator gases.

Energy complex based on BUCG is mobile-modular enterprise for intensive productivity, quality and variety of the fossil fuels products, which enables dynamic and flawed without reorienting the release of the final product, in the form of heat, electricity, chemicals and chemical products due to the flexible shifts technological parameters with regard to the conditions of dynamic change of mining and geotechnical conditions.

Profitability and efficiency of these enterprises is obvious, rising oil prices and gas, as well as rising prices for coal, which is associated with the costs of production, transportation, processing, environmental protection and the depletion of commercial reserves of energy raw materials.

These tasks require a comprehensive approach, which fully meet the concept of radical technological schemes of BUCG developed in the National Mining University in the Department of Underground Mining.

2 EXPERIENCE, DEVELOPED AND IMPLEMENTED PROJECTS IN NATIONAL MINING UNIVERSITY

Establishing the school of underground coal gasification in the State Mining Institution (Dnipropetrovsk Mining Institute, now the National Mining University) refers to the 20-th. A famous scientist, professor A. Terpigorev proposed basic principles of underground gasifiers and actively participated in the commercialization of the first stations "Pidzemhaz" in Ukraine.

Several members of the Underground Mining Department in National Mining University (Dnipropetrovsk Mining Institute) since 1968 take active part in research and development the technologies of BUCG on the stations "Pidzemgaz" in Russia (Shatska, South-Abenska station), Uzbekistan (Anhrenska station), on lignite coal deposits Synelnykovska research area (Ukraine) with scientific and research institutes from Russia (IGD Skochynskoho, "Uhlyehaz"), Uzbekistan (Tashkent Polytechnic Institute), Ukraine (Lviv Heolohorozvidochnyi Research Institute, Donetsk NDIHazu, Dniprodiproshaht, Krivyi Rig Mining Institute), Poland (Central Mining Institute in Katowice, Krakow Mining and Metallurgical Academy, Research Institute of radical technologies in Warsaw).

National Mining University under the state budget plan the work of the Ministry of Education and Science of Ukraine for the period 1991 – 2000 was performed research theme: GP-57 "To conduct research, develop recommendations and specifications for the establishment and development of industrial production of artificial energy through the controlled process of underground coal gasification" state registration number 01910054149; PRO-1 "Development and application the technologies and equipment for underground coal gasification and complex processing of its products in order to obtain highly efficient energy and chemicals" state registration number 0194U008928, VF-3 – "Research the underground coal gasification", under a contract with the firm JCQAL, Japan Center for Energy and coal. 2007 – 2010; the realization of the joint project "HUGE: Hydrogen Oriented Underground Coal Gasification for Europe" (Figure 1) with the creation of experimental in-situ gasifier and its testing in mine "Barbara" (funding by the Fund "Coal and Steel", contract No RFCR-CT-2007-00006). From 2010 up to 2013 at support the Ministry of Education and Science of Ukraine and companies "Donbass Fuel-energy company", "Donetsksteel" was conducted analytical and laboratory research on the experimental union for determination the opportunities of thin coal gasification in the Central Donbass, Western Donbass (Ukraine) and Kuzbass deposits (Russia).

The complexity of the experiment allow to confirm a number of analytical solutions and bring them to the real recommendations un order to designe the power-chemical enterprises for production and complex processing the coal seam s in the place of occurrence (Falshtynskyy et al. 2012)

As a result of experience, taking into account the critical analysis of the achievements and limitation in the operation of industrial plants "Pidzemhaz" in the Soviet Union, experiments conducted on experimental underground gasifier in the United States, United Kingdom, Poland, Belgium, Czech Republic, Germany, China, Australia, Japan and during research the processes borehole underground coal gasification on laboratory unit, using mathematical modeling, determination and developed radical technologies BUCG that provide comprehensive processing of coal seams with account the changing in pressure integrity, controllability of a system, safety and quality productivity of mobile complex.

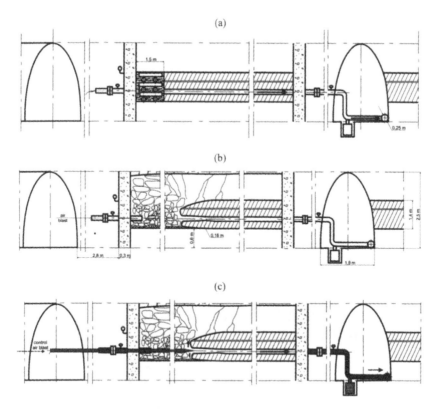

Figure 1. Technological schemes of experimental in-situ gasifier with air blast supply: (a) start of gasification (coal ignition); (b) coal seam gasification; (c) coal seam gasification by controlled retraction injection point.

On the basis of scientific results was developed three working project of "Pidzemhaz" stations for lignite coal "Techno-economic justification of building the station of "Pidzemhaz" Synelnykove deposits (Figure 2, a), "The pilot project of underground gasifier" Monastyrschynske deposit (Figure 2, b) and "Project of experimental station "Pidzemhaz" for "Donbass Fuel-energy company" (Figure 3).

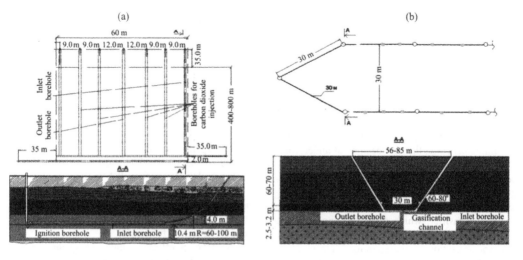

Figure 2. Technological schemas of opening and preparation the underground gasifier for: (a) Synelnykove field and (b) Monastyrschynske coal deposit.

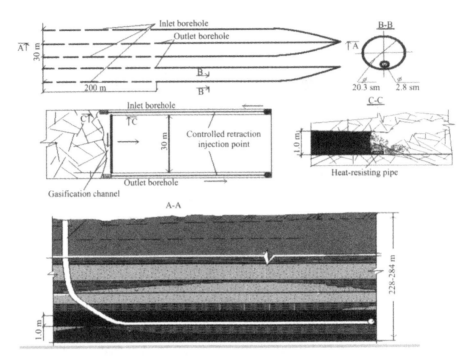

Figure 3. Technological schemas of opening and preparation the underground gasifier pilot station for Donbass Fuel-energy company "Pavlogradvugillia".

Improving technology BUCG associated with the implementation of surface complex power and chemical complex in underground part that are copyrighted by series of patented inventions in Ukraine and abroad (96 patents). The main researches are publishing in 204 publications (articles, monographs, textbooks etc.). For the conditions of Ukrainian energy and power sector the research and justification the possibility of alternative mining technologies are essential. Substantial deposits of coal can be converted to generators gas turn-on commercially reasonable level, to solve problems of providing specific kind of energy and political aspects of the energy security (Falshtynskyi et al. 2014, Lozynskyi et al. 2014).

The essence of the "Know-How" of modern developments in the field of underground coal gasification, that was developed on the Department of Underground Mining (National Mining University) resides:

– implementation the new designs of underground gasifier, technology and technical solutions in processing complex, control of the thermo-chemical gasification process, taking into account the geological conditions and technological parameters of underground gasifier, which gives the opportunity to develop technical documentation for the construction and operation of the stations "Pidzemgaz" in preparation of underground gasifier from the surface and in mining conditions, with gasification commercial and non-commercial reserves in the conditions of high rock massif fracturing;

– development technological schemes of BUCG and constructions of underground gasifier with artificial rocks hermetization, directed and selective blast injection, material and thermal balance control of the gasification process.

3 KEY FINDING

Development concept of technology borehole underground coal gasification allows:

– to substantiate the fitness criteria of sufficient impermebility of underground gasifier in the real conditions with account mining and geological conditions, technical and technological parameters of coal seam gasification;

– to expand area of application the technology BUCG on shallow workings (20 – 370 m) with productive bed thickness from 0.4 to 30 m;

– to develop layer by layer coal reserves gasification in descending and ascending order with injected stowing of deformed rocks of roof and goaf with account rock massif fracturing;

– to develop effective and safe system of coal seam through the horizontal production wells;

– to ensure activation of the oxidizing and reducing zone in gasification channel of underground gasifier with variation of air injection in time and space by combining injection and exhaustion of air-flue gas mixture for activation the gasification processes in gasifier channel;

– to build adaptive scheme of underground gasifier with the preparation from the surface and mine conditions, taking into account geological and man-made conditions, technical and technological parameters of the gasification process and changes in geomechanical factor;

– to provide technological and environmental safety of BUCG through controllability of material and thermal balance during thermochemical conversion process of the coal seam in an oxidizing and reducing zones in gasifier channel, which is achieved by providing the stowing operations, controlled retraction injection point on a combustible face through flexible pipes in opening inclined borehole, supply pulsing blast injection and reverse zones of underground gasifier;

– to increase the pressure in the gasification channel up to 5.7 MPa (upon conditions injection stowing of rock massif and goaf in underground gasifier) for intensifying thermochemical process by effectively selecting blast mixtures including carbon dioxide (CO_2) which ensures its utilization and improves combustible composition;

– to improve the efficiency and effectiveness of the process by reducing the cost value of gasification products on $33.6 - 45.1\%$, while the cost of $1000\ m^3$ of combustible gases will be $389 - 878$ UAH; cost of 1 kWh of electricity – $0.24 - 0.31$ UAH, given the cost of underground gasifier (underground part) – $12.5 - 44.1$ mln \$. The length of the extraction panel, respectively ($250 - 500$ m), the cost of the station "Pidzemhaz" provided performance from 55 to 100 thousand m^3/h of gaseous products that include $4 - 5$ underground gasifier, mobile-block complex of coal gas purification system on the industrial site – $135.2 - 409.6$ mln \$. Economic playback time – $1.3 - 2.7$ years;

– to reduce the loss of air and gasification products on $9.8 - 28.6\%$, increase the calorific value of combustible gases to $46 - 61\%$ using to produce heat and electricity free-piston gasifier with efficiency coefficient – $0.72 - 0.85$.

4 CONCLUSIONS

Technological scheme of borehole underground coal gasification, that are described allows to increase range of application the technology BUCG and in an environmentally safe and closed cycle with chemical products, energy and heat generation.

The technology is effective at commercial and non-economic coal gasification in mine, do not requires a significant investment in deposit opening, mine development, utilization and waste storage, extends the long life operating of mines, solves a major social problem of employment.

REFERENCES

Bondarenko, V., Tabachenko, M, Wachowicz, J. 2010. *Possibility of production complex of sufficient gasses in Ukraine.* New Techniques and Technologies in Mining. The Netherlands: CRC Press/Balkema: 113 – 119.

Falshtynskyi, V., Dychkovskyi, R., Lozynskyi, V., Saik, P. 2014. *Some aspects of technological processes control of in-situ gasifier at coal seam gasification.* Progressive Technologies of Coal, Coalbed Methane, and Ores Mining. The Netherlands: CRC Press/Balkema: 109 – 112.

Falshtyns'kyy, V., Dychkovs'kyy, R., Lozyns'kyy, V., Saik, P. 2013. *Justification of the gasification channel length in underground gas generator.* Annual Scientific-Technical Colletion-Mining of Mineral Deposits 2013. The Netherlands: CRC Press/Balkema: 125 – 132.

Falshtynskyy, V., Dychkovskyy, R., Lozynskyy, V., Saik, P. 2012. *New method for justification the technological parameters of coal gasification in the test setting.* Geomechanical Processes During Underground Mining – Proceedings of the School of Underground Mining. The Netherlands: CRC Press/Balkema: 201 – 208.

Khomenko, O., Kononenko, M., Myronova, I. 2013. *Blasting works technology to decrease an emission of harmful matters into the mine atmosphere.* Mining of Mineral Deposits 2013. The Netherlands: CRC Press/Balkema: 231 – 235.

Khomenko, O.Ye. 2012. *Implementation of energy method in study of zonal disintegration of rocks.* Naukovyi Visnyk Natsionalnoho Hirnychoho Universytetu, No 4: 44 – 54.

Lozynskyi, V.G., Dychkovskyi, R.O., Falshtynskyi, V.S., Saik, P.B. 2015. *Revisiting possibility to cross the disjunctive geological faults by underground gasifier.* Naukovyi Visnyk Natsionalnoho Hirnychoho Universytetu, No 4.

Theoretical and Practical Solutions of Mineral Resources Mining – Pivnyak, Bondarenko & Kovalevska (eds)
© 2015 Taylor & Francis Group, London, ISBN: 978-1-138-02883-8

Analytical, laboratory and bench test researches of underground coal gasification technology in National Mining University

V. Falshtynskyi, R. Dychkovskyi, V. Lozynskyi & P. Saik
National Mining University, Dnipropetrovsk, Ukraine

ABSTRACT: In the article results of analytical, laboratory and bench test researches presented. The present paper deals with the creation of generator gas from coal seams in-situ. It is proposed to use received results to intensify underground coal gasification process. The results of laboratory and bench test experimental researches are presented in the form of graphs and tables. The conclusions are drawn at the end of the paper with perspective trends for future studies.

1 INTRODUCTION

The analysis of coal deposits development under modern conditions shows the necessity of new solutions for a line of problems to provide safety of mines exploitation, complex development of mineral resources and protection of the environment. One of these problems is a mine methane utilization recovered to the surface with various degassing methods, and methane taken onto the surface by ventilation current (Ovchynnikov et al. 2013).

Development of the underground coal gasification (UCG) technology with artificial hermetization of rock massif with effective and mobile adaptation to changing geological conditions and technological parameters required to elucidate the dependence of rock massif behavior during coal seam gaification that was need to substantiate the rational parameters of the rock massif and goaf backfill injection technology (Falshtynskyi 2014).

Underground coal gasification can be examined as one of examples of generator gas reception which carry out new principle – combination of coal extracting with its simultaneous processing in the unique technological process (Bondarenko et al 2010).

From each and every factors that affect to substantiation fitness criteria of coal deposits for borehole underground coal gasification is difficult to distinguish the principal, so the choice of deposits is carried out on the total impact of various alternative factors. They are defined by the presence in the gasification zone – combustion source, which supports by the achievement of the desired temperature, the formation of artificial gas, as well as features of the geological environment, hydrogeological conditions, forecast water inflows, abatement techniques, leaks of blast injection and gas, quality of coal seam.

2 ANALYTICAL, LABORATORY AND BENCH TEST RESEARCHES OF THE UCG TECHNOLOGY

As a methodological approach on the Underground Mining Department "National Mining University" were substantiated the fitness criteria of coal deposits to underground coal gasification, study guide: "Development and substantiation the fitness criteria to assess the feasibility of underground coal gasification" Dnipropetrovsk: 1987. This technique has passed industrial approbation and received confirmation of its effectiveness in the development of technical documentation and underground gasification project feasibility study for the conditions Sinelnikovske and Tarnavske deposits and mines in the Western Donbass.

For the calculation of materially-thermal balance of BUCG, the program MTBalanse SPGU was utilized. It was designed by the employees from the underground mining department from the National Mining University. The calculation algorithm includes thermo chemical conversions of solid fuel into gas and condensed fluid in the conditions of elementary composition of coal seams, external water inflow and the thermal balance of in situ gasifier.

The program for calculating material and thermal balance parameters of BUCG processes takes into account the following conditions: changes of anthropogenic situations in rock layers that contain in situ gasifier qualities taking into account mining-geological conditions and technological parameters of the process; the peculiarity of the composition of air blast mixtures and their influence on coal seam gasification processes; the change of qualitative and quantitative indices of BUCG gas with grades of coal seams and air blast mixture; the influence of geometrical parameters of oxidizing and the restoration zone of gasifier reactions channeled on the balance of kinetic indices of chemical reactions and physical rates; the influence of coal seam degassing efficiency on thermal balance; the influence of gasification process ballast gases on the qualitative indices of in situ gasifier; the practicality of the substantiation of balance calculation parameters for the prediction of production indices manageability for the "Pidzemgas" station. For the effective design of mine workings is often resorted to the use of mathematical modeling as nature experiments are not always available (Vladyko et al. 2012).

With the purpose to provided reliability and efficiency of underground gasifier was developed the technology of sufficient hermetization of rock massif around the gasifier by injection backfill material in deformed rocks of roof and goaf according load distribution in rock layers (Figure 1).

To solve this problem, the mathematical models of rock massif behavior during coal seam gasification was created, that takes into account the immediate rocks of roof and bottom, exposed to high temperatures to swelling, filling goaf and features of coal seam gasification depending on seam thickness.

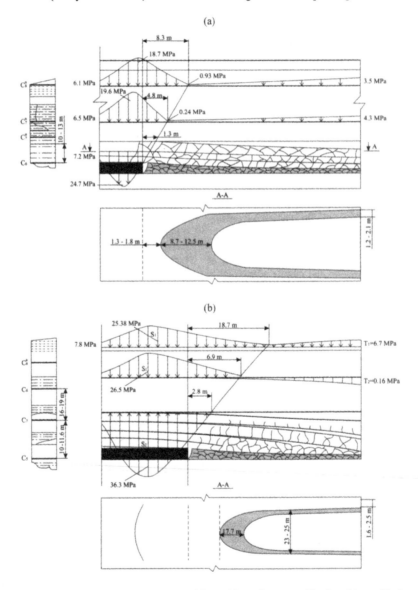

Figure 1. Load distributions in rock layers above the gasifier at thin coal seam gasification: (a) area No 1, coal seam C_5; (b) area No4; coal seam C_6.

Based on the mathematical model the proposed algorithm for calculating the parameters of behavior and stratification of rock massif are developed, based on characteristics of coal seam gasification process (Falshtynskyy et al. 2012).

The specific geological conditions and technological parameters of gasification process during the calculation are entered in personal computer. After completion of calculation the geometrical and physical stress parameters, subsidence and horizontal displacement of rock layers in the table form, graphs and diagrams are displayed on the screen. Deformations and rocks subsidence in an underground gas generator take a place under the action of two factors. A primary factor is mining pressure. The second factor is a high temperature of gasification process at gas generator exploitation (Falshtyns'kyi et al. 2013).

Analyzing Figure 1 we can say that bearing pressure zone at the level of coal seam C_7 during gasification of coal seam – C_5 with the gasification advance 1 m/day, is located in rock massif at a distance 2.8 m from the gasification channel above underground gasifier. During coal seam C_5 gasification above goaf on the distance 10 – 16 m is formed stratification cavity.

During analytical substantiation of heat transfer in the system "gasifier-rock massif" was considered quasi-stationary regime, non-isothermal flow of high-temperature products, taking into account their heat with rock massif in the conditions of coal seam gasification. Rock massif has isotropic thermal conductivity. On the rocks surface take heat exchange with the gaseous products that under pressure flow deep into the rocks of roof by vertical cracks and stratification, occurring under strain in coal seam.

Analyzing the calculations that were performed to substantiation the research work it should be noted that the convection heat transfer and temperature change of migratory gaseous products in rock massif is sensitive to significant changes in the thermal conductivity of rocks. Thus, the reduced thermal conductivity in 1.5 times and 2 times increased specific heats of gases (3.2 – 6.0 m distance from the coal seam) temperature is changed to 222^0C ($318 – 96^0$C). In convection heat transfer in rock massif accounts for 54 – 65% of all heat loses, the second part is related to heating rocks by conductive heat exchange (Figure 2 and 3).

Mass and thermal balances are substantiated for analysis and efficient use the technological process of BUCG. By dint of mass and thermal balances establish the actual yield, the efficiency of energy use, costs, loss of raw materials, fuel and other materials. From digital data of both balances set organizational and technical solutions to improve the operation of the equipment, make the best use, disposal or recovery of material and energy resources.

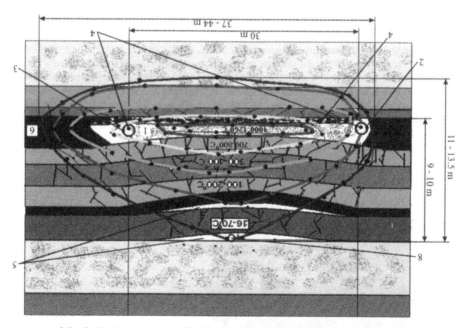

Figure 2. The parameters of distribution the temperature field in the rock massif during borehole underground coal gasification across stratification; 1 – reaction channel of underground gasifier; 2 – air injection borehole; 3 – gas production borehole; 4 – ash and slagged ash; 5 – stratification of rocks of roof; 6 – coal seam; 7 – flexible pipe; 8 – stowing pipe.

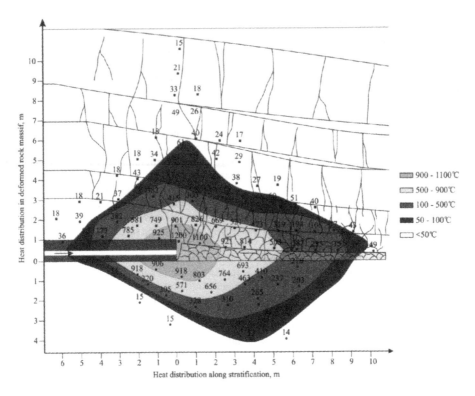

Figure 3. The parameters of distribution the temperature field in the rock massif during underground coal gasification along stratification.

Conducting experiments in laboratory and bench units focused on simulation process of BUCG in terms of geological modeling and technological parameters (Ovchynnikov 2013). To simulate on laboratory and bench unit the process of underground coal gasification was developed a mathematical model of scale factors and similarity conditions. Laboratory model has dimensions 100×100×12 cm (Figure 4). In the model pack the rock layers for simulation rocks of roof (direct and main roof) and coal seam.

To simulate the combustion face thermal element was installed in the hole. The main task was modeling the thermal field during gasification in the terms of compliance with the laboratory model similarity of field conditions. Based on the laboratory results the graphs that characterize the heating of rock massif, during coal gasification, were built (Figure 5 and 6).

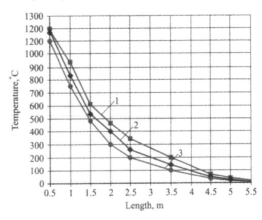

Figure 5. Heating the rocks of roof: 1 – pressure in combustion face 0,3 MPa; 2 – pressure 0,6 MPa; 3 – pressure 0,4 MPa and oxygen injection O_2 – 31%.

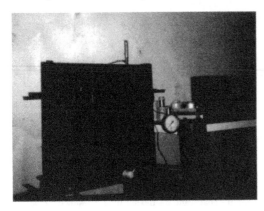

Figure 4. Laboratory model of gasifier.

100

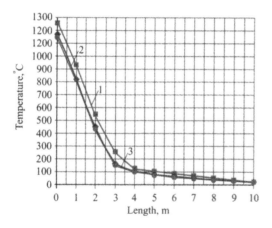

Figure 6. Heating the bottom rocks: 1 – pressure in combustion face 0,3 MPa; 2 – pressure in combustion face 0,6 MPa; 3 – pressure in combustion face 0,4 MPa and oxygen injection O_2 – 31%.

Experimental study of rock massif behavior and study the temperature field in the rocks that contain gasifier, held on bench units in 1988 at the mine No1 "Hostra" colliery group "Kurahovske" SC "Selidovovygillia" and in 2002 at the mine "Ternivska" SC "Pavlogradvugillia". Bench unit represents a simplified model of underground gasifier. General views of the main part of experimental test unit are presented on Figure 7 and 8.

Figure 7. Experimental bench unit.

Analyzing the experimental results, we can say that at a distance of 1.5 m (6 m in naure) from the coal seam, the rocks of roof heating by gaseous products of underground gasification due to their migration through cracks and stratified rock massif. At a distance of 1.5 – 2.3 m layer (6 – 9.2 m in nature) intensity of heaing reduce trough decrease the size of irregularities in the stratified rock massif. The heatming degree of rocks of roof, depending on the composition of injected (air, oxygen) and pressure integrity of rock massif are shown on Figure 9.

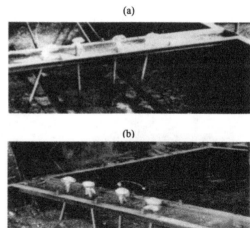

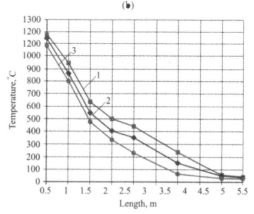

Figure 8. Formation of coal seam and rock massif in the gasifier model: (a) bottom and wells; (b) rock massif.

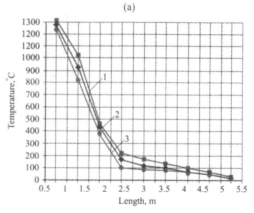

Figure 9. Heating the rocks of roof (a) and bottom rocks (b): 1 – air injected enriched oxygen (O_2 – 30%); pressure P = 0.4 MPa; 2 – injected air, pressure P = 0.6 MPa; 3 – injected air, pressure P=0.4 MPa.

101

In 2009 – 2010 the experimental unit for simulation the technology of borehole underground coal gasification was constructed and patented in National Mining University (Falshtynskyy et al. 2012).

Modeling the processes of coal seam gasification in bench unit with account mining and geological conditions for SC "Pavlohradvugillia" was conducted in 2011 (Figure 10).

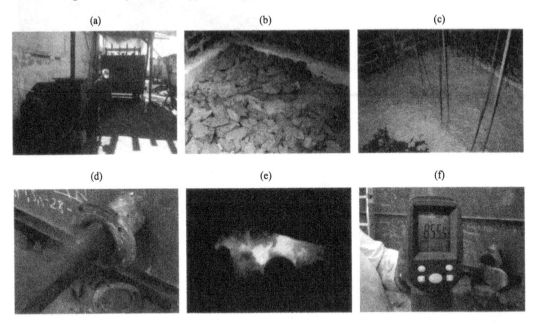

Figure 10. Installation of experimental unit No 1: (a) general view of experimental unit; (b) formation of rock massif; (c) installation of ranging mark and thermocouples in rocks of roof; (d) coal ignition; (e) formation of gasification channel; (f) fixation of the temperature in gasification channel.

Parameters of ignition, gasification and starvation during underground coal gasification in the experimental laboratory unit at the different regimes of the thermal processing with fixation temperature conditions and output of combustible generator gases are presented on Figure 11.

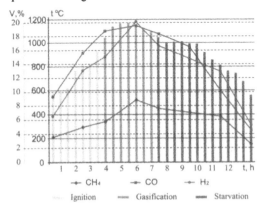

Figure 11. Temperature condition ($t°$) and output of combustible components (V) dependently to time from the beginning of experiment (t).

The results of analysis of liquid products coal gasification that was gotten during an experiment on the laboratory unit are presented in table 1.

Table 1. Results of runback analysis.

No	Name of indexes	Amount, mg/m³	Amount, g/m³	kg/612kg of coal	g/kg of coal
1	Phenols	7500	7.5	4.60	7.53
2	Hydrogen sulfide	11900	11.9	7.31	11.95
3	Volatile ammonia	3100	3.1	1.90	3.11
4	Fixed amonia	2580	2.58	1.58	2.59
5	Resinous substance	5450	5.45	3.34	5.47
6	Aromatic compounds	100	0.10	0.06	0.10
7	Ammonium: thiosulphate	112000	112	68.80	112.43
7	Ammonium: sulfate	27000	27	16.58	27.10
	At all	142630	142.63	87.62	143.18

As a result of experiment on the experimental unit was gotten the information about rocks, which contain gasifier, composition of generator gas, parameters of coal seam ignition, burning of gasification channel with application the combined mode of coal combustion, rock massif subsidence above gasifier.

A runback is a liquid with the presence of hard matters and not characteristic for a coke-chemical enterprise smell which reminds the smell of tar. At the protracted contact with air a test acquires a black. Power indexes of gasification process, during an experiment on the laboratory unit presented in Table 2.

Table 2. Power indexes of gasification process.

Amount of coal, kg	Output of generator gas, m³		Thermal power of the experimental unit, GKal		Power of the experimental unit , MVt	Q, MJ/m³	Output of chemical products g/kg of coal
	generator gas	(CH₄, CO, H₂)	for 1 h	for 15 h			
612	143.1	529.4	2.89	43.35	3.35	2.13	0.143

Subsidence of rocks of roof in an underground gasifier (Falshtyns'kyy et al. 2013) and traditional mining (Koshka et al. 2014) is related to the mine and geological conditions, technological parameters of underground gasifier and process of coal seam gasification. Lines which characterize the changes of the rocks of roof subsidence during coal seam gasification in experimental laboratory unit are presented on Figure 12.

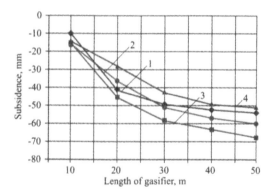

Figure 12. Subsidence of rocks of roof at gasification in experimental laboratory unit: 1 – 4 – ranging marks in the direct roof.

The methods of laboratory researches included a choice and preparation of equivalent materials, formation and preparation of models, carrying out of experiment (Kononenko & Khomenko 2010).

Analysing the results of experiments, it is possible to establish, that in the distance 0.5 m (2 m in nature) from a seam, the rocks of roof are added the intensive warming up by the gaseous products of borehole underground coal gasification due to their migration on cracks and stratifications in rock massif.

Subsidence of rocks of roof in a maximum made 67.2 mm, that it is related to the well-organized lowering and ash presents, which make 15 – 18% from coal seam. Rocks swelling above a combustion face was calculated, coming from the coefficient of swelling $K_{sw} = 1.27 – 1.42$ (depending the chemical compositions of rock massif).

In 2012, the experimental research on the bench experimental unit for modeling the underground coal gasification process was continued in a specially equipped room inside the territory of Donetsk electrical plant (Lozynskyi et al. 2014). It is one of the core companies of the group "Donetsksteel", in order to establish the parameters of gasification process in respect to coal seams of Solenovsk deposit (Figure 13).

(a)   (b)   (c)

Figure 13. Installation of experimental unit No 2: (a) general view of experimental unit; (b) formation of rock massif with installation of ranging mark and thermocouples in rocks of roof; (c) coal ignition with temperature control.

As a result of the experiment on the test bench unit of the BUCG were obtained: the heating parameters of temperature field around the gasifier, ignition and burning of the gasification channel in combined cycle. The changes in the energy balance of the reaction zones of the gasifier depending on the channel type of the blast injection and natural processes of generator gas composition are presented (Figure 14 and 15). The energy balance of the reaction zone of the gasification channel allows us to estimate the balance between chemical reactions and physical processes of gasification rates.

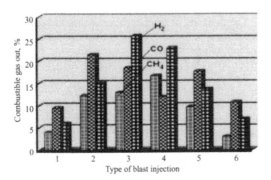

Figure 14. Combustible gas out (generator gases CH_4, CO, H_2) with account the similarity factor depending on the type of blast injection and physical processes.

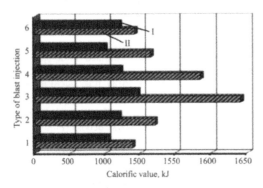

Figure 15. Energy of chemical zones in gasification channel depending on the type of blast injection and physical processes: I – oxidizing zone (exothermic reaction); II – reducing zone (endothermic reaction).

In Figure 15 and 16 number 1 – blast injection; 2 – blast injection enriched with O_2 – 25% post-reverse regime of gasification; 3 – blast enriched with O_2 – 30% and 20% steam (T = 128°C); 4 – blast enriched with O_2 – 27% and 15% steam (T = 128°C); 5 – blast enriched with O_2 – 25% (O_2 supply in a pulsed regime) and 12% steam (T = 128°C); 6 – steam-air blast injection (in pulsed regime) O_2 – 21%, 10% steam (T = 128°C).

Results of experiments showed that at a distance – 0.4 m (1.6 m in nature) from the rocks of roof above underground gasifier subjected to intense heating of the gaseous products of BUCG (P = 0.25 – 0.3 MPa) due to their migration cracks and in percarbonic rock massif. Varying pressure in the oxidizen zone (exothermic processes) of gasification channel and taking of generator gases from the restoration zone (endothermic processes) in a gasgenerator with providing equilibrium of physical speeds and kinetic reactions, provided the combined serve of blowing mixture in the pulsating mode (Falshtyns'kyy et al. 2013).

A temperature was controlled by the temperature-sensitive elements which were fixed in the TERA "Devices Systems" Firebird 2.1 database. The temperature balance and combustible gases outlet depending on the different blowing components during the time on the experimental unit is shown in Figure 16. The intensity of the warming of roof falls at the reducing expense of the sizes of breaking rock layers, the conditions of conductive heat exchange is observed.

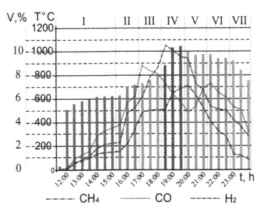

Figure 16. Temperature balance (t°) and combustion gases outlet (V) depending of different blowing mixture in a time: I – Ignition; II – air blast injection; III – blast injection enriched with O_2 – 25% post-reverse regime of gasification; IV – blast enriched with O_2 – 30% and 20% steam; V – blast enriched with O_2 – 27% and 15% steam; VI – 25% (O_2 supply in a pulsed regime) and 12% steam; VII – steam-air blast injection (in pulsed regime) O_2 – 21%, 10% steam.

On Figure 16 is shown, those during oxygen injection the temperature balance and output of combustible gases were increased very fast. This is particularly evident after the reversal air injected, when the injected borehole is becoming a gas production borehole and inversely (gas production borehole is becoming injected borehole). At the same time during steam-air injection is performed extinction of coal burning particles. Then, we need to provide re-activation of the gasification process by oxygen injection.

The distribution of temperatures on length of the reactionary channel is connected with the length of the channel, its section, the quantitative and qualitative structure of the blowing mixture and received gases, the extent of deformations and the temperature indicators of rocks.

The methods describing adequately the phenomena of rock behavior cannot be described on the base of existing concepts of physics (Khomenko 2012).

Rocks of roof subsidence above the gasifier were fixed by the Monitor QB program. Differences between general results following investigation make up 1 – 8 marks. The curves characterizing the rocks of roof subsidence are presented in Figure 17.

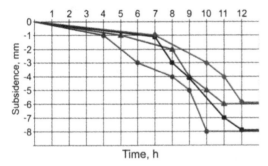

Figure 17. Rocks of roof subsidence in the process of goaf formation: 1 – 4 number of ranging mark.

With the help of smoke exhaust in gas-pipeline formed estuary discharge point (discharge), which could provide additional waterproofing of combustion channel, due to redistribution of static and dynamic groundwater of overlaying rock massif (Falshtynskyi et al. 2014).

As we can see from the subsidence schedule (Figure 17), the rocks of roof subsidence by a maximum 8 sm, this is connected with the ordered subsidence on an ash residue (17 – 22% from the thickness of the gasified coal seam) and rocks swelling above the combustion face with account the factor of swelling – $K_{sw.} = 1.4$. The error of conducted results makes between 9 to 27%, with an average – 24.7%.

4 CONCLUSIONS

Technological scheme of borehole underground coal gasification, that are described allows to increase range of application the technology BUCG and in an environmentally safe and closed cycle with chemical products, energy and heat generation.

The technology is effective at commercial and non-economic coal gasification in mine, do not requires a significant investment in deposit opening, mine development, utilization and waste storage, extends the long life operating of mines, solves a major social problem of employment. Authenticity of model on the laboratory unit is provided by criteria of similarity to the mine and geological conditions of nature.

The combined system of air injection and removal of generator gases allows to reduce time on ignition for 40% and to increase the speed of thermal preparation of the gasification channel of a gas for 32% at the reduction of expenses on blast injection for 15 – 18%. The variation of injection blast (air, oxygen-aerial, steam-oxygen-aerial and steam-air) gives the chance to derive the gasification process on necessary level of power gases. Reversive modes of oxidizing and reducing zones of underground gasifier is obligatory for maintenance of thermal balance of gasification process and provides a mode of kinetics of endothermic reactions during coal seam gasification. At a stable mode of gasification the outlet of generator gases make $1.82 – 2.49 \ m^3/kg$ of coal with calorific value $2.05 – 2.36 \ MJ/m^3$ and composition of combustible components in generator gas CH_4 – 3.6 – 15.2%, CO – 9.5 – 22.3%, H_2 – 6.2 – 25.8%, output of chemical matters 0.14 kg/kg coal.

Stability of coal gasification at rocks instability at the increase of goaf in gasifier are provided: by sufficient impermeability of underground gasifier, varying of temperature and pressure in blowing, by point of transfer, and also support and control the thermal and adiabatic mode of zone in reactionary channel.

Control of temperature and pressure regime during gasification process allow to define the parameters of injection blast pressure and mode of inlet and outlet streams, allowing to provide stability of gasification during discontinuity of the rocks of roof and goaf growth. Warming the rocks of roof during gasification process carried out by convection – 56.7% and conductive – 43.3% of heat transfer.

Upper limit of heating the rocks of roof above underground gasifier make 8.4 – 9.2 m and is located between oxidizing and restoration zone. Heating the rock massif in the direction of stratification is in 1.2 times less than normally and distributed in rock massif on distance 3.9 – 4.7 m from the combustion face. The volume of heated rocks of roof up to the temperature 100°C make 19.6%.

In goaf rocks collapsed in the direction of stratification are heated to 50°C on the distance 7.6 – 11.5 m from the combustion face. Volume of deformed and collapsed rocks that are heated up to 100°C above underground gasifier make 35.2%.

Thermal conductivity of disturbed rock around underground gasifier is characterized by the influence of convection (rock formation in the roof) and conduction (bedrock) in heat transfer and stored in per-cubic dependence on the intensity of fracturing rocks size of cracks and.

REFERENCES

Bondarenko, V., Tabachenko, M, Wachowicz, J. 2010. *Possibility of production complex of sufficient gasses in Ukraine*. New Techniques and Technologies in Mining. The Netherlands: CRC Press/Balkema: 113 – 119.

Falshtyns'kyy, V., Dychkovs'kyy, R., Lozyns'kyy, V., Saik, P. 2013. *Justification of the gasification channel length in underground gas generator*. Annual Scientific-Technical Colletion – Mining of Mineral Deposits 2013. The Netherlands: CRC Press/Balkema: 125 – 132.

Falshtynskyi, V., Dychkovskyi, R., Lozynskyi, V., Saik, P. 2014. *Some aspects of technological processes control of in-situ gasifier at coal seam gasification*. Progressive Technologies of Coal, Coalbed Methane, and Ores Mining. The Netherlands: CRC Press/Balkema: 109 – 112.

Falshtynskyy, V., Dychkovskyy, R., Lozynskyy, V., Saik, P. 2012. *New method for justification the technological parameters of coal gasification in the test setting*. Geomechanical Processes During Underground Mining – Proceedings of the School of Underground Mining. The Netherlands: CRC Press/Balkema: 201 – 208.

Khomenko, O.Ye. 2012. *Implementation of energy method in study of zonal disintegration of rocks*. Naukovyi Visnyk Natsionalnoho Hirnychoho Universytetu, No 4: 44 – 54.

Kononenko, M., Khomenko, O. 2010. *Technology of support of workings near to extraction chambers*. New Techniques and Technologies in Mining. The Netherlands: CRC Press/Balkema: 193 – 197.

Koshka, O., Yavors'kyy, A., Malashkevych, D. *Surface subsidence during mining thin seams with waste rock storage*. Progressive technologies of coal, coalbed methane, and ores mining. The Netherlands: CRC Press/Balkema: 229 – 234.

Lozynskyi, V.G., Dychkovskyi, R.O., Falshtynskyi, V.S., Saik, P.B. 2015. *Revisiting possibility to cross the disjunctive geological faults by underground gasifier*. Naukovyi Visnyk Natsionalnoho Hirnychoho Universytetu, No 4.

Ovchynnikov, M., Ganushevych, K., Sai, K. 2013. *Methodology of gas hydrates formation from gaseous mixtures of various compositions*. Annual Scientific-Technical Colletion – Mining of Mineral Deposits 2013. The Netherlands: CRC Press/Balkema: 203 – 205.

Vladyko, O., Khomenko, O., Khomenko, O. 2012. *Imitating modeling Stability of mine workings*. Geomechanical processes during underground mining. The Netherlands: CRC Press/Balkema: 147 – 150.

Results of applying drilling and injection technologies for strengthening soil massif

Ye. Tsaplin
Geological and Prospecting Expedition, Antratsyt, Ukraine

V. Rastsvietaiev, K. Dudlia & T. Morozova
National Mining University, Dnipropetrovsk, Ukraine

ABSTRACT: Results of work performance as for strengthening watered loess soils by cement and silicate solutions through inclined wells applying technology of pressure injection solution inside the floor are presented. Given drilling and injection technologies allowed to provide safe operation of compressor station.

1 INTRODUCTION

At present, well drilling is widely used in industrial and civil constructions for erecting protective and support structures providing required bearing capability of buildings and their safe operation. Piles of various applications, supporting walls, dowel and anchor supporting of mining and foundation area slopes are constructed using wells. Problems of prevention of building sagging at unstable floor and collapse of foundation area slopes are solved by the drilling and injection technologies. Landslides on the slopes are minimized.

2 RESEARCH RESULTS

Massif strengthening was performed in the mode of pressure injection of cement solutions inside the floor to stabilize the base of piston compressor foundation at Kremenchug petroleum processing plant. According to (Tielnikh et al. 2007) this work was performed by using vertical and inclined wells equipped with perforate injector.

The necessity of work performance as for foundation base strengthening of piston compressors has arisen due to the equipment vibration increase caused by the floor watering at the depth of up to 1.5 – 2 meters and possible sagging of foundation and walls of compressor station. Recorded water inflow was within the range of 0.08 – 1 meters per day. The depth of foundation laying is 2 meters.

Test exploratory well with the depth of 12 meters was drilled to refine current hydrogeological conditions at a distance of 50 meters from compressor station. It was stated that the soils at foundation base are sagging loess located at a depth of 5.3 meters. Water

bearing level is at a depth of 8 meters. There was no water while drilling test exploratory well at a depth of 8 meters. Furthermore, chemical composition of underground water of the water bearing level and the water which was found out near compressor station was different.

Thus, the main reason of soil watering near foundation of compressor stations is unknown. To find out soil watering analysis of geological and geophysical information which was obtained by the method of electric sounding was performed. Hence, anomalous area corresponding to the geodynamic structure of the north-eastern extension with high activity rate (the width is 14 – 16 meters) near which compression station is located is highlighted. It was determined that anomalous area is natural drain for the surface and technogenic water of the large territory. It has become the reason of soil watering in the base of compressor foundation. Watering depth directly under the foundation of compressor station within sagged loess is not more than 3.5 meters. According to the results of study (Dudlya et al. 2011) connection with water bearing level is not found out. Furthermore, performed geophysical study allows to choose rational method of soil stabilization in the foundation base and to optimize work volume.

Recovery of soil bearing capability at the foundation base was performed by the pressure injection of the cement and silicate solutions inside the floor with injectors which are within inclined wells drilled around foundations.

Perforate metal pipe which is reinforcing element of the pile was used as an injector. According to the recommendations (Dudlya et al. 2011) soil stabilization was performed within the range of the depths of collapsing soil 0 – 5.3 meters (Figure 1).

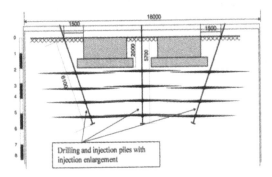

Figure 1. Layout chart of well location during soil strengthening at the foundation base of piston compressors.

Well drilling was performed directly inside compressor station by underground drilling machine BSK-2M2-100. Solution of bentonite clay with CMC was used as a drilling fluid. Water loss was $4 - 6$ cm^3 per 30 minutes. According to the design solutions floor-by-floor injection of 19.1 m^3 of cement and silicate solutions was performed within fifteen wells. Injection pressure was $4 - 10$ atmospheres. It is necessary to point out that numerous returns of cement solution to the daylight surface near wells as well as beyond buildings at a distance of 8 meters from the well where the solution was injected was recorded. While injecting cement solution to the test wells between foundations (inside the soil consolidated by injection) pressure injection has greatly increased (in some cases it was 25 atmospheres). It has become evident sign of sufficient massif consolidation and increase of its bearing capability.

Furthermore, repetitive geophysical studies show the absence of anomalous water inflow at the foundation base of piston compressors. Data comparison of results of geophysical study before and after solution injection under the foundations of compressors shows that the area of the upper layer of ground water extension was considerably reduced and was shifted north-west outside construction. Electromagnetic field change with the radius of $20 - 25$ meters within the area of the plants is due to soil consolidation by pressure injection of cement and silicate solutions of the floor. It is given in Figure 2.

It is necessary to point out that test vibration diagnostics together with geophysical methods of quality control of performed operations was carried out to estimate technical state of piston compressor foundations in the process of soil reinforcing. Diagnostic cargo canting with the weight of 2000 kilograms within the area neighboring to the unit was performed as an impact percussion effect on the inspected units.

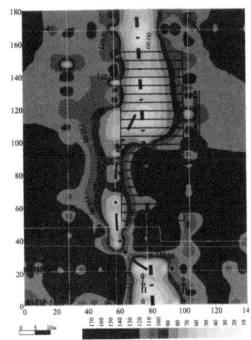

■·■ – axial line of water bearing geodynamic structure before solution injection

■·■ – change of location of water-saturated soils after solution injection

Figure 2. The map of water bearing soil extension within the area of compressor station after performing cement job.

Figure 3 shows the results of performed investigations. It is necessary to stress that overstretched front of foundation vibrations if response duration is 1.2 seconds is clearly outlined.

The first mode of foundation characteristic vibrations is dominant. There are eight cycles per second. They correspond to the frequency of the basic resonance equal to 8.05 Hz and have a great effect on high vibration initiation in operation process as they are close to vibration frequency of the compressor pipe which is 8.43 Hz. Time of the unit vibrations after excitation confirms the presence of the high humid soil directly under foundation bottom. After work completion as for injection floor reinforcing within the foundation of the piston compressors and strength development of cement and silicate solutions repetitive experimental investigations concerning impulse generation of the given units with data storage were performed.

Figure 3 also shows the results of these studies. Time of the response after performed strengthening was 0.6 seconds. It is practically two times less than initial value. But it is necessary to point out that the first mode of foundation characteristic vibrations.

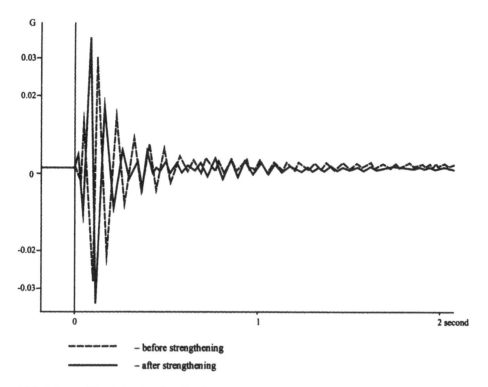

G

0.03 —

0.02 —

0

-0.02 —

-0.03 —

| 0 | 1 | 2 second |

------- – before strengthening

———— – after strengthening

Figure 3. Mode shape of foundation decaying vibrations.

(there are four cycles per second) which corresponds to the frequency of basic resonance dominates. It is 4.15 Hz. Decreasing curve of response reaction is not the smooth function. It shows the presence of the other modes of characteristic vibrations. Increase of the basic resonance frequency from 8.05 Hz to 4.15 Hz provides decreasing piston compressor vibration.

Cutting foundation response time of piston compressors of the percussion impulse excitation is two times as much in comparison with initial value. It is the result of factual improvement of physical and chemical soil properties within considered area.

Furthermore, surveying of the foundation vibration and the sagging shows that there were no changes from the moment of finishing injection operations.

3 CONCLUSIONS

1. Determining massif state by geophysical methods allows controlling and changing its structural and mechanical properties using drilling and injection technologies.

2. Wells of collapsing loess sediments are necessary to drill using drilling fluids water loss of which is not less than 7 cm^3 per thirty minutes.

3. Soil massif strengthening in the mode of pressure injection by floor-by-floor injection of cement solution through perforate injector in complicated conditions of technogenic anomalies allowed increasing foundation strength of buildings, providing safe operation and improving ecological situation.

REFERENCES

Dudlya, N., Popov, A., Telnikh, N., Tsaplin, Ye. & Korotkov, A. 2011. *Ecological problem solution using drilling and injection technologies.* Scientific papers DonNTU. Mining and geological series, No 14: 92 – 96.

Tielnikh, M., Popov, O., Smolii, P., Tsaplin, Ye. & Veremchuk, S. 2007. *Drilling and injection friction piles with cement enlargement.* Patent for utility model. Antratsyt, January 15.

Theoretical and Practical Solutions of Mineral Resources Mining – Pivnyak, Bondarenko & Kovalevska (eds)
© 2015 Taylor & Francis Group, London, ISBN: 978-1-138-02883-8

Sorbents of purify mine waters

O. Svetkina, O. Netiaga & H. Tarasova
National Mining University, Dnipropetrovsk, Ukraine

ABSTRACT: The technology of sorbents based on coal combustion products through a variety of methods was researched. It is shown that these sorbents are distinctive because their structure has non-localized π-electrons of the graphite-like networks of crystallites of carbon. This circumstance determines not only the uniqueness of electro-physical properties of coal but also adsorption, redox, chemisorption processes on the border of coal-slurry. The listed circumstances allow you to use the original methods of chemical and mechanochemical modification of the surface chemical and coal, due to introducing desired donor and acceptor atoms in carbon frame, which increases the absorption capacity and selectivity carbon sorbents.

1 INTRODUCTION

Ecological problems of mining, metallurgy, energetic and other industries include cleaning up of polluted waters from oil and other petroleum products, salts of polyvalent metals, acids, etc. There are many sorbents, which differ according to the principle of action, however, each of them works in certain circumstances and is not universal. This causes the need of a universal sorbent, serving in different conditions and with different dirt.

Very promising porous adsorbent is activated carbon. Owing to a combination of valuable qualities – highly developed porous structure, diversity of the chemical nature of the surface, special electrophysical properties as well as chemical, thermal and radiation resistance, activated carbon of various types is widely used for absorbing gaseous and dissolved substances. The uniqueness of carbon adsorbents is that the solutes substances can be absorbed by coal by different mechanisms, their sorption may be due to the different nature of substances.

Knowledge of the mechanism of sorption of solutes allows you to develop efficient methods for extracting the latter from solutions of complex composition, the separation of the mixture components in the sorption process, search, and synthesized carbon materials with high sorption capacity and high both individual and group selectivity.

A distinctive feature of activated carbon as sorbents is that in their structure there are non-localized π-electrons included in the graphite-like networks of crystallites of carbon. This fact explains not only the originality of the electrical properties of the coal, but also adsorption, redox, chemisorption processes on the boundary of the coal-solution. These circumstances allow the use of original meth-

ods of chemical surface modification of coal by introducing the desired donor and accepting atoms in the carbon frame and thereby increase the absorption capacity and selectivity of carbon sorbents.

2 FORMULATING THE PROBLEM

The aim of this work is the development of technology for production of sorbents from coal combustion products.

3 MATERALS UNDER ANALYSIS

Active coals represent one kind of microcrystalline carbon, which is characterized by high values of specific surface area, resulting in wide usage of activated carbon in the adsorption of various gases, vapors and solutes.

The proportion of crystalline form during metamorphism of coal increases, which is associated with the growth of flat aromatic and enshrined fragments (lamellae) by dehydrocyclization "bridges", the dehydrogenation of acyclic rings and condensation reaction and a gradual relative orientation of these fragments, forming a pack of several parallel layers of carbon.

The organic matter of coal is heterogeneous. It consists of bitumen, humic acids and residual carbon. The macromolecules of the so-called transitional forms of carbon contains carbon atoms in different states of hybridization of the valence electrons $[C\,(sp^3, sp^2, sp)_m]$.

When considering the structure of the skeleton of the transitional forms of carbon, it is possible to allocate two basic elements:

– crystal graphite-like part consisting of carbon atoms with sp^2-hybridization;

– atom molecular component of and a disordered (amorphous component, including the carbon atoms in different states of hybridization of valence electrons sp^3–, sp^2–, nsp–, atomic type frame.

In Figure 1 the structure of the carbon frame of mixed type is shown.

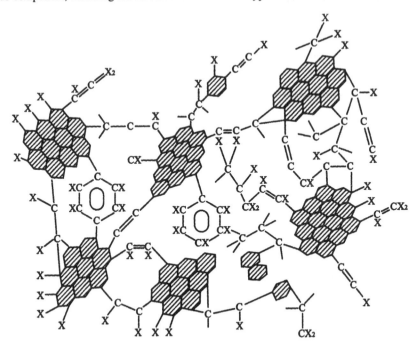

Figure 1. The structure of the carbon frame of mixed type.

The frame of atomic-molecular structure is formed of crystalline and disordered fragments of structures with different types of carbon bonds chemically linked and combined in the spatial-polymer structure.

The peculiarity of this structure is that this picture of the complex structure is implemented not only within samples, but also on its surface.

At the vibro-loading as a result of local temperature increase at the moment of impact there appears possibility of transition of disordered parts in graphite structural organization, resulting in the reorganization of the framework of the atomic and molecular structure to molecular diagram $[C\,(sp^3,\,sp^2,\,sp)_m]{\rightarrow}[C\,(sp^2)_m]$

Thus, the coals consist of mainly of carbon atoms organized into a condensed aromatic fragments with trigonal hybridization, which are connected in the space-polymeric structure carbon frag ments of diagonal (chain), trigonal (planar) and tetrahedral (frame) structure, forming a mixed type core – atomic-atomicmoleculer.

In this regard, it is interesting to study the residual coal (RC). RC – this is the part that remains after extraction of the solvent and treatment with alkali.

It is important to note that the active sites in the coal will be located on the carbon atoms belonging to the surface of the frame: $S_{act} = S_{gen} = S^x_{act} + S^y_{act} + S^c_{act}$, where S – the specific surface of coal. These centres will form a series of activated atomic compounds of frame type.

At vibro-loading influence reactivity of the surface of the coal will depend on the degree of carbonization of the carbon substances and graphitization, and for carbonized centers $S_{act} = S_{gen} = S^x_{act} + S^y_{act} + S^c_{act}$, and graphite for maximum centres $S_{act} = S_{gen} + S_{inact}$ forming rows of atomic-atomicmolecular and atomic-molecular compounds, respectively.

As is known (Balaz 2008), by mechanical dispersion of coal mechanical destruction occurs, i.e., rupture of valence bonds. This leads to the fact that in place of the old destroyed factions carbon-oxygen and carbon-carbon new atomic grouping and links appear. End groups and decomposed fragments of the structure of coal serve as the sources of volatiles and soluble products, as well as parallelomania of condensed aromatic nuclei takes place. Physical and physico-chemical properties of coals are closely related to their molecular structure, which was formed during metamorphic transformations in coal body.

The research objects were coals of different degrees of metamorphism, the residual coal (RC) B brand, as well as sandstones. Residual coal (RC) represents a hydrocarbon frame of the organic mass of coal (OMC), obtained by exhaustive extraction of coal grade B chloroform. The feature of this carbon (RC) is that after the extraction "mobile molecular phase" is removed.

Vibration activation was carried out in two regime grinding: continuous and periodic. The restructuring of the coal studied by IR spectroscopy and chemical functional analysis (Svetkina 2013).

Physical and technical characteristics of coals used in the process of vibro-impact loading is represented in Table 1.

Table 1. Physical and technical characteristic of used coals.

Brand of coal	A^a, %	W^a, %	S^r_t, %	V^a, %
G (gas)	4.6	2.1	3.1	45.1
LF (long-flame)	20.0	13.3	2.3	42.2
B (bold)	26.1	3.0	0.9	32
C (coking)	13.4	1.0	3.9	19.4
S (skinny)	7.2	1.6	1.25	12.5
OC (otoschenno caking)	36.9	2.6	4.85	18.9

During vibro-loading we determined the content of functional groups (K^g) on the surface of the coals according to the analysis technique.

After vibro-loading in the periodic regime grinding of coals of different stages of metamorphism evaluation of the energy output of mechanical degradation was determined by carbon-oxygen and carbon-carbon bonds.

The relationship between the vibroactivation material and energy output was carried out according to the formula (Danielyan & Torosyan 1991):

$$G = W/I , \qquad (1)$$

where G – the energy yield, mol/MJ; W – the amount of product formed per unit time per 1 kg of crushed product, mol/(kg·s); I – energy intensity of the mill, MWt/kg.

Experimental data of vibro-loading on the coals in the periodic regime grinding is shown in Table 2.

Table 2. The process of destruction of coal at vibro-loading.

The destruction of chemical bonds	Coal					
	LF	G	B	C	OC	S
Bond $C - C$	0.04	0.43	0.19	0.09	0.1	0.11
Bond $C - O$	0.12	0.12	0.10	0.08	—	0.03

As seen from Table 2, mechanochemical transformations of chemicals of coal depends on the stage of metamorphism.

Increasing of time of vibro-loading the solubility of RC increases, as well as the degree of aromaticity, and the atomic ratio H/C on the contrary decreases. In addition, as shown by functional analysis there is the destruction and changing the following atomic groups: $-CH_3$, $\equiv C-$, $-OH$, $-CH_2-$, $=C=O$, C_{ar}.

In Figure 2 presents the kinetic curves of formation and loss of functional groups from the time of vibro-loading.

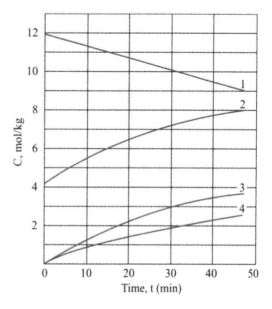

Figure 2. Kinetic curves of the formation of chemical groups and their destruction at vibroactivation: 1 – $-CH_2-$; 2 – C_{ar}; 3 – $\equiv C-$, $=C=O$; 4 – $-CH_3$.

Thus, the content of functional groups is determined by the competing of processes of formation and destruction. Maximum concentration (C mol/kg) of atomic groups formed during vibro-loading are presented in Table 3.

Table 3. The maximum concentration of functional groups.

The process	Loading time, min	Concentration of atomic groups [C], mol/kg
The formation of $-CH_3$	5.0	4.6
Formation $=C=O$	5.0	2.4
Formation CON	5.0	2.0
Increasing C_{ar}	5.0	38
The loss $-CH_2-$	1.0	10.3

X-ray diffraction analysis of the products of the activation by the method of the vibro-loading showed that with increasing of time of loading the processes of structuring takes place we see increases of the size of the layer and the height of the cluster, the number of associated layers in it, and the distance between layers is decreased.

Analysis of experimental data shows that in the process of vibro-impact loading in the periodic mode, there is a gap bridging $-CH_2-$ and $-CH_2-CH_2-$ linkages between condensed aromatic fragments in the layer, and the formation of $-CH_3$ groups (the period of the γ-band 4.44 – 4.70 Å).

In connection with the degradation of methylene bridges, a reduction of residual structural units of coal relative from $-OH$ to 520 a.e.m takes place.

Vibroactivation of RC was carried out in a continuous mode of grinding, passing the material through the grinding chamber a certain number of times to set the path of the grinding of L equal to 1, 2 and 3 meters. The results of the experimental data are presented in Table 4.

Table 4. The experimental dates of vibroactivation.

№	Length of grinding path L, m	Specific surface area S, m^3/g	Angle of wetting Θ^0
1	0	0.25	40
2	1	0.69	41
3	2	0.75	41.5
4	3	0.75	41.8

Functional chemical analysis showed that crushing of coal in continuous mode, the interaction of coal with oxygen takes place, resulting in the formation of surface and bulk oxides of the basic or acidic nature. Thus, as a result of the vibro-loading of the shelter various oxygen-containing groups and layered compounds of introduction are formed on the surface various. Deformation of carbon layers due to internal source stress and a wedging effect of the introduced foreign phase in the form of microcrystalline grafite like compounds lead to a reduction in the size of crystallites to a complete amorphization RC.

Surface properties of coal have been studied for continuous and periodic modes, using IR spectroscopy and thermography. The appearance in the IR spectro absorption bands 1740, 1780, 1880 cm^{-1} indicates the presence on the surface of activated carbon carboxyl and carbonyl functional groups.

It should be that at the vibroactivation of materials the adsorption capacity of the crushed particles increases. In this regard, experiments were conducted on determination of adsorption isotherms of activated coals.

Analysis of experimental data of adsorption of some inorganic electrolytes (in particular of aluminum sulfate, the concentration of which is equal to $10^{-1}\%$) from the time of the vibro-loading shows that there is a change in adsorption capacity of aluminum sulfate. The formation of active oxygen on the coal surface in the form of carbonyl, carboxyl and hydroxyl groups leads to an increase of negative charge on the activated material. In this regard, there is an amplification of the electrostatic interaction between the cation of aluminum and the "active" oxygen of coal surface, and as a result, the increase in adsorption.

To give active coals selective properties, for example, with respect to certain metal ions for their selective extraction from aqueous solutions of complex salt composition they are usually subjected to oxidation at relatively low temperatures (300 – 450°C),as a result in the weight content of oxygen in coal can reach 13 – 14%.

Surface chemistries activated coals do not represent a new phase on coal, but are functional groups of the particular nature associated with the peripheral carbon atoms in the graphite-like nets of crystallites of coal. Despite the large number of papers devoted to the chemistry of the surface of the coal, the question about the nature of surface chemical compounds of activated coals can be considered one of the most complex and least understood. In this regard, it should be noted that most researchers link as acid-base and redox properties of active coals with the existence on them of certain functional groups of differences of nature. Since the study of the chemical nature of the surface compounds of coal, included the study of their sorption capacity against acid and alkali, we introduced the idea of basic and acidic oxides on the surface of the coal. Taking into consideration that the amount of chemisorbed carbon oxygen is always higher than its content in basic or acidic groups, we introduced the idea of nonionic surface oxygen compounds to explain this discrepancy.

If the coal after activation comes in contact with oxygen or air at a low temperature, for example 0 to 100°C, then compounds of the main character form on coal surface. Although the basic nature of oxides still remains insufficiently understood, it is established that the basic oxides are products of chemisorptions of oxygen on coal, which correspond to high values of adsorption energy of oxygen on the more active parts of the surface of coal. The oxygen of these oxides in contact with water or aqueous solutions goes into solution in the form of hydroxyl ions, charging the surface of the coal positively.

At temperatures of 400 – 500°C the interaction of oxygen with coal is different. In this case, the coal hemosorbents 13 – 15%, and by some data up to 22 – 25% oxygen, which firmly binds the coal in the form of various surface oxygen-containing compounds. About 1/5 of chemisorbing in these conditions, oxygen is a member of various protogenic groups.

Carbon sorbents can be obtained by high-temperature prossesing of coal and, in particular, by burning. The optimum carbon content in the sorbent can be achieved by regulating the combustion process of coal combustion, or by separation of the resulting products. The separation process can occur in air (electric and pneumatic separation) and water (flotation, selective flocculation, etc.) environment.

To obtain the sorbent raw coal is subjected to ruined grinding to a particle size of 10 – 500 mkm when the content of coarse fraction is 90 mkm 10 – 70% and burn in the boilers of thermal power plants (TPP) at 1200 – 1500°C. As a result of thermal processing of carbonaceous raw materials at such the specified temperature the output of the sorbent increases and the obtained product at 45 – 60% consists of a carbonaceous fraction and 20 – 55% of particles of silicates of spherical shape.

The sorbent obtained by the proposed method, in contrast to the known ones allows purification of mine waters from viscous oil, the content of which may vary from a few milligrams per liter to tens of grams per liter.

Table 5 shows that the decrease in the content of the coal fraction less than 45% reduces the cleaning efficiency by reducing the sorption capacity of the mixture and worsens the conditions of its subsequent combustion. The increasing content of coal by over 80% does not affect the completeness of purification of waters.

Table 5. The influence of content of carbon fraction on the degree of purification of mine water.

The content of components, wt. %		Content of kerosene in water, mg/l		The degree of purification, %
The coal fraction	Alumi-nosili-cates	Before filte-ring	After filtering	
25	75	17.2	4.1	76.3
35	65	16.4	2.0	87.3
45	55	15.8	0.9	94.3
55	45	19.5	0.5	97.4
65	35	18.9	0.4	97.8
75	25	15.2	0.3	98.2
85	15	16.7	0.3	98.2

The content in the filter mixture is of silicates less than 20% reduces the rate of filtration of polluted water. The increasing of the content of aluminosilicates more than 45% reduces the degree of purification due to the decreasing in the sorption capacity of the mixture. The sorbent saturated with apolar surface-active substances is not exposed to regeneration as it is a cheap material and is burned in boilers.

The mineral component of the sorbent is represented by the particles of aluminosilicates of spherical shape the existence of which allows to increase the speed of filtering the purified water through a lay of sorbent and to improve the effectiveness of their puring. The presence in filtering mixture coal particles and silicates with different hydrophobic properties, increases the sorption capacity of the filter.

Products of high temperature processing, obtained by the method described above are effective sorbents of apolar surface-active substances (SAS), including petroleum products. At mixing of the sorbent with sewage, polluted, for example, with fuel oil and viscous engine oils, oil agglomeration occurs, i.e. the formation of agglomerates as granules size 1 - 5 mm, consisting of particles of the sorbent.

Thus, the sorbent obtained on the basis of waste of Pridneprovskaya GRES after high-temperature activation can be applied for waste water purification from various contaminants: metals, electrolytes, superficially surfactants substances and organic substances. For wider application of ashes of power plants as sorbents and utilization of waste, polluting the environment, studies have been conducted to study the properties of these sorbents.

4 CONCLUSIONS

At vibroactivation of waste of thermal power plants the activation of the surface takes place which leads to additional active centers.

It's found that the most effective are sorbents obtained by special burning of certain grades of coal under the WES.

It is shown that the obtained sorbents with their sorption properties are not less than known amphoteric the adsorbents.

REFERENCES

Balaz, P. 2008. *Mechanochemistry in Nanoscience and Minerals Engineering.* Berlin: Springer : 413.
Danielyan, N.G. & Torosyan, A.R. 1991. *Energy yield under high pressure and shear strain. Mechanochemical synthesis works in inorganic chemistry.* Novosbirisk: Science: 56 – 59.
Svetkina, O. 2013. *Receipt of coagulant of water treatment from radio-active elements.* Mining of Mineral Deposits. The Netherland: CRC Press/Balkema: 227 – 230.

Theoretical and Practical Solutions of Mineral Resources Mining – Pivnyak, Bondarenko & Kovalevska (eds)
© 2015 Taylor & Francis Group, London, ISBN: 978-1-138-02883-8

Use of geographic information systems at creation three-dimensional models of mine objects

D. Kirgizbaeva, M. Nurpeisova & Zh. Shakirov
Kazakh National Research Technical University named after K.I. Satpayev, Almaty, Kazakhstan

E. Levin
Michigan Technological University, Houghton, United Stated

ANNOTATION: The article presents results of work on 3D modeling for underground workings and above mine objects, carried out on the basis of ArcGIS, with the use of stored materials as initial data, as well as newest materials of ground and air scanning. Selected modes of laser scanning provided a high level of accuracy of the representation of objects in the horizontal and vertical position. 3D model has a specific purpose. It is intended to assess the visibility of designed mine objects. The use of three-dimensional modeling techniques in geographic information systems greatly increases the effectiveness of a variety of tasks of Geodesy and Cartography.

1 INTRODUCTION

Geographic information systems (GIS) for the collection, storage, processing, access, display and distribute spatially coordinated data (spatial data) are an indispensable tool in the resource inventory, analysis, evaluation, monitoring, management and planning and serve as support in making solutions. Three-dimensional modeling of any object with the use of GIS include materials in the form of vector maps, satellite images, photographs, three-dimensional models, diagrams, plans, diagrams, graphs, tables, databases, multimedia, and text documents. In connection with the formation of new independent states of the former Soviet republics, from 1 January in 1992 stopped the existence a single cartographic and geodetic survey of the USSR. However, the needs of the political and economic development have necessitated the establishment of national cartography and geodesy services.

At the present time in the Republic of Kazakhstan in connection with the global changes taking place in the whole of its territory, the rapid development of the construction, development and updating of topographic maps has become an urgent task in the field of geodesy and cartography.

Contemporary development of information technology in the field of geodesy, cartography and integration with adjacent fields of science broke new ground in the principles of production, processing and storage of spatial information. In recent years the market for geographic information systems took the leading position.

All GIS systems are based on information technology, creation, processing and complex analysis of difficult structured digital cartographic and geodetic information. That is one of the important issues in the creation with the latest facilities, which would lead to a rapid, completely new type of data collection and data quality, reduce costs and reduce turnaround time. The transition from traditional methods to modern technology with the use of methods and means of satellite geodesy led to the development and wide application of global navigation satellite systems (GNSS). Along with it, recently brought to the forefront the problem of control of the environment and predict potential of multi-temporal changes in natural conditions in different regions. This explains the special attention to the study of the dynamics of geo systems, the processes occurring in the Earth's crust and the creation of dynamic maps of natural phenomena.

2 RESEARCH OF THE RECENT RESEARCH AND PUBLICATIONS

Relevance of article due to the need to geodetic studies and the national economy in the operational study of the dynamics of natural objects and phenomena in order to prediction their development, as well as the widespread introduction of satellite Positioning of coordinates (ASPC) computer mapping techniques, allowing to reflect the complex geo systems more adequate. In the development of methods for solving problems of cartography had contributed:

A.M. Andreev (Andreev 1995), A.M. Berliyant (Berliyant 1986), A.S. Vasmut (Vasmut 1983), A.V. Koshkarev, I.S. Tikunov, A.M. Trofimov (Koshkarev 1990, Koshkarev et al. 1991), S.N. Serbreniyk (Serebreniyk 1999) and others.

Currently, despite of the definite achievements to carry out work applied differing modern methods and means. But all of the methods and equipment that used to achieve these goals, don't always economically and technically compensated. Application of 3D laser scanning due to the high cost of equipment, excessive density and accuracy of measurements is a promising method. Stereo topographical method has disadvantages when taking pictures with high density of building and dense woody vegetation, requires certain rules and time limits execution skilled personnel and is not rational in terms of cost of works and equipment. Modern methods of satellite technology have found great application. These technologies are developing dynamically in the integration with the region of high computer technology and can solve a wider range of surveying and mapping tasks. According to the results of the comparative analysis in the first stage we can say that the built environment is the most preferred method of combined Survey, using tachometric and satellite as the most operative maneuverable and economically effective method for large-scale Surveying. Works are almost independent of the time of day and weather conditions and can be carried at any time of the year. Maximum economic efficiency is practically achievable at use of modern electronic tachometers and integrated apparatus of Satellite Positioning of coordinates (IASPC).

Therefore, further technological processes considered of using modern appliance as hardware and software.

This study would be incomplete if you do not see at least the basic and most common hardware and software. To date, these programs have become an integral part of the process in creating, maintaining and updating of dynamic maps of nature. During the last century there were constantly changes in the state technical and software used to perform field surveying, creating cartographic information. These changes are associated with the development of IT-technologies, automation requirements of all types of surveying and introduction of the latest advances in science and technology. Actual tasks performed in real time by surveyors, require the use of new technologies and high-tech equipment, the use of integrated devices and software tools that allow solving them as soon as possible with a maximum economic efficiency. Now widely are used geodetic instruments of a new generation that would quickly solve the problems of cartography in an automated mode

with a completely new principle of collecting information on the spatial locality. Such measuring devices are equipped with integrated computing facilities, full-color screens and storage devices, creating the possibility of recording and storing the measurement results for further use in the technological process. At cameral processing of the results of field work it is necessary a high performance and multifunctional software to translate raster information in digital form and the actual formation of a digital map.

Currently, Integrated Automated Positioning System (IAPS) of Geodesy and Cartography has great importance. Processing of the results of field measurements is made using special software. To consider in detail the programs in this study has no meaning, because each manufacturer for the processing of measurement results obtained by appropriate devices offers its program solution.

Therefore at presentation of the materials are considered GIS, focused on the work with IASPC and joint use of raster and vector materials (photo plans, maps, digital terrain models), which provides: recoding and converting the array, "raw" data; translation of starting materials into a vector form; binding to vector objects databases of semantic information; organization of relationships between objects; with extensive functions on analysis jointed and separated metric and semantic information. Software tools can process large amounts of data. Certain preference, simplicity of learning and use in specific works in Kazakhstan bring to the fore "MapInfo". However, long-term use and world experience of using GIS technology ArcGIS, in countries with a high level of informatization allow us to speak of its advantages over the others. The fact that demonstrates the quality of this program is the use of this product by the RK Agency for Land Management, Department of Defense and cartographic factory in Kazakhstan. Those facts allow us to say more advanced and modern ArcGIS system in the creation, renewal and store dynamic maps using GIS technology and (IASPC).

To summarize the analysis of this stage we can conclude the following: the main criterion for the selection of instrumentation and hardware is the opportunity to work with a lot of information, simplicity of use and compatibility with GIS packages. As a result, it is clear that existing technologies and methods do not allow talking about the introduction of new technology in production. According to the results of the analysis with existing devices, hardware and software necessary to develop a universal method and technology for joint use and application of modern surveying methods, modern technical equipment and GIS. This tendency is dictated by the requirements and timing of works is an important task in the field of geodesy and cartography. It is extremely important from the point of view of economic efficiency, the use of modern combined method for the process of creating, processing, storage, distribution and update digital maps (Nurpeisova & Kirgizbaeva 2010a).

3 RESULT OF INVESTIGATION

Architecture of the system of universal technologies for creating products using IASPC and GIS technologies, consistent with the principles and requirements of its construction, identified from the analysis of the domain in the first section focuses on the following tasks:

– minimization of expenses and labor;
– operative monitoring and updating of topographic maps and plans;
– versatility and simplicity of use of the entire geospatial information;
– maximum automation of carrying out all processes;
– compatibility of formats and conversion of information;
– exceptions of gross errors and quality control of finished products.

Given universal technological scheme involves the use of modern methods of surveying and hardware and software (Figure 1).

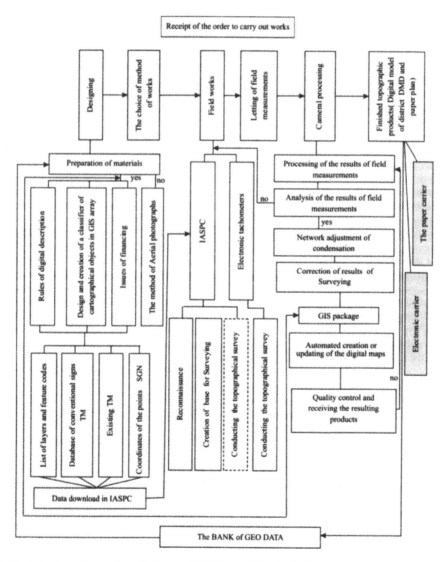

Figure 1. The technology of surveying with the use of IASPC and GIS technologies.

Technological process includes four main stages: 1) preparatory; 2) field; 3) cameral; 4) quality control and acceptance of finished products.

In the first stage of the bank's existing geospatial data collected all available information about the area in which it is supposed to conduct work.

In the second stage work is carried out in the field. At the satellite and tachometric method brigade may consist of two people, which greatly reduces the cost of wages. Inspection of the initial frames is carried out in the field and determined the changes since the last update of the plan or map.

The next step is the creation of Surveying bases. With the advent of satellite methods to create justification for the Surveying has become easier and faster. A new level of surveying equipment allows to produce maximum integration with the use of GIS.

Apart from of geodetic measurements receiver enables to collect semantic information in the field, which is encoded using the selected and loaded classifier. In carrying out field work additionally collects a variety of photos and videos, which can also be tied to the project and put in the database. Besides, any picket point information can be written in text or audio format. Minimum of two devices requires for field works with high accuracy: one (base station) works on a point with exactly known coordinates, and the other (the rover) – a t the point to be surveyed. Most of the static observations can be carried out in an automated way.

Mapping methodology on models, processing and drawing facilities are usually automated, entry of new GIS data begins after the selection of the classifier and the creation of the survey project. At the third stage carried out the analysis and processing of field measurements and their conversion into a GIS area for: creation, filling, storage, updating and preparing of maps for printing (Nurpeisova & Kirgizbaeva 2010b).

Thus, three-dimensional modeling is an integration of GIS data from various sources into a single system.

The use of three-dimensional modeling techniques in geographic information systems greatly increases the efficiency of a wide range of applications in urban planning, architectural planning, in creating of navigation systems and so on.

A three-dimensional Geographic Information System, first of all, develops a modern tool of management of cities and regions, based on the use 3D-technologies for sustainable development of the state; increases the efficiency of established and newly created information systems of state significance. The distinctive advantage of using 3D-GIS is the rapidity of assessment the situation and accuracy accept the decision, the integration of GIS data from various sources into a single system. Possibilities of 3D-GIS consist in the following: monitoring of means of transport with navigation systems; the use of a unified address-information database with a multifactorial; providing access to information, which is difficult to obtain from other sources; low expenses for modeling processes related to a significant change in appearance of the mine objects; visibility of situational and analytical information; the use of a unified address-information database with the multifactorial searching.

3D-GIS applications are widely used in the field of safety. The system is a set of specialized modules for solving a wide range of complex problems related to the management and maintenance of the property, from the security of a larger complex to manage and accounting for leases separate facilities. A distinctive feature is that the system is based on web-technologies and can run on any computer from anywhere in the world through a standard web-browser (Internet Explorer, Mozilla Fire Fox, Google Chrome and etc.), does not require installation on your computer, making the decision universal, works on almost all operating systems and platforms. Thus the system has a high degree of protection against unauthorized access. All information systems are stored in a single database that allows multiple employees to work in real time with single and up to date information. Moreover the module enables optimal and efficient to choose space for installation of cameras.

Providing module of the security object easily complemented into the system. For this case, the camera indoor and outdoor observations "attached" to the three-dimensional interactive model of the office building (and its surroundings). So the camera images are not displayed as a mosaic of images, and are projected directly into the 3D-model, which significantly increases the visibility and perception of the degree of the video stream.

The Main System is based on multi-functional interactive three-dimensional model of object and surrounding territory. The three-dimensional model is built with a high degree of detail on the basis of the set of accurate and reliable data (from floor plans and schemes of construction of communications, to sketches and photographs). This approach allows us to create a three-dimensional model with the level of detail that is needed to solve specific problems of controlling object. At the same time the model can be created and still being built, reconstructed or even a developed building. The functional interactive three-dimensional model allows you to display an object with any point of inspection from the outside of the building, and from the inside. It is possible to view the floor repeater and cross sections of building and location diagrams of communications, security systems, firefighting, etc. Three-dimensional model can also reflect the characteristics of the interior layout and decoration. Thus, 3D-GIS allow us to represent the described object in as close to real form in the most visible manner and with the necessary details.

While working with mining elaboration models implemented the following idea. During the design phase of mining operations are creating capital model project of mine elaborations, which are located in the corresponding horizon. At the stages of implementation of these projects move into a group model of the actual elaborations (Figure 2).

Figure 2. 3D model of the infrastructure capital mine elaborations.

In process of occurrence data about the drifting of the workings made gradual replacement of sections on actual project with binding to sections of the date of their amount of drivage. In the 3D model is possible to realize the full information on each site of mining operations.

One of the major works of surveying service is orientation through one vertical shaft. Modeling of the process of contiguity by connecting triangle is shown in Figure 3.

Figure 3. 3D modeling of process orientation in 1 shaft.

Three-dimensional modeling of the processes used in solving problems of forecasting and risk assessment of emergency situations, he definition of flood zones and pollution, drainage area, avalanche areas, the level of soil erosion, etc.

4 CONCLUSION

In such a way three-dimensional models are effective for using in various fields of human activity. Therefore, for today one of the main trends of the world market in the design is the transition from two-dimensional to three-dimensional modeling design, as well as the introduction of modern three-dimensional GIS and come to the fore.

REFERENCES

Andreev, A.M. 1995. *Computer technology of visual information processing and their application in the publication of maps.* Geodesy and Cartography, No 8: 47 – 49.

Berliyant, A.M. 1986. *The use of maps in Earth Sciences.* Outcome of science and technology. Cartography, Vol. 12. M: VINITI AN SSR: 175.

Falshtynskyy, V., Dychkovskyy, R., Lozynskyy, V., Saik, P. 2012. *New method for justification the technological parameters of coal gasification in the test setting.* Geomechanical Processes During Underground Mining - Proceedings of the School of Underground Mining. The Netherlands: CRC Press/Balkema: 201 – 208.

Koshkarev, A.V. 1990. *Cartography and geoinformatics.* News of academy of Sciences of the USSR, No 1: 27 – 37.

Koshkarev, A.V., Tikunov, V.S. & Trofimov, A.M. 1991. *Theoretical and methodological aspects of geographic development information systems.* Geography and natural resources, No 1: 11 – 16.

Nurpeisova, M.B. & Kirgizbaeva, D.M. 2010b. *Creation of multi-temporal maps of the Aral region.* Materials of Scientific and Practical Conference "Innovative technologies for the collection and processing of spatial data for natural resource management". Ust-Kamenogorsk: 2010.

Nurpeisova, T.B. & Kirgizbaeva, D.M. 2010a. *Technology creation of dynamic maps.* Materials of International Scientific and Practical Conference "Stosowane naukowe opracowania-2010". Praga: 67 – 70.

Serebreniyk, S.N. 1999. *Cartography and Geoinformatics and their interaction.* M: Mir: 159.

Vasmut, A.S. 1983. *Modeling of cartography by COMPUTER application.* M: Nedra: 200.

Theoretical and Practical Solutions of Mineral Resources Mining – Pivnyak, Bondarenko & Kovalevska (eds)
© 2015 Taylor & Francis Group, London, ISBN: 978-1-138-02883-8

The results of gas hydrates process research in porous media

V. Bondarenko, K. Sai, K. Ganushevych & M. Ovchynnikov
National Mining University, Dnipropetrovsk, Ukraine

ABSTRACT: The results of the gas hydrates formation process research are received using various porous media. A series of laboratory experiments directed to receiving gas hydrates samples in climate regulation chamber KTK-3000 is conducted. The change of parameters of gas hydrates formation process at rock sand presence as porous medium is established.

1 INTRODUCTION

High prices for hydrocarbons and depletion of traditional hydrocarbon deposits are pushing humanity to search for alternative sources of fuel, which are gas hydrates, shale gas, coal bed methane and technologies of coal seams development (Ovchynnikov et al. 2013, Ganushevych et al. 2014, Falshtynskyy 2012, Lozynskyi 2015). Gas hydrates are considered to be one of alternative and potentially extensive sources of gas. Preliminary estimates of the world gas hydrates reserves indicate they considerably exceed existing reserves of natural gas. A number of leading countries of the world, such as the US, Canada, Japan, India, Norway, Germany, shows great interest in creation of technology of gas hydrate deposits development.

Natural gas hydrates present a metastable mineral, formation and dissolution of which depend on pressure and temperature, gas and water composition, and properties of the porous medium they are formed in, too.

The major part of hydrates can be found in pore spaces of sedimentary rocks but not in a free state (in the water). If hydrate deposits locate at shallow depths, they form a layer of several hundred meters thickness, and, if they happen to be found on larger depths (several kilometers below sea level), the thickness goes up to 1 km. To ground rational parameters of the systems of natural gas hydrates deposits development, it is necessary to explore in details characteristics of the gas hydrates stable existence in pore spaces.

2 STUDYING THE PROCESS OF GAS HYDRATES PROPERTIES IN POROUS MEDIA

The first information on the gas hydrates formation conditions in porous media was received by Y.F. Makogon in 1974. The experiments were conducted on quartz sand and natural core samples with usage of methane. It was ascertained that comparing to free volume the temperature of the beginning of hydrate formation decreases by $2 - 4.5°C$, and impact of the porous media on the conditions of the beginning of hydrate formation weakens while pressure rises. Also, he introduced analytical description of the conditions of gas hydrates formation in porous media. That description is based on interconnection of the bond water content and vapor pressure above it. This implies that with decrease of particles' diameter, which form the porous medium, and the pore water capillaries the temperature of the beginning of gas hydrates' formation in porous media considerably reduces (Makogon 1974).

Many researchers (S. Dallimore, CoUett T. Field, E.M. Chuvilin, V.S. Yakushev, S.E. Grechishchev, V.I. Ponomarev) hypothesize about existence of clusters of gas hydrates at depths more than calculated by baric and temperature depth conditions. Under equal conditions of gas saturation level and the temperature regime of hydrate formation, to a greater extent, will be determined by characteristics of the soil, and namely, mineral and granulometric composition and primary moisture. Movement of unfrozen water can occur during the formation of gas hydrates in ice-containing porous rocks, and this phenomenon might play an important role in nature (Dallimore et al. 1996).

In their work, J.F. Wright, F.M. Nixon, T. Uchida presented laboratory results of examination of thermobaric conditions of gas hydrate stability in waste rocks from the JAPEX/JNOC/GSC Malik 2L-38 borehole off the stratums at depths of $897 - 952$ m. Together with that, influence of the dusty fraction content and moisture mineralization in pores on decay of gas hydrates in the ground was investigated. The authors also noted that in the absence of the dusty fraction and low salinity (4‰) there is no significant impact on change in the temperature and pressure conditions for hydrate stability. However, with salinity increased to 40‰ measures of these conditions move to lower temperatures and higher pressures, while the shift of the temperature reaches 2°C.

When forming gas hydrates in salt deposits, increase of the mineralization level of pore fluids may eventually move thermobaric conditions of gas hydrates' constancy to some extent at which their further growth stops. On this basis, there is a possibility of a very high content of gas hydrates in non-saline sandy deposits while fine-dispersed rocks may not contain hydrates at all (Wright et al. 1999). In fact, these data do not fully consist with the results that have been obtained by V.P. Melnikov and A.N. Nesterov. The authors conducted experiments on the hydrate formation process from brine and it was established that the amount of gas hydrates formed from mineralized water is bigger than off pure water (Melnikov & Nesterov 1997).

3 STATEMENT OF THE PROBLEM OF THE RESEARCH

In connection with quite different opinions and hypotheses, it was decided upon reasonability of conducting a series of experiments to explore the process of hydrate formation in porous media. Hereby,

the aim of this research is to create models of the host rocks of bottom marine sediments and carry out a series of experiments on the production of gas hydrates under conditions similar to the natural ones.

4 MAIN MATERIAL AND RESULTS

Methods of experimental researches of the hydrate formation process are based on physical modeling and natural experiments in the KTK-3000 climate regulation chamber. This methodology is built on usage of the experimental NPO-5 plant (Figure 1) which allows to simulate specified parameters of the process of gas hydrates' production in a wide range of temperatures and pressures. In general, the technique of experimental research includes preparation of the samples for physical modeling, conducting a series of experiments and processing of the experimental data.

For the most of the previous studies on the hydrate formation process in porous media, artificial media were selected as for the hydrate-forming environments that were modeled with glass beads, silica gel, substance (Tohidi et al. 2001, Smith et al. 2002, Uchida et al. 2002).

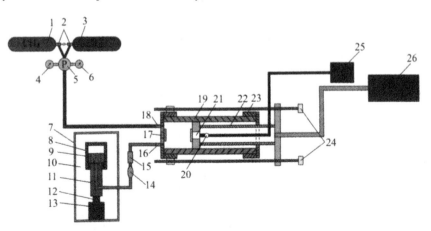

Figure 1. Laboratory unit for gas hydrate receive: 1 – methane tank; 2 – cylinder valve; 3 – cylinder with carbon dioxide; 4 – high pressure gauge; 5 – adjustable gear; 6 – gauge of low pressure; 7 – rigid frame; 8 – cylinder of water pressure unit; 9 – water; 10 – unit to supply water under pressure; 11 – rod of the water pressure unit; 12 – rod of the hydraulic jack; 13 – hydraulic jack body to produce 5 MPa pressure; 14 – water pressure gauge; 15 – water feed control valve; 16 – water input choke; 17 – cylinder transparent window; 18 – gas input choke; 19 – device for the formation of gas hydrates; 20 – block of LEDs; 21 – transparent window; 22 – rod; 23 – guide flange; 24 – coupling bolts with nuts; 25 – power battery; 26 – tank for removal of substituted methane.

For experiments carried out in this research natural sand with varying grain size and kaolinite clay were used.

During the experiments, as gas hydrate-forming gas methane was being used because it is its hy-

drates which are the most commonly spread in nature and form gas hydrates' deposits in the Black Sea. The actual volume fraction of methane is 94.5%, the rest accounts for homologues of methane (Table 1).

Table 1. Composition of the hydrate-forming gas used in experimental studies.

Composition of gas	Percentage, %
Methane	94.5
Ethane	2
Propane	2.5
Butane	1

In the experiments, five samples, which simulated porous media for formation of gas hydrates, were used. The temperature of the hydrate-formation process was maintained constant in the range of +8...+9°C for all of the experiments. The start pressure for the hydrates' formation of each experiment was obtained by injecting gas into the reactor and subsequent determination of a sharp drop of the pressure in the reactor which indicated dissolution of the gas molecules in the water molecules, and as a consequence – the beginning of the gas hydrates' crystals' formation. After reducing the pressure, it was artificially augmented to a value determined experimentally for each case using a sample of a rock, and kept then constant until the end of the process of hydrate formation. Completion of the gas hydrates' formation was defined by visual inspection through a window in the reactor.

As the control experiment, a hydrate sample was obtained by free contacting of water and gas under appropriate thermobaric conditions as simulating the conditions of formation of the gas hydrate deposits in the Black Sea (Figure 2).

Stages of the gas hydrate' formation are shown in the Figures 2 – 7, wherein: I – the stage preceding the hydrate formation (pressure buildup); II – the stage of the formation of the gas hydrates' crystallization centers; III – the stage of the process of the active hydrate formation; IV – the stage of the hydrate consolidation.

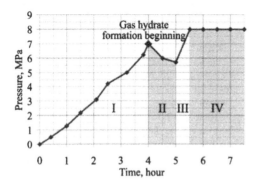

Figure 2. Stages of the formation of a gas hydrate when free contacting of gas and water.

Figure 2 shows that the process of hydrate formation with the free mixing of gas and water started after 4 hours when having reached a pressure of 7 MPa. The beginning of nucleation of a gas hydrate is marked with a pressure surge to 5.5 MPa, indicating that the gas has dissolved in the water. Further formation of the gas hydrate was occurring at a pressure of 8 MPa.

For obtaining the second hydrate sample, coarse-grained sand was selected as a porous medium; to produce the third specimen, middle-grained sand was chosen; the fourth case was examined by using some fine-grained sand; for the fifth one – coarse-grained sand with the addition of clay particles (10% of the particles based on the weight of the sand) was picked; and the sixth test was carried out with use of middle-grained sand with the addition of clay particles (10% of the particles based on the weight of the sand).

Parameters and stages of the hydrate formation process of these experiments are shown in the Figures 3 – 7.

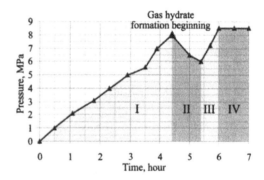

Figure 3. Stages of the formation of a gas hydrate in the sample of coarse-grained sand, gas and water.

The conducted experiments showed that parameters of the methane hydrate formation applying coarse-grained sand are almost identical to the parameters of the hydrate formation process when there is a free gas-water contact (Figure 3). Thus, deviation of barometric scale constituted ≈1 MPa towards increasing the pressures of the hydrate formation beginning when a coarse-grained sand sample used and which reached 8 MPa.

When applying middle-grained sand, hydrate formation pressure increased by 1.5 MPa comparing to the first laboratory experiment, and reached the value of 8.5 MPa (Figure 4). Nucleation process of the hydrate crystals started after 4.5 hours.

Throughout the experiment, the temperature was being kept constant (as it was in all of the previous experiments) and equal to +8...+9°C.

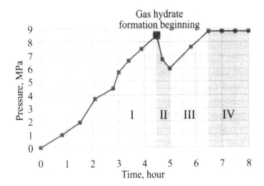

Figure 4. Stages of the formation of a gas hydrate in the sample of middle-grained sand, gas and water.

The experiment with using fine-grained sand yield the following results: pressure at the beginning of the hydrate formation increased by 2 MPa in comparison to the free mixing of gas and water when a gas hydrate is being formed (Figure 5).

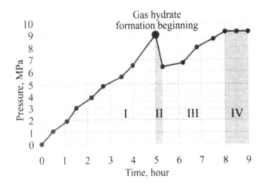

Figure 5. Stages of the formation of a gas hydrate in the sample of fine-grained sand, gas and water.

Analyzing the Figure 5, it can be stated that the increase of pressure of the gas hydrate formation in fine-grained sand comparing to free volume indicates that the smaller diameter of the particles which make up the porous medium is, the higher pressure is needed for gas hydrate formation then.

Laboratory researches of the hydrate formation process with coarse-grained sand with clay particles added (10% of the particles based on the weight of sand) as a rock-forming medium used showed that the pressure of the beginning of gas hydrates' formation increased by 3 MPa comparing to the first experiment and it reached 10 MPa (Figure 6).

Having added a clay fraction to middle-grained sand at the rate of 10% of its weight, gas hydrates' formation pressure increased by 12 MPa and at the end of the experiment amounted to 14 MPa (Figure 7).

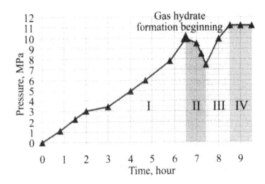

Figure 6. Stages of the formation of a gas hydrate in the sample of coarse-grained sand with addition of clay particles, gas and water.

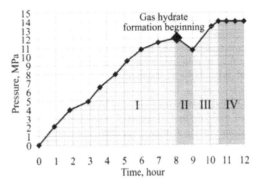

Figure 7. Stages of the formation of a gas hydrate in the sample middle-grained sand with the addition of clay particles, gas and water.

Analyzing the results, it can be concluded that a change in the structure of the rock-forming medium varies the pressure necessary for the gas hydrates' formation. Revealed a displacement of the equilibrium line of hydrate formation in a higher-pressure zone. This could be explained with the fact that with decreasing size of particles, composing the rock-forming medium, formation of the bond water films on grains of the rock takes place. The binding energy of the water in such films is greater than that of free water. This means that for the conversion of the skin water into the hydrated condition it is necessary to increase the pressure of the hydrate formation.

Beginning of the gas hydrate nucleation process in the reactor was observed visually through a window in climate regulation chamber. It should be noted that for the nucleation of a gas hydrate's crystallization centers in various porous media, an important condition is a sufficient amount of gas dissolved in the water, since if the amount is small or insufficient gas hydrates do not form.

5 CONCLUSIONS

1. The leading developed countries consider natural gas hydrates as an alternative source of hydrocarbons due to enormous reserves of methane gas in a hydrate form. Many researchers believe that the amount of hydrocarbon in the gas hydrate deposits noticeably exceeds existing reserves of natural gas. This caused large-scale studies of gas hydrates in order to develop technology to extract methane from them.

2. Since almost all of the gas hydrates' reserves reside in the pore space of sedimentary rocks under oceans and seas, the authors analyzed scrutiny of the gas hydrates' formation process in porous media. Systematized basic regularities of the hydrate formation obtained by renowned scientists such as Y.F. Makogon, V.S. Yakushev, E.M. Chuvilin, V.P. Melnikov, A.N. Nesterov, S. Dallimore, CoUett T. Field, T. Uchida formulated the tasks of further research.

3. The laboratory research of the gas hydrates' formation process in porous media using a variety of sandy and clay-grained rocks on the KTK-3000 plant was conducted. As the hydrate-forming medium, natural sand with varying grain size and kaolinite clay were chosen.

4. It is found that the parameters of the gas hydrates' formation process in vitro in the presence of coarse-, middle- and fine-grained sand are shifted on the barometric scale upward increasing pressure of hydrate formation: when using a sample of coarse-grained sand, the pressure increased by ≈ 1 MPa, using middle-grained sand resulted in 1.5 MPa growth, applying fine-grained sand made up 2 MPa increase. Modification of the pressure parameters of the phase transitions of methane and water into a gas hydrate can be explained in that in sand, water is less associated with particles' surface and, substantially, is in a capillary form.

5. While examination of the coarse-grained sand sample with the addition of clay particles, the pressure of hydrate formation increased by 3 MPa relatively to the free mixing of gas and water, and resulted in 10 MPa. When adding to the middle-grained sand some clay fraction of 10% of its weight, the hydrate formation pressure increased to 14 MPa. An increase in pressure of the hydrate formation in the sandy-clay samples comparing to clean sands can be explained by the Van der Waals' bond water powers on clay particles.

6. So, development of the technologies and establishment of their rational parameters should take into account the effect of the granulometric composition of porous medium.

REFERENCES

Dallimore, S., Chuvilin, E., Yakushev, V., Grechischev, S., Ponomarev, V., Pavlov, A. & T. CoUett. 1996. *Field and laboratory characterization of intrapermafrost gas 44JL hydrates, Mackenzie Deila, NWT, Canada.* Proceeding of the 2nd International Conference on Natural Gas Hydrates. Toulouse: 525 – 531.

Falshtynskyy, V., Dychkovskyy, R., Lozynskyy, V. & Saik, P. 2012. *New method for justification the technological parameters of coal gasification in the test setting.* Geomechanical Processes During Underground Mining - Proceedings of the School of Underground Mining. The Netherlands: CRC Press/Balkema: 201 – 208.

Ganushevych, K., Sai, K. & Korotkova, A. 2014. *Creation of gas hydrates from mine methane.* Progressive Technologies of Coal, Coalbed Methane, and Ores Mining. The Netherlands: CRC Press/Balkema: 505 – 509.

Lozynskyi, V.G., Dychkovskyi, R.O., Falshtynskyi, V.S., & Saik, P.B. 2015. *Revisiting possibility to cross the disjunctive geological faults by underground gasifier.* Naukovyi Visnyk Natsionalnoho Hirnychoho Universytetu, No 4:

Makogon, Y.F. 1974. *Hydrates of Natural Gases.* Moscow: Nedra: 208.

Melnikov, V.P. & Nesterov, A.N. 1997. *Water migration during gas hydrate formation in porous media.* Rotterdam: Ground Freezing 97, Knutsson (ed.): 391 – 394.

Ovchynnikov, M., Ganushevych, K. & Sai, K. 2013. *Methodology of gas hydrates formation from gaseous mixtures of various compositions.* Annual Scientific-Technical Colletion – Mining of Mineral Deposits 2013. The Netherlands: CRC Press/Balkema: 203 – 205.

Smith, D.H., Wilder, J.W., Seshadri, K. & Zhang, W. 2002. *Equilibrium pressure and temperature for equilibria involving SI and SII hydrate, liquid water, and free gas in porous media.* Proceeding of the 4th International Conference on Natural Gas Hydrates. Japan: 295 – 300.

Tohidi, B., Anderson, R., Clennell, M.B., Burgass, R.W. & Biderkab, A.B. 2001. *Visual observation of gas hydrate formation and dissociation in synthetic porous media by means of glass micromodel.* Geology, Vol. 29, No 9: 867 – 870.

Uchida, T., Ebinuma, T., Takeya, S., Nagao, J. & Narita, H. 2002. *Effects of pore sizes on dissociation temperatures and pressures of methane, carbon dioxide and propane hydrates in porous media.* J. Phys, Chem., Vol. 106: 820 – 826.

Wright, J.F. Dallimore, S.R., Nixon, F.M., Uchida, T. & CoUett, T.S. 1999. *Influences of grain size and salinity onpressure-temperature thresholds for methane hydrate stability.* Scientific Results from JAPEX / JNOC / GSC Mallik 2L-38 Gas Hydrate Research Well. Mackenzie Delta, Northwest Territories, Canada: Geological Survey of Canada, Bulletin 544: 229 – 240.

Theoretical and Practical Solutions of Mineral Resources Mining – Pivnyak, Bondarenko & Kovalevska (eds)
© 2015 Taylor & Francis Group, London, ISBN: 978-1-138-02883-8

Methodological approach to the development of gas hydrate deposits

E. Maksymova
National Mining University, Dnipropetrovsk, Ukraine

ABSTRACT: Complex methodological approach to the development of underwater deposits of natural gas hydrates using circular well system with thermal medium circulation by counter flows is proposed. Introduction of the given production parameters as well as new technological operations will allow economically expedient, controllable, and continuous gas extraction both at shallow depths and deep ones with minimum effect on underwater ecosystem.

1 INTRODUCTION

As world reserves of power resources are being exhausted, a number of major countries are engaged in search of both alternative and additional power sources. It is known that almost half of Earth's carbon being a part of numerous organic and non-organic compounds is in the form of hard gas hydrates (mostly in the form of methane hydrate). Today the resource is one of the most promising additional power sources in the world. Methane molecule surrounded by water molecule and creating so-called clathrate structure is that very gas hydrate being studied by many scientists and researches during last four decades. Conditions of their formation and stable existence period are researched (Byakov & Krugliakova 2001, Shnyukov & Ziborov 2004, Makogon 2010, Bondarenko et al. 2012). It depends on intensive exploitation, complication of its development, and depletion of such mineral deposits as oil, gas condensate, gas, and coal ones. World oil prices grow constantly. Hence, other energy resources become more expensive as well. Necessity to transfer to alternative power sources or additional ones is obvious. Scientists from different countries believe that gas hydrate is the additional power resource not extracted before but with great potential reserves. Today physics, chemists (supramolecular clathrate theory), and ecologists pay much attention to the issues. Mining specialists should be involved in the research work connected with methods of its development to solve technical problems concerning this unique resource. From the viewpoint of mining practice and oil-producing branch the development of methods and techniques of gas extracting from natural deposits of gas hydrates it is required to have clear idea of future mineral deposit mining conditions. It is also necessary to take into account occurrence features, water and physical properties of enclosing rocks, hydrostatic pressure and deposit temperature while opening and developing the deposits. The research is carried out by a team of mining engineers at the Underground Mining Department of the National Mining University (Bondarenko et al. 2012, Ovchynnikov et al. 2013, Bondarenko et al. 2013). Thus, the issue of underground / underwater development methods as well as methods for gas extraction from natural gas hydrate deposits is rather topical.

The article is aimed at the analyses of available methods of gas dissociation from gas hydrates to propose a method of methane extraction from natural gas hydrate deposits in terms of deep gas hydrate rock masses.

2 METHODOLOGICAL PRINCIPLES OF THE DEVELOPMENT

Clear system methodology for the development of deep gas hydrate deposits is not available today. As for modern areas of gas hydrate deposits location, their industrial development is carried out only in Western Siberia and Alaska; however, these deposits are associated with continental permafrost zones. In 2013 Japanese ship of Japan Oil, Gas and Metals National Corporation (JOGMENC) was the first to extract gas from gas hydrates of ocean floor. According to the information by the Ministry of Economy of Japan, JOGMANC bored wells fifty kilometers from Atsumi peninsula. Japanese experts state that these wells contain almost 1.1 trn m^3 of hydrates. Such fuel reserves can provide power resources of the country for the next eleven years. According to information by Bloomberg with reference to JOGMENC data, total hydrate reserves within the shelf of Japan can cover its fuel requirements for 100 years.

At the end of October 2012, the Ministry of Energy of Russia recognized that just gas hydrates rather than shale gas (as it was stated before) were main potential competitor for Russian "Gazprom" in the fuel market. According to information by the Ministry of Energy, reserves of gas hydrates may be more than reserves of other fuel resources taken together. Only in the Black Sea these reserves are almost 30 trn m^3.

Techniques of continental development of gas hydrate deposits differ greatly from underwater extraction conditions. It turns out that extraction from underground deposits in permafrost zones is not very profitable to compare with the traditional oil and gas extraction. It is quite possible that neither scientists nor investors have paid much attention to this fact. It is probably the reason why today the development for extraction techniques of such deposits delays. Maybe it also depends on the fact that inland deposits are minor share of the world deposits of gas hydrates which major part is in seas and oceans. Thermodynamic conditions for such compounds formation are optimum at the depth of 200 down to 1000 m below sea level (Shnyukov & Ziborov 2004).

Four basic principal approaches to gas extraction from natural underwater deposits of gas hydrates are available today. These are: pressure decrease below phase equilibrium condition; heating of hydrate-bearing rocks above equilibrium temperature; combination of the techniques; and inhibitor injection immediately into gas hydrate seam. The technics are based upon one fundamental idea – to shift phase equilibrium towards hydrate dissociation with its following disintegration into gas and water. As for inhibitor injection into seams, in this case, taking into account global occurrence of gas hydrates, negative effect of such a technique on ecosystem is obvious; that is why we do not consider it. Any selected extraction technique depends on geological and structural features of specific deposit and the adopted development system should meet optimum technical and economic specifications as well as the environmental requirements. Scientific sources describe several techniques of heat source effect on gas hydrate to extract methane gas from it. However, the mistake of many experts is that it is proposed either to extract ice-like masses to the surface or to extract water saturated with gas to the surface using air-lift columns with the following gas separation and extraction. The described techniques are very energy-consuming and they are not attractive from economic viewpoint. For example, there are many methods of gas extraction from gas hydrate of bottom sediments by means of dissolution of upper

layer of gas hydrates accumulations with the water having natural reservoir temperature delivering it into a bottomed cone with water and gas mixture with methane and its homologues formation in it with following pumping-out of the mixture to the surface using air-lift facilities. Such techniques are rather expensive requiring considerable amount of metallic structures subjecting to active corrosion due to sea water action; besides they engage transportation of huge water amount with low gas content being useful only at shallow depths and within limited areas.

Moreover, solid hydrocarbons can be developed using well drilling-out operations with the system of closed horizontal side sections with following formation of non-controllable thermal field within lower seam and hydrocarbon extraction from upper gas hydrate seam. Operational process involves continuous effect of hot water on the seam with its constant circulation along the formed closed channels. It is proposed to extract water and gas mixture using air-lift approach with its following separations at the surface. Such methods require huge amounts of hot water and their transportation. They are complicated by multilevel development procedure and necessity of following separation. In addition such methods can result in non-controllability of the whole process of gas hydrate mass disintegration.

We propose to perform dissociation within a seam itself preserving natural hydrodynamic equilibrium, to leave free water in the development level, to captures gas with the help of gas-handling well, and to get pure gas from the pipe at the surface of the floating platform. The technique proposed by the authors involving introduction of the specified extraction parameters as well as new technological operations will give the possibility of economically expedient controllable constant gas obtaining at both shallow and great depths with minimum effect on underwater ecosystem.

Stage to select specified parameters includes following substages. Before deep-water drilling operations start it is required to determine complexity of geological structure of gas hydrate productive levels. While selecting extraction procedure for deep-water deposits of gas hydrates it is quite important to take into consideration the complexity of mining and geological conditions. Deposits may be of complex geological structure – disturbed tectonically into the series of blocks and zones with variable nature of productive levels – lithological composition, collector properties etc. and of simple one – productive seams within such deposits are characterized by relative persistence of lithological composition, collector properties, and productive levels within the whole deposit area. Then the range of extraction depth and number of lithological differences of enclosing rocks of cut of the deposit to be developed as well as their thickness are determined. If it is possible one seam is specified to be the development object. Such parameters as its thermal conductivity, specific heat, temperature mode of rocks according to their depth, porosity, density, and specific weight are estimated. Degree of deposit saturation with gas hydrates depends on these parameters. The next stage determines possible operating production rate (for example, extra low-output are up to 25000 m³/day; low-output are 25000 – 100000 m³/day; medium-output are 100000 – 500000 m³/day; high-output are 500000 – 1000000 m³/day; super high-output are over 1000000 m³/day). Next stage determines a range of allowable seam pressures within which the deposit development is possible.

Thus, before specific development parameters are selected the deposit should be explored in detail, reserves of gas hydrate within selected boundaries should be estimated, full-scale testing of the deposit in terms of different depths of future development should be performed, and laboratory tests of all basic parameters should be carried out. Development plan of each specific deposit involves core sampling from the productive levels to provide ra-

ther complete characteristic of physical properties of both productive seams and enclosing rocks. Design of future well cluster should involve possible contacts of rocks and various lithological differences, and structural and geometrical features of the deposit structure.

Full-range of industrial geophysical research including determination of well shaft curvature and azimuth should be performed in each exploratory well. Advance without logging, measurements of curvature and azimuth should not be more than 200 m. Geological and engineering program approved in accordance with projects of prospecting and pilot development of the deposit determines research efforts and types of geological and field study while drilling wells.

Geothermal gradient is special wells should be determined for each potential area.

It is proposed to develop deposit as follows: deposit is bored-out by means of circular well system; at the depth of 1200 m under bottom it transfers to horizontal plane with following approaching to a bottom surface to transform some part of the deposit into closed cylindrical cavity for heat effect (Figure 1).

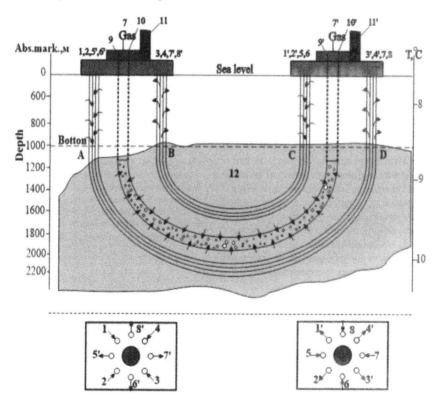

Figure 1. Plan of gas hydrate deposit development:1 – 8 and 1'– 8' – deep-water wells with thermal medium, 9, 9' – development well; 10, 10' – a gas-collecting tank; 11, 11' – a hydrating reactor; 12 – gas hydrate deposit.

Shift of phase equilibrium of methane containing thickness stability is fundamental criterion to extract methane from gas hydrate deposit. Equilibrium of temperature and pressure is the key parameter for gas hydrate stability. As the considered plan specifies constant depth of the deposit then pressure is taken as a constant value.

For example deposit temperature may be within $+8.5...+9.5°C$, then $+9.0°C$ is taken as the initial one. In summers water temperature in the upper layers varies from $+15.0$ to $+20.0°C$.

It is proposed to make heat effect by means of sea water injection (for example, at $+18°C$ temperature) from one platform ($1 - 4$ and $5 - 8$ wells) and its spouting at the second platform from the opposite openings ($1' - 4'$ and $5' - 8'$ correspondingly). Such technology helps avoid formation of cones/flowing under floating platforms. While injecting thermal medium being water from upper sea layers, $+1...+2°C$ shift of temperature mode will take place due to poor heating of ABCD working zone (Figure 1). Any variations of these parameters will effect on dissociation rate and as a result on a production rate of gas obtained at (9 and $9'$) openings which can be controlled by the operation of wells with thermal medium. As soon as gas starts its dissociation from the seam, the plan provides its capturing with a perforated pipe of (9 and $9'$) development well getting into (10 and $10'$) gas-collecting tank of the platform. As free water is in the seam, natural hydrodynamic mode is preserved. The process will be reflected in simultaneous change of production rate of 9 and $9'$ well. According to the adopted development procedure if the specified production rate parameters are violated, then either decrease or increase in operation of circular well system with thermal medium is possible. Hence, such development technique means the use of natural processes of exothermal catalytic reactions within a zone of gas hydrate occurrence, i.e. it allows using emitted heat to decompose gas hydrates. It is proposed to perform drilling from two floating platforms with (10 and $10'$) gas-collecting tank and (11 and $11'$) facilities being reactors for gas recoversion into gas hydrate for its further optimum transportation.

3 CONCLUSIONS

1. The work proposes innovative method for the development of gas hydrate deposit occurring at various depths on the basis of complex approach and estimation of the initial parameters of mining and geological conditions of a deposit.

2. Shift of gas hydrate phase equilibrium towards its dissociation with water and gas liberation with complete water and enclosing rock leaving in a seam is the basic principle of the proposed development system for gas hydrate seam.

3. Owing to the technique features, the development method makes it possible to achieve controllable process in terms of commercial extraction of natural gas from gas hydrates.

4. All the principal advantages of the technique minimize negative environmental effect of this development.

REFERENCES

Bondarenko, V., Ganushevych, K. & Sai, K. 2012. *Substantiation of technological parameters of methane extraction from the Black Sea gas hydrate*. Materials of the conference "Szkoła Eksploatacji Podziemnej". Krakow, February 20 – 24.

Bondarenko, V., Maksymova, E. & Koval, O. 2013. *Genetic classification of gas hydrates deposits types by geologic-structural criteria*. Mining of Mineral Deposits. The Netherland: CRC Press/Balkema: 115 – 119.

Byakov, Y.A. & Krugliakova, R.P. 2001. *Gas hydrates of sedimentary strata of the Black Sea – hydrocarbon raw of the future*. Exploration and protection of the interior, No 8: 14 – 19.

Makogon, Y.F. 2010. *Gas hydrates as additional source of energy for Ukraine*. Oil and gas industry, No 3: 47 – 51.

Ovchynnikov, M., Ganushevych, K. & Sai, K. 2013. *Methodology of gas hydrates formation from gaseous mixtures of various compositions*. Mining of Mineral Deposits. The Netherland: CRC Press/Balkema: 203 – 205.

Shnyukov, E.F. & Ziborov, A.P. 2004. *Mineral wealth of the Black Sea*. Kyiv: Scientific publication of The National Academy of Sciences of Ukraine: 283.

Theoretical and Practical Solutions of Mineral Resources Mining – Pivnyak, Bondarenko & Kovalevska (eds)
© 2015 Taylor & Francis Group, London, ISBN: 978-1-138-02883-8

Stress-and-strain state of rock mass around the working behind the longwall face

I. Sakhno, A. Nosach & L. Beletskaya
Krasnoarmeisk Industrial Institute of Donetsk National Technical University, Ukraine

ABSTRACT: The most important task facing Ukraine's coal industry is to ensure the operational state of mine workings. The aim of the investigation presented in this paper is to study the changes in geomechanical environment conditions around the development working at the site of its transition to the longwall face supported area. The investigation was carried out on the basis of the finite-element method. The simulation results proved that in case of supporting the developments behind the longwall, localization of maximum stress and the type of the volumetric strain state in the regular roof, hanging up at the boundary with the worked out areas, is determined by the protective face side pack rigidity. The area of maximum stress formation in the main roof indicates the place of its destruction and defines the protected working stability. The investigation determined the application sphere for induced caving in of the main roof.

1 INTRODUCTION

For many recent years one of the main problems of Ukraine's coal mines has been insufficient operational reliability of mine workings. The current situation with exploiting and maintaining mine workings has led to disparities in the distribution of workers for certain types of operation. The number of drifters is nearly equal to the number of productive breakage face workers, and the number of workers engaged in subsidiary operations, at the expense of the quantity of repairmen, has exceeded 70% of the total number of workers participating in the process of coal extraction (Babiyuk 2012).

Therefore, the most important task facing the coal industry of Ukraine, along with improving productivity and upgrading the mine stocks, traditionally is providing mine workings operational condition and raising their reliability. The greatest mine rock displacements are known to occur in development workings. The results of assessment of development workings conditions show that some 14 – 17% of them are in unsatisfactory conditions. Moreover, the workings, supported behind the longwall, are repaired at least once.

The rigidity of mine workings is primarily determined by the stress-strain state (SSS) of the surrounding mine rock mass. Alterations in SSS of the rocks surrounding the mine working at great depths have become the reason why most of the known methods of supporting have exhausted their potentialities in the line of ensuring the workings rigidity. In modern mines, in the capacity of protective structures are most commonly used solid bands of different materials set up behind the longwall (gob stowing, constructions made of artificial elements, cast blocks). The variations of the protective constructions material's deformation modulus are up to three orders of magnitude. The aim of the investigation was the study of the change in SSS of the rock around the development working at the site of its transition to the supported area behind the breakage face and determination of the affect of the protective construction rigidity on the stability of the development working behind the longwall.

2 RESEARCH METHODS

A characteristic feature of investigation into geomechanical processes is a wide application of physical and mathematical simulation. Methods of physical simulation, commonly applied to equivalent models, allow to receive a clearly cut picture of displacements and destructions in the rock mass. However, realization of these methods is associated with a great labour-consuming process of constructing models, their calibration, considerable time consumption and a rather scarce amount of the information obtained. In addition, by using this method it is difficult enough to provide precision in conditions for preparation of the equivalent material and loading the model while carrying on a series of similar experiments.

Mathematical simulation, realized by means of analytical and numerical methods, is one of the basic up-to-date tools which make it possible to investigate into the stress and strain state of the rock mass.

Currently, numerical simulation methods are widely used in geomechanics. Of these, a leading position is occupied by the finite-element method (FEM) (Zenkevich 1975), which was used in solving this problem. The implementation of the FEM was realized in software complex ANSYS.

3 THEORETICAL PART

An arched working supported behind the longwall and exploiting a coal seam of 1.5 m thickness at the depth of 800 m was simulated. The sheet deposit is arbitrarily specified as horizontal. The immediate roof is presented as siltstone 2.5 m thick, with uniaxial compression strength of 40 MPa. The main roof is 60 m thick, 70 MPa strength sandstone. The ground rock is mudstone of 40 MPa uniaxial compression strength. According to the described structural core of the seam, for each layer a special deformation modulus (MPa), a lateral strain factor (Poisson), internal friction angle, adhesion factor and dilatation factor were set. Initial data for simulating were taken from the inventory of the mine rocks physical properties (Melnikov et al. 1975) for the conditions of Donetsk-Makeyevka coal-bearing region. The materials, simulating mine rocks, were described with reference to Druker-Prager's basic isotropic model. A composite problem was solved in non-linear formulation. The geomchanical state formed in the rock mass behind the longwall after the coal extraction and the subsequent caving in of the immediate roof is considered. In this case, we proceeded from the assumption that the immediate roof, which has a relatively low strength, caves in without appreciable hanging up over the open goaf.

The SSS of rocks which embed the development working behind the longwall, depending on the type of the protective construction was simulated. A solid face side pack is adopted as a means of protecting the working. As far as protective packs may be of different widths, shrinkage and mechanical compliance, their integral index is adopted as specific rigidity of the pack normal to the seam bedding strike. This feature allows to assess the ability of the component to resist deformation under external impact, being, in fact, the magnitude of the reverse compliance. The specific rigidity will be determined as the product of the deformation modulus by the unit of the pack cross-section. The immediate entry into the operation of the protective pack with the specified mode of deformation is assumed in the model. The simulated means of protection are shown in Table 1.

Table 1. Simulated protective constructions.

Means of protection	Specific rigidity of the protective construction, c_p GN
Cast pack of quick-setting materials	12
Concrete block with wooden gaskets	6
Coal pillar (before breakage)	2
Band of rock semi-blocks	1.2
Filled cribs	0.6
Gob stowing (pneumatic filling)	0.2
Gob stowing (mechanical filling)	0.02

At present, there is no consensus as to the most appropriate theory of mine rocks strength in a volumetric strain area; the existing classical strength theories give underrated values of extreme strains by 3 – 8 times, depending on the type of a volumetric strain state. In this case, the strength limit depends on the ratio of the tensor strain components. Therefore, the analysis of strains around the working was carried out with reference to the patterns of algebraically greatest main tension strains – S1 (σ_1) (I strength theory), to the equivalent strains – SEQV (IV strength theory or the criterion of maximum reshaping power) and to the strain intensiveness – SINT (maximum tangent strain theory, III strength theory). According to the above-mentioned hypotheses the calculated strains are to be compared to the strength limit (yield) at an ordinary tension.

4 RESULTS AND DISCUSSION

Figures 1 – 3 show the distribution patterns of strains around the protected working when the length of the hanging up cantilever is 30 m and calculated by different strength theories at the protective pack rigidity 2 GN.

From Figure 1 it is evident, that in the top part of the main roof two areas of maximum main strains S1 (σ_1) are formed, whose strains reach the extreme values. At the same time, over the protective pack this area tends to be of a greater size and the strains emerging in it are also bigger. This may indicate the most likely place of the cantilever failing over the protective face side pack.

The analysis of Figures 2 and 3 shows that in the roof in the same places maximum strains SINT and SEQV are formed, although the absolute value of these strains is greater over the coal seam. In the area of maximum strains the excess over the extreme factors is more than two times. It may indicate the cantilever falling at its seam end.

Changing the protective construction rigidity may lead to displacement of the maximum tensions area in the main roof relatively to the protected development working, that can be clearly seen in Figure 4. Thus, in case, when the working is protected with gob stowing, set up by pneumatic filling, the area of maximum tensions S1 (σ_1) shifts to the side of the virgin seam.

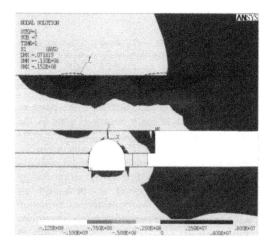

Figure 1. Main strains S1 (σ_1) distribution around the mine working supported behind the longwall at the face side pack specific rigidity 2 GN: 1 – area of maximum tensions.

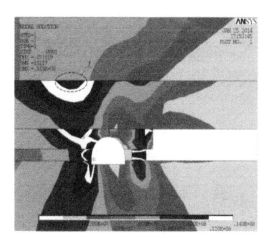

Figure 2. Strains SINT distribution around the mine working supported behind the longwall at the face side pack specific rigidity 2 GN: 1 – area of maximum strains.

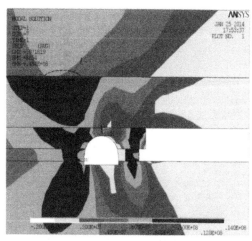

Figure 3. Strains SEQV distribution around the mine working supported behind the longwall at the face side pack specific rigidity 2 GN: 1 – area of maximum strains.

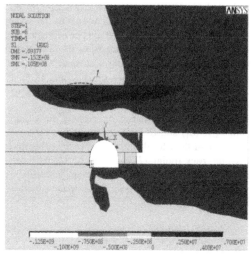

Figure 4. Main strains S1 (σ_1) distribution around the mine working supported behind the longwall at the face side pack specific rigidity 0.2 GN: 1 – area of maximum tensions.

To evaluate the affect of the protective construction rigidity on the maximum strains area localization in the main roof which is, accordingly, the most likely place of the roof caving in, let's analyze the strains formed in the top part of the hanging cantilever of the main roof. The points of strains location are shown in Figure 5. The initial points of the coordinates shown in Figure 5 are associated with the centre of the arched support vault.

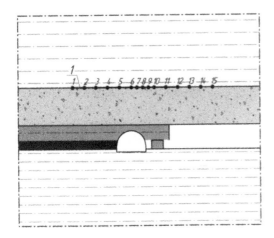

Figure 5. Model diagram with the points of strains location: 1 – the line in the top part of the main roof hanging cantilever.

Figures 6 and 7 show the diagrams of the changes in the calculated strains S1 (σ_1) and SEQV in the model along line *1*, drawn in the middle of the upper face of the main roof layer. The axis of ordinates (y-axis) in the diagram corresponds to the transverse axis of the protected working's cross-section. The positive direction of the axis of abscissas (x-axis) coincides with the direction from the working toward the open goaf of the longwall.

The graphs show that two areas of maximum strains (S1 (σ_1) and SEQV) are formed in the roof. The place of their localization is over the protective face side pack (point *11*, Figure 5; horizontal coordinate *6*, Figure 6, 7) and over the coal seam (between points *3* and *4*, Figure 5; horizontal coordinate – *5*, Figure 6, 7).

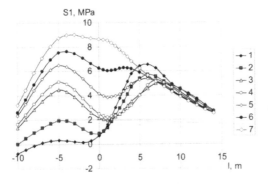

Figure 6. Calculated strains S1 (σ_1) along line 1, drawn on the upper face of the main roof layer, at the protective face side pack rigidity respectively: 1 – 12 GN, 2 – 6 GN, 3 – 2 GN, 4 – 1.2 GN, 5 – 0.6 GN, 6 – 0.2 GN, 7 – 0.02 GN.

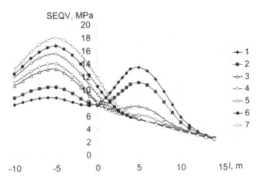

Figure 7. Calculated strains SEQV along line 1, drawn on the upper face of the main roof layer, at the protective face side pack rigidity respectively: 1 – 12 GN, 2 – 6 GN, 3 – 2 GN, 4 – 1.2 GN, 5 – 0.6 GN, 6 – 0.2 GN, 7 – 0.02 GN.

The comparison shows, that from the point of view of the first strength theory (Figure 6) the failure over the protective pack and over the seam at the pack rigidity 1.2 GN is equally probable, because the maximum tensions are equal and constitute 5.15 MPa. According to the strain energy theory of strength (Figure 7), the probability of failure is equal over the protective construction and the seam at the pack rigidity 6 GN, maximum strains about 10.5 MPa. Thus, the difference in values is by 5 folds depending on the strength theory applied. So far, as the rocks are in a volumetric strain area, in order to choose a more suitable strength theory for subsequent analysis, let's consider the diagrams of dependence of Lode-Nadai μ_σ strain state parameter for the main roof along line *1* (Figure 8) and in the points of maximum strain (Figure 9) upon the protective pack rigidity. The least power consuming type of volumetric strain state is known as "generalized shear" (Norel 1982).

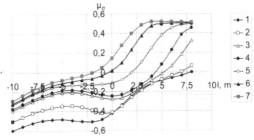

Figure 8. Dependence of the strain state μ_σ type in the main roof along line 1, drawn on the upper face of the main roof layer, at the protective face side pack rigidity respectively: 1 – 12 GN, 2 – 6 GN, 3 – 2 GN, 4 – 1.2 GN, 5 – 0.6 GN, 6 – 0.2 GN, 7 – 0.02 GN at the cantilever length 30 m.

136

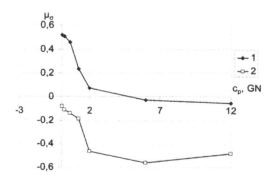

Figure 9. Dependence of the strain state μ_σ type in points 11 (1) and between points 3 and 4 (2) along line 1 from the protective face side pack rigidity

The diagrams show that the main roof rocks in the area of maximum strains over the protective face side pack (point *11*, Figure 5) at the pack rigidity of 12 to 2 GN, are in the state approaching the generalized shear; lowering the pack rigidity to less than 2 GN leads to transition of the strain state nearer to the generalized tension, which is the most power-intensive state for destruction. At the same time, the area of the main roof rock formation over the seam (between points *3* and *4*, Figure 5) at the face side pack rigidity of 12 to 2 GN is in the state between generalized strain and generalized shear, while further lowering rigidity to less than *2* leads to approximation of the strain state with the generalized shear. From Figure 9 it is possible to conclude that when similar strains emerge in the areas of maximum factors, with the range of pack rigidity 12 – 2 GN, a failure over the protective pack is more likely. Taking into account the ratio of tensor components in the area of maximum strains in the main roof and the mechanism of the cantilever failure and, also, that the mine rock tensile strength limit is less than its compressive strength limit approximately by one order, the first strength theory is to be adopted as a criterion for further analysis.

Thus, for the case considered the following conclusion can be made: at the protective face side pack rigidity of 12 to 2 GN the area of maximum strains is formed in the main roof rocks over the protective construction, in the upper part of the hanging rock cantilever (point *11*, Figure 5), in this case the rocks in the area of maximum factors are in the state of generalized shear. The formation of maximum stress area is due to the main roof bending in the open goaf after coal extraction. The initial destruction of the main roof rocks is most likely in this place. This will ensure the hanging main roof failure into the open goaf that will lead to the reduction of loading on the protective pack and on the outline of the pro-

tected development working. This will allow for the further stability of the working.

At the face side pack rigidity of 2 GN to 1.2 GN the emergence of primary destruction is equally probable both over the protected construction and over the seam, that will be determined by the rock's structural composition.

At the pack rigidity less than 1.2 GN the maximum strains area is formed in the main roof rocks over the seam, in the top part of the hanging cantilever (between points *3* and *4*, Figure 5). The rocks are also in the state approaching the generalized shear. The primary main rock destruction is likely to occur over the seam. After the main roof failure in this case, the weight of the rock cantilever will partly concentrate on the outline forming rocks and partly on the protective construction, causing an asymmetric loading on the development working support and its deformation; a further progress in geomechanical processes around the working will lead to reduction in the protected working stability.

It's apparent, that the obtained results are determined not only by the face side pack rigidity c_p, but also by the coal seam rigidity c_c, which may vary; so, it's more correct to make conclusions with no regard to the absolute values of the protective pack rigidity, but making use of its ratio to the coal rigidity. In the problem being solved the coal seam rigidity was defined as 2.0 GN. Thus, the results obtained can be reduced to the conclusion, that at the face side pack rigidity exceeding that of the coal seam the destruction of the main roof will take place over the protective pack, that will allow to safeguard the working stability. In case, when the pack rigidity is less than coal stiffness, the destruction takes place over the coal seam that leads to a significant reduction of the working's cross-section. While at equal rigidity of coal and protective pack the probability of destruction is equally likely both over the protective construction and over the seam and it will be determined by the rocks structural composition.

From the outcomes of the investigation it follows that the most efficient methods of protecting the workings behind the longwall are rigid constructions: reinforced concrete bands and cast packs. Provided such constructions are promptly set up after coal extraction and they are immediately engaged in the acting deformation mode of the main roof, at the boundary with the open goaf conditions for its caving in over the protective pack are created, that may ensure the workings stability behind the longwall. In case of insufficient protective pack rigidity or its belated engagement in the operation of supporting the workings behind the longwall, extra measures should be taken for induced caving in of the main roof cantilever on the side of the open goaf.

5 CONCLUSION

1. In supporting the workings behind the longwall the distribution of strains and the type of volumetric tension state in the main roof hanging over the boundary with the open goaf is, to a large extent, determined by the protective face side pack rigidity.

2. The area of maximum strains formation in the main roof indicates the site of its failure and the protected working stability.

3. The results obtained allow us to conclude, that at the pack rigidity exceeding the coal seam rigidity the main roof failure will take place over the protective pack that will safeguard the working stability. In the case, where the face side pack rigidity is less than the coal stiffness, the main roof failure occurs over the coal seam, that leads to a considerable reduction in the working's cross-section. While at the similar rigidity of coal and the protective pack the failure is equally likely over the protective construc-tion and over the seam and it will be determined by the rock structural composition.

4. Hence, it follows that there is a need of further developing protection methods aimed at induced caving in of the main roof over the protective face side pack in order to ensure stability of the workings behind the longwall, and the field of application of these methods is pointed out.

REFERENCES

Babiyuk, G. 2012. *Dependability of mine workings control*. Donetsk: Svit knyhy: 420.

Melnikov, N., Rzhevskiy, V. & Protodiakonov, M. 1975. *Handbook (cadastre) of mine rocks physical properties*. Moscow: Nedra: 279.

Norel, B. 1982. *Change in coal seam mechanical properties in massif*. Moscow: Nauka: 128.

Zenkevich, O. 1975. *The finite-element method used in technology*. Moscow: Mir: 539.

Current condition of damaged lands by surface mining in Ukraine and its influence on environment

I. Gumenik & O. Lozhnikov
National Mining University, Dnipropetrovsk, Ukraine

ABSTRACT: The volume of extraction minerals by surface mining in Ukraine has significant indexes due to the large mineral deposits potential of the country. That's why the problem of natural resources rational use and environment protection at the minerals extraction and processing gained national importance. Solving of these problems inseparably linked with monitoring of comprehensive utilization of mineral resources at the mining and processing, decreases of minerals losses and damaged lands area. The results of conducted researches allow to substantiate necessity of development the rational technological charts and method for man-caused deposits forming and repeated mining of dumps, tailings and sludge dumps.

1 INTRODUCTION

Human activity associated with all stages of surface mining leads to a change in the natural condition of environment. The various mining stages of deposit exploitation have different influence on land surface and subsurface, the air basin, surface and groundwater, and especially on flora and fauna. The result of intensive man-caused impact on the environment is the violation of the ecological balance on the mining region territory.

Man-caused impact of surface mining on the ecosystem at different stages of the deposit exploitation can be divided on an influence and changes that in the many cases can't be reclaimed to its original condition.

1. Preparation of the land surface for mining include deforestation, drainage of lakes and rivers, replace of the roads, power lines, communication facilities, demolition of buildings and structures. Main influence sources are dust, noise, vibration, exhaust gases. As a result mining area has change in the landscape, flora and fauna, surface water.

2. Deposit drainage from surface water includes creation of special drainage excavation, dewatering hole systems, creation of drainage ditches. On this stage dominant effects are noise, exhaust gases. As a result are changes in the landscape and land interior, the systems of surface and groundwater.

3. Pit preparation and construction consist in capital, haulage and working tranches building. Effects on environment at this stage are dust, noise, vibration, exhaust gases. The changes deal with landscape and subsoil.

4. Exploitation stage includes preparation for deposits mining, excavation of overburden and minerals, haulage works, storage of overburden in the dumps, and minerals into temporary storage. Main influence sources are dust, noise, vibration, exhaust gases. On this stage main changes in the landscape and the natural rock mass, including hydrogeological system, its composition and properties, qualitative and quantitative characteristics of mining rocks.

2 MAIN DIRECTIONS OF SURFACE MINE DISTURBANCE ON ENVIRONMENT

Influence of mining on the natural environment can be divided into direct and indirect. First is an impact immediately in time, second is a result of subsequent ecosystem response from a direct effect.

As an example of direct impact considered destruction of the rock massive during the blasting, excavation and loading operations, haulage of the rock mass, minerals storage and overburden dumping. The indirect influence may be occurred by certain species of flora on the dumps surface at the relevant conditions.

Thus, the upper layer of the lithosphere – geological environment (rocks, soil, surface and groundwater, all organisms and vegetation that live in soil and rocks) is one of the main objects of environment on which surface mining influenced at the all stages of the pit life cycle. All components of the geological environment in nature are in dynamic equilibrium (Drebenshtedt 2013).

Artificial intervention in condition of any part geological structure at process of surface mining causes geological, physical, chemical and biological processes, which leads to activation or decay of natural phenomena and produce new man-caused processes that have not met previously in natural condition. Among them should be allocated primarily disturbance of soil and land. Distinguish the physical, chemical and mechanical disturbances.

Physical disturbance of soil associated with the transformation of the landscape, the deformation of the surface, changing the soil structure and development of erosion processes.

Chemical disturbance of soil concerned with contamination of the process waste and emissions from sludge and tailings, dumps and minerals store, dumps with off-grade ore and man-caused deposits.

Mechanical disturbances caused during haulage of rocks, dumping, erosion of dumps and tailings, as well as dust emissions at the mining processes.

Essential and integral part of human activity at the surface mining on environment is violations on the composition and mode of surface and groundwater.

The main sources of pit waters pollution are destroyed parts of overburden and minerals in faces; internal and external piles; erosion of non-working surface in pit. Pit water include dust, oil and other substances that enter to it's during the haulage in mine excavation, as well as those that fall into them during haulage through the rock surface.

For reduce damage and low pressure on environment at the mining activities implement complex of rehabilitation work on mining sites due to the interests of territory society.

3 SURFACE MINING IMPACT ON THE ENVIRONMENT IN UKRAINE

At present the problem of natural resources rational use and protection of the natural environment at the minerals mining and processing in Ukraine gained national importance. This is due to the fact that the environment conditions in all mining regions characterized as critical.

The reasons are not only in a high concentration of mining production on a limited area and large volume of mining production waste, but are related primarily to the specifics of the territorial distribution of productive forces, as well as nature use model of mining regions economic development. There, as a rule, concentrated enterprises of mining, metallurgy, chemical industry, energy sector and others.

As a result, Donetsk, Dnipropetrovsk, Lviv and other areas formed as regions of raw material orientation. In this regard, according to the low production technological level and ecologically reasonable accommodation of the enterprises in these regions, there were areas of environmental pressure. First of all this influence observe in the Dnipropetrovsk region. With a relatively small land area (31900 km^2) in the region works more than 120 pits and mines for the extraction of iron and manganese ore, coal, rare metals, refractory clay and kaolin, raw materials for the production of building materials (Gorova et al. 2013).

In the old mining regions of Ukraine such as Donetskiy, Krivorogskiy, Lviv-Volynskiy, Carpathian, long-term and large-scale mining production led to a delay of the dynamics of natural processes, create a dangerous man-made disasters and to degradation of the whole components of natural environment. The main factors of influence on the environment are: numerous mining excavations, intensive pumping of groundwater, waste accumulation, blasting.

Since the beginning of the mining industry restructuring and the general closure of unprofitable pits and mines in this regions under the influence of geotechnical, hydrogeological and hydrochemical factors began to form a new ecological and geological conditions, which is currently characterized as a worsening of the existing and the emergence of additional threats.

The best example that confirms the importance of the problem is the situation that prevailed in the sulfur-bearing plants Carpathian Basin, in particular, Javoriv Industrial Merger "Sera". Those enterprises are designed and built as a raw materials source for the production of chemical fertilizers.

During the deposit exploitation was produced 20 million tons of commercial ore. In the period from 1990 to 1993 world prices for sulfur decreased from 200 to 30 dollars. This led to the fact that the production of sulfur became unprofitable, not only in Ukraine, but also in other countries. Therefore, since 1993, production of sulfur by surface mining on Javoriv IM "Sera" completely finished.

The situation is complicated by the necessity of large work volumes for maintain environmentally hazardous facilities and systems in a safe state. In addition, it is necessary to eliminate the consequences of long-term enterprises operation.

The main objects that are now threatened for environmental catastrophe of Javorovsky region and the basin of the Vistula River are pit residual spaces and the rivers system that was forced removed from the deposit territory.

According to preliminary estimates the scales of the open pit mining negative impact on the environment are:

– increment in pit residual space 40 – 50 million tons per year of hydrogen sulfide, saline drainage water from violation of the pressure horizon by the mining operations;

– as a result of long-term pumping drainage water around the quarry formed depression crater area about 100 square kilometers, that bring to intensified cave phenomenon that offer a real threat to human safety;

– created and operates the forced removal system of rivers and surface water from the territory of the deposit with a total pumping volume 77.2 million tons per year;

– emission to the atmosphere 180 tons of hydrogen sulfide per year due to degassing.

A similar situation exists in other mining regions of Ukraine (Kryvbas, Dniprovsky lignite basin, Marganecky basin and others).

In Ukraine coal is mined at the three major basins Donbas, Lvov-Volynskiy basin and Dneprobas (lignite). The impacts on ecological situation in coal mining regions are intensified through the fact that every third mining company operated more than 50 years. At the mine closure the best ability for the resumption of natural indicators has a surface groundwater level. This lifting at the partial or complete filling of mining excavation is a major factor in the deterioration of mining regions environmental conditions.

Complex influence of brown coal cast on the Pridneprovsk lignite region landscape shown in Figure 1.

Figure 1. Complex influence of lignite open cast mining on the Pridneprovsk region landscape.

As a result of large-scale mining operations the areas of disturbed lands account up to 8.2% of the Lugansk and 7.8% of the Donetsk regions. The territories under waste dumps employ about 7200 hectares and 2400 hectares under the coal preparation tailings.

In Kryvyi Rig mining region extensive mining of iron ore led to catastrophic damage of environmental-economic situation: withdrawn and contaminated large areas of arable land, man-caused landscape complexes formed, surface waters and groundwater polluted. Iron ore mining was carried out on 10 pits with depths ranging from 100 to 300 m or more, and 23 underground mines with a maximum depth 1400 m. The result of mining activity was a violation of 700 square km of high quality fertile land, including under dumps nearly 70 square km.

Open pit mining has a very different and specific effect on the lithosphere. This is due primarily to the penetration into the crust and remove from the bowels a large rock masses, violation of the land surface, and others.

With atmospheric flows replaced some heavy metals – mercury, lead, cadmium, copper and zinc. Especially dangerous is the accumulation of toxic metals (mercury, lead, cadmium) in tissues and organs of commercial fish.

As a result of mine drainage systems operated by pump to the surface lift large quantities of groundwater with different chemical composition containing a variety of trace elements and man-made substances. This has a negative impact on the hydrosphere.

Natural landscapes at the open pit mining are usually partially or completely damaged. The formation of mining-technical relief characterized by the elements is developed pits residual space, dumps, tailings, etc. Nature of the initial landscapes change depends on the used technology.

The situation is exacerbated by the fact that much of the mining deposits located in areas of active farming.

Due to the production of mining, exploration, construction and other work in 2014 registered 155.1 thousand hectares of disturbed land (Figure 2). Damaged lands in technical reclamation stage up to 3000 ha, in the process of shrinkage up to 2000 ha, biological reclamation up to 1300 ha. Land reclamation performed in 2011 on an area of 0.9 thousand ha, which is on 0.2 thousand ha less than had been violated in the same year – 0.6 thousand ha (Table 1).

According to the National Scientific Center "Institute of Soil Science and Agricultural Chemistry" reclaimed lands are characterized by considerable heterogeneity and diversity of ground cover. The humus contented in the reclaimed layer is lower than in the zonal soils, almost 1.5 – 2 times and equal 1.5 – 2% (in some cases, 2.0 – 3.0%), at least – 4%.

Table 1. Violation and land reclamation, thousand hectare (Statelandagency of Ukraine, 2013).

Type of economic activities	1995	2000	2005	2010	2011	2012
Disturbed land	17.0	1.9	2.0	1.3	1.2	1.1
Mined land	16.4	2.8	1.8	1.0	0.9	0.3
Reclaimed land	8.4	3.7	2.1	0.5	0.6	0.7
Including arable land for agricultural	3.0	2.0	0.8	0.3	0.4	0.6
Reclamation direction:						
– arable land	1.5	1.5	0.3	0.2	0.2	0.4
– forest (shrub) plantations	3.0	1.2	0.7	0.1	0.1	0.4
– reservoirs	0.6	0.2	0.1	0.04	0.04	0.02
– building	0.3	0.0	0.0	0.0003	0.008	0.003
– recreational and others	1.5	0.3	0.5	0.02	0.09	0.03

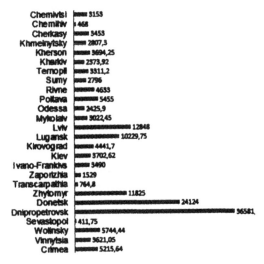

Figure 2. The presence of disturbed land in Ukraine on areas and regions (2012), ha.

According to the Ministry of Agrarian Policy of Ukraine in 2012 compared to 1990 the area of arable land in Ukraine has decreased by 1.18 million hectares (1.5%) and amounted 31.12 million hectares (79.5%).

Ecosystems destabilization and negative impact of large-scale mining operations on the environment negatively affects on the health.

The contribution of anthropogenic factors on the health deviations 10 – 57%. There is a close relation between the number of newly registered pathologies and anthropogenic pollution in Ukraine:

– significantly increased the number of congenital anomalies by 2.12 times;

– by 1.28 times increased the number of pathologies associated with diseases of the skin and tissue;

– by 1.72 times increased indicators of diseases related to the circulatory system;

– number of tumors increased by 1.2 times;

– by 1.16 times became more pathologies at the pregnancy.

The same disease may be caused or triggered by various environmental factors. According to different scientists, due to arbitrary mixtures of air emissions, solid waste, industrial waste water may form up to 60 thousand of persistent toxic compounds. Therefore, even small displays of new pollution dramatically increase morbidity.

Today in Ukraine the average life expectancy is 62.4 years opposed 69.3 years in 1991. This is largely due to the poor quality of the environment (air and water quality, the food quality and safety, exposure of ionizing radiation).

To improve the environment needs enormous funds. However, in Ukraine one person had only $ 3 on environmental expenditures, at that time in Germany – 150, Poland – 30; Lithuania and Russia – 15.

According to the World Bank with the inclusion in the cost of the projects the cost of overcoming the possible negative impacts on the environment (including the environmental constraints) the greening of production pays an average of 5 – 7 years. The inclusion of environmental factors in the decision-making process at the design stage costs 3 – 4 times cheaper than the subsequent installation of additional pollution control equipment. The costs on responding to the use of non-environmental technology and equipment by 30 – 35 times higher than the cost required for the development of environmentally-oriented production technology.

Conducted researches shown dangerous situation that is exacerbated each year, in addition to the implementation of a unified state policy in the field of environmental management necessity to organize targeted financing the development and implementation of resource-saving and environment-friendly technologies for the extraction, processing and mineral resources comprehensive utilization (Gumenik 2012). Reclamation of disturbed territories at the mining activities should be solving with economic mechanisms as a priority task.

4 CONCLUSIONS

Problems of damaged lands stipulates for low level of mineral resources extraction at the mining of rock mass (30 – 40% or less). The rest of the byproduct is stored in the dumps. Meanwhile its have various types of conditioned raw materials – construction, refractory, foundry, fluxing and others.

Conducted researches show large losses of minerals at the mining and processing. Loss at the coal production 19 – 21%, iron ore – about 15%, refractory raw materials up to 10%, mining and chemical raw materials up to 60%. Dilution factor amount 10%. As a result, reserves of stored iron quartzite exceeded 500 million tons, manganese ore raw materials – 27 million tons.

A lot of problems consist in low level of mineral resources comprehensive utilization (2 – 5%). Many deposits were mined unclustered that's leading not only to minerals loss, but also to the alienation and significant land disturbance.

Comparative high land involve in mining (from 6 to 35 hectares at the production 1 million tones of minerals) bring to appear hundreds hectares of damaged lands.

In the dumps, tailings and sludge ponds accumulated near 1 billion cubic meters of mining waste, which have significant human impacts on environment.

REFERENCES

Drebenshtedt, K. 2013. *Modern environmental and economic approach of mining industry.* Economy of Region, No 1: 114 – 122.

Gorova, A., Pavlychenko, A., Borysovska, O., Krupska, L. 2013. *The development of methodology for assessment of environmental risk degree in mining regions.* Annual Scientific-Technical Colletion – Mining of Mineral Deposits. The Netherlands: CRC Press/Balkema: 207 – 209.

Gumenik, I., Lozhnikov, V., Maevskiy, A. 2012. *Methodological principles of negative opencast mining influence increasing due to steady development.* Geomechanical Processes During Underground Mining. The Netherlands: CRC Press/Balkema: 45 – 51.

Theoretical and Practical Solutions of Mineral Resources Mining – Pivnyak, Bondarenko & Kovalevska (eds)
© 2015 Taylor & Francis Group, London, ISBN: 978-1-138-02883-8

Effective technology of stripping operations in deep coal opencasts with railway and auto truck transport

B. Rakishev, S. Moldabaev & E. Kuldeev

Kazakh National Research Technical University named after K.I. Satpayev, Almaty, Kazakhstan

ABSTRACT: There are given scientifically grounded technologies of stripping operations for the conditions of coal strip cut "Vostochnyi" of Ekibastuz field using railway and auto truck transport. Constrained conditions of stripping only from one flank of opencast field predetermined usage of daylight area for excavation and shipment of stripped soils up to the depth of 60 m by draglines to railway transport. There is shown the scheme of stripping of coal layers using draglines also to the railway transport. There was proposed technology of mining of the bottom part of the stripping zone by excavator-auto truck complexes with a backlog recovery of stripping operations. Its feature is two-level mining of benches by cross-section panels with the construction of stripping grooves on the flanks of opencast fields. At the boundary of the application of railway and auto truck transport before coming of crushing and conveyor complexes to design capacity it is offered to organize a haulage level with a overloading of the part of auto stripping to railway transport. Implementation of these measures will allow transferring mining of the part of stripping operations volumes to a later date, reducing the amount of capital mining operations for the construction of stripping excavations that by increasing of the exposed coal reserves will improve the productivity of opencast on 25%.

1 INTRODUCTION

Implementation of continuous technology in opencasts with angled positioning of coal layers has its own peculiarities. In this regard, opencast mining of coal layers of Ekibastuz field in Kazakhstan on the cut "Vostochnyi" is interesting. Excavation front with length of 2.8 km is stripped by three steep trenches. In these trenches are constructed four lifting conveyors: in the central – two, and one in both the northern and southern (Belik et al. 1992). Through a central conveyor are organized coal cargo flows from the upper mining bench, and through two flanking conveyors – from the bottom mining bench. The front of the bottom bench is completely divided in the center of opencast field, where are installed connecting conveyors.

Coal mining is carried out by bucket-wheel excavator $SR_s(k)$-2000 using coal face loaders $BR_s(k)$-2000.65. Cutting of a new coal bench with height of 25 m is mainly completed by this bucket-wheel excavator in combination with interbench $AR_s(k)$-5000.95 and above mentioned coal face loaders.

On the cut there is implemented advanced one-bench (two-bench) technological scheme with one transport level with division of front of mining operations into four blocks of equal lengths. With the advance of face conveyor to a new position advanc-

ing of connecting conveyor is carried out, that eliminates their reassembling to change the direction of transportation of coal.

Dosed continuous uploading of wagons is carried out on the surface blending-and-loading complex. Stability of coal quality is ensured by a uniform mining of low-ash layers 1, 2 and high-ash layer 3, respectively, at the top and bottom benches (subbenches).

According to the general scheme of development of Ekibastuz field in the second phase of reconstruction of cuts of the Southern Group it was planned to implement continuous and cyclical-continuous technologies of complement of stripping operations using, respectively, conveyor and combined auto truck-conveyor transport, including on the cut "Vostochnyi". Due to the increase of the hardness of stripping soils in depth the area of application of continuous technology is limited.

Usually stripping operations with using of power shovels are conducted in constrained conditions with narrow working sites. Placement of a powerful bucket-wheel excavator of type ERP-6500 and systems of face conveyors requires a significant spacing of the working side and investment.

Working experience on the stripping benches of Ekibastuz field shows that in the next few years in the upper part of the stripping zone it is necessary to still focus on the well-proven railway transport.

Increasing of rational insertion depth of railway transport at the present stage is possible by reducing the number of transport levels and simplifying the construction of scheme of track system. The studies of LLP "Bogatyr Komur", Institute of Mining, named after A.A. Skochinskiy, LLP "Karaganda-giproshaht and Co" on the changeover at deep levels on cyclical-continuous technology using both the conveyor and the combined auto truck-conveyor transport deserve attention.

Performance of stripping operations is characterized by: the increasing complexity of the transport scheme structure because of the increasing number of transport levels; fine pitch shifting of railway lines (maximum 20 – 24 m); long time arrangement of face track on a new route; an increase in the weighted average lifting height of the mined rock from the opencast to the daylight surface due to the gradual displacement of the main volumes of the extraction to the area of deep levels; constrained conditions in a confined space.

2 APPLICATION OF DRAGLINES WITH RAILWAY TRANSPORT

One example of decrease of the weighted average lifting height of the mined rock from the opencast could be providing transshipment of rocks of the second stripping bench from the daylight surface on its upper site and unloading them by draglines with different linear dimensions when loading in the means of railway transport.

At the daylight surface of the upper stripping bench front could be divided into the required number of sites of excavation and loading operations (ELO) due to the simplified schemes of placement of railway lines over top opencast outline and to ensure each ELO site with autonomous access roads. There appears a possibility of maximum load of each face track of ELO site by mining thick mass of covering and adjacent rocks of the upper zone of the working side. The upper and underlying benches with approximate total height of 50 – 60 m are mined with draglines of various types by downward digging. The most powerful dragline mines rocks of underlying bench with backfilling them in pile (Figure 1).

To complete this ahead working dragline prepares platform by drifting stope on the upper stripping bench and completes unloading of rocks pillar directly to the means of railway transport. Pile is disassembled by third dragline at this ELO site and as well shipped in means of railway transport (Rakishev et al. 2013).

Independent functioning of ELO sites is provided by permanent laying of access roads at angles of aproximately 60 – 80° to the front of advanced bench. Simultaneous loading of two locomotive wagonages at each ELO site is carried out through bilateral supplying of empty cars at the first half of the site where access roads are adjacent and at the second half – through construction of second parallel face track. An example of implementation of the proposed technological solution is as follows.

The front of works of the upper zone of the working side is divided into excavation-and-loading sites with length of 1170 m (based on standard performance of draglines and provision of annual advance of stripping benches on 100 m at sites 7, 8 of the cut field "Vostochnyi"), in each of which operates three different types of dragline. Upper bench with height of 23 m is mined with dragline ESh-13/50 with immediate loading of rocks into means of railway transport. The underlying bench with height of 37 m is mined with powerful dragline ESh-20/90. At that the rocks are placed in formed with dragline ESh 13/50 container. Pile is filled in ring to the maximum dump height of 37 m. The loading of pile rocks into means of railway transport is carried out with dragline ESh-20/65.

Organization of working of the draglines. After drifting the stope on platform with width of 40 m, ESh-13/50 No 1 bypassing the deadlock of face track of its ELO site returns to its original position and carries out mining of another stope. Loading rocks in means of railway transport is carried out on a face track of the following stope. At the same time the previous track up to half of the site must be disassembled before mining of a new stope begins. Inclining to the second half of the site, as well as empty cars supply to the first half for the ESh-13/50 No 1 dragline is carried out through a face track of the following stope, adjacent with the previous in the center area of the ELO site.

Dragline ESh-20/90 No 2 after mining its stope on pile (on platform of also 40 m wide) stops for current repairs till the end of the pile shipment with dragline ESh-20/65 No 3 and only then returns to the starting position and starts drifting new stope on debris of underlying bench.

Face trucks advancing of draglines ESh-20/90 and ESh-20/65 is slightly more intense than advancing of the ESh-13/50 (the last one is uploaded to the limit of standard performance), so by the end of drifting another stope the first two approach on the minimum distance allowed by the rules of safety.

To ensure the design capacity of the cut "Vostochnyi" it is necessary to organize three ELO sites. The total annual amount of mining of stripping soils in them will be 21 million m^3 with a length of the operation front of 3510 meters.

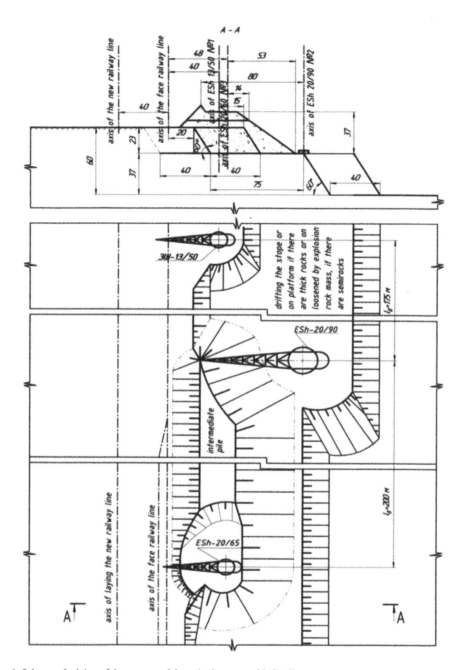

Figure 1. Scheme of mining of the top part of the stripping zone with draglines.

Considered technology and organization of stripping operations in the upper zone of the working side on example of the cut "Vostochnyi" will allow:
– to reduce the number of transport levels on four units;
– to reduce the volume of reconstruction of movable railway tracks in 11 times or more, and their length to at least 10...11 km;

– to transport stripping with lower power consumption at relatively high speeds of locomotive wagonages;
– to achieve high performance of equipment complexes, comparable on less investment with the continuous technology.

Increase in angle of working side by nearly 8 degrees, stop carrying out stripping operations in the upper zone for the period of reconstruction and intense new completion of advanced benches during transition to a new phase of mining development will allow to stabilize the mode of mining operations with rational technical and economic characteristics and more flexibly allocate the quantities of stripping by years. The increase in the volume of planning works is compensated by an increase of turnover of rock "trains" for removal of stripping from the underlying benches and decreasing of weighted average lift height of the mined rock.

Limiting fact of timely formation of operation front to mining excavators is significant backlog of stripping operations in the bottom part of the stripping zone. We have considered the option of simultaneous usage of dragline type ESh-13/50, at dressing of the back of coal layer both for non-transport and for transport technology (Figure 2). Waiting for rock "train" it can operate also with the transshipment of stripped rocks in pile, shipped then by excavator with large linear dimensions to the means of railway transport (such as EKG 6.3y).

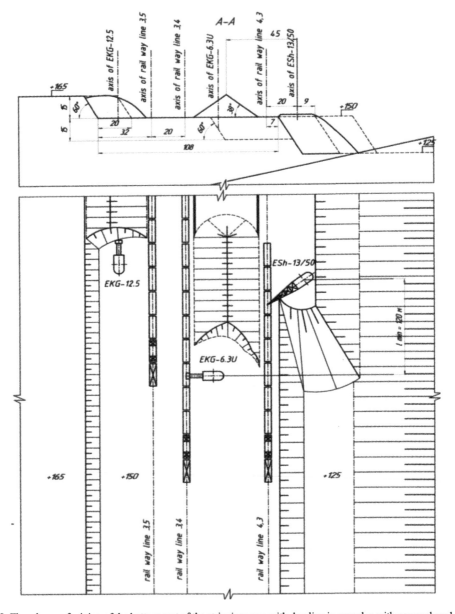

Figure 2. The scheme of mining of the bottom part of the stripping zone with dragline in complex with power shovel.

3 TECHNOLOGY OF STRIPPING OPERATIONS WITH USAGE OF RAILWAY AND AUTO TRUCK TRANSPORT

Usage of cyclical-continuous technology with conveyor lifting of crushed hard rock is still one of the major ways to solve the transport problem of deep ore opencasts (Stolyarov 2003, Melnikov et al. 1995, Drizhenko et al. 2009, Malgin et al. 2002). Therefore, in 2010, on the cut "Vostochnyi" were completed precommissioning works on commissioning crushing and conveyor complex with the replacement of railway transport into combined auto truck-conveyor transport (Figure 3).

Figure 3. Commissioning of the crushing and conveyor complex on the cut "Vostochnyi".

First time in the world curved conveyor lines with an elevation angle of 18 degrees were tested (Figure 4).

Figure 4. The conveyor lift for stripping operations of the cut "Vostochnyi".

The first crushing and conveyor complex with an annual capacity of 10 million m^3 stripped rocks consists of two hydraulic excavators R994B and 9350 of company LIEBHERR (straight power shovels) with an electric drive and a bucket capacity of 18 m^3, 12 dump trucks with carrying capacity of 90 tons of company COMATSU, two two-roll crushers BSW, a system of lifting, main and dump conveyors with belt width of 2 m and console spreader AR$_s$-B.45.50. Crushing complex of stationary type is located in the end of the cut.

Recommended mining technology on the cut "Vostochnyi" for the period of implementation of the crushing and conveyor complex is shown in Figure 5.

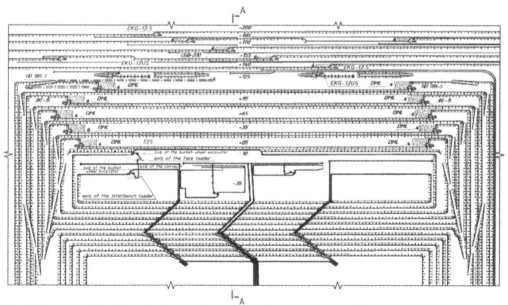

Figure 5. Recommended mining technology on the cut "Vostochnyi" for the period of implementation of the crushing and conveyor complex.

The high efficiency of the usage of excavator-auto truck complexes with reduction of current stripping volumes is achieved when switching to a two-level technology of mining of benches of the bottom part of the stripping zone by cross-section panels (Figure 6) and construction of temporary ramps at the ends of coal cut (Rakishev et al. 2012).

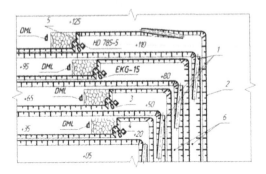

Figure 6. Two-level technology of mining benches with cross-section panels: 1 – temporary ramps at the ends of coal cut; 2 – excavator; 3 – dump truck; 4 – mining of the upper parts of benches.

The main volume of auto stripping is moved to crushing and loading unit (Figure 7), and part of auto stripping is transported to external piles by railway transport through transshipment points of the haulage level (Figure 8).

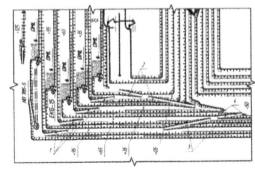

Figure 7. Development of the system of ramps at the ends of the cut "Vostochnyi": 1 – temporary ramps; 2, 3 – permanent ramps are correspondingly above and below the level of placement crushing and transshipment points(CTP); 4 – the direction of transporting flows of auto stripping to crushing and transshipment points (CTP).

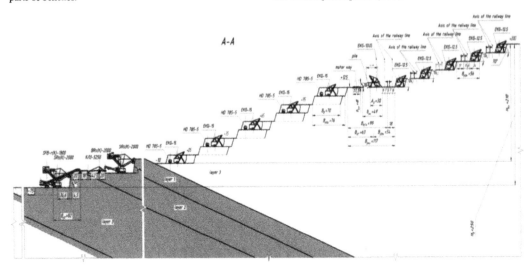

Figure 8. Organization of the haulage level at the boundary of usage of railway and auto truck transport.

By the time of the implementation of a crushing and conveyor complex 9 stripping benches with an average height of 15 m, and with a width of work sites around 45 m were in operation. Only railway transport was applied on stripping operations. The angle of slope of stripping zone at boundary with the hanging wall of coal layer 1 was 15 degrees. According to the project 5 upper benches will be mined using railway transport, and the lower stripping zone – using combined auto truck-conveyor transport.

To move to the regulated technological zones it is necessary initially with mining of cross-section panels with a width equal to the width of the work sites, complete extraction of stripping with excavator-auto truck complexes through one bench. On this mining begins from the bottom bench leaving only safety berms. Subsequently, when reaching the designed capacity in each technological zone initially will be mined upper parts of benches (Figure 6). For this purpose temporary ramps will be constructed in each of the ends of opencast. When mining the lower parts of the benches only safety berms will be left. This will allow completing in practice the transition into intensive mining of high benches with separate equipment complexes in each technological zone from both flanks of the opencast field.

Placement of stripping workings for movement of auto truck transport at the ends of the cut is shown in Figure 7. For this purpose, transportation berms are left every 30 meters. On the level, +50 m from the both flanks of the opencast are placed crushing and transshipment points (CTP) of crushing and conveyor complexes. Cargo movement connection of working levels of the upper and lower parts of the benches with CTP placement site is performed sequentially through the transport berms and a system of permanent ramps. When mining the upper parts of the benches temporary ramps are constructed on each flank of the opencast field to perform communication with transport berms.

Permanent ramps to the downhill service dump trucks from transport berms on the levels +95 and +65 m in the freight hauling direction. Newly established transportation berms every 30 m (levels +35, + 05, etc.) and cut upper parts of new benches are provided with connection with the CTP placement site by consistently extensible and also permanent ramps. They service dump trucks on lifting in the freight hauling direction. The length of permanent ramps with a slope of 80‰ between the benches is 375 m, and subbenches adjacent to the CTP site and re-cut – 187.5 m.

Mining of cross-section panels with excavator-auto truck complexes allows almost eliminate the breaking of blast holes in the direction of mined-out space. Usage of multi-row short-delay blasting to shortpile toward the mined-out panel improves the quality of hard rock crushing, significantly reduces the yield oversize that will positively affect the reliability of the of crushing and transhipment points operation.

According to the project crushing and transshipment points are located at a depth of about 150 m on both flanks of opencast field. At operating cycle of excavators to the center of opencast from the both flanks of the opencast field weighted average distance of transportation by dump trucks in the reporting period (2011 – 2025) will grow from 1.1 to 1.8 km. Since the rational height of the parts of the benches became equal to 15 m, then it is recommended to use an excavator EKG-15 or its modification – EKG-18P as the excavation and loading machine, depending on the desired intensity of development of stripping zone.

4 OBJECTIFICATION OF TWO-LEVEL TECHNOLOGY

For mining conditions of the cut "Vostochnyi" (horizontal capacity of layers 600 m, the angle of slope is 19 degrees) with an annual output in coal of 25 million tons the advance rate of operation front of stripping zone should be at least 60 meters per year. When mining benches with height of 30 m with excavators EKG-15 from both flanks of the opencast the advance rates of their operation work is found to be equal to 64 meters per year. In this case, 5 top stripping benches with height of 15 m should be simultaneously mined with 7 excavators EKG-12.5 with the railway haulage (Figure 5). Calculated advance rate of stripping benches with excavator-railway complexes is nearly 70 meters, which will provide the proportionality of development of stripping zone and ensure reaching the design capacity of the cut in coal of 25 million tons.

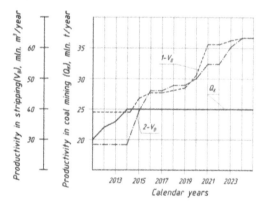

Figure 9. On the objectification of the time schedule of mining operations: 1-V_B, 2-V_B, – performance of the cut on stripping correspondingly according to the project and recommended schemes; Q_k – productivity of the cut in coal.

Analysis of the time schedule of mining operations shows that even in case of overloading the part of volumes of auto stripping through transshipment points (warehouses) of the haulage level on the boundary of usage of excavator-railway and excavator-auto truck complexes to the means of railway transport in the amount of up to 5 million m³ per year there is no guarantee of completing the planned volumes of excavation of stripped rocks on the cut "Vostochnyi". Therefore, during the period of transition to crushing and conveyor complex one of the alternative ways to eliminate the stripping backlog is to increase the slope angle of stripping zone, while simultaneously forming temporary haulage level and applying the combined auto truck-railway transport.

5 CONCLUSIONS

Based on the identified patterns of development of the opencast space, connections between elements of different technological opencast mining complexes there are given entirely new scientific-based solutions to improve the efficiency of operation of coal cuts. As a result of the completed research was established methodology of designing resource-saving technologies in open development of coal deposits, including:

– a two-level technology of mining of the bottom part of the stripping zone with excavator-auto truck complexes with cross-section panels and construction of stripping workings on the flanks of the opencast field that will allow to reduce by 19% the current stripping volumes and improve productivity on coal mining for 25% with reduction of the term of achieving design capacity at the cut "Vostochnyi";

– organization of temporary haulage level between zones of functioning of excavator-railway and excavator-auto truck complexes for the period of reaching design capacity of crushing and conveyor complexes;

– method of optimization of the operating area position in stages of mining of the angled coal layers that allows to substantiate the construction of stripping zone at the continuous technology of coal mining and carrying out stripping operations on cyclical and cyclical-continuous technology;

– technique of optimal parameters objectification of a two-level technology of mining of the bottom part of the stripping zone with cross-section panels in connection with the rates of coal mining and the transfer pitch of the transshipment points.

REFERENCES

Belik, N.M., Fedotov, I.P. & Dzhaksybaev, S.I. 1992. *Coal of Ekibastuz*. Moscow: Nedra.

Drizhenko, A.Y., Kozenko, G.V. & Rykus, A.A. *Opencast development of iron ores in Ukraine: status and prospects: monograph*. Dnipropetrovsk: Publishers Poltavsky literator.

Malgin, O.P., Shemetov, P.A., Lashko, V.T. & Kolomnikov, S.S. 2002. *Improvement of cyclical-continuous technology of mining in deep opencasts: monograph*. Tashkent: Fan.

Melnikov, N.N., Usynin, V.I. & Reshetnyak, S.P. 1995. *The cyclical-continuous technology with mobile crushing and transshipment complexes for deep opencasts: monograph*. Apatity: Mining Institute of the Kola Scientific Center of Russian Academy of Sciences.

Rakishev, B.R., Moldabaev, S.K., Samenov, G.K. & Nurgaliyeva, M.S. 2013. *Development of the working area of coal cuts when switching to a cyclical-continuous technology*. Moscow: Moscow State Mining University, Journal Mining informational and analytical bulletin., No 6: 127 – 142.

Stolyarov, V.F. 2003. *The problem of cyclical-continuous technology of deep opencasts: monograph*. Ekaterinburg: Ural Branch of the Russian Academy of Sciences.

Theoretical and Practical Solutions of Mineral Resources Mining – Pivnyak, Bondarenko & Kovalevska (eds)
© 2015 Taylor & Francis Group, London, ISBN: 978-1-138-02883-8

Mathematic model for parameter matching in evaluators group model with application of the distributed information processing principle

A. Tsekhovoy & Zh. Sultanbekova
Kazakh National Research Technical University named after K.I. Satpayev, Almaty, Kazakhstan

ABSRACT: Parameters matching in the applied evaluators group model as per the distributed information processing principle in tractive, operational and velocity calculations for transport should be considered in a form of mathematic model through final values averaging and use of the matching factor defined on base of the model.

1 INTRODUCTION

To computerize a routine designer's work in selecting type of transport and its parameters the program runs a database (DB). Duration of matching process of desired values received from the evaluators and duration of the process of calling to program blocks and database has an influence upon calculation period. Possible PC pending should be taken into consideration too.

The methods studied have been implemented in the form of program tools and series of tests have been completed.

Complex structure of program tool was developed together with the distributed information processing model enabling to select the project execution criteria to speed up the design process significantly while maintaining a high quality project or to execute the project at low material costs without reducing the design period.

Calculation of performance index of open pit transport used in the course of minerals recovery as a sample for implementing the distributed information processing approach during the opencast mining design is observed further. Calculations relating to open pit dump trucks provided sequentially (cascaded) include the following phases:

– defining actual carrying capacity of dump trucks, carrying capacity use factor and body capacity;

– calculation of velocity of dump truck travel subject to the open pit road elements;

– checking the conditions of loaded dump truck travel at ruling gradient;

– defining the value of interval between the trucks;

– calculation of operating dump trucks and inventory stock.

Three independent calculation modules based on different methods determining the production rate of open pit transport have been generated on a basis of five-phase calculation of duration and content for the open pit transport. These calculation modules were implemented as a part of the evaluators' group model represented in the form of individual program blocks executed by the coordinator or with the help of one evaluator.

Coordinator or chief designer who assembles a final design on a basis of results of individual modules calculations serves as a match link between individual blocks.

2 MATHEMATIC MODEL

Mathematic model of operational calculations for the open pit transport serves as a base of sequential algorithm when developing the Auto software complex implemented in object-oriented environment of MS Visual Basic (Evgenev 2006, Kuzmin et al. 1991, Share point service 2005) programming.

The developed software complex is used as basic parameter for checking the adequacy of the developed distributed information processing model when providing the calculations relating to open pit transport.

Since the work assumes application of the evaluators group for independent execution of all three specified modules the interrelation of all parameters in the evaluators group model (CGM) was examined. When studying such interrelation between the open pit transport calculation parameters in the distributed information processing model the parameters involved in the complex of open pit transport calculations must be adjusted.

Every evaluator in the EGM defines the value of shift-based production rate for dump trucks by different simplified methods which are to be approved by the coordinator.

Shift-based production rate of dump trucks should be defined as per the following formula:

$$Q_{sh} = \frac{60 P_a T_{sh}}{T_f} K_u,\qquad(1)$$

where P_a – actual carrying capacity of a dump truck; T_{sh} – duration of a shift; T_f – full time of a trip; K_u – use factor of shift time.

However, in reality the calculation of the parameters included in the main formula (1) is usually provided by different methods depending on qualification of the evaluator executing this procedure. In virtue of the established procedure, especially on design works and operations plans calculations level the value P_a actual carrying capacity of a dump truck, T_{sh} – full time of trip, K_u – use factor of shift time are usually different from case to case. Within the frame of this task we are to check the probability of getting a reasonable result by using a formula (1). To see the work of every independent evaluator the methods applied for calculation of designed parameters on basis of existing practice have been selected.

Out of four parameters involved in calculation of this parameter the value T_{sh} is provided by the customer, i.e. duration of shift will remain as a constant. The values of rest three parameters (P_a, K_u, T_f) may differ since calculated by specific method or accepted by independent evaluators.

Actual carrying capacity of dump truck P_a in module one should be defined as per the given formula subject to values of the following indicators: number of buckets required for loading the vehicle body; volume of rock mass in excavator bucket. This indicator in module two and three should be based on standards.

The value T_f should be defined through exact values related to dump truck travel time along the roads, time required for loading/offloading, manipulations and waiting time for loading. In the second computation block the value of this indicator should be defined as per the given calculation method and in first and third block the value is defined by the evaluator on grounds of length of the sites and roads surface. Use factor for the shift time K_u should be defined as a travel time-shift length ratio. Therefore, this ratio in third block should be defined as correlation and the same in first and second blocks is to be defined in accordance with the designers' opinion.

As a result three computation blocks work we have three different Q_{sh} values. Coordinators task is focused on the results matching. The Q_{sh} value is a random value dependent on different organizational, technological and technical factors (Figure 1).

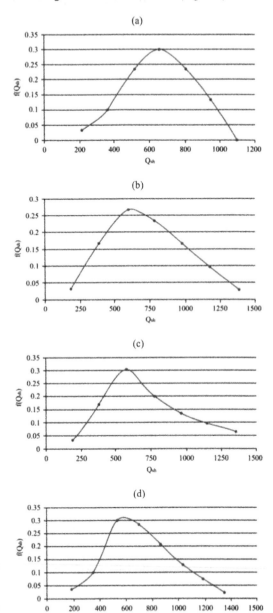

Figure 1. Distribution of experimental indicators: (a) distribution of the first computation module results; (b) distribution of the second computation module results: (c) distribution of the third computation module results; (d) distribution of experimental indicators.

154

Indicators of dump truck shift-based production rate defined by three different methods for 30 samples provided a list of random indicators that serve as a base for Q_{sh} value distribution function. The Figure 1 shows the distribution of experimental indicators functions in three computing blocks (a, b, c) and distributing experimental data function (d).

Mathematic expectation of the shift-based dump truck production rate indicator under study is $M = 645.595$, mean square deviation $\sigma = 489.38$ and dispersion $D \approx 239494.3$. Mathematic expectation of the first evaluator $M_1 = 635.336$, second evaluator $M_2 = 641.3333$, third evaluator $M_3 = 660.116$ and therefore, the mean square deviation $\sigma_1 = 247.680$, $\sigma_2 = 287.1356$ and $\sigma_3 = 279.215$.

As it is known the density function distribution β is limited on the left and the right and looks as follows:

$$E(t) = \bar{t} = \frac{1}{6}(A + 4M + B),\tag{2}$$

where A and B – lower and upper border of random value; M – the most probable value t or mode.

To process results of the calculations made by the evaluators group in order to get the value which is the closest to a basic indictor the two approaches for matching of the parameters received from the evaluators were developed.

3 CONCLUSIONS

Option one of matching model.

Step 1. Define arithmetic mean of three values of shift-based production rate as per the following formula:

$$P_1 = \frac{Q_{sh1} + Q_{sh2} + Q_{sh3}}{3}.\tag{3}$$

Step 2. Define a relative error of mean value of shift-based production rate against the reference value of this parameter. Find dispersion and mean square deviation of the indicators on the base of error as per the following formulas:

$$D_1 = \frac{\sum_{i=1}^{n}(P_1 - Q_{sh\_et})^2}{n-1},\tag{4}$$

$$\sigma_1 = \sqrt{D_1}.\tag{5}$$

Step 3. Define ratio of values of shift-based production rate received from every evaluator to a reference value:

$$k_1 = \frac{Q_{sh1}}{Q_{sh\_et}}, \quad k_2 = \frac{Q_{sh2}}{Q_{sh\_et}}, \quad k_3 = \frac{Q_{sh3}}{Q_{sh\_et}}.\tag{6}$$

Step 4. Coefficient amount is to be divided by number of tests and we get an averaged value of the coefficients as follows:

$$\bar{K}_1 = \frac{1}{\sum_{i=1}^{n}k_1}, \quad \bar{K}_2 = \frac{1}{\sum_{i=1}^{n}k_2}, \quad \bar{K}_3 = \frac{1}{\sum_{i=1}^{n}k_3}.\tag{7}$$

Step 5. Calculate an averaged value of shift-based production rate:

$$P_2 = \frac{Q_{sh1} \cdot \bar{K}_1 + Q_{sh2} \cdot \bar{K}_2 + Q_{sh3} \cdot \bar{K}_3}{\bar{K}_1 + \bar{K}_2 + \bar{K}_3}.\tag{8}$$

Step 6. Define relative error of the mean value of shift-based production rate P_2 against the reference value of this parameter. On a basis of error value find dispersion and mean square deviation similarly to the formulas (5) and (6):

$$D_1 = \frac{\sum_{i=1}^{n}(P_2 - Q_{sh\_et})^2}{n-1},\tag{9}$$

$$\sigma_1 = \sqrt{D_1}.\tag{10}$$

Step 7. Define arithmetic mean of four shift-based production rate values as per the following formula:

$$P_3 = \frac{Q_{sh1} + Q_{sh2} + Q_{sh3} + P_2}{4},\tag{11}$$

Step 8. Similarly to the formula (7) define the following coefficient:

$$k_4 = \frac{P_1}{Q_{sh\_et}}.\tag{12}$$

Step 9. Get a mean value of the following coefficient:

$$\bar{K}_4 = \frac{1}{\sum_{i=1}^{n}k_4}.\tag{13}$$

Further repeat this iteration similarly to the formulas (6) – (11) and get the values of matched parameters P_4, P_5, P_6, P_7 of the dump truck shift-based production rate subject to the matching ratio $\bar{K}_1, \bar{K}_2, \bar{K}_3, \bar{K}_4, \bar{K}_5, \bar{K}_6, \bar{K}_7, \bar{K}_8$ which is to be calculated as per the formulas (13) – (7).

As per this method the optional values of P_4, P_5, P_6, P_7 should be defined as adjusted values.

This approach is ineffective for matching the parameters received from three independent evaluators, i.e. averaged and adjusted values of the parameter under study diverge with the reference indicator.

155

REFERENCES

Evgenev, G.B. 2006. *Systemiology of Engineering Knowledge*. Information Science in Technical Institutes. Moscow: Publishing House of the MHTU after N.E. Bauman.

Kuzmin, I.A., Putilov, V.A. & Filchakov, V.V. 1991. *Distributed Information Processing in Scientific Researches*. Saint Petersburg: Publish House-PGU.

Share point service. 2005. *Step by Step Official Curriculum*. Microsoft Windows. Moscow: Nauka: 240.

Theoretical and Practical Solutions of Mineral Resources Mining – Pivnyak, Bondarenko & Kovalevska (eds)
© 2015 Taylor & Francis Group, London, ISBN: 978-1-138-02883-8

Improving enhancing methods of instrumental monitoring of rock mass geodynamics at mining ferruginous quartzite in open pits

Ye. Babets
State Research Ore Mining Institute, SIHE "Kryvyi Rih National University", Kryvyi Rih, Ukraine

ABSTRACT. The article shows that the relation between electric and elastic properties of crystalline rocks enables enhancing methods and devices for monitoring rock mass geodynamics at mining ferruginous quartzite in open pits.

1 INTRODUCTION

Kryvyi Rih Iron Ore Basin is one of the world's largest iron ore basins and main raw material base of Ukraine's ferrous metallurgy. The deposits have been developed for over 130 years. However, proven ore reserves enable many decades of iron ore mining.

At present Kryvbas iron ore is mined by the combination of open pit and underground methods as well as the open pit method in areas influenced by underground mining. The movement of strata rock slides influenced by underground mining is not finished yet – the open pit face is entering craters, benches and cavities fractures, movement troughs are still developing.

This is caused by the fact that for a long time the same deposits in Kryvbas have been mined by both underground and open pit methods almost simultaneously. Underground and open pit methods are often applied at the same part of the deposit without matching technological cycles that causes certain problems, deteriorates engineering-and-economical performance of the mining enterprises, mine safety and environmental situation in the region.

The complex use of both mining methods will enable mining iron ore to a fuller extent, decreasing capital expenditures and investments and improving engineering-and-economical performance of the mining enterprises, mine safety and environmental situation in the region.

Along with positive ones, there are also negative factors of simultaneous application of the both methods. There occur anthropogenic geostatic pressure disturbances in the rock mass worked by underground mining. Pit crater outbreaks under pit benches and bottom or dumps can considerably complicate mining in pits. Mass underground blasting can decrease bench stability.

Combined open pit and underground mining requires thorough coordination. However, the both methods being properly ordered, the negative factors can be minimized. To do that, it is necessary to work out flow sheets of safe and highly efficient methods of mining and a system of monitoring rock geotechnics and geodynamics.

The present situation in Kryvbas requires working out and implementing measures to prevent the negative geotechnics and anthropogenic influence of open pit and underground mining on the geological terrain and the environment. The primary task of detecting structural changes in the rock disturbed by underground mining is solved by applying the geophysical method of registering Natural Earth Impulse Electromagnetic Field (NEIEMF) parameters.

It should be noted that NEIEMF method is mainly based on the empiric knowledge of genesis and temporal progress of potential rock movement slide centers influenced by underground mining. This fact slows down both the development of methods of monitoring the above objects dynamics and, in particular, the prediction of potential movement slide processes.

2 MAIN PART

The widely used geophysical method of studying the rock mass geodynamics influenced by underground mining is known as the method of registering Natural Earth Impulse Electromagnetic Field (NEIEMF) parameters. It has some disadvantages and its empiric base restricts its application.

Firstly, as mentioned above, formation of local electromagnetic radiation centers in regions with developed open pit or underground mining is caused by exclusively anthropogenic factors; therefore, this is the nature's response to invasion of geostatic pressure. This phenomenon does not exist every time and everywhere, but Earth's magnetic field does (Barechar 1981). Secondly, and very importantly, some scientists think that it is not NEIEMF momentums that are registered but individual non-linear magnetic waves (Sena 1977, Filippov 1990).

So, considering the existing physical phenomenon, the geophysical qualification of the NEIEMF method requires certain adjustment. Thirdly, any natural physical process is obligatory mathematically formulated that reveals its content and stimulates its proper use.

It should be noted that the NEIEMF method is empirical; this comes from the current understanding of the physical phenomenon. In view of the above it is necessary to analyze the dynamics of the discrete non-linear magnetic field in the continuous static magnetic field. There is every reason for this as the principal genesis of the physical phenomenon to be identified is as follows.

In high load conditions at certain parts of a solid environment (sedimentary and crystalline rock influenced by underground mining) an unbalance of the magnetic field of the inner structure of rock is formed. Resulted from the magnetic field unbalance, some of the most vulnerable in terms of stability rock areas influenced by underground mining are "stimulated", centers of stable non-linear magnetic wave emission, including those caused by the energy of the local gravitational field that sustains and develops high external load of centers of magnetic field dynamics under conditions of industrial dynamism, e.g. industrial seismicity.

Discrete magnetic nonlinearity dynamics occurs in the continuous static magnetic field that evidently offers certain resistance to intrusion of magnetic nonlinearity. Due to that nonlinearity intrusion decreases to zero. Intrusive nonlinearity movement is of certain length depending on the initial capacity power of intrusive nonlinearity and energy of the external static magnetic field that contains intrusive nonlinearity movement until zero.

Crystalline rock has both magnetic and electric ordering. Interrelation of magnetic, electric and elastic properties results in cross effects connecting magnetic and electric characteristics of the material. Applying the external electric field to such structure causes magnetization changes, and applying the external magnetic field causes polarization changes.

This effect called magnetoelectric (ME) enables designing fundamentally new solid-state electronic devices. Rather small ME-effect values in monocrystalline rock confine significantly the use of the devices. In rocks consisting of mechanically coupled magnetostrictive and piezoelectric components the ME-effect value is considerably greater.

This allows designing ME-effect based devices, e.g. magnetic-field sensors, with much greater sensitivity than that of Hall ones. ME-effects are conditioned by connection of magnetostriction and piezoelectricity. The in the alternating magnetic field due to magnetostriction there appear mechanical stresses that are transferred into the piezoelectric phase; and due to the piezoelectric effect polarization changes and electric pressure potential occurs.

As piezoelectricity is a linear function of the electric field intensity and magnetostriction is a non-linear function of magnetization, in general both linear and non-linear ME-effects appear. Many studies dealt with the linear ME-effect that consists in the electric potential occurring in, for example, a capacitor (with a dielectric made of magnetostrictive-piezoelectric material) when it is placed into static continuous (bias) and alternating magnetic fields. Frequency dependence of the effect is resonant, at the antiresonance frequency ME-factor shows peak increase.

The effect value depends on the static magnetic field and there is a pronounced maximum on the so called field dependence. In the weak fields the effect value is proportional to the bias field value. This is because of the fact that in the area far from saturation magnetostriction is a quadratic function of magnetization. So, when there is a bias field, the ME-effect value is proportional to the product of static and alternating fields' densities intensities as there is a linear in the alternating magnetic field ME-effect. However, along with the linear one there is also a non-linear ME-effect with the value inversely as the square of the alternating magnetic field.

In case of large bias fields its value is far less than that of the linear one and can be neglected. But in case of weak fields its value is commensurable or more than the linear effect value. This fact should be considered when designing magnetoelectric sensors for measuring weak fields.

With the quadratic effect of the magnetic field the frequency of mechanical oscillations vibrations in the magnetostrictive phase will make twice the frequency of the applied alternating magnetic field. Being transmitted into the piezoelectric phase through mechanic interaction, these oscillations vibrations will change the piezoelectric polarization and, as a result, the doubled frequency electric voltage tension appears on the capacitor plates. When the alternating magnetic field frequency equals half the antiresonance frequency, resonant effect step-up will occur. Unlike the linear effect, this resonance will take place at a zero bias field and its value will be quadratic of the alternating magnetic field intensity.

In the low-frequency spectrum region due to superposition of signals from linear and non-linear effects on the temporary dependence of a signal there appears difference of amplitudes of two neighbor maximums. The value of the difference is proportional to the static magnetic field intensity. This effect can be used for measuring a static magnetic field in the area of fields far from a saturation field of a magnet. A plate of a magnet (thickness ^mt) and a piezoelectric (thickness ^Pt, polarized perpendicular to the plate plane – axis Z) can serve a model of the theoretical description of the ME-effect (Venevtsev et al. 1982, Belovm 1987).

Let us assume that the direction of a static and alternating magnetic fields with the frequency ω' and the direction of polarization coincide. The movement equation for the x-projection of the environment displacement vector ^au_x considering nonuniformity of vibrations in the direction perpendicular to the interface horizon contrast is as follows:

$$^a\rho\frac{\partial^2 {}^au_x}{\partial t^2} = \frac{\partial {}^aT_{xx}}{\partial x} + \frac{\partial {}^aT_{xz}}{\partial z}, \quad (1)$$

where a equals m for magnetostrictive and p for piezoelectric layers, ^ap – ferrite or piezoelectric density, $^aT_{ij}$ – stress tensor.

Considering the plate long and narrow inhomogeneity, heterogeneity along the plate width, taking into account the fact that vibrations oscillations in the magnetostrictive phase are transferred into the piezoelectric phase through shear deformations, equations for the tensor of deformations of the magnetostrictive $^mS_{ij}$ and piezoelectric $^PS_{ij}$ phases and z-component of the electric displacement vector are as follows:

$$^mS_{xx} = \frac{1}{{}^mY}{}^mT_{xx} + g_{xx,z}(H_z)^2, \quad (2)$$

$$^mS_{xz} = \frac{1}{{}^mG}{}^mT_{xz}, \quad (3)$$

$$^PS_{xx} = \frac{1}{{}^PY}{}^PT_{xx} + d_{z,xx}E_z, \quad (4)$$

$$^PS_{xz} = \frac{1}{{}^PG}{}^PT_{xz}, \quad (5)$$

$$D_z = \varepsilon_{zz}E_z + d_{z,xx}{}^PT_{xx}. \quad (6)$$

where ^mY, ^mG, ^PY, ^PG – moduluses of elasticity (Young modulus) and magnetostrictive and piezoelectric phase displacement correspondingly; $g_{xx,z} = \delta\gamma_1(\delta(M_z)^2)\chi^2$ – magnetostrictive factor; γ_1 – magnetostrictive deformation in the direction perpendicular to the magnetic field; x – magnetic susceptibility; $d_{z,xx}$ – piezoelectric tensor; ε_{zz} – dielectric tensor of the piezoelectric; $H_z = H_m \exp(i\omega't)$ – intensity of the magnetic field with frequency ω'; E_z – z-projection of the electric vector in the piezoelectric.

Solution of equation (1) is as follows:

$$^au_x = {}^au(x,z)\exp(i\omega t), \quad (7)$$

where $\omega = 2\omega'$ – vibration frequency.

Solving the equation of movement for x-projection of the environment displacement vector and inserting the received expression into equation (6) using the condition of an open circuit for the potential difference on capacitor plates, the following expression is obtained:

$$U(t) = \frac{{}^PYd_{xx,z}g_{xx,z}}{\varepsilon_{zz}\Delta_\alpha} \cdot \frac{{}^mY^mt}{{}^mY^mt\frac{th({}^mk)}{{}^mk} + {}^PY^Pt\frac{tg({}^Pk)}{{}^Pk}} \cdot \frac{tg(k)}{k}\frac{th({}^Pk)}{{}^Pk}{}^Pt(H_z(t))^2. \quad (8)$$

Here the following symbols are introduced:

$$\Delta_\alpha = 1 - K_p^2\left(1 - \frac{{}^PY^Pt}{{}^mY^mt\frac{th({}^mk)}{{}^mk} + {}^PY^P\frac{tg({}^Pk)}{{}^Pk}t} \cdot \frac{tg(k)}{k}\frac{tg({}^Pk)}{{}^Pk}\right), \quad (9)$$

$$K_p^2 = \frac{{}^PY(d_{xx,z})^2}{\varepsilon_{xx}} - \text{squared} \quad \text{electro-mechanical}$$

coupling factor; $^mk = {}^m x^m t$ and $^Pk = {}^P x^P t$ – dimensionless variables; $^mx^2 = 2(1+\nu)\left(k^2 - \frac{\omega^2}{{}^mV_L^2}\right)$,

$$^Px^2 = -2(1+\nu)\left(k^2 - \frac{\omega^2}{{}^PV_L^2}\right) - \text{Z variables that determine discontinuity, heterogeneity; inhomogeneity}$$

environment displacement slide shift along axis Z;

$$^mV_L = \sqrt{{}^mY/{}^m\rho}, \quad {}^PV_L = \sqrt{{}^PY/{}^P\rho} - \text{dilatational}$$

159

acoustic wave velocity; v – poisson ratio that is to be the same for magetostrictive and piezoelectric phases; $k = kL/2$ – dimensionless parameter; k – wave number.

Dispersion relation that connects the wave number and angular frequency, is determined based on consistency of the system of equations (1) – (5):

$$^m x^m Ytg\left(^m k\right) = ^P x^P Yth\left(^P k\right). \qquad (10)$$

Equation (10) for thin layers is considerably simplified and turns into:

$$k = \sqrt{\frac{^m \rho^m t + ^P \rho^P t}{^m Y^m t + ^P Y^P t}}\,\omega. \qquad (11)$$

Equation (8) describes frequency dependence of ME-effect. It is determined through the dependence of a wave number on frequency. Equation (8) shows that at so called antiresonance frequencies when $\Delta_\alpha = 0$, ME-factor shows peak increase. It should be noticed that antiresonance frequencies are near resonance frequencies that are determined by the condition $k = (\pi/2)(2n - 1)$ or $L = (\lambda/2)(2n - 1)$, where λ – acoustic wave length, $n = 1,2,\dots$ – an integer. Then, considering (10) frequency of the first and main resonance will occur near the frequency determined by:

$$f_{res} = \frac{1}{2L}\sqrt{\frac{^m Y^m t + ^P Y^P t}{^m \rho^m t + ^P \rho^P t}}. \qquad (12)$$

Undoubtedly, when mining ferruginous quartzite in open pits, the mentioned dynamics of the ME-effect enables improvements of methods and devices for monitoring geodynamics of rock mass influenced by underground mining.

Unlike the existing empiric one, the suggested approach to timely preventing the destruction of surface rock areas influenced by underground mining allows effective monitoring the temporal progress of destructive energy of deformation of potential movement slide of surface rock areas influenced by underground mining as well as informs about the time of their destruction in areas of large-scale underground and open pit mining.

3 CONCLUSION

In Kryvbas intensive open pit and underground mining have been carried out for a long time results in anthropogenic destruction the natural geological geostatic pressure in rock mass influenced by underground mining.

Anthropogenic influence on the environment causes landscape and social problems.

Scientific studies of the process of accumulating destructive energy at areas with potential deformation of rock mass influenced by underground mining are carried out by the State Research Ore Mining Institute of SIHE "Kryvyi Rih National University".

The current empiric level of knowledge of genesis and temporal progress of potential rock movement slide centers influenced by underground mining slows down development of geophysical methods of monitoring open pit dynamics and, in particular, prediction of potential disturbances deformations of the above objects.

The currently applied method of studying rock geodynamics called the method of registering Natural Earth Impulse Electromagnetic Field (NEIEMF) parameters is mostly empiric. It has some disadvantages and its empiric base restricts its application.

The article provides grounds for improving methods and instruments of monitoring rock mass geodynamics at mining ferruginous quartzite in open pits within areas influenced by underground mining.

REFERENCES

Barechar. 1981. *Magnetysm, what is it.* K: Naukova dumka: 208.

Sena, L.A. 1977. *Physics units of inhomogeneity.* M: Nauka: 336.

Filippov, A.T. 1990. *Many faces soliton.* M: Nauka: 288.

Venevtsev, Yu.N., Gagulin, V.V. & Lubimov, V.N. 1982. *Ferroelectromagnetics.* M: Nauka: 224.

Belovm, K.P. 1987. *Magnetostrictive effects and their technical applications.* M: Nauka: 160.

Theoretical and Practical Solutions of Mineral Resources Mining – Pivnyak, Bondarenko & Kovalevska (eds)
© 2015 Taylor & Francis Group, London, ISBN: 978-1-138-02883-8

The monitoring of earth surface displacements during the subsoil development

G. Kyrgizbayeva, M. Nurpeisov & O. Sarybayev
Kazakh National Research Technical University named after K.I. Satpayev, Almaty, Kazakhstan

ABSTRACT: The results of studies were summarized in the article, which conducted by the authors in the course of carrying out scientific studies on the vertical movements of the earth's surface under the influence of the development of the Tengiz field. Nowadays, in many oil and gas basins of the world recorded strong, and even catastrophic geodynamic phenomena related to oil and gas extraction. These phenomena are realized in the form of earthquakes, fault activation in the sediment and intensive subsidence in the earth's surface. It is proposed the technique of repeated observations points geodynamic polygon using total stations, laser levels, satellite systems, which increases the accuracy and efficiency of determining the earth's surface subsidence, and the effectiveness of monitoring through field and office computerization of surveying and geodetic works.

1 INTRODUCTION

The subsoil of the western region of the Republic of Kazakhstan is rich by hydrocarbon fields. A large scale development of oil and gas resources, leads to intense movements of the earth's surface, both within the local area, and in the individual structural elements, resulting in bending occur boreholes tear gas and water pipelines, disabling the railways and roads, underground utilities and engineering structures, which in turn leads to a considerable economic loss. All this is a direct consequence of the change of the geodynamic regime of the geological environment under the influence of large-scale development of the subsoil, which conclusively proved by the results of experimental studies of the motion of the Earth's surface by the example of nature-technological system "Caspian zone".

2 MAIN CONTENT

For safe and efficient oil and gas development is necessary to investigate the impact of natural and anthropogenic factors on the development of deformation processes that will assess the possibility of controlling their effects on the rock mass, the earth's surface and engineering facilities. There is also the unsolved question of operational monitoring of ground deformation, processing and analysis of the received information. Therefore, one of the actual problems of intensive mining, especially in seismically active regions, is the study technological

movements of the earth's surface. Various conditions of occurrence of oil reservoirs, their sizes and shapes, as well as the technological features of conducting mining operations require individual, technical solutions based mining subsoil protection, environmental protection and industrial safety.

All these issues are regulated by the decree of the President of the Republic of Kazakhstan having the force of law from 29.01.96 №2828 "On Subsoil and Subsoil Use" and the Law "On Oil" and "On Environmental Protection", where oil and gas company is responsible for monitoring state the array undermining objects and utilities to ensure the normal operation of the enterprise.

Perfection of a technique of surveying and geodetic observations of the land surface subsidence and prediction of their parameters in the management of subsurface monitoring is a fundamental factor in improving the safety and efficiency of development of hydrocarbon fields.

The work contains research results obtained by the authors in the course of carrying out research as direct contractors "Study of vertical movements of the earth's surface under the influence of the development of the Tengiz field" (NCI No 1-14-01 from 01.01.2004) and "Bookmark geodynamic polygon and conducting instrumental observation on the field Botakhan" (NCI № 2.771.08).

Currently accumulated extensive material for traffic on earth, both within the local areas – geodynamic polygons (GFC) and over large areas. In most cases, GFC observations are complex: at the same time to study the motion of the Earth's surface and conducted a wide range of geological, geomorphological, geophysical, seismological and geodetic studies mainly, the possibility of deformation of the earth's surface caused by the engineering activities of people (technological movement).

For a correct prediction of subsidence over the earth surface (SES) and the adoption of appropriate measures to prevent the harmful effects of oil and gas, it is necessary to know the technological component of the total amount of vertical SES, or measures to prevent subsidence of these lead to unnecessary material costs and will be ineffective. In this regard, the special importance of reliable and timely forecast technological SES, which is understood scientifically, based judgments about possible states of the object in space and time.

Research in the field of study of the motion of the earth's surface in large-scale oil and gas development based on many scientific research organizations and individual researchers (Rock burst...2005, Kondratyev 2003, Lyubushkin 2003, Simmons 2004, Viktorov et al. 2005, Owen 2005, Mora & Keipi 2006, National Center..., Kozyrev et al. 2005, Nurpeisova & Bekbasarov 2006). However, despite some progress in addressing changes in geodynamic regime of the geological environment for large scale development of mineral resources, some questions remain insufficiently studied.

At the present stage of development of applied geodesy no scientifically based methodology for monitoring land surface deformations developed oil and gas fields. Regulatory framework is also missing. Traditionally used scheme in such work is thickening of the state geodetic network in the area of oil and gas lines, leveling of class II. In this calculation is not carried out the required measurement accuracy is not justified by the required frequency of observations, not a principle of reasonable minimum amount pledged survey markers within the area of potential subsidence. In connection with this unnecessarily increases the cost price of the work and informative study is lost.

Thus, the organization of monitoring the movement of the earth's surface (MES) in the area of oil and gas experience the following specific requirements:

– improving the information content of the results;

– increasing the efficiency of observations;

– cost-effectiveness of research.

It was our main aim of our research.

Kazakh National Technical University made theoretical and experimental surveyor – geodetic research examining the movements of the earth's surface in the Tengiz and Botakhan.

Tengiz – it is the largest oil field in Kazakhstan, opened in 1979 and located in the Caspian oil and gas province 160 km southeast of Atyrau. Producing horizons are in the depth interval 3.8 – 5.4 km from the earth's surface. The essential feature of the deposit is the presence of large amounts of hydrogen sulfide which is under abnormally high reservoir pressure. The reservoir is represented sophisticated constructions carbonate crack or fracture-vacuum incollector and salt confining beds. Features of the structure reservoir- available separate hydraulic unit, the connection between them is difficult or absent.

Abnormally high reservoir pressure, complex geological structure, as well as a significant field life with ever increasing volumes of recoverable oil yield suggests the possible drawdown of anthropogenic origin, and that was the basis for this study.

In solving the problem of determining the technological SES along with geodetic methods widely used methods of theoretical calculations of SES. In determining the SES caused oil and gas extraction, solves two problems: the calculation of sedimentation seam roof (SSR) and the calculation of the PCR from the known value of SSR.

The accuracy of the calculated value of the SES depends on the accuracy of the calculation of the vertical compression of the skeleton oil reservoir (or subsidence seam roof). The results of theoretical calculation of anthropogenic SES give less accurate results than the actual repeated geodetic measurements. However, it seems promising to use them both at the design stage of geodynamic polygon (GDP), so to compare the calculated values of the SES with the same values obtained by re-leveling.

State oil and gas reservoir in the process of its development depends on natural and anthropogenic factors. Natural factors include the geological features of the structure and properties of petroleum reservoir rock from which the reservoir is. Among the technological factors include: the ways of placing production and injection wells, the rate of production and injection fluids, and others.

In this regard, besides the Tengiz and Botakhan also just studied a number of oil fields in Kazakhstan sector of the Caspian Sea. Areas of development of many fields are located in areas with intense tectonic movements. According to the geological documentation and field observations identified the main characteristics of tectonic disturbances.

Efficiency and safety of mining operations during extraction of minerals, also during exploitation of facilities, which are placed in the operations site of these actions, depends on geodynamic state of thickness of minerals and tectonic and techno genic processes occurring in those minerals. Meanwhile, projection of mining companies are usually based on poor information about the state of rock mass, because it is impossible to make observations on the construction site of future facility before the beginning of mining operations.

These problems and complications are dealt with the assistance of geodynamic monitoring, the main goal of which is obtaining operative information about geomechanical processes and consequences caused by them occurring in the earth formation and earth surface, which are needed for taking well-timed preventive measures.

According to the regulation about the geological and surveying provision of industrial safety and protection of natural resources in the Law of natural resources and their usage No 2828, Law about the natural and techno genic emergencies, Regulation about Governmental monitoring of natural resources of Republic of Kazakhstan, functions of surveying service include monitoring of natural resources, including processes of rock mass and earth surface movements, geomechanical and geodynamic processes in natural resources usage to prevent deleterious effect of mining operations on capital mining operations site, surface facilities and environment.

Based on the analysis of geology and tectonics of region, numerical modelling and experimental evaluation of the state of strain in the massive, energetically loaded zones could be pointed out, which determine boundaries of geodynamic monitoring zones. Afterwards, monitoring of danger zone is done, which mainly includes control of deformation and parameters of geophysical fields.

Reliable information about deformations of rock mass could be obtained through direct geodetic observations on geodynamic polygons. Earth surface movements (ESM) are learnt through method of repetitive leveling based on first class technique, when measurements of excess in sections between adjacent frames are analyzed. Technology of leveling must correspond to requirements of accepted technique.

The expected structure of learning technique and prediction of dangerous states in NTS considering the whole complex of geophysical and geodetic measurements and their joint processing are represented on Figure 1.

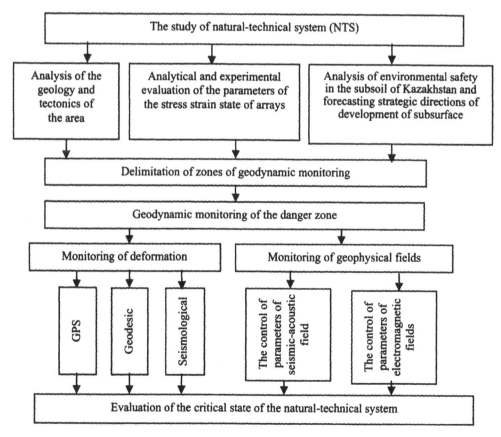

Figure 1. Scheme of learning method and prediction of geodynamic processes in NTS.

The first leveling work in the Tengiz field were conducted in 1989 – 1991 by ukrainian "Gipro-NIIneft" together with the expedition of 105 Hooke in Kazakhstan. Performing at that time consisted of 48 geodetic leveling characters.

Analysis of materials devoted to the study of the SES developed oil and gas fields, showed that the use of repeated leveling by trunk moves leveling SGN can not get enough information about vertical movements of the earth's surface (VMES) deposits. The scheme of special geodetic constructions planned on the GFC not always lead to the desired result, particularly in areas that coincide with areas prone to active anthropogenic influence, where there is a problem common VMES division into separate components.

In order to study technogene phenomena, and seismotectonic movements earth's surface, caused by the development of mineral resources, the management of "TCO" decided to resume in this area surveillance array. In this regard, after a detailed geological, geodetic surveying-study areas and technological parameters of the Tengiz field was recreated geodynamic polygon by the proposed technique by us in 2005, placing a quadrilateral polygon in the center of the expected displacement trough with radial profile lines within the deposit.

Given that the primary major faults Caspian region have the latitude, profile lines were mainly oriented meridionally. Tengiz GFC consists mainly of seven specialized lines, as well as additional leveling networks of about 1000 points, including 5 – fundamental, 20 – (reference and working) GPS network, 5 – seismological, 5 – light ranging and 10 – deep. At 407 points set security labels (Nurpeisova et al. 2003).

Long-term observation of deformations over 1000 benchmarks GFC showed the complexity of the field work, especially the transfer from one point to another set of devices (the device itself, stand, rack, etc. In this regard, developed the device circuit is shown (Figure 2), where the upper part of the center is equipped with forced centering table, for the installation of instruments and measuring the efficiency of operations. The invention relates to a geodetic center for installation of new devices and signals. Such items on GFC Tengiz-25 pieces. The purpose of the invention – to increase the accuracy of centering, speed measurement without using a tripod in points standing and watching. The new device allows fast and precise alignment, and eliminates the use of tripods.

In the study of modern movements of the earth's surface of the vertical component is determined based on the results of repeated precision leveling. At GFC "Tengiz" leveling class II – produces digital laser level firm Leika WILD NA 3003 with invar

strips by the method of double leveling in the forward and reverse direction. For leveling accuracy tolerance was set ±0.4 mm, which corresponds to the leveling tolerances I – class (Nurpeisova & Zhardayev 2004).

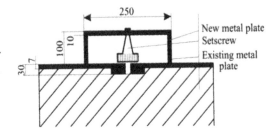

Figure 2. New installation for appliances.

The principle of leveling is based on processing the encoded signal (obtained from the rack), and then send the data to the computer and processing of a special program. The advantage of this system is the ease of measurement, no error with reading and writing, automatic calculation of height during the measurement and data logging.

Included in the GFC "Tengiz" points form a framework region network. On these points was organized geodynamic monitoring of the condition of pipelines and engineering structures.

The observations were made by three satellite receiver GPS, which allowed to measure in some areas three vectors at the same time. Processing of satellite observations are made with a special program that is included with satellite receivers, the result will be horizontal coordinates of all points of the network and elevations.

To assess emerging and developing at GFC ground deformation we analyzed the horizontal and vertical displacements 9 working points of the pipeline profile line. Stage coordinates of points determined from satellite measurements.

To analyze the accuracy of measurements made satellite receiver GPS, measurements were made of the control network of precision electronic tachometer TS 1200. The circuit control network is a geodesic quadrangles with measurements of angles and distances. Comparative analysis of the measurements (Table 1) showed the accuracy of satellite measurements practically coincides with the total station measurement accuracy.

Years of experience surveying instrument observations of the state of the rock mass in a number of fields in Kazakhstan helped to develop and implement a high-technique observations using modern electronic equipment (Figure 3).

Figure 3. Modern technology of high-precision observations using GPS.

Principles of leveling is based on processing of coded signal (received from hub through lens). Microprocessor computes readings of hub and appropriate horizontal distance between hub and level. Measurement results are saved in REK module of WILD GRM 10 with following data transfer into computer and data processing in special program. Advantages of such system are simplicity of measuring, absence of computation and reading errors, automatic computation of height during measurement and data logging.

With increasing accuracy and efficiency of geodetical measurements expands the list of problems, major contribution to solve those problems could come from modern geodesy. One of such problems is detailed study of technogenic processes. Solution of this problem could lead to creation of constantly functioning geodynamic technogenic polygons, which rely on regularly performing complex of geophysical and geodetic observations with geodetic interpretation of its results. Highly detailed geodetic observations on geodynamic polygons are held by surveying services. Increased volume and intensity of mineral resources development lead to the occurrence of technogenic disasters, which are related to certain shifts and subsidence of earth surface.

Among modern methods and research tools of dislocations and deformations of earth surface the technology of satellite system (GPS system) proved to be effective. Satellite GPS systems(global positioning) are needed to create calculation basis of deformations and geo informational systems, which allow to predict parameters of geodynamic processes. Along with GPS technologies, systematic observations of CEP are conducted with electronic tachometers and digital levels for different regions of intensive extraction of solid mineral resources (Nurpeisova et al. 2007, Nurpeisova & Turdahunov 2006).

Principles of leveling are based on processing of coded signal (received from hub), with following data transfer to computer and processing of received data through special program. Advantages of such system include simplicity of measurement, absence of computational and recording errors, automatic calculation of heights during measurement and data logging. Precise repeated leveling operations were conducted in two cycles (fall and spring periods). Results of leveling attributes of 2008 are shown in Figure 4, which indicates about:

– maximum speeds of deformations related to zones of fracture.

– regularity of deformation processes of earth surface, besides deformations occur highly irregularly in terms of space and time (subsidence has positive and negative signs);

– maximum deformation speeds, which occurred in period of 2001 – 2008, related to fracture zones;

– tectonic nature of earth surface subsidence (according to positive signs of dislocation of frames in height).

Repeated geodetic measurements were also conducted with electronic tachometers TS110, TS 120 of Leica Corporation and results frames subsidence were compared with results of repeated leveling. Evaluation of height measurement accuracy of these frames with electronic tachometer TS 1200, where SKP of 1 km was equal to $\mu = 2.51$ mm, on station $\mu = 0.79$ mm.

On GDP "Tengiz" in period from 2000 to 2005, 8 cycles of following GPS measurements were conducted:

– transmission of coordinates from p. ITRF to supporting point No 2;

– determination of coordinates of all three supporting points of GPS network;

– determination of dislocations of working points from pipelines locations;

– average measurement time on one working point was equal to six hours depending on length of lines and conditions of receiving radio signal on point.

Processing of satellite observations were conducted with the standard program SKI (Leica Corporation, Switzerland), which is included in GPS receivers kit, as a result horizontal coordinates of all network points in local system of coordinates and height marks of frames were received. To analyze the accuracy of measurements, conducted with GPS satellite receivers, measurements of main satellite network with electronic tachometer TS 1200 were conducted. Comparative analysis of conducted measurements (Table 1) showed that accuracy of satellite measurements almost matches with accuracy of measurements of electronic tachometer. Thus, development of geodynamic polygon's network for precise observations using electric and satellite GPS receivers allowed to decrease time spent on determining coordinates in recalculation of one point 10 – 15 times and increase accuracy of determining coordinates at least 2 times.

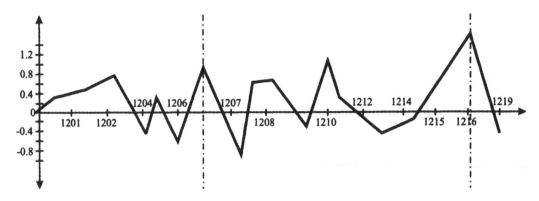

Figure 4. Chart of dislocations of frames GDP "Tengiz".

Table 1. Comparative analysis of satellite and linear measurements.

From the point	To the point	S-GPS (netwok scheme)	TS 1200	dS, m	Relative error
1	2	2359.266	2359.265	0.001	1/1947000
2	4	2606.720	2606.714	0.006	1/429000
2	3	1220.430	1220.428	0.002	1/663000
1	4	2276.461	2276.465	-0.004	1/625000
1	3	2840.789	2840.796	-0.007	1/386000
4	3	1962.898	1962.896	0.002	1/1002000
		Note – average quadratic error		0.005	

On Figure 4 under numbers 1201, 1202, 1203 are indicated working frames. It was determined that supporting GPS points were chosen correctly based on comparison data of leveling works, because in period frame of 204 – 2008 they had dislocations of 2 to 8 mm, besides frames 1207 and 1219 ($\eta_{max} = 0.5 - 0.8$ mm) confined to tectonic fractures (Nurpeisova 2010).

In Figure 5 is shown a graph of displacement of benchmarks leveling network GFC "Tengiz" for the period 2001 – 2005 and 1992 – 2008. 1 – 3 on a profile consisting of 25 benckmarks. In the period of

2001 – 2005, benchmark had a displacement of 2 mm to 8 mm, and for the period 1992 – 2008 $\eta_{max} = 2.9$ sm.

Thus, the organization of further research organization deserves greater attention on GFC geodetic monitoring, based on the use of satellite methods in those areas of engineering structures on the territory of the Tengiz field, which are subject to technogenic processes (Nurpeisova et al. 2013, Nurpeisova et al 2014).

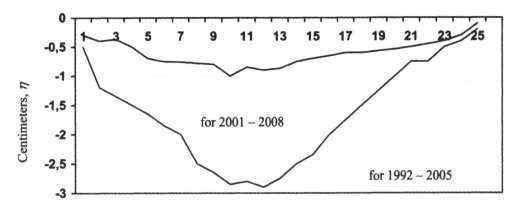

Figure 5. Chart of dislocation of frames by profile 1 – 3.

Afterwards, all information on legitimacy of process of system displacements and parameters of critical condition proceeds into expert system, where assessment of natural-technical system (NTS) is done on the basis of integration of database and knowledge, and justified by appropriate management decisions. Final goal of these decisions is to provide either adapting functioning of NTS or controlled withdrawal of its critical state.

3 CONCLUSIONS

Analysis of results of monitoring of the earth surface displacements that occurred over the Tengiz field from 2001 to 2008 and registered in the processing of observational data showed the presence of accelerating time-subsidence surface. Geodetic measurements are discrete, they do not provide a complete picture of the deformation processes in time. This can only be done with comprehensive research, including seismological, geophysical methods, and space radar monitoring of displacements.

REFERENCES

Kondratyev, O.K. 2003. *Forecast of earthquakes*. Reasons and solutions of a problem. M: PFUR, Vol. 1: 148 – 152.
Kozyrev, A.A., Panin, V.I. & Maltsev. V.A. 2005. *Parameters of seismic process as harbingers of technogenic catastrophes*. Development of subsoil and environmental problems – look in XXI of eyelids. M: Publishing house Academy of mountain sciences of the Russian Academy of Sciences: 111 – 121.
Lyubushkin, A.A. 2003. *Geodinamic monitoring*. Geophysics problems in XXI an eyelid. M: Science: 70 – 94.
Mora, S., Keipi, K. 2006. *Disaster risk management in development projects: models and checklists*. Bulletin of engineering geology and the environment, V 65, No 2: 155 – 165.
National Center of information on earthquakes, American Geological service [Electronic resources]. http://neic.usgs.gov/neis/eqlists/eq-stats.html.

Nurpeisova, M.B., Aitkasinova, Sh.K. & Kirgisbaeva, G.M. 2014. *Geomechanical monitoring of the massif of rocks at the combined wave of development of fields*. Geodesy & Mine Surveying 17 – 26 June: 279 – 192.
Nurpeisova, M.B. & Bekbasarov, Sh.Sh. 2006. *Environmental problems of development subsoil*. Almaty: Messenger Kazakh national technical university: 53 – 57.
Nurpeisova, M.B., Kasimkanova, H.M. & Kurgizbaeva G.M. 2010. *Methods of the separation of the active systems of the rifts with using GPS-technologies*. Materialy VI – Miedzynardowej naukowi praktvcznej konferencji "Stosowane naukowe opracowania 2010". Prague: 57 – 59.
Nurpeisova, M.B., Kasymkanov, H.M. &, Kyrgizbayeva, G.M. 2007. *Improvement of ways of tool supervision on mines*. Materials XVII international scientific and practical conference "Deformation and destruction of rocks and developments". The Crimea, Alushta, Taurian national university: 235 – 238.
Nurpeisova, M.B., Kyrgizbayev, G.M. & Aytkazinov, Sh.K. 2013. *The forecast of technogenic danger of the territory according to geodetic and satellite measurements*. Karaganda: Marksheyderiya zhne Geodeziya, No 3: 18 – 19.
Nurpeisova, M.B. & Turdahunov, M.M. 2006. *Regulation of f regime of mining operation at opencast*. The 15th international symposium on mine planning and equipment selection Torino. Italia, September 20 – 22: 1459 – 1464.
Nurpeisova, M.B., Zhardayev, M.K. & Beck, A.Sh. 2003. *Forecasting of technogenic catastrophe of development subsoil at large-scale*. Almaty: Geodesy. Cartography. Geographic information systems, No 3: 72 – 76.
Nurpeisova, M.B. & Zhardayev, M.K. 2004. *Results of natural supervision on the geodynamic Tengiz ground*. Almaty: The Industry of Kazakhstan, No 4: 120 – 124.
Owen, M.L. 2005. *Calibrating a semi-guanti-tative seismic risk model using rock burst case studies from underground metalliferous mines*. Australia: Controlling seismic risk. Australian Centre for Geomechanics: 191 – 204.
Rock burst and seismicity in mines proceedings. 2005. Australia: Australion Centre for Geomechanics.
Simmons, J.V. 2004. *Geotechnical risk management in open pit coal mines*. Australian Center for Geomechanics Newsletter, No 22: 1 – 4.
Viktorov, S.D., Iofis, M.A. & Odintsev, V.N. 2005. *Destruction of rocks and risk of technogenic accidents*. Mountain magazine, No 4: 30 – 35.

Theoretical and Practical Solutions of Mineral Resources Mining – Pivnyak, Bondarenko & Kovalevska (eds)
© 2015 Taylor & Francis Group, London, ISBN: 978-1-138-02883-8

Determination of borehole parameters for the degassing of underworked coal seams

I. Pugach & O. Mukha
National Mining University, Dnipropetrovsk, Ukraine

ABSTRACT: Developed a mathematical model to determine the parameters of laying degasification wells drilled on undermined flat and inclined coal seams. The essence of the method consists in determining the coordinates of the mouth of hole and borehole bottom using the apparatus of analytical geometry. The method of calculation of rational parameters of degasification wells under the determination of the place and the amount of accumulation of methane by the stress-strain state of rock strata is developed.

1 INTRODUCTION

The problem of methane from gassy coal mines today is a complex problem that represents both industrial and environmental hazard. If wrong calculate the parameters of degassing this create a threat of an emergency and reduced coal mining.

As a result of the formation of mined-out space the wall-rock are set in motion. Layer the immediate mine roof above the mined-out space from the high-roof is separated and flexed along the normal to the formation. Then the overlying rock layers successively are exfoliated and bended after the bottom layers. Under the bend between layers with different physical and mechanical properties the voids of ply separation can be formed.

When drilling the boreholes towards working face and pillar system development, the increases of working face speed is deteriorating connection of these wells with methane zone and reduced their effectiveness. In addition, the angle of the total displacement of the roof rocks depends from the working face speed. Getting started of the borehole and the effectiveness of its actions depend on the angle of displacement of the roof rocks.

When the borehole crossing the plane of displacement of rocks gets into the zone of the loading-out from the rock pressure therein starts flowing methane. When the borehole approaching the zone of disorderly rockfall its efficiency is reduced due to intake air from the goaf. Angle of displacement with increasing the rate of stope is reduced (Bondarenko 2013). The mouth of hole can be destroyed before she starts to give methane. With this in mind, the calculation of topological parameters of the well with the mining and geological factors is an urgent task.

2 CALCULATION THE PARAMETERS OF THE LOCATION OF THE DEGASIFICATION BOREHOLE

The position of the borehole defined by three parameters (Degassing of coal ... 2004): l_b – length of the borehole, m; β – the angle between the borehole and the horizontal plane, deg; φ – the angle between the projection of the borehole and the horizontal plane perpendicular to the axis of the output in the same plane, deg.

In the article the analytical substantiation of calculating the parameters of degasification borehole in the mining and geological conditions is suggested.

For effective the degassing of overlying rock borehole must to cross the degassed layer into the zone of the loading-out from the rock pressure. When drilling is necessary to consider the angle of unloading rock from rock abundant pressure ψ. And it must take account of as from the line stope, and from the mine working, from which the borehole is drilled.

Methane drainage borehole prior to disconnecting from the degassing network should not cross the uncontrolled rock caving zone. The height of this zone is usually taken as $h_{rcz} = (4-6) \cdot m$, where m – extracting seam thickness, m.

The borehole cannot pass through the intersection of two planes that limit zone disorderly collapse of rocks and their unload from the rock pressure at an angle ψ from the line of the stope. The borehole must intersect degassed layer on the line of maximum gas release (y_{max}) from the chitter into the methane drainage borehole. This line is parallel to the mine working, from which boreholes are drilled.

When the distance between the layers $M = 25 \div 60$ m, then $y_{max} = (0.4 + 0.5) M$. If the advance rate of stope is $1.0 \div 1.5$ m/day and $y_{max} = (0.6 + 0.8) M$ if the advance rate of stope is 2.5 m/day and more.

The scheme of degassing the overlying flat-lying seams by the boreholes from the liquidated mine working, when the extraction pillar mining across the pitch is represented. Topological parameters of the methane drainage borehole are determined at the time of disconnection from the degassing network and disposition the wellhead at a distance l_{disc} from the line of the stope.

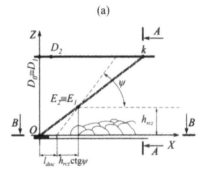

(a)

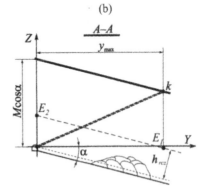

(b)

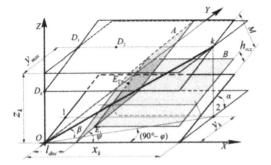

Figure 1. The calculation scheme for determining the parameters of methane drainage borehole drilled at working mining across the pitch from the ventilation drift.

Line with the coordinates of the borehole passes through the origin, the line of intersection of planes A and B and lines lying at a distance parallel to the axis drift (y_{max}).

Shall find the equation of the plane which contains the origin of coordinates, and a line of intersection of the planes A and B. Shall present this plane through the coordinates of three points that do not lie on a straight line:

– the origin of coordinates $O(x_0, y_0, z_0)$;

– two points of the line of intersection of the planes A and B: $E_1(x_1, y_1, z_1)$ and $E_2(x_2, y_2, z_2)$, where

$x_0 = y_0 = z_0 = 0$; $x_1 = l_{disc} + h_{rcz} ctg \psi$; $y_1 = y_{max}$;

$z_1 = h_{rcz} \mp y_{max} tg \alpha$; $x_2 = l_{disc} + h_{rcz} ctg \psi$; $y_0 = 0$;

$z_2 = h_{rcz}$, where l_{disc} – distance from a main point turned-off of the methane drainage borehole from the pipeline to the line stope, m; α – angle seam inclination, deg; M – the distance between the layers, m (Figure 1).

To determine the coordinates of the points we construct a projection borehole on the planes XOZ, XOY, YOZ (Figure 2).

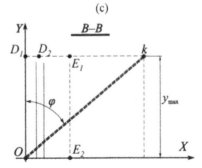

(c)

Figure 2. The projections of borehole on the planes: (a) XOZ (b) YOZ, (c) XOY in developing the coal bed across the pitch.

The upper sign (plus or minus), here in after adopted during drilling towards the dip, and the bottom – to the uprising.

The flat surface passing through the three points that do not lie on a straight line is described by the equation:

$$\begin{vmatrix} x - x_0 & y - y_0 & z - z_0 \\ x_1 - x_0 & y_1 - y_0 & z_1 - z_0 \\ x_2 - x_0 & y_2 - y_0 & z_2 - z_0 \end{vmatrix} = 0. \qquad (1)$$

170

Substituting the values of the coordinates of points of the plane OE_1E_2 in the equation (1), we find the determinant of the matrix:

$$\begin{vmatrix} x & y & z \\ l_{disc} + h_{rcz}ctg\psi & y_{max} & h_{rcz} \mp y_{max}tg\alpha \\ l_{disc} + h_{rcz}ctg\psi & 0 & h_{rcz} \end{vmatrix} =$$

$$= x \begin{vmatrix} y_{max} & h_{rcz} \mp y_{max}tg\alpha \\ 0 & h_{rcz} \end{vmatrix} -$$

$$- y \begin{vmatrix} l_{disc} + h_{rcz}ctg\psi & h_{rcz} \mp y_{max}tg\alpha \\ l_{disc} + h_{rcz}ctg\psi & h_{rcz} \end{vmatrix} +$$

$$+ z \begin{vmatrix} l_{omк} + h_{oбp}ctg\psi & y_{max} \\ l_{omк} + h_{oбp}ctg\psi & 0 \end{vmatrix} = 0, \qquad (2)$$

$$xh_{rcz} \mp (ytg\alpha - z)(l_{disc} + h_{rcz}ctg\psi) = 0. \qquad (3)$$

The coordinates of the borehole bottom are located at the intersection of three planes:
 – the plane which described by the equation (3);
 – the plane of the superimposed seam;
 – a vertical plane passing through the line of maximum gas release y_{max}.

We define three points which are belong to the plane of the superimposed seam $D_0\left(0,0,\dfrac{M}{cos\,\alpha}\right)$;

$D_1\left(0,1,\dfrac{M}{cos\,\alpha}\mp tg\alpha\right)$; $D_2\left(1,1,\dfrac{M}{cos\,\alpha}\mp tg\alpha\right)$ and

find the equation of the plane $D_0D_1D_2$:

$$\begin{vmatrix} x & y & z-\dfrac{M}{cos\,\alpha} \\ 0 & 1 & \mp tg\alpha \\ 1 & 1 & \mp tg\alpha \end{vmatrix} = x\begin{vmatrix} 1 & \mp tg\alpha \\ 1 & \mp tg\alpha \end{vmatrix} -$$

$$- y\begin{vmatrix} 0 & \mp tg\alpha \\ 1 & \mp tg\alpha \end{vmatrix} + + \left(z - \dfrac{M}{cos\,\alpha}\right)\begin{vmatrix} 0 & 1 \\ 1 & 1 \end{vmatrix} = 0, \qquad (4)$$

$$\mp ytg\alpha - z + \dfrac{M}{cos\,\alpha} = 0. \qquad (5)$$

Vertical plane passing through the line of maximum gas release y_{max} is described by the equation:

$$y = y_{max}. \qquad (6)$$

Coordinates of the borehole bottom is the solution of equations (3 – 5):

$$\begin{cases} xh_{rcz} \mp (ytg\alpha - z)(l_{disc} + h_{rcz}ctg\psi) = 0 \\ \mp ytg\alpha - z = -\dfrac{M}{cos\,\alpha} \\ y = y_{max} \end{cases} \qquad (7)$$

Let us find the determinant of the system:

$$\Delta = \begin{vmatrix} h_{rcz} & \mp tg\alpha \cdot A2 & -A2 \\ 0 & \mp tg\alpha & -1 \\ 0 & 1 & 0 \end{vmatrix} = h_{rcz}, \qquad (8)$$

where $A2 = (l_{disc} + h_{rcz}ctg\psi)$.

Determinant is not zero, therefore, the system of equations (5) has a unique solution: $x_k = \dfrac{\Delta x}{\Delta}$, $y_k = \dfrac{\Delta y}{\Delta}$, $z_k = \dfrac{\Delta z}{\Delta}$.

Replacing in the system (6) elements of the first column of free terms of equations (5), we obtain the determinant Δx, similarly – Δy and Δz:

$$\Delta x = \begin{vmatrix} h_{rcz} & \mp tg\alpha(l_{disc} + h_{rcz}ctg\psi) & 0 \\ 0 & \mp tg\alpha & -\dfrac{M}{cos\,\alpha} \\ 0 & 1 & y_{max} \end{vmatrix} =$$

$$= \dfrac{M}{cos\,\alpha}(l_{disc} + h_{rcz}ctg\psi), \qquad (9)$$

$$\Delta y = \begin{vmatrix} h_{rcz} & 0 & -A2 \\ 0 & -\dfrac{M}{cos\,\alpha} & -1 \\ 0 & y_{max} & 0 \end{vmatrix} = y_{max}h_{rcz}, \qquad (10)$$

$$\Delta z = \begin{vmatrix} h_{rcz} & \mp tg\alpha(l_{disc} + h_{rcz}ctg\psi) & 0 \\ 0 & \mp tg\alpha & -\dfrac{M}{cos\,\alpha} \\ 0 & 1 & y_{max} \end{vmatrix} =$$

$$= \dfrac{h_{rcz}}{cos\,\alpha}(M \mp y_{max}sin\,\alpha). \qquad (11)$$

The system of equations (7) has a solution:

$$\begin{cases} x_k = \dfrac{M}{h_{rcz}cos\,\alpha}(l_{disc} + h_{rcz}ctg\psi) \\ y_k = y_{max} \\ z_k = \dfrac{(M \mp y_{max}sin\,\alpha)}{cos\,\alpha} \end{cases} \qquad (12)$$

The length of the borehole is defined as the unit vector d_k – the distance between two points: $O(0,0,0)$ – mouth of the methane drainage borehole and $k(x_k, y_k, z_k)$ – borehole bottom. Coordinates downhole can be found by solving the system of equations (12):

$$|d_k| = \sqrt{x_k^2 + y_k^2 + z_k^2} \, ,$$
(13)

$$l_{bor} = \left(\frac{1}{\cos^2 \alpha} \left(\left(\frac{M}{h_{rcz}} \right)^2 (l_{disc} + h_{rcz} ctg\,\psi)^2 + \right. \right.$$

$$\left. \left. + (M \mp y_{max} sin\,\alpha)^2 \right) + y_{max}^2 \right)^{0.5} .$$
(14)

Angles of inclination β and rotation φ of the borehole can be set according to the formulas:

$$\beta = arctg \left(\frac{z_k}{(x_k^2 + y_k^2)^{0.5}} \right) ,$$
(15)

$$\varphi = arctg \left(\frac{x_k}{y_k} \right) ,$$
(16)

and substituting the coordinates of the borehole (12) in (15) and (16) we obtain the expression:

$$\beta = arctg \left(\frac{h_{obr} (M \mp y_{max} sin\,\alpha)}{\left(M^2 (A2)^2 + h_{obr}^2 y_{max}^2 \cos^2 \alpha \right)^{0.5}} \right) ,$$
(17)

$$\varphi = arctg \left(\frac{M (l_{disc} + h_{rcz} ctg\,\psi)}{h_{rcz} y_{max} \cos \alpha} \right) .$$
(18)

Consider the scheme of degassing of undermined flat seams by the boreholes drilled from the liquidated mine working, when mining down-dip (up-dip). The parameters of degassing boreholes thus define the same as in the previous case. Projections of this borehole are shown in Figure 3.

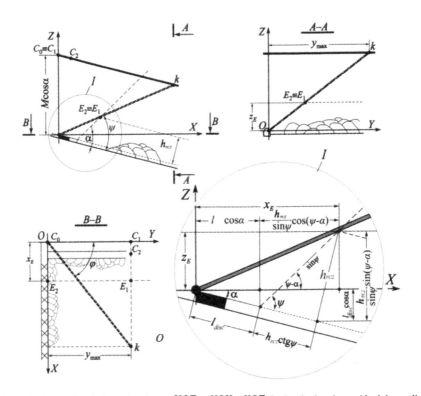

Figure 3. The projections of borehole on the planes XOZ, XOY, YOZ in developing the coal bed down-dip (up-dip)..

According to the scheme shown in Figure 3:

$$x_E = \frac{h_{rcz}}{sin\,\psi} \cos(\psi \mp \alpha) + l_{disc} \cos \alpha ,$$
(19)

$$x_E = y_{max} ,$$
(20)

$$z_E = \frac{h_{rcz}}{sin\,\psi} sin(\psi \mp \alpha) \mp l_{disc} sin\,\alpha .$$
(21)

Coordinates of the borehole bottom:

172

$$\begin{cases} x_k = \dfrac{-x_E M}{\left(x_E(\mp \operatorname{tg}\alpha) - z_E\right)\cos\alpha} \\ y_k = y_{max} \\ z_k = \dfrac{-z_E M}{\left(x_E(\mp \operatorname{tg}\alpha) - z_E\right)\cos\alpha} \end{cases} \qquad (22)$$

The length of the borehole for these conditions will be:

$$l_{bor} = \left(\frac{M^2\left(x_E^2 + z_E^2\right)}{\left(x_E(\mp \operatorname{tg}\alpha) - z_E\right)^2 \cos^2\alpha} + y_{max}^2 \right)^{0,5}, \quad (23)$$

the angle to the horizontal plane:

$$\beta = \operatorname{arctg}\left(\frac{-z_E M}{\left(x_E^2 M^2 + y_{max}^2\left(x_E(\mp \operatorname{tg}\alpha) - z_E\right)^2 \cos^2\alpha\right)^{0,5}} \right), \quad (24)$$

the angle of rotation:

$$\varphi = \operatorname{arctg}\left(\frac{-x_E M}{y_{max}\left(x_E(\mp \operatorname{tg}\alpha) - z_E\right)\cos\alpha} \right). \quad (25)$$

Parameters of the boreholes can be defined by the formulas (23), (24) and (25) and the formulas (13) and (15), (16) using coordinates (22) of the borehole bottom.

Calculation of the amount of methane produced from the wells, its concentration in the mixture, and the parameters of the degassing system (vacuum in wellhead and vacuum pumps, rational diameters of the degassing pipes) may be determined by the method described in (Kremenchutskiy 2012).

3 CONCLUSIONS

The method of determination of topological parameters of the degasification boreholes bottom coordinates allows, in contrast to the existing, consider the height of the uncontrolled rock caving zone and the angle of unloading rock from rock pressure. This reduces the likelihood of increasing the flow of air into the boreholes and the destruction of the wellbore when it approached to the stope.

REFERENCES

Bondarenko, V., Kharin, E., Antoshchenko, M. 2013. *Basic scientific positions of forecast of the dynamics of methane release when mining the gas bearing coal seams.* Naukovyi visnyk Natsionalnoho Hirnychoho Universytetu, No 5: 24 – 30.
Degassing of coal mines. *Requirements for methods and schemes of degassing.* SOU 10.1.00174088.001-2004. Kyiv: Mintopenergo of Ukraine: 162.
Kremenchutskiy, N., Muha, O. & Pugach, I. 2012. *Degassing systems rational parameters selection at coal mines.* The Netherland: CRC Press/Balkema: 87 – 93.

Theoretical and Practical Solutions of Mineral Resources Mining – Pivnyak, Bondarenko & Kovalevska (eds)
© 2015 Taylor & Francis Group, London, ISBN: 978-1-138-02883-8

Analysis of current trends in development of mine hoists design engineering

K. Zabolotny, O. Zhupiev & A. Molodchenko
National Mining University, Dnipropetrovsk, Ukraine

ABSTRACT: This article highlights the key issues of hoisting and identifies the current trends in the design of mine hoists.

1 INTRODUCTION

Hoists are considered to be the most crucial link in mining among all the machinery used in the ore mining industry. Emergency situations in the process of operation of hoisting machines do not only lead to significant financial losses, but also endanger people's lives. For this reason, high requirements to mine hoists (MH) shall comply with the special requirements with regard to the technical level, quality of manufacture, reliability and safety.

The practice of using MH provides the constant improvement of their performance characteristics, in particular, weight-carrying capacity, speed and hoisting depth. The reducing of the equipment material content by increasing its efficiency, performance quality and control system reliability is very important as well. At the same time, it is necessary to maintain and even to improve the reliability of the whole hoisting. The solution of this complex problem is possible only through improving the parameters of mine hoisting functional systems, including the emergency operation.

2 ANALYSIS OF CURRENT TRENDS IN THE DEVELOPMENT OF MINE HOISTS DESIGN ENGINEERING

Currently the main way to improve the drum hoists design is to reduce their size while increasing the rope winding surface on the drum. Therefore, M.G. Kuzhel, V.A. Kravtsov, I.P. Kovalevsky, V.V. Roizen, the developers of new mine hoists design, proposed the following technical solutions: to place the brake gear within a drum; to downsize the reversible drum by replacing one of the hubs with thrust support rollers and equipping the front, shell and flanges with box-shaped supports; to use the non-metallic removable lining that allows to increase the drum winding surface during operation and improve the rope function; to ensure the variable pitch of the rope winding (Zabolotny 1997).

Implementation of the above listed suggestions is impossible without proper scientific justification. All this can have negative consequences, such as unreasonable uprating of the unit metal content, unbalancing of the structural solidity, and difficulties in the mine hoists performance in some cases. The latter happened due to the fact that the haulage distance of the characteristic points on the surface of narrow reversible drums under the internal brake exceeded the acceptable distance extremely. Besides the thrust support rollers blocking and a rope in the gap occurred. In addition, an increase in the brake actuator upstroking, cracks in flanges and forehead took place.

We will consider the efficiency of the proposed technical solutions and the problems having arisen.

Thus, the desire to combine downsizing and reducing of a mine hoist mass with its high reliability and productivity contributed to the development of a new class of DHE (Drum-type Hoist Engine) units. The brakes are placed inside the drum in this type of machines.

As opposed to mine hoists with external post brakes, where the cylindrical surface of the drum includes a winding part and brake fields, the design of the proposed hoisting machine allows to use the whole cylindrical surface of the drum for rope winding.

Specialists of NMBP (Novokramatorsk Machine Building Plant) decided to make significant changes in the design of the drums to implement this idea. For instance, whether preliminary the uncoupler devise was arranged outside the resettable drum, now its tooth wheel rims are attached directly to the resettable and fixed drums. At the same time, they managed to avoid combined torque impact on the shaft of the hoisting machine under the maximum static stress of the rope.

When hoists with a split-type drum and external brakes are used, the resettable drum is arranged on two shaft bearings that enlarged its winding surface. In the new design, the narrow resettable drum is placed on the one self-aligning bearing, and thrust support rollers are mounted along the perimeter of the pipe shell's forehead.

Described hoist design modifications were successful. In particular, the machinery foundation was simplified, brake installation conditions were improved, equipment performance was advanced, and the usable width and coiling length were increased. At the same time, metal consumption of the uncoupler and the size of a drum were reduced, the shaft stress was shrunk, the number of assembly units for installation on a foundation was cut, the construction and installation work volume and hoisting down time reduction while replacing a worn out machine. Everything mentioned gives an idea of the reality to obtain the significant economic benefit.

We admit that despite a number of advantages some significant faults in the parameter calculation and in the hoist drums with internal brakes design were made at the very beginning. Thus, the calculation was made with respect to the rope load only, while the load under the influence of the brake in general motion may reach 60%. It is necessary to consider that during the operation of the drum mine hoists with external brakes the load emerging under their influence is taken by the forehead in the radial direction, i.e. the structure deformation due to the surge is equal to the forehead radial deformation. In this regard, the designs do not provide for the braking field points transposition on a substantial distance.

In the design of a drum with the internal brakes, the load emerging under their influence applies to the pipe shell extension. Since the value of the pipe shell and the forehead bending stiffness is many times lower than the forehead radial stiffness, the brake load becomes the key parameter in the calculation of the braking field points motion distance. Under such conditions, the expected drum points motion distances exceeded the predicted ones significantly that is peculiar to narrow resettable drums under the influence of the internal brake. Consequently, thrust support rollers freezing and the increase of the brake actuator stroke occurred, cracks in the rim and forehead appeared. Quite often the designers had to increase the drum rigidity by means of various types of constructional reinforcements emergently. Thus, the foreheads with reinforcements in the form of radial fins with the curved shape of outer edge and triangular gussets were installed on the drum initially. Afterwards the design was changed as follows: a box-like section

partition was attached to the drum edge, and a plate stiffener was welded to the rim. Further, the design modifications were made. In one case, the stiffening ring connecting the pipe shell with the rim disc was installed; under the stiffening ring of the drum the fins with the width of 50 mm were placed. In the other variant the plate stiffeners near to the central part of the forehead applied.

Thus, taking into account the effect of the drum points motion on the substantial distances under the hoists with internal brake operation, the designers were forced to increase the constructional drum stiffness by means of introducing the additional reinforcements, such as box-like section edges, a partition, a plate rim stiffener, a stiffening pipe shell ring, fins under the stiffening ring on the pipe shell, and plate stiffeners near to the forehead central part. Unfortunately, these actions were performed without a clear understanding of how this complex structure would work. In addition, the construction's parameters were not calculated properly being the evidence of chaotic attempts to reinforce the drum. Under these circumstances, the constructional modification was performed on the hoist assembled under the ground. This approach helped to achieve the necessary drum stiffness, but this construction with the overestimated weight was irrational.

Considering the use of non-metallic removable rope tread aimed at increasing the winding surface of the drum, we should take into account the usage of the hoist unit in difficult geological conditions often requires an increase in the drum rope length. This can be achieved through the increase of a rope diameter and the winding drum surface, and the rope pitch. A metal wear plate with spiral slots fitting the new parameters of the winding surface is used traditionally. The plate is assembled and welded to the hoist drum.

The disadvantages of using the metal plate are as follows:

– reduction in the drum rigidity due to the welded joints of the metal plate especially in the maximum cross-bending stress zone;

– absence of coordination in the drum pipe shell and the metal plate operation leading to additional shear stress in the drum;

– significant increase in loads that affects the hoist joints under the influence of the additional weight of the metal plate;

– difficulties in replacing of the previously mounted plate in the operational need.

In addition to the metal plate, the old design hoists were also equipped with the ones made of high-impact valuable sorts of timber that is now banned. Wooden plates are now used abroad. However, polymeric materials are used for this purpose in Sweden.

Domestic designers suggested using compressed wood, secondary polyamide and other plastics for production of the plates for sheaves and drums.

For example, V.V. Franchuk, candidate of engineering sciences, developed the combined removable plate made of elastic elements that have different values of the elasticity coefficient (polyamide and rubber).

Operating experience of the drums equipped with the elastic plate helped to formulate the basic technical requirements to the material. It must be wearproof (immediate plate shall not break the mine hoist operations down), sufficiently stiff and sound (to ensure high quality laying of the rope into the drum lining groove), lightweight, simple for mechanical processing, low cost, providing permissible contact stress together with the rope that increases its lifetime.

There are two trends in the design of rope drums in contemporary mine hoists, namely:
– foreign manufacturers produce drums with a heavy pipe shell (up to 160 mm) and without reinforcements;
– domestic companies produce drums with a thin pipe shell (55 mm maximum), but use different types of reinforcements.

Due to the small thickness of the pipe shell, the drums of a domestic production are more competitive. Their weight and metal content are approximately 2 – 2.5 times lower. However, the necessity to use reinforcements in the hoist drum degrades its performance characteristics.

M.A. Rutkovsky and K.S. Zabolotny developed a semi-empirical method that provides the use of an analytical model of the axisymmetric shell during calculating the profiled pipe shell bend of the mine hoist drum to solve this problem. This model has equivalent mechanical characteristics. In particular, there are factors of coordination with empirical data. Simulation of the machine design as a discrete-continuous interacting system allows defining the dependence between its parameters that allows achieving the optimal design (Zabolotny & Rutkovski 2012).

Upgrading of the braking system is another current trend in the designing of mine hoists, since braking is the most dynamically dangerous mode of mine hoists operation. This applies to both the operating and safety shutdowns.

A.P. Nesterov, E.S. Traube, A.G. Stepanov, V.I. Belobrov, N.G. Garkusha, A.N. Shatilo, L.V. Kolosov, A.A. Belotserkovski and other native scientists made a great contribution to the solution of this problem. They developed regulatory requirements and provided the means to insure the operational safety of a hoisting unit in case of a breakdown based on the results of the hoisting dynamics represented as a multimass system with elastic couplings study. Moreover, the traditional calculation methods and theoretical study of the brake system dynamics to determine their rational parameters fall behind practice significantly. Currently the dynamics of a hoisting unit under operating and safety breakdown is not yet sufficiently studied yet, and the methods of calculation of the brake heating up and wearing are not sufficiently developed (Belobrov, Abramovskiy, Samusya 1990).

Traditional radial mine hoists brakes with the force lever transmission were widely replaced with the hydraulic disc brakes in foreign mining companies since 1970s. It is not a tribute to fashion, but numerous undoubted advantages of the disc brakes (Traube & Naydenko 1980).

Some of them are as follows:
– an opportunity to achieve a high-rated automation of hoisting and, consequently, increasing of its productivity;
– hoisting safety improvement through the use of automatically regulated safety braking and high stability of braking characteristics;
– significant reduction in the hoisting downtime caused by the brake system breakdowns due to its high reliability;
– improvement of the braking quality and associated reduction in dynamic loads that ultimately increases the lifetime of the main hoisting unit elements (ropes, grips, a mechanical part of the hoist, etc.);
– significant increase in performance by reducing the idle time (approximately by 3 times) and the action time of the device;
– improvement in the indicator of the overall braking device reliability in case of failure form 50% (typical for the post brake) to 75% (typical for the disc brake) that increases the operational safety of the hoisting unit in general;
– obvious increase in accuracy of a vehicle shutdown as a result of good brake performance, and a consequent accuracy of the hoisting cycle execution;
– wide standardization of numerous elements of this brake type that makes them suitable for different types of a hoisting machines. It is very convenient for major mining companies possessing several hoists;
– higher maintainability due to the presence of many elements;
– opportunity to implement the gravitational descent mode in conjunction with the automation system that allows to lower the vehicle to the bottom level through partial braking under gravity in case of a dump in the actuator of a hoisting unit;
– higher performance index, ease of setting-up and regulation;

– mutual compensation of forces pressing the brake shoes to the disc that helps to avoid the impact of radial forces on the drum and the forehead;

– higher compactability and lower weight of the disc brake compared to the radial brake (in terms of equality of braking torques), and less response time, high performance and regulation accuracy of a unit;

– availability of flat surface of the brake shoe that is less susceptible to thermal deformation, and therefore, the braking field is cooled well through convection and emission;

– almost complete avoidance of the brake disc thermal expansion influence on the braking effect.

At the same time, the use of the disc brake cannot ensure safe braking, if the single-stage systems are used, so it is reasonable to provide more complex braking systems subjected to automatic control (Vasil'yev 2012).

3 CONCLUSIONS

Design parameters of the mine hoists have undergone significant changes. To solve certain tasks connected with safe hoisting, designers have implemented the following technical ideas:

– the placement of braking devices inside the drum;

– the downsizing the reversible drum by replacing one of the hubs with thrust support rollers, and using foreheads, pipe shells and rims with a box-shaped reinforcements;

– the use of a removable non-metal rope tread (plate) to increase the winding surface of the drum aiming at improving the rope performance while hoisting;

– the use of a semi-empirical method to determine the parameters of the mine hoists for their optimal design;

– the ensuring the variable pitch of the rope winding;

– the replacement of the radial drum brakes with the disc brakes.

The hoists designing analysis shows a tendency of increasing the winding speed and load-carrying capacity as well as reducing the metal content and downsizing of these installations resulting in improvement of operating and safety braking. Currently a lot of foreign and domestic mine hoists manufacturers look to the development and implementation of disc systems that provide automatically regulated safety brake.

Determination of the modern trends in the development of hoists control systems being the complex electromechanical systems suggests the priority of integrated drive control systems as well as application of the principles of fuzzy logic to ensure the certain modes of safety brake.

REFERENCES

Belobrov, V.I., Abramovskiy, V.F., Samusya, V.I. 1990. *Brake Systems of mine winders*. Kyiv: Naukova Dumka: 174.

Traube, E.S. & Naydenko, I.S. 1980. *Braking devices and safety of mine winders*. Moscow: Nedra: 256.

Vasil'yev, V.I. 2012. *Substantiation of rational dynamic parameters of the safety braking of mine hoist plants*. Ph.D. thesis in Engineering Science. Sumy: SumDU: 211.

Zabolotny, K.S. 1997. *Scientific substantiation of technical-ray solutions to improve the cable length and reduce the size of mine hoist machines with cylindrical drums*. Thesis Doctor in Engineering Science. Dnipropetrovsk: NMU: 325.

Zabolotny, K.S. & Rutkovski, M.A. 2012. *Semi-empirical method of constructing generalized parametric model shell drum mine hoist machines*. Naukovyi Visnyk Natsionalnoho Hirnychoho Universytetu: 88 – 92.

Environmental aspects of waste management on coal mining enterprises

A. Gorova, A. Pavlychenko & S. Kulyna
National Mining University, Dnipropetrovsk, Ukraine

O. Shkremetko
St. Petersburg Energy Institute, St. Petersburg, Russia

ABSTRACT: The peculiarities of coal mining enterprises influence on the state of environment objects was analyzed. The results of comprehensive analysis of the levels of waste dumps ecological danger are given. The complex of measures targeted on the efficiency improvement of coal wastes treatment on mining enterprises and also on the reduction of their negative influence on the environment.

1 INTRODUCTION

The issues of accumulation and recycling of coal mining enterprises wastes become more and more urgent with every passing day. The average content of kibbled waste rock is $110 - 150$ m^3 per ton of coal (for deep mining). Each thousand tones of cleaned coal is associated with $100 - 120$ m^3 of waste dumps. As a result of long-term coal mining, 8 milliard tons of wastes was accumulated as lying on the Earth surface and representing danger for the environment. There are more than 900 waste dumps, and a quarter of them is burns actively or has the hotbeds of fire. The waste dumps embrace an area of over 150 000 ha (The National... 2012, Petrova 2002, Phillip Pack 2009, Saksin & Krupskaya 2005).

Most waste dumps are located on open surface and thus the bed is affected by the environmental factors, particularly by atmospheric precipitation, pressure and temperature difference, air mass circulation etc. As a result, the gullies, chasms and gaps appear, and the waters containing solution of chemical components stream down the waste dump surface. The climatic changes cause the processes of physical alteration, decomposition and biological weathering of beds. The loss of solid matter as fine-dispersed particles is observed. The waste dump is gradually transformed from the form of separate pile of bed into the environmental component, a source of negative influence on the behavior of physical and chemical, biological and ecological processes on the site (Koskov et al. 1999, Krupskaya et al. 2014, Lubchik & Varlamov 2002).

The active release of deleterious substances from the waste dumps causes the upset of the ecological balance on waste dump territories. The uncontrolled changes of the waste dumps state, caused by external and internal factors, are the serious problem too. We should point out that about 80% of waste dumps passed the stage of spontaneous ignition and burning or are currently burning. In case of waste dumps open burning the carbon oxides and carbon dioxides, nitrogen oxides, sulfur oxides and other contaminants are emitted into the atmospheric air. (Zborshchik & Osokin 1990, Matveeva 2007, Zborshchik et al. 1989).

The water streaming down the waste dumps causes the chemicals migration, so the content of the elements in the surface layer of waste dumps is mainly homogenious and includes the following elements: manganese, copper, zinc, lead, nickel, and cobalt. Their quantity differs significantly in different coal mining regions. The water, streaming down the waste dumps, percolates through the soil and penetrate the ground waters, and change their chemical composition.

It should be noted that coal mining wastes can be mentioned as the anthropogenic deposit, which is suitable for industrial development. The utilization of industrial wastes enables their quantity decrease and additional extraction of useful components, as well as minimizing the demand for further mining of the definite types of mineral raw materials (Zubova et al. 2012).

The waste dumps on each stage are a source of increased ecological danger for the environmental objects and population health. In addition, the low level of utilization of mining wastes and implementation of recycling technologies are the obstacle in the way of sustainable development of coal industry in Ukraine (Pivnyak et al. 2005). That is the reason for the necessity of the development and implementation of the technologies aimed at minimizing of the waste dumps negative influence on the state of environment components and population health.

2 FORMULATING THE PROBLEM

The storage of coal mining wastes goes with the sequential physical and geochemical processes as well as environmental discharge of polluting substances. In such a case the contamination of the atmospheric air, soils, surface and ground water, plants etc. takes place.

The estimation of influence of coal mines waste dumps on the environment has some milestones with the different approaches to the assessment of the environmental threat.

To prevent rocks from spontaneous ignition and control the temperature conditions the temperature survey is carried out. Firstly, the technical documentation is analyzed to study the parameters and indicators of waste dumps. After the project development, the proving grounds are determined and superimposed on the topographic plan. The temperature of the rock surface and beds is measured by means of pyrometers and drill-rods at the depth of $0.5 - 2.5$ meters. If the temperature at the depth of 2.5 meters is higher than 80°C, the waste dump burns, and the extinguishing measures should be taken in this case.

The concentration of carbon oxide (CO), sulfuric anhydride (SO_2), nitrogen oxide (NO_2, NO_3), and hydrogen sulfide (H_2S) is also measured.

In addition to the methods mentioned above, the estimation of the mining wastes toxicity level is carried out. The procedure is based on the ecological and hygienic assessment of the wastes composition and properties. The study of the normality of the rock properties from the point of ecological and hygienic standards is a background of the process. This requires the following analyses:

– the estimation of the waste dumps radiation hazard;

– the laboratory examination of the acute toxicity of pulverized coal on rats and the determination of the hazard levels by means of the indicators of peroral and lethal effect;

– the substantiation of the danger class of coal mining wastes based on the maximum concentration limit of polluting substances in the soil;

– the qualitative assessment of the possible presence of high-toxic substances, notably compounds of heavy metals moving fractions, which are able to move in the environment;

– providing conclusions about the toxicity level of coal mining wastes.

But such an analysis of the rocks toxicity properties sets aside the reaction products in the waste dumps and appearing in the environment in the form of toxic gases, toxic dust, hot steams, acid flow from the descents surface, as well as the increase of radionuclide activity and moving activity of heavy metals.

It should be pointed out that the waste dumps are the dangerous objects of anthropogenic origin with the complicated internal transformations. The nature and intensity of the chemical reactions in the waste dumps and on their surface reason the danger level of coal mining wastes as related to the environment and population living not far from the place of storage. The matters is complicated by the absence of clear and reasoned criteria of assessment of the ecological risks levels for the environmental objects (Gorova et al. 2013, Kolesnik et al. 2012, Borisovskaya & Fedotov 2014).

Thus there is a need for the development of the integrated methodological and theoretical research base, which could enable the high degree of assessment reliability of the environmental objects pollution levels in the mining regions. Thus, the specific researches are also needed to develop the new and affordable technologies, providing the decrease of environmental pollution in the areas of mining wastes storage without the substantial increase of prime cost of coal mining.

3 GENERAL PART

The study was conducted on the territory of Chervonograd mining region (ChMR). The coal has been worked for 50 years in the region. Although the coalfield occupies the small territory, there are seven operating coal mines and two closing ones. Central Concentrating Plant (CCP) is also situated in the region. The anthropogenic load caused the transformation of the landscapes and their components, the formation of anthropogenic land forms and transfer of large fertile land territories for the storage of coal mining enterprises wastes. Three mines are currently closed in the region (No 1 "Chervonohradska", "Visean" and No 5 "Velykomostivska"). "Zarichna" mine is going into liquidation. So, the ecological conditions in the ChMR, as well as in other mining regions in Ukraine are very difficult.

Almost 561.5 ha of lands in region are influenced by mining enterprises. Waste dumps cover 137.5 ha, which is about 25% of mining lease territory. But it should be noted that the waste dumps of the liquidated mines of No 5 "Velykomostivska" and "Bendiuzka" were not taken into consideration.

Visual examination of waste banks enabled the determination of yellow, white and heat source zones. As usual, it is possible to distinguish the local zones of rocks oxidation on the waste banks edges as there is no temperature rise here. The heat source zones are characterized by vapor liberation comprising water, sulfuric acid, carbon acid gas, and nitrogen dioxide. The combustion process changes the rocks mineralogical makeup both in the heat sources and out of them. There are specific zones around the sources, which contain rocks of the colors form natural to brown, covered with yellow sulphate deposit. In the zones distant from primary rocks there were the white areas representing the progress of white sulphate mineralization.

During 50-year coal mining in ChMR the mining enterprises accumulated more than 37 m tons of wastes as waste dumps, which cause considerable environmental pollution.

As a result of study of the soils, adjoining the waste dumps, it was defined that the soils acidity was between 4.6 and 7.4. The pH level means that the soils vary from acidic to alkaline types in different points of sight. It is known that in acidic recrements with pH 4.0 – 5.5, the minerals, containing manganese, zinc, chromium, lead and other elements, are dissolved by sulphuric acid and move to the adjoining territories via surface flow. The high level of soil acidity has a bad influence on plants, as such elements as iron, aluminum and manganese, move into readily accessible for plants fractions, and their concentration can reach the toxicity level. The excess of such metals breaks carbohydrate and protein metabolism of plants, which reduces their crop capacity.

The high level of soil acidity deteriorates its filterability, capillarity and penetrability. As for the other results of qualitative assessment of soils state, it should be noted that the soils, adjoining the waste banks, are also contaminated by sulphates. The concentration of water-soluble sulphate-ion was between 518 and 4219 mg per kg, thus, the maximum permissible concentration was 3.2 – 26.4-fold exceeded. It should be noted that the manganese and cobalt content was according to the standard for the mine "Velykomostivska" except some points of sight. The chloride content in the samples from the mines denoted the salinity, which has also a bad influence on plants.

Analyzing the results of radioassay of equivalent dose rate of external gamma radiation on the ChMR waste dumps, it was stated that permissible level was not exceed.

The process of waste dumps spontaneous ignition is a serious threat for the region. According to the official statistics the waste dumps in the ChMR do not burn, but the temperature survey, carried out in 2014 (December), showed the heat sources in waste dump of "Chervonohradska" mine; the average temperature at a depth of 0.5 m was 75°C; the atmospheric temperature was +2°C. In recent years, the zones with a temperature of 128°C were observed in the inactive waste dump of the closed "Visean" mine.

The region accumulates not only the coal mining wastes, but also the coal cleaning wastes. The waste dump of CCP occupies the largest territory among the waste dumps; it has an area of 89 ha and is 68 m in height. The waste dump is harmful for the environment as the heavy metals content exceeds the maximum permissible value (MPV) significantly. E.g.: the excess for lead is 45.5-fold (MPV is 6 mg per kg), for copper – 81.3-fold (MPV is 3 mg per kg), for nickel – 19.8-fold (MPV is 4 mg per kg), and zinc – 31-fold (MPV is 2 mg per kg (Baranov 2008). Moreover, the heat sources are observed too.

The deformation processes, observed on the waste dumps of mines, are characterized by the formation of gullies in different waste bank zones; the rocks move down through them at a distance of up to 6 m; the width of gullies range from 2 to 4 m; the depth – from 1 to 3.5 m.

The average composition of waste dumps is as follows: 67% of argillites, 20% of siltstone, 10% of sandstone, 3% of coal and coaly rocks (sapropelites). The rock is represented by fragment and grained components of red, grey and black rock with fragment size of 150 – 200 mm. The 61% of waste dumps of red and grey colour is comprised of noncombustible rock (Knysh 2006).

More than 70% of wastes in waste banks are comprised of argillites with the property of heavy metals entrapping (*Li, V, B, P, Zn, Pb, Bi, Co*), as well as *Hg* and *As* (as a result of high content of sulphide sulphur (pyrite). The high content of sulphur in the waste dumps promotes the formation of aureoles of acidic waters at the foots of waste banks. In addition, the waters, accumulating at the foots of waste banks, are comprised of calcium, magnesium, iron, fluorine, manganese etc (Knysh 2006).

The results of chemical analysis of water from the foot of the waste bank show its unsatisfactory state. The studies demonstrate that the migration of harmful substances from the waste bank surface takes place; further it is preceded by groundwater flow and causes the contamination of soils and groundwater.

To evaluate the toxicity of water from the foot of the waste bank we used the growth analysis with the application of Allium cepa L bioindicator. The advantage of the test is that it allows the assessment of water-soluble components of water assays. The test is easy to carry out and responsive to the general water toxicity. The level of inhibition of Allium cepa L. rootlets growth is an indicator of toxicity, as it known that the process is inhibited in case of less concentration of toxicant than plants germination (Rudenko et al. 2003).

Twelve test tubes were prepared for each water sample, and then they were filled with 25 ml of water specimen. The prepared and peeled onions were placed on the top of each test tube in such a manner that the bulb stem touched the liquid. The water in the test tubes was changed every day. After two days of the experiment the two bulbs with the shortest rootlets were removed from each variant. The experiment lasted for 72 hours. Then the rootage length of 10 bulbs was measured (the shortest and the longest rootlets were not taken into consideration). So, 12 bulbs were couched for each sample; from 4 to 30 rootlets of each onion were measured. The truth of each trial was approved by triple reproduction for 48 test tubes on all study stages.

The results of the research were processed by means of mathematical-statistical analysis; the arithmetical mean accuracy and Student's coefficient were also calculated.

Phytotoxic effect (*PE*), namely degree of oppressing of growth processes was determined in percents in relation to control (by mass and by sprouts or roots length of test-culture) according to the formula (Bilyavskiy & Butchenko 2006):

$$PE = \frac{m_o - m_x}{m_o} \cdot 100\ \%, \tag{1}$$

where m_o – mass or length of sprouts (root or above-ground part) in control; m_x – mass or length of sprouts in variants of research.

The following estimation scale was used for the determination of water samples toxicity by means of growth analysis, Table 1 (Rudenko et al. 2003).

Table 1. Scale of assessment of water toxicity level.

Phytotoxic effect, %	The toxicity level
0 – 20.0	No or weak toxicity
20.1 – 40.0	Average toxicity
40.1 – 60.0	Above average toxicity
60.1 – 80.0	High toxicity
80.1 – 100.0	Maximum toxicity

The results of study of morphological changes of phytoindicators under the conditions of water toxicity assessment in 2014 are tabulated in the Table 2.

Table 2. The indicators of phytoindicator Allium cepa L rootage length, which was raised on the water assays sampled at the foots of waste banks, 2014.

Sampling place	Average length of rootage, $x \pm m$	The truth of trial, t
"Velykomostivska" mine	1.75 ± 0.1	3.55
"Chervonohradska" mine	1.59 ± 0.08	4.98
"Lisova" mine	0.33 ± 0.02	16.3
"Zarichna" mine	1.76 ± 0.09	3.61
"Nadiya" mine	2.18 ± 0.1	0.78
Mine No 5 "Velykomostivska"**		
– waste dump with fused rock (red)	0.61 ± 0.03	13.98
– waste dump with unfused rock (black)	0.36 ± 0.05	15.06
CCP	*	*
Check (settled tap water)	2.3 ± 0.12	—

* the growth processes are not observed for all samples; ** closed mine

The obtained data demonstrated the proven toxic effect of water from the foot of the waste bank on the growth processes of Allium cepa L rootlets for all samples. The samples from "Nadiya" are the exception to the rule. It is established that all the processes of growth of bioindicator rootlets are inhibited in all points of sight; this denotes that water from the foot of the waste bank has toxical properties. The highest levels of rootage inhibition were observed for the samples from the foot of CCP waste dump, as there were no growth processes in the samples. As for the other samples, we can state that the levels of rootlets growth inhibition of the bioindicator were 7 – 1.1-fold different from the

check sample ones. So the highest level of growth processes was observed for the samples of water from the foot of the waste bank from "Nadiya" mine, and the lowest – for "Lisova" and No 5 "Velykomostivska" mines. As for the samples from "Nadiya" mine, they were not different from the check samples, as the growth rate in the samples was the same as for the check samples.

It should be noted that the development of herbage of indicator plants from samples was almost not observed. The results of the determined toxicity of water from the foot of the waste bank, sampled at the foots of waste dumps from Chervonograd mining region in 2014 are tabulated in the Table 3.

Table 3. The toxicity levels of water from the foot of the waste bank, 2014.

Sampling place	The levels of rootage inhibition of bioindicator, %	The toxicity level
"Velykomostivska" mine	24	average
"Chervonohradska" mine	31	average
"Lisova" mine	86	maximum
"Zarichna" mine	18	low
"Nadiya" mine	5	low
Mine No 5 "Velykomostivska"		
– waste dump with fused rock (red)	73	high
– waste dump with unfused rock (black)	85	maximum
CCP	100	maximum

As a result of data analysis (from Table 3), it was stated that the value of phytotoxic effect range from "low" for "Lisova" and "Nadiya" mines to maximum for "Lisova", No 5 "Velykomostivska" mines and CCP. The toxicity level is proved by the chemical diagnosis of water from the foot of the waste bank.

So, the waste dumps belong to the most hazardous anthropogenic objects of the territories of mining regions in Ukraine. They cause permanent contamination of the natural environment objects. The study enables us to distinguish a range of ecological risks for the biota, including humans, on the territories of mining wastes storage (Figure 1).

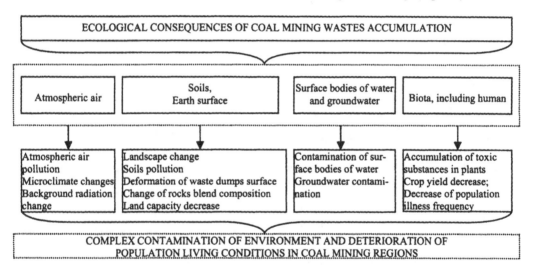

Figure 1. Ecological consequences of coal mining wastes accumulation on the earth surface.

It should be noted that wastes of coal mining enterprises are not only a source of negative influence on the environment, but also the considerable raw material for different branches of national economy. That is the reason for the researches of the ways of wastes utilization in the production process of construction materials for industrial engineering, reconditioning of roads and for goaf stowing of mining enterprises (Buzylo et al. 2014, Knysh 2006). As a result of waste dumps recycling, the coal, chemical elements and other associate mineral deposits can be attained.

In order to increase the efficiency of management of coal mining enterprises wastes the following items should be provided:

– the count of the inventory and certification of coal mines and dressing plants wastes;

– the development and implementation of innovative assessment for contamination risks of environment for the purpose of the protection of population health in the coal mining regions;

– the improvement of sanitary rules for the treatment of coal mining wastes;

– the decrease of the environment objects contamination risks of the territories of wastes storage by means of technological land restoration with the application of naturally occurring materials;

– the creation of safety shields for the protection of the territories, adjoining the waste dumps, from the leakage of the contaminated water into the soils and groundwater;

– the improvement of the technology of ecological danger class assessment of the coal mining wastes by the application of highly-sensitive methods of bioindication and biotesting;

– the determination of danger class according to the standard and innovative methodology;

– the development of electronic passports of wastes location, which contain the information about the waste dumps parameters, useful components content, as well as the levels of environment objects on the adjoining territories;

– the development of wastes detoxing technology for wastes of coal mining industry using the naturally occurring materials;

– the processing of waste dumps, as anthropogenic deposits, by means of useful components extraction.

4 CONCLUSIONS

As based on the carried out researches, the peculiarities of external and internal physical and chemical transformations in the waste dumps of coal mines are studied. It was distinguished that the waste dumps comprise high-level danger for environment objects and population health without reference to the stage of existence.

As a result of bioindicative tests the levels of the toxicity for the water sampled at the foot of waste dumps of coal mines, both operating and liquidated.

The complex of measures targeted on the decrease of negative influence of waste dumps on the environmental state and population living conditions in the coal mining regions.

The results of long-term studies enables the decrease of truth of trial for the assessment of contamination levels of environmental objects in coal mining regions and the decrease of the efficiency of the treatment system of coal mining enterprises wastes.

REFERENCES

Baranov, V. 2008. *Ecological scope of rock dump coal mines CCM CJ-SC "Lvivsistemenergo" as subject to plant trees and grass*. Bulletin of L'viv University. Biological Issue, No 46: 172 – 178.

Bilyavskiy, G.O. & Butchenko, L.I. 2006. *Elements of ecology: the theory and the practice*. School-book. Libra: 368.

Borisovskaya, Ye.A. & Fedotov, V.V. 2014. *Improvement of the method of danger class definition of coal-mining*

solid wastes, Naukovyi Visnyk Natsionalnoho Hirnychoho Universytetu, No 3: 130 – 137.

Buzylo, V., Savelieva, T., Saveliev, V. & Morozova, T. 2014. *Utilization of the waste products for stowing*. Progressive Technologies of Coal, Coalbed Methane, and Ores Mining. The Netherlands: CRC Press/Balkema: 31 – 34.

Gorova, A., Pavlychenko, A. & Borysovs'ka, O. 2013. *The study of ecological state of waste disposal areas of energy and mining companies*. Annual Scientific-Technical Collection – Mining of Mineral Deposits. The Netherlands: CRC Press/Balkema: 169 – 171.

Knysh, I.B. 2006. *Prospect of utilization of wastes of coal industry of Lvov region as a new mineral raw material*. Bulletin of L'viv University. Geology, No 20: 111 – 123.

Kolesnik, V.Ye., Fedotov, V.V. & Buchavy, Yu.V. 2012. *Generalized algorithm of diversification of waste rock dump handling technologies in coal mines*, Naukovyi Visnyk Natsionalnoho Hirnychoho Universytetu, No 4: 138 – 142.

Koskov, I.G., Dokukin, O.S. & Kononenko, N.A. 1999. *Conceptual Foundations of Ecological Security in the Regions of Mines Closure*. The Coal of Ukraine, No 2: 15 – 18.

Krupskaya, L.T., Zvereva, V.P., Bubnova, M.B., Chumachenko, E.A. & Golubev, D.A. 2014. *Environmental monitoring of ecosystems under the impact of gold and tin mining wastes in the Far Eastern Federal District*. Russian Journal of General Chemistry. Pleiades Publishing Ltd, No 13: 2616 – 2623.

Lubchik, G.L. & Varlamov, G.B. 2002. *Resource and environmental issues of global and regional energy consumption*. Energy and Electrification, No 9: 35 – 47.

Matveeva, N.G. 2007. *The possibility of application of international experience in the processing of rock dumps of coal mining-processing industry in the Donbass coal region, Ecology*. Visnik of the Volodymyr Dahl East Ukrainian National University, No 1: 35 – 40.

Petrova, L.O. 2002. *The environmental impact of coal mining and coal waste processing*. Geological Journal, No 2: 81 – 87.

Phillip, Pack. 2009. *Risk Assessment in Donieck Basin: Closure of Mines and Waste Heaps*. Arendal: 21 – 22

Pivnyak, G.G., Pilov, P.I., Bondarenko, V.I., Surgai, N.S. & Tulub, S.B. 2005. *Development of coal industry: The part of the power strategy in the Ukraine*. Mining Journal, No 5: 14 – 18.

Rudenko, S.S., Kostyshyn, S.S. & Morozova, T.V. 2003. *General ecology: a practical course*. Ruta: 320.

Saksin, B.G. & Krupskaya, L.T. 2005. *Regional estimation of the effect of mining production on the environment*. Mining Journal, No 2: 82 – 86.

The National Report about the State of Environment in Ukraine in 2011. 2012. Kyiv: Ukraine Ministry of Ecology and Natural resources. LAT&K: 258.

Zborshchik, M.P. & Osokin, V.V. 1990. *Prevention of spontaneous combustion of rocks*. Kiev: Technique: 176.

Zborshchik, M.P., Osokin, V.V. & Panov, B.S. 1989. *Mineralogical features of sedimentary rocks that are prone to spontaneous combustion*. Kiev: Development of mineral deposits, No 83: 92 – 98.

Zubova, L.G., Zubov, A.R., Verekh-Belousova, K.I. & Oleinik, N.V. 2012. *Production of Metals from the Waste Piles of Donbass Coal Mines*. Lugansk: VNU im. V. Dalya: 144.

Theoretical and Practical Solutions of Mineral Resources Mining – Pivnyak, Bondarenko & Kovalevska (eds)
© 2015 Taylor & Francis Group, London, ISBN: 978-1-138-02883-8

Research and development of software and hardware modules for testing technologies of rock mass blasting preparation

A. Kopesbayeva & A. Auezova
Almaty University of Power Engineering & Telecommunications, Almaty, Kazakhstan

M. Adambaev & A. Kuttybayev
Kazakh National Research Technical University named after K.I. Satpayev, Almaty, Kazakhstan

ABSTRACT: Training of highly qualified personnel at Higher Educational Institutions will be incomplete if students (bachelors, masters and doctors) don't have the opportunity to test modules of technological and production processes, as they would not be able to determine the structure and parameters of these processes via control channels and estimate the equation of relationship of input and output coordinates graphically/visually, consequently, such personnel cannot operate (manage) production, much less create and design new more efficient technologies and control systems thereof. Modern supervisory control and data acquisition systems (SCADA) – allow developing virtual models of different complexity most closely resembling actual technological and production processes, controller management of such processes, visualization and opportunity of testing parameters changes of technological processes in real time mode for use in corresponding training researching in major technical specialties at HEI. This article considers the research of complex structured blocks in quarry conditions, development of software hardware complexes for determination of internal structure of debris and their rock processing characteristics, parameters of drilling and blasting operations, and visualization in real time mode of rock mass with different blasting parameters based on modern SCADA systems.

1 INTRODUCTION

Many works are dedicated to the research of various aspects of open cut complex structured mining of non-ferrous metal ore, noble and rare metals. They emphasize the role of blasting operations in achievement of required breaking quality and arrangement of aggregated ore in debris. However the relationship between initial geological and morphological data, parameters of blasting charges location, technology of blasting operations and their final result in these sources is not yet disclosed which requires corresponding research.

During the blasting operations in quarries fragmentation of rock mass takes place which leads to the transformation of internal structure of the section. Depending on parameters of blast-hole charges placing within the mass, sequence of their initiation and blasting method different results can be achieved in the specified aspect.

Effective exploitation and rational utilization of mineral resources is based on identification of a consistent pattern of natural resources disposition within the mass and debris, operating control of processes that ensure stable quality of ore mining. In connection therewith development of software

hardware modules with integrated SCADA system, based on parameters of charges placing within the complex structured block, aggregated rock disposition of complex structured blocks within the debris taking into account their mine processing characteristics remains relevant (Rakishev et al. 2013).

2 MAIN PART

2.1 *Method description*

Today the issue of development of an automated system for the calculation of blasting operations parameters and engineering of large-scale blasts, as well as automated data analysis systems on the structure and mechanical properties of the breakable medium is quite relevant. For that purpose the system of computer modeling of mining technology facilities has been developed in the Unity Pro software environment and on the basis of this system, a computer technology of automated calculation and designing of open large-scale blasts of any complexity, based on modeling of mining technology facilities, interactive graphics, computerized means of processing results modeling and generation of design drawings and technical documentation. Main advantage of this technology is the systematic approach in solving geological, surveying and technological tasks, providing implementation of software designing means within the consolidated information area of mining enterprise.

Software package can be a part of automated control system of technological process, scientific experiment, production process etc. Software can be installed on computers. Programmable logic controllers PLCs, input-output drivers or OLE (Object Linking and Embedding) for Process Control/Dynamic Data Exchange (OPC/DDE) servers are used for connection with the facility. Program code can be written in programming language, as well as generated within the designing environment.

This software package solves the following tasks:
– data exchange between "facility connection devices", (i.e. with controllers and input/output boards) in real-time via drivers;
– data processing in real-time;
– logic control;
– information display on monitor in a simple and user-friendly form;
– real time database management with technological information;
– alarm signaling and alarm messages control;
– preparing and generation of reports on the technological or production workflow;
– providing communication with external applications (database control systems, electronics worksheets, word processors etc.).

In the enterprise management system such applications most often are applications referred to the level of Manufacturing Execution System (MES) – specialized application software intended to solve tasks of synchronization, coordination, analysis and optimization of product output within a production framework (Adambaev & Auezova 2013a).

2.2 The analysis of recent researches

The performed analysis of scientific and technical literature and experience of foreign mining enterprises show that during exploitation of complex structured deposits by open-cut mining not only high duty work of excavation and loading, transporting and breaking machinery depend on the results of blasting preparation for the excavation but also quantitative and qualitative loss of natural resources. Blasting preparation of complex structured blocks for excavation should meet the following requirements:
– provide required technological parameters of shot pile;

– provide even qualitative shattering of blasted rock mass, which predetermines the scope and quality of complex selection;
– provide minimal shattering mixing during the blast of various sorts of ore and with barren;
– to shape the ore body (inclusions) within the debris suitably for isolated excavation incurring the least qualitative and quantitative loss.

Therefore, the objective of this work is development of software and hardware complexes to design parameters of drilling and blasting operations, visualization of debris models of blasted rock mass under different blasting methods in real-time; forecasting distribution of aggregated rock of complex structured block in debris. Working out of recommendations on quantitative and qualitative loss minimization during development of complex structured blocks. Moreover, as a rule, it is inaccessible to watch real experiment at production filed, while there is an opportunity to restart the same experiment several times on a virtual stands in real-time with changed parameters and compare the results.

Despite unyielding searching of scientists and production workers no adequate alternative to the drilling and blasting method of preparation for half-rock and rock formations extraction at mining enterprises have yet been discovered. First of all this is stipulated by its indisputable technological and cost effectiveness.

At the same time the issue of shattering quality control of rock mass under the conditions of more common vertical bore-hole charges remains relevant. A need for scientifically based choice of parameters for placing of explosive within the blasting mass, and technologies of their blasting fully corresponding to the natural properties of the rock mass, and forecasting technological parameters of blasted rocks is arising. The above specified tasks can be solved with the help of gradual rock bench breaking method when blasting bore-hole charges, developed by B.R. Rakishev (Rakishev 2006).

Explosive calculation is carried out based on the maximum filling up principle of the bored hole volume with the explosive. Construction of charges and its shape significantly affect the duration of explosive pulse effect on a mass. When applying decked charge with air gap, blocking of detonation products of the main charge takes place, which increases the duration time of blast effect on to a mass. However there are no scientifically based recommendations on this matter.

Application of charges of variable diameter allows for more even distribution of explosive energy within the blasting mass. However no working constructions of such drilling means have yet been developed.

During delay-action blasting bore-hole charges are blasted in a sequential order with set delay intervals.

Effectiveness of delay blasting application depends on the adequate choice of delay interval, parameters of bore-holes spacing and selection of initiation pattern. The interval of decked charge blasting is one of the most crucial parameters of blasting operations. Adequately selected delay interval allows improving rock breaking, decreasing seismic impact of the blast, obtaining desired profile of rock mass debris, preserving geological structure of the mass to minimize loss and contamination of ore, lowering the air blast and cutter break into the rear area of the mass. The interval of delay action blasting effect on different blasting indices is interdependent. Thus, when the blasting energy release into the seismic wave is lower, the quality of breaking is higher (Rakishev et al. 2014).

Empirical dependencies contain coefficients, application of which gives only approximate values of the interval. The disadvantage, to our opinion, is that many authors define delay intervals in isolation from blasting results, i.e. scope of destruction, energy release into the seismic wave etc., which, first of all, depend on the parameters of the pattern of holes, physical and mechanical properties of the rock.

The performed analysis proves that the problem of parameters substantiation of bore-hole charges disposition within the mass requires thorough theoretical working out. Practically it has been discovered that for higher efficiency, convenience and accuracy of experiment conduction it is more reasonable to use possibilities of virtual stand, as virtual experiments allow replacing physical experiments, while at the same time keeping virtual process in line with the real technological or production process to the limit.

Taking drilling and blasting (D&B) parameters calculation method as a basis, based on accounting of physical and mechanical rock properties and detonation characteristics of explosive, rational parameters of D&B can be established without resort to real experiment. Whereas this will allow improving training of highly qualified personnel at Higher Education Institutions (Rakishev et al. 2013).

Integrated SCADA-system includes a complex of software means for organization and conduct of automated measurements, processing and presentation of measurement data, storing measurement results in the database, generation of reports on research results, visualization of the process under research in real-time.

Software and hardware complex of automated design system of D&B calculation is demonstrative, easy and has a user-friendly interface. Program of the automated D&B design system allows quick calculation of parameters for D&B for bore-holes drilling passport/certificate and designing of large scale blasts under certain production conditions, as well as to visualize D&B results and generate recommendations on minimization of qualitative and quantitative loss.

Opportunities of integrated SCADA-systems are vast, except for libraries of different devices applied directly to a software environment; they also enable creating facilities randomly of any complexity, programing principles of various process operation and virtual launching in real-time, taking into account all factors of the process under research.

2.3 Algorithm description

Block-diagram of the program shown on Figure 1.
Program specification:
1. Start.
2. Input of initial data: ρ_0 – rock density, kg/m^3; c – sound velocity in rock, m/s; v – Poisson's ratio; σ_c – ultimate compressive strength, Pa; σ_t – ultimate tensile strength, Pa; ρ_{ex} – explosive density, kg/m^3; D – detonation velocity, m/s; H – blasting block height, m; d_0 – borehole diameter, m; p – capacity of borehole length unit, kg/m.
3. Calculation of: P_c – environment strength characteristic, Pa; P_i – initial pressure of detonation products, Pa.
4. Calculation of: $\overline{r_{lim}}$ – relative maximum radius of cavity; r_{lim} – maximum radius of cavity, m; r_2 – radius of fine breaking area, m; r_1 – radius of radial fracture area, m.
5. Calculation of: W – line of resistance along bench toe, m; a – distance between holes, m; a_d – distance between columns of holes, m; h_l – loaded length above grade of bench, m; l_2 – length of uncharged section of hole, m; l_h – length of hole sub drilling, m.
6. Calculation of: l_1 – charge length in a borehole, m; l_d – borehole depth, m; h_{int} – length of interval between parts of the charge, m; n – number of charge units.
7. Calculation of: Q – mass of charge in a hole, kg; u – borehole walls travelling velocity, m/s; τ – slowing down time, s; q_c – specific charge of explosive, kg/m^3.
8. Calculation results and bench visualization are displayed.
9. End.

2.4 Program interface description

Designing of Automated process control systems using integrated SCADA-system Unity Pro (Schneider Electric), supposes use of project browser, which is intended for developing project's structure and mathematical base for data processing and control.

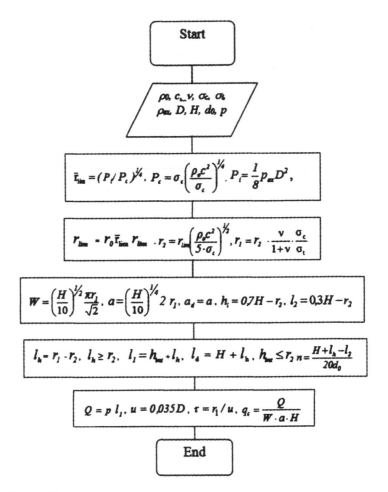

Figure 1. Block-diagram of the program.

Use of data presentation processor is designed to develop graphical component of the project. Application of latest information technologies and program products, which enable successful use of applied approaches in industrial facilities identification of various classes with algorithms of their implementation for obtaining not only qualitative but quantitative models of facilities under research, increases control effectiveness of complex industrial facilities.

Unity Pro – is a software environment for configuring, programming, debugging and diagnosing of executive system of Modicon industrial controllers produced by Schneider Electric: Modicon M340, TSX Premium (including Atrium) and Quantum. Uniform software environment – "all in one", 5 languages of International Electrotechnical Commission standard (IEC) 61131-3, embedded adaptable Derived Function Block (DFB), simulator of Programmable Logic Controller PLC on a personal computer for programs debugging, embedded diagnosis, complete set of online services (Adambaev & Auezova 2013b).

2.5 Computer aided automated design of D&B program

A project in the Unity Pro environment should be created, then a configuration of Modicon M340 controller with PLC BMX P34 2020 processor, a discrete module DDM 3202K should be added into the 1 slot, discrete module DDO 3202K – into slot 2, analogue module AMI 0410 – into slot 3, analogue module AMO 0410 – into slot 4. Controller configuration window is shown on Figure 2.

Reset of words on internal memory %MWI at "cold" start.

Number of internal bits % M – 512.

Number of words on internal memory %MW – 1024.

Number of constant words %KW – 256.

188

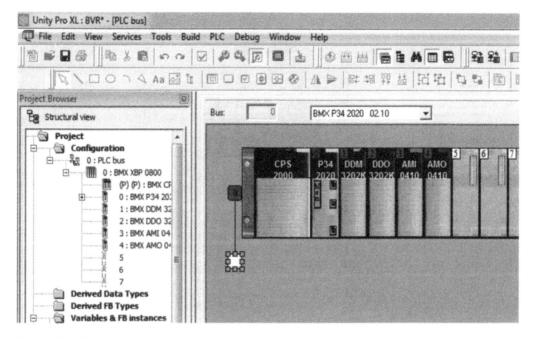

Figure 2. Controller configuration window.

Calculation is made based on example of diorite porphyry of Sarbaiskoe minefield.

All data is entered into the symbolic table of variables shown on Figure 3.

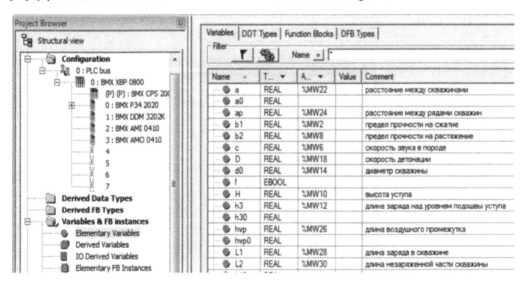

Figure 3. Symbolic table of variables.

Next window on Figure 4 shows working filed of the program of automated D&B designing.

Once the program is started next window is opened and corresponding data entry should be performed: rock density ρ_0, kg/m^3; sound velocity in the rock c, m/s; Poisson's coefficient v; ultimate

compression strength σ_c, Pa; ultimate tensile strength σ_t, Pa; density of applicable explosive; ρ_{ex} kg/m^3; detonation speed D, m/s; bench height H, m; borehole diameter d_0, m; capacity of borehole length unit p, kg/m. Data input window is shown on Figure 5.

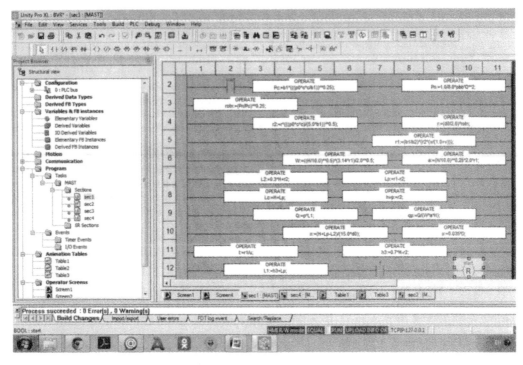

Figure 4. Working filed of the program in LADDER (LD) language.

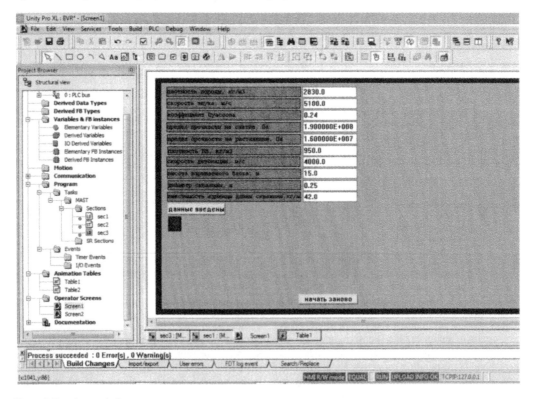

Figure 5. Data input window.

Once all data entry is complete it can be changed, or calculated, by pressing corresponding button "Data entered". As a result we see a program window displaying all results; contours of the mass are drawn in compliance to the calculating parameters. After that calculation can be made again by pressing "start again" option or any data can be changed, shown on Figure 6.

Data can be entered via animation table. Displayed results are also registered in animation table and are changed in real-time as shown on Figure 7.

Figure 6. Results display window.

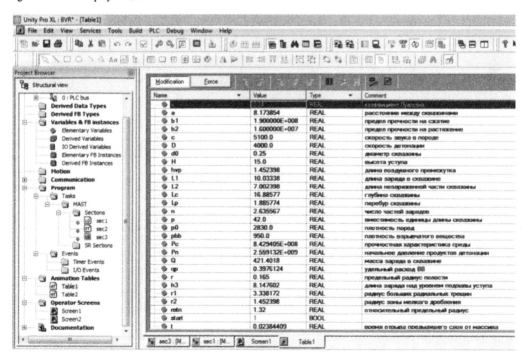

Figure 7. Animation table of results display.

Test results agree with the results of control experiment which proves program operating efficiency and its applicability.

Numerous scientific and technical literature addresses mostly only some specific issues and is limited by general reasoning based on complex mathematical arguments and calculations, which are difficult to follow by the students, and do not provide any specific recommendations regarding practical application thereof, as well as any demonstration of real production processes and impact of some parameters changes in real-time onto the end results (product). In this regard, development of virtual models in the integrated SCADA-system, simulating real technological and production processes, controller management thereof, visualization in real-time is a necessary requirement when researching modern processes.

3 CONCLUSIONS

1. Research and development of software package of drilling and blasting operations parameters designing, visualization of rock mass models under various blasting parameters in real-time.

2. Forecasting displacement of aggregated rocks of a complex structured block in the debris.

3. Recommendations development on minimization of qualitative and quantitative loss when mining complex structured blocks.

REFERENCES

Adambaev, M.D. & Auezova, A.M. 2013a. *Use of the Programmable Logical Controllers in Studies Technical Specialties of Higher Education Institutions.* Materials of the International Conference "Modern Science: Problems and Perspectives". Las Vegas, April 15. International Center for Education & Technology: 337 – 339.

Adambaev, M.D. & Auezova, A.M. 2013b. *Programming controllers and visualization in the software environment Unity Pro.* Materials of the 11th International Scientific Conference "Information Technologies and Management". Riga, April 18 – 19: 74 – 76.

Rakishev, B.R. 2006. *Automated designing of parameters and large scale blasts results in quarries.* Almaty: KazNTU: 110.

Rakishev, B.R., Auezova, A.M., Kuttybayev, A.Y. & Kozhantov A.U. 2014. *Specifications of the rock massifs by the block sizes.* Naukovyi Visnyk Natsionalnoho Hirnychoho Universytetu, No 6 (144): 22 – 27.

Rakishev, B.R., Rakisheva, Z.B., Auezova, A.M. & Daurenbekova, A.N. 2013. *The regression models of different block-sizes rock massifs.* Bulletin of Kazakh National Technical University, No 6 (100): 104 – 110.

Theoretical and Practical Solutions of Mineral Resources Mining – Pivnyak, Bondarenko & Kovalevska (eds)
© 2015 Taylor & Francis Group, London, ISBN: 978-1-138-02883-8

The level of atmospheric pollution around the iron-ore mine

I. Myronova
National Mining University, Dnipropetrovsk, Ukraine

ABSTRACT: Calculation of ecologically harmful substances diffusion into the atmosphere from ventilation shaft of iron-ore mine that situated within city is executed. Isolines of near-the-ground concentrations of total impact of ecologically harmful substances around contamination source that change with distance increasing from ventilation shaft on exponential dependencies are given. Empiric formula that allows to determine near-the-ground concentration of total impact of ecologically harmful substances with taking into account explosive ratio and distance from emission source is established.

1 INTRODUCTION

Mining industry of Ukraine has powerful mining potential and take one of the leading place between European Union countries on iron-ore extraction. Ukraine is entering to the number of leading mineral raw countries of the world, that occupies only 0.4% of land and has 5% of mineral raw potential in own interiors. In Ukraine surveyed 49 iron-ore deposits with total stocks of 28 billion tons that is equal to 8 – 9% of the whole world ones.

The most significant iron-ore deposits of Ukraine are following: Kryvorizhskyi iron-ore basin, Kremenchuhskyi iron-ore region, Bilozirskyi iron-ore region, Prydniprovskyi iron-ore region, Odesko-Bilotserkivskyi iron-ore region and Kerchenskyi iron-ore basin. At present time, iron-ore deposits of Kryvorizhsko-Kremenchuzka zone (basin) are mined by underground and open-pit methods and Bilozerskyi iron-ore region is mined by underground method.

With the purpose of decreasing negative influence on environment during blasting, all quarries are transited on emulsion explosives. It is well-known that during blasting of one kilogram of emulsion explosives into the air release only 20 liters of gas, that is almost in 14 times less than during usage of trotyl-contain explosives. On underground operations in iron-ore mines usage of without trotyl-contain explosives is equal to 3%, that because of development of technology of underground mining operations conducting and charging machines for emulsion explosives usage. That is why nowadays on Ukrainian iron-ore mines is still using trotyl-contain explosives. During blasting of 1 kilogram of trotyl precipitates 275 liters of toxic gases in the air (Pozdnyakov &

Rossi 1977). After blasting conducting, that connected with underground mining operations, waste air releases into the atmosphere through ventilation shafts without clearing. It is related to the fact that at present moment does not exist effective devices and treatment facilities for gas collecting and cleaning, that release onto the surface in significant volumes. Depending on mines and ventilation shafts situation, return ventilation air negatively impacts on environment objects, human health and biota that grow on contiguous to enterprise area (Saarkoppel 2007, Dzhuvelikyan 2002).

Deterioration of nature quality condition in the places of mining enterprises location causes necessity of search ways and method for negotiation of negative consequences of people intrusion in nature systems functioning. That is why for increasing ecological safety level of underground mining processes of iron-ore extraction it is necessary to establish regularities of dispersion of iron-ore releases dispersion into environment.

The main purpose of work is examination of atmosphere air condition in places of underground extraction of iron-ore enterprises within the city. That is why industrial site of mine "Nova" LLC "Vostok-Ruda" and contiguous to it area is actual polygon for air condition research conduction.

2 MEASUREMENT OF ECOLOGICALLY HARMFUL SUBSTANCES CONCENTRATION

Zhowtorichenske iron-ore deposit, where working "Nova" mine LLC "Vostok-Ruda" situated in the northern part of Kryvorizhskyi iron-ore basin within the city Zhovti Vody, Dnipropetrovsk region

(Khomenko et al. 2011). The main sources of atmosphere air pollution are following: grinding-sorting factory, backfilling, auto transporting, railway shops, boiler house and mine. In 2000 together with enterprise "Ecouniversal" was provided inventarization of all emission sources. It is allowed to decrease releases on 1165 tons per year into the atmosphere. On grinding-sorting factory has been installed dust cleaning installation (DCI) of wet cleaning. For cemented dust collecting has been installed hose filters in backfilling complex, boiler house has been transited from oil fuel to gas. That is why significant quantity of ecologically harmful substances brings by used airflow from ventilation shaft of mine.

During iron-ore mining on "Nova" mine LLC "Vostok-Ruda", the used airflow releases into the atmosphere through the ventilation shaft "Severnaya-Drenazhnaya". Absence of effective equipment and treatment facilities for ecologically harmful substances collecting from the used airflow of mine atmosphere which releases onto the surface in big volumes with significant velocity lead to harmful substances can easily come out into the environment and cause disprove it quality.

Measurement of ecologically harmful substances concentration release from ventilation shaft of mine has been executed by express-method of physics-and-chemical analysis. Sampling of analyzed air realized in inter-shift break after blasting conducting at mine. Through the gas-analyzer "Paladii-3M" and gas detector GH-M measured carbon and nitrogen oxides concentrations.

Gained results of ecologically harmful substances concentrations in collected samples of mine atmosphere during 2009 – 2013 are represented in Tables 1 and 2.

Table 1. Results of ecologically harmful substances concentration measurement in outgoing airflow of main fan channel.

Gas	Harmful substances concentration, mg/m³				
	2009	2010	2011	2012	2013
CO	3.5	12	5	4	13
NO_x	0.5	3	1	1	2

Table 2. Calculation results of emission intensity.

Year	Fan productivity, m³/s	Air motion velocity, m/s	Emission intensity			
			CO		NO_x	
			g/s	kg/h	g/s	kg/h
2009			0.684	2.46	0.098	0.35
2010			0.492	1.77	0.123	0.44
2011	131.16	5.69	0.680	2.45	0.136	0.49
2012			0.542	1.95	0.135	0.49
2013			1.705	6.14	0.262	0.94

For establishment of ecologically harmful substances concentration interaction in collected air samples and indices on mine over a period of 2009 – 2013 were collected data on annual productivity, total and specific charge of explosives. Information is represented in Table 3.

Table 3. Annual productivity, total and specific charge of explosives on "Nova" mine over a period of 2009 – 2013.

Year	Annual productivity, t/year	Annual overall consumption of explosives, kg	Specific charge of emissions, kg/ton
2009	82000	33700	0.41
2010	390000	160884	0.41
2011	540000	248941	0.46
2012	645000	364465	0.56
2013	670000	327800	0.49

Harmful substances concentration in outgoing airflow of main fan channel allowed to establish empiric formulas of ecologically harmful substances concentration from annual specific charge of explosives.

Carbon oxide concentration:

$$C_{CO} = 1395 \cdot q^{6.7}, \tag{1}$$

with the approximation at $R^2 = 0.903$, where q – specific charge of explosives, kg/m³:

$$C_{NO_x} = 405 \cdot q^{7.6}, \tag{2}$$

with the approximation at $R^2 = 0.9725$.

3 RESEARCH OF ATMOSPHERE AIR CONDITION ON THE TERRITORY OF IRON-ORE MINE

For determining specialties of ecologically harmful substances dispersion into the atmosphere from ventilation shaft was used automatization calculation system of atmosphere pollution "EOL-2000[h]". In the base of software laid rates that has been settled by methods of harmful substances concentrations calculation in atmosphere air "OND-86" (Calculation methods… 1986).

In calculations were used following initial data: coefficient of atmosphere stratification $A = 200$, coefficient of land topography $\eta = 1$, average maximum air temperature in the most hot month of the year is equal to 26.9°C, average maximum air temperature in the most cold month of the year is equal to -5.6°C, annual average wind speed is equal to 10.5 m/s. Calculation polygon is like a square with side size 8000 m, in the middle of which is situated emission source. In calculation square all objects are marked by topographic signs. Emission source: ventilation shaft "Severnaya-Drenazhnaya" – height 5 m, diameter of exit hole 4.7 m, airflow rate 131.16 m³/s and air temperature is equal to 26.9°C. Harmful substances: carbon oxide – maximum allowable concentration (MAC) is equal to 5 mg/m³, hazard class – 4, settlement factor 1, exponentiation coefficient is equal to 0.9; nitrogen oxide – MAC is equal to 0.085 mg/m³, hazard class – 2, settlement factor 1, exponentiation coefficient is equal to 1.3.

According to calculations results of were built isolines that characterized near-the-ground concentration of total impact of harmful substances from ventilation shaft (Figure 1).

Data analysis, which represented on Figure 1, a allowed to establish that main isolines of near-the-ground concentration of total impact of ecologically harmful substances around ventilation shaft of mine in a radius of 355 m and is equal to 0.12 unit fractions (u.f.) from MAC and decrease to 0.02 on a distance of 2450 m from emission source. Decreasing of near-the-ground concentration happens after each 200 – 750 m on 0.01 – 0.04 unit fractions of MAC.

The main isolines of near-the-ground concentration of total impact of ecologically harmful substances around ventilation shaft in a radius of 355 m and is equal to 0.14 u.f. from MAC and decrease to 0.02 on a distance of 2450 m from emission source (Figure 1, b). Decreasing of near-the-ground concentration happens from 200 to 750 m on 0.01 – 0.04 u.f. of MAC.

Isolines of near-the-ground concentration of total impact of ecologically harmful substances around ventilation shaft in a radius of 355 m and is equal to 0.16 u.f. of MAC and decrease to 0.02 on a distance of 2450 m from emission source (Figure 1, c and d). Decreasing of near-the-ground concentration happens from 200 to 750 m on 0.01 – 0.05 u.f. of MAC.

The main isolines of near-the-ground concentration of total impact of ecologically harmful substances around ventilation shaft in a radius of 355 m and is equal to 0.32 u.f. from MAC and decrease to 0.04 on a distance of 2450 m from emission source (Figure 1, b). Decreasing of near-the-ground con-

centration happens from 200 to 750 m on 0.03 – 0.1 u.f. of MAC.

Total picture of near-the-ground concentration value changing of ecologically harmful substances with increasing distance to emission source can be observed on changing its concentration in u.f. of MAC (Figure 2).

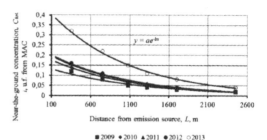

Figure 2. Isolines of near-the-ground concentrations of total impact of ecologically harmful substances around ventilation shaft.

As a result of maximum values approximation conducting were received empiric equations of dependencies u.f. of MAC of total impact from distance to emission source.

Value of near-the-ground concentration of total impact:
– in 2009:

$$C_{tot.i} = 0.15 \cdot e^{-0.001 \cdot L}, \tag{3}$$

with the approximation at $R^2 = 0.9602$, where L – distance to emission source, m;
– in 2010:

$$C_{tot.i} = 0.19 \cdot e^{-0.001 \cdot L}, \tag{4}$$

with the approximation at $R^2 = 0.9751$;
– in 2011:

$$C_{tot.i} = 0.22 \cdot e^{-0.001 \cdot L}, \tag{5}$$

with the approximation at $R^2 = 0.9958$;
– in 2012:

$$C_{tot.i} = 0.25 \cdot e^{-0.001 \cdot L}, \tag{6}$$

with the approximation at $R^2 = 0.9871$;
– in 2013:

$$C_{tot.i} = 0.45 \cdot e^{-0.001 \cdot L}, \tag{7}$$

with the approximation at $R^2 = 0.9964$.

Executed analysis of near-the-ground concentrations values of total impact of ecologically harmful substances allowed to establish that with increasing distance to 2450 m from emission source concentration values decrease in 6 – 8 times.

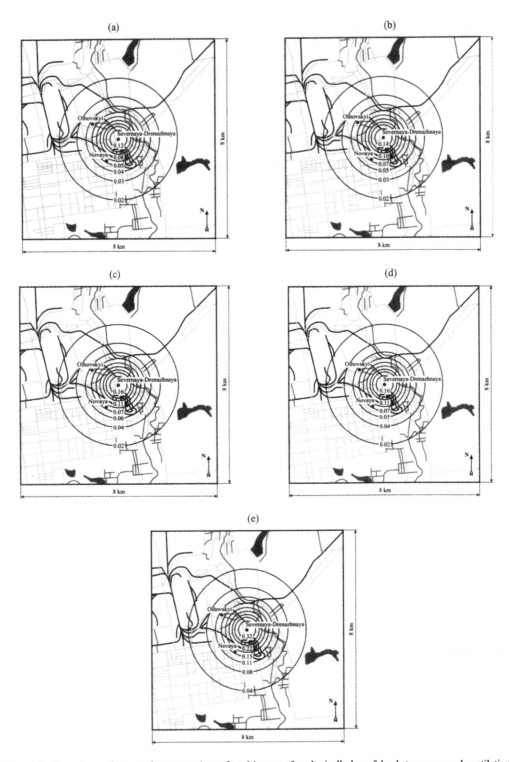

Figure 1. Isolines of near-the-ground concentrations of total impact of ecologically harmful substances around ventilation shaft of mine in 2009 (a), 2010 (b), 2011 (c) 2012 (g) and 2013 (e).

The main factor that influences on concentration values of ecologically harmful substances is annual specific charge of explosives.

Therefore, near-the-ground concentration of total impact $C_{tot.i}$ with increasing distance to emission source changes on exponential difference:

$$C_{tot.i} = a \cdot e^{-b \cdot L}, \tag{8}$$

where a and b – given numbers that contain known values.

Each value changes from specific charge of explosives on following regularities:

$$a = 3 \cdot q^{3.4}, \tag{9}$$

$$b = 0.001. \tag{10}$$

To substitute equations (3) – (7) in expression (8) and executed necessary transformations we will get empiric formula that determined near-the-ground concentration of total impact of ecologically harmful substances with taking into account specific charge of explosives and distance to emission source:

$$C_{tot.i} = 3 \cdot q^{3.4} \cdot e^{-0,001 \cdot L}, \tag{11}$$

with the approximation at $R^2 = 0.9034$.

4 CONCLUSIONS

Determining of near-the-ground concentration values of total impact of ecologically harmful substances allowed to establish that near-the-ground concentration of harmful substances has been influenced by specific charge of explosives and distance to emission source.

Further research allowed to establish that near-the-ground concentration of total impact of ecologically harmful substances with distance increasing to ventilation shaft "Severnaya-Drenazhnaya" "Nova" mine LLC "Vostok-Ruda" and specific charge of explosives change on exponential dependence.

Therefore, established regularities prove necessity of development and implementation of measures complex with direction on decreasing of harmful substances concentration in mine atmosphere of iron-ore mines.

REFERENCES

Calculation methods of concentration in atmosphere air of harmful substances that contain in enterprise emissions. "OND-86". 1987. Leningrad: Gidrometeoizdat: 76.

Dzhuvelikyan, Kh.A. 2002. *Role of iron-ore industry in environmental pollution by hard metals.* Ecology and industry of Russia: 26 – 29.

Khomenko, O.E., Kononenko, M.M., Vladyko, O.B. & Maltsev, D.V. 2011. *Ore mining in Ukraine.* Dnipropetrovsk: National Mining University: 288.

Pozdnyakov, Z.G. & Rossi, B.D. 1977. *Reference book on industrial explosives and explosive items.* Moskow: Nedra: 253.

Saarkoppel, L.M. 2007. *Comparison estimation of health condition of ore industry workers.* Occupational medicine and industry ecology, No 12: 17 – 22.

Theoretical and Practical Solutions of Mineral Resources Mining – Pivnyak, Bondarenko & Kovalevska (eds)
© 2015 Taylor & Francis Group, London, ISBN: 978-1-138-02883-8

Integrated sustaining of technogenic mine structures

M. Nurpeisova, D. Kirgizbaeva, K. Kopzhasaruly & A. Bek
Kazakh National Research Technical University named after K.I. Satpayev, Almaty, Kazakhstan

ABSTRACT: The problems of controlling the stability of man-made mine structures presented in the form of open pit and underground workings in the development of mineral deposits in Kazakhstan. Recommendations to improve their resistance to the development of activities and production monitoring tool. Relevance of research is characterized by an increase in depth of field development and the transition to the combined production of minerals from the open method on underground.

1 INTRODUCTION

Problems of dynamics of the Earth's crust due to the engineering activity, one way or another connected with the extensive development of mining operations and increase their depth. In the short term, up to 70% of ore quarries will have a depth of more than 250 m, and in some cases from 300 to 500 m.

The overall evaluation of the stability of benches, pit, waste dumps should be borne in mind that in the real world they are exposed to a large number of factors that can be grouped into two categories: natural and mining engineering. The combination of these fac tors determines the strength of the rock mass and the conditions of its deformation.

Existing methods of assessment and justification of the nature of stress-strain state of based on traditional classical basis, but with increasing depth of mining operations engineering-geological situation is complicated and is accompanied by a qualitatively new manifestations of violating the safety of mining operations and the environment.

In recent years, a significant step in addressing the sustainability of mining slopes. But despite the significant achievements of domestic and foreign scholars in the field, the problem is still not solved because of the complexity and diversity of geological features of deposits at this stage.

2 MAIN CONTENT

In assessing the interaction of rocks with different structures, become important structural features of the rock mass, as fracture strength and the forecast of their properties. The work is focused on the study of patterns of change in the strength properties of rock masses with their depth and development on this basis of new technological means to ensure the sustainability of buildings.

Office of the stability of rock slopes of various engineering structures including quarries have been engaged as separate domestic and foreign researchers, and the whole organization (Fissenko 1965, Turintsev et al. 1984, Galustyan 1980, Popov & Okatov 1980, Mashanov 1961).

But despite the significant achievements of domestic and foreign scholars in the field, the problem is still not solved because of the complexity and diversity of geological features of deposits. Summarizing the analysis of issues related to the stability of the pit walls, it is possible to chart the research slopes of mine ranges in order to develop techniques and methods for the assessment and management of their stability (Figure 1).

It is from this position of the goal, substantiated the idea, formulated research problems. Experimental facilities and the introduction of the study of objects and the challenges of sustainability quarry slopes were deposits of the Republic of Kazakhstan. Methods of study of fracturing of the rock mass are divided into direct and indirect. Among the indirect methods for prediction of fracture array become widespread methods.

On the basis of the known theoretical propositions developed reliable ways of assessing the disturbance of rocks that will resolve the issue of operational control of their condition and sufficient for the solution of problems of geomechanics detail perform zoning career fields by a factor of disturbance. For large-scale studies of disturbance of rock masses, the definition of the boundaries and sizes of homogeneous and identification of potentially unstable areas benches and pit is proposed to use thermal-metric method comprising registering the intensity distribution of the thermal radiation of the controlled object, and based on the relationship we have identified the parameters of the thermal radiation of the rock mass and the characteristics of its violation.

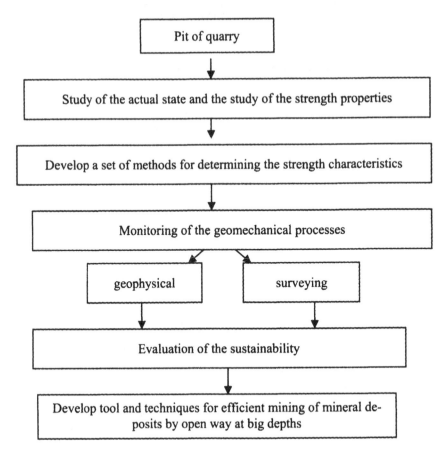

Figure 1. Scheme of studies.

The method differs from similar other ways to improve accuracy and speed measurement with simplicity and usability, while providing multi-drop and simultaneous measurement processes. For registration of the radiation intensity and determination of thermal-metric properties of rocks we have developed a device, a block diagram of which is shown in Figure 2.

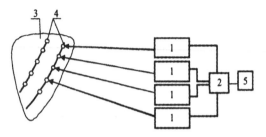

Figure 2. Functional diagram of the device for certain violations of the array: 1 – Thermometers; 2 – microprocessor; 3 – mine range; 4 – hole; 5 – output device registration.

The results of measurements in the quarries Akzhal deposit showed that the observed feature of the difference in the nature of the cooling sections, depending on the fragmentation of rocks, in general, conserved (Figure 3). In the initial period of fractured land cools faster than the monolithic array. Then fractured portion cools slower monolithic, which is due to the fact that the deep layers of the fractured portion of the array retain more heat due to the smaller coefficient of thermal conductivity of the surface layer. Thereafter, the area under the curves from the reference point to the line of intersection of the curves and the temperature difference of the areas. The maximum value of the difference of areas served to establish the potentially unstable areas of rock mass (Patent No 20700).

The test results of integrating a number of schemes to determine the areas under the curves 1, 2, 3 have shown that the use of them in a manner convenient to automate the processes of measuring and machining data obtained. The value of the crack opening is set by the developed ultrasonic method comprising sounding controlled object and the definition of the transmission coefficient (A) of the elastic waves through the object.

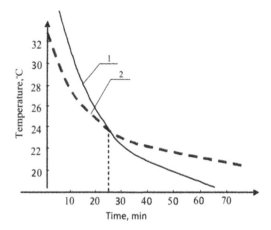

Figure 3. Graph of temperature changes in the rock mass during the cooling: 1 – the process of cooling small cracks area array; 2 – the process of cooling strong cracks area array.

The value of A is determined by a monotonic increase in the amplitude of the emitted ultrasonic signal to A. In Figure 4 is shown a functional diagram of an ultrasonic method.

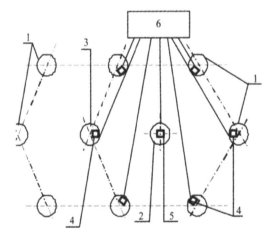

Figure 4. Functional diagram of an ultrasonic method: 1 – wellbore; 2 and 3 – the top corner of the hexagon; 4 – receivers of ultrasonic waves; 5 – emitter of ultrasonic waves; 6 – multi-microprocessor.

Application of the method makes it possible to simultaneously six dimensions, which consequently accelerates research massif and completeness of the study.

According to the results of the acoustic sounding rock samples obtained dependence shown in Figure 5.

When the ultrasonic wave passes through a crack having a width comparable to the wave length, then the change in amplitude of the transmitted wave.

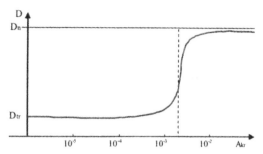

Figure 5. Change the value of the transmission coefficient of elastic waves (D) on the amplitude of the transmitted signal (A).

By an abrupt increase in the transmission coefficient D ultrasound trial of a crack in the controlled object, and the amplitude of the emitted ultrasonic wave A corresponding to an abrupt increase in the transmission coefficient, determine the magnitude of the crack opening (λ) by the formula:

$$\lambda = A \cdot e^{-\delta l_0}, \tag{1}$$

where A – amplitude of the radiated ultrasonic waves; δ – the coefficient of the ultrasonic attenuation in the sample; l_0 – the distance from the point of excitation of the ultrasonic wave to crack.

The value of the crack opening λ can be expressed in units or in units of length. In this case, A is determined by the product of the peak reading voltmeter U and K coefficient conversion radiator. The novelty of the developed methods confirmed by the patents REPUBLIC Kazakhstan (Patent No 2007/1031).

One of the main stages of the study the stability of slopes quarries was to study the physical and mechanical properties of rocks.

The interrelation between the velocity of longitudinal ultrasonic waves V_0 by a factor of rock strength on M.M. Protodyakonov f. To determine the velocity of propagation of longitudinal waves in the laboratory, we have prepared samples of a separate, different types of rocks, taken from different horizons at the studied fields. Based on a detailed analysis based on observations, as well as literary sources other ore deposits we have established graph-analytic dependence of the velocity of longitudinal ultrasonic waves V_0 with the strength of rocks f and this relationship is expressed by the following formula:

$$f = 0.45 V_0^2, \tag{2}$$

where: V_0 – velocity of longitudinal ultrasonic waves is expressed in km/s.

Figure 6 shows the dependence for determining the coefficient of the fortress of the greater speed of 1.5 km/s for rocks of any composition.

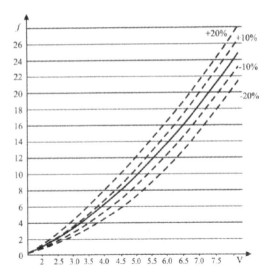

Figure 6. Graphic analytical dependence to determine the coefficient of the fortress f rocks velocity V of ultrasonic waves.

The postulated relationship f and V_0 puts on a real basis for the decision cottages associated with the study of the spatial distribution of the strength properties of rocks, the dynamics of their changes under the influence of natural factors and the impact of Engineering.

The level of stress R_m, corresponding to the ultimate strength of the rocks in the array is an important strength characteristic of rocks. Tensile strength of the rocks in the array is determined by the formula:

$$R_m = \sigma_c \cdot \lambda_c, \tag{3}$$

where: σ_c – rock strength in compression in the sample; with λ_c – coefficient of structural weakening.

With the aim of finding common patterns of variability of the strength properties of rocks in the array method of mathematical statistics and correlation analysis summarizes the properties of the host rocks and deposits Akzhal set of graph-analytical relationships between density, adhesion, strength and depth of rocks.

Figure 7 shows the results generalize the experimental work on the definition of change of the linear dimensions of the structural units of the depth of their occurrence in the bowels by quarries, analytical dependence of which are shown in Table 1.

To find the common patterns of variability in strength properties of rocks, summarizes a number of fields, and also established Graphic analytical relationship between the average density, adhesion, strength of rocks and their depth (Figure 8 and Table 2).

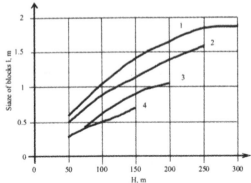

Figure 7. Change in the average size of the building blocks of rocks with their depth by quarries: 1 – Akzhal; 2 – Sayak; 3 – Zherek; 4 – Rodnikovoe.

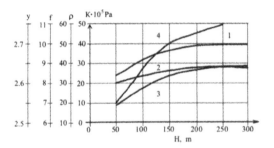

Figure 8. Dependence of the strength properties of massive limestone with their depth at Akzhalskom career. 1 – clutch, k; 2 – angle of internal friction, ρ, 3 – strength, f, 4 – medium.

Curves of changes in the properties of rocks carried by the average indicators in depth in 50 m. Qualification and reliability dependency determination made by the formulas of mathematical statistics. To create a rational observation station geomechanical analysis was performed, which included zoning and prediction of stress-strain state of the massif. Their stability is analyzed by solving the problems of the theory of elasticity, obtained by numerical methods, mainly using the finite element method.

Table 1. The relationship between the size of the structural changes in the wall rocks with the depth of their occurrence.

Quarries	Equation	Value of reliability	The limits of the aleph
Akzhal	$l_{cp} = 0.025 + 0.012\,H - 0.00002\,H^2$	$r_k = 0.75$	$300 < H < 50$
Sayak	$l_{cp} = 0.23 + 0.006\,H - 0.000002\,H^2$	$r_k = 0.87$	$250 < H < 50$
Zherek	$l_{cp} = 0.12 + 0.005\,H - 0.00001\,H^2$	$r_k = 0.79$	$200 < H < 50$
Radnikovsky	$l_{cp} = 0.11 + 0.004\,H$	$r_k = 0.90$	$200 < H < 50$

202

Table 2. The constraint equations properties of rocks with their depth.

The study of quantity	Equation	Value of reliability	Limits of action
Grip on crack k, Pa·10^5	$k = 14.5 + 0.2\,H - 0.0004\,H^2$	0.88	$300 < H < 50$
The angle of the internal its friction r, degree	$\rho = 25.5 + 0.1\,H - 0.0002\,H^2$	0.90	$250 < H < 50$
Castle rock	$f = 6.15 + 0.018\,H - 0.00003\,H^2$	0.89	$300 < H < 50$
Density g, t/m^3	$\gamma = 2.36 + 0.0038\,H - 0.000008\,H^2$	0.88	$250 < H < 50$

Zoning massif with the layering, fracturing and faulting of the array revealed six stretches of (three on the north, and three on the south side) with similar conditions of stability. Length of all the breaking strain on the front, as a rule, exceeds the height of the deformed shoulder 2 – 5 times. This indicates that the effect (clamping) on the stability of the lateral ledges of rocks that are in the limit position, is effective only as long as the length of the disturbed portion exceeds its height. Therefore, the inception of the workers of frames along the contour career allows you to monitor the stability of the beads throughout their length, and established network is used as a basis for the expansion of the observations.

Investigation of the distribution of isolines durability massif career "Akzhal" in the observed depth shows (Figure 9), that it really is possible to allocate the prism of the collapse, in which the most actively occurring deformation and disintegration processes. With increasing depth increases the risk of caving in ledges. Forming surface starts sliding upwards.

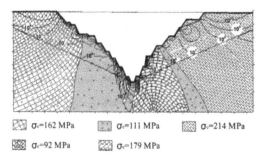

σ$_c$=162 MPa   σ$_c$=111 MPa   σ$_c$=214 MPa

σ$_c$=92 MPa   σ$_c$=179 MPa

Figure 9. Distribution contour longevity massif on the career of "Akzhal".

Of course, that the findings are specific to career "Akzhal". Time education and the sliding surface configuration depends primarily on the nature (strength of the rocks composing with bead array of fracture, the presence of faults and the others). And mining-technical factors (depth of reference works, the angle of slope of career, technology of open cast mining and etc.).

The application of this technique will allow at least a first approximation to assess the stability massif,

given the depth of mining and associated time elapsed since the beginning of the outcrop of rock mass.

Instrumental surveying observations are the primary means of obtaining information about deformations pit and the most reliable basis for the prediction of their stability. Observation, analysis and interpretation of the results of observations can be used: to determine the values of displacements, strains, velocities of deformation process and boundaries of the deformation; set the type of breaking strain massif rocks; establish the relationship between the factors that determine the stability massif, and the process of deformation of the pit and to determine the quantitative relationships between them; determine the critical value of deformation prior to the beginning of the active phase of deformation for various geotechnical rock units; exercise control over the conduct of mining operations on the deformed sections of boards and dumps; determine the effectiveness of against deformation events.

The observations were made with the strong points of the geodetic network laid on the edge of the pit. To improve the accuracy and efficiency of alignment measurement, we made improvements vantage point, namely, the surface of the existing metal plate was welded new item other metal plate with tripod screw, which allowed for fast and accurate alignment, as well as to exclude the use of tripods (Figure 10).

Observations of the absolute deformation boards on the objects under investigation were carried out on the profile lines observation station devices of new generation. Repeated geodetic measurements were made tachometers company Leika TS110 and TS1206 in combination with reflectors installed on a permanent basis, as well as laser tape measures in the inaccessible areas of career (Nurpeisova & Kopzhasaruly 2013). For each profile line drawn sheet vertical and horizontal displacements of the frames, as well as graphs of displacement.

To automatically receive the proposed software package (software package CREDO running operating systems from Microsoft: Windows 95/98, Windows 2000 because at the moment it is the most popular and available operating systems, as well as the program has a modern graphical user interface).

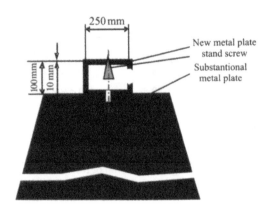

Figure 10. Advanced vantage point.

Results of field observations of the deformation of slopes quarries have shown that this process has a certain period of time within which the behavior of this process remains unchanged. This allows you to select the specified time intervals in separate stages, to formulate their definitions and set the parameters by which they are characterized.

Effective methods for controlling the stability of slopes associated with hardening of the rock mass and dusty surfaces. Created solution to strengthen fractured rocks with a low cost, sufficient fluidity for filling small cracks and adhesion to rocks, high strength. The solution contains cement, filler and water.

The filler used tails OF BGMK. Parallel has been researched and prepared a new composition for strengthening reinforcement strongholds nab lyudatelnoy station in wells, allowing the wastes as well as the mining industry and to increase the strength and frost resistance of the resulting material (Kazakhstan (Patent No 14023, Patent No 14475, Patent No 14476, Patent No 19861).

3 CONCLUSION

Developed new automated methods for isolating weak tectonic disturbances portions of pit and methods of allocation parameters joint systems based on cluster analysis and display them on rectangular diagrams that provides zoning pit according to their degree of disturbance.

Effective methods for controlling the stability of slopes associated with hardening of the rock mass and dusty surfaces. Created solution to strengthen fractured rocks with a low cost, sufficient fluidity for filling small cracks and adhesion to rocks, high strength.

Results of the study to clarify the parameters of stable boards and steps to limit and developed "Guidelines for assessing the stability of the pit walls Akzhal's field" (Bek 2002), which have been introduced into production in the form of regulations.

REFERENCES

Bek, A.S. 2002. *Guidelines for assessing the stability of pit walls Akzhal.* Almaty: KazNTU: 23.

Fissenko, G.L. 1965. *Stability of pit walls and waste dumps.* M: 378.

Galustyan, E.L. 1980. *Management geomechanical processes in the quarries.* M: 237.

Mashanov, A.Z. 1961. *The mechanics of the rock mass.* Alma-Ata: 166.

Nurpeisova, M.B. & Kopzhasaruly, K. 2013. *Automatic deformation monitoring system.* Materials of 10th international conference of young scientists and specialists "Problems of development of mineral resources in the XXI century through the eyes of the young generation". Moscow: IPKON RAN: 116 – 120.

Patent No 14023, RK. *Way to strengthen the shoulder Career.* Nurpeisov, M.B., Bek, A.S., Kyrgizbaeva, G.M. Pub. 05.12.2003.

Patent No 14475, RK. *Composition for strengthening reinforcement in wells at hardening array.* Nurpeisov, M.B., Bek, A.S., Beck, A.S., Kasymhanova, H.M. et al. Pub. 13.04.2004.

Patent No 14476, RK. *The solution is to strengthen the fractured rock.* Nurpeisov, M.B., Bek, A.S., Bek, A.S & Kasymhanova, H.M. Pub. 13.04.2004.

Patent No 19861, RK. *Composition for fixing dusty surfaces of tailings and other objects.* Nurpeisov, M.B. & Bek, A.S. Pub. 26.02.2007.

Patent No 2007/1031, RK. *A method for determining fracture array.* Nurpeisov, M.B., Bek, A.S., Kasymhanova, H.M. et al. Pub. 02.10.2008.

Patent No 20700, RK. *A method for determining rock mass disturbance.* Nurpeisov, M.B., Bek, A.S., Kasymhanova, H.M. et al. Pub. 15.01.2009.

Popov, I.I. & Okatov, R.P. 1980. *Fighting landslides in the quarries.* M.: 239.

Turintsev, Y.I, Polovov, B.D., Gordeyev, V.A. et al. 1984. *Geomechanical processes in open cast mining.* Sverdlovsk: 56.

The development of idea of tunnel unit design with the use of morphological analysis

K. Zabolotny, A. Sirchenko & O. Zhupiev
National Mining University, Dnipropetrovsk, Ukraine

ABSTRACT: Solid-state models of perspective units of tunnelling system complexes were developed, and their mass properties were identified in the study. Sequence diagram of basic operations involved in underground railway tunnelling was composed. Optimum alternatives of units as part of tunnel complexes that require the installation of additional removable modules for hole drilling and tunnelling were selected by means of morphological analysis. Modules can transform into each other and adapt to changes in the rock hardness within wide limits. The unit designed can perform tunnelling by means of shearer mining under the risk of a roof collapse and ground-based facilities destruction.

1 INTRODUCTION

The most efficient method of the underground railway tunnelling is a shield tunnelling method. However, if the linking of the parallel tunnels or the continuation of an unfinished tunnel is required, the mentioned method does not apply. Rock or shearer mining tunnelling applies in the above cases.

Tunnel-boring complexes (TBC) produced by the domestic industry subjected for the work in specific mining-and-geological and economic conditions of metro engineering are not competitive to the foreign counterparts due to poor performance or inability to apply to the drill and blast tunnelling method in soil with low solidity (when $f \leq 6$ according to the Protodiakonov Scale) as well as by the shearer mining method due to the risk of buildings or mine rock collapse.

The suggested technical solution for this problem is as follows:
– to improve the design of TBC separate machines;
– to unitize, i.e. to place several devices on a single trolley.

Scientific research mission is to justify optimal parameters for TBC separate machines and select the rational schemes for their unitization.

The research objective is to improve the efficiency of the underground railway tunnels construction through the development of the conception of tunnel unit design by means of morphological analysis of its characteristics.

To achieve the objective the following scientific missions were set and solved:
– to develop solid-state models of TBC units per-spective in the practical use, and to determine their mass properties.
– to compose sequence diagrams of basic operations involved in underground railway tunnelling.
– to estimate the priority in the application of operations and selection of the parameter types; to determine the numerical values of the indicators in each case of breaking and lining.
– to perform the morphological analysis of models and to define the most perspective TBC units for production.
– to develop the design of TBC capable of working in the tunnels with the rock hardness to vary within wide limits.

The idea of the research is to use the advanced methods of computer simulation and morphological analysis of the devices developed.

The object of the study is the mechanical processes that accompany tunnel construction.

The subject of the study is the determination of the relations between TBC specifications and its design parameters.

2 RESULTS OF THE STUDIES

The efficiency of the construction of the underground tunnels main line directly depends directly on the tunnelling method and specifications of the equipment used. Technologies used to construct tunnels are normally classified considering the actual rock hardness of available and possible limitations in their use (Table 1).

In cases when both rock and shearer mining tunnelling are possible, it is advisable to choose the latter as more efficient.

Table 1. Tunnel construction technology characteristics.

Name of the Technology Based on Restriction of its Application			
Soil Hardness according to Protodiakonov Scale	Risk of Roof Collapse	Risk of Damage to Buildings	No Limitations
0 – 6	Rock tunnelling method	—	Rock tunnelling method Shearer mining method
6 – 10	Rock tunnelling method	Shearer mining method	Rock tunnelling method Shearer mining method
10 – 20	Rock tunnelling method	—	Rock tunnelling method

Construction of main line underground tunnels involves such basic operations as breaking, loading and transportation of soil, and lining elements mounting.

TBC designed for blast hole drilling (BHD) applies in the rock tunnelling. In this case, first face drilling with boreholes by means of the drilling rigs is performed and powder charges are loaded. Then explosion destroying mine rock follows. Afterwards the rock mass is transported with a rock loading machine into trollies and lifted to the surface. Further, permanent lining is mounted by means of special tunnel erectors. Shear mining tunnelling method is similar to the one described above, but the soil excavation is carried out not through blasting but using tunnel-boring machines, where operative parts of various types can be installed.

Two types of TBC are used in the construction of the Moscow and Dnipropetrovsk underground. The first one is the rock tunnel-boring complex (RTBC). It includes separate machines, namely, the 3DB-5D drill rig, the STA-2 tunnel lining erector, the 1RLM-5 loading machine. The shearer tunnel-boring complex (STBC) consists of the following units: the TOM-42 tunnel-boring machine, the STA-2 erector, the 1RLM-5 loading machine. We consider the named complexes to be the basic ones in our study (we have chosen them as prototype). By carrying out morphological analysis of their parameters, we describe an algorithm for creation of a conceptual model of the most efficient complexes below.

The algorithm suggested in the study with a slight adjustment of the tables with the results of the morphological analysis of the tunnelling parameters can be used for the construction of other undergrounds.

Scientists of the National Mining University (Zabolotny et al. 2013) substantiated rational parameters of tunnel-boring complexes that minimize the weight of such erector components as an undercarriage; a manipulator swing drive; guide beams that support the lining elements during installation; a portable arch; a tier with movable platforms by 12%. Another study (Sirchenko et al. 2013) proved the rationality of the two manipulators erector actuator usage that enhances its performance by 1.8 times on the basis of one cycle of operation. The material

of the research performed was the basis of the conceptual scheme of lightweight upgraded erector – MTPM of improved efficiency developed by the authors of this article.

In this scheme, determining the rational TBC parameters we confine to the breaking and lining operations exclusively, and the existing 1RLM-5 loading machine is suggested for transporting rock.

Algorithm of breaking and lining operations in both rock and shearer tunnelling can be used either for two independent machines, or for a single unit application.

It is known that the basic unit of the tunnel erector is a trolley with other objects such as turning devices and manipulator extension, hoods and movable platforms, etc., installed on it.

The authors of this article propose to upgrade the trolley (erector basic unit) with a blast-hole drilling module or a tunnelling device.

A previously created experimental TSTE 5.5 tunnel machine model (Figure 1) is able to destroy the rock which hardness does not exceed the value of $f \leq 6$, with a cutter 1 mounted on manipulator 2 – lining mount actuator, mechanically.

The tunnelling unit CMPM-1 (Figure 2) with the boom 1 and the drill bit 2 mounted on the frame of the upgraded tunnel erector MTPM (these devices are also used in a tunnel-boring machine), has been developed to dig harder rock ($f \leq 10$) at the National Mining University (NMU). In addition, the manipulator 3 for the tunnelling placement is located on the boom.

On the basis of the work the NMU experts have designed the CMPM-2 tunneling unit (Figure 3) capable of digging even harder rock ($f \leq 10$). The latter is an advanced design of the CMPM-1 unit that requires the installation of the boom with the drill bit on the upgraded tunnel erector MTPM.

Another tunnelling unit DMPM (Figure 4) developed at the NMU is used for breaking the rock with the hardness $f \leq 20$ Two drill rigs 1 are mounted on the frame of the upgraded tunnel erector MTPM in this construction. This allows to drill holes directly from the erector trolley.

To determine the rational parameters of TBC we will use the morphological analysis method.

(a)

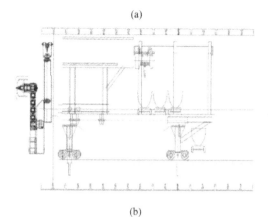

(b)

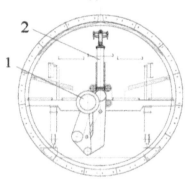

Figure 1. Diagram of TSTE-5.5 experimental tunnelling unit: (a) sectional view; (b) view from the front: 1 – cutter; 2 – manipulator.

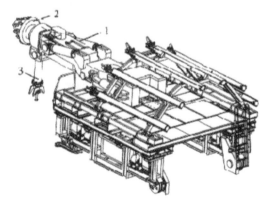

Figure 2. Model of the CMPM-1 tunneling unit for shearer mining tunneling: 1 – boom; 2 – drill bit; 3 – manipulator.

We will consider six above-mentioned options to implement the breaking and lining operations (the selection of the devices: STA-2 + 3DB-5D DMPM; STA-2 + TOM-42; CMPM-1; TSTE-5,5; CMPM-2).

Figure 3. General view of the CMPM-2 tunneling unit for shearer mining tunnelling.

(a)

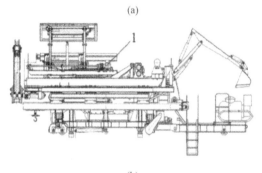

(b)

Figure 4. View of the tunnel unit UTMB for rock tunneling: (a) diagram: 1 – drill rigs; (b) general view.

Analysing the work of existing complexes as part of the STA-2 erector, the 3DB-5D drilling rig or the TOM-42 tunnel-boring machine we conclude that the final parameters of every option implemented are as follows: the specific quantity of metal m and the time of execution t of the operations within one operating cycle; the minimum f_{min} and maximum f_{max} allowable hardness of the rock being destroyed during digging; the probability of the ground-based facilities destruction p due to the rock mining tunnelling.

Since each option of the shearer mining tunnelling requires the loading coincides with the rock breaking, the time t shall mean the duration of the operations performed by all machines and units constituting TBC.

To carry out the morphological analysis we use such dimensionless parameters when the minimal value is considered to be the best one. For this purpose, we take the values of weight and time respectively to the STA-2 + 3DB-5D option ($m_0 = 45$ t, t_0), and the reciprocal value of hardness ($1/f$) according to the Protodiakonov Scale.

Priority values of each option of the basic operations to be implemented selected according to the experts' estimation are summarized in Table 3. Since it was necessary to solve the issue of increasing the TBC performance efficiency, the parameter of time as opposed to the priority of weight was the priority. Since the implementation of each digging operation is

possible only within a certain range of the rock hardness (Table 1), the priority of this very indicator was accepted the highest one.

Mass properties of actual equipment were determined on the basis of existing TBC according to datasheet specifications while indicators of the construction developing were determined by means of solid simulation with the use of the software package SOLIDWORKS.

To select the values of the dimensionless parameter of time we compose a sequence diagram of the basic operations that correspond to the selected operations to be implemented (Table 2). The results of the morphological analysis were summarized in Table 3.

Table 2. Sequence diagram of basic operations of tunnelling in different options of its implementation, h.

Name of an option	Name of an Operation										
	Arrival of 3DB-5D or TOM-42	Blast-hole Drilling	Hole Charging	Explosion and Ventilation	Breaking and Loading of Soil	Departure of TOM-42	Arrival of IRLM-5	Loading of Soil	Lining Mounting	Track Laying	Total Time
STA-2 + 3DB-5D	1	3.5	1	0.5	0		1	5	8.5	2.5	23
DMPM	0	3.5	1	0.5	0		1	5	5.5	2.5	19
STA-2 + TOM-42	1	0	0	0	2.35	1	0	0	8.5	2.5	15.35
CMPM-1	0	0	0	0	2.35	0	0	0	8.5	2.5	13.35
TSTE-5.5	0	0	0	0	2.35	0	0	0	8.5	2.5	13.35
CMPM-2	0	0	0	0	2.35	0	0	0	5.5	2.5	10.35

Table 3. Results of Morphological Analysis of Dimensionless Parameters of Tunnelling Implementation Options.

Name of an Option	Dimensionless Parameters of Options to be Implemented					
	Weight (m/m_0)	Time (t/t_0)	Reciprocal Value of Hardness		Destruction Probability (p)	Consolidated Figures
			($1/f_{min}$)	($1/f_{max}$)		
			Priority			
	1	5	50	50	1	—
STA-2 + 3DB-5D	1	1	0	0.05	1	9.50
DMPM	0.56	0.82	0	0.05	1	8.20
STA-2 + TOM-42	2.22	0.66	0.16	0.1	0	18.89
CMPM-1	0.91	0.58	0	0.1	0	8.81
TSTE-5,5	0.55	0.58	0	0.16	0	11.79
CMPM-2	0.93	0.45	0	0.1	0	8.18

As the data in Table 3 shows the most optimal implementations of the tunnelling (they correspond to the minimum values of the consolidated figures) are the structures mounted on the basic erector unit. They have additional removable modules for blast-hole drilling, or for tunnelling, in particular, DMPM

is applicable in the rock mining tunnelling while CMPM-2 is applicable in the shearer mining tunnelling. Module unification and the opportunity of rapid units transformation will improve the tunnelling efficiency in case of any changes in the rock hardness within wide limits.

3 CONCLUSIONS

Solid models of TBC units, promising in the practical use were developed, and their mass properties were determined.

On the basis of morphological analysis the best options of TBC units mounted on the basic erector unit equipped with additional removable modules for blast-hole drilling or tunnelling were selected. Besides the modules can transform into each other, and adapt to changes in the rock hardness within wide limits.

The unit CMPM-2 capable to perform shearer mining tunnelling under the risk of roof collapse and ground-based facilities destruction.

REFERENCES

Sirchenko, A., Zabolotny, K. & Panchenko, E. 2013. *Features double lever manipulator simulation tunnel stacker*. Materials of the scientific-practical conference "Mathematical problems of technical mechanics". Dniprodzerzhynsk: Vol. 2: 72 – 73.

Zabolotny, K., Sirchenko, A. & Panchenko, E. 2013. *Theoretical and Computational study of stress-strain state of the elements of single-lever tunnel stacker UTC-2*. Materials of the scientific-practical conference "Mathematical problems of technical mechanics". Dniprodzerzhynsk: Vol. 2: 74 – 75.

Theoretical and Practical Solutions of Mineral Resources Mining – Pivnyak, Bondarenko & Kovalevska (eds)
© 2015 Taylor & Francis Group, London, ISBN: 978-1-138-02883-8

Technical, economic and environmental aspects of the use of emulsion explosives by ERA brand in underground and surface mining

T. Kholodenko, Ye. Ustimenko & L. Pidkamenna
State Enterprise Research-Industrial Complex "Pavlohrad Chemical Plant", Pavlohrad, Ukraine

A. Pavlychenko
National Mining University, Dnipropetrovsk, Ukraine

ABSTRACT: Technical, economic and environmental efficiency factors were analyzed for application of ERA emulsion explosives containing the solid propellant processing products. Applicability of ERA emulsion explosives for mining enterprises using both underground and surface mining operations was proved.

1 INTRODUCTION

Currently in Ukraine, as in other countries with developed mining industry, there is a steady tendency to refuse TNT containing explosives and change to the commercial emulsion explosives (Yefremov 2010, Kholodenko et al. 2014, Stupnik et al. 2013, Khomenko et al. 2013).

The use of emulsion explosives (EE) allows reducing the costs for blasting and enables control and optimization of the parameters of explosion depending on the application. At the same time it provides higher safety of blasting operations and creates favorable conditions for use of effective mechanized transportation and loading of boreholes (Efremov et al. 1996, Stupnik et al. 2013).

Production of emulsion explosives has changed the concept of arrangement and carrying out of blasting in mining enterprises and promoted the development of new types of raw materials and equipment for the production of explosives and loading equipment.

The main advantages of emulsion explosives are that their components are not assigned to explosive materials prior to mixing and are safe for handling, transportation, exposure to electrical discharge and mechanical effects. The non-explosive components take on the explosive properties only upon mixing in the borehole soon thereafter. If the broken rock keeps a part of unexploded emulsion explosive, then due to destructive processes such explosives lose their explosive properties with time (Shyman & Ustimenko 2009).

It should be noted that the minimum emission of toxic gases when using emulsion explosives is observed only when they have a certain balanced composition. Thus, to avoid the formation of nitrogen oxides and to minimize the carbon monoxide content in the composition of the explosion products, the oxygen balance of emulsion explosives must be in the range of -0.2 to -2.0%. The latter is sometimes contradicted by the need to obtain explosives with high energy efficiency, which increasing requires the addition of high energy additives into emulsion explosives such as: aluminum powder, pyroxiline or ballistit powder, ANFO, etc (Efremov et al. 1987, Kozlovskaya & Chebenko 2010, Ustimenko et al. 2009).

An important advantage of EE application is the possibility to significantly reduce the environmental damage due to release of toxic gases during blasting operations, as compared to other types of explosives (Doludareva et al. 2012, Zakharenkov 2010, Krysin et al. 2004, Krysin & Novinskij 2006, Saksin & Krupskaya 2005, Mironova & Borysovs'ka 2014).

At present at the premises of SE RIC PCP the set of studies are carried out in order to produce new types of emulsion explosives, which formulations comprise the solid propellant (SP) and the products of its high-level processing (Kirichenko et al. 2012, Ustimenko et al. 2009, Kirichenko et al. 2012).

2 FORMULATING THE PROBLEM

Currently the various types of emulsion explosives under trademarks "Ukrainit" and "Emonit" (Kryvbasvzryvprom), "Anemix" (Intervzryvprom) and "ERA" (SE "RIC PCP") are produced and used in Ukraine. ERA emulsion explosives include a wide range of various explosives of Division 1.1, including ERA-I, ERA-II and ERA-III which contain the products of SP high-level processing.

The development of the above types of emulsion explosives and their blasting technologies was carried out with regard to balanced formulations and processes of blasting that allow achieving the required EE energetic and explosive parameters with minimum formation of harmful substances in the explosion products.

The study of the environmental impact of drilling and blasting operations using emulsion explosives took into account a number of factors, which can be divided into two groups. The factors of the first group characterize the so-called EE "internal" formulation features and its adaptability to the different changes under application conditions in blasting operations. The factors of the second group characterize the so-called "external" conditions associated with the special production and loading of emulsion explosives into boreholes and blastholes, special initiation of borehole charge blasting, as well as mining and geological conditions of blasting operations.

It should be noted that the factors characterizing EE application "external" conditions and parameters in blasting operations, are essentially determinant for the formation of gas-and-dust mixture of the explosion products and therefore, for description of the environmental impact. Ignoring the impact of these factors can lead to deviations from and non-conformance with the conditions for which the emulsion explosives are applicable, as well as to deterioration of their explosive properties and development of deflagration processes within charge. As a consequence, this leads to the generation of large amounts of such toxic products of EE thermal decomposition as CO and NO_x.

Deterioration of explosive properties may occur in case of violations of parameters and manufacturing process of emulsion explosives, because of failures or malfunction of dosing units and mixers in mix-pump trucks as well as due to lower control on the part of personnel. As a result the balanced mixture ratio is not maintained, the EE structural stability is not generated or prematurely lost. It results in break in detonation over a column of EE charge and its transition to deflagration processes. Such processes occur also in case of "flowing" of emulsion explosives to cracks and cavities around boreholes where there are no conditions for their explosive transition. Destruction of EE structure from premature initiating pulses and effects of borehole water environment, as well as insufficient capacity of boosters contribute to deterioration of their explosive properties and release of products of incomplete oxidation and explosive transition of the formulation components.

To perform the ecological and technical and economic feasibility report of possible application of ERA emulsion explosives in the mining enterprises it is necessary to carry out the theoretical and practical evaluation of their formulation features and the results of the explosion products control.

Therefore, the objective of study is an ecological and technical and economic feasibility of possible EE application with SP using in underground and surface mining operations.

3 GENERAL PART

In development of formulations of ERA emulsion explosives the computational and experimental methods were used, allowing to optimize technological, operational and ecological parameters and properties of specific types of explosives. As the basic criterion for choice of EE formulation, its balance for parameters of energetic characteristics and minimization of generation of explosive transition harmful products has been chosen.

The experimental researches used the methods developed by SE RIC PCP. The measurements were carried out by means of differential gravimetric thermoanalyzer Diamond TG/DTA Perkin Elmer (USA), laser light scattering particle size analyzer SALD 30IV (manufactured by Shimadzu, Japan) and calorimeter C-2000 (Germany).

The experimental investigations used the methods developed by SE RIC PCP. The measurements were carried out by means of differential gravimetric thermoanalyzer Diamond TG/DTA Perkin Elmer (USA), laser light scattering particle size analyzer SALD 30IV (Japan) and calorimeter C-2000 (Germany).

Set of studies performed includes the investigations of EE properties with addition of SP from all ICBM SS-24 stages as high-energy components and EE properties with addition of products of high-level processing.

In order to evaluate the safety ecological impact, with regard to influence of the so-called factor of "internal" formulation conditions, the choice and the analysis of components application for ERA emulsion explosives is carried out. The basic components of ERA emulsion explosives of Division 1.5 are ammonium nitrate and calcium nitrate used as dry components and (or) solutions in quantity of 75 – 95%, and industrial oil mixed with hydrocarbonic emulsifier in quantity of 4.5 – 7.0%. The crumb of products of SP processing of up to 10% is added to some compositions as energetic additives increasing EE pushing action. In order to allow the required processing properties and EE stable structure and to achieve EE sensitizing characteristics, their compositions contain the appropriate processing aids and gas-generating reagents in total up to 2% (Table 1).

Table 1. Components of emulsion explosives, Hazard Class with regard to toxicity of substances to be used.

Component	Hazard Class	ERA-I (II) (IIIn)	Ukrainit	Anemix	Emonit N	Content, %
Ammonium nitrate	4	+	+	+	+	
Sodium nitrate	3	—	—	—	+	75 – 83
Calcium nitrate	3	+	+	—	—	
Industrial oil	3	+	+	+	+	
Emulsifier	3	+	+	+	+	4.5 – 7.0
SP processing product	3	+	—	—	—	< 10.0
Processing aids	2	+	+	+	+	1.0 – 2.0
Water	4	+	+	+	+	10 – 15

Table 2 represented the data on elemental constituents of ERA emulsion explosives suitable for composition conditional formula per 1 kg taking into account the manufacturing tolerances, show that the molar content of carbon (C), hydrogen (H), nitrogen (N), oxygen (O) and sulfur (S) elements for such emulsion explosives is about the same as for other EE brands.

Table 2. EE elemental constituents and characteristics.

EE brand	Elemental constituents, mole/kg								Characteristic			
	C	H	O	N	Ca	Cl	S	Al	Ox. Bal., %	T, K	k	δ_{ee},%
ERA-I (II) (IIIn)	4.4 – 6.1	52.0 – 55.0	35.0 – 37.5	17.0 – 18.5	0.85 – 1.25	<10¹	<10¹	0.25 – 0.56	–0.2... –0.8	2370 – 2470	1.298	89.8
Ukrainit	4.5 – 4.6	52.5 – 53.0	37.5 – 38.0	15.5 – 16.0	1.65 – 1.75	—	<10¹	—	–0.2... –2.2	2080 – 2150	1.274	87.4
Anemix	3.4 – 3.6	63.2 – 64.0	38.0 – 38.5	20.0 – 20.5	—	—	<10¹	—	–0.85... –1.9	2030 – 2100	1.276	87.1
Emonit N	4.8 – 4.9	58.0 – 58.5	37.0 – 37.5	17.0 – 17.5	1.80 – 1.90 (Na)	—	<10¹	—	–0.2... –0.8	1920 – 2000	1.254	85.2

Taking into account the technology ability to manufacture the emulsion explosives using mix-pump trucks, the computational and experimental evaluation was held concerning the compositions balance with regard to permissible variations of content of any component of ERA EE formulation that does not result in a deterioration of its characteristics, including the composition of the explosion products.

The presence of calcium and sodium elements in EE compositions is caused by the introduction of their nitrates in EE composition, providing the best processing properties and stability of the emulsion matrix structure. Also, the presence of these elements makes it possible to neutralize the generation of acidic products of explosive transition of EE components, such as CO, NO_x, HCl, H_2S and thus to reduce the harmful explosion products. The presence of the aluminum and chlorine elements is caused by the introduction of the SP high-energy components in EE composition, increasing the temperature of the explosion products and their pushing action. The presence of a sulfur element is caused by the use of the fuel phase hydrocarbons and processing aids containing this element in the EE com-

positions. It should be noted that the molar content of sulfur and chlorine elements is small and being at the level as for impurities which accompany the technical raw materials for various EE brands.

In case of elemental composition of ERA I emulsion explosives containing SP components shown in Table 2, the oxygen balance falls within the limits of –0.2 – –0.8% that proves the balance of compositions within allowable variance (not above 1.5%) in respect of one or another component content.

The formulation composition of ERA emulsion explosives allows obtaining 10 – 15% higher level of thermodynamic temperature (T) during explosion as compared to other emulsion explosives. Besides, explosion products of ERA-I (ERA-(II) (IIIn) emulsion explosives have higher value of specific heat ratio (k) and a better explosion heat availability factor (δ_{expl}) of up to 90% has been obtained for them respectively.

The Brinkley-Wilson model of chemical transition taken for explosives with a small negative oxygen balance was used to assess the thermochemical ability of emulsion explosives compositions to form certain explosion products. When performing thermochemical calculations for the products of explosive transition of explosives, substances for which there are appropriate possibilities for formation are specified. Such possibilities are determined by means of chemical and kinetic principles of the processes when atoms combine into molecules of only those substances, the formation of which results in release of the maximum amount of energy, and when a number of simple substances increase. In addition, when assessing the possibility of formation of certain substances during explosion, it is necessary to take into account such parameters as enthalpy of formation of substances that should be negative, entropy and Gibbs energy that should tend to decrease. All these requirements are met by the products of explosive transition of ERA emulsion explosives containing products of SP processing specified in thermochemical equation.

The detailed assessment of the possibility of formation of organochlorine and dioxin substances in the explosion products shows that there are no thermochemical bases for their formation. This is due to the fact that, firstly, there are no such substances in the composition of emulsion explosives, and secondly, organochlorine and dioxin substances have a complex molecular composition (for example, $C_{12}H_4O_2Cl_4$), a property of maintaining condensed state at a temperature of more than 400°C at extremely high positive enthalpy of formation (more than 323 kJ/mol) and increasing Gibbs energy.

Thermodynamic evaluation of the ability of emulsion explosives compositions containing products of SP high-level processing to form certain explosion products under various conditions, including furan ($C_aH_bO_c$), organochlorine ($C_aH_bCl_d$) and dioxin ($C_aH_bO_cCl_d$) ones is made based on the results of thermodynamic calculations shown in Table 3.

Apart from substances shown in Table 3, the substances of intermediate state and partial carbon and hydrogen oxidation with concentrations of less than 10^{-4}, as well as molecular nitrogen are thermody-

namically present in the products of explosive transition of emulsion explosives.

The received thermodynamic calculations data confirms correspondence and correctness of performed thermochemical assessment with regard to the composition of products of explosive transition of ERA emulsion explosives containing products of SP processing. In addition to the above, thermodynamic calculations showed that the products of chemical and explosive transition of emulsion explosives do not contain not only furan, dioxin and organochlorine substances, but also simple chlorohydrocarbons or fragments thereof, which may be the source of formation of mentioned substances.

Disposable SP and the products of its processing are high-energy materials, the addition of which into emulsion explosives allows improving their energy efficiency. At the same time, these additives alter the oxygen balance of explosives and under certain conditions are capable to increase the output of toxic gases unusual for emulsion explosives. Therefore, while developing emulsion explosives containing SP and products of its processing most of the focus is on selection of the composition that ensures high energy potential and the minimum content of toxic gases in the detonation products (Kirichenko et al. 2012).

Tables 4 and 5 show calculated physical-chemical and explosive parameters for emulsion explosives containing stage I and II SP characterizing a change of amount of emissions of carbon oxide and nitrogen oxides depending on concentration of SP of these stages.

The analysis of the data in Tables 4 and 5 showed that with the increase of SP content in the formulation of emulsion explosives the explosion energy increases and the oxygen balance decreases. Nitrogen oxides content in emissions decreases simultaneously. As for carbon oxide, the marked increase in its concentration is observed when the content of SP in emulsion explosives is 10% and more.

Table 3. The content of products of explosive transition of ERA emulsion explosives under various pressures, mole/kg.

Explosion products	ERA-AM			ERA – I (II) (IIIn)		
	under ambient pressure, MPa			under ambient pressure, MPa		
	1000	100	0.1	1000	100	0.1
CO	0.39	0.39	10^{-6}	0.34	0.34	10^{-6}
CO_2	4.40	4.40	3.46	4.14	4.14	3.64
H_2O	25.3	25.3	26.7	26.2	26.3	27.2
HCl	0	0	0	0.12	0.11	0
H_2S	0.01	0.01	0	0.01	0.01	0
H_2	0.44	0.45	0.03	0.27	0.26	0.02
Al_2O_3	0	0	0	0.12	0.12	0.12
$Ca(OH)_2$	1.33	1.33	0	0.91	0.91	0
$CaCl_2$	0	0	0	10^{-4}	10^{-3}	0.06
$CaCO_3$	0	0	1.32	0	0	0.85
$CaSO_4$	0	0	0.01	0	0	0.01

Table 4. Physical-chemical and explosive parameters of emulsion explosives containing stage I SP.

Parameter	Parameter value when the content of stage I SP in emulsion explosives is:				
	8%	9%	10%	11%	15%
Heat of explosion, kJ/kg	3560.8	3673.0	3746.2	3755.7	3769.1
Volume of gases, l	821.58	817	815	813.3	812
Oxygen balance, %	1.22	0.67	0.02	−0.43	−2.82
Composition of explosion products, %:					
CO	—	—	0.5	1.33	5.7
CO_2	13.89	14.43	14.28	13.4	8.8
H_2O	44.98	44.85	44.75	44.6	44.1
N_2	22.24	22.55	22.7	22.5	21.7
NO_x	1.84	0.75	—	—	—
Al_2O_3	2.87	3.23	3.59	3.9	5.4
$CaCl_2$	1.78	2.0	2.2	2.44	3.3
$CaCO_3$	12.42	12.23	12.0	11.83	11.

Table 5. Physical-chemical and explosive parameters of emulsion explosives containing stage II SP.

Parameter	Parameter value when the content of stage II SP in emulsion explosives is:				
	8%	9%	10%	11%	15%
Heat of explosion, kJ/kg	3620.1	3736.5	3814.0	3826.9	3873.0
Volume of gases, l	821.28	816.7	813.7	813.3	811.0
Oxygen balance, %	0.99	0.44	−0.14	−0.74	−3.02
Composition of explosion products, %:					
CO	—	—	0.38	1.45	5.5
CO_2	13.12	13.54	13.38	12.2	7.4
H_2O	44.76	44.6	44.45	44.3	43.7
N_2	23.11	23.51	23.74	23.64	23.3
NO_x	1.75	0.69	—	—	—
Al_2O_3	3.17	3.57	3.94	4.36	6
$CaCl_2$	0.5	0.61	0.68	0.74	1
$CaCO_3$	—	—	0.38	1.45	5.5

Dependence of toxic gases emissions on stage I'm and III SP content in emulsion explosives has a similar nature. Studies have found that the addition of SP as high-energy components in emulsion explosives allows reducing emissions of harmful gases such as nitrogen oxides and improving the energy characteristics of explosives. At the same time, pure SP is an explosive and is not safe to handle. Thus, their use is limited when preparing emulsion explosives directly on site where blasting operations and mechanical charging of holes are carried out.

The calculated physical-chemical and explosive parameters for emulsion explosives containing the products of SP high-level processing (HMX, ammonium perchlorate, polymer matrix) are given in Tables 6 – 10.

The analysis of the data in Table 6 has found that with the increase of HMX content in the formulation the explosion energy increases. The total amount of toxic gases during explosion decreases with the increase of HMX content to 15% in emulsion explosives. Further increase of its content results in the increase of carbon oxide emissions.

With the increase of ammonium perchlorate content in the formulation the explosion energy decreases and the amount of toxic gases during explosion increases.

Consequently, in order to minimize negative impact of blasting works on the environment, it is necessary to use explosives with minimum content of toxic substances found in products of explosive transition of emulsion explosives. In order to eliminate generation of nitrogen oxides and to minimize the content of carbon monoxide as part of explosion products, the oxygen balance of explosives must be within −0.2…−2.0%, and the content of energetic additives shall not exceed 5%, products left from SP high-level processing are recommended for use as such additives (polymer matrix containing aluminium powder and plasticized butyl-rubber or polybutadiene rubber). In this case, high explosion heat (over 3700 kJ/kg) of explosives is ensured, nitrogen oxides formation is eliminated and minimum content of carbon monoxide in detonation products is ensured.

Table 6. Physical-chemical and explosive parameters for emulsion explosives containing HMX.

Parameter	Parameter value when the content of HMX in emulsion explosives is:			
	10%	11%	13%	15%
Heat of explosion, kJ/kg	3424.1	3500.3	3652.7	3696.2
Volume of gases, l	835.6	833.4	829	828.6
Oxygen balance, %	1.47	1.05	0.22	−0.61
Composition of explosion products, %:				
CO	—	—	—	1.0
CO_2	15.6	16.2	17.4	16.9
H_2O	43.9	43.7	43.3	42.9
N_2	23.68	24.1	24.9	25.1
NO_x	2.79	2	0.45	—
$CaCO_3$	14	14	14	14

Table 7. Physical-chemical and explosive parameters for emulsion explosives containing ammonium perchlorate.

Parameter	Parameter value when the content of ammonium perchlorate in emulsion explosives is:		
	8%	10%	15%
Heat of explosion, kJ/kg	2633.6	2628.2	2610.6
Volume of gases, l	848	845.8	839.7
Oxygen balance, %	6.75	7.03	7.73
Composition of explosion products, %:			
CO	—	—	—
CO_2	11.1	11.5	12.4
H_2O	45.5	45.3	45.0
N_2	17.4	16.8	15.3
NO_x	11.7	11.9	12.6
$CaCl_2$	3.8	4.7	7.08
$CaCO_3$	10.6	9.7	7.6

Table 8. Physical-chemical and explosive parameters for emulsion explosives containing polymer matrix from SP of stage I.

Parameter	Parameter value when the content of stage I SP polymer matrix in emulsion explosives is:				
	5%	7%	8%	9%	10%
Heat of explosion, kJ/kg	3704.0	3707.0	3709.7	3712.6	3713.9
Volume of gases, l	821.6	823.0	823.7	824.5	825.0
Oxygen balance, %	−0.19	−2.52	−3.68	−4.85	−6.01
Composition of explosion products, %:					
CO	0.3	4.4	6.4	8.5	10.5
CO_2	13.3	8.6	6.3	3.9	1.53
H_2O	45.1	44.7	44.6	44.4	44.2
N_2	23.8	23.5	23.3	23.2	23.0
NO_x	—	—	—	—	—
Al_2O_3	3.4	4.7	5.4	6.1	6.76
$CaCO_3$	14	14	14	14	14

Table 9. Physical-chemical and explosive parameters of emulsion explosives containing polymer matrix from SP of Stage II.

Parameter	Parameter value when the content of Stage II SP polymer matrix in emulsion explosives is:				
	4%	5%	6%	8%	10%
Heat of explosion, kJ/kg	3671.7	3715.6	3718.1	3724.8	3603.8
Volume of gases, l	822	820.8	821.6	822.7	839.9
Oxygen balance, %	0.19	−1.13	−2.51	−5.25	−7.92
Composition of explosion products, %:					
CO	—	2.0	4.4	9.2	13.84
CO_2	13.2	11.0	8.1	2.4	—
H_2O	45.3	45.1	44.9	44.6	44.2
N_2	23.6	23.6	23.3	22.9	22.4
NO_x	0.4	—	—	—	—
Al_2O_3	3.5	4.3	5.2	7.0	8.69
$CaCO_3$	14.0	14.0	14.0	14.0	10.8

Table 10. Physical-chemical and explosive parameters of emulsion explosives with aluminium.

Parameter	Parameter value when the content of aluminium in emulsion explosives is:				
	2%	3%	4%	5%	6%
Heat of explosion, kJ/kg	3261.7	3562.7	3862.5	4162.7	4295.9
Volume of gases, l	825.9	810	793.9	777.8	768
Oxygen balance, %	3.45	2.36	1.27	0.18	−0.91
Composition of explosion products, %:					
CO	—	—	—	—	1.6
CO_2	9.6	9.6	9.6	9.6	7.2
H_2O	45.1	44.6	44.2	43.7	43.3
N_2	21.0	21.6	22.2	22.8	22.6
NO_x	6.5	4.5	2.4	0.4	—
Al_2O_3	3.8	5.7	7.6	9.4	11.3
$CaCO_3$	14.0	14.0	14.0	14.0	14.0

In the process of manufacturing and while using emulsion explosives that contain products from SP reprocessing, factors of safe use of the given type of explosives were assessed, also the influence was assessed of so called "operational" factors on their safe use, as well as formation of toxic explosion products, in particular, preserving stable ratios of ingredients of emulsion explosives during mechanized manufacturing and loading into boreholes.

When loading emulsion explosives having a critical detonation diameter close to the borehole diameter, detonation of emulsion explosives may be interrupted, and deflagration processes may begin, being accompanied with generation of toxic products due to incomplete oxidation of fuel components, and explosive properties are lower. The similar situation is found when loading strong fracture boreholes with emulsion explosives of low viscosity and low thixotropy – in this case emulsion explosive in fractures burns out to generate a large amount of toxic products. All this results in poorer characteristics of an explosion and, conse-

quently, in poorer rock mass fragmentation quality.

The performed thermodynamic calculation proved that ERA emulsion explosives containing SP reprocessed products are characterized with low content of harmful substances in products of explosive transition. The next important stage of development of the process of manufacturing of bulk emulsion explosives containing rocket propellant designed for mechanized borehole loading using mix-pump trucks is experimental evaluation of emulsion explosives' impact on the atmospheric air, underground water and soil.

The ecological survey done in the areas of blasting works has determined parameters in terms of presence and concentration of: in the atmospheric air (carbon oxide, nitrogen dioxide, hydrogen chloride); in the surface soil (nickel, zinc, copper, lead, nitrates, ammonium); in surface and underground water (nitrates, nitrites, ammonium nitrogen, ammonium perchlorate).

The ecological monitoring at the stage of acceptance tests of ERA brand emulsion explosives containing SP was performed during blasting works at mining enterprises developing mineral deposits using open-cut and underground mining methods. These deposits were characterized with various structural, geological, physical and chemical properties of rocks, and side by side with ecological problems, a series of problems related to industrial applicability of ERA emulsion explosives containing SP were solved.

The results of studies performed in the years 2007 through 2015 showed that, during blasting of ERA emulsion explosives containing SP reprocessing products, in the air environment on blasting sites there was neither hydrogen chloride nor organochlorine substances. In soil or underground waters, no ammonium perchlorate or organochlorine substances were found, and the content of nitrate ions and chloride ions remains on the level of background concentrations and does not exceed maximum-allowable concentrations. The analysis of samples of air medium done $30 - 45$ minutes after an explosion shows that, due to dispersal, these parameters turn back to their initial values as before the explosion.

The analysis of properties of ERA emulsion explosives components containing SP reprocessing products demonstrated that they contain no toxic substances. The used components are traditional in compositions of industrial explosives. The analysis of characteristics of these emulsion explosives formulas proved both their equilibrium in terms of oxygen balance and absence of thermo-chemical and thermo-dynamic pre-conditions for generation of toxic organochlorine and dioxin substances in the products of explosive transition of emulsion explosives.

The results of ecological monitoring prove that the formula, conditions of use and manufacture of ERA brand emulsion explosives containing up to 10% of SP reprocessing products ensure completeness of chemical reactions during explosion processes without generation of toxic explosion products.

The structural stability of emulsion explosives with SP reprocessing products and relevant rheological characteristics ensuring the continuity of an emulsion explosive charge column have been confirmed. All this makes it possible to realize in full the energy potential of emulsion explosives containing SP high-level processing products with the consumption rate of 0.6 to 1.0 kg/m^3 for rock mass blasting (depending on rock hardness), with content of toxic substances in detonation products being low (less than 1%).

Cost-and-saving factors for the use of various explosives are compared in Table 11.

Table 11. Economic effectiveness of the use of emulsion explosives.

Type of explosives	Name of explosives	Volume of BA per 1000 boreholes		Sell price, UAH/kg (VAT excl.)	Cost of BA per 1000 boreholes, UAH	Economic effectiveness of the use of ERA brand EE	
		Pcs.	kg			UAH	%
Blasting agents (BA)	TNT blocks ZTP 800	2000	1600	27.00	43200	—	—
	Packaged ERA-R brand EE 070-90 mm (1 kg each)	2000	2000	12.33	24660	18540	42.9
Packaged explosives	Packaged Ammonite 6ZhV	5000	1000	18.00	18000	—	—
	Packaged ERA-R brand EE 032-90 mm	5000	1000	11.00	11000	7000	38.9
Bulk emulsion explosives	Bulk ERA-N EE	—	1100	6.00	6000	12000	66.7

The analysis of the readings of Table 11 evidence that the use of ERA brand emulsion explosives is advantageous not only ecologically but also economically.

4 CONCLUSIONS

ERA brand emulsion explosives containing SP reprocessing products having minimum impact on the environment have been developed, and the study of effect of the developed emulsion explosives on ecosystems in blasting areas has been conducted.

The effect of various high-energy additives (including SP of all stages and SP high-level processing products) on energetic characteristics and toxicity of explosion products from emulsion explosives has been studied.

The most practical range of SP high-level processing products content in emulsion explosives, which ensures energetic characteristics with a minimum impact of explosion products on the environment, has been determined.

The results of performed works have determined optimal parameters for the manufacturing process of ERA brand emulsion explosives containing SP high-level processing products that ensure safe performance of process operations at all its stages and allow the negative impact onto environmental conditions to be eliminated when using this type of explosives.

REFERENCES

Yefremov, E.I. 2010. *Experience of use of simple explosives in open-cast mines of Ukraine.* Ukrainian Union of Mining Engineers. Information Bulletin, No 4: 9 – 11.

Kholodenko, T., Ustimenko, Ye., Pidkamenna, L. & Pavlychenko, A. 2014. *Ecological safety of emulsion explosives use at mining enterprises.* Progressive Technologies of Coal, Coalbed Methane, and Ores Mining. The Netherlands: CRC Press/Balkema: 255 – 260

Stupnik, N.I., Kalinichenko, V.A., Fedko, M.B. & Mirchenko, Ye.G. 2013. *Prospects of application of TNT-free explosives in ore deposits developed by underground mining.* Naukovyi Visnyk Natsionalnoho Hirnychoho Universytetu, No 1: 44 – 48.

Khomenko, O., Kononenko, M. & Myronova, I. 2013. *Blasting works technology to decrease an emission of harmful matters into the mine atmosphere.* Annual Scientific-Technical Colletion – Mining of Mineral Deposits. The Netherlands: CRC Press/Balkema: 231 – 235.

Shyman, L. & Ustimenko, Y. 2009. *Disposal and destruction processes of ammunition, missiles and explosives, which constitute danger when storing.* NATO Security through Science Series C: Environmental Security: 147 – 152.

Efremov, E.I., Beresnevich, P.V. & Petrenko, V.D. 1996. *The problems of ecology massive explosions in quarries.* Dnepropetrovsk: Sich: 179.

Efremov, E.I., Kravtsov, V.S. & Myachina, P.N. 1987. *Destruction of rock explosion energy.* Kyiv: Naukova Dumka: 264.

Kozlovskaya, T.F. & Chebenko, B.N. 2010. *Ways to reduce the level of environmental danger in regions of the open pit mining.* Transactions of Kremenchuk Mykhailo Ostrohradskyi National University, No 6: 163 – 168.

Ustimenko, Ye.B., Shyman, L.N. & Kirichenko, A.L. 2009. *Peculiarities of water-emulsion explosive features for safe use during blasting.* Transactions of Kremenchuk Mykhailo Ostrohradskyi National University, No 2: 86 – 89.

Doludareva, Ya.S., Kozlovskaya, T.F., Lemizhanskaya, V.D. & Komir, A.I. 2012. *Influence of the surface-active substances implementation in the rock failure area on the intensity of rock crushing by means of the pulse loads.* Naukovyi Visnyk Natsionalnoho Hirnychoho Universytetu, No 4: 93 – 97.

Zakharenkov, Ye.I. 2010. State of blasting in Ukraine. State supervision in the treatment of industrial explosives. *Ukrainian Union of Mining Engineers.* Information Bulletin, No 4: 3 – 8.

Krysin, R.S., Ishchenko, N.I., Klimenko, V.A., Piven, V.A. & Kuprin, V.P. 2004. *Explosive ukranit-PM-1: Equipment and fabrication technology.* Mining Journal, No 8: 32 – 37.

Krysin, R.S. & Novinskij, V.V. 2006. *Model of the explosive rock crushing.* Dnipropetrovsk: Art-press: 144.

Saksin, B.G. & Krupskaya, L.T. 2005. *Regional estimation of the effect of mining production on the environment.* Mining Journal, No 2: 82 – 86.

Mironova, I. & Borysovs'ka, O. 2014. *Defining the parameters of the atmospheric air for iron ore mines,* Progressive Technologies of Coal, Coalbed Methane, and Ores Mining. The Netherlands: CRC Press/Balkema: 333 – 339.

Kirichenko, A.L., Ustimenko, Ye.B., Shyman, L.N. & Politov, V.V. 2012. *Study of detonation characteristics of blast-hole charges of packaged emulsion explosives.* Naukovyi Visnyk Natsionalnoho Hirnychoho Universytetu, No 6: 37 – 41.

Kirichenko, A.L., Ustimenko, Ye.B., Shyman, L.N., Podkamennaya, L.I. & Politov, V.V. 2012. *Method optimization of loading and initiation of blast-hole charges of packaged emulsion "ERA" explosives during the drifting operations in coal-bearing massifs.* Transactions of Kremenchuk Mykhailo Ostrohradskyi National University, No 2: 84 – 87.

Theoretical and Practical Solutions of Mineral Resources Mining – Pivnyak, Bondarenko & Kovalevska (eds)
© *2015 Taylor & Francis Group, London, ISBN: 978-1-138-02883-8*

Research of drainage drift during overworking of adjacent coal seam C_5 under conditions of "Samarska" mine

V. Sotskov, V. Russkikh & D. Astafiev
National Mining University, Dnipropetrovsk, Ukraine

ABSTRACT: The scrutiny degree of the determining problem of rational parameters of the development roadway support construction in overworking conditions were analyzed. The expediency of conducting extensive scientific researches using the computer simulation by finite element method. A series of calculations on the approaching the working face to the development roadway in elastoplastic formulation, taking into account the physical and mechanical properties of rocks and layered rock mass were conducted. On the basis of the obtained data, rational parameters of fastening system for this type of mine working that is confirmed with additional researches are developed.

1 ACTUALITY

Characteristic of conducting mining operations on mines of the Western Donbas is development of the suite consisting of three-six couples pulled together layers with parting height less than 12 m. In such layers about 40% of balance stocks of mines that are presented by mainly thin and very thin seams with thickness 0.6 – 1.2 m with seam inclination 3 – 5 degrees are concentrated (Sally et al. 2014).

One of the most indicative examples are the difficulties which arose with working off 3 blocks of Mine Management "Samaraska" PJSC "DTEK Pavlohradvuhillia" including C_5 and C_4^2 seams with distance between 8 – 12 m.

For assignment of strong water inflow during collapses of the main roof (30 – 60 m³/hour) from an extraction site it was carried out drainage ventilation roadway on C_4 layer at a depth of 8 – 9 m from coal seam C_5 (Figure 1).

Drainage ventilation roadway (DVR) will serve not only for branch of water inflow during mining of seam C_5, but also as a drainage roadway during mining of seam C_4. Therefore, preservation of its cross section for mine is extremely important.

A number of technology factors has impact on stability of development. First, water, which on drainage wells arrives in roadway from the overlying horizon, promotes a slaking of rocks of the soil and provokes a heaving, secondly, the part of the extraction columns fulfilling coal layer passes directly over roadway that leads to hit of development in a abutment pressure zone ahead of a face.

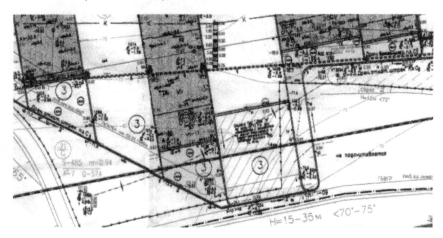

Figure 1. Relative position of drainage ventilation roadway (DVR) of seam C_5 and DVR of seam C_4.

As a result, there is a situation when the choice of rational parameters of system of fastening of a drainage roadway can't be carried out according to recommendations for maintenance of development mine workings. It is necessary to develop the new parameters, which are completely answering to specifics of carrying out development in difficult mine-and-geological conditions taking into account influence of an overworking and the acquired water content.

2 TASK ASSIGNMENT

During mine supervision of a condition of a drainage ventilation roadway of seam C_4^2 (DVR) mine "Samarska" it is established that during long use of mine working in difficult hydrogeological conditions, and also intensive rock pressure during mining considerable reduction of section of development happened longwall of overlying C_5 layer. On the specified site of the drainage ventilation roadway support lost the bearing ability, the area of cross section doesn't correspond to the demanded parameters, a condition of a railway line the unsatisfactory. It was necessary to make refastening of its separate sites (Figure 2) for further safe operation of development.

On the basis of data on refastening of development it is possible to draw a conclusion that the initial system of fastening couldn't provide effective maintenance of development throughout all service life. However, despite compliance of the passport to normative documents, even within a straight section of development in different places different schemes of installation are applied fix. The installation step changes, type fix, are selectively applied an anchor. Obviously, there is no uniform approach to a choice of rational parameters of fastening of this kind of developments.

On Figure 3 the condition of a drainage ventilation roadway at the time of pass over it a stoping faces presented. In general, development kept the functions for the admission of air and reception of water from the overlying horizon. However, use of a rolling stock, and also pass of people it is complicated. Influence of mountain pressure upon development led to deformation of a contour, situation ally established racks fix strengthenings significantly didn't correct a situation. As a result of inrushes of rock the grid is deformed. Strong water inflow, which provokes a slaking of rocks of the soil that leads to loss of integrity of a railway line, and also a flash of racks frame has the greatest impact fix.

(a)
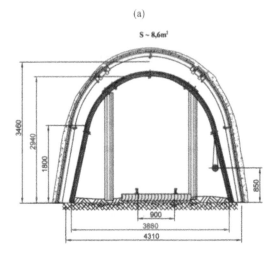

(b)
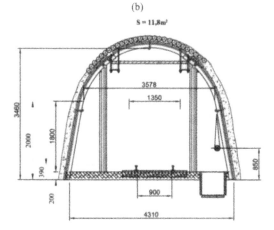

Figure 2. A condition of the DVR in the course of refastening: a – before; b – after.

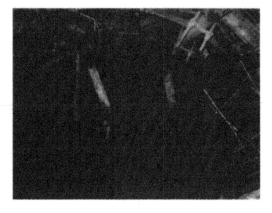

Figure 3. Condition of DVR during re-bolting.

The analysis of the existing approaches to carrying out developments in the conditions of an overworking showed that completely there are no regulations at the choice of location of mine working in the massif of rather overlying coal layer.

Decisions on a venue of mine working are made on the basis of rational proposals of technical officers of the enterprise. The unforeseen emergencies connected with discrepancy of system of fastening to the demanded parameters for maintenance of stability of development during all term of work result that in turn causes relevance and carrying out additional researches of the presented problem.

3 COMPUTING EXPERIMENT CONDUCTING

Very ample opportunities for the solution of similar tasks has a program ANSYS complex (a product of ANSYS Inc.), having method of finite elements method as a mechanical and mathematical basis and possessing expanded computing opportunities. The complex allows to solve a wide range of engineering tasks, including definition of the intense deformed condition of designs (Fomychov et al. 2014).

The considered model (Figure 4) consists of 25 rock layers, thus height of model is equal to 48.2 m. Power and physic mechanical properties of each layer completely correspond to the geological forecast for the 547th longwall of seam C_5.

Width of model id equal to 290 m that includes directly a longwall (230 m), extraction entries, and also sites of earlier fulfilled longwalls on 25 m from each party. In a roof over the fulfilled longwalls, and also round extraction mine workings of joint-block zone displacement with splitting rock layers into blocks with various sizes is modelled. At a depth of 9 m from coal layer and on removal 18 m from a first line of model it is simulated Drainage ventilation roadway. According to the passport of fastening of a drainage roadway it was simulated frame and anchor support. Depth of model made 55 m. The mechanized complex was modelled in the form of the continuous rectangular block by width 5 m and height 1 m. Worked-out area is filled with the brought-down rocks simulated in the form of the continuous block with the strength characteristics corresponding to the destroyed massif (Russkikh et al. 2013).

At approach of a stoping face there is an association of a frontal abutment pressure zone ahead of a face and in development sides. Therefore, there is a distribution of the squeezing tension in the range of 7 – 10 MPa on the considerable square of the massif.

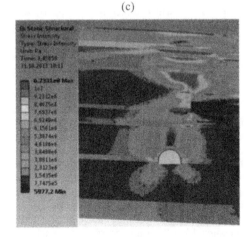

Figure 4. Diagram of tension intensity σ of the rock massif during overworking of DVR: (a) 14 m from a stoping face before mine working; (b) 7 m from a stoping face before mine working; c) stoping face is at the level of the central axis of mine working.

This area occupies all space between coal layer and development and reaches 10 – 12 m in width and approximately as much in height. Thus, the increase in intensity of a heaving of rocks of the soil in development that is characterized by increase in concentration of the stretching tension is noted.

Diagram of tension intensity (Figure 4) reflects the fullest picture of stress-strain state (SSS) of rock massif. The frontal abutment pressure zone ahead of a face extends on height to 10 m and width 5 – 7 m with a characteristic bias in a roof towards the worked-out area, in connection with a flash of rocks of a roof in the formed cavity. Concentration of tension increases from 20 to 30 MPa and above, depending on a site.

In the main roof over the worked-out area there is a formation of block structure on height to 12 capacities of the taken-out layer. It is followed by increase of tension over 25 – 30 MPa that exceeds strength for these rocks. In sides of mine working abutment pressure zones reach to 5 – 7 m in height and 3 – 5 m in width, at concentration of tension in the range of 15 – 20 MPa. Thus, there is an association to a frontal abutment pressure zone ahead of a face in this connection, lying between coal layer and mine working, layers are affected by tension intensity σ about 10 MPa. In a roof and the bottom of mine working unloading zones were formed.

For receiving a full picture of SSS influence on a massif of mine working, it is necessary to analyze a condition of the frame and anchor fastening established in a roadway. For this purpose, we use diagram of tension intensity σ for three main stages of calculation (Figure 5).

For start we will consider frame support, which racks, at all three stages of calculation, are considerably loaded. It is connected with pressure of side rocks, and also in connection with transfer of pressure of a cap board through a pliable element. Tension is in racks on limit of strength of steel of 270 MPa. A cap board fix at the first stage of calculation (Figure 5, a) is exposed to tension which isn't exceeding 200 MPa, thus the area with higher concentration which is localized with shift aside, opposite to approach of a face is already formed now. Asymmetry of distribution of tension in a cap board with each subsequent calculation only amplifies, thus the sizes significantly don't change, occupying 2 – 2.5 items, unlike tension which increased to 270 MPa.

Side an anchor are loaded rather poorly, tension exceeding 100 MPa isn't recorded. An anchor, established in a roof, participate in development maintenance process more effectively. At distribution of tension, there is an asymmetry, similar with a frame, shift of tension to the opposite side of approach of a clearing face.

(a)

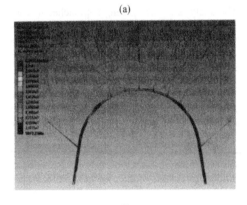

(b)

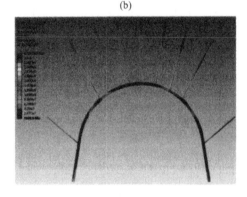

(c)

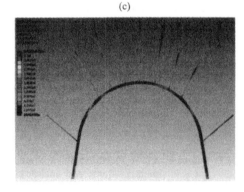

Figure 5. Diagram of tension intensity σ fastenings of mine workings in the conditions of water content: (a) 14 m from a stoping face before mine working; (b) 7 m from a stoping face before mine working; (c) stoping face is at the level of the central axis of mine workings.

In the field an anchor at 80% of the length are exposed to tension in the range of 180 – 220 MPa. In separate places, it is obvious in places of formation of longitudinal cracks in the massif and rock displacement relatively each other, there is an increase of tension to the maximum 270 MPa that lead to possible deformation of a steel midstream in this site. Three anchors from approach of a face are loaded rather poorly, at the level with side anchors, tension doesn't exceed 120 MPa.

The received results allowed to get the fullest picture of SSS of the massif and fastening of development in the conditions of an overworking. On the basis of experiments it is possible to define shortcomings of the applied system of fastening that will be the basis for development of the rational scheme of fastening the overworking mine workings with taking into account all parameters with high degree of objectivity.

4 TECHNOLOGICAL PART

Recent trends of development geomechanics researches directed on maintenance of mine workings, which are in a zone of influence of stoping operations are directed on the accounting of all factors influencing stability of mine working. It is unambiguously established that such approach to maintenance of mine workings when the massif is considered separately, and fastening elements separately, has no right for serious consideration. A number of theoretical works and mine experiments confirmed that frame support doesn't carry out the direct functions until the contour of development doesn't start putting pressure upon a frame. As a result, maintenance of development begins only after the beginning of active process of displacement of breed in development that reduces overall performance most fix that could constrain this process even at the time of its development. This shortcoming partially decreases by increase of culture of production, qualities of installation fix, and uses of wooden, nevertheless, it is rather half measures which don't yield necessary result.

Use of anchor fastening in the basic is also directed on correction of a shortcoming frame fix also its addition. Drilling of shots directly in the surrounding massif with installation in them of steel fittings, and also use the resin-grouted materials showed the efficiency in practice. Creation by means of anchors, so-called "support plate" in a development roof, allows to reduce partially pressure upon a frame support, and also interferes with a chaotic collapse of rocks in development.

In the conditions of the soft flooded rocks of the Western Donbas an optimum condition is use of a combination of two main types of fastening. Thus the technique that allows to unite anchor and frame support in the uniform cargo bearing system by means of rope communications (Sotskov & Saleev 2013) perfectly proved. As a result of an anchor, fixed in the massif are used for increase of effect of resistance of a frame to pressure of rocks, relying on strength properties of the massif. Completely closed structure, which is most effectively supporting a development contour, turns out.

Not less important the factors influencing fastening work are the installation step, number of a special profile and type frame fix, use of pliable elements, length and type of anchors, and also the scheme of their arrangement. Successful operating by the listed factors, allows to increase significantly overall performance of fastening and to reduce metal consumption of a design that is considerably reflected in cost of fastening of running meter of mine working.

Each change in system of fastening underwent testing by carrying out computing experiment by results of which it was possible to give the corresponding assessment to efficiency of its work. Depending on result calculation repeated taking into account amendments, or there was a transition to consideration of the following element in case of positive result.

Some schemes of an arrangement of anchors (Figure 6) were developed for search of optimum parameters of the unit of anchor fastening. When developing schemes were considered not only results of calculation of the VAT of the used system, but also feature of operation of overworking mine working which is exposed to influence of passing of a clearing face, atypical for preparatory developments, in a roof. According to calculation of the SSS for these schemes, the comparative analysis by results of which it is established was carried out that the most effective is use of the scheme in Figure 6.

Use of a combination the resin-grouted and rope anchors allows to maintain more effectively rocks of the main roof for decrease in influence on beam frames. There was an opportunity to refuse side anchors, which were poorly loaded, by installation the resin-grouted anchors under rope as the main load of fastening occurs at the time of passing of a stoping face that provokes intensive pressure in a development roof, thus sides are loaded less.

5 CONCLUSIONS

Analysis of system of drainage ventilation roadway fastening of Mine Management "Samaraska" during mining of working by stoping face. By results of computing experiments series, the analysis of SSS of the used system of fastening is carried out. Shortcomings of this system work are established. On the basis of the obtained data, rational parameters of fastening system for this type of mine working that is confirmed with additional researches are developed.

225

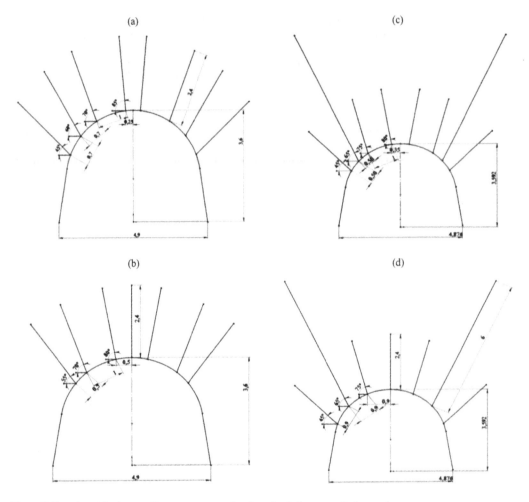

Figure 6. Experimental schemes of an arrangement of anchors for drainage ventilation roadway.

REFERENCES

Fomychov, V., Lapko, V. & Fomychova, L. 2014. *Modern technologies of bolting in weakly metamorphosed rocks: experience and perspectives.* Progressive Technologies of Coal, Coalbed Methane, and Ores Mining. The Netherlands: CRC Press/Balkema: 347 – 350.

Russkikh, V., Demchenko, Yu., Salli, S. & Shevchenko, O., 2013. *New technical solutions during mining of seam C₅ under the complex hydrogeological conditions of western Donbass.* Mining of Mineral Deposits. The Netherlands: CRC Press/Balkema: 257 – 260.

Sally, S., Pochepov, V. & Mamaykin, O. 2014. *Theoretical aspects of the potential technological schemes evaluation and their susceptibility to innovations.* Progressive Technologies of Coal, Coalbed Methane, and Ores Mining. The Netherlands: CRC Press/Balkema: 479 – 483.

Sotskov, V. & Saleev, I. 2013. *Investigation of the rock massif stress strain state in conditions of the drainage drift overworking.* Mining of Mineral Deposits. The Netherlands: CRC Press/Balkema: 197 – 201.

Theoretical and Practical Solutions of Mineral Resources Mining – Pivnyak, Bondarenko & Kovalevska (eds)
© 2015 Taylor & Francis Group, London, ISBN: 978-1-138-02883-8

First metals discovery and development
the sacral component phenomenon

H. Haiko
National Technical University of Ukraine "Kyiv Polytechnic Institute", Kyiv, Ukraine

V. Biletskyi
National Technical University named after Yuri Kondratyuk, Poltava, Ukraine

ABSTRACT: This article is accentuated on the civilizational significance of the first industrial metals discovery, it reveals the prerequisites for the development of ores and the main stages of metals mastering connected with centuries-old experience of the previous mining activities of mankind. The authors suggest the hypothesis that the metallurgy was born in the depths of sustainable communities of archaic miners, which were motivated by not only utilitarian but also sacral factors. A new hypothesis is examined in comparison to the established versions of accidental discovery of metals. Key phrases: history of mining, industrial archeology, the birth of metallurgy, primal mining communities, sacral factors, mining cults, irrational traditions.

1 THE PROBLEM AND THE STATUS OF ITS STUDY

The phenomenon of discovering and development of the first metals has always been considered as one of the most important stages of the mankind civilizational growth. However, the continuous absence of any archaeological artifacts and the inability to use written sources (the birth of metallurgy outstripped and perhaps defined the conditions of the invention of writing) led to the hypothetical explanations for the processes of metals and metal appearance, which were based solely on the concepts and logic of their authors. This basic thesis was that the early humans randomly were witnesses of an unexpected melting of metal from ores in a home fire or forest fire and after this hint of nature they started a self-dependent metallurgical activity. At the beginning of the 20th century these hypotheses were set out in the fundamental papers (Borhers et al 1901, Kramer 1904) and the history books. From the second half of the 20th century archeology of metals begins to accumulate numerous artifacts that detailed stages of development and historical geography of metals (Chernykh 1972, Hauptmann 2000). Some archaeological works marked the frequent use of sacral factors among the primary miners (Chernykh 1972, Hauptmann 2000, Babel 2008, Kargala 2002). The authors made an attempt to explain the close relationship between the formation of sustainable mining communities, sacred worship and a birth of mining and metallurgical

technology in several previous works that explored the phenomenon of primitive mining activity (Gaiko 2012, Gaiko 2009, Gaiko & Biletskyi 2013). Formed on their attempts hypothesis stands to the attention of the reader.

The purpose of this article is an analysis of possible ways of discovering the first metals, the assessment of the impact of sacral components on mining and smelting activities motivation and organization, the reasoning of the new metallurgy birth hypothesis, which answers some questions that could not be solved within the framework of other hypotheses.

2 THE PRESENTMENT OF THE MAIN PART

It is hard to overstate the importance of a comprehensive meaning of the metals discovery in a human civilization progress. The ore mining supplied it with fundamentally new materials that had previously unknown properties (ductility, melting) and important advantages (ability to change its shape, high toughness, operational durability). That not only effectively improved the available types of stone tools, but also created entirely new types, which opened opportunities for a technological progress. A famous American anthropologist L.H. Morgan said: "When a barbarian, moving forward step by step, discovered the native metals, began to melt them in a crucible and cast in a mold; floated native copper with tin and created bronze and ultimately, even through the greater effort of thought he invented the furnace and pulled out iron ore – nine-tenths of the struggle for civilization was won".

Simulating ancient production processes, the researchers compared the effectiveness of using stone and copper tools. As a result, they concluded that productivity increases in case of using copper tools for felling trees – 3 times, for cutting – 6 – 7 times, for drilling – 22 times (Semenov 1968). Metal tools had the particular importance in agriculture, where they greatly accelerated the cultivation, harvesting and construction of irrigation systems. More properties of metals were needed for the manufacturing process of weapons, including its new type – swords.

The emergence of wars, conquering tribes and peoples, the establishment of government became possible largely due to the advantages of copper weapons. One of the results of the mining and metallurgical industry formation process had become an international division of labor and an extensive development of ancient world communication system on its basis (transport of ores, metals and products of them). Search and development of new fields promoted the migration, population and development processes of distant lands. Metals also had a considerable impact on trade, as they were the first universal equivalent for commercial exchange. Civilization changes caused by the mining and metallurgical activities had a global nature and combined both technical and social components of society development (Gayko & Biletskyi 2013).

The unique phenomenon of learning and mastering ores and their metallurgical processing is one of the highest and mysterious collective manifestations of human genius. "Why did this happen? Why do human beings do not live today as they did in the Mesolithic period?" – raised the question a researcher of primitive man R. Braidwood. The answer on that question, which is still formed by historical science, is the cornerstone of human history.

An example of "inertia of thinking" according to the authors is the centuries-old hypothesis of the random nature of the discovery of metals, which claimed the dubious (on our opinion) idea that the birth of metallurgy is connected with observing of rare random events of copper ore melting that suddenly landed in a primitive man's fireplace. The discoverers of metal, according to this version, could be ancient hunters or ranchers who unexpectedly witnessed an accidental copper melting. Most of the modern historical reconstructions are still revealing the version of "how a shepherd became a miner".

The authors consider that this version cannot answer the basic question of R. Braidwood, why for tens of thousands years of Homo sapiens activity there were no "cases" of an accidental copper melt-

ing till Neolithic. Finding the answer on a difficult question "Why?" we have tried to study the problem from a different angle and set in the middle of the research another question, "Who?" – Who could be a discoverer of metallurgy?

The importance of this issue substantially growth considering the limited number of possible centers of metallurgy origin. Archaeological evidences of the recent decades localized the occurrence era of metallurgy (IX – VIII millennium BC) to the few centers in the eastern and central parts of Middle East. Later (VI – V millennium BC) a powerful copper development center formed in the Balkans, and a thousand years later – in the South Caucasus. This "historical geography" divided researchers in mono- and poly-centrists. The first ones consider that the development of copper occurred in a single center from which the mining and metallurgical knowledge was distributed ("diffused") on a broad band around the Black Sea and the Middle East. The second ones defend the hypothesis of generation of metallurgy in three or four separate centers with the following spread on the nearby and remote areas.

As we can see, only inhabitants of a several regions could participate in the initial development of metals. The presence of the rich copper ore deposits near the surface was a necessary but no a sufficient condition for their discovery.

It is also noteworthy that the termination of the ancient copper deposits operation was almost always connected not with the exhaustion of stocks, but with the resettlement (or destruction) of the population, which owned a mining and metallurgical expertise. The most typical example is the development history of a powerful Karhalynsky deposit in the Southern Urals (Kargala 2002). During III – II millennium BC it was one of the richest copper mining centers in Eurasia (on the territory of size 50 by 10 km researchers discovered about 30 thousand mine shafts, most of which was considered as primitive mines). At the end of the II millennium BC for unknown to us reasons experienced in ore mining tribes abandoned these numerous mines. After them for three thousand years all nations that lived in these lands were no longer able to master the development of local ores, despite the fact that thousands of mine shafts basins indicated rich deposits. The complexity of this task show the sheets of Ural mining plants head G.V. de Gennin to Peter I, in which he points out that Saxon masters cannot smelt Uralian copper ores advises to invite masters from Hanover and Mansfeld, "which are able to smelt the shiver ore". Despite the participation of renowned specialists and personal control of top officials "ore-smelting industry in Russia started incredibly hard and strained, it was new and very complicated business" (Pavlenko 1962).

This example shows clearly enough how difficult it was to develop the ore deposits and to get a metal even for the experts of eighteenth century and how far from scientific truth is an illusion that the copper business opening was rapidly and easily accessible for primitive people who lived in the copper-rich area.

To find out who were the possible "authors" of metal discovery, we should assess the probability of random events of copper-melting and their role in the problem of development of metal. Unfortunately for "lovers of simple solutions" the melting probability of even a fusible metal copper ore in the usual fire is very low (insufficient temperature). The researchers conducted hundreds of experiments that proved impossibility of metallurgical copper ore smelting without a purposeful air blowing into the fire. In exceptional cases where the ore was caught in the kiln for firing ceramics or in a huge forest fire, the required temperature (700 – 800°C) could be achieved, but it is not sufficient condition of metal smelting. An important factor is a presence of a special environment that demanded a comprehensive contact with charcoal.

Let us suppose that an extraordinary case brought together all the necessary conditions. In this case in the fire would have been discovered not bars or lumps of metal but only small drops of copper, inseparable from the ore, and even under a layer of ash. Even if they were found, they could not say much to an ignorant man and an attempt to replicate the melting in normal fire would probably ended as a failure. These considerations give a rise to the motivated doubts about random nature of copper ore discovering, which serves as an established version of the metal discovery. The hypothesis of successful ore searches by ignorant people also causes doubts.

It is also important to consider that in addition to ore the native metals could be found in the mountains and valleys by anyone. Most researchers connect the primary use of the metal with the native copper. However a native cooper in a separate "stone" could be found in a huge quantity. The oldest articles thereof are small decorations (charms, beads, etc.). With high probability it can be argued that a primitive man who accidentally found a piece of copper on the earth's surface would never find the other one for the second time. With such a limited amount of metal, it is unlikely for a primal man to ponder about new production opportunities (if such considerations were possible at all for that time "inhabitants").

As you can see, the role of a "random person" in the process of metal discovering could not be decisive. Moreover, the arguments lead to the idea of targeted long-term development of metallurgical processes. As L. Pasteur said – "In the fields of observation chance favors only the prepared mind."

To our opinion, there are a lot of evidences that the discovery of metals is connected with the centuries-old experience of the previous mining activities of mankind, distinguishing processes of some specific communities of archaic miners ("Hunters of Stone) in the Neolithic (and possibly earlier), accumulation and development of first knowledge about metal in these professional societies. Centuries-old experience of communities that mined and processed flint, obsidian, pyrite, raw materials for mineral paints, etc., formed a special search archaic worldview of miners that saw in stones the presence of capabilities that were sacred and hidden to a human being. At a certain stage of development of mining experience (collecting samples of native metals and ores, identifying their search attributes, styding their properties, the formation of the sacred traditions) appeared a creative idea of a new material – "malleable stone" and later "fluid stone" (metal). It has been formed and implemented by mining communities, which passed it from generation to generation for achievement of a rock art in the system of special sacral worship. The most ancient pit developments of mineral raw materials for the manufacture of paints (hematite), steel (pyrite), etc. are dated by the period of 35 – 40 thousand years ago. The underground flint mining started much later (there was enough of it on the surface), but in IX – VIII millennium BC appeared the ancient flint mines that solved a range of complex technical problems already in Neolithic times.

It is important to note the scale and the focus of mining operations of Neolithic era. Almost every large deposit was opened by several thousands of mine shafts (depth 20 m) with a network of horizontal workings (usually – by petal pattern). The continuous development of such deposits dates back several hundred (sometimes – thousands) of years, and there were identified tens of old mining centers (we may assume that archaeologists discovered only a small part of them), which gives a reason to believe in existence of a special sustainable community of miners, who were sedentary, were separated from the other tribes with their activity and specificity of original sacral culture.

Indirect evidences of the miners' connection to their predecessors who developed stone raw materials can be observed in similar construction techniques of Neolithic pits (the usage of "fire way" to destroy rocks) in almost identical mining tools usage, even in mandatory filling of created space. The last typical factor for flint and copper mines in the vast areas of Eurasia is especially significant, since the very labor-intensive production technology is not life-essential (especially filling trunks), it is more a reflection of a cultural tradition. Perhaps there was a taboo for "wounding" the earth's surface, which required "curing" it by returning everything to the original state (filling cavities with wasted rock). If ore mining was carried out by people who would not have anything to do with flint mining (societies of random hunters or ranchers) this tradition would be certainly interrupted.

In favor of the copper extraction concept as a single mining complex testified numerous archaeological excavations of ancient mines that bind ore treatment, metal melting, and even metal fabrication tools with mining activities. A placing of the oldest metallurgical plants near mines, miners settlements, mutual placing of metallurgical and mining tools, treating and smelting areas give a reason to believe that the miners and metallurgists lived for a long time the one community that dates back to the Neolithic. As an eloquent example of the mining experience continuity can serve one of the oldest human settlements – Catal-Hüyük (VIII millennium BC, southern Turkey). Close to two extinct volcanoes, it has been a center of obsidian development (volcanic glass) – the best weapon material of his time. Archaeological excavations have revealed deposits of obsidian in many buildings of the city, as well as numerous obsidian workshops. Exactly in this small village were found copper products, copper slag and scale, what confirms the possibility of the first copper smelting by the same miners, which developed stone.

Describing the first organizational and ideological factors of copper development, it should be noted that the interference into the metal's nature was perceived by a primitive man as a mystic, wonderful thing. It was filled with symbolism of sacral forces. The archaic miners felt themselves in a constant and direct contact with the other, invisible world which was to them no less apparent than a real one. There are many evidences of extremely high value of sacral and magic factors in an ancient miners' activity, which are indicating a significant impact of ancient magical cult's servants.

An objective basis for the emergence and spreading of these cults was a significantly higher (compared to other productive activities) role of uncertainties in the work results of miners, metallurgists. It was never known for certain whether the excavation (the result of hard, long and dangerous work) encounters on ore deposits, the ore quality is fine, a collapse of the mine won't happen to the worker under the ground, a flooding or fumes happen. It was unknown if the metal's smelting will be successful, if miners will be able to get a desired quantity and quality of metal.

Among the typical examples of sacral worship of miners called irrational tradition must be notified a filling the mines with a waste rock (as already mentioned); left in the Neolithic mines solar signs as mystical appeal to patrons; mining tools cult, which "helped" the successful conduct of underground works and so on. Also significant are many burial places of small ungulates in the mines and on the industrial areas (perhaps as a gratitude to the other-worldly forces for obtained mineral resources). An ancient burial miner tradition is known for burring metallurgists with working tools and a rich copper ore, which probably was a kind of "calling card" of the dead man, when he traveled to another world. Archaeologists have reconstructed some magical actions (for example, the construction of sacred labyrinth of trenches, the use of divination bones, etc.) that archaic miners performed to detect deposits for search shafts digging. Ethno-archaeologists, who studied the lifestyle of modern human societies in conditions close to their original state (the tribes of Africa, Polynesia) report that metal melting is a subject of many magical rituals: the choice of "destined by spirits" smelting time; a sanctification of melting furnace; anvil worship; numerous taboos were always respected by metallurgist and blacksmiths. Only a few men in the tribe knew the secret art of metallurgy (by R. Forbes, "the given by fire spirits power to turn stones in metal").

We believe that there was a high degree of interaction between archaic miners and archaic priests of magical cults. Presumably, this relationship was stronger among miners than in other activity fields, what leaded to more rapid formation of the ruling tribe elite (suspended – priestly caste), since the presence of minerals largely determines the richness and power of the tribe.

It is logical to assume that the first mining and metallurgical knowledge that was considered as magical could be persisted and developed by the ancient priestly caste, which was related to mining communities, and passed them down from generation to generation. Their authority and ceremonial activities created an effective consolidation in mining community to do especially labor-intensive and dangerous underground works for numerous repetitions of complicated attempts to smelt metal (most of which did not give the desired result). We do not exclude that the invention of air blowing in the center of fire of ore, which opened up the possibility of melting steel, was associated with magical rituals of giving the ore the power, i.e. high-temperature regime could be initially invented through "ideological" and not "technological" ideas, but in an environment where both of these factors were very close.

The factor of sacred worships and their priests' activities in mining communities should be considered as one of the prerequisites for the birth of metallurgy. If the fact that the sacred practice of medieval alchemists helped discover many properties of ores, metals and their alloys and get new metallurgical technologies (Rabinovich 1979) is perceived as obvious, why we expect from the primitive communities not sacral, but purely rational way of opening of metals' secrets? Perhaps this methodological error stood in the way of accurate historical reconstruction of the beginnings of metallurgy for a long time?

From the written sources of III millennium BC is known that miners of Sumer was owned and operated directly by temple priests. This practice also took place in ancient Egypt, where existed a cult of the goddess Hathor, patroness of miners. Presumably, the projection of such relations can be reasonably widespread on the earlier period of social development. Factors of sacral worldview of a primitive man and the role of ancient cults' priests should be considered as part of the necessary conditions for the metallurgical activities establishment.

3 CONCLUSIONS

Prerequisites for metallurgy's genesis (according to the author's version) lied in the progressive development of mining technology of Stone Age and have been associated with the formation of a stable cultural community of "mountain people" already in the Neolithic who developed the non-metallic minerals mining and accumulated experience for centuries. In the depths of this community began a deliberate, long-term, and associated with the cult of sacral miner's activity of metals exploration.

Factors of sacral and magical character and activities of magical cults' priests among archaic miners should be considered as a necessary condition for the birth of metallurgy, assuming that there were not only rational, but also sacral practices aimed at the discovery of the mysteries of metal (similar to the mystical practices of alchemical historic times).

It is possible that the invention of air blowing, which provided an opportunity of metal smelting process discovery, was associated with magical rituals of ore testing with the fire of huge power, i.e. high-temperature regime could be initially created through "ideological" and not "technological" motivation.

Achievements of archaeometallurgy of recent decades add efficiently new evidences to the modern idea of the development of the Neolithic revolution; allow to separate a well-established community of miners, metallurgists and to put it next to the husbandry and pastorally communities as an important component of the Neolithic revolution.

REFERENCES

Babel, J. 2008. *Krzemionki Opatowskie. The earliest beginnings of modern mining.* New Challenges and Visions for Mining 21st World Mining Congress.

Borhers, V., Vyust, V. & Treptov, E. 1901. *Mining and Metallurgy: translation from German.* Industries and Technology (Encyclopedia of industrial knowledge). Association "Education": 677.

Chernykh, E.N. 1972. *Metal – man – time.* M: Nauka: 208.

Gaiko, G.I. 2009. *The history of the development of the Earth's interior.* Donetsk: East Publishing House: 292.

Gaiko, G.I. 2012. *Birth of metallurgy – why it all happened?* Technique – Youth, No 11: 16 – 21.

Gayko, G. & Biletskyi, V. 2013. *History of Mining: Textbook.* Kyiv-Alchevsk Publishing House Kyiv-Mohylanska Academy of Publishers "Lado" DonSTU: 542.

Hauptmann, A. 2000. *Zur fruhen Metallurgie des kupfers in Fenan Jordanien.* Veroffentlichungen aus Deutschen Bergbau-Muzeum, No 87.

Kargala, T.I. 2002. Languages Slavic culture.

Kramer, G. 1904. *The universe and humanity. The history of the study of nature and the application of its forces in the service of humanity.* Association "Education": 460.

Pavlenko, N.I. 1962. *The History of Metallurgy in Russia XVIII century.* Plants and their owners. M.: Publishing USSR Academy of Sciences: 567.

Rabinovich, V.L. 1979. *Alchemy as phenomenon of Middle Ages culture.* M: Nauka: 392.

Semenov, S.A. 1968. *The development of technics in the Stone Age.* M: Nauka: 362.

Theoretical and Practical Solutions of Mineral Resources Mining – Pivnyak, Bondarenko & Kovalevska (eds)
© 2015 Taylor & Francis Group, London, ISBN: 978-1-138-02883-8

State support of stability in coal-producing Donbas regions

S. Salli
National Mining University, Dnipropetrovsk, Ukraine

ABSTRACT: This article represents a methodological approach to restore the capacity of mines and enrichment plants of Donbas, damaged by military operations. It was offered matrix models of the fuel supply for power plants for energy security of Ukraine and reduction of social tension in the coal-producing regions.

1 SETTING PROBLEM

Military operations in the eastern regions of the country are making their adjustments to the level of recovery costs in general and the Donbas coal industry in particular. The government estimates the loss of several billion hryvnias, and the extent of the devastation continues to grow. EU plans to send a special mission to Ukraine for inventory and assessment of damage and render possible support. Obviously, in general plan of returning east Ukraine to life, should include measures to restore the coal industry. It should be emphasized that at the coal mines was seen a serious outflow of young staff of leading professions. If the coal industry is recognized as a guarantee of energy security, then the need for training also will require corresponding budgetary costs. Resumption of enterprises in Donbas and, above all, the coal industry provides resolution of social issues that are caused by:

– first, a significant share of social expenditure in total expenditure on repair of mines corrupted by war state;

– second, the large difference between the actual financing and needs of the industry to support each ton of production capacity;

– third, the negative impact of past restructuring the socio-economic situation of the mining monofunctional towns, manifested in increasing unemployment, lower industrial production and low income in population of these cities.

Today in Ukraine – 450 cities, 344 of them are classified as small; their population is less than 50 thousand residents. In general, they are about 6.5 million people, or 13.5% of the total population. Only 7% of small towns feel more or less comfortable. All others belong to the so-called depressed because they are classified as monofunctional settlements whose lives provide one or two industrial enterprises. Turning to specifics, the main problems that concern citizens of Shakhtarsk, Dzershynsk, Ukrainsk, Torez, Snizhne and many others, are low income (50.3% of respondents) and unemployment (Vagonova 2005).

Employment still remains a major unresolved issue in the Donbas. If we solved it, possibly, we would avoid many problems. Thus, to the question of what most cities are missing to improve their lives, the vast majority of respondents (48.6%) indicated that they do not have the financial resources for social support and investment in the city's economy (22%). Today depressed areas include settlements, which since 1996 have coal mining and processing enterprises liquidated or are being liquidating. A striking example of such territory in the Donetsk region may be Snizhne city, where since the dawn of independence operated seven mines. Today, only one left, and unemployment reaches 75%. The same situation exists in Torez, where out of 13 mines are only three, and they were stopped after being damaged by military action (Without coal...).

Thus, state support for coal-producing Donbas region, where most of the population is connected with the work of coal mines, will be directed to achieve full employment. Moreover, due to shortage of budget funds targeted investment is needed, means the choice of the region, assessments of its decline and promising coal mining assets in terms of quality coal reserves.

2 ANALYSIS OF RECENT RESEARCH AND PUBLICATIONS

Analysis of researches, related to the social problems of the coal industry of Ukraine, contains mainly statistical data on mines of Donetsk coal basin without a comparative analysis of the social situation of workers of coal enterprises within particular coal-producing regions. It should be admitted that in the works from experience analysis of the restructuring of the coal industry are no recommendations for modeling processes of mines in the area of expanded reproduction (Amosha et al. 2002, Pivnyak et al. 2004,). Several studies suggested methods of predicting mobility of mine workers, but no calculations not only for individual mines, but also for cities or regions (Buts & Gerasymova 2007, Sally et al. 2014). The last makes it possible to predict the level of employment potential immediately at the regional level and will promote interregional industrial units, which is prospective for the economy of Ukraine. Thus, determining the cause and effect of relationships forming the depression state of the regional economy and mining towns is the basis for evaluating the effectiveness of the reconstruction process of the coal industry.

Several concepts of impact on social state of Donbas coal-regions in terms of interference consequences of the mines, enrichment plants and power plants are presented in this work.

3 THE WORDING OF ARTICLE PURPOSE

The aim of this work is a generalization and development of scientific and methodological foundations, development tools and algorithm for modeling interference of population living standards in coal-producing regions and stability of mines and power plants work.

4 THE MAIN MATERIAL

In the current situation thousands of people, lost their jobs, are forced to seek refuge in other regions of Ukraine or abroad, need state protection to improve social status. More than others suffered workers of coal mines in the eastern regions.

Unfortunately, for many years during the planning of development of individual regions dominated industrial approach to the assessment of mines when all 130 companies were considered as objects of industrial production. This resulted in unilateral solutions to many issues related to the distinction of regional social policy. Today we need a new approach, connected with European integration, meaning adjusting existing terms of the assessment of coal mining assets, which always considered in isolation from the social and demographic block issues. It should be noted that the degree of social location of Ukraine mines is very different: from locations in big cities, to separate, remotely located villages. In other words, the mine should be considered also as socio-demographic feature (Amosha & Yashchenko 1999).

With regard to the situation of restoration of, impaired by war, social stability and coal-Donbas region potential a multi-layered distributing problem should be considered. The essence of this problem: by the end of the planning period in spatially separated points of a given region it should be restored the company to be considered as a source of energy and as electricity consumers. There are options for recovery mines' potential and their maximum possible power is known.

We introduce the following notation: X_{1i}^{2j} – volume of deliveries of coal i-th mine to the j-th power plant; X_{2j}^{s} – electricity supply from the j-th power plant; s-th consumer; X_{2j}^{li} – unused capacity l_i-th mine; f_1 – the cost of producing 1 thousand tons of coal in the i-th mine; f_2 – the costs of producing 1 million kWh/h of electricity j-th by power plant; u_i – the cost of transporting 1 ton of coal from i-th mine to the j-th power plant; λ_i – power losses and costs of transmission of the j-th power plant to the s-th user; M_i, M_j, M_{2i} – the maximum possible capacity respectively of mines, power plants and processing plants; Ds – demand of s-user; γ_i – need of i-th mine in electricity; l_{1i}, l_{2i} – respectively productivity at the i-th mine and enrichment plant; L_R – the working-age population in the region.

The problem is described by the following system of relations.

To minimize:

$$F = \sum_{i=1}^{m}\sum_{j=1}^{m2}\left[f^{1i}+u_{1i}^{2j}+\lambda_{1i}^{2j}\right]X_{1i}^{2j}+$$

$$+\sum_{j=1}^{m2}\sum_{s=1}^{m3}u_{2j}^{s}+\sum_{j=1}^{m2}\sum_{i=1}^{m}u_{2j}^{li}X_{2j}^{li}=min, \qquad (1)$$

with the conditions:

$$\sum_{j=1}^{m2}X_{1i}^{2j}+X_{1i}=M_{1i}, \quad \sum_{i=1}^{m}\lambda_{1i}^{2j}X_{1i}^{2j}+X_{2j}=M_{2j} \qquad (2)$$

$$\sum_{s=1}^{m}X_{2j}^{s}+\sum_{i=1}^{m}X_{2j}^{li}+X_{2j}=M_{2j}, \qquad (3)$$

$$\sum_{j=1}^{m2}\lambda_{2j}^{s}X_{2j}^{s}=D_{s}, \qquad (4)$$

$$\sum_{j=1}^{m2}\lambda_{2j}^{s}X_{2j}^{s}=D_{s}, \qquad (5)$$

$$\sum_{i=1}^{m}\frac{M_{1i}}{l_{1i}}+\sum_{i=1}^{m2}\frac{M_{2i}}{l_{2i}}\leq L_{R}, \qquad (6)$$

The best indicator of a region is the living level of its population. Living level is complex. The core of it is the level of cash income. However, this level is also heterogeneous, as it includes income from production and not production activities. Ultimately, they are directed to necessity to ensure all members of society with a certain standard of living no worse than in other regions. In this sense we should talk about balanced development of coal-regions. These requirements are implemented in production of each region, which is a specific territorial economic complex. It provides the most efficient development of productive forces in order to best meet the needs of the country in the finished coal products, produced in the region, and meet local needs. These highlights of planning regions' economic, differ by monofunction – coal mining.

Built on the model (1) – (6) matrix (Figure 1) allows us to formulate a number of important principles of stabilizing the situation in a particular region.

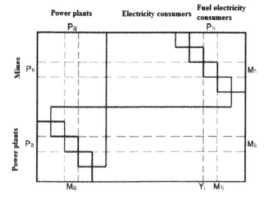

Figure 1. Matrix of incidents "mine – power plant".

The essence of these principles is to replace positive feedback (the larger the mine produces coal, the more electricity it needs) with two consecutive negative bounds (the larger the mine produces coal, the less unused capacity and the less unused capacity of the mine, the more power it needs).

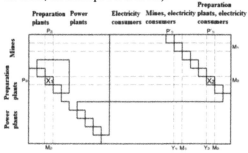

Figure 2. Matrix of incidents "mine – preparation plants – power plants".

On the Figure 2 is considered a more complex case, where the scheme included along with mines enriching plants, which are provided with electricity from the same power. Unused capacity of these plants should play a dual role here: first, to ensure consistency between the number of unenriched coal, supplied to the factory, and enriched coal shipped from the factory; secondly, to reflect the real energy needs of plants.

Solutions of this type of distribution problems with additional constraints like type $X_i = X_2$, usually performed by block methods (Takha 1985).

5 CONCLUSIONS

Restoration of the coal industry requires extraordinary measures by the state of social policy and improvement the structure of mine stock through developing exactly profitable reserves, a category of which is still limited in Ukraine.

Any approach to the assessment of coal-producing regions of Donbas production should ensure full employment. Currently, this situation is particularly fundamental because decline in Ukraine economic security is unacceptable due to deficiency of energy.

A better indicator of any coal-region is the living level of the population that live in there, meaning population cash income from productive and non-productive activities.

REFERENCES

Vagonova, A.G. 2005. *Economic problems connected with Ukrainian coal mines maintenance and investment.* Dnipropetrovsk: National Mining University: 287.
Without coal... 2014. Facenews. [Electronic resource]: http://www.facenews.ua/articles/2014/248922/.
Pivnyak, G.G., Amosha, A.I., Yashchenko, Y.P. et other. 2004. *Reproduction of mine fund and investment processes in coal industry of Ukraine.* Kyiv: Naukova Dumka: 331.
Amosha, A.I., Ilyashov, M.A. & Salli, V.I. 2002. *System analysis of a mine as investment object.* Donetsk: IEP NAS of Ukraine: 68.
Buts, Y.V. & Gerasymova, I.Y. 2007. *Labour efficiency in coal mines.* Dnipropetrovsk: National Mining University: 156.
Sally, S., Pochepov, V. & Mamaykin, O. 2014. *Theoretical aspects of the potential technological schemes evaluation and their susceptibility to innovations.* Progressive Technologies of Coal, Coalbed Methane, and Ores Mining. The Netherland: CRC Press/Balkema: 479 – 483.
Amosha, A.I. & Yashchenko, Y.P. 1999. *Regions are the most important component of the structure of market economy control.* Donetsk: IEP NAS of Ukraine: Industrial economy, No 2(4): 3 – 9.
Takha, H. 1985. *Introduction into operational research.* Books one and two. Moskva: Mir: 479.
Bondarenko, V., Symanovych, G. & Koval. O. 2012. The mechanism of over-coal thin-layered massif deformation ofweak rocks in a longwall. Geomechanical processes during underground mining. The Netherlands: CRC Press/Balkema: 41 – 44.

Theoretical and Practical Solutions of Mineral Resources Mining – Pivnyak, Bondarenko & Kovalevska (eds)
© 2015 Taylor & Francis Group, London, ISBN: 978-1-138-02883-8

On influence of additional members' movability of mining vehicle on motion characteristics

K. Ziborov & S. Fedoriachenko
National Mining University, Dnipropetrovsk, Ukraine

ABSTRACT: The paper is devoted to mining rail vehicle motion stability while motion on the rail track of steady and transient curvature. Several kinematical schemes of mining locomotives are studied. Assumed, that the most dynamically stable is locomotive with pin-joint coupling design. In addition, additional local movability of wheel set can increase safety factor as well as sectional conception of locomotive chassis. Substantiated, that local movability within locomotive chassis' coupling can increase safety factor and reduce the dynamical forces acting on the rail track. The research of mining motion stability provided in the paper. Substantiation of additional kinematical movability of mining vehicle elements is studied.

1 INTRODUCTION

The tendency of increasing an adhesion weight of locomotives up to 10 – 14 (28) ton (Bilichenko et al. 2005) in order to haul more heavy mining tub induced significant growth of static loads on the rail track. Because of the fact, that existing mining rail tracks have been design for much lower locomotives' weight, increased axial loading on the rail track elements rocketed up to 1.5 – 2.5 times and for mining tub 7 times more. However, while so huge increase of masses, axial loadings, motion speed and freight flow, modernization of rolling stock have not been provided.

The force interaction character and coupling class of mining rollingstock elements suppose the usage of the lowest kinematical pairs. At the same time, the dimensions of machines' elements can vary owing to wear and gap adjusting within kinematical pair, elastic deformations, heat expansion, mistakes while mounting and repair, etc. Thus, it is obligatory to choose such mechanisms' scheme when the requirements for accuracy will be not so high.

The statically determined mechanisms requires the properties where member are self-excited, without odd joints. Such member linkage induces reduction of dynamical loading and growth of motion safety. It happens owing to additional movability of construction elements, while the linkages, with appear a kinematical pair, takes stable space position. They trace the variable trajectory and do not induce additional force disturbance.

As a output element of the mining rolling stocks' majority used to account a wheel, which develops a frictional pair with rail while motion. The wheel set parameter must meet the requirements of optimality both in wheel function as a support rod to take the weight of the locomotive and transmit it on the rail, and the tractive element for generating the traction force to overwhelm the motion reaction. These requirements are contradictory, that measures the design parameters both the chassis and mining vehicle in general (Garg & Dukkipaty 1984).

At the other side, the wheel as an support rod, might trace the trajectory while motion on the rail track, which can not be linear in vertical surface, that induces dynamical load component in all vehicle components. An additional loading factor is that the wheel must trace horizontal trajectory either that causes constant lateral interaction (Garg & Dukkipaty 1984). While the motion on the curvilinear rail track, the wheel roll on the rail can cause a significant lateral forces, which may result in stability loss of the mining vehicle. At the other side, the tractive effort realizes at the point of wheel-rail contact through friction, and it is limited by frictional properties of bonded surfaces and pressing force (Ziborov 2014). All these factors evoke unstable motion regime of mining vehicle.

The modern design methods (Ziborov et al. 2013), which base on the scientific approach of mining machines simulation and research, facilitate to define the location and character of arising dynamical loading and prevent their growth while forming at the mining vehicle chassis. This prevents the following dynamical loads transmission on the bolster structure. Thus, the structure selection and selection of mining machines parameters, which bases on the detail analysis of running processes, might be an essential part of energy-mechanical system and its scheme development on the design stage (Ziborov et al. 2013).

The study purpose is determination of qualitative and quantitative influence of mining vehicle coupling on the stability while motion on the ail track with constant and variable radius.

2 THE RESEARCH RESULT

It is well known, that odd coupling are such connection type, which removal do not increase the total mechanism movability (Levinskiy 1990). The limited line displacement causes the necessity of force transmission within the kinematical pair between member, and limited angular displacement – the torque between members of the pair. The kinematical pair must be designed for these forces and torques. To check the odd linkages within the mechanism we can use constructional formula. We might take into account the quantity of known odd couplings. The quantity p_i of kinematical pairs of i-th class, that applies ip_i constraining conditions, and all kinematical pairs $\sum_{i=1}^{i=5} ip_i$ constrained conditions.

There can be general or local (passive) mechanism movability. The local movability is such one that does not influence on the total mechanism movability. The local movability has rollers (because of the possible slipping), pulley, band wheel, bushing and pin, cylindrical sliding bar with ball end. The kinematical pair coupling defines without clearances, if kinematical pair produced like ball bearing, where the clearance is very small or even can be tightness.

While clearance existence an additional kinematical movability arises, which can be used in machine operation. These clearances can occur because of frictional wear of the coupled kinematical members, and grow uncontrollable. Such growth evokes additional dynamical components of operational loadings and reduces machine's exploitation indexes. However, it is possible to revise the machine design and additional kinematical movability either to reduce the duration of nonstationary motion regime. This is essential for mining conditions, which is marked by lots of unfavorable factors of numerous nature (Ziborov 2014). To provide the smooth wear of coupled kinematical members a coupling with local movability can be applied (Ziborov & Fedoriachenko 2014).

The authors of the paper proposed and implemented on the industrial enterprises new technical solutions of mining vehicle components and machines. For example, locomotive of the module scheme, that includes a few sections. It allows for development of the vehicle with different trailing weight, energy supply system and necessary exploi-

tation indexes. The distinguish feature of such locomotives is kinematical coupling between bogie and tractive section (Figure 1).

(a)

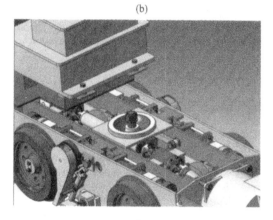

(b)

(c)

Figure 1. Pin joint locomotive (a); locomotive joint (b); mining wagon's wheel set (c).

Such connection provides necessary relative movability and transmits vertical loading from frame to bogie, horizontal lateral forces – centrifugal force, reaction of overrunning rail, which has geometrical imperfections in all surfaces. Movability around the vertical axis is necessary for tractive bogie turn and in order to avoid odd couplings, because the pin does not carry the chassis weight; around lateral axis – for correct weight distribution between locomotive axles and reduction influence on the rail track; longitudinal movability is absent because the tractive effort transmits in this direction.

Additional kinematical wheel movability is realized in cylindrical joint of mining wagon's axle box. It allows for reduction of the angle of attack of the wheel on the rail without additional force disturbance (Figure 2) (Ziborov et al. 2011, Patent No 96497).

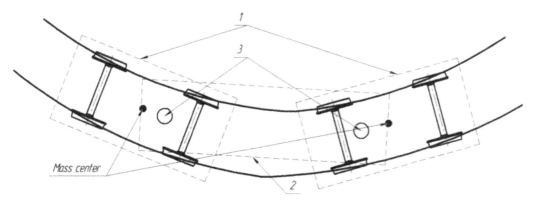

Figure 2. General scheme of tractive bogie rotation in relation to mass center while wheel climbing on the: 1 – tractive bogie; 2 – middle section; 3 – pin joint.

To determine the relations between kinematical and dynamical characteristics of mining vehicle we need to provide an analysis of force interaction relations in dependence on rail track parameters and subject to mining vehicle output members. The obtained data allows for assessment of the safety index, which is used to describe by safety coefficient (Ziborov et al. 2009):

$$K_y = \frac{tg\,\chi - \mu}{1 + \mu \cdot tg\,\chi}\left(\frac{P_v}{P_l}\right) > 1, \qquad (1)$$

where χ – angle of wheel flange; μ – friction coefficient; P_v – normal rail reaction under ongoing wheel, N; P_l – guiding force on the ongoing wheel, N.

Local and regular rail imperfections lead to additional growth of guiding force P_b, that can cause the derailment at some certain critical value. Reduction of guiding force can improve stability and predict derailment.

The most complex motion regime is driving through curvilinear rail track with wheel flange climbing by both rear and front axles. This induces the rotation of tractive bogie in relation to mass center (Figure 2). Simultaneously, the middle section rotates around pin joint. At axial displacement of the wheels, a reaction force arises at the point of flange contact, which acts flatwise to motion direction. A sudden growth of these forces appears while wheel misalignment. To reduce reactive forces an additional local movability of kinematical pair coupling is necessary.

The usage of mathematical simulation facilitates the designing and dynamical interconnection definition (Ziborov 2013). The study of mining vehicle dynamics is provided via developed system of differential equations.

Thus, we have obtained several relations of dynamic forces and safety factor (SF) indexes (Figure 3).

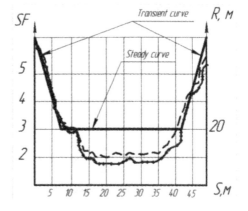

Figure 3. Safety factor relation to track curvature subject to structural scheme. V=4 m/s - - - - sectional locomotive; —— conventional locomotive.

Analysis of the dependences (Figure 3) shows that the rigid connection of the traction bogie section with a body provokes unsteady hunting and lateral displacement, which amplitude continuity growths until the wheel flanges touch the rail. If locomotive's connection is rough, then energy of the impact will be more significant, due to traction bogie mass influence and mass of the middle section partially. In order to avoid this, it is more appropriate to apply kinematical coupling with local movability, which allows for reduction of the middle section inertia influence on the rail track through bogie.

As mentioned above, the characteristics of contact surfaces, and the pressing force define friction properties at the contact point. When the position of the wheelset in the rail track cannot be achieved through the friction forces, there is a two-point contact appears and lateral forces on the flange, which protects the wheelset from derailment. At the same time, an additional resistance force arises. However, the forces on the flange is connected with frictional components, which may lead to force reduction in the contact area. Thereby it facilitates the wheel climbing on the rail, especially on curved track sections of small radius (Figure 4).

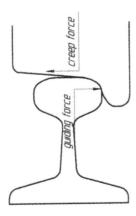

Figure 4. The scheme of wheel and rail interaction.

One of such forces is moment resistance. For wheel with rough wheel bandage the torque will be defined by the formula (Figure 5):

$$M_c = R\int_{\varphi_n}^{\varphi_c} p(\varphi)\mu R\varphi d\varphi = \frac{2Q_z R\mu}{3\pi\varphi_n^2}\sqrt{\left(\varphi_n^2 - \varphi_c^2\right)^3} \quad (2)$$

where Q_z – normal load on the wheel, N; φ_n – contact surface angle of ongoing wheel, rad; φ_c – contact surface angle of offgoing wheel, rad; φ – angular coordinate, rad; R – wheel radius, m.

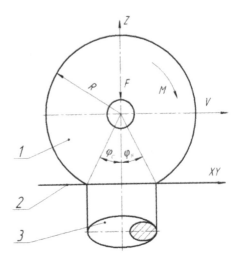

Figure 5. Ongoing wheel and rail interaction scheme: 1 – wheel; 2 – rail; 3 – contact area.

Normal component of full reaction Q_z compensates the pressing force F. The component Q_{xy} appears wheel rolling resistance that predicts wheel sliding along the rail.

The relations of force resistance in dependence on bogie characteristics and exploitation characteristics are depicted on the Figure 6.

It is obvious that increasing the curvature radius of the rail track the load on the chassis elements both locomotive and wagon will decline.

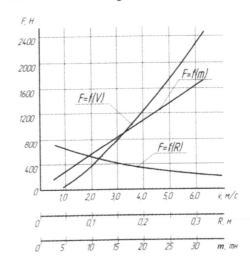

Figure 6. Force resistance of mining rolling stock: V – velocity; m – mass; R – turning radius.

Decrease of motion resistance can be achieved by using mining vehicles with short wheelbase or using a new wider rail track. However, all these measures will reduce the performance of locomotive haulage or increase operating costs. The most appropriate design changes due to the economical meanings is development a new vehicle design with additional local mobility in the kinematic pairs (Ziborov 2013).

3 CONCLUSIONS

To enhance the stability and safety, reduce load on the vehicle's chassis and on the track, reduction motion resistance is possible while the usage of a new kinematical design where the kinematical pairs will have an additional local movability. Thus, it will reduce the number of redundant links with shortage the unnecessary weight. To determine the appropriate value of mobility, providing the necessary performance, we can use modern means of computer simulation interoperability of mine transport and track.

REFERENCES

Bilichenko, M., Pivnyak, G. & Rengevich O. 2005 *Mining Transport*. Dnipropetrovsk: National Mining University: 636.

Garg, V. & Dukkipaty, R. 1984. *Dynamics of Railway Vehicle Systems*. New York: Academic Press.

Patent No 104207 UA. *Mining wagon*. Ziborov, K.A., Fedoriachenko, S.A., Vanzha, G.K. National Mining University. Declared 19.03.12, Published 10.01.14, No 1.

Patent No 96497 UA. *Mining locomotive*. Ziborov, K.A., Fedoriachenko S.A., Protsiv, V.V., Litvin, V.V. National Mining University. Declared 21.03.10, Published 18.10.11, No 21.

Ziborov, K. & Fedoriacheko, S. 2014. *The frictional work in pair wheel-rail in case of different structural scheme of mining rolling stock*. Progressive technologies of coal, coalbed methane and ores mining. The Netherland: CRC Press/Balkema: 517 – 521.

Ziborov, K. 2010. *Frictional pair characteristics subject to kinematical and force imperfections of the rail track*. Mining machines and equipment, No 100: 26 – 32.

Ziborov, K., Blokhin, C. & Litvin, V. 2009. *Motion stability of mining sectional locomotive in curves of steady and transient segments*. Hoisting-transportation equipment No 29: 67 – 76.

Ziborov, K., Protsiv, V. & Fedoruachenko, S. 2013. *Application of computer simulation while designing mechanical systems of mining rolling stock*. Naukovyi Visnyk Natsionalnoho Hirnychoho Universytetu, No 6: 55 – 59.

Ziborov, K., Protsiv, V., Blokhin, S. & Fedoriachenko, S. 2013. *On formation of kinematical and dynamical parameters of output elements of the mine vehicles in transient motion*. Naukovyi Visnyk Natsionalnoho Hirnychoho Universytetu, No 4: 65 – 70.

Theoretical and Practical Solutions of Mineral Resources Mining – Pivnyak, Bondarenko & Kovalevska (eds)
© 2015 Taylor & Francis Group, London, ISBN: 978-1-138-02883-8

The development of deep pits steep slope layers

O. Anisimov
National Mining University, Dnipropetrovsk, Ukraine

ABSTRACT: The development of mining operations sloping layers with a sizeable angle to the horizon is kindly task of mining operations. Formation edges' pit with sizeable angles of slope is an important task, as it allows largeness of overburden postpone to a later period of the development of the field. Considered technique and dependencies are determined the height of the layer in the formation of open pit by steep slopes layers.

1 INTRODUCTION

Currently, many deep pits of Ukraine, along with mineral extraction, there are a significant slippage of development of overburden from the intended technical project. Formed edges deep pits have steep position. Due to the impact of the seismic effects from industrial explosions, the dynamic impact of the mining equipment and other factors, they are practically absent berms security. Sizeable amount of breakaway rocks have accumulated on existing areas. Mechanical cleaning of them is impeded by the decreasing width areas compared with the design values.

In modern conditions during developing deep pits is very important sequence of the open pit formation in space. Since last century, the depth of pits change, which were considered deep. Development of geotechnical sciences has led to opportunities of sizeable angles edges pits and to they formation by laying sizeable angle. Currently, the design (limiting) angles of slopes edges of deep pits have a value from 35 to 42°. Developing of field's the open pit provides for the formation of areas with the development of the front of mining operations in the horizontal direction with the movement of the upper horizons for possible mining development on the lower. Because of the desire to extract more minerals many deep open pits of CIS countries, including Ukraine, have come to a situation where it is necessary to review the technological schemes of mining operations. Opened reserves decrease, and the development of the front excavations of waste rock in the upper levels is become late.

Maintaining all working ledges in overburden working area of open pit in regulations with working areas correspond to pointed graphics mode of mining operations. Consequently, the productivity of working excavators in each of ledges is small, their number is quite large, and the total volume of overburden developed when entering the boundary contours of open pit on the surface is maximized. To improve the organization of the excavations waste rock at increase of the current ratio of stripping is proposed to change the direction of mining ledges instead of horizontal of steep slope.

Increase the speed of preparation for minerals extraction is possible only with the timely development of overburden in edges of the pit. To accelerate the development of overburden is necessary to use the technological scheme of mining operations with the formation of areas edges of the pit at sizeable angles of slope. Theoretical bases of the formation of steeply slope layers were considered by prof. Drizhenko in publications (Drizhenko 2011, 2014). Prof. Drizhenko offers developing of overburden to produce steep layers with sizeable angles of laying edges of the pit. This should be respected the technology of conducting mining operations, safety and developing of the formation of individual sections of the open pit, which provides transport links between faces and the surface.

2 THE MAIN OBJECTIVES

1. Develop a technological scheme of conducting mining operations in the formation of edges of the pit with sizeable angles relative to the horizontal level.

2. Set basic parameters of the formation of open pit space steep slope layers and establish the dependence between the volume of excavation of the overburden and a height of some steep sloping layers.

3 THE SOLUTION OF THESE OBJECTIVES

In the development of the extended field of deep pits the formation worked out section is carried out stepwise. At the first stage of the formation of the open pit of the first turn is limited the depth and then is worked out the next turn of open pit. One way to solve the task of developing the first phase of deep open pit is the transition to the new scheme with the minimum necessary extraction of overburden and maximal load of the excavator, the stage development with ground movement faces from top to down overburden zone.

Technology of the formation of steeply slope layers provides the execution for all kinds of works on preparation, excavation and transportation of rock mass but is not differ from the technology of formation edges in the usual way.

In this case, the working area in a steeply slope layer is oriented along the front edge of the pit at an angle of 20 – 30 to the strike and worked through in the direction from the end of uncovering descents to the opposite. Changing the direction of the movement front of mining operations on steeply slope within the stable state of naked rock mass allows significantly increase the angle of slope of the working edge. At the same time mining operations are conducted with the using of road transport in faces at work berms, and railway transport with the overload of excavation of the rock mass – on non-working area at the ends of the open pit.

Steeply slope layers should be formed by the rock overburden on the front edges, beginning from the output of minerals. Layers are formed on the lying and hanging edges of the open pit in the cross section. At each end of the open pit are provided descents for hauling rocks at transfer points, and the width of descents at the extra base includes catching area. Between working areas in the layer, in accordance with the conditions of safety, are left temporarily closed ledges with berms security. Deepening of mining operations is produced in the end on the open pit with constant transport communications. Also transfer points are recommended to be placed in the end of the open pit's area formed on the non-working edges of the pit. In the application of the railway transport their communications are organized constantly, the open pits end with their location and the surrounding areas of the open pit are rebuilt in the designed position in the first turn. In each cross section of the open pit the slope angle of frontal edges on the overburden is formed on the maximum possible considered stability of rocks.

Formation of the open pit steeply slope layer allows to move sizeable volumes of overburden at a later period of mining development. Layers are formed from top to down by one working area. Development works are conducted with the movement of the front of mining operations along the strike of the deposit. Usually, the upper horizon for overburden has a bigger extension, and the lower has a less (Figure 1).

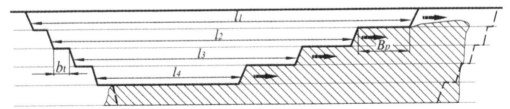

Figure 1. Formation of open pit area along the strike of the deposit.

The height of a steep layer is one of the important parameters that affect the consistency and speed of mining development of the overburden to prepare the mineral for excavation. At the height the steeply slope layer is influenced by the physical and mechanical properties of rocks, their stability, mining technology, the type of used drilling equipment, excavator and vehicles. In this article are considered theoretical conditions of determination of the height of a steep layer.

The development of steep layers could be produced with the same height of each layer (variant 1). The volume of overburden excavation in the upper levels will exceed the volume of the lower. The second option of the development is carried out for a excavation of equal volumes of overburden and at the same time will change the height of individual steeply slope layers. Considering that the width of steeply slope layer in two versions are identical and corresponds to the width of the working area. The width of the working area is calculated by the usual method. On Figure 2, bold arrows indicate the direction of the conducting of mining operations. Each layer is fulfilled horizontal areas with a certain height of the ledge.

During the consideration of the first variant (Figure 2, a), the height of each steeply slope area is identical, and according to Figure 1, the length of each horizon varies depending on the position relative to the bottom of the open pit or surface. All calculated dependences are shown in formulas (1) – (3):

$$h_1 = h_2 = h_3 = ...h_n , \tag{1}$$

$$V_1 > V_2 > V_3 > ...V_n , \tag{2}$$

$$\begin{cases} V_1 = h_1 \cdot Bp \cdot l_1 \\ V_2 = h_2 \cdot Bp \cdot l_2 ...V_n = h_n \cdot Bp \cdot l_n \end{cases} . \tag{3}$$

Development works with the same volume at each area of the steeply slope layer (Figure 2, b) makes it necessary to calculate the height of each individual layer. The height of the layer will vary taking into account the different lengths of individual horizons of conducting of the mining operations. Upper horizons will have a shorter length, and the lower will havelarger. Dependences between the main parameters are shown in the formulas (4) – (5):

$$V_1 = V_2 = V_3 = ...V_n , \tag{4}$$

$$h_1 = \frac{V_1}{Bp \cdot l_1}, \quad h_2 = \frac{V_2}{Bp \cdot l_2}, \quad h_n = \frac{V_n}{Bp \cdot l_n} . \tag{5}$$

(a)

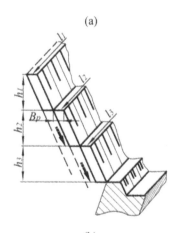

(b)

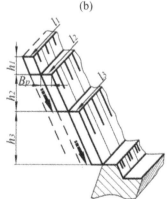

Figure 2. Formation of open pit steep layers: (a) layers of the same height; (b) layers of different heights.

4 RESULTS

Were performed preliminary calculations for the deposit that have the depth of 370 m, the horizontal thickness of the deposit is 100 m, uncovered resources at the bottom of the open pit is 100 m, width of the transport areas $(bt) - 35$ m. The width of working areas, and thus the width of the layer (Bp) are – 70 m. The formation of working areas provides a zone of conducting excavation and loading operations (breast excavator) within the area of blasted rock, the area of the movement of transports and the area foe equipment and people safety from the possible collapse of rocks from the overlying sections of edges of the pit. The edges height of the separate area plays the important role in the formation of the entire edge and possible conducting of mining operations.

On the basis of demonstrated variations (Figure 2, a and b) were obtained graphs (Figures 3, 4). The graph in Figure 3 shows that at a constant height of the steep layer change volumes of overburden's excavation. This is due to the different lengths of developed horizons. As closer to the surface the working horizon, the longer is its working area.

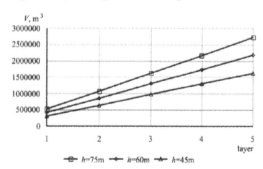

Figure 3. Graph of changes of volumes of overburden's excavation during the development of mining layers (1, 2, 3, 4, 5) of the same height.

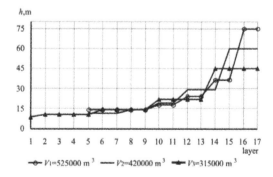

Figure 4. Graph of changes of the height of layers during the development of equal volumes of overburden at some horizons (1 – 17).

245

The graph may be linear, as shown in Figure 3, or it acquires a nonlinear shape during the changing of the length of individual horizons (upper, lower or medium).

The graph in Figure 4 illustrates the change of the height of some steep layers when a constant volume of overburden is recovered. The volume of the overburden has been calculated with taking into account the required width of the prepared area and terms of working off the lower horizon for minerals. During the period of the development of the lower horizon for minerals, the bottom of the steep layer of overburden should be worked out. Initially accepted volume of the overburden corresponds to the maximum height of the layer calculated by the variation 1, and then is calculated the height of individual layers. According to the graph, it is shown that the curve is changed for the decrease of lower horizons, which correspond to the position height of 8 – 10 m. If the EKG-8 (bucket capacity – 8 m^3) is taken, then at the conduction of mining operations the height of rock edges can be 15 m. From this, it could be concluded that upper horizons might worked out by two or three excavators with a height of the layer 30 m.

5 CONCLUSIONS

1. The formation of the working area is carried out steep slope layers, depending from the height of the group of edges forming the phase that determines the position of mining operations at dropping the bottom of the open pit on some edges. In the theoretical work on the graphs are shown the results of the determination of the height of group berms that allowed to work out volumes of overburden and carried out the excavation of mineral. The height of the layer could be taken the same while changing the volume of the excavation of overburden, which requires the insertion of the additional equipment for their exploitation. If it is assumed that each layer has the same volumes of the overburden, which are excavated, then the change of the height of steep slope layers according to the graphs is shown.

2. At many Ukrainian iron open pits, the most part of edges to overburden has temporarily deactivated with a large angles of the slope, the proposed method of the conduction of mining operations allows to normalize the production of iron ore in planned volumes without significantly increase of resources for the stripping in compliance with safety regulations. Technological scheme of conducting works of ground steep layers is ambitious for many deep pits where formed steep slopes of temporarily closed pit's edges.

REFERENCES

Drizhenko, A. 2011. *Milestone developing of overburden iron open pits steep slope extractive layers*. Mining Journal, No 2: 25 – 28.
Drizhenko, A. 2014. *Open pit mining: a textbook*. Dnipropetrovsk: SHEI "NMU": 2nd Edition: 590.

Theoretical and Practical Solutions of Mineral Resources Mining – Pivnyak, Bondarenko & Kovalevska (eds)
© 2015 Taylor & Francis Group, London, ISBN: 978-1-138-02883-8

The new method of extraction of poor and extremely poor ores in underground conditions of Vatutynske deposit (Ukraine)

O. Vladyko & D. Maltsev
National Mining University, Dnipropetrovsk, Ukraine

ABSTRACT: Extraction of uranium ore at the Vatutynske deposit (Ukraine) with the possibility of using the leaching technology from underground workings are considered. The main stages of obtaining minerals using two different technologies of production are proposed. The algorithm for calculating the leaching parameters of poor and extremely poor uranium ore in stoping chambers and pumping the resulting solution to the surface is presented. The layout of movement of leaching and productive solutions in the chamber and the surface respectively with the safety of the works are proposed. Layout of blasthole-injectors, waterproofing stratum and likely direction of movement of solution sulfuric acid is given. Preliminary calculations of leaching time consumption depending on the physicochemical properties of rocks, consumption of reagents and additional work are obtained. The effectiveness of the proposed combined technology of receiving uranium oxides is proved, technological schemes for leaching on situ for Vatutynske deposits (Ukraine) are shown.

1 INTRODUCTION

Oil, gas, coal and uranium are the most common in the world of energy. Ukraine has sufficient amount of coal and uranium only, and supplies of the latter makes 4% of the world reserves. This ensures that our country occupies its place in the top ten ranking leaders of international uranium ore owners. With the current level of provision, balance of using uranium energy in other energy sources is 44 to 56%, which is not in favor of uranium. Extraction of uranium ore in Ukraine is performed by the State Enterprise "Eastern Mining Processing Plant" (SE "EMDP"), owning three mines. It is proposed to improve technology of obtaining uranium from Vatutynske deposit where there is a complicated form of the deposit and increasing number of extremely pure ores, which requires the use of more flexible technology of uranium mining in comparison with classical ways. One way to reduce costs is to develop a deposit using the combined method of extraction of uranium containing elements of underground development and drilling technology. Thus, the search for optimal parameters of uranium mining is relevant to development in this field. Production of uranium ores is associated with high costs of production. This is due to the use of development with laying out space, dangerous working conditions of staff and increased costs for the safe conduct of manufacturing, which significantly reduces the technical and economic parameters of produc-

tion. A partial solution is to use geotechnological means of uranium mining. They are more effective compared with underground mining and allow one to use fewer people at work.

The enterprises that produce uranium ore, now apply chamber system design with the backfilling technology. Such technology development systems include loosening of rock mass with its subsequent release by cone raise on traversing the horizon and the subsequent hoisting it up to the surface. After this, it is transported to the concentration mill for further processing. After that, the chamber is backfilled. In such a large part of the technological chain of production, the cost mainly includes ore output to the surface, transportation costs to the hydrometallurgical plant complex, backfill works, and the concentrating mill – processing (crushing and leaching). So finding ways to reduce the cost of obtaining uranium ore is an important issue not only for the company but for the country as a whole. The solution is to use the geotechnological methods of production that enable to reduce transport costs and backfilling and so on.

2 ANALYSIS OF RECENT RESEARCH AND PUBLICATIONS

Geotechnological ways of uranium mining are comprehensively highlighted in many papers by Prof. V.Zh. Arens (Arens 2010). A huge contribution in the development of geotechnology was made by academicians V.V. Rzhevsky and E.I. Shemyakin, Prof. D.P. Lobanov, Y.D. Dyadkin and I.I. Sharovarov (Sharovar 1999, Korotkov et al. 2001). They proved the reality of the use of geotechnological methods of production directly in a rock mass and made preliminary calculations of technological possibilities of production (Fazlullin 2001, Tolstov et al. 2002, Vladyko 2013).

3 MAIN PART OF THE RESEARCH

The analysis of uranium mining conducted by the authors has shown that many scientific works are devoted to underground mining of uranium, but technologies using elements of underground mining are very rare. Based on this method of production and its parameters, additional investigation is necessary.

The main aim is to develop a method of extracting the poor and extremely poor uranium ores using the underground and geotechnological methods at the operating enterprise. To reach the goals, the following tasks will be set and resolved:

1. Analyze the existing technologies and basic parameters of uranium mining.

2. Develop the algorithm of calculating technological parameters in obtaining uranium oxides.

3. Perform preliminary calculation for obtaining uranium oxides conditions of Vatutynske deposits.

Let's define the stages that will be used in carrying out the work and schematically depict these stages in two technologies: the first one is the current technology of uranium ore processing and further obtaining uranium oxides at a concentration mill, and the second one is the proposed technology of obtaining uranium oxides at the enterprise without transportation and processing units at the concentration mill (Figure 1).

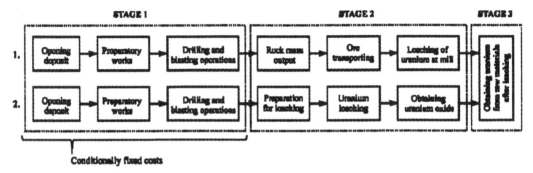

Figure 1. Stages of obtaining uranium for two technologies.

Stage 1. At this stage, the cost of the first and the second technology is approximately the same and can be attributed to the relatively fixed costs.

Stage 2. When comparing the two technologies, in the first one, the major cost is output and transportation of ore, while in the second one, it is obtaining uranium oxides. Significant difference between the two technologies is obvious, so it we will be considered in more detail.

Stage 3. Obtaining uranium metal from oxides and final processing for using in nuclear power plants. This phase of the current study is not decisive, as such processing is not carried out in the country and is only possible abroad.

The current technology of uranium ore is already known and described in more detail in the sources (Maltsev 2013, Chernova et al. 2001), that's why current research does not deal with this issue. Instead, we turn to a detailed presentation of the recommended technologies.

The theoretical consideration of the proposed methods of extracting extremely poor uranium ore shows that the most effective way of extraction is

geotechnological one (Korotkov et al. 2001). It shows greater efficiency in comparison with the traditional underground mining technology. Under the concentration of uranium up to 0.3%, by geotechnological method is 20% more efficient than underground technology, but with increasing concentration of uranium ores up to 0.5% the production costs tend to be equal (Korotkov et al. 2001).

It should be noted that the main processes in stage 2, significantly affecting the cost of extracted ore, is the hoisting up of minerals on the surface and transportation to the mill, material costs during the subsequent backfilling and payroll costs for further processing and more. The recommended geotechnology of extraction (leaching at an enterprise) solves a substantial part of these issues, and removes some issues altogether.

Let us consider separately the basic processes of leaching in chamber and hoisting up (productive) solution saturated with uranium compounds to the surface. For this, we give the general layout of geotechnological complex of uranium oxide mining by mining methods in the leaching chambers (Figure 2).

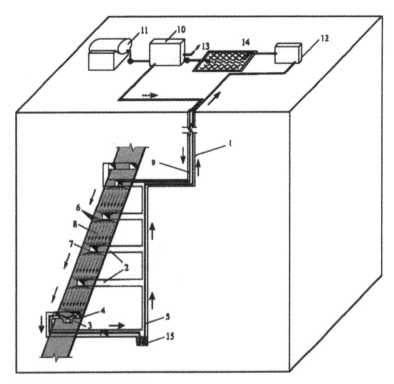

Figure 2. Flowchart of uranium leach using geotechnology in underground conditions and traffic productive (solid line) and leachable (dotted line) solutions: 1 – shaft; 2 – crosscut; 3 – deep pump; 4 – tank for solution saturated with uranium oxide; 5 – saturated solution transport network; 6 – blasthole-injectors; 7 – cavity in the opening floor for collecting solution; 8 – leachable chamber; 9 – pipes to supply the solution in the chamber; 10 – technical oxidation unit; 11 – the composition of sulfuric acid; 12 – pumping station for pumping productive solution; 13 – discharge of enriched solution after leaching process; 14 – surface settling tank for fine cleaning; 15 – contour tank.

In Figure 2, value *12* shows a pump station for pumping the productive solution. The network *5* transports productive solution enriched with uranium oxide. Further productive solution enters the tank *14*. The surface settling tank *14* purifies the solution from impurities by settling and sedimentation. Then the solution is fed to the acidification technical unit *10*. At this stage, the under-enriched solution is injected to a network *9* after the addition of sulfuric acid from the tank *11*, which is saturated to the desired concentration. The solution supplied to the mine over the network enters the productive horizon, where the uranium is leached by acid solutions. It is at this stage that the solution containing uranium is formed. After reaching the desired concentrations of productive solution, it is withdrawn from the circulation by discharge *13*.

To define the efficiency of the technology in the circumstances it is important to calculate the solution loss in situ uranium leaching, this loss reaches about 20% of the total volume of solution by spreading through cracks, bypassing the waterproofing layer. Under each chamber, 4 to 5 vacuum pumps are mounted. From the contour tank *15* solvent is pumped to the surface through pipes in the pumping station *12* (Figure 2). The cycle ends when the productive solution of uranium falls below the minimally acceptable level.

For a more detailed understanding of how the solution has to circulate in the chamber and the mines, Figure 3 shows the schematic diagram of solution circulation. The motion of solution is shown by dotted line from the surface to the upper drilling horizon, which runs across the top-down chamber. Productive solution that has passed through the rock mass is specified by solid line and moves bottom-up, pumped through the corresponding workings on the upper horizon, and then the solution is transported to the surface.

Let us define the motion of solution in chamber and basic technological parameters of blasthole-injectors at the top sublevel, within the circle in the section range A-A (Figure 4).

After the processing cycle is finished, the productive solution is pumped to surface for further processing.

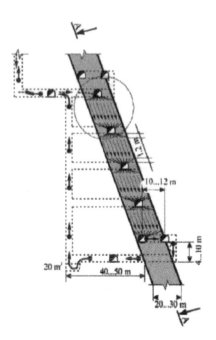

Figure 3. Principal diagram of obtaining productive solution from uranium ore by in situ leaching with an indication of the main technological parameters.

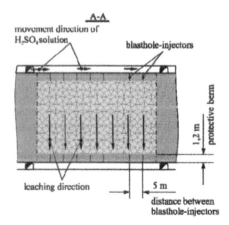

Figure 4. Principal layout of blasthole-injectors with the distance between them, waterproofing layer and the direction of the sulfuric acid solution.

The total time of the leaching process according to preliminary calculations is 260 days, where the watering of the chamber takes 30 – 50 days, acidification – about 90 days, and sorption – about 130 days. The losses of productive solution are about 8%. Extraction of uranium into productive solution from the rock mass makes about 50% for the described conditions, and under the conditions of obtaining uranium from the solution (in enrichment) is around 60 – 80%.

To determine the parameters of loosening rock mass for its further leaching, we use the generally acceptable calculation of loosening in restricted area.

1. For the drilling and blasting operations in the rock mass in restricted environment, it is necessary to define the line of minimal resistance (LMR) charges W and the distance between the ends of the wells a (Maltsev 2013):

$$W = K_{fract} \cdot \frac{90}{\sqrt{f}} \cdot \sqrt{\frac{L}{L' \cdot m}} \cdot d_{bhole} , \qquad (1)$$

where K_{fract} – coefficient that accounts fracturing of rock mass; f – strength of rocks; L' and L – volume and concentration of the energy of standard explosives; m – coefficient of charges convergence; h – the maximum height of sublevel; B – width of the chamber; d_{bhole} – blast-hole diameter; k – factor dependent on the h/b ratio and other mining-technological factors; P – charge density.

$$a = m \cdot W . \qquad (2)$$

Under the change of the factor of charges convergence, one keep in mind that the product $m \cdot W$ should be kept constant. Then let us perform the calculation of leaching geotechnological parameters.

Let us choose key geotechnological indicators depending on mining method parameters (Tabachenko et al. 2012):

– total solutions V_0 required for leaching;
– average metal content in the resulting $C_{aver.}$ productive solution;
– the number of boreholes n_{pump} required for pumping solution into the chamber.

2. Total volume of solution V_0, which is required for leaching, is determined by the formula:

$$V_0 = Q_e \cdot f_o , \qquad (3)$$

where Q_e – volume of ore in the chamber; f_o – consumption of solution per ton of ore.

3. Volume of uranium which in theory can be converted into productive solution, V_U:

$$V_U = \alpha_U \cdot Q_e \cdot k_U , \qquad (4)$$

where k_U – probable value of extraction of uranium compounds in the total weight of the chamber; α_U – content of useful component in the ore.

4. Average content of metal $C_{aver.}$ in productive solution that is received from the chamber:

$$C_{aver.} = \frac{V_U}{V_0} . \qquad (5)$$

5. Duration of working off the volume of ore in the chamber, T:

$$T = t_{mount.} + t_{acid.} + t_{dismant.} , \qquad (6)$$

250

where $t_{mount.}$ – equipment   mounting   time; $t_{acid.}$ – ore acidification time; $t_{dismant.}$ – equipment dismantling time.

6. Total number of boreholes N, which are necessary for stoping chamber:

$$N = 2(n_h - 1) \cdot n_{pump.}, \qquad (7)$$

where n_h – number of drilling horizons in the chamber; $n_{pump.}$ – number of input or output blast-holes on the horizon.

4 PRELIMINARY CALCULATION OF LEACHING PARAMETERS FOR CONDITIONS OF VATUTYNSKE FIELD

LMR for loosening rock mass without changing existing equipment will be $W = 1.8$ meters, the distance between boreholes – $a = 2.1$ meters. Calculation of the number of blasthole-injectors is performed based on a statistical average size of the chamber 25×70×55 m (L×H×W).

The distance between injectors can vary from 1 to 7 m for our conditions, then we need 4 blastholes as the distance between the ends of the injectors will be up to 5 m. From this, we can determine the number of injectors on one blasthole-injection sublevel, based on the length of the chamber. The number of rows of the blasthole-injectors equal 10, on the basis that the distance between rows is 5 m. Total number of blasthole-injectors for sublevel equals 20 units. According to the technology, the number of input blasthole-injectors is the number of output blasthole-injectors. Then, the total number serving one sublevel will be 200 units. Assuming that the average chamber has five sublevels, we can determine the total number of blastholes per chamber, which is 1000 units. Value of the acidification solution will be 2500 kg and obtained productive solution will be 3128 kg, as is calculated by these formulae. Acidification time of rock mass will be 90 days.

5 CONCLUSIONS

The result of the research is the extraction technology that takes into account parameters of leaching uranium in situ. Development of technology using geotechnology proved that it can be effectively used for mining in Vatutynske field. Its implementation will allow extracting even those ores that were not previously suitable for production.

REFERENCES

Arens, V.Zh. 2010. *Physico-chemical geotechnology*: textbook for high schools / under the general editorship of V.Zh. Arens. Moscow: Moscow Mining University: 575.

Chernova, A.P., Babak, M.I., Koshyk, Y.E. & Andreev, O.K. 2001. *Mining and processing of uranium ore in Ukraine*. Kyiv: ADEF-Ukraine: 238.

Fazlullin, M.I. 2001. *Heap leach of noble metals ed.* Izd. AGN.

Korotkov, V.V., Lobanov, D.G., Nesterov, Yu.V. & Abdulmanov, I.G. 2001. *Mining and chemical geotechnology of uranium mining*. Moscow: Geos.

Maltsev, D.V. 2013. *Substantiation of blast-hole drilling parameters during destruction of massif that contains uranium ore*. Ph.D. Thesis. Dnipropetrovsk: National Mining University: 177.

Sharovar, I.I. 1999. *Geotechnical development methods of layered deposits*. Moscow: Moscow Mining University: 242.

Tabachenko, M.M., Vladyko, O.B., Khomenko, O.E. & Maltsev, D.V. 2012. *Physico-chemical geotechnology*: Textbook for high schools. Dnipropetrovsk: National Mining University: 310.

Tolstov, E.A. & Thick, D.E. 2002. *Physicochemical geotechnology field development of uranium and gold in the Kyzylkum region*. Moscow: Geotechnology Geoinform Center.

Vladyko, O.B. 2013. *Technological parameters of cutoff curtains, created with the help of inkjet technology*. Geomechenikal Processes During Underground Mining. The Netherland: CRC Press/Balkema: 135 – 140.

Theoretical and Practical Solutions of Mineral Resources Mining – Pivnyak, Bondarenko & Kovalevska (eds)
© 2015 Taylor & Francis Group, London, ISBN: 978-1-138-02883-8

Environmental aspects of the use of overburden rocks for hazardous waste detoxification

O. Borysovs'ka & A. Pavlychenko
National Mining University, Dnipropetrovsk, Ukraine

ABSTRACT: The features of industrial waste impact on the environmental components are analyzed. Possibilities of overburden rocks use for detoxification of hazardous wastes are defined. The technological scheme of waste detoxification using clay materials is proposed.

1 INTRODUCTION

Many years of energy and raw material specialization and low technological level of industry of Ukraine set it among the countries with the highest absolute volume of waste generation and accumulation. Negative side of industrial production growth as well as scale of natural resource use is very high volumes of production and accumulation of producing and consuming waste, what results in large areas of land occupied by solid waste.

Waste is one of the most important factors of environmental pollution and negative impact on virtually all of its components. The area of land that is given under the sites of accumulation waste in slime storages, dumps, landfills etc. is increasing every year (The National... 2012, Lubchik & Varlamov 2002, Saksin & Krupskaya 2005).

Technologies of recycling are outdated, the cost of energy resources and all kinds of transportation is dramatically increased, and the activities of procurement organizations are significantly narrowed. There are no tools of effective influence on the companies and they continue to use only such wastes that give a quick profit. The costs of storage and waste disposal are almost 20 percent of the total production costs.

The existing level of recycling of waste as secondary resources has no effect on improvement of the environment. This is due to the fact that the recycling involves mainly large-tonnage mining waste and some other waste – mostly low-toxic or neutral (inert) (Buzylo et al. 2014, Zubova et al. 2012, Matveeva 2007). Therefore, environmental impact of waste recycling is negligible.

So far, Ukraine has no properly organized system of collection, storage and disposal of toxic waste. Insignificant use of technologies, machinery and equipment, along with the accumulated experience of waste disposal of coal, fuel-and-energy, metallurgical, chemical and other industries, as well as secondary raw materials considerably worsens the situation. Whereas the waste of production and consumption may play the important role in supply of industry with raw materials, replacing primary resources (ore concentrates, natural non-metallic materials, fuel, etc.).

The important areas of sustainable development of Ukraine are comprehensive utilization of minerals, industrial waste reduction, implementation of recycling technologies and more. Only recycling processes are able to complete former imperfect cycles of movement of materials, to ensure their sustainable use and to limit the volumes of raw materials extraction from the earth's crust to the necessary minimum (Pivnyak et al. 2013).

A significant number of mining wastes can also be used in the environmental practice. The most promising is the use of natural materials that are capable of stopping the harmful substances and preventing their further migration. One of possible approach to address this problem is the use of widely available clays from overburden rocks of open mining as the most affordable, low-cost and highly efficient materials (Gumenik et al. 2012).

2 FORMULATING THE PROBLEM

Especially critical situation in Ukraine exists in the field of municipal solid waste (MSW) management. Millions of tons of waste are stored on overloaded landfills without sorting and recycling of valuable components. Thermal processing of MSW in the conditions of national incineration plants is ineffective neither economically nor environmentally because in the result there is a need for neutralization of products of solid waste combustion – slag and fly ash (Ablaeva & Borisovskaya 2011).

Solving the problem of hazardous waste management requires the creation of scientific principles of increasing ecological safety of waste's placement areas. It is necessary to assess the real environmental hazard from technogenic objects and components of waste, to predict changes in the environment and to develop ways to detoxify waste (Gorova et al. 2013, Kolesnik et al. 2012).

Comparative analysis of recycling and disposal methods for waste of MSW combustion discovered that the costs of measures to reduce the waste toxicity are disproportionate to economic effect of their implementation through ultra-low environmental tax rates for waste disposal.

From the environmental point of view the use of clay to protect and improve the quality of the environment is the most interesting approach. Clay is a unique natural mineral that has exceptional properties such as dispersion, hydrophilic, flexibility, ability to adsorption and ion exchange with the environment. It is widely used for treatment of wastewater, gas and dust emissions, for creation of buffer zones around storage of toxic waste, etc. (Demenko & Borysovs'ka 2014).

Therefore, the purpose of this work is to develop technology of hazardous waste decontamination through the use of clays from overburden rocks of mining production.

3 GENERAL PART

Western part of the Nikopol manganese ore deposit is developing by Ordzhonikidze Mining and Processing Plant (MPP). The average ratio of overburden is $17 \, \text{m}^3/\text{t}$, annually over 100 million tons of overburden rocks are ejected in developing. A significant part of sedimentary rocks of Western part of basin is clay rocks. The properties of several types of clays (green and black) that belong to the overburden rocks of Chkalovsky career No 2 of Ordzhonikidze MPP were investigated: their chemical, spectral, grain and mineral composition, mineral type and other features were determined.

The composition of clay largely determines their properties, although direct quantitative relationship between composition and properties has not yet been established. Therefore, mostly the material composition is the clay classification indicator that allows simultaneous to judge about some of their properties and conditions of occurrence.

According to the data of spectral analysis is established that the studied clay contains such elements as copper – $3 \cdot 10^{-3}\%$, manganese – $(5 - 7) \cdot 10^{-4}\%$; zinc – $(7 - 15) \cdot 10^{-3}\%$; molybdenum – $(0.3 - 1.5) \cdot 10^{-3}\%$; cobalt – $(0.2 - 2.0) \cdot 10^{-3}\%$ and others.

As a result of grain size determination is established that in terms of the content of particles smaller than 0.001 mm green clay belongs to dispersed clays and black clay belongs to coarsely dispersed clays. In addition, the high content of particles smaller than 0.01 mm characterizes green and black clays as highly plastic clay (with a plasticity number of at least 15).

Determination of the concentration of hydrogen ions in the aqueous extract of the studied clays showed that an extract of green and black clay has a slightly alkaline reaction – pH = 7.5...8.

Assessment of mineral composition of clay was carried out by its wetting. A sample of clay in air-dry state ground into powder and sieved through a sieve with holes in 0.5 mm diameter. Sifted powder with mass of 40 – 50 g were placed in a porcelain cup with a diameter of 7 cm, than compacted well by flat pestle and aligned to the horizontal surface. Then 1 ml of 4% solution of sodium pyrophosphate was injected drop by drops from pipette to the centre of studied powder surface. The absorption time t counted from the moment of the first drop descent to the moment of the last drop of solution tidying up. It was determined accurate to 0.1 minutes. After tidying up the last drops spot diameter d formed by wetting the powder was determined in mm. Time of tidying t when moistening the green clay powder was 3.6 min.; spot diameter d – 29 mm, hence the index $K = 104$ ($K = t \cdot d$). Thickness of wet clay layer in the central part was about 4 mm, swelling were weakly expressed. According to the results of observations green clay can be classified as kaolinite-hydromicaceous.

In the case with black clay time of tidying t was 2.5 min.; spot diameter d – 24 mm, hence $K = 60$. Thickness of wet clay layer in the central part was about 8 mm, swelling was almost imperceptibly, and the surface of moistened layer was slightly concave. So, black clay refers to hydromicaceous.

Determining the mineral type of clay minerals was carried out by the nature of the drying of fraction smaller 0.001 mm in size. Collection of fraction smaller 0.001 mm was performed by clay decantation. The resulting suspension of clay particles was evaporated in porcelain cups in a water bath, and after reaching the density of fluid mass it was dried at room temperature.

The suspension, which was dry, covered thinly the bottom and edges of cups, and the nature of the drying of fraction smaller 0.001 mm was different for different types of clay mineral. At green clay drying the matte surface was formed, broken with radial cracks that indicate a preference of kaolinite clay minerals in the composition. Black clay at drying gave glossy surface, divided on individual plates with mutually intersecting cracks. This suggests that black clay refers to hydromicaceous clay again.

Absorption capacity of clays depends directly on their mineral composition. Kaolinite-hydromicaceous clays, which include green clay, have absorption capacity of 20 to 40 meq per 100 g clay and hydromicaceous clays – about 15 (Avidon 1968).

As a result of studies it found that green clay has a higher dispersion, absorption capacity and physic and chemical activity.

To investigate the possibility of decontamination of products of MSW combustion in laboratory conditions samples of pre-crushed slag and fly ash of Dnipropetrovsk incinerator were processed by green clay, the concentration of which was 10, 20, 30 and 40%. Then wastes were moistened with distilled water in the ratio 1:1 and dried to air-dry state for 5 days. Each version of the study was presented by samples with mass of 15 g. Then weights were selected from mixed samples in which the mass of waste was 5 g, taking into account dilution factor, to consider the waste impoverishment effect by clay and to determine the concentration of heavy metals (HM) in 1 mg of researched waste instead of 1 mg of mixed samples.

Determination of water-soluble and mobile forms of heavy metals was performed by atomic absorption. Determination of gross content of HM in raw waste and processed by clay waste was performed by spectral analysis.

It was established that all studied heavy metals are contained in the combustion products in a mobile form, and metals of the first hazard class are most intensively removed with ammonium acetate buffer – zinc, cadmium and lead. As for metals in water-soluble form their leaching percentage from wastes is lower than for the mobile forms. This represents a danger to the environment, given the accumulation of significant amounts of slag and fly ash.

As a result of determination of gross concentrations of HM in the treated waste (Table 1) no significant changes in the waste after processing by clay were found.

The results of determination of HM content in water-soluble and mobile form in the treated waste presented in Table 2 and 3, respectively.

Table 1. The results of spectral analysis of products of MSW combustion before and after clay treatment.

Sample	I hazard class			II hazard class			III hazard class
	Zn	Cd	Pb	Cu	Ni	Co	Cr	Mn
Clay	70	—	20	30	30	10	70	500
Fly ash	4000	60	550	300	35	13.5	400	700
10% of clay	3000	50	200	1000	50	10	1000	1000
20% of clay	1500	0	500	300	50	15	500	700
30% of clay	700	0	500	100	20	7	200	500
40% of clay	700	0	500	100	30	7	300	700
Slag	1500	5	400	700	85	15	400	750
10% of clay	500	0	500	150	15	7	200	500
20% of clay	300	0	500	150	10	7	200	500
30% of clay	1000	10	500	500	30	10	300	700
40% of clay	300	0	200	100	15	15	200	300

Table 2. Content of heavy metals in the aqueous extract, mg/kg of investigated waste.

Sample	I hazard class			II hazard class			III hazard class
	Zn	Cd	Pb	Cu	Ni	Co	Cr	Mn
Clay	0	0	0	0.43	0	0.17	1.3	0.43
Fly ash	0.29	1.19	3.57	0.29	1.29	0.57	2.57	0.59
10% of clay	0.29	1.19	2.70	0.14	1.29	0.54	1.08	0.54
20% of clay	0.28	0.72	1.67	0.14	1.11	0.44	0.83	0.41
30% of clay	0.20	0.41	1.35	0.14	0.95	0.38	0.81	0.27
40% of clay	0.14	0.35	1.35	0.14	0.95	0.38	0.81	0.27
Slag	0.81	0	0.25	2.25	0.50	0.20	0.75	0.50
10% of clay	0.50	0	0.13	1.88	0.50	0.20	0.54	0.50
20% of clay	0.14	0	0	2.03	0.54	0.14	0.50	0.54
30% of clay	0.07	0	0	1.45	0.53	0.11	0.40	0.53
40% of clay	0.06	0	0	1.45	0.53	0.05	0.26	0.54

Table 3. The content of heavy metals in ammonium acetate extract, mg/kg of investigated waste.

Sample	I hazard class				II hazard class			III hazard class
	Zn	Cd	Pb	Cu	Ni	Co	Cr	Mn
Clay	0.5	0	6.0	0	2.0	0.4	0	6.0
Fly ash	3400.0	56.0	650.0	33.0	14.0	4.4	60.0	47.0
10% of clay	3100.0	52.0	630.0	34.0	14.0	4.4	56.0	47.0
20% of clay	2800.0	44.0	600.0	31.0	14.0	4.0	53.0	47.0
30% of clay	2600.0	36.0	540.0	33.0	15.0	3.6	50.0	44.0
40% of clay	2400.0	36.0	470.0	42.0	14.0	4.4	48.0	42.0
Slag	1400.0	2.8	240.0	115.0	12.0	2.4	14.0	180.0
10% of clay	1400.0	2.8	240.0	73.0	7.0	2.0	13.0	40.0
20% of clay	1150.0	2.0	240.0	75.0	8.0	2.0	13.0	38.0
30% of clay	1200.0	1.6	240.0	74.0	8.0	2.0	13.0	36.0
40% of clay	1150.0	1.2	230.0	61.0	5.0	1.6	12.0	35.0

Experimental studies indicate that the content of virtually all the water-soluble forms of heavy metals in waste is reduced under the clay processing, except nickel and manganese in the case with slag. As for mobile forms of heavy metals, their content in the waste also is reduced by the clay processing, with the exception of lead in the case of slag and nickel in the variant of the study with fly ash.

For water-soluble forms of the HM that are absorbed from the fly ash, the following dependences of the absorption efficiency on the concentrations $Me(x)$ were defined:
– for heavy metals of the first hazard class:

$$Pb(x) = -0.07\,x^2 + 4.83\,x - 16.18 \qquad R^2 = 0.99, \quad (1)$$

$$Zn(x) = 0.04\,x^2 - 3.10\,x + 3.45 \qquad R^2 = 0.99, \quad (2)$$

$$Cd(x) = -0.09\,x^2 + 6.69\,x - 58.61 \qquad R^2 = 0.99; \quad (3)$$

– for heavy metals of the second hazard class:

$$Cu(x) = 51.7\cdot[1 - exp(-x)] \qquad R^2 = 0.99, \quad (4)$$

$$Co(x) = -0.013\,x^2 + 1.45\,x - 2.51 \qquad R^2 = 0.99, \quad (5)$$

$$Ni(x) = -0.035\,x^2 + 2.66\cdot x - 23.64 \qquad R^2 = 0.99, \quad (6)$$

$$Cr(x) = 67.5\cdot[1 - exp(-x)] \qquad R^2 = 0.98; \quad (7)$$

– for heavy metals of the third hazard class:

$$Mn(x) = -0.011\,x^2 + 1.98\,x - 3.54 \qquad R^2 = 0.94. \quad (8)$$

For water-soluble forms of the HM that are absorbed from the slag, the following dependences of the absorption efficiency on the concentrations $Me(x)$ were defined:
– for heavy metals of the first hazard class:

$$Pb(x) = -0.11\,x^2 + 6.75\,x - 1.94 \qquad R^2 = 0.98, \quad (9)$$

$$Zn(x) = -0.08\,x^2 + 5.22\,x - 2.36 \qquad R^2 = 0.99; \quad (10)$$

– for heavy metals of the second hazard class:

$$Co(x) = 2.4\,x - 22.5 \qquad R^2 = 0.99, \quad (11)$$

$$Cu(x) = 0.90\,x + 1.429 \qquad R^2 = 0.81, \quad (12)$$

$$Cr(x) = -0.008\,x^2 + 1.79\,x + 3.28 \qquad R^2 = 0.96. \quad (13)$$

For mobile forms of HM that are absorbed from the fly ash the efficiency of metals absorption from waste is described by the following equations:
– for heavy metals of the first hazard class:

$$Pb(x) = -0.07\,x - 2.77 \qquad R^2 = 0.94, \quad (14)$$

$$Zn(x) = 0.74\,x + 1.18 \qquad R^2 = 0.99, \quad (15)$$

$$Cd(x) = -0.01\,x^2 + 0.41\,x - 2.04 \qquad R^2 = 0.96; \quad (16)$$

– for heavy metals of the second and the third hazard class:

$$Cr(x) = 0.50\,x + 1.0 \qquad R^2 = 0.99, \quad (17)$$

$$Mn(x) = -0.53\,x - 10.28 \qquad R^2 = 0.99. \quad (18)$$

Under the further study of the absorption capacity of green clay in relation to mobile forms of certain HM by methods of physical and chemical modelling is established that green clay effectively absorbs HM from solution. In order of increasing the limit capacity of separate absorption metals are as follows: cobalt > nickel > manganese > chromium > cadmium > zinc > copper > lead. Green clay absorbs heavy metals of the first hazard class the most effectively.

At research of the HM absorption of clay from polycomponent mixtures of eight toxic elements in concentrations corresponding to their concentration in the slag, found that green clay is able to absorb all investigated metals from polycomponent mixtures, and increasing weight of clay enhances the absorption of each metal from the solution.

In order of increasing the efficiency of absorption from polycomponent mixtures HM can be placed in the next row: cadmium > nickel > zinc > cobalt > manganese > lead > copper > chrome. Thus, differences between absorptive capacity of clay in relation to the separate heavy metals and metals in complex were specified and the feasibility of using of kaolinite-hydromicaceous green clay from overburden rocks of Nikopol manganese ore deposit in environmental protection purposes was confirmed.

Analysis of the results allowed determining optimal concentration of clay from an environmental and economic point of views that provides most complete absorption of harmful substances (Table 4).

Table 4. The calculation of the optimal concentration of clay for the toxic waste treatment.

Dependence	Concentration of clay, %			
	10	20	30	40
The efficiency of absorption of HM water-soluble forms from fly ash, %				
$Pb(x) = -0.0721\,x^2 + 4.8305\,x - 16.176$	24.9	51.6	63.8	61.7
$Zn(x) = -0.0345\,x^2 + 3.1034\,x + 3.4483$	31.0	51.7	65.5	72.4
$Cd(x) = -0.0861\,x^2 + 6.6849\,x - 58.613$	0.0	40.6	64.4	71.0
$Cu(x) = 51.7\cdot[1 - exp(-x)]$	51.7	51.7	51.7	51.7
$Co(x) = -0.0125\,x^2 + 1.4486\,x - 2.5063$	10.7	21.5	29.7	35.4
$Ni(x) = -0.035\,x^2 + 2.66\,x - 23.64$	0.0	15.6	24.7	26.8
$Cr(x) = 67.5\cdot[1 - exp(-x)]$	67.5	67.5	67.5	67.5
$Mn(x) = -0.0109\,x^2 + 1.9782\,x - 3.5351$	15.2	31.7	46.0	58.2
Average	25.1	41.5	51.7	55.6
The efficiency of absorption of HM water-soluble forms from slag, %				
$Pb(x) = -0.1057\,x^2 + 6.7486\,x - 1.9429$	55.0	90.7	100.0	98.9
$Zn(x) = -0.0785\,x^2 + 5.522\,x - 2.3633$	45.0	76.7	92.6	92.9
$Co(x) = 2.4\,x - 22.5$	1.5	25.5	49.5	73.5
$Cu(x) = 0.9022\,x + 1.4222$	10.4	19.5	28.5	37.5
$Cr(x) = -0.0076\,x^2 + 1.7981\,x + 3.2762$	20.5	36.2	50.4	63.0
Average	26.5	49.7	64.2	73.2
The efficiency of absorption of HM mobile forms from fly ash, %				
$Pb(x) = 0.6923\,x - 2.7692$	4.2	11.1	18.0	24.9
$Zn(x) = 0.7353\,x + 1.1765$	8.5	15.9	23.2	30.6
$Cd(x) = -0.0102\,x^2 + 1.4082\,x - 2.0408$	11.0	22.0	31.0	38.0
$Cr(x) = 0.5\,x + 1$	6.0	11.0	16.0	21.0
$Mn(x) = 0.5319\,x - 10.284$	0.0	0.4	5.7	11.0
Average	5.9	12.1	18.8	25.1
The efficiency of absorption of HM mobile forms from slag, %				
$Cd(x) = 1.8571\,x - 14.286$	4.3	22.9	41.4	60.0
$Zn(x) = -0.0357\,x^2 + 2.2857\,x - 17.857$	1.4	13.6	18.6	16.5
$Ni(x) = 0.0063\,x^3 - 0.3929\,x^2 + 7.1726\,x + 0.5952$	39.3	37.3	32.3	62.1
$Cu(x) = 0.0041\,x^3 - 0.2776\,x^2 + 5.7867\,x + 0.472$	34.7	38.0	34.9	50.2
$Cr(x) = 0.0012\,x^3 - 0.0714\,x^2 + 1.3095\,x - 7E-13$	7.2	7.2	7.4	14.9
$Co(x) = 0.0028\,x^3 - 0.1667\,x^2 + 3.0556\,x - 2E-12$	16.7	16.8	17.2	34.7
$Mn(x) = 77.8\cdot[1 - exp(-x)]$	77.8	77.8	77.8	77.8
Average	25.9	30.5	32.8	45.2
Average for slag (water-soluble and mobile)	26.2	40.1	48.5	59.2
Average for fly ash (water-soluble and mobile)	15.5	26.8	35.2	40.3

Based on the positive results of the model experiments a method of reducing the concentration of the mobile forms of HM in the products of MSW incineration and preventing their migration from the waste to the environment was developed.

Analysis of clays absorption of HM revealed that rational concentration of clay for decontamination of MSW incineration products is the concentration 30 ± 5% of the weight of the sample. In processing the researched waste with this clay amount the decreasing of content of HM mobile forms was observed in slag in average of 50%, in fly ash in average of 35%. Further increasing the amount of clay for processing these wastes is unnecessary, because the addition of 40% clay increased in the extent of absorption of heavy metals by only 3 – 4%.

To protect the environment from combustion products of municipal solid waste by reducing the amount of mobile forms of HM and to prevent their migration into the environment appropriate environmental technologies were developed.

The technology is implemented as follows: MSW incineration slag is subjected to sifting to remove large inert fractions such as stones, broken glass, ceramics, etc. Sifted out slag is ground to obtain a uniform fractional composition and the green clay is ground in the same way. Because the technological scheme of thermal processing of solid waste at domestic plants involves wet removal of slag from waste heat boilers the additional moisture o slag is unnecessary. In the case with fly ash there is no need for the previous grinding and sifting because 70 – 90% of fly ash has particles with a diameter less than 50 microns. But fly ash requires moistening before mixing with clay. Then clay is added to waste in specific percentage and mixed them until a homogeneous mass. At the interaction of clay with products of MSW combustion ions of heavy metals are linking and fixing by the whole mass of clay as a result of ion exchange, which reduces the environmental hazard class of studied waste.

After drying processed in such a incineration wastes are suitable for storage and the further secondary use (e.g. road construction) without threat to the environment.

Taking into account modern disparity of payments for waste disposal and its real potential danger for the environment and significant costs for the purchase and transportation of clay, designed method will have an economic effect only when standards of environmental tax for environmental pollution at the state level will be viewing.

4 CONCLUSIONS

Green clay of Nikopol deposit of manganese ores mined at Ordzhonikidze MPP as passing raw material is dispersed kaolinite-hydromicaceous clay that contains minerals of montmorillonite and hydromica and has great absorption capacity in relation to the heavy metals. Established that the limit HM absorption capacity of given clay rocks varies from 10 to 110 mg/d, in descending order of the limit capacity heavy metals can be arranged as follows:

lead > copper > zinc > cadmium > chrome > manganese > nickel > cobalt.

Also found that clay is able to absorb heavy metals not only individually, but also from polycomponent mixtures, while as decreasing the efficiency of absorption heavy metals can be arranged in the following order: chrome > copper > lead > manganese > cobalt > zinc > nickel > cadmium. Thus differences between the absorption capacity of clay in relation to the separate heavy metals and metals in the complex are concretized. The obtained results allow recommending the use of clay as an effective sorbent for environmental purposes.

The method of reducing environmental hazard of MSW combustion products that consists in the processing of wastes with clay sorbent at a concentration of 30 ± 5% is developed. As a result of waste treatment with clay the content of water-soluble forms of heavy metals in the slag is reduced by 64%, mobile form of HM is reduced by 33%. In the fly ash the amount of HM in water-soluble form is reduced by 52%, in the mobile form – by 19%. The implementation into practice of the proposed method of neutralization of MSW incineration products will reduce the level of heavy metals migration from waste into the environment and ensure the environmental safety of such waste storage areas.

Thus, the use of overburden rocks of opencast mineral deposits reduces the toxicity of industrial waste. We believe that the overburden rocks can also be used to detoxify other hazardous waste due to its high absorbing properties.

REFERENCES

Buzylo, V., Savelieva, T., Saveliev, V. & Morozova, T. 2014. *Utilization of the waste products for stowing.* Progressive Technologies of Coal, Coalbed Methane, and Ores Mining. The Netherlands: CRC Press/Balkema: 31 – 34.

Demenko, O. & Borysovs'ka, O. 2014. *Ways of soils detoxication that are contaminated by heavy metals using nature sorbents.* Progressive Technologies of Coal, Coalbed Methane, and Ores Mining. The Netherlands: CRC Press/Balkema: 261 – 266.

Gorova, A., Pavlychenko, A. & Borysovs'ka, O. 2013. *The study of ecological state of waste disposal areas of energy and mining companies.* Mining of Mineral Deposits. The Netherlands: CRC Press/Balkema: 169 – 171.

Gumenik, I., Lozhnikov, A. & Maevskiy, A. 2012. *Methodological principles of negative opencast mining influence increasing due to steady development.* Geomechanical Processes During Underground Mining. The Netherlands: CRC Press/Balkema: 45 – 49.

Kolesnik, V., Fedotov, V. & Buchavy, Yu. 2012. *Generalized algorithm of diversification of waste rock dump handling technologies in coal mines.* Naukovyi Visnyk Natsionalnoho Hirnychoho Universytetu, No 4: 138 – 142.

Lubchik, G.L. & Varlamov, G.B. 2002. *Resource and environmental issues of global and regional energy consumption*. Energy and Electrification, No 9: 35 – 47.

Matveeva, N.G. 2007. *The possibility of application of international experience in the processing of rock dumps of coal mining-processing industry in the Donbass coal region. Ecology*. Visnik of the Volodymyr Dahl East Ukrainian National University, No 1: 35 – 40.

Pivnyak, G., Kovrov, O. & Cherep, A. 2013. *Modern role of resource universities for ensuring sustainable environmental development of mining regions*. Naukovyi Visnyk Natsionalnoho Hirnychoho Universytetu, No 1: 77 – 83.

Saksin, B.G. & Krupskaya, L.T. 2005. *Regional estimation of the effect of mining production on the environment*. Mining Journal, No 2: 82 – 86.

The National Report about the State of Environment in Ukraine in 2011. 2012. Kyiv: Ukraine Ministry of Ecology and Natural resources: LAT&K: 258.

Zubova, L.G., Zubov, A.R., Verekh-Belousova, K.I. & Oleinik, N.V. 2012. *Production of Metals from the Waste Piles of Donbass Coal Mines*. Lugansk: VNU im. V. Dalya: 144.

Ablaeva, L.A. & Borisovskaya, E.A. 2011. *Perspective directions of natural clays for urban areas cleaning*. Kyiv: Reports of the National Academy of Sciences of Ukraine, No 3: 187 – 192.

Avidon, V. 1968. *Preliminary tests of clays in field conditions*. Moscow. Nedra: 168.

Theoretical and Practical Solutions of Mineral Resources Mining – Pivnyak, Bondarenko & Kovalevska (eds)
© 2015 Taylor & Francis Group, London, ISBN: 978-1-138-02883-8

Test load envelope of semi – premium O&G pipe coupling with bayonet locks

V. Protsiv, K. Ziborov & S. Fedoriachenko
National Mining University, Dnipropetrovsk, Ukraine

ABSTRACT: The most prospective (amongst the existing premium and semi premium) pipe couplings for O&G production and industry are defined. Stress, which arises within the thread coupling elements while different offset and preload, subject to loading regimes according to standard ISO/FDIS 13679 is calculated by the usage of the finite element analysis (FEA). The stress in 9" pipe coupling with buttress thread is simulated. The simulation provided for pipe couplings, with bayonet locks on the pipes ends. Substantiated, that such connection is enough hermetical and preserves the failure of the box caused by overstress and possible breaking owing to too high torque, which arises during assembling of pipeline. Denoted situation can occurred while assembling a new pipe segment or under influence of external factors, which act on a pipe during exploitation in underground or underwater environment.

1 INTRODUCTION

There are many pipe's threaded couplings, which are used for casing or exploitation pipes, drill rods for oil-extracting wellsites or analogical, such as geothermal one (Super-sealed tubular...2004). There are also several threaded pipe couplings, which are used on G&O casings, known as Riser and are used for bores connection with marine oil platforms on the shelf (Meeterns et al. 2010). Such pipe couplings are subjected to numerous different loads (axial tension – compression, internal or external pressure of liquid or gas substance, bending, torsion) which can act in multiple combinations (e.g. torsion and internal pressure) with lasting variable intensity. Hermetical (or "semi premium", also known as "standard") and extra hermetical ("premium") threaded pipe couplings might have not only high tensile strength, but to stay hermetical (in particular it concerns gas pipelines) in spite of the stress influence including combined loads. Above all, the boxes, which connect separate pipe segments, are prone to rupture. On the Figure 1 a pipe and a box are depicted (pipes are represented as threaded nipples).

The study purpose. Provide strength calculation of the semi-premium coupling with conical buttress thread and bayonet locks on the connecting pipe ends.

2 THE RESEARCH RESULTS

The most perspective semi-premium pipe coupling is to be read a buttress inch thread (Buttress, API spec. 5B USA) (Specification for Treading...2008). It has incline of side thread edges in cross-section equals to 3° and 10°, which while the conicity 1:16 that allows to provide self-braking. Moreover, the symmetrical profile of the pipe and box thread, tiny thread element bending radii (from 0.2 to 0.76 mm) provide higher pressure in the coupling, that is the result of screwing pipe in the box with predefined torque that is necessary to obtain certain preload. Thus, linear overlap of initial thread contour of pipe and box generates tensile and compression stresses that must not exceed the proportional limit of the material.

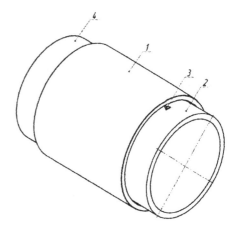

Figure 1. Threaded pipe coupling: 1 – box; 2 – the first connecting pipe; 3 – check control mark on the first pipe; 4 – the second connecting pipe.

This parameter (proportional limit) thought to be maximal, because pipe coupling trials according to CAL I, CAL II, CAL III and CAL IV predict to provide six serial hand tight screwing – unscrewing with no thread damage. Only after seventh screwing with maximal necessary torque the pipe comes to the next trial stage with full exploitation loads. As a material for coupling boxes and drill casings could be used quenched and tempered steels P110 and Q125 according to API 5 CT Specification. Their proportional limits are 758 and 860 MPa correspondingly (Specification for Casting…2008). Therefore, the first stage of computational experiment is devoted to finite element analysis of pipe and box coupling of the first production stage at the enterprise.

After the pipe and the box are produced, they are getting coupled. For this reason, the box *1* (Figure 1) screws on the pipe end *2* while the pipe end coincide with check mark *3*, situated on the depicted pipe. It allows to screw the pipe near the box center. As it stands, the pipe set delivers to consumer (if required, another procedures can be done, e.g. colouring, oiling, dressing an end cap etc.).

In order to study stress-strain state of the coupling the geometric preload 0.45 mm has been included (the preload of thread contour of the pipe and the box). Geometrical way to put the preload is more efficient in comparison with the usage of default in-built functions of Ansys R15, because the contact is non-linear and the preload value can vary across the edges.

The majority of using stress-strain simulation technologies includes generation of 3D bar, that resulted as the cross section of two parallel surfaces, which are situated on some distance between them along the pipe coupling axis. To improve the quality of the research we have designed 3D couplings which let us take into account the compensation of internal pressure of finite elements massive with minimum assumptions. Furthermore, it let describe a frictional contact of nonlinear treaded contact. Thus, we have obtained a maximal convergence with physical model.

The disadvantages of denoted method are significant hardware and software requirements: necessity to provide big RAM and high performance CPU. To increase convergence and cut off the processing time, the contact rigidity of a turn of the tread has been reduced, that, by the way, did not reflected on the results.

The stress, which arises in box-pipe coupling after assembling in industrial facilities and following shipping to a supplier is depicted on the Figure 2.

As we can see on Figure 2, the maximal stress in this case exceeds both proportional limit and yield stress (862 and 930 MPa correspondingly for chosen steels), that, naturally, will cause thread destruction and box blowout, as the weakest element of the coupling.

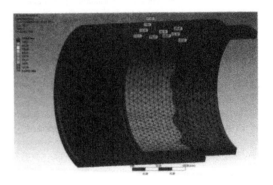

Figure 2. The stress in threaded pipe coupling with 0.45 mm preload.

Hence, it will be reasonable to reduce the preload up to 0.25 mm, for instance. Such case of a coupling is depicted on the Figure 3.

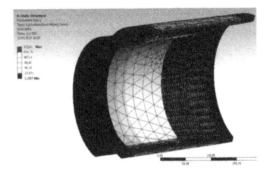

Figure 3. The stress in pipe coupling with 0.25 mm preload.

In this case the maximal stress does not exceed the proportional limit, thus such pipe coupling can be run for the following research under the load. For example, let's apply an internal pressure of 80 MPa that imitates liquid or gas pressure.

Before the next step, necessary to provide the second production step – to screw the second pipe end till both ends will be in tight frictional contact inside the box. In this case, too big torque will cause the skid of the pipe ends. Inasmuch that friction coefficient of a steel by steel with obligated lubricant between contacting surfaces can be miserable (0.05 – 0.10), the friction force will be insufficient to couple the pipes in this case. Even extra pipe end treatment for bigger roughness and grooving (including wave, tooth or combined profiles) will not come up with necessary result, because along the "handmade" assembling the pipe's ends will be damaging, that is unacceptable.

The picture of the stress, the arises in pipe coupling with 0.25 preload and 80 MPa internal pressure is brought in the Figure 4.

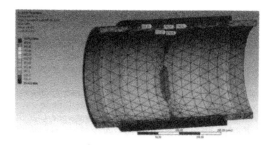

Figure 4. The stress in pipe coupling with 0.25 preload and internal pressure 80 MPa.

Because the maximal stress in the thread elements does not exceed even the proportional limit, we can assume that such coupling is able to work, although it can not predict box blowout when extra torque will be applied to a new pipe.

To predict the jacking of a new pipe over another, lets use the perspective bayonet pipe coupling (Pipe Threaded Couplings 2014).

If necessary to add another pipe segment during drilling a well or a pipe line building, into the free pipe box 1 of pipe line is screwing threaded pipe end of the next pipe. The pipe screws until, for instance, the conical pipe coupling will be hermetical (Figure 5; here brought the same element indexes as on the Figure 1).

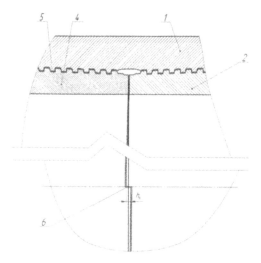

Figure 5. The finished coupling of two bayonet pipes: 1 – box; 2 – the first coupling pipe; 4 – the second coupling pipe; 5 – preload in threaded coupling; 6 – the interacting line of two bayonet locks of coupling pipes.

In order to get that done to pipe 4 necessary to apply such torque, that is needed to evolve certain preload 5 (the dash lines on the pipe and box section

form cross grid) By the way, the same preload has been formed on first production stage (while screwing box 1 on the first pipe 2 in the plant facilities). The coupling process of the pipe 4 with box 1 will finish, when bayonet locks on the pipes' ends 2 and 4 will coincide and joint together along the line 6 inside the box. The length of this line will be equal to overlap height h_n of side bayonet snug. The bayonet snugs are brought on the Figure 6 under the positions 7 and 8 for spiral and rectangular notches, and 9 – for embedded into a lengthwise canals pivots.

The picture of the stress, allocated in pipe threaded coupling with bayonet locks and 0.25 mm preload, internal pressure 80 MPa is brought on the Figure 7.

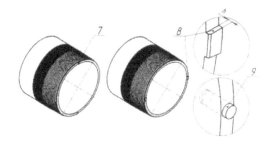

Figure 6. The drill pipes with bayonet locks of different types: 7 – spiral notch; 8 – rectangular notch; 9 – pivot on the pipe end.

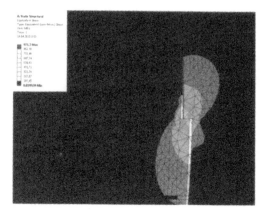

Figure 7. The stress in pipe threaded coupling with bayonet locks with 0.25 mm preload and 80 MPa internal pressure.

Because the maximal stress does not exceed the proportional limit we can assume, the developed pipe coupling is able to work (hermetical and rigid). An additional screwing of a new pipe in relevance to another is predicted, so the stress growth is impossible, and, correspondingly, box blowout.

3 CONCLUSIONS

1. Semi premium hermetical pipe coupling can be produced basing on the conical buttress thread with bayonet locks on the pipes' ends.

2. The sufficient preload in the thread elements equals to 0.25 mm, ignoring production allowance.

3. The stress simulation using the finite element method have to take into account the frictional contact of interacting threaded elements with preload; the geometrical overlap while non-linear contact with variable preload along the edges is reasonable to define in CAD software.

REFERENCES

API Spec 5B, *Specification for Treading, Gauging and Thread Inspection of Casing, Tubing and Line Pipe Threads*. 50th edition. 04.01.2008. American Petroleum Institute.

API Spec 5CT, *Specification for Casting and Tubing*. 9th edition. 04.01.2008. American Petroleum Institute.

Meertens, B., De. Baets P. & Van Wittenberghe J. 2010. *Fatigue behavior of threaded couplings – experimental research*. International Journal "Sustainable Construction & Design". Belgium, February 20 – 21.

Patent 77473, MKB F16L 15/00. *Super-sealed tubular threaded joint comprising at least one threaded element with end edge*. Vallourec Mannesmann Oil & Gas France. Sumitomo Metal Industries ltd. No 20040705325. 16.08.04, 15.12.06 (FR).

Patent MKB F16L 15/00. *Pipe threaded couplings*. Protsiv V.V. & Fedoriachenko S.A. No 2010407920. 30.07.14 (UA).

Theoretical and Practical Solutions of Mineral Resources Mining – Pivnyak, Bondarenko & Kovalevska (eds)
© 2015 Taylor & Francis Group, London, ISBN: 978-1-138-02883-8

Analytical modeling of the backfill massif deformations around the chamber with mining depth increase

O. Khomenko, M. Kononenko & M. Petlovanyi
National Mining University, Dnipropetrovsk, Ukraine

ABSTRACT: Using thermodynamic analytical research method, the investigation of strains formation in the backfill massif around the chamber of second stage was performed. It was established that the horizontal and vertical dimensions of areas vary with increasing deformation depth of the development by logarithmic and exponential law with relevant empirical equations.

1 INTRODUCTION

During the development of steeply dipping ore deposits using chamber systems with solid backfill the quality of extracted minerals is determined by the stability of backfilling mass. Use the chamber mining with a solid backfill mainly carried out on the "chamber-pillar" method with the mining from the center of the mine field to its flanks. The technology of mining involves practicing inter-chamber ore pillars in the second stage, which are located near artificial rock massif. With this order of ore deposits mining, there is contamination of ore with waste rock and contaminated backfill. Contaminated ore with collapsed backfill causes significant harm to the economic activity of the enterprise, manifested in the reduction of product quality and technical and economic performance. For example, getting only 1% of the backfill material in the ore is accompanied with decrease value $1.7 - 2\%$ (Kuzmenko et al. 2014).

Deformation characteristics of backfill significantly inferior ore-rock massif, so the increased concentration of the stress increases with mining depth increase that contributes the collapse and fall out of rock massif (Falshtynskyy et al. 2012, Falshtynskyi et al. 2013). In this regard, the assessment of deformations occurring in backfilling mass adjacent to the mining chambers by increasing the depth of the development is an important scientific and practical problem, the solution of which will allow forecasting the stability of artificial rock massif.

2 THE MAIN PART OF THE ARTICLE

The object of the research is the technology of mining in Pivdenno-Bilozerske rich iron ore deposit, which is developing by Private Joint Stock Company "Zaporizhskyi iron ore plant" (PJSC "ZIOP") using the chamber system development with solid backfilling. Developed and implemented recommendations of Mining Research Institute of State Higher Educational Institution "National Kryvyi Rih University" a new form of chambers for the conditions of the Pivdenno-Bilozerske deposits has reduced the amount of roadway construction and the time of delivery equipment in blocks. This has been achieved through the formation of a high sloping rock at the bottom of the hanging wall deposits. Changing the system design development led to intensification of rock pressure increase on the contours of the working space, where there are rock inrush, adjacent strata and backfill.

The research of destructive strains that have been developed in adjacent strata of backfill massif around chambers of second stages was examined. This is due to the fact that at the location of chambers on the second stage of contact with hanging wall rocks, the influence of the space and time factor of their exposure to the stability of the rock massif, so the impact of backfilling in this case is very difficult.

Contamination of ore during ore extraction in the footwall has an extraordinary character (Kuzmenko 2014). During iron ore extraction in the chamber of the first stage contamination of ore makes up to $0.5 - 1.8\%$ of the rock. Then from chambers of the second stage such indicators make up to $3 - 5.1\%$ of the rock. This confirms the negative impact of backfilling mass during mining of chamber stock in contact with the footwall side of rocks. Increased ore contamination is the result of complex geomechanical processes occurring in the rock massif, consisting of ore, adjacent strata and backfill. Establishing of contamination character is necessary for the production of the cutter-loader and in general for mining science. Similar studies have been carried out previously for mining ore deposit for the floor of 940 – 1040 m (Khomenko et al. 2014), but in this paper we consider the change of destructive strains with mining depth increase.

Under condition of Pivdenno-Bilozerske deposit physical-mechanical properties of ore, rocks and backfill are represented in a wide range. Strength of ores ranges from 60 to 80 MPa, the adjacent strata 80 – 140 MPa, backfill 50 – 60 MPa (by a factor of stability). Bulk weight of ore is $0.39 – 0.4$ MN/m^3 and adjacent rocks varies from 0.21 to 0.29 MN/m^3. The strength of rocks and backfill significantly affect the rock massif deformation around the chamber of the second stage. To assess the spread of strain fields around the extraction chamber in the rock massif was selected one of the modern analytical method. Analysis of theoretical research methods allowed us to determine the most appropriate method – thermodynamic (Lavrinenko & Lisak 1993), which gives the highest accuracy of (85 – 90%) calculated and directly measured elastic deformations.

Assess the impact of destructive deformations on backfilling massif formed in the zone of chambers influence is possible with mining depth increase (Lozynskyi et al. 2015). Ore extraction in footwall chambers surrounded backfilling massif strike on ore deposit shows a wide variation of contamination indicators from 2 to 7%. This is due to the variety of influencing factors such as hardness of rocks, fracturing, area and the time of exposure, varying elements of the block parameters (height, width, length of the chamber). Therefore, to determine the influence of mining depth is useful to consider ore contamination during mining with the same chamber parameters for a particular area of the deposit. For this we consider abandoned place – 4s deposits at a depth of 465 – 840 m, chamber 1/4s (level 465 – 580 m), 1/4s (level 548 – 640 m), 2/4s (level 640 – 740 m) 1/4s (level 740 – 840 m) with an average chamber parameters: height – 110 meters, width – 30 m, length – 45 m. Ore contamination indicators and volume of reserves are shown in Table 1. According to the data presented in Table 1 was plotted diagram (Figure 1) the influence of mining depth to backfill collapse during ore extraction along the axis 4s.

Table 1. Contamination of ore extracted in second stage chamber along the axis 4s.

Chamber	Level, m	Reserves in chamber, thousand tons	Contamination of ore, thousand tons/%
1/4s	465 – 580	323	7.9/2.47
1/4s	548 – 640	313	13.2/4.23
2/4s	640 – 740	303	16.6/5.55
1/4s	740 – 840	590	18.6/3.16

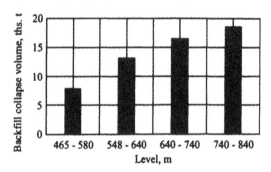

Figure 1. Influence of mining depth to backfill collapse during ore extraction along the axis 4s.

From Figure 1 is shown that there is correlation between the depth and the volume of backfill massif caving in the working chamber. The trend shows that with mining depth increase from 465 to 840 m volumes of caving increased by 2.3 times. For more reliable results need to be exploring more of the number of chambers located in different axes along the strike of the deposit. The aim of further research is the study of the nature of destructive deformations in the rock massif surrounding ore reserves in the chamber of the second stage with mining depth increase. Main objectives of the analytical modeling are: the study of destructive deformations areas around the secondary chamber; the influence of the depth of the secondary chamber on the rock massif; to identify the law of destructive deformation changes in the rock massif.

Increasing the depth of the chamber is accompanied by an increase in the destressed zone and rising tensions of surrounding rocks. Deformation occurs in places where the real stress arising within rock massif destressed zone, exceed the maximum allowable tensile or shear. The difference between the current and ultimate strain reflects the safety factor. For the same geological conditions analytical modeling of the influence of second stage chamber on the rock massif and backfill was made to the depth of exploitation 840 – 1040 m. As an example, we saw the formation of deformation areas in the rock massif and backfill of the second stage chamber, located across the strike of the deposit at different laying depth (Figure 2).

Figure 2 shows that the area of backfill deformation around the second stage chamber across the strike of the deposit becoming elongated shape resembling an ellipse and arranged in a backfill massif of the primary chamber and the chamber roof. On 2/3 the height of the chamber in backfill massif of the primary chamber, the deformation area with increasing depth of the chamber from 840 to 1040 m spreading towards the rock outcrop hanging wall at a distance of 27 – 30 m.

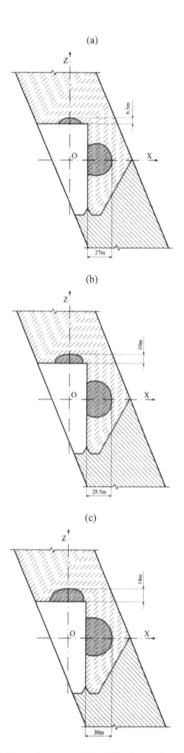

The width of this strain region increased from 30 – 35 m up to 38 – 42 m and has an effect on the power required reinforcement layer of backfill. In the roof of the chamber, at a depth of 840 m of mining operations in backfill massif of overlaying layer, the deformation extends to the height of 6.5 m. With depth increase of the chamber to 1040 m, the area of deformation increases up to 14 meters. The width of this area with mining depth increase from 840 to 1040 m increases from 20 – 26 m to 27 – 32 m. It should be noted that the destruction of the roof and the chamber is more likely caused by the gravitation forces, while caving out of the chamber sides occur in its path. At the same time, in the depth of the backfill massif occurs fracturing and discontinuity.

Gross findings of the stress-deformed state of the backfill massif can be seen from changes of areas of possible destructive strains (Figure 3).

(a)

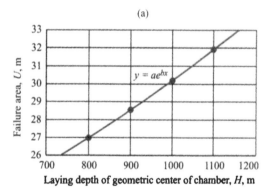

(b)

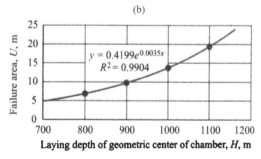

Figure 3. Change of failure area in the backfill massif in chambers of the first stage (a) and in the top of chambers of the second stage (b) with mining depth increase.

After approximation of the maximum values using Microsoft Excel 2013 empirical law equations of the failure area value U on depth H was obtained. For the backfill massif, surrounding secondary chamber along the strike of deposits the empirical law is written as:

Figure 2. Formation areas of possible destructive strains in the unloading area of second stage chambers across the strike of the deposit at a depth of their location: (a) 840 m; (b) 940 m; (c) 1040 m.

– failure area in backfill massif of the first stage chamber mining:

$$U = 17.2 \cdot e^{0.0006 \cdot H} \text{, m, } R^2 = 0.9974, \tag{1}$$

where H – depth of the geometric center of the chamber, m; R – authenticity of approximation.

– failure area in backfill massif in roof rocks of the second stage chamber mining:

$$U = 0.42 \cdot e^{0.0035 \cdot H} \text{, m, } R^2 = 0.9904. \tag{2}$$

As an example the formation of strain fields in the backfill massif, surrounding chambers of the second stage located along the strike of the deposit at different laying depth (Figure 4).

(a)

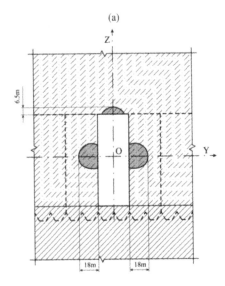

(b)

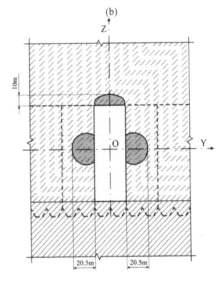

(c)

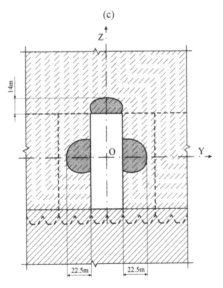

Figure 4. Formation of the failure area around the chambers of the second stage located along the strike of the deposit at a depth of: (a) 840 m; (b) 940 m; (c) 1040 m.

Figure 4 clearly shows that the area of backfill massif deformation around the chamber has an elongated shape resembling an ellipse and arranged in sides of the chamber. On 2/3 the height of the chamber the deformation area with increasing depth of the chamber from 840 to 1040 m spreading towards the rock outcrop at a distance of 18 – 22.5 m. The width of this strain region increased from 20 – 25 m up to 28 – 31

Changes in the stress-strain state in the backfill massif of chambers of the second stage, which are located along the strike of the deposit with mining depth increase, can be observed on areas of possible destructive strains (Figure 5).

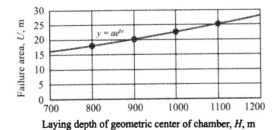

Figure 5. Change of the failure area in the backfill massif in chambers of the first stage located at sides of chambers of the second stage with mining depth increase.

After approximation of the maximum values using Microsoft Excel 2013 empirical law equations of the failure area value U on depth H was obtained. For backfill massif, surrounding secondary chamber along the strike of deposits the empirical law is written as:

$$U = 7.7 \cdot e^{0.0011 \cdot H} \text{, m, } R^2 = 0.9953. \quad (3)$$

Further studies will allow changes in width of the failure area in backfill massif in the chamber of second stage with mining depth increase (Figure 6).

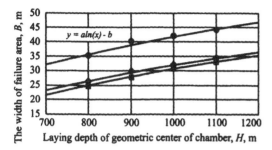

Figure 6. Change of the width of failure area in the backfill massif in chambers of the second stage with mining depth increase.

After approximation of the maximum values using Microsoft Excel 2013 empirical law equations of the failure area value U on depth B was obtained. For the backfill massif, surrounding secondary chamber along the strike of deposits the empirical law is written as:
– failure area in the backfill massif of the first stage chamber mining:

$$B = 27.6 \cdot ln(H) - 148.7 \text{, m, } R^2 = 0.9587; \quad (4)$$

– failure area in the backfill massif in rocks of the roof of second stage chamber mining:

$$B = 24.7 \cdot ln(H) - 138.4 \text{, m, } R^2 = 0.9796; \quad (5)$$

– failure area in the backfill massif in rocks of a workings of the second stage chamber mining:

$$B = 25.5 \cdot ln(H) - 145.4 \text{, m, } R^2 = 0.9972. \quad (6)$$

Research results of conducted investigation in the context of failure area in the backfill massif around the chamber of second stage with mining depth increase can be described by the exponential law and its width by the logarithmic law. This empirical law of the deformation and its width for different mining laying depths allow defining the size and shape of the failure area, which significantly affect the stability of backfilling massif in chambers. The studies confirm the importance of mining in the areas of destructive deformations that must be followed with accurate backfilling massif deformation.

3 CONCLUSIONS

During the chamber mining with a hardening backfill with mining depth increase by 1.6 times the volume of backfill massif caving on the contour of the second stage chamber mining increased by 2.3 times. Analysis of the caving formation should be carried out down-dip of ore deposits by ones geological section.

The size of failure deformations in the backfill massif in the chamber of first stage, as well as sides and roof of the second stage chamber exponentially depend on the mining depth. The most intense deformations developed in roof rocks of the chamber, with mining depth increase by 1.6 times. Their value increased by 2.1 times, and in the backfilling massif in chambers of the first stage and the sides of the chambers of second stage only by 1.1 – 1.2 times.

With mining depth increase the width of failure area develops more intense than in the horizontal plane by 1.3 times and change by logarithmic law, which leads to additional costs for forming the hardened backfill layer.

REFERENCES

Khomenko, O., Kononenko, M. & Petlyovanyy, M. 2014. *Investigation of the stress-strain state of the rock massif around secondary chambers*. Progressive Technologies of Coal, Coalbed Methane, and Ores Mining. The Netherlands: CRC Press/ Balkema: 241 – 245.

Kuzmenko, A. Petlyovanyy, M & Heylo, A. 2014. *Application of fine-grained binding materials in the technology of hardening backfill construction*. Progressive Technologies of Coal, Coalbed Methane, and Ores Mining. The Netherlands: CRC Press/Balkema: 465 – 469.

Kuzmenko, A.M. 2014. *An impact of the structure of the rock massif and the order of chamber mining on ore contamination*. Collection of research papers NAS of Ukraine, IGTM, No 118: 37 – 45.

Falshtyns'kyy, V., Dychkovs'kyy, R., Lozyns'kyy, V. & Saik, P. 2013. *Justification of the gasification channel length in underground gas generator*. Annual Scientific-Technical Colletion – Mining of Mineral Deposits 2013. The Netherlands: CRC Press/Balkema: 125 – 132.

Falshtynskyy, V., Dychkovskyy, R., Lozynskyy, V. & Saik, P. 2012. *New method for the justification of technological parameters of the coal gasification in the test setting*. Geomechanical Processes during Underground Mining – Proceedings of the School of Underground Mining. The Netherlands: CRC Press/Balkema: 201 – 208.

Lavrinenko, V.F & Lisak,V.I. 1993. *Physical processes in the rock mass in the disequilibrium*. News of Higher Educational Institution, Mining journal, No 1: 1 – 6.

Lozynskyi, V.G., Dychkovskyi, R.O., Falshtynskyi, V.S., Saik, P.B. 2015. *Revisiting possibility to cross the disjunctive geological faults by underground gasifier*. Naukovyi Visnyk Natsionalnoho Hirnychoho Universytetu, No 4.

The questions of main drift protection in Western Donbas coal mines

O. Gayday & N. Gurgiy
National Mining University, Dnipropetrovsk, Ukraine

ABSTRACT: The estimation of application of coal pillars of rectangular form is given for the protection of main drifts on mine "Zahidno-Donbaska" Open Corporation of "DTEK "Pavlogradugol". The losses of coal are evident on a particular example, as a result of pillar abandonment, and also the issues of stability, related to the estimation of the protection of main drifts, are considered.

1 INTRODUCTION

At implementation of mining of coal fields the main production link is the mining site providing mining of coal with the set quality indicators. Successful work of a site depends on many factors. One of the main factors is the condition of mines workings, but not only preparatory, but also main which are necessary links in mineral transportation, and also for implementation of ventilation of mine in general.

So far the general principle of protection of mine workings from influence of mining works by means of pillar of coal was its arrangement out of a zone of basic pressure. The mine workings in this case settle down perpendicularly to the main drift therefore the left protective pillar get a rectangular shape (Bondarenko et al. 2012). At approach to the main drift of mining works from a mine working column occurs a "reeling" wave of mining pressure in the direction of action of the maximum stretching tension on horizontal axis that leads to easing of the bearing ability of a pillar and deformation of a contour of the main drift. Inefficient protection of the main mine workings is result of this action, to this fact the huge statistics of refastening of mine workings. The unsatisfactory condition of mine workings does impossible normal mining of coal, and especially the successful.

So, for synthesis of the correct technological decision, what parameters the left protective pillar of coal for the main workings have to possess, it is necessary to have regularities of change of a condition of a massif in specific conditions taking into account the predominating jointing.

The jointing of rocks breaks integrity of rocks weakens their stability. Distinguish born (natural) jointing (cleavage), the tectonic jointing and cracks caused by mining pressure (Falshtyns'kyy et al. 2013). Duration of operation of the main drifts assumes need of their maintenance for operating state. For this purpose the way of protective pillars of coal of mine workings is mainly applied. By means of such way protect about 90% of extent of capital and main mine workings of coal mines of Ukraine.

Coal pillar has large dimensions both along the strike and on descent or ascension of coal seam. Along the strike they are limited to workings of extraction columns, and on descent (ascension) – borders of the goaf. The transport, conveyor and ventilating mine workings that adjoin the main drifts usually meet at the angle 90°. In this case coal pillar get the form close to a rectangle.

To ensure sustainability of mine workings they are constructed with taking into account the mining-and-geological parameters such as depth of their arrangement, coefficient of structural easing of rocks and so on. According to Instructions (Instructions on a rational arrangement... 1986) dimensions of protective coal pillar of the preparatory mine working which are carried out in the massif by the narrow-web-mining should be accepted according to requirements of subsection 5.7.

2 MAIN PART

The coefficient of structural easing for rocks of the Western Donbas lies within 0.6 – 0 75 (soapstone's and siltstones). Taking into account that stress value in calculation of rocks resistance with applied monoaxial compression doesn't exceed 20 MPa.

Mine working, as a rule, cross numerous zones of plicative and disjunctive dislocations and therefore the established depth of their arrangement should be increased in comparison with actual, approximately by 1.2 times. It means that width of a protective coal pillar of main drifts which are 450 – 500 m deep, has to be 96 m.

The specified width, has been determined by calculations of "Dniprogiproshakht" institute in coordination with the National Mining University, establishes the minimum protection requirements of the main drifts in mines of Western Donbas.

The maximum width of protective pillar reaches 150 m that leads to unfairly big losses of coal.

After mining out 50 – 65 m of extraction columns using the protective pillar of the main drift there is such amount of left out coal that it exceeds its annual design capacity.

Despite it, stability of the main drifts on considerable extent more than is unsatisfactory for normal functioning of mine working.

Considering the aforesaid, while designing a protective pillar of sheeted mine working it is important to choose its form correctly.

The general principle of protection of the main drifts from influence of mining activity by means of pillar is locating them out of a zone of bearing pressure. However, due to the lack of reliable ways to determine the stress in the rock massif, the size of a zone of basic pressure usually has been established according to indirect characteristics of rock pressure. Therefore at measurement of different characteristics of manifestation of bearing pressure (shift, deformation, etc.) the received parameters of a bearing pressure zone will differ. The analysis of publications shows that the big difference in estimations of width of a zone of bearing pressure (from 18 to 240 m) is substantially caused by distinction in methods of an assessment of zone size (Vasilyev & Malinin 1960).

We will discuss the issue of unjustified losses of coal, and also sustainability of the main mining working on a concrete example in conditions of mines of Western Donbas; in particular, of "Zahidno-Donbaska" mine of "DTEK "Pavlogradvugillia".

At the mine some coal seams are developed, in particular coal seam C_8^1 the block No 1. The main drifts, capital cross-cuts and other mine workings are necessary for transportation and ventilation, and therefore particularly important mine areas count tens of kilometers. The unsatisfactory condition of development owing to influence of mining works in the 833rd stoping face becames result of actions for protection of the southern main tramming drift (losses of coal in a pillar made 21384 t).

In the 833rd stoping face (length of the stoping face is 160 m, coal seam capacity C_8^1 on this site is 0.9 m) extraction works were finished at distance of 110 m from the southern main tramming drift (Figure 1).

The face is parallel to a longitudinal axis of a drift therefore for its protection it is left pillar squared coal. On a site of pickets (Pic) of the 30 – 50 in three years is needed for refastening arose. Changing of shift of side rocks from time of supervision is given in Figure 2.

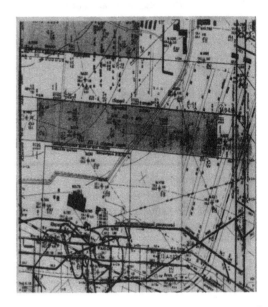

Figure 1. The Fragment of the plan of mining works of seam C_8^1.

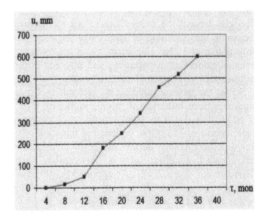

Figure 2. Nature of shifts u side rocks from time τ supervision.

The southern main tramming drift on a piece from picket 1 to 81 was retimber by ring lining that considerably improved stability of mine working for further use. Results of coalmine tool measurements at control of the content of mine working are specified in Table 1.

Experience of conducting mine workings in Western Donbas shows, that leaving of protective pillar doesn't provide a satisfactory condition of mine working during operation time.

It is established by the researches conducted on Zahidno-Donbaska mine that the reason for that – moving of a zone of the increased mining pressure (IMP) to depth of a pillar after a stoping face stopage.

Table 1. Results of mine tool measurements.

The installation location of metering stations	The height of the mine workings, mm	Time of supervision, days	Shifts in the roof and the soil, mm	Distance from a drift to a stoping face, m
Pic 1	3481	391	253	300
Pic 5	3416	2343	662	250
Pic 10	3420	1264	1015	200
Pic 15	3475	1662	923	150
Pic 20	3600	2343	1442	100
Pic 24	3720	1581	1289	50
Pic 30	3180	1447	644	10
Pic 40	3345	1551	452	10
Pic 50	3287	1802	422	10
Pic 60	3480	240	90	10
Pic 70+6m	3325	293	336	50
Pic 75+6m	3239	324	127	100
Pic 81	3320	229	103	150

Thus, when performing further researches, there is a need of justification of rationally effective way of protective pillars of coal the main drifts, namely their forms as the rectangular doesn't satisfy to questions of stability of excavations during the long period (Kravchenko 1970).

Variable width of a pillar promotes attenuation of loadings on support when conducting mining works near the main drift and to the best perception of internal tension of the massif of rocks (Kolokolov & Gayday 2004).

As it is known (Grishkova 1947), the main indicator of a form of cross section of a pillar is the hydraulic radius, that is the relation of the area of cross section of a pillar to its perimeter:

$$R = S/L,$$

where S – the area of cross section, m²; L – perimeter length, m.

The best ratio at a triangular form of cross section pillar of coal was reached.

Besides, change of a form of the protective pillar on triangular will allow reducing losses of coal by 35 – 45% in comparison with a rectangular shape. Thus strength properties of a pillar don't decrease, and deformation fix in a drift decreases.

In view of that the coal massif near the main drift is weakened by an unloading zone, mining works are stoped at its border. It gives to a pillar the form of a trapezium (Figure 3).

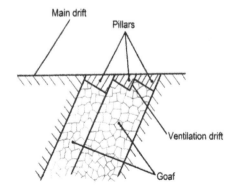

Figure 3. Scheme of an arrangement of local mine workings at an angle to the predominating jointing.

At an arrangement of an extraction column at an angle $\beta < 90\,^\circ$ the zone of basic pressure of a stoping face moves in the direction of action on a horizontal axis of the main squeezing tension. It promotes decrease in probability of manifestations of the gas dynamic phenomena at clearing dredging and reduction of number of inrushes of rocks of a roof.

3 CONCLUSIONS

1. For ensuring stability of the main drift on mines of Western Donbas pillar 100 – 150 m wide is leaved that leads to unfairly big losses of coal.

2. Leaving of pillar at main mining workings provides temporary maintenance of their initial stability.

3. The main reason for the created real situation is insufficient attention to an assessment of mining-and-geological conditions in consequence of which quite often form pillar of coal of the insufficient sizes.

4. It is necessary to reconsider a question concerning a form of the left pillar of coal, to correct the general principle of a way of protective pillar of the main drifts and to continue researches in this direction.

REFERENCES

Bondarenko, V., Symanovych, G. & Koval. O. 2012. *The mechanism of over-coal thin-layered massif deformation ofweak rocks in a longwall.* Geomechanical processes during underground mining. The Netherlands: CRC Press/Balkema: 41 – 44.

Falshtyns'kyy, V., Dychkovs'kyy, R., Lozyns'kyy, V &, Saik, P. 2013. *Justification of the gasification channel length in underground gas generator*. Annual Scientific-Technical Colletion - Mining of Mineral Deposits 2013. The Netherlands: CRC Press/Balkema: 125 – 132.

Grishkova, N. 1947. *Determination of mechanical properties and elastic constant rocks of Donbas*. Works of the commission on management of a roof and reports at the I All-Union meeting of GONTI of Ukraine. Donetsk: 55 – 67.

Instructions on a rational arrangement, protection and maintenance of mine workings on coal mines of the USSR. 1986 Leningrad: VNIMI: 222.

Kolokolov, O. & Gayday, O. 2004. *Influence of the predominating massif jointing on a condition of a stoping face*. Materials of the Ukrainian-Polish forum of miners "Mining industry Ukraine and Poland: actual problems and prospects". Yalta, September 13 – 19: 32 – 37.

Kravchenko, V. 1970. *Prevention of blockages in stoping faces*. Moscow: Nedra: 210.

Vasilyev, P. & Malinin, S. 1960. *Influence of the major geological factors on behavior of rocks in mine workings*. Moscow: Gosgortekhizdat: 94.

Rationale for the parameters equipment for rope dehydration of mining hoisting installations

K. Zabolotny, S. Zinovyev, A. Zupiev & E. Panchenko
National Mining University, Dnipropetrovsk, Ukraine

ABSTRACT: The interaction of air current with water film on a surface of moving hoist rope of mining hoisting installation is investigated. The mathematical models of carrying water by hoist rope and removing water from hoist rope surface are developed. The conformity of setting mathematical models is proved by experiments in the laboratory and mine. The parameters of the equipment for rope dehydration are grounded. Use of this parameters guarantees complete deleting of water from rope surface with probability of 95%. The methodical recommendations for ground of rational parameters of equipment for rope dehydration are developed and introduced on the production amalgamation "Krasnoarmiyskvugillya" and into design works of the state project institute "Kryvbasproekt". The idea of water-deleting device is used in the methodical recommendations. The coming economic effect from introduction of equipment for rope dehydration is the 600 thousand hryvnias for public corporation "Kryvbaszalizrudkom" mine "Gvardiyska" due to running costs reduction.

1 INTRODUCTION

An important role belongs to the vertical (hoist) transport in the mineral resources mining technology by underground method. For example, a mining hoisting installation (MHI) is a very responsible link in the production process. In the case of any MHI accident, the extraction of mineral resources may stop, causing huge losses. Thus, there are high demands to the reliability and technical level of MHI, the quality of their design and production. According to the research results, 85% of mine shafts in Donbas, Kryvbas and 75% of non-ferrous metals mining are watered, and each rope of MHI is capable to bring to the surface from 10 to 20 m³ of groundwater per day. During the machine operation water brought up by hoist ropes (HR) inevitably fall on the brakes and the traction sheaves, reducing the operating security of the entire mining hoisting complex. To remove water from the surface of the ropes (dehydration) use Different devices are used to remove water from hoist rope surface (dehydration) are used. Specialists of the National Mining University have developed the design of the dehydrating equipment, in which the air current to remove the water is formed by underpressure in a chamber of the device (Figure 1). The equipment is installed in the tower below the angle sheaves, consisting of a device to remove water by air current (WR device), a water separator, a fan with a drive and a pipeline system.

Figure 1. Model of the equipment to remove water from ropes.

Removing water from the rope occurs in the chamber of the WR device. There, fan operations cause underpressure, which generates air current in the gap between the rope and the sealing ring, which is directed along the rope surface. The resultant water-air mixture through nozzles of the WR device enters the water separator, in which it is divided into phases.

Operation of the device to remove water is accompanied by complex hydrodynamic processes of the interaction between air current and water film, which is sprayed on the hoist rope surface.

Currently, there are no models to describe the process of removing water by air. This fact does not allow a reasoned approach to the selection of performance of the equipment capable of removing water from the ropes. Thus, the rationale for the design parameters of the equipment to remove water from the ropes based on the determination of occurring operating procedure regularities is a relevant scientific mission.

The objective of the research is to improve the operating efficiency of the equipment to remove water from the ropes of the mining hoisting installations, using air current.

2 RESULTS OF RESEARCHES

Complete dehydration of water from the hoist rope surface is possible if the removal intensity In of water exceeds the its carry-over intensity Q, i.e.

$$In \geq Q .$$ (1)

Processes, which occur in the dehydrating equipment, can be divided into the following: water carry-over by a rope, water spraying by air current, air current flow in a device chamber, water removal from a spraying area, division of the two-phase flow in a water separator.

To explore the process of water carry-over by a hoist rope, we used the physical model described below. The flow of water with the thickness h flows off the surface of the vertical rope with the diameter d_k, it moves at the constant velocity v_k vertically upwards. Air current at the constant velocity v_{AN} moves in the cylindrical neighbourhood of the rope. In this case, the following assumptions apply: the water flow is laminar, uniform along the rope and consistent over time, the air ring cylinder with the thickness $h_0^{\bullet}$ moves at the velocity v_{AN} without tangent interaction with water; the water and air flow is axisymmetric, removal of water by the hoist rope is proportional to its perimeter.

For mathematical description of the process, let us introduce the coordinate system $oxyz$, where the z-axis is directed along the generator of a surface of

a solid cylinder, the y-axis along the arc, and the x-axis along the outward normal, wherein when $x = 0$, the axis will correspond to the rope surface.

According to the results of solving the equations of motion (Navier-Stokes and continuity) together with the boundary conditions, an expression that describes the distribution of water velocity over the film thickness has been obtained, namely:

$$v_z^b = v_k + \frac{g}{2 \cdot v} \cdot x^2 - \frac{g \cdot h}{v} \cdot x .$$ (2)

Let us denote the specific intensity of water removal Q_y to determine the maximum intensity of water removal by the rope, which corresponds to the maximum of the function Q_y, as follows:

$$Q_y = \int_0^h v_z(x) \cdot dx = v_{cp} \cdot h = v_k \cdot h - \frac{g \cdot h^3}{3 \cdot v} ,$$ (3)

where v_{cp} – shall mean the average velocity of the water film flow, taking into account the thickness of the flow.

With increasing velocity of the rope motion the specific intensity of water removal is increasing; and each value of this velocity corresponds to such film thickness at which the intensity of water removal is at a maximum, it is defined as follows:

$$h_Q = \sqrt{\frac{v}{g} \cdot v_\kappa} .$$ (4)

As seen, the thickness of water film removed by the hoist rope is proportional to the square root of the rope velocity. The intensity of water removal by the rope is defined as its flow rate, taking into account cross-section of film by the following formula:

$$Q = \int_P Q_y(v_\kappa, h) \cdot dy ,$$ (5)

where P – shall mean the rope perimeter.

Designating the ratio of the rope perimeter to the cylinder perimeter of the corresponding diameter as the rope perimeter coefficient K_p, and the ratio of the maximum water removal intensity (depending on the film thickness h_Q) to water removal intensity based on the arbitrary film thickness in the form of the water removal coefficient K_w, and also taking into account the dependence (4), we conclude that the intensity of water removal by the hoist rope is determined from the following expression:

$$Q = \frac{2}{3} \cdot K_b \cdot K_p \cdot \pi \cdot d_\kappa \cdot \sqrt{\frac{v}{g}} \cdot v_\kappa^{3/2} ,$$ (6)

where v – shall mean the viscosity of water.

Coefficient K_b characterizes the real intensity of water removal by HR (in the existing conditions of the rope operations) in relation to the maximum possible theoretical, and therefore, determination of its value can be done only experimentally in a study of the hoisting installation, which is operated in the most watered mine shaft. The experiment has been performed at mine "Tsentralna" of the production amalgamation "Krasnoarmiyskvugillya". Here, the main shaft is considered one of the most watered among the mines in Ukraine.

The experimental installation to remove water has been placed in the tower below the angle sheaves (see its diagram in Figure 2. It includes the following objects,: 1 – a MHI drum; 2 – angle sheaves; 3 – a hoist rope; 4 – a device to remove water (Zabolotny et al. 2013); 5 – a pipeline; 6 – U-shaped water manometer; 7 – a network resistance regulator; 8 – fans; 9 – a water separator.

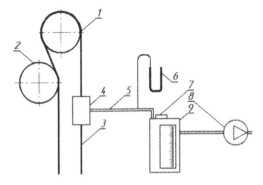

Figure 2. Diagram of the mining installation to remove water from the hoist rope surface.

Based on the experimental data values of the coefficient K_b have been calculated, and the hypothesis of their distribution normal law has been verified. Figure 3 shows a histogram of experimental and a polygon of theoretical distribution of K_b values, and Figure 4 shows deviation of the experimental data obtained in the study of the rope with the diameter of 42 mm (indicated by dots) from the intensity curve of water removal by the rope (shown by a solid line).

With probability of 95% water removal intensity coefficient $K_w = (6.23 \pm 0.92) \cdot 10^{-1}$; with the deviation of confidence limits from the mathematical expectation amounting to 14.8%. The adequacy of the expression (6) to the experimental data has been proved by the Fisher's ratio test. To study the process of removing water from the hoist rope surface, a physical model has been developed, which lies in the fact that water flow loses its stability at a certain velocity of air current, thus causing the wave flow. Influence of air current on water film increases the amplitude of wave oscillation, resulting in irregularity of their symmetric shape. It is proved that the rapid increase in the am-

plitude of wave oscillation contributes to its significant deformation. Thus, the wave becomes unstable and droplets are detached from the surface of its tipping ridges. In solving tasks with the application of this model we use these assumptions: the hoist rope is a cylindrical body; film motion over its surface is axisymmetric, laminar; wave oscillation amplitude is small compared to its length that corresponds to the beginning of unstable wave formation; tangent stress is absent on the water film surface; the water spraying intensity is proportional to the area, where water interacts with air current.

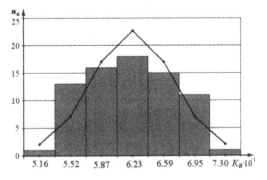

Figure 3. Distribution histogram of values of water removal intensity coefficient K_b.

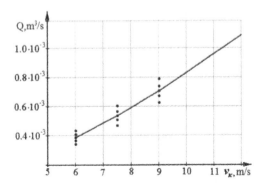

Figure 4. Dependency graphs of water removal intensity on HR velocity.

Application of the variable separation method in solution of linearized motion equations (Navier-Stokes and continuity) and representation of the wave flow as sum of two currents (unperturbed and of small perturbations) allows us to formulate the eigenvalue problem as a system of dispersion equations that describe the wave flow, that is:

$$\begin{cases} \rho \cdot \alpha \cdot v \cdot K^d + \rho^* \cdot (\alpha + i \cdot k \cdot v_{AN})^2 + \sigma \cdot k^3 = 0 \\ \alpha = v[\ell^2 - k^2] \end{cases}, \quad (7)$$

where ρ, $\rho*$ – shall mean the density of water and air, respectively; α – shall mean characteristics of the time-variable wave flow; v shall mean the water viscosity; k – shall mean the wave number;

ℓ – shall mean the wave oscillation damping coefficient in the radial direction; v_{AN} – shall mean the gas flow velocity; σ – shall mean the water surface tension. In this system:

$$K^d = \frac{2 \cdot \ell \cdot k^2}{k \cdot sh(\ell \cdot h) \cdot ch(k \cdot h) - \ell \cdot ch(\ell \cdot h) \cdot sh(k \cdot h)} + (\ell^2 + k^2) \cdot \frac{k \cdot th(\ell \cdot h) \cdot th(k \cdot h) - \ell}{k \cdot th(\ell \cdot h) - \ell \cdot th(k \cdot h)}. \qquad (8)$$

As shown in the diagram of changes in the velocity and pressure distribution in water film (Figure 5), wave flow is exponentially damped in the direction from the interface of gas and liquid phases; at a distance from said surface, which is equal to one-sixth wavelength, the oscillation amplitude is reduced by 95%; potential flow of liquid does not affect its rotational motion; potential flow of air determines the nature of rotational motion; the rope surface shape does not affect the process of water rotational motion.

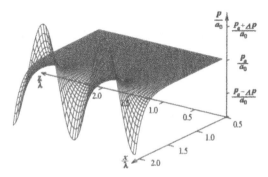

Figure 5. Diagram of pressure distribution in water film, covering the MHI rope.

Considering the rapid damping of wave oscillation amplitude in the direction from the phases interface, let us assume that $K^d = (\ell^2 - k^2)$. Due to the fact that air density as compared to water density is much lower, and hence the value of the second term is small compared to the others, the system of dispersion equations (7) can be simplified to a single equation, and the following dependence can be further obtained:

$$\alpha_0 = k \cdot \sqrt{\frac{\rho^* \cdot v_{AN}^2 - \sigma \cdot k}{\rho}}, \qquad (9)$$

from this equation we determine the wavelength, corresponding to the maximum increase in oscillation amplitude, i.e.:

$$\lambda_0 = \frac{3 \cdot \pi \cdot \sigma}{\rho^* \cdot v_{AN}^2}, \qquad (10)$$

According to the results of flow parameters determination, based on the spraying model

V.G. Levich, also considering the designation of the airflow velocity integral function, we come to the following conclusion:

$$J = \int_0^L v_{AN}(z) \cdot dz, \qquad (11)$$

where L – is the length of a chamber in the WR device.

Then, we obtain the expression for calculating the intensity of water removing from the HR surface, namely:

$$In = \frac{8}{3\sqrt{3}} \cdot \frac{\pi^3 \cdot K_k^3 \cdot K_n \cdot K_p}{ln(M)} \cdot \sqrt{\frac{\rho^*}{\rho}} \cdot d_\kappa \cdot J, \qquad (12)$$

where K_k – shall mean the coefficient of proportionality between the droplet diameter and the wavelength; K_n – shall mean the irregularity coefficient of water film thickness; M – shall mean the enlargement factor of wave oscillation amplitude, at which droplet detachment occurs.

Next, let us discuss the influence of design parameters of the WR device on the water removal process, and determine the dependence of the airflow velocity integral function on the characteristics of the chamber in the device and the pressure drop.

Above it was noted that the intensity of water removal from the hoist rope surface is directly proportional to the airflow velocity integral function (11), which is used in this case. At the same time, due to the fact that the air flow generates the WR device, the integral function of this flow velocity depends on the structural features of the device. Obviously, it is necessary to study this dependency.

To solve the problem set, it was decided to apply the computational experiment, the results of which are updated using the tests of a laboratory prototype of the WR device. Whereas the formula for calculating the water removal intensity includes coefficients that take into account the properties of liquid, which ultimately affects the value of a parameter in question, then, in our study, these coefficients are generalized under the name of the water removal intensity coefficient. Its numerical value was determined in the laboratory and mine experiments.

In the numerical experiment the equations k-ε models of turbulent motion of incompressible air flow, which were solved using an application software package of ANSYS 6.1/ED system. At the same time, we were focused on such boundary conditions:

on the hoist rope surface and on the chamber walls of the device the flow velocity is zero; pressure is equal to atmospheric (p_a), and it corresponds to the depression of a fan (Δp^*) in the exit outlet.

As shown by the results obtained by solving the mentioned equations, changing the air flow velocity along the chamber of the device and the value of the function J enter the complicated dependence on a number of factors. For ease of the analysis, we express the indicated function through equivalent data, the values of which can be varied in the research process.

Suppose $J = \Lambda \cdot v_B$, where v_B – is the air flow velocity, which is determined by using Bernoulli's equation; wherein, if $v_B = \sqrt{2 \cdot (p_a - p_{AN}) \big/ \rho_*}$, then, the value of Λ corresponds to the length of an area equivalent rectangle (Figure 6), it can be considered the effective length of an interaction area of air flow with water on the rope.

Influence of the following parameters on the value of the function J was examined: length and gap of the opening forming the flow, underpressure in the chamber of the device and its length (i.e., the four-factor experiment was conducted). As a result, we have such a linear regression equation $J = \Lambda \cdot v_B$, wherein $\Lambda = 0.175$ m.

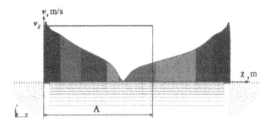

Figure 6. Distribution diagram of air flow axial velocity over the rope surface along the length of the chamber of the WR device.

Thus it is proved that during the operations of the WR device two annular air jets, which are formed by underpressure in the chamber, move through the gaps between the rope and the sealing ring along the rope surface toward each other. Figure 7 shows a graphical interpretation of the computational experiments results in the conditions where the exit outlets are located in the middle of the chamber of the WR device. Thus, it was possible to determine a change in the parameter value (the effective length of the flows interaction area) depending on the location and number of pipes that remove water. It will be appreciated that the pipes were shifted symmetrically and asymmetrically in small increments in the direction from the middle of the chamber.

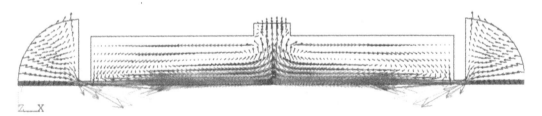

Figure 7. Air flow velocity field in the chamber of the WR device in turbulent flow ($\Delta p^* = 0.05 \cdot p_a$).

As a result, the adequacy of the regression equation to the experimental material was confirmed. It follows from this equation that the influence of the opening parameters, i.e. the values of the gap between the rope and the sealing ring, ranging from 2 to 6 mm, and the length of the opening, which was enlarged in the range of 5...15 mm, causes change in the valued of the function J not more than 2.3%.

It was established that the intensity of water removal from the rope surface is directly proportional to its diameter multiplied by the square root of the value of fan depression, i.e.:

$$In = \frac{8\sqrt{2}}{3\sqrt{3}} \cdot \frac{\pi^3 \cdot B \cdot K_p}{\sqrt{\rho}} \cdot \Lambda \cdot d_\kappa \cdot \sqrt{\Delta p} , \qquad (13)$$

where B – shall mean the water removal intensity coefficient.

The coefficient B characterizes the operation of the equipment to remove water from ropes. Its value was determined in the laboratory experiment and was specified during the mine experiment.

The laboratory experiment was full factorial. The hoist rope was simulated as a cylinder of the appropriate diameter and lengths of the round-strand rope (manufactured in accordance with GOST 7669-80). In this case, water was delivered to the fixed vertically rope that flowed down over its surface. The experiment involved such input values (factors): the rope diameter, the pressure drop in the chamber of the WR device, the chamber length, and the output intensity indicator of water removing from the rope surface.

These experimental results were tested for the presence of gross errors, homogeneity of variance of the results by the Kochren's criterion and influence of each factor by the Fisher's ratio test. Processing of the experimental results revealed that with probability of 95% the influence of the length of the device chamber on the water removal intensity is absent, indicating the reliability of the calculations done using the application software package ANSYS 6.1/ED. At the same time, the influence of other factors proved to be very significant.

After determining the value of B, let us test the hypothesis on the correspondence of experimental data distribution to the normal law. In this case, we analysed the values of the water removal coefficient from the cylinder surface (B_c) and the same value in relation to the round-strand rope (B_k). A slight discrepancy between the empirical and theoretical frequency of falling of the results in a confidence interval was determined according to the Pearson's test. Thus, the values of both data are characterized by the mathematical expectation and the mean-square deviation. So, a confidence interval of values of the coefficient $B_c = (5.45\pm0.36)\cdot10^{-4}$ and the coefficient $B_k = (5.42\pm0.39)\cdot10^{-4}$. Deviation of the confidence limits from the mathematical expectation with a probability of 95% was found to be 6.6 and 7.2%, respectively.

In the mine experiment the coefficient B was determined within five fixed ranges of fan depression provided the maximum velocity of HR motion. As established by the Pearson's test, the distribution of the coefficient B corresponds to the normal law.

Comparison of values of the coefficients B, B_c and B_k identified with a probability of 95% of their falling into the confidence interval shows that the range of values of the coefficient B, found according to the results of the mine experiment, with a deviation of the confidence limits from the mathematical expectation amounting to 13.7%, covers the same parameter values obtained in the laboratory tests, the range is equal to $(5.34 \pm 0.73)\cdot10^{-4}$. In view of the Fisher's ratio test the adequacy of the expression (1) to the experimental data was proved.

Using the condition (1) and expressions (6) and (13), we come to the following dependence:

$$\Delta p^* = \frac{3}{32} \cdot \frac{K_e^3}{B^2} \cdot \frac{\rho \cdot v}{\pi^4 \cdot g \cdot \Lambda^2} \cdot v_\kappa^3 . \qquad (14)$$

We consider this equation the basic in determination of parameters of the equipment to remove water from the ropes on the basis of the WR device. To comply with the inequality (1) it is necessary that the minimum water removal intensity exceeds the maximum value of its removal in the process of the hoist rope motion. For this reason, in the determination of the coefficient of fan depression ξ, where water film removal becomes possible with a probability of 95%, we use the maximum values of the water carry-over intensity coefficient K and the minimum values of the water removal intensity coefficient B. Reasoning from this fact, we come to the following conclusion:

$$\Delta p^* = 0,001 \cdot \xi \cdot \frac{\rho \cdot v}{g \cdot \Lambda^2} \cdot v_k^3 , \qquad (15)$$

where $\xi = 2.41\cdot10^6$.

3 CONCLUSIONS

The research results of the process of water removal by the hoist rope using a mathematical model in the form of the Navier-Stokes equations indicate that the water removal intensity by the hoist rope of MHI is proportional to its diameter and is in the power-law dependence with indicator 3/2 on the rope velocity.

Research of the process of removing water from the hoist rope surface ensures that the intensity of water removal from the rope surface is proportional to the air flow velocity integral function and the diameter of this rope, and depends on the ratio of the density of water and air.

According to the results of numerical experiments using the finite-element k-ε-models of turbulent motion it was proved that where the length of the chamber of the device is changed in the range of 300...600 mm, and the pressure drop there is between 1.0 to 15.0 kPa, then the effective value of the length of the flows interaction area with an accuracy of 5.3% is constant and equal to 175 mm. Statistical analysis of the regression model for the integral velocity function showed its adequacy to the experimental data. The integral velocity function, and thus the intensity of removing water from the rope surface are proportional to the effective value of the length of the flows interaction area multiplied by the square root of the fan depression value.

The laboratory experiments confirmed that the value of the water removal intensity from the surface of a cylindrical and round-strand hoist ropes is directly proportional to the rope diameter multiplied by the square root of the fan depression value. The water removal intensity coefficient was also determined: in the calculation of a model of the locked-coil rope it is equal to $(5.45\pm0.36)\cdot10^{-4}$, and in relation to the round-strand rope (GOST 7669-80) is equal to $(5.42 \pm 0.39) \cdot 10^{-4}$.

At full water spraying over the rope surface of the device chamber we obtained an equation based on which it is possible to determine the parameters of the equipment to remove water from the ropes, which suggests that depression of a fan is directly proportional to the velocity cube of the hoist rope motion. Statistical processing data of a physical experiment conducted at mine "Tsentralna" of the production amalgamation "Krasnoarmiyskvugillya" confirmed that with probability of 95% the value of the water removal intensity coefficient will amount to $(5.34 \pm 0.73) \cdot 10^{-4}$, and the water carry-over inten-

sity coefficient will amount to $(6.23 \pm 0.92) \cdot 10^{-1}$. Moreover, the complete removal of water from the hoist rope surface is possible, when depression in the water removing device will be 5.7 kPa at its velocity of 9 m/s, and 13.2 kPa at the velocity of 12 m/s.

REFERENCES

Zabolotny, K.S., Zinovyev, S.N., Zhupiiev, A.L., Panchenko, Y.L. 2013. *The theory of working processes of the parameters of equipment for rope dehydration of mining hoisting installation.* Monograph. Dnipropetrovsk: National Mining University: 151.

Theoretical and Practical Solutions of Mineral Resources Mining – Pivnyak, Bondarenko & Kovalevska (eds)
© 2015 Taylor & Francis Group, London, ISBN: 978-1-138-02883-8

Applying the distributed information processing principle in opencast mining design

Zh. Sultanbekova & S. Moldabayev
Kazakh National Research Technical University named after K.I. Satpayev, Almaty, Kazakhstan

ABSTRACT: Possibilities to use a model of the evaluators group in the course of designing the open cast mining, in particular during estimation of open pit transport. The methods reducing a design period with maintenance of high quality project include application of new IT trends in the course of works.

1 INTRODUCTION

Application of the distributed information processing principle was studied for the first time for engineering estimations on a basis of calculations of performance indicators for open pit transport involved in mineral extraction process.

The issues relating to implementing the distributed information processing are reflected in the work of the Russian Academician, E.V. Evreinov. The operations as per the established practice relating to opencast mine design consisting of multiple sections, are to be completed sequentially (cascaded). The target and idea, tasks and structure of research were defined in the course of opencast mine designing allowing to complete the calculations in parallel, on a basis of this principle

2 MATHEMATIC MODEL

The Open Pit Transport Section was selected as the main object of research since it is the most difficult and significant in the opencast mine design and interconnected with all engineering processes.

The existing types of program products structure (Scheme 1 and 2) and the proposed structure (Scheme 3) shown in the Table 1 with evaluation of computational procedures duration in different software schemes as well as the costs required for development and implementation are provided bellow.

For purpose of the task assigned a complex structure of program tools on a basis of the distributed information processing use including a sequential algorithm, complex software with subroutines, co-ordinator software and individual computing blocks, program blocks interacting with database, was developed. Time for computing procedures completed by the coordinator in general is to be defined as per the following formula:

$$T_{B.P.} = \sum_{k=1}^{R} t_{ful.k} + \sum_{n=1}^{N} t_{o.sub} + t_{main} + \sum_{l=1}^{L} t_{r.bl.l} +$$

$$+ \sum_{m=1}^{M} t_{ev.m} + t_{DB} + \sum_{m=1}^{M} t_{adj.m} + t_{co} + t_{call} + t_{pen}, \quad (1)$$

where $t_{ful.k}$ – time spent for fulfilling the section; K – maximum number of project sections, $k = \overline{1, K}$; $t_{o.sub}$ – operation time of subroutine; n – a number of subroutines, $n = \overline{1, N}$; $t_{r.bl.l}$ – time required for completing the l program block, $l = \overline{1, L}$; $t_{ev.m}$ – operation time of the m evaluator, $m = \overline{1, M}$; $t_{adj.m}$ – time for adjustment of results of the m evaluator; t_{DB} – operation time of database; t_{main} – operation time of the main program; t_{coor} – operation time of the coordinator; t_{call} – time of call to subroutine; t_{pen} – time of possible computer pending.

When operating this scheme the following limitations have been defined:

– when developing the software under the sequential algorithm a computational process runs sequentially, i.e. the section is executed after completing the previous section, K – number of the evaluation sections;

– when operating the software with subroutine the computational process is to be executed by the main program which, if required, calls to the subroutine, N – number of subroutines;

– when operating the software with individual subroutine blocks the computational process is controlled by the coordinator possessing the L program blocks;

– when operating the software with evaluators group the process is executed by the coordinator, individual evaluators have different program blocks, in this case the M means number of evaluators involved in the project process.

Table 1. Main types of software product structures.

Schemes	Duration of computational procedure	Costs for development and implementation
1. Sequential algorithm. 	$$T_{B.P.} = \sum_{i=1}^{k} t_{dev.i} \cdot$$	$$Z_{qbn} = Z_{dby} + Z_{sd} + Z_{pay} \cdot$$
2. Program with subroutines. subroutine	$$T_{B.P.} = \sum_{i=1}^{n} t_{dev.i} + t_{main} + t_{call},$$ t_{call} – time when the main program calls to the subroutine, $$t_{call} \approx 0.$$	$$Z_{gen} = (Z_{dev} + Z_{devi}) +$$ $$+ (Z_{cm} + Z_{pay}).$$
3. Model of the distributed information processing. subrout Last calculat. pr.bl	Option 0 – sequential program: $$T_{B.P.} = \sum_{i=1}^{k} t_{dev.i} \cdot$$ Option 1 – with subroutines: $$T_{B.P.} = \sum_{i=1}^{n} t_{SDi} + t_{BD} + t_{main.pr} + t_{call} \cdot$$ Option 2 – with program blocks: $$T_{B.P.} = \sum_{i=1}^{l} t_{prbli} + t_{BD} + \sum_{i=1}^{l} t_{adji} \cdot$$ Option 3 – with evaluators: $$T_{B.\Pi.} = \sum_{i=1}^{m} t_{evi} + t_{DB} + \sum_{i=1}^{m} t_{adji} + t_{co} \cdot$$	Option 0 – sequential program: $$Z_{gen} = Z_{dev} + Z_{cm} + Z_{pay} \cdot$$ Option 1: $$Z_{gen} = (Z_{main} + Z_{zd} + Z_{db}) +$$ $$+ (Z_{cm} + Z_{pay}).$$ Option 2: $$Z_{gen} = (Z_{main} + Z_{pt} + Z_{db}) +$$ $$Z_{cm} + Z_{pav} \cdot$$ Option 3 $$Z_{gen} = (Z_{dv} + Z_{co} + m \cdot Z_{cm}) +$$ $$+ m \cdot Z_{pay} \cdot$$

A cost for development and implementation of software generally is to be defined as follows:

$$Z_{gen} = Z_{dev} + Z_{SD} + Z_{pr.bl.} + Z_{ev} + Z_{DB} +$$
$$+ Z_{CM} + Z_{pay.} + Z_{co} + Z_{main},$$ (2)

where Z_{dev} – costs for development of software to execute the project sections; Z_{SD} – the costs for subroutines development; $Z_{pr.bl}$ – the costs for development of program blocks; Z_{ev} – the costs for development of programs for the evaluator; Z_{main} – the costs for development of program under sequential algorithm; Z_{coor} – the costs for coordinator development; Z_{DB} – the costs for database development; Z_{CM} – the costs for using the computing machinery; Z_{pay} – the payroll costs.

All designations shown in the formulas (1) and (2) and in further formulas have the same meaning.

The first and second structural schemes of the programs are applied for solving not sophisticated tasks. While for operational decision on current tasks a prompt reaction to different demands will be required with the changes to the program structures.

Requirements diversity includes multiple tasks scheduling, designing and management on a basis of different criteria such as qualitative assigned tasks implementation within short time, existing program tools review and new relevant algorithms development due to shift from the computing machines to personal computers (Kuzmin et al. 1991, Evgenev 2006, Burlakin & Voronov 1998).

Due to this the 3-rd scheme contains the distributed information processing model proposed by us which operates on a basis of level of complexity, volume and targets of the tasks.

The 3-rd scheme of software products with application of the distributed information and database processing principle may be implemented by four options: sequential program, subroutines and program blocks as well as with the help of evaluators. 4 options of the proposed model work are described below.

Option 1, Scheme 3: the program is implemented on the grounds of the sequential algorithm with use of database and the computing process runs in linear manner.

Occasionally, the program calls to database for required parameters, the duration of the computational procedure should be defined as per the formula (1) and defined through summation of all sequential sections execution time and database operation time. Other parameters in this formula will be equal to zero since the specified blocks are not involved in the computational process. The costs for software development and implementation include the costs for individual sections and database development as well as for the computing machines complex (CMC) use and for remuneration of designer's labor.

Option 2, Scheme 3: the computation process is executed by the main program consisting of n subroutines and database. The calculations are made through call to subroutines based on simplified method defining the desired parameters. Time for calculations is defined as follows:

$$T_{B.P.} = \sum_{i=1}^{n} t_{SDi} + t_{DB} + t_{main.pr} + t_{call} a .$$ (3)

Option 3, Scheme 3: complex task maybe solved and to do so the individual program blocks may be applied. Results of the program blocks calculations should be processed by the coordinator through their matching. Then individual project sections or chapters may be executed through involvement of 1 program blocks and computation process may be executed by one person, duration of execution and costs should be defined as per the following formulas (1) and (2):

$$T_{B.P.} = \sum_{i=1}^{l} t_{pr.bli} + t_{DB} + t_{co} .$$ (4)

Costs for development and implementation of this type of program should be defined by the following method:

$$Z_{gen.} = (Z_{co} + Z_{pr.bl} + Z_{DB}) + Z_{CM} + Z_{pay} .$$ (5)

Option 4, Scheme 3: m evaluators or designers will be involved, the project may be executed by the coordinator and one evaluator. Fulfillment period depends on professional level and qualification of the personnel, their quantity as well as level of busyness of the

specialists. Therefore, costs for development and implementation of this model will be increased since the part-time specialists (0.25 or 0.5 fte) from other works or projects may be involved. For these evaluators work the work places should be equipped with the relevant personal computers.

Duration of execution should be defined by the following formula:

$$T_{B.P.} = \sum_{i=1}^{m} t_{evi} + t_{DB} + \sum_{i=1}^{m} t_{adj.nd} + t_{co} .$$ (6)

Design process duration as well as prime cost of this program tools depends on number and level of busyness of the specialists involved (so called operative efficiency), i.e. material costs may be as follows:

$$Z_{gen.} = (Z_{ev} + Z_{co}) + m Z_{CM} + m Z_{pay} .$$ (7)

3 CONCLUSIONS

Design process duration with selection of program tool structure subject to the customer's requirements and capability may be defined by the following two criteria: duration of execution and project development costs. If project is to be executed within short time without material costs limited the design process should involve individual evaluators whose number should be limited or defined depending on busyness and qualification level of the personnel and program blocks availability. When the customers requires to execute the low cost project with no design period limits the design process should be executed by one designer owing to implication of individual program blocks included in the coordinator's structure.

REFERENCES

Burlakin, A.N. & Voronov, E.M. 1998. *Parallel Implementation of Optimisation Algorithms for Multisite and Multicriterion Systems*. Moscow: MGTU.

Evgenev, G.B. 2006. *Systemiology of Engineering Knowledge*. Information Science in Technical Institutes. Moscow: Publishing House of the MHTU after N.E. Bauman.

Kuzmin, I.A., Putilov, V.A. & Filchakov, V.V. 1991. *Distributed Information Processing in Scientific Researches*. Saint Petersburg: Publish House-PGU.

Theoretical and Practical Solutions of Mineral Resources Mining – Pivnyak, Bondarenko & Kovalevska (eds)
© *2015 Taylor & Francis Group, London, ISBN: 978-1-138-02883-8*

Efficient land use in open-cut mining

T. Kalybekov, K. Rysbekov & Y. Zhakypbek
Kazakh National Research Technical University named after K.I. Satpayev, Almaty, Kazakhstan

ABSSTRACT: The article deals with methods of efficient use of land allotted for open development of mineral deposits basing on a timely mine reclamation.

1 INTRODUCTION

Open mining of mineral deposits includes allotting vast lands on organization of external and tailing dumps, transport communications and other various industrial facilities. Open cut mining results in land and natural landscape disturbance, pollution of air, surface and ground water adjacent to mining area and deterioration of sanitary-hygienic environment of local population. In this regard, it is possible to provide impact mitigation caused by mining by efficient use of lands allotted for open mining by timely carrying out mine reclamation.

The paper deals with the following tasks of efficient land use in open mining:

– to choose rational mine reclamation directions based on sustainable development of mining areas;

– to justify criteria of efficient mined land use;

– to study regime of land disturbance at open mining for mined land reclamation management.

The solution of the said issues during mine development will contribute to effective carrying out mined land reclamation and efficient land use for a sustainable development of mining areas.

2 MAIN PART

Open mining disturbs natural terrains, changes groundwater composition, pollutes air and deteriorates soil productivity due to groundwater lowering. It badly changes the development area i.e. the original ecological system is changed into a new "field-environment system". Depending on capacity of a mining enterprise the environmental quality and productivity of adjacent lands deteriorates. So impact mitigation at open mining requires timely mined land reclamation during a field development. In the coming years, open mining remains to be the main engineering method of field development and a cause of land disturbance increase. The paper investigates various aspects of efficient use of lands allotted for open mining.

The use of excavated area for storing overburden will reduce the cost of minerals by reducing overburden transportation distance and external dump area. In determining a land disturbance regime it is desirable to use a graphical method that allows to take into account the features of a pool distribution, excavation methods and systems of a pool development (Tomakov & Kovalenko 1984). The paper describes various methods for upgrading process technologies in terms of efficient land use and basic graphical method elements. A pit extension will change a current stripping ratio up to 30 – 50 t/t of mined ore. So, disposal of large volumes of waste will require immense territory, for example, an open deposit being 500 – 1000 m deep the waste dump area is expected to exceed the open-mined deposit by 4 – 7 times (Mikhailov 1981). That is why a study of land disturbance and reclamation regimes will encourage the efficient land use in open mining.

The environmental protection progress and a further pit extension face new environmental challenges connected with deterioration of ecological and geochemical characteristics of enclosing rocks. Given this in designing extension works at the "Jubileiny" deposit there was used a technology of a maximal overburden disposal inside the excavated pit i.e. use of a stripped area for reducing environmental impact caused by external dumps (Wolpert & Martynov 2011). Organization of internal dumps in the Sarbaisky deposit has reduced the volume of mining waste stored at external dumps. To reduce mining waste quantities we for example practice using carbonate rocks for the production of fractional crushed gravel and tails of dry magnetic separation for the construction of temporary roads in pits and dumps (Bondarenko 2014).

All these works aim to reduce external dump areas. Open mining necessitates the organization of overburden disposal in other worked-out quarries for the formation of a man-made landscape that would correspond to selected mined land reclamation purposes (Fomin et al. 2014).

These measures promote mined land reclamation and impact mitigation. The open mining research allowed offering (Ryazantsev 2012) the following impact mitigation methods like a simultaneous development of all dump layers, accounting the coefficient of a landscape for reducing the size of plots allotted for dumps, reduction of the area of allotted land due to a compact disposal of mining facilities.

These activities must be based on analytic dependences that allow making the analysis of concrete field development conditions. A study of how a deposit area is allotted into separate plots for main mining and processing objects revealed that a significant part of land is allotted for external dumps i.e. it speaks about a technogenic disturbance of natural environment (Kalabin 2012). It is noted that a degree of environmental risks grows if processing plants and tailings dumps are located close to quarries and reaches 0.55 – 0.65. A brief review of research speaks about the necessity of a comprehensive analysis of efficient land use for finding the best open field development option.

A selection of efficient mined land reclamation direction in open mining should be based on an in-depth analysis of economic and physic-geographic features of a deposit area. Selection of the most rational and efficient mined land reclamation methods depends on many factors that influence on the development of mined area and utilization of available natural resources of the region (Figure 1).

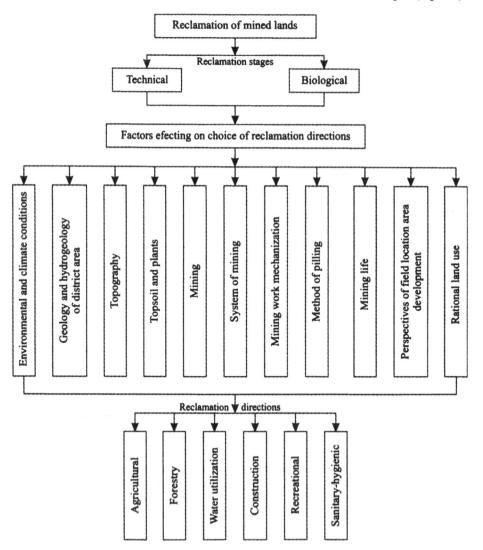

Figure 1. Efficient land reclamation direction selection diagram.

A correct choice of a mined land reclamation direction enables to solve in future the issues of external dump formation in accordance with soil reclamation requirements. A rational mined land reclamation direction in open mining usually does not pretend to restore the original state of the environment but it aims to solve important environmental and social issues of human life and activities.

The paper proposes to justify the criteria of efficient land use in open mining by using analytical formulas (Kalybekov 2014, Chulakov et al. 1994) derived under the assumption that an open-pit bottom is of a round, square or rectangular shape. The formula includes such evaluation criteria as a size of the area to be disturbed in open mining, waste dump capacity and land disturbance degree.

Evaluation of open mining parameters at planning a system of field development ensures efficient land use. That is why, if it is a rectangular shape dump then a specific size of open mine area which is equal to remove material-field space relationship should seek to a minimum:

$$V_{prk} = \frac{H_k(a + H_k ctg\gamma_a)(v + H_k ctg\gamma_a)}{(a + 2H_k ctg\gamma_a)(v + 2H_k ctg\gamma_a)} \to min, \quad (1)$$

where a – rectangular pit bottom width, m; v – pit bottom length, m; H_k – pit depth, m; $ctg\gamma_a$ – average inter-ramp pit slope angle.

Waste dump parameters evaluation is based on calculation of a total remove rock volume. A rational use of dump areas is characterized by the volume of stored overburden, so a specific capacity of square dumps which is equal to overburden volume-dump area relationship should strive for a maximum:

$$\frac{\frac{H(L - HCtg\beta)^2}{K_p}}{\left(\sqrt{\frac{VK_P}{H}} + HCtg\beta\right)^2} \to max, \quad (2)$$

where V – volume of stored overburden, m³; H – dump height, m; L – dump wall, m; β – a resulting dump slope angle, degree; K_p – overburden fragmentation index.

A producer should carry out timely mined land reclamation in order to minimize a degree of land disturbance. Then a land disturbance degree in open mining should seek to a minimum:

$$S_{n.k} + S_{n.o} = S_c \to min, \quad (3)$$

where $S_{n.k}$ – quarry disturbed area, m²; $S_{n.o}$ – dump disturbed area, m².

The paper suggests to use mining and geometrical analysis and an analytical method for sizing up to

be-disturbed area in open mining. For this purpose, we accept a target quarry depth and agree that a quarry bottom shape may be round, square or rectangular. In this context, a mining-geometrical analysis for evaluation of land disturbance is calculated as follows.

Mining area specific size criterions enable to evaluate the total land disturbance area. The paper suggest to evaluate an average pit slope angle at close of operations by the following formula:

$$\gamma_a = \frac{L_h \cdot \gamma_{v.c} + L_f \cdot \gamma_h}{L_h + L_f}, \text{ degree,} \quad (4)$$

where $\gamma_{h.c}$ – and inter-ramp pit slope angle from the side of a hanging wall at close of mining operations; γ_f – inter-ramp pit slope angle from the side of a footwall; L_h – pit wall length from the side of hanging wall; L_f – pit wall length from the side of a footwall.

The increase in average inter-ramp pit slope angle at close of mining operations results in land disturbance decrease on the land surface level. Therefore, evaluation of a rational value of this slope angle is very important in designing open field development.

Further, to evaluate land disturbance size during a field development it is needed to calculate an average inter-ramp pit slope angle. Then, there for each average inter-ramp angle is evaluated annual land disturbance area at a pit in-depth extension. The average inter-ramp pit slope angle is set according to the formula:

$$\gamma_a = \frac{L_h \cdot \gamma_{v.c} + L_f \cdot \gamma_h}{L_h + L_f}, \text{ degree,} \quad (5)$$

A diagram dependence of disturbed lands on pit extension at an inter-ramp angle equals to $10 - 30^0$ (Figure 2).

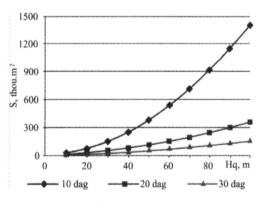

Figure 2. Land disturbance-quarry depth area dependence diagram.

Mining-geometrical analysis results at a different pit in-depth extension rate shows that land is disturbed before a certain state of a field development i.e. topsoil stripping from a quarry area is carried out at the beginning of a running period. It is connected with the formation of quarry working zone when operating angles of high wall slope are flatter than high wall slope angles at close of mining operations.

Within a certain time the area of land disturbance is calculated as difference between adjacent pit extension rates $h = h_{i+1} - h_i$ at the adopted average high wall slope angle. For example, for evaluating the size of land disturbance it is sufficient to restore perpendiculars from respective pit extension depths up to curves (Figure 3) that indicate average $10 - 30^0$ high wall angle slopes and from these points to the line that points out a disturbed area size. The difference between these two land disturbance values will show a target size of land to be disturbed for top topsoil stripping during a planned period by taking into account that pits may be of a different shape. Field development at 70 m level will result in land disturbance of a pit at different values of average high wall slope angles and the size of land disturbance is expected to be as follows: $S_{10}^0 - 141.000 \text{ m}^2$; at $S_{20}^0 - 35.000 \text{ m}^2$; at $S_{30}^0 - 16.000 \text{ m}^2$.

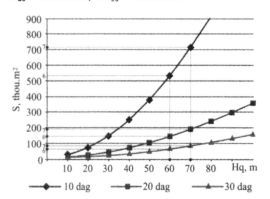

Figure 3. Field area land disturbance evaluation diagram.

The paper proposes to control analytically a land disturbance regime at dump the allocated sites depending on volume of stored overburden, dump height and a resulting slope angle for various shape dumps taking into account a waste dump capacity criterion. By using analytical formulas, one can set parameters of various form dumps and storage of certain amount of overburden in those dumps. This analytical method enables to predict and justify the size of area expected to be disturbed under dumping various overburden volumes.

At external dumping for the purpose of efficient land management one should strive for a simultaneous development of the front of dump layers in a ground plan and in a profile. Open mining provides for overburden remove and storage in at external dumps of various forms. Figure 4 presents a diagram for evaluation of land disturbance size depending on annual waste volume stored and different height of a dump.

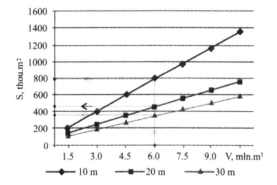

Figure 4. Dumping land disturbance evaluation diagram.

Establishment of land disturbance regime at dumping enables to schedule topsoil stripping and efficiently control mined land reclamation in strip-pits.

3 CONCLUSION

A timely mined land reclamation and creation of a productive and updated landscape badly depends on a correct recording of factors enlisted in the selection diagram of efficient reclamation direction at the stage of field development.

The paper proposes to use new analytical formulae as rational land use criteria for establishing a minimum size of land to be allotted for open mining operations and waste dump capacity maximum for minimizing land disturbance degree at open mining. The said formulae differ from the known ones by pit parameters adopted at the time of establishing rational land use criteria.

The analytical method proposed in the paper for justification of land disturbance regime at open mining enabled to determine the following dependencies: land disturbance-vertical advance and land disturbance-adjacent vertical advance; land disturbance size – annual overburden volume stored. The planning of topsoil excavation, stripped soil storage and dump formation are based on obtained dependencies.

It is recommended to continue further studies in the following directions:

– control on topsoil stripping and its transportation on reclamation sites in accordance with land disturbance regime in open mining;

– selective overburden disposal in external dumps in accordance with the adopted land reclamation direction.

Calculation of land disturbance size at open mining serves as a basis for controlling topsoil excavation works. A selection of a rational land reclamation direction serves as a basis for combining the overburden selective storage in the body of a dump and acceleration of mined land reclamation.

REFERENCES

Bondarenko, A.V. 2014. *Organization of environmental work at JSC "SSGPO"*. Mining Journal. No 6, 96 – 98.

Chulakov, P.Ch., Begalinov, A. & Kalybekov, T. 1994. *Recultivation intensification of lands disturbed by open mining activities*. Almaty: Gylym: 272.

Fomin, S.I., Vedrova, D.A. & Doan, V.T. 2014. *Restoration of disturbed mining territory*. Mine Survey and subsoil resources management, No 6 (74): 67 – 70.

Kalabin, G.V. 2012. *Typification of a general plot plan of quarries and assessment of their ecological compatibility*. Mine Survey and subsoil resources management, No 3 (59). 61 – 65.

Kalybekov, T. 2014. *Establishment of rationale optimization criteria of reclamation works at open-cast deposits*. Mine Surveying and subsoil, No 4. 44 – 50.

Mikhailov, A.M. 1981. *Environmental protection at open mining*. M: Nedra: 184.

Ryazantsev, S.S. 2012. *On criteria of land allotments use at open development of mineral deposits*. Mining informational and analytical bulletin, No 4: 403 – 406.

Tomakov, P.I. & Kovalenko, V.S. 1984. *Rational land use at open cast mining*. M: Nedra: 213.

Wolpert, Ya.L. & Martynov, G.A. 2011. *Main directions of minimizing environmental impact of diamond mining in Yakutia*. Mining Journal, No 1: 100 – 102.

Theoretical and Practical Solutions of Mineral Resources Mining – Pivnyak, Bondarenko & Kovalevska (eds)
© 2015 Taylor & Francis Group, London, ISBN: 978-1-138-02883-8

Development of methodology for assessing geospatial variability of primary kaolin

R. Sobolevskyi, O. Vashchuk & O. Tolkach
Zhytomyr State Technological University, Zhytomyr, Ukraine

ABSTRACT: The article focuses on analysis of the instruments which were used to determine whiteness of kaolin. Experimental studies of the whiteness based on the proposed method, approximation and statistical processing of the data were made. The optimal parameters of the scanner settings for maximum performance and qualitative determination of whiteness were determined. Based on the studies the program for determination of kaolin whiteness was made on the basis of scanned image of the sample by the RGB color coordinates. The influence of the deposits exploration degree and individual parts of deposit on the evaluation of spatial variability of kaolin whiteness of Velyko-Gadominetsky deposit was investigated. The method of variogram analysis was used in this investigation.

1 INTRODUCTION

Whiteness is the main quality indicator for the majority directions of uses of primary kaolin. Accordingly, this fact determines the cost of raw materials and the cost-effectiveness of mining. Therefore, in planning mining operations the spatial variability of whiteness of primary kaolin should be taken into account. Analysis of executed researches in this area and the results of mining of Velyko-Gadominetsky, Glukhovetsky and Zhezhelevsky primary kaolin deposits showed that the significant problems are the relatively high cost of determining the whiteness of individual samples of kaolin and the complexity of the geostatistical prediction of variability values of whiteness in the limits of individual sites of deposit or deposit in general.

D.O. Sorokin developed a gradation scale of whiteness Wiso, which is used for assessing the level of porcelain quality and he also proposed the technology of constructing expert systems of products identification (Sorokin et al. 2009).

N.P. Belov proposed to leave in the opto-spectral sensor for whiteness measurement the only one channel at the wavelengths in the blue-violet areas of spectrum (Belov et al. 2009).

A.D. Yakymchuk decreased the determination error of the whiteness of fabric samples on 4 – 7% by the proposed computer method in comparison with a conventional method determining whiteness of the fabric samples (Yakymchuk & Petelsky 2001).

Works by S.S. Hordyuhina, R.M. Bulguchev, D.I. Rebrikov, L.D. Lozhkin focus on the possibility of using the colored coordinate systems to determine various industrial applications (Hordyuhina & Grigoriev 2011, Bulguchev et al. 1998, Rebrikov et al. 2008, Lozhkin & Suvorov 1979).

L.D. Lozhkin proposed a device based on the spectral colorimetric method of measuring color coordinates and chromaticity.

A.M. Gotra proposes to avoid losses of information due to the fragmentation testing, by optimizing the scheme of sampling soil samples (Gotra & Meshalkina 2003). A.V. Isaeva developed a complex of programs which implement the modification of Kriging method proposed in the thesis, the morphological algorithm, methods for solving nonlinear equations, algorithms, computer simulation of random fields (Isaeva & Serdobolskaya 2011).

V.A. Sidorova showed the effectiveness of using geostatistics techniques for the study of spatial patterns of soil characteristics, characteristics of crop yields and crop quality due to the tasks of precision farming (Sidorova & Krasilnikov 2007).

M.A. Fishman reviewed a mineral aggregate as grains combination which allowed to introduce a probability space which can be represented by autocorrelation of mineral species closely spaced grain, and therefore in this case the method of spatial variation can be applied (Fishman 2000).

2 RESULTS AND DISCUSSION

Most of the devices used for determining the whiteness at the enterprises are often characterized by low speed determination of the sample whiteness, the error in determining the whiteness caused by the human factor, minimum possibility of automating the measurement process and low cost. Modern devices are characterized by high accuracy in determining the whiteness, high speed, but at the same time by high cost. Based on the principle of existing technologies for determining the whiteness a possibility of using a flatbed scanner with the subsequent processing of color coordinate points of the image for determining the sample whiteness was investigated in this work.

In processing the scanned image a selected color coordinate system is an important factor determining the efficiency of this process. Nowadays there is a large number of color models, such as: CMYK, RGB, Lab, HSB, XYZ, CMY, xyY, LCH_{ab}, Luv, LCH_{uv}. Based on the analysis of theoretical resources the RGB color model was selected for processing the scanned image. The benefits of this model are explained by visualization, wide color scope, simplicity of hardware implementation, the image in the computer's RAM and a minimum size of the output file. Also worth noting is the fact that in the computer graphics the main device of output

information works in this system. This model is constructed on the basis an eye structure.

In this paper the whiteness of image scanned sample is proposed to be calculated using formula:

$$W = \left(\frac{\sum\limits_{i=1}^{n}\left(\frac{1}{3}R+\frac{1}{3}G+\frac{1}{3}B\right)\cdot k}{\sum\limits_{j=1}^{k}(R_{max}+G_{max}+B_{max})\cdot n} \right)\cdot 100\,\%, \qquad (1)$$

where R, G, B – values of the color coordinates in the RGB system, R_{max}, G_{max}, B_{max} – maximum values of color coordinates; k – the elements' number of color coordinate system; n – the number of pixels for which the whiteness is determined.

Relationship between the determined whiteness values and actual values is confirmed by the correlation coefficient 0.88. Analytically, this dependence is described by the pair of linear regression:

$$y = 2.0388x - 110.1439. \qquad (2)$$

This correlation is shown in Figure 1.

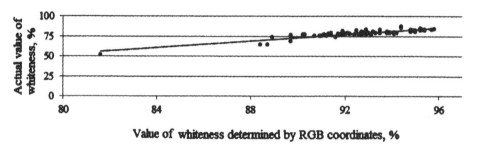

Figure 1. Regression model of relationship between actual value of whiteness and value of whiteness determined by RGB coordinates.

An important parameter in the evaluation of image quality is a resolution, therefore study of the impact of the resolution to the accuracy and productivity of determining whiteness of model was done. Samples of kaolin in the form of compressed tablets of 50 mm in diameter in an amount of 20 pieces were placed on a scanning surface of a scanner. Further the same kaolin samples with resolution from 300 to 1800 pixels per inch with a spacing of 100 were scanned. After that, from the scanned image of A4 size were cut out individual samples of kaolin and saved as a separate image. On the next step the images of samples were downloaded in the program "The whiteness", where the value of their whiteness by RGB color coordinates was defined as well as the actual values of whiteness using previously calculated the regression equation.

The effectiveness of the proposed methodology for determining the kaolin whiteness will be deter-

mined by productivity, which is recommended to be determined by the formula:

$$P = \frac{60\cdot N_s}{t_1+t_2+t_3+t_4+t_5}, \left(\frac{pcs}{min}\right), \qquad (3)$$

where N_s – the number of samples, pcs; t_1 – a necessary time for laying samples on the scanner, sec, t_2 – a time of scanning the samples, sec; t_3 – the time required to cut images of samples with the scanned image, sec, t_4 – the time required for loading into the program, sec, t_5 – the time spent on image processing in the program, sec.

Table 1 shows that the value of defined whiteness if scanned with a task from 300 to 1200 pixels per inch, and 1300 dots per inch is increased by 0.1 – 0.2% while the tendency to growth the values of whiteness with increasing values of whiteness is not observed.

Table 1. The results of determination the samples' whiteness.

The number of pixels per inch (ppi)	Scan time, sec	Whiteness RGB, %	Actual whiteness, %	One sample size, Mb	The maximum number of samples for one scanning, pcs	The processing time of one sample in program, sec	Productivity, pcs/min
300	11	93.7	80.9	0.5	20	0.9	7.0
400	14	93.8	81.1	0.9	20	1.5	6.5
500	26	93.7	80.9	1.4	20	2.3	5.6
600	26	93.7	80.9	2.1	20	3.4	5.1
700	72	93.8	81.1	2.7	20	4.3	4.0
800	76	93.7	80.9	3.5	20	5.6	3.6
900	96	93.7	80.9	4.9	20	7.8	3.1
1000	96	93.7	80.9	5.4	20	8.6	2.9
1100	96	93.7	80.9	6.3	20	10.0	2.7
1200	96	93.7	80.9	8.3	20	13.1	2.4
1300	237	93.9	81.3	9.8	15	15.5	1.6
1400	237	93.9	81.3	10.7	15	16.9	1.5
1500	236	94	81.5	13.6	15	21.5	1.3
1600	237	93.9	81.3	15.1	15	23.8	1.3
1700	322	93.9	81.3	16.4	15	25.9	1.1
1800	322	93.9	81.3	17.9	10	28.2	0.9

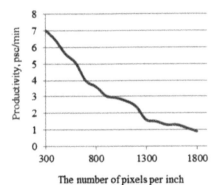

Figure 2. Dependence for productivity of determining whiteness and number of pixels per inch.

Analyzing the Table 1 and Figure 2, the optimum resolution in the scanner settings was chosen as 600 ppi, because scanning with this value of resolution takes a little time, the scanned image takes up little space on the disk, the number of pixels in the image of a scanned sample of more than 720000 are enough for statistical processing. In this case the productivity of determining whiteness is 5.1 samples per minute. At the moment of choosing a scanner settings it should be noted that parameters such as sharpening, removing raster, color restoration, background correction, removal of defects and other parameters must be turned off, in case if they change the original image.

The methodology described above became the basis for creating the program algorithm "Whiteness" for determining kaolin whiteness using scanned image of the sample. Software source code was written in high-level general-purpose programming language Python.

An important part in planning calendar of mining operations is reliable prediction of values whiteness in some areas and the field in general. One of the main approaches to the analysis and processing of spatially distributed data is a geostatistical modeling. According to the data of geochemical analysis of selected samples from 628 welles of Velyko-Hadomynetsky deposit the variogram analysis was made (Figure 3, a) and Kriging map for whiteness was built (Figure 3, b).

Analysis of Kriging map indicates that a zone of maximum whiteness is located in the northern part of the deposit, and variogram shows that the value of whiteness has autocorrelation in azimuth of 90° at a distance of 435 m. In geostatistical prediction important parameters are number of data and places of taking samples. Therefore the influence extent of exploration deposit on result of prediction values whiteness was investigated in this paper (Figure 4).

Variogram analysis for whiteness considering the extent of exploration and their basic statistical indicators showed that for operational exploration for whiteness with azimuth of 315° characterized by minimum variance (23.5%) in the range of 43 autocorrelation, because this part of deposit is characterized by maximum density of exploratory wells and the minimal variability values of whiteness (Figure 4, a).

The value of the variance (41%) corresponds to consolidated results of operating and previous exploration, which is in 1.7 times more than in the operational exploration on a range of autocorrelation of 140 m with azimuth 157.5°, that is obviously explained by a large spatial variability of whiteness on the deposit (Figure 4, b). Simultaneously, for preliminary exploration results the mean value of dispersion (30%) in the range of autocorrelation 435 m with azimuth 90° is typical, which is explained by the influence of uniformity network and much wider distance between exploratory wells compared to the consolidated results (Figure 4, c).

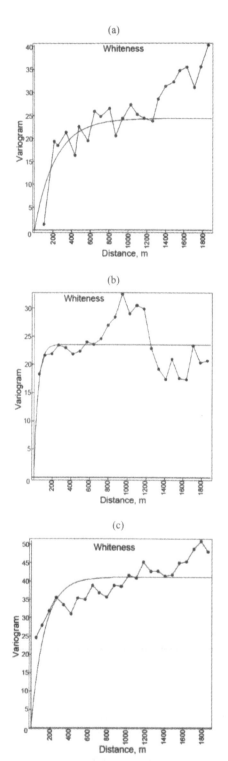

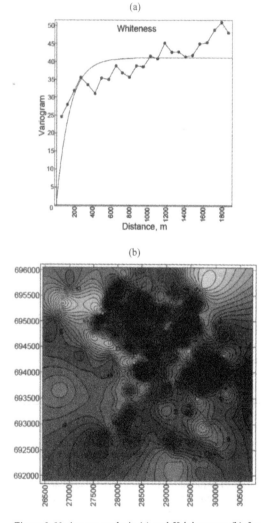

Figure 3. Variogram analysis (a) and Kriging map (b) for whiteness.

Figure 4. Results of variogram analysis taking into account extent of exploration: (a) previous exploration; (b) operational exploration; (c) consolidated results.

3 CONCLUSIONS

The methodology of using a flatbed scanner with further processing of color coordinates pixel of scanned image for determine the whiteness of kaolin sample was proposed.

The experimental studies for determine whiteness were made based on the proposed methodology. The approximation and statistical analysis of the data was done. A regression model of relationship for actual value of whiteness and value of whiteness determined by RGB coordinate was defined and is expressed by the following equation: $y = 2.0388x - 110.1439$.

The optimal scanning resolution for maximum performance and qualitative determination of whiteness was determined, which is 600 pixels per inch. The productivity of determining whiteness for a given speed is 5.1 samples per minute.

Based on the studies the program for determining the kaolin whiteness using scanned sample image for the RGB color coordinates was designed.

A geostatistical prediction for whiteness samples according to the extent of deposit exploration was made.

REFERENCES

Belov, N.P., Gaydukova, O.S., Panov, I.A., Patyaev, A.Yu., Smirnov, Yu.Yu., Sherstobitova, A.S. & Yaskov, A.D. 2009. *Laboratory Spectrophotometer for Ultraviolet Spectral Region*. Proceedings. St. Petersburg State University for Information Technologies, Mechanics, and Optic: 81 – 87.

Bulguchev, R.M., Gogol, A.A. & Black, V.J. 1998. *Variants of construction the optical color sensors*. 51th STC prof. prep. St. Petersburg: 80.

Fishman, M.A. 2000. *Adaptation kriging procedure for using in the rocks space*. Petrography at the turn of the XXI century: results and prospects. Proceedings of the Second All-Russian petrographic meeting. Syktyvkar Geoprint: 103 – 104.

Gotra, O.M. & Meshalkina, Y.L. 2003. *Description of the distribution of humus in the range of one zero on the example of the arable layer of sod-podzolic soil*. Bulletin of Moscow University. No 4: 3 – 8.

Hordyuhina, S.S. & Grigoriev, A.A. 2011. *The definition of specific coordinates of physiological systems' color using a statistical model of color vision*. Semiconductor lighting, No 1: 16 – 19.

Isaeva, A.V. & Serdobolskaya, M.L. 2011. *The hypothesis of local stationary in the problem of the stochastic forecast using kriging*. Vestnik MGU, Ser. No 3: 14 – 19.

Lozhkin, L.D. & Suvorov, G.A. 1979. *Questions of the spectral color measurement*. Technique of Film and Television, No 3: 35 – 39.

Rebrikov, D.I., Bityukov, V.K., Khvostov, A.A. & Ponomareva, E.I. 2008. *Formation the color spectrum of the surface by the color models of digital images*. Bulletin of the Voronezh State Technological Academy, No 2: 40 – 44.

Sidorova, V.A. & Krasilnikov, P.V. 2007. *A geostatistical analysis of the spatial structure of acidity and organic carbon content of soil zone of the Russian Plain*. Geostatistics and soil geography. Moscow: Science: 67 – 80.

Sorokin, D.A., Platov, Yu.T. & Platova, R.A. 2009. *Colorimetric identification of porcelain by material type*. Glass and Ceramics. Kluwer Academic Publishers-Consultants Bureau: 125 – 128.

Yakymchuk, O.D. & Petelsky, M.B. 2001. *The methods of mathematical planning of the experiment in the optimization of the composition and conditions of using detergents*. St. Petersburg: Sixth St. Petersburg Assembly of Young Scientists and Specialists: 13 – 14.

Theoretical and Practical Solutions of Mineral Resources Mining – Pivnyak, Bondarenko & Kovalevska (eds)
© 2015 Taylor & Francis Group, London, ISBN: 978-1-138-02883-8

Stochastic model of rock mass strength in terms of random distance between joints

O. Sdvyzhkova, S. Gapeiev & V. Tykhonenko
National Mining University, Dnipropetrovsk, Ukraine

ABSTRACT: Impact of scale effect upon rock strength while transferring from the scale of laboratory sample to the one of rock mass is studied. Rock strength is examined as a random value which magnitude in statistical sampling is distributed according to a certain distribution law. It is shown that logarithmic-normal law is the best one to describe strength dispersion. Effect of macrodefects (joints) upon the strength during laboratory sample testing cannot be taken into account directly; that is why definite number of virtual "defective" samples with the strength being underestimated but not equal to zero are added into the sampling. In this context, distance between the joints is also considered as a random value distributed according to Rayleigh distribution law.

1 INTRODUCTION

While developing deformation model of rock medium it is necessary to take into account the fact that rock is heterogeneous medium as for its nature and texture. Rock heterogeneity is manifested in terms of variableness of its physical properties and substance composition within the space. These variations can be the reason of anisotropy resulting in random variations of the values of physical field parameters.

Main characteristic to determine the model of medium deformation is its resistance to internal loads. As for rock, its basic resistance characteristic is compressive strength being determined using rock samples according to standard testing technique. Transfer from the testing results to the strength of the whole rock mass is a complex task. It is explained by the fact that rock samples have limited sizes, moreover they cannot simulate complex rock mass structure, i.e. they cannot show heterogeneity of various scale levels being characteristic for rock medium.

Difference of rock mass strength from rock sample strength is called scale effect in rocks that means available regularities of changes of mechanical processes within an object in terms of changes of its dimensions.

Statistic interpretation is one of hypothesis explaining scale effect; according to this interpretation, actual solids always have internal defects in form of vacancies, dislocations, joints, inclusions of different strength microvolumes distributed within the solid randomly. The more solid volume is, the more number of defects it contains, consequently, the less its strength is.

Structural factor, k_c, being considered as the ratio of the value of specific strength characteristic in a mass to its value obtained while testing the samples of standard linear dimensions, is a quantitative scale effect estimator. As a rule, this is the ratio of compressive strength within R_m mass to the average compressive strength of $\overline{R_c}$ rock sample, i.e. $k_c = (R_m)/(\overline{R_c})$. Since ultimate strength level and parameters of rock elastoplastic state around mine workings are connected with this characteristic, determination of objective value of structural factor is an important and complex task connected with rational design of underground facilities.

A great number of theoretical studies based on statistic explanation of solid strength nature are devoted to determination of this value (for example, Volkov 1960, Sedrakian 1968, Rats 1968, Brady 1970, Koyfman 1963 etc.). Dependences of different complexity levels have been obtained to express qualitative pattern of strength decrease of large samples. However, it is difficult to estimate strength decrease quantitatively due to the difference of initial idealized physical models from real rock masses.

2 PROBABILISTIC ROCK MASS MODEL BASED ON NORMAL STRENGTH DISTRIBUTION OF STRUCTURAL ELEMENTS

Papers by A. Shashenko and E. Sdvizhkova (Shashenko et al. 2008), based on statistic theory of strength, consider rock mass as the aggregate consisting of structural elements which physical and mechanical properties are distributed randomly according to a certain distribution law with $f(R)$ distribution density and $F(R)$ integral distribution function.

Difference of R_m mass (aggregate) strength from mathematical expectation of the strength of its $M(R)$ structural elements (laboratory samples) is estimated using k_c structural factor. In this context, mass strength should be estimated using such R_m value so that the strength of its structural elements with the specified reliability would not be less than this value.

Following expression is used to determine probability of such occurrence:

$$p(R \geq R_m) = 1 - F(R), \tag{1}$$

where $F(R) = \int_{-\infty}^{R} f(x)dx$ is integral function of R value

distribution. This inequation should be solved relative to R_m value, i.e. $R_m = t$, where t is the argument of $F(R)$ function if its value is equal to $1 - P$. Assume such designation as $t = arg\, F(1 - P)$. Then structural factor will be determined using expression:

$$k_c = \frac{arg\, F(1 - P)}{M(R)}. \tag{2}$$

Definite form of expression (2) will depend upon the selection of $F(R)$ probability distribution function of R random value; in this case, strength of mass structural elements is represented by this very value.

As a rule, distribution law is selected on the basis of both physical nature of a random value and analysis of statistic information. Most often, especially in case when the volume of such information is minor, researchers select normal distribution law as a probabilistic model of quantitative criterion under study.

Papers by A. Shashenko, N. Surgai, and L. Parchevskiy (Shashenko et al. 1994) obtain structural factor value in assumption that the strength of mass structural elements is distributed just according to a normal law. In this case inequality (1) is as follows:

$$P(R \geq R_m) = 1 - F_0\left(\frac{R_m - a}{\sigma}\right), \tag{3}$$

where $F_0(t) = \frac{1}{2\sqrt{\pi}} \int_{-\infty}^{t} e^{-\left(\frac{t^2}{2}\right)} \cdot dt$ is normalized integral function of normal distribution.

While solving equation (3) relative to R_m value and dividing both equation parts by $M(R) = a$ mathematical expectation, the authors of the paper (Shashenko et al. 1994) obtained following expression:

$$k_c = \eta \cdot arg\, F_0(1 - P) + 1, \tag{4}$$

where $\eta = \sigma / a$ is relative variation of structural elements strength.

Thus, structural factor is obtained as the value depending upon η relative variation characterizing level of medium heterogeneity: the more number of internal defects structural elements contain, the higher value of relative variation is.

It should be noted that normal distribution law describes adequately only the values which variations are not more than 33% (it comes from three sigma rule). As a rule, considerable data variation is not characteristic for the results of laboratory testing; variation of sample strength is usually not more than 25 – 30%. Thus, large number of samples of coal-bearing Donbass rock has been tested in the National Mining University (Shashenko et al. 2008). Testing analysis has shown that the values of η relative variation, β_1^2 asymmetry, and β_2 kurtosis say for normal distribution law which has been selected by the authors of the paper (Shashenko et al. 1994) as probabilistic-statistical model of the strength of mass structural elements to determine structural factor in the form of (4).

However, while studying strength of rock mass structural elements being represented in reality by the samples made of selected ones attention should be paid to following facts: does the sampling obtained as the result of rock sample testing really express the nature of general totality? Is basic requirement for the sampling (equality of probabilities for all the elements of general totality to be added to the sampling) met?

It is known that rock features natural jointing. While preparing samples, the ones with jointing are broken until testing begins. Thus, structural elements containing microdefects are not used for standard testing. Only "good", i.e. unbroken mass elements, are taken as sampling parts. Nevertheless, as a result of such samples testing according to standard techniques there is scatter of strength values with η variation within 25 – 30%. Consequently, in terms of standard testing, especially when there is standard determination of sampling characteristic, i.e. statistic distribution moments, η variation expresses only internal heterogeneity of mass structural elements at microlevel (defects of crystalline lattice etc.). Then, formula (4) means that structural factor takes into account rock heterogeneity at microlevel only.

Actually, rock is essentially heterogeneous medium with numerous defects of various levels – from hairline fractures being almost invisible up to discontinuous faults of various amplitude and orientation. Macrodefect features of rock medium have definite effect upon distribution law of its structural elements strength stipulating its deviation from normal distribution.

3 RECORD OF MICRODEFECTS IN STATISTIC MODEL OF ROCK MASS STRENGTH

To take into account heterogeneity of macro- and megalevels available in rock mass (for example, natural joints) it is necessary to put joint-containing elements into testing statistics. Thus, real sampling and real variation series are corrected by introduction of additional elements which strength is well below the one of unbroken elements. It is obvious that artificial sampling supplementation will change all he distribution moments including variation characteristics: dispersion, standard, and relative variation. Then, according to Pearson diagram, law of distribution of structural elements strength will vary as well.

The paper (Shashenko et al. 1994) shows techniques to introduce tested elements broken with microdefects into statistics. It is supposed that strength of such elements is close to zero and their number in sampling depends upon the distance between joints in real mass. As a result "corrected" initial first order moment (average sampling) and central second order moment (sampling dispersion) as well as relative strength variation connected to them are determined for the sampling into which n_m elements with zero strength are introduced. According to (Shashenko et al. 2008) η' relative strength variation for "corrected" sampling is determined by means of η standard testing variation and l_m distance between joints within real rock medium:

$$\eta' = \sqrt{\frac{l_m + l_0}{l_m}(\eta^2 + 1) - 1}.$$ (5)

where l_0 – characteristic size of the sample being tested for strength under laboratory conditions.

Physical significance of the problem shows that the distance between joints cannot be less then linear size of the sample otherwise it cannot be made. It is seen that artificial introduction of "defective" elements with zero strength into sampling increases strength value variation. Thus, if $\eta'^2 + 1 = 1.1$ (that corresponds to the value of initial variation of laboratory testing $\eta = 0.31$) and distance between joints being five times more than sample size, then mass strength variation is $\eta = 0.58$, i.e. it increases by 1.8 times comparing with the initial variation. Such sharp variation increase is explained by excessive model idealization meaning that the strength of broken elements is equal to zero.

Professor E. Sdvizhkova has proposed to take into account the fact that the strength of the elements with macrodefecs is not reduced down to zero but

decreased comparing with unbroken elements. Thus, R_i $(i = 1...n_e)$ general totality of strength values contains structural elements which strength is less than the strength of unbroken samples being estimated by certain $f(\alpha)$ function of strength reduction depending upon α angle of joint inclination to horizontal plane. Then n_m of broken samples should be added to the initial totality of n_e samples; strength of n_m of broken samples is equal to:

$$R_{im} = f(\alpha)R_i \qquad (i = 1...n_m),$$ (6)

where $f(\alpha)$ function characterizing decrease of structural element strength depending upon inclination angle of reduction plane has been obtained by E. Sdvizhkova using simulation by boundary element method of breaking rock sample with joint in terms of its various orientation relative to load axis and having the form of parabolic dependence:

$$f(\alpha) = 7 \cdot 10^{-4} \alpha^2 - 0{,}0386\alpha + 0{,}9359.$$ (7)

In the paper (Shashenko et al. 2008) technique of macrodefects consideration taking into account function of strength reduction was fully generalized. Concept of joints effect coefficient was introduced being taken into account while correcting all the moments of statistic distribution.

It can be shown that all the initial moments of the kth order of the "corrected" and initial series are interconnected by the ratio:

$$m_k' = K_k m_k,$$ (8)

where $m_k = \frac{1}{n_e}\sum_{i=1}^{n_e} R_i^k$ is the initial moment of the kth order for standard sampling; and coefficient of joints for the moment of the kth order is determined by the expression:

$$K_k = \frac{\dfrac{l_m}{l_0} + f^k(\alpha)}{\dfrac{l_m}{l_0} + 1}.$$ (9)

In a particular case, if strength of broken elements is assumed to be equal to zero, then coefficient of joints effect is similar for all the initial moments.

Dispersion of the "corrected" variation series is equal to:

$$D' = K_2 m_2 - K_1^2 m_1^2.$$ (10)

Then taking into account that the dispersion of the "corrected" variation series is equal to $D' = K_2 m_2 - K_1^2 m_1^2$ relative variation of the "corrected" series in the assumption that defective elements have certain strength $(f(\alpha) \neq 0)$ will be determined by the expression:

$$\eta' = \frac{\sqrt{D'}}{m_1'} = \sqrt{\frac{K_2 m_2 - K_1^2 m_1^2}{K_1^2 m_1^2}} = \sqrt{\frac{K_2}{K_1^2}(\eta^2 + 1) - 1}. \quad (11)$$

If broken elements preserve their partial bearing capacity, then strength variation is not so high as in case of total breaking of "defective" elements. In particular, if $\eta^2 + 1 = 1.1$, that corresponds to the value of initial variation of laboratory testing of $\eta = 0.31$ and the distance between joints being five times more than sample size, mass strength variation now is $\eta' = 0.42$, i.e. it grows by 1.3 times comparing with the initial variation. Such dispersion of rock mass strength is considered to be more realistic comparing with the hypothesis for zero strength of broken elements being used by E. Sdvizhkova to develop more adequate probabilistic-statistical model of rock medium strength based on logarithmic-normal law of random distribution (strength of rock samples). She used this law as the basis to obtain following expression to estimate structural factor:

$$k_c = \frac{exp\left(arg\, F(1-p) \cdot \sqrt{ln\left(\eta'^2 + 1\right)} \right)}{\sqrt{\eta'^2 + 1}}. \quad (12)$$

In this case, k_c factor is calculated by means of η' (11) "corrected" relative strength variation expressing degree of rock medium heterogeneity stipulated by the available microdefects (joints).

4 EFFECT OF DISTANCE VARIABILITY BETWEEN JOINTS UPON STRENGTH DISTRIBUTION OF ROCK SAMPLES

The studies having been noted above show that the distance between joints is a constant value though it is not the case actually. Distance between joints is the main fissility parameter; this value is developed under the effect of large number of stochastic events being random itself. Consequently, being the result of manifestation of mass stochastic events it follows certain probability distribution law.

S. Batugin (Batugin 1988) gathered a body of statistic information considering distances between joints. To develop empirical distribution, measurement results were selected within the sites of mining enterprises of Kuzbass being homogeneous in their fissility. The author points out that according to Pearson criterion in terms of 5% level of significance several statistic hypotheses are not rejected

(that is characteristic for moderate-sized samplings). However, by reference to physical nature of the given random value it is necessary to take Rayleigh law as the probabilistic model of distance distribution between joints.

Following the selected statistic model, determine probability of the fact that the distance between joints of a certain system will be not less than a certain l^* boundary value:

$$p(l \geq l^*) = 1 - \int_0^{l^*} \frac{l}{\sigma} exp\left(-\frac{l^2}{2\sigma^2}\right) dl = exp\left(-\frac{l^{*2}}{2\sigma^2}\right), \quad (13)$$

where $\sigma(l) = 0.52m$ is mean-square deviation, m is mathematical expectation of Rayleigh random value.

Having solved this equation relative to l^* we obtain formula for boundary value of random distance between joints expressed by means of mathematical expectation:

$$l^* = 0,8m \sqrt{ln\frac{1}{p^2}}. \quad (14)$$

It is obvious that if probability tends to a unit, then boundary value tends to zero so it is necessary to work out definite practical certainty to determine value dispersion of random value under consideration relative to average. For example, with $p = 0.9$ probability all the values of distances between joints will be more or equal to $l^* = 0.25m$ value, and with $p = 0.8$ probability all the distances between joints will be not less then $l^* = 0.53m$ value etc.

Taking into account the obtained correlations while determining coefficient of joints effect in formula (9) it is necessary to replace l_m determinate value with l^* random value being determined depending upon m average value and p specified probability of boundary value of the distance between joints:

$$l_m = l^* = 0,8m \sqrt{ln\frac{1}{p^2}}, \quad (15)$$

i.e. the distance between joints should be represented not as a constant value equal to average one but as a random value varying within 50 – 52% according to Rayleigh distribution law.

Having united formulas (9), (11), (12) and (15) we obtain system of congruencies representing stochastic model of rock mass strength which takes into account the fact that the distance between joints is a random value characterized by dispersion relative to its m mathematical expectation according to Rayleigh law (see below – (16)). Here K_1 and K_2 are coefficients of joints effect for the first and second order moments respectively.

Table 1 represents the results of the proposed model being used to estimate structural factor of rock mass and calculated strength value comparing with normal values to determine the same values.

As calculations show according to normative standards, decrease of the distances between joints from 0.5 m down to 0.2 m does not effect the value of rock mass strength. According to the calculation using the proposed technique, this value decreases by 20%. It means that the proposed technique makes it possible to be more precise while determining rock mass strength within the area of mine working development.

Thus, based on the results of the studies having been carried out it is possible to state that the difference of rock mass strength from the strength of laboratory samples is determined by the variation of distance between joints within 50 – 52% according to Rayleigh law; it decreases exponentially depending on this parameter that allows increasing accuracy while estimating rock resisting strength and reliability of mine working design.

$$
\begin{cases}
k_c = \dfrac{exp\left(arg\, F(1-p)\cdot \sqrt{ln\left(\eta'^2+1\right)} \right)}{\sqrt{\eta'^2+1}}, \\[4mm]
\eta' = \sqrt{\dfrac{K_2}{K_1^2}\left(\eta^2+1\right)-1}, \\[4mm]
K_1 = \dfrac{\dfrac{l^*}{l_0}+f(\alpha)}{\dfrac{l^*}{l_0}+1},\; K_2 = \dfrac{\dfrac{l^*}{l_0}+f^2(\alpha)}{\dfrac{l^*}{l_0}+1}, \\[4mm]
l^* = 0{,}8m\sqrt{ln\dfrac{1}{p^2}}
\end{cases}
\qquad (16)
$$

Table 1. Values of structural factor and calculated value of strength limit.

Average distance between joints, l_m, m	Structural factor, k_c		Calculated strength value, R, MPa	
	Technique	Normative standards	Technique	Normative standards
0.2	0.21	0.4	8.40	16.0
0.5	0.30	0.4	12.0	16.0

5 CONCLUSION

1. Statistic interpretation is one of hypothesis explaining scale effect; according to this interpretation, actual solids always have internal defects in form of vacancies, dislocations, joints, inclusions of different strength microvolumes distributed within the solid randomly. The more solid volume is, the more number of defects it contains, consequently, the less its strength is.

2. To take into account possible heterogeneities of macro- and megalevels available in rock mass, it is necessary that sampling statistic contains nominally elements with joints. Thus, real sampling and real variation series are corrected by means of introduction of additional elements which strength is considerably lower than the one of unbroken elements but not equal to zero.

3. Distance between joints is the basic fissility parameter; this value is developed under the effect of large number of stochastic events being random itself. Consequently, being the result of manifestation of mass stochastic events it follows certain probability distribution law.

4. The difference of rock mass strength from the strength of laboratory samples is determined by the variation of distance between joints within 50 - 52% according to Rayleigh law; it decreases exponentially depending on the parameter that allows increasing accuracy while estimating rock resisting strength and reliability of mine working design.

REFERENCES

Batugin, S.A. 1988. *Rock mass anisotropy*. Novosibirsk, Siberian Division: Nauka: 83.

Brady, B.T. 1970. *A mechanical education of state for brittle rocks*. Int. I. Rock Mech. and Mining Sci, Vol. 7: 485 – 421.

Koyfman, M.I. 1963. *Basic scale effect in terms of rocks and coal*. Problems of mining mechanization. M.: Publishing House of the Academy of Sciences of the USSR: 39 – 56.

Rats, M.V. 1968. *Heterogeneity of rocks and their physical properties*. M: Nauka: 106.

Sedrakian, L.G. *Elements of statistic theory of deformation and breaking of brittle materials*. Yerevan: Aistan Publishing House: 247.

Shashenko, A.N., Sdvizhkova, E.A. & Gapieiev, S.N. 2008. *Deformability and strength of rock mass*. Dnipropetrovsk: National Mining University: 224.

Shashenko, A.N., Surgai, N.S. & Parchevskiy, L.Ya. 1994. *Probability theory methods in geomechanics*. K.: Tekhnika: 209.

Volkov, S.D. 1960. *Statistic theory of strength*. Sverdlovsk: Mashgiz: 115.

Theoretical and Practical Solutions of Mineral Resources Mining – Pivnyak, Bondarenko & Kovalevska (eds)
© 2015 Taylor & Francis Group, London, ISBN: 978-1-138-02883-8

Thermodynamics of multiphase flows in relation to the calculation of deep-water hydraulic hoisting

E. Kyrychenko, V. Samusya, V. Kyrychenko & A. Antonenko
National Mining University, Dnipropetrovsk, Ukraine

ABSTRACT: Concentrated in depths of World Ocean significant stocks on volume of polymetallic ores represent commercial interest by way of industrial production of scarce nonferrous metals. Results of the executed geological researches create objective preconditions for intensive development of new branch of mining-sea mining. Therefore development on creation of the unique deep-water mining equipment get a special urgency. One of the basic technological operations is transportation of the extracted minerals on a surface. Among known ways experts allocate airlift variant of hydraulic hoisting. Owing to high parameters of reliability of its work in difficult conditions of greater depths. On a line with it, the basic lack of airlift is high power consumption of installations. Therefore the tendency of development of airlift hydraulic hoisting consists in development of ways and means of raising installation's efficiency.

1 INTRODUCTION

Nowadays creation of deep-water mining complexes on the basis of hydraulic system of rise of mineral raw material differs high probability of technical realization.

In airlift's variant of hydraulic hoisting the basic electromechanical equipment settles down on a surface of water pool, that considerably simplifies its maintenance service and reduces the periods of time necessary for it. The given circumstance increases working capacity of hydraulic hoist of such principle of action, however high power consumption of compressors causes low coefficient of efficiency (COE) of installations. Nevertheless, according to dynamics of patenting, at a choice of the equipment for rise of nodules on a surface of sea water the preference is given to airlift installations that specifies perspective of their use in new area of mining.

It is, determines necessity of increase of accuracy of used methods of calculation of parameters of heterogeneous streams. Let's bring the brief review of known analytical models of two-phase gas-liquid currents (Uoliss 1972).

The model of homogeneous current describes movement of separate phases in a stream on the basis of concepts of speed of a mix and relative speed of a phase. Such approach is expedient for using in cases when relative movement insignificantly depends on charges of phases, and it is determined basically by other parameters. For example, at vesicular structure of current in vertical pipes of the big section with low speeds of a stream. In given conditions relative movement between phases is defined by balance of forces of pushing out a bubbles and resistance to their movement, olumetric concentration, instead of charges of phases.

For practical calculations the continuous models which are being, as a matter of fact, by a version of model of a stream of drift are widely used. Continuous models operate with parameters of a mix, averaged-out on the area of cross-section section of a pipe, and suppose the correct account of sliding of phases. Their advantage consists in an opportunity of an effective utilization of an available empirical information.

2 MAIN PART

For model of separate current consideration of movement of phases with the description of their own properties and speeds is characteristic. In model the equations of indissolubility, movement and energy which enter the name for each phase separately are used. Interaction of phases among themselves and with walls of a pipe is considered by the separate equations. Complexity of model depends on quantity of equations containing in her. At most simple statement of a problem discrepancy only on one of parameters of phases is considered, and the equation of preservation is made for a mix as a whole. In case of excess by number of variables of quantity of the equations simplifying assumptions are used or correlation dependences are entered.

Let's spend a comparative estimation of models of current of a biphase mix.

Distinction of speeds and temperatures of phases causes a mutual exchange of quantity of movement and heat. Often these processes proceed quickly and it is possible to make an assumption about achievement of balance. In this case the most convenient method of research is the theory of homogeneous current, for example, for research disperse and vesicular structures. However the homogeneous model becomes inexact in conditions of sharp acceleration and change of pressure. For the description of such current use of more exact models, for example, based on the theory of a stream of drift or separate current is required.

The difference between model of a stream of drift and the theory of homogeneous current consists in an opportunity of the account of sliding of phases of a mix. For vertical vesicular structure of current the opportunity of reception of more exact results gives model of a stream of drift, however in this case at the decision of the equation of movement of a mix it is necessary to use numerical methods. That is, increase of accuracy of a method of researches is reached by its complications. The theory of a stream of drift is widely used at studying vesicular, whizbang, scummy-turbulent and disperse structures of current gas-liquid mixes, and also suspensions of firm particles in a liquid. This model is the basic point for distribution of one-dimensional theoretical methods on streams in which changes of speed or density on section of the channel are essential. She also is a basis for the decision of some non-stationary problems

In opinion of authors of given article the most perspective theory is separate current theory which tendency of development consists in specification of tangents of pressure on border of section of phases is. Such approach yields comprehensible results at research of ring and dispersive-ring structures, but demands use of a plenty of empirical parities.

There is a vertical stream of a three-phase mix in sea airlift (a liquid, gas and firm particles). Presence in a mix of a firm phase significally changes not only structure of the differential equations, but also ideology of construction of mathematical models. In (Kyrychenko et al. 2014) it is developed refined method of calculation of parameters of deep-water airlift hydraulic hoistings, in which connection to each structure of current of a multiphase stream there corresponds the mathematical model. The author's concept of studying of multiphase streams is based on synthesis of theoretical models and methods of various complexity.

However lack of the given method is that it is devoted exclusively to questions of hydrodynamics when being free from thermodynamic processes proceeding in installations.

Results of the state-of-the-art review of the researches, devoted to airlifts have shown theories (Kyrychenko et al 2014) that in all settlement techniques one of the basic assumptions is postulation of isothermality of process of current in an elevating pipe (EP). Such assumption is justified for mine airlifts with rather short elevating pipes and speeds of current of a mix of $10-15$ km/s.

At the same time the analysis of technological schemes of functioning deep-water airlift hydraulic hoisting (DAH) has shown, that the degree of expansion of air in an elevating pipe at movement of a hydromix from the amalgamator to an air separator makes hundreds times. This circumstance describes a new scope of airlift's usage, puts under doubt reliability of an assumption about isothermal expansion of air.

The purpose of given article consists in development of innovative technical decisions, on increase of efficiency DAH, based on for the first time established thermodynamical effects connected with new area of operation of installations by development of deep-water deposits of minerals.

Let's pass to consideration three-phase models of the hydraulic fluid, containing two discrete phases: firm particles and gas bubbles. Movement we shall consider with continual points of view, including, that each phase moves with the speed. We shall assume, that movement one-dimensional, we neglect forces of direct friction of particles about walls of the pipeline, interference of firm and gaseous particles, and also we consider, that own speeds of phases are small in comparison with speed of a sound in a mix.

At the made assumptions the mathematical model of current of a three-phase hydraulic fluid can be presented by following system of the differential equations of indissolubility $(1) - (3)$ and movements $(4) - (6)$, written down accordingly for liquid, firm for liquid, firm and gaseous phases:

$$(1 - C_1 - C_2)\frac{\partial p}{\partial t} - \rho_0 a_0^2 \frac{\partial C_1}{\partial t} - \rho_0 a_0^2 \frac{\partial C_2}{\partial t} +$$

$$+ \rho_0 a_0^2 (1 - C_1 - C_2)\frac{\partial V_0}{\partial x} = 0, \tag{1}$$

$$C_1 \frac{\partial p}{\partial t} + \rho_1 a_1^2 \frac{\partial C_1}{\partial t} + \rho_1 a_1^2 C_1 \frac{\partial V_1}{\partial x} = 0, \tag{2}$$

$$C_2 \frac{\partial p}{\partial t} + \rho_2 a_2^2 \frac{\partial C_2}{\partial t} + \rho_2 a_2^2 C_2 \frac{\partial V_2}{\partial x} = 0, \tag{3}$$

$$\left(1 + \frac{C_1 k_1 + C_2 k_2}{2}\right)\frac{\partial V_0}{\partial t} - \frac{C_1 k_1}{2}\frac{\partial V_1}{\partial t} -$$

$$-\frac{C_2 k_2}{2}\frac{\partial V_2}{\partial t}+\frac{(1-C_1-C_2)}{\rho_0}\frac{\partial p}{\partial x}=\phi_0 , \tag{4}$$

$$\left(\frac{\rho_1}{\rho_0}+\frac{k_1}{2}\right)\frac{\partial V_1}{\partial t}-\left(1+\frac{k_1}{2}\right)\frac{\partial V_0}{\partial t}+\frac{1}{\rho_0}\frac{\partial p}{\partial x}=\phi_1 , \tag{5}$$

$$\left(\frac{\rho_2}{\rho_0}+\frac{k_2}{2}\right)\frac{\partial V_2}{\partial t}-\left(1+\frac{k_2}{2}\right)\frac{\partial V_0}{\partial t}+\frac{1}{\rho_0}\frac{\partial p}{\partial x}=\phi_2 , \tag{6}$$

where:

$$\phi_0=(1-C_1-C_2)g\sin\alpha-\frac{\lambda}{2D}\frac{\rho_c}{\rho_0}|V_c|V_c-\frac{3}{8}\times$$

$$\times\left[\frac{C_1 Cf}{R_1}|V_0-V_1|(V_0-V_1)+\frac{C_2 C_b}{R_2}|V_0-V_2|(V_0-V_2)\right], \tag{7}$$

$$\phi_1=\left(\frac{\rho_1}{\rho_0}+\frac{k_1}{2}\right)\frac{\partial V_1}{\partial t}-\left(1+\frac{k_1}{2}\right)\frac{\partial V_0}{\partial t}+\frac{1}{\rho_0}\frac{\partial p}{\partial x}, \tag{8}$$

$$\phi_2=-\frac{\rho_2}{\rho_0}g\sin\alpha+\frac{3}{8}\frac{C_f}{R_1}|V_0-V_2|(V_0-V_2), \tag{9}$$

$$\frac{1}{a_1^2}=\frac{\rho_1}{K_1}+\frac{\rho_1}{F}\left(\frac{\partial F}{\partial p}\right), \tag{10}$$

$$\frac{1}{a_2^2}=\frac{\rho_2}{K_2}+\frac{\rho_2}{F}\left(\frac{\partial F}{\partial p}\right), \tag{11}$$

$$\frac{1}{a_0^2}=\frac{1}{a_f^2}+\frac{\rho_0}{F}\left(\frac{\partial F}{\partial p}\right), \tag{12}$$

$$\alpha_f^2=\frac{K_f}{\rho_0}, \tag{13}$$

$$K_1=\frac{E_1}{3(1-2v_1)}, \tag{14}$$

$$\rho_c=\rho_0^*+\rho_1^*+\rho_2^*=(1-C_1-C_2)\rho_0+C_1\rho_1+C_2\rho_2 , \tag{15}$$

$$V_c=\frac{1}{\rho_c}\left(\rho_0^* V_0+\rho_1^* V_1+\rho_2^* V_2\right). \tag{16}$$

where K_1, E_1, V_1 – the module of volumetric compression, Junga's module and Puasson's factor of firm particles; K_2 – the module of volumetric compression of gas bubbles; K_f– the volumetric module of elasticity of a liquid; a_f– speed of a sound in a pure unlimited fluid; R_1, R_2 – equivalent radiuses of firm particles and gas bubbles; k_1, k_2 – the factors considering influences of nonsphericity and also concentration of firm particles and bubbles on attached weights; g – force of gravity; a – corner of an inclination of the pipeline to horizon; D – diameter of the pipeline; λ – Darcy's factor; t – time; C_f, C_b – factors of resistance of firm particles and bubbles;

C_i – volumetric concentration of a phase; p – pressure; ρ_i – true density of a phase; ρ_i^* – the-resulted density of a phase; V_i – speed of a phase, F – the area of interior cross section of the pipeline.

On the basis of the developed method of calculation of airlift hydraulic hoistings (Kyrychenko et al. 2014) and the corresponding software researches of transportation polymetallic nodules (PN) for installation with constant diameter of the elevating pipeline are executed. In calculations variants of hydraulic hoisting of various monodisperse structures of firm particles were examined. For approach of the received results plants productivity on dry mineral raw material is accepted 77.8 kg/with (1000000). The sizes of particles varied in a range 0.001...0.15 m. Density of particles in each considered monodisperse structure of a firm material was considered as a constant. Diameters of bringing and elevating pipelines in the given calculations made $D_{und}=D_p=0.46$ m. Results of calculations are selectively brought below.

On Figure 1 dependences of change of absolute speeds of firm, liquid and gas phases (corresponding curves 1, 2 and 3) on extent of elevating pipeline DAH are resulted at transportation PMN in diameter of 0.07 m and by density 1250 kg/m^3, and also of 1900 m. Depth of development of a deposit corresponds to depth of immersing of the amalgamator 6000 m. Zones I, II, III, IV denotes vesicular, shell, ring and disperse structures of current. On extent of vesicular and shell structures of current speed of a liquid exceeds speed of firm particles. For ring structure of current essential expansion of air attracts transportation of the basic part of a firm material to volume of a gas phase, moving with a continuous stream in the central part of the pipeline. As a result speed of firm particles in a gas phase exceeds speed concentrated at a wall of the elevating pipeline of a liquid that speaks increase of speed of gas concerning a liquid phase. In disperse structure of current speed of liquid drops again exceeds speed of particles (Goman 2008).

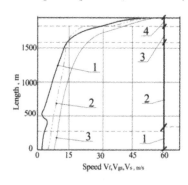

Figure 1. Dependence of change of speeds of firm, liquid and gas phases on length of the elevating pipeline.

307

Apparently from the resulted schedules at ring structure of current in EP firm particles advance water since it is transported basically by air in the centre of the pipeline while the liquid phase concentrates on periphery pipes. Thus relative speed of gas and liquid phases is great enough. In this plan the certain interest is represented with the research problem of parameters of a ring mode of current in view of thermodynamic processes. With this purpose attempt to develop thermodynamic model based on use of the equations of indissolubility, movement and energy separately for each phase has been undertaken. However, bulkiness of model and complexity of a choice closing empirical dependences have not allowed to receive correct decisions in a wide range of change of initial data. But, the analysis of results of some settlement cases has allowed to establish, that by virtue of significant relative speeds of air power interchange between a gas and liquid phase is complicated, therefore really existing in a stream polytrophic process is closer to adiabatic, than to isothermal. Calculations of parameters of current with use of an effective parameter of an adiabatic curve $k = 1.4$ have shown, that increase of settlement density of compressed air at isentropic process leads to reduction of its relative speed, redistribution of structures of current on height of an elevating pipe and, as consequence, to change of a hydraulic mode of transportation in comparison with isothermal process.

Let's stop on applied aspects of thermodynamics of DAH. For this purpose we shall execute estimated calculation of final temperature fulfilled in airlift air.

The settlement scheme is presented on Figure 2 (Patent No 30168).

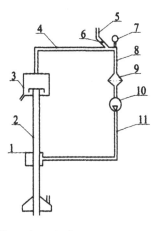

Figure 2. The settlement scheme.

The problem is solved at following basic assumptions: the density of water is considered a constant; external heat exchange – is absent; on an output from airlift water, air and firm particles are in thermal balance (it is the most adverse from the point of view of decrease in temperature of air a case); influence of phase transitions is a little. The equation of the first law of thermodynamics (the law of conservation of energy) for considered system looks like:

$$I_{fen} + I_{gsen} + I_{sen} + m_f \frac{V_{fen}^2}{2} + m_{gs} \frac{V_{gsen}^2}{2} +$$

$$+ m_s \frac{V_{sen}^2}{2} + m_f g(H + H_n + h) +$$

$$+ m_{gs} g(H + H_n + h) + m_s g(H + H_n + h) -$$

$$- \left(I_{fbg} + I_{gsbg} + I_{sbg} + m_f \frac{V_{fbg}^2}{2} + \right.$$

$$\left. + m_g \frac{V_{gsbg}^2}{2} + m_s \frac{V_{sbg}^2}{2} + m_{gs} g H_n \right) = 0, \quad (17)$$

where I_i – full enthalpy.

Indexes: "f", "gs", "s", "c" define accordingly parameters of water, air, firm and mix particles on an input (an additional index "bg") and on an output (an additional index "en") an elevating pipe of airlift.

Enthalpy of a stream of water it is defined under the formula:

$$I_f = C_f T m_f + P_f \frac{m_f}{\rho_f}, \quad (18)$$

where C_f – a thermal capacity of water; T – absolute temperature of water.

Enthalpy of air:

$$I_{gs} = C_p T m_{gs}, \quad (19)$$

where C_p – isobaric thermal capacity of air.

Absolute pressure of water at the input of airlift:

$$P_{fbg} = P_o + \rho_f g(H + H_n), \quad (20)$$

where P_o – atmosphere pressure upon a free surface of water.

Substituting $(18) - (20)$, in (17), after transformations we shall receive the equation (21).

For example, for installation with parameters; $h = 2000$ m, $H = 2000$ m, $V_{fen} = 55$ m/s, $V_{gsen} = 55$ m/s, $V_{fbg} = 0$ m/s, $V_{gsbg} = 3$ m/s, $C_f = 4150$ J/(kg·K), $C_{pgs} = 1000$ J/(kg·K), $T_{bg} = 275$ K, the final temperature makes $T_{en} = 274$ K.

$$T_{en} = T_{bg} - \frac{m_f\left(V_{fen}^2 - V_{fbg}^2\right) + m_g s\left(V_{gsen}^2 - V_{gsbg}^2\right) + m_s\left(V_{sen}^2 - V_{sbg}^2\right)}{2\left(C_f m_f + C_{pgs} m_{sg} + C_s m_s\right)} - \frac{m_f g h_n + m_{gs} g(H + h) + m_s g(H + h + H_n)}{C_f m_f + C_{pgs} m_{gs} + C_s m_s}. \quad (21)$$

Dismissed has estimated character since it is executed within the limits of homogeneous equilibrium model of two-phase current and allows to determine the top border of possible temperature of air on an output from an elevating pipe. Actually, in top sections of HP as it was specified above, the thermodynamic mode of current is close to adiabatic, therefore the valid temperature of air will be below settlement.

In view of this fact the way according to which fulfilled in airlift cold air on the tight pipeline again moves to the input of the compressor together with atmospheric air (Figure 3) that results both in increase in productivity of the compressor, and to decrease specific power consumption is developed. Really, productivity of the compressor is defined on volume of sucked in air. Geometrical volume V it will not change, and the quantity of air "m" with other things being equal, as is known depends on its temperature T as follows:

$$m = \frac{pV}{RT}, \tag{22}$$

where R – a gas constant.

Change of productivity δ_n, % of the compressor can be estimated for various conditions of operation on dependence:

$$\delta_n\% = \frac{\left(m_e + m_x\right) - m_x'}{m_x} \cdot 100\,\%, \tag{23}$$

where: $m_e = \dfrac{PV\alpha_e}{RT_e} \cdot n$ – quantity acted in the compressor fulfilled in airlift air; $\alpha_e = a$ volume fraction of the fulfilled air which has acted in the compressor; $m_x' = \dfrac{pV(1-\alpha_e)}{RT_{bg}} \cdot n$ – quantity of atmospheric air which has acted in the compressor; $m_x = \dfrac{pV(1-\alpha_e)}{RT_{bg}} \cdot n$ – quantity of atmospheric air which has acted in the compressor during the work of the compressor under the traditional scheme; T_e and T_{bg} – temperatures of air which has fulfilled in airlift and in an environment.

Thus, dependence (23) will become:

$$\delta_n\% = \frac{\dfrac{\alpha_e}{T_e} + \dfrac{1-\alpha_e}{T_{bg}} - \dfrac{1}{T_{bg}}}{\dfrac{1}{T_{bg}}} \cdot 100\% = \alpha_e \frac{T_{bg} - T_e}{T_e} \cdot 100\,\%, \tag{24}$$

For example, $\alpha_e = 0.6$, $T_{bg} = 303$ K and $T_e = 274$ K $\delta_n = 6.8\%$.

Besides as on an entrance of the compressor cooled air specific work of the compressor due to decrease in work of compression of air will decrease, at its intensive cooling moves.

As a first approximation this decrease in power consumption can be estimated on parameters of the first step of the compressor, including, that operating modes of the subsequent steps have remained constant.

It is well-known that, specific (falling a unit of quantity of air) display work of a step (section) of the compressor is determined by expression:

$$l = \frac{z}{z-1} R(T_n - T_\kappa), \tag{25}$$

where z – a parameter of process; T_{in} and T_{fin} – initial and final temperatures of air of a step.

Thus, relative decrease specific power consumption δ_1, % of the compressor can be estimated by dependence:

$$\delta_1\% = \frac{l_o - l_e}{l_o} \cdot 100\,\%, \tag{26}$$

where l_e and l_o – specific display works at use of fulfilled air and at use only atmospheric air.

Accepting T_{en} and m constants, and temperature T_a of air acting in the compressor proportional to mass fractions of fulfilled and atmospheric air and substituting (25) in (26), we shall receive:

$$\delta_1\% = \frac{T_{bg} - T_a}{T_{bg}} \cdot 100\% = \alpha_e \frac{T_{bg} - T_e}{T_{bg}} \cdot 100\%, \tag{27}$$

where: $T_a = \alpha_e T_e + (1 - \alpha_e) T_{bg}$. $\tag{28}$

For example, at $\alpha_e = 0.6$, $T_{bg} = 303$ K and $T_e = 272$ K, $\delta_n = 6.14\%$

The analysis of results of regular calculations with use of the above-stated mathematical device (1) – (28) considering interference hydrodynamic and thermodynamic effects, has allowed to formulate following scientific result: at ring structure of current in the certain range of change of initial data the firm particles transported in the centre of elevating pipe of DAH in a mode close to pneumatic transport, advance the liquid phase concentrated on periphery of the pipeline. By virtue of significant relative speeds of gas and liquid phases interphase interchange it is complicated and really existing polytropic process is closer to adiabatic than to isothermal.

The given result specifies expediency of study of essentially new "closed" constructive schemes of DAH directed on knocking over of pneumatic energy and allowing effectively to use cool potential of deep waters. According to our reckoning development of the specified schemes in many respects defines strategy of designing deep-water airlift hydraulic hoistings with the improved power characteristic. The given reasons are put in a basis stated below the invention (Figure 3).

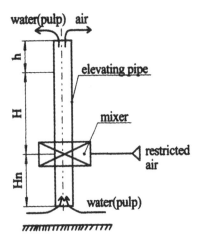

water(pulp) air

elevating pipe

mixer

restricted air

water(pulp)

Figure 3. Scheme of DAH of the closed type.

The invention concerns to area of rise of minerals from a bottom of the seas and oceans and can be used at designing and operation of airlift installations in other branches.

The purpose of the invention consists in increase of an overall performance of airlift installations by increase in productivity, reduction specific power consumption of process and protection of an environment.

The object in view is reached by that fulfilled in airlift air possessing in low temperature potential, is again used in the compressor of airlift installations.

The essence of the technical decision consists that separeted on an output of airlift cool air by heat sealed pipeline moves to the soaking up pipeline of the compressor in front of the air filter where for indemnification of losses of air in airlift in addition, through the adjustable gate, air from an atmosphere moves. Thus mass submission of the compressor as a result of mixing fulfilled in airlift cold air and atmospheric air on an input of the compressor air with temperature below an ambient temperature moves increases, that in addition leads to decrease in a specific power consumption on compression of air in the compressor and excludes emission fulfilled in airlift air in an atmosphere.

On Figure 3 it is schematically represented, an offered design.

Airlift installation contains the elevating pipe 2 connected through the amalgamator 1 and air pipe of a high pressure 11 with output of the compressor 10 Output of an elevating pipe 2 through air separator 3 with drainage branch pipe, the pipeline-air line of 4 compressors 8 connected to the soaking up pipeline 10. In the soaking up pipeline 8 the filter 9 is established and the manometer 7 Soaking up

pipeline 8 is connected to an atmosphere through the soaking up branch pipe 5 equipped by the adjustable gate 6.

The device works as follows. Before start of installation the gate 6 is completely open, providing the minimal resistance of a soaking up line 8. After start of the compressor the compressed air on an air pipe 11 acts through the amalgamator in an elevating pipe therefore there is a start of airlift. Air getting into airlift, has reference temperature close to temperature deep layers of water (+1°C...+ 2°C). During the movement of air in airlift with the big relative speed (up to 50 km/s), as a result of intensive thermodynamic processes occuring there, the temperature of air goes down so, that in an air separator the temperature separeted air will be below its reference temperature. As at work of the compressor in a tight air line 4 the vacuum leaving from airlift, air continues to extend is created moves on heat sealed pipeline to an airline 4 in soaking up pipeline 8 where mixes up with acting on a soaking up branch pipe 5 of an atmosphere air and cools it. As a result of it density 10 air acting in the compressor increases, that leads to increase in productivity of the compressor and its decrease specific to power consumption. During work of installation pressure in the soaking up pipeline of the compressor is supervised by a manometer 7 and by means of the gate 6 established on the soaking up branch pipe, supported equal nominal.

3 CONCLUSIONS

The new approach to studying processes in airlift, based on the joint analysis of hydrodynamical and thermodynamic parameters of high-gradient three-phase compressed stream is offered.

For the first time it is established, that interference of hydraulic and thermodynamic fields in high-gradient of three-phase compressed stream leads to occurrence is abnormal low temperature potential of fulfilled air on an output from elevating pipe of DAH.

The closed constructive schemes of DAH directed on knocking over of pneumatic energy in a running cycle of installations, using features of thermodynamic process of expansion of compressed air and a cold of deep waters are developed. Use of such schemes allows raising essentially productivity of hydraulic hoistings, alongside with reduction of their power consumption.

REFERENCES

Goman, O. 2008. *Development of multipurpose dynamic model of the multiphase environment with reference to airlift hydraulic hoisting.* Scientific Bulletin of National Mining University, No 8: 89 – 93.

Kyrychenko, E., Goman, O., Kyrychenko, E., & Evteev, O. 2014. *Designing of systems of hydraulic*

transport of polymetalic ores of the World ocean. Nikopol: FOP Feldman O.: 611.

Patent No 30168, UA, IPC F04F1/00, F04F1/20. *The way of management of airlift and airlift installation.* Kyrychenko, E., Vyshnjak, E. & Nakidaylo, A. Pub. 15.02.2002.

Uoliss, G. 1972 *One – dimensional biphase currents.* Moskva: Mir: 440.

Theoretical and Practical Solutions of Mineral Resources Mining – Pivnyak, Bondarenko & Kovalevska (eds)
© *2015 Taylor & Francis Group, London, ISBN: 978-1-138-02883-8*

Assessment of groundwater rise in urban areas:
Case study the city of Dnipropetrovsk

T. Perkova & D. Rudakov
National Mining University, Dnipropetrovsk, Ukraine

ABSTRACT: The numerical groundwater flow model has been developed for the city of Dnipropetrovsk in order to assess the risk of groundwater rise (GWR) caused by natural and man-made impacts. The 3D model takes into proper account the heterogeneity of aquifer properties as well as flow budget components in permeable and confining layers. The model allows more accurate long-term predicting the groundwater level in built-up urban areas based on the evaporation-precipitation balance, and evaluating the total area under high risk of flooding due to groundwater rise up to the critical depth. Assessment of groundwater rise in urban areas is considered as the useful tool for decision making in water management and geotechnical stability in large cities affected by floods.

1 INTRODUCTION

At the time being, groundwater table exceeds the natural level in more than 900 cities and towns in Ukraine with the total area of about 4000 km² (Information yearbook 2012). High groundwater level triggers landslides, suffusion, land subsidence and may intensify karst development under specific geological settings. Groundwater rise is believed to pose a very risk to public safety and can cause significant damages and economic losses (Yakovlev et. al. 2002). One of those cities is Dnipropetrovsk located in Dnipro valley where the trend of rising groundwater has been observed since the end of the nineteenth century.

High groundwater level in urban areas is most likely to occur in lowlands underlain by permeable rocks like the left bank and lower terraces of right bank of Dnipro River. Besides, this process may also occur in the areas with higher ground surface elevation underlain by low-permeable loess and loamy-sandy soils. For the case of Dnipropetrovsk city, a number of man-made factors contributed to higher flood risks which are:

– leaky sewer and municipal pipeline systems;
– obstructed runoff in depth;
– extensive building;
– large sealed areas, f. i., covered with asphalt.

By 1990s artificial recharge in some city areas due to leaks from pipelines became comparable with the mean annual precipitation rate of about 460 mm. In some cases these may reach 70 percentage of total water supply (Statistical Yearbook 2011).

Likely, the leaks has triggered a catastrophic landslide in June 1997 that destroyed one nine-story block of apartments, one school, two kindergartens, and several one-story buildings on the area of 4.5 ha.

Engineering solutions proposed and developed to protect of flooding proved to be insufficient. A number of measures to prevent flooding, including sedimentary control, upgrade of municipal pipeline systems, and water drainage have been taken till 2011 (Comprehensive Programme 2002). Regrettably these measures had no significant effect.

To predict and control groundwater rise in challenging urban environment with various types of ground surface ground water flow models have to accurately consider surface runoff and precipitation-evaporation balance. At present, ground water flow models are not systematically applied to large urban areas. Commonly, evaporation in these models is input without any reference to ground water level, which lowers the prediction validity, especially when groundwater occurs at shall depth (Været et. al. 2009, Rudakov et. al. 2014). Thus, adequate assessment of recharge and evaporation is getting of critical importance for groundwater flow models and valid assessments of flow budget and ground water rise.

Therefore, the study aims to assess the risk of groundwater rise in the city of Dnipropetrovsk by evaluation of surface and underground runoff, taking into proper account relief, unsaturated zone thickness, evaporation, leakages, and hydraulic conductivity for soils in different districts of the Dnipropetrovsk city.

2 SITE DESCRIPTION

Dnipropetrovsk is located at the connection of two major geomorphologic elements that are the right pediment plain bank which is part of the Dnipro upland and left lowland banks. These provide the great originality of terrain and features of modern exogenous processes. Geological settings of the terrain are the complex of Quaternary and Paleogene soils, and crystalline Proterozoic rocks.

The main part of Dnipropetrovsk is located on the right bank of Dnipro River dissected by ravines and gorges deeply incised in sedimentary rocks (Figure 1).

In the highly heterogeneous ground water flow domain within the city area three layers are discriminated which are:
– unconfined aquifer of loess loam and sandy-clay soils;
– confining low-permeable clay bed;
– confined-unconfined sandy-clay aquifer.

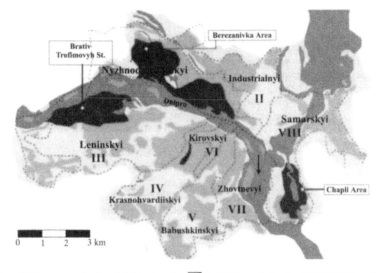

Figure 1. Locations of the areas under flood risk in the city: ▭ district boundaries; I...VIII district numbers, ▨ areas of raising groundwater.

The risk of flooding is higher, particularly, for Amur-Nyzhnodniprovskyi (AND), Leninskyi and Samarskyi districts. Some residential areas there have been flooded in February 2012. Melted snow water has covered large territory having affected 20 houses and 33 homesteads in AND district. Five houses and 21 homesteads were flooded on Brativ Trofimovyh Street (Figure 2) as well as two homesteads in Chapli Area.

More than 22000 m^3 of water were pumped out and 200 m of drainage channels were cleaned by the Emergencies Ministry and municipal services during February, 26 to urgently dewater this area.

3 DEVELOPMENT OF THE CONCEPTUAL GROUNDWATER FLOW MODEL

The groundwater flow domain has been digitized by assigning hydraulic conductivity K, aquifer thickness m, and groundwater depth h for each layer (Table 1).

Figure 2. Groundwater flooding of homestead in Brativ Trofimovyh Street, February 2012.

The impact of precipitation seasonal changes on ground water level variation, the leaks from water and heat supply and sewer systems as well as building density were assessed by time series analysis for ground water recharge. These adjustable parameters varied between extreme values while testing the numerical model. Infiltration rate was assumed to be much higher for lowlands, foots of slopes and in gullies because of collecting more surface runoff.

Table 1. Groundwater flow model parameters.

Layer #	Aquifer / confining bed	Thick-ness, m, m	Hydraulic conductivity, K, md⁻¹	Ground water depth, h, m
1	Loess loam, sandy-clay	2 – 42	0.2 – 12	1 – 20
2	Clay	10 – 15	0.0001	–
3	Sandy-clay	40	1.2	15 – 75

Empirical formula derived by F. Averianov was used to estimate annual evaporation from the ground water surface (Averianov 1978). In case of high elevation of ground water table near the soil surface the evaporation rate E (mm/year) was calculated as follows:

$$E(H) = E_0 \left(1 - \frac{H}{H_{cr}}\right)^n, \qquad (1)$$

where E_0 – evaporation rate from the water surface, mm/year; H – the ground water depth, m; H_{cr} – the critical depth of ground water level, m; n – the empirical factor depending on soil properties that ranges from 1 to 3. If the ground water level falls below the critical depth H_{cr}, no evaporation is assumed to occur.

The critical depth H_{cr} has been estimated following the formula of V. Kovdy (1947):

$$H_{cr} = (170 + 8T), \text{ cm}, \qquad (2)$$

where T – the mean annual temperature assumed to be 9°C for the site area; $H_{cr} = 2.8$ m.

The mean annual head variation ΔH for an arbitrary planar area within the flow domain can be evaluated on the base of the water balance equation as follows:

$$F_{ev} \cdot E(H) + F_{ev} \cdot w_p + Q_t - Q_s -$$
$$- Q_{lk} + Q_{sd} = F_0 \frac{\Delta H}{\mu \, \Delta t}, \qquad (3)$$

where F_{ev} – evaporation area, m²; $F_{ev} = \alpha \cdot F_0$; α – the fraction part of unsealed surface of the total area that is able to evaporate; F_0 – the total area, m²; w_p – precipitation rate, m/d; Q_t – artificial recharge, m³d⁻¹; $Q_t = F_0 \cdot w_t$, m³d⁻¹; w_t – artificial recharge rate, m/d; Q_s – surface runoff, m³/d, $Q_s = F_0 \cdot w_p \cdot \alpha_s$; α_s – surface runoff fraction in precipitation flow; Q_{lk} – vertical discharge through the underlying clay bed, m³d⁻¹; Q_{sd} – groundwater discharge through flow domain boundaries, m³d⁻¹; μ – specific yield; t – time period, d.

Ground water depth H was evaluated as a result of numerical solution of equation (1) and (3) for

steady state condition, providing parameter ΔH is equal to 0. The iterative procedure coded in software MathCad and used for calculations allowed correlating ground water head and recharge rate for different fraction of sealed surface, which is important factor for urban areas (see Section 4).

The numerical groundwater flow model is based on the equation:

$$\frac{\partial}{\partial x}\left(K_{x,s} m_s \frac{\partial H_s}{\partial x}\right) + \frac{\partial}{\partial y}\left(K_{y,s} m_s \frac{\partial H_s}{\partial y}\right) +$$
$$+ \frac{K'_{s,l}}{m'_{s,l}}(H_s - H_l) + W_s - E_s = n_{0,s} \frac{\partial H_s}{\partial t}, \qquad (4)$$

where "s" and "l" – layer indices; $K_{x,s}$, $K_{y,s}$ – hydraulic conductivities along the x and y coordinate axes in layer "s", m/d; H_s, H_l – the potentiometric head in layers "s" and "l"; $K'_{s,l}$ and $m'_{s,l}$ – average hydraulic conductivity and layer thickness, respectively, md⁻¹ and m; W_s – volumetric flux per unit volume, representing sources of water inflow, m³d⁻¹; n_0 – effective porosity.

Numerical simulation of groundwater flow has been performed using software Visual Modflow 2009.1 that applies 3D block-centred finite difference procedure to solve equation (4) (Harbaugh et al. 2000).

The software Visual Modflow provide some advanced options for accurate simulation of 3D groundwater flow, particularly, dewatering and rewetting modes both for aquifers and confining layers. This is especially important under strong impact of time-dependent parameters that quantify various man-made and natural factors.

The flow domain covers residential and industrial areas in both parts of the city. It takes into account space variations of recharge, groundwater discharge and surface-ground water interaction. The flow domain area of 341 km² has been divided by the grid with node resolution of 500×500 m onto 1363 cells. The grid spacing has been agreed with the geological monitoring scale that allowed trustworthy evaluation of representative model parameters. This spacing enabled reproducing the ground surface elevation (Figure 3) with the accuracy sufficient for modelling, taking into proper account spotty patterns of evaporation and precipitation over the ground surface in urban areas.

The flow domain is bounded by water divides and lowlands near the river in the south-western and northern parts of the city area. The long-term mean annual water level is assumed on the boundaries.

The ground water flow domain includes two unconfined and confined-unconfined aquifers separated by the low-permeable bed of heavy loams and clays.

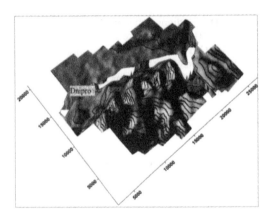

Figure 3. Ground surface elevation of the site.

The shallow Quaternary aquifer occurs on the right river bank on the watershed area and in ravine bottoms as well as throughout the left riverbank. The deeper Neogene aquifer occurs within the watershed area and on the ravine slopes.

Layer bottom elevation of the Quaternary aquifer has been assigned to each block of the grid according to geological surveying data. The Neogene aquifer has been reproduced in the model using the available data on rock/soil permeability, flow velocity, and head elevation at the external model contours.

The heterogeneity of soil/rock permeability in the model was simulated by 10 different zones according to the ranges of conductivity and storage (Table 1) and available geological cross-sections.

4 RESULTS AND DISCUSSIONS

To evaluate quasi-steady ground water depth depending on the evaporation rate by equation 3 we varied the input data of ground water recharge, surface runoff, and the fraction of evaporation area. The other parameters were defined according to typical conditions of Dnipropetrovsk: $w_p = 450$ mm/year; $E_0 = 0.8$ m/year; $H_{cr} = 2.8$ m; $\mu = 0.2$. Descending vertical and horizontal flows were assumed insignificant.

Reduction of evaporation area under intensive artificial recharge w_t of 150 mm/a (at the mean value w_t for the city of 91 mm/a) will result in groundwater rise or decrease of groundwater depth from 1.4 m to zero, i.e. to the earth's surface (Figure 4).

Reduction of unsealed surface inhibits evapo-ration, which likely causes the convex form of the curves.

Recharge rate was evaluated by the procedure described above regarding to the available data on leaks from water supply and sewer systems, surface runoff and evaporation.

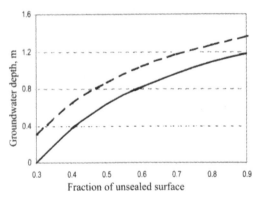

Figure 4. Calculated groundwater depth H versus the fraction of unsealed surface α for various recharge rates: $w_{r1} = 100$ mm/a (dashed curve); $w_{r2} = 150$ mm/a (solid curve).

Increasing of recharge rate from 0 to 200 mm/a contributes groundwater level rise; groundwater depth demonstrates almost linear correlation with surface runoff for in the range from 1.2 – 2 m to 0.5 – 1 m. Reduction of the surface runoff results in rising groundwater level up to 0.5 – 0.8 m (Figure 5).

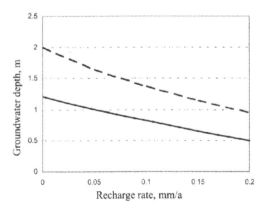

Figure 5. Calculated groundwater depth H versus recharge rate w_t, for different fraction of surface runoff: $\alpha_s = 0.6$ (solid curve); $\alpha_s = 0.9$ (dashed curve).

This parameter ranging within 0.05 – 160 mm/a was distributed over the city area using 15 zones with different values.

The developed model has been validated by inverse modelling under conditions of steady-state flow. The calculations showed good correlation (80%) to groundwater monitoring data.

According to modelling, the water budget of the shallow Quaternary aquifer depends primarily on infiltration and descending water flow to the underlying aquifer (Figure 6, a). However, both natural and man-made recharges (over 13000 m³/d total) contribute the most to groundwater rise and heighten the risk of flooding in lowlands.

These budget components discharge primarily to the river (roughly 80%), and the rest water comes to the deeper Neogene aquifer. Groundwater and, partially, rain water may enter the Neogene aquifer through the zones of relatively high permeability in the confining bed encountered mainly in the central part of the right bank city area (Figure 6, b).

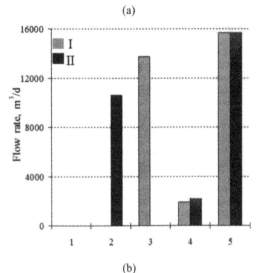

(a)

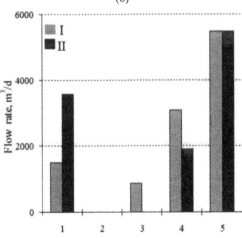

(b)

Figure 6. Inflows (I) and outflows (II) of the Quaternary aquifer (a) and Neogene aquifer (b): 1 – discharge through boundary; 2 – groundwater inflow and outflow to the river; 3 – recharge; 4 – water flow exchange between adjacent layers; 5 – total water balance.

Ground water flow is assumed to have backup water level within the Neogene aquifer toward the river bank, where discharge decreases due to hydraulic gradient fall.

Groundwater flow will likely remain stable till 2018 on the left riverside. Thus, groundwater level in Industrial and Samara districts is expected to rise insignificantly (Figure 7).

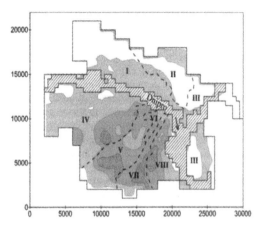

Figure 7. Ground water rise map of Dnipropetrovsk city for the next five years based on modelling results: I...VIII – district numbers; ☐ natural groundwater level area; groundwater raising domain: ▨ 0 – 1 m; ▨ 1 – 2 m; ▉ over 2 m.

At the same time, groundwater level will rise another 1 m to in a small area within Industrial district in the northern city area on the distance 200 m far from the river. The same trend refers to the lowlands in Samara district and residential area Chapli (South-East of the city). Groundwater level variations may have the tendency to rise 1 m to in the North-Western part of the city; however, this trend will be restrained by higher evaporation and evapotranspiration within the green zones on the border of Frunzenskiy residential area (Figure 7).

Throughout the right riverside city part except most of its western districts, ground water rise of 1 m to 2 m is expected.

The increases of ground water level up to 2 m may occur more likely in South-West of the city; under the highest risk of groundwater rise are the central part of city, the districts to South from downtown and the suburb Krasnopolie where groundwater may rise over 2 m above the current levels. The areas with critical groundwater depth of less than 2 m below the surface will likely to appear within central and south-eastern districts till 2018 if no preventive measures will be taken.

317

Therefore, the model highlighted the most problematic areas of the city in terms of ground water rise risk. Figure 7 can be interpreted as the groundwater rise susceptibility map that visualizes the size of flood risk zones depending on groundwater and surface parameters, evaporation, recharge, and various leaks.

5 CONCLUSIONS AND RECOMMENDATIONS

For the first time the numerical model of 3D ground water flow has been developed and validated for the area of Dnipropetrovsk city.

As a result of increasing infiltration and leaks from municipal pipeline and sewage system, the natural groundwater balance of the city changed dramatically. Prevailing water input mostly from man-made sources resulted in the levels much higher than those observed natural groundwater flow.

Troublesome groundwater rise in urban areas occurs mostly in slopes and lowlands with relatively high permeability aquifer within a right bank of Dnipropetrovsk city. Hydraulic gradient in these areas is very low, which leads to forming of large stagnant flow zones.

The total area affected by groundwater rise over 2 m to the critical depth has been evaluated numerically; it may reach above 15 km^2 in the next five years.

The preventive measures to be taken urgently should be focused on the drainage in flood susceptible residential areas. The measures have to include installation of radial drainage systems and intake wells that allow self-draining of groundwater and lowering ground water head in the sites of its maximum rise (Inkin & Rudakov 2012). Besides, water supply and sewer networks need urgent modernization and repair to avoid or minimize significant leaks.

Further rise of groundwater level within the city is expected to be limited. The only exception is the built-up areas with dense population where leakages of water supply and sewer systems may frequently occur in the future.

The living map based on the ground water flow model helps to highlight the critical points for more thorough monitoring and early warning systems. These can alert the authorities on possible groundwater flood events in advance. Such early warning implies that authorities can be prepared in advance to respond properly to flood event threats.

REFERENCES

Averianov, S.F. 1978. *Salinity control of irrigated lands.* Moscow: Kolos: 288.

Comprehensive Programme of consequence management of flooding areas in cities and settlements of Ukraine (approved by the Cabinet of Ministers of Ukraine No 160 dated February 15, 2002).

Harbaugh, A.W., Banta, E.R., Hill, M.C. et al. 2000. *Modflow-2000, the U.S. Geological Survey modular groundwater model – user guide to modularization concepts and the ground-water flow process.* Open-file report 00-92 "U.S. Geological Survey": 210.

Information yearbook on intensification of hazardous exogenous processes in Ukraine according to EGP monitoring. 2012. Kyiv: The State Service of Geology and Mineral Resources of Ukraine, State Scientific and Production Enterprise "State Geological Information Fund of Ukraine": 105.

Inkin, O.V., Rudakov, D.V. 2012. *Estimation of measures efficiency of flooded areas transfer to self-draining regime.* Materials of the International Scientific and Technical conference "Sustainable development of industry and society". Kryvyi Rig, May 22 – 25: 4 – 7.

Rudakov, D.V., Sobko, B.E. Perkova, T.I. et. al. 2014. *Forecast efficiency of clay screen creation below the bottom of the tailing facilities of titanium and zirconium ore treatment waste.* Metallurgical and Ore Mining Industry, No 1: 73 – 76.

Statistical Yearbook of Ukraine for 2010. 2011. Kyiv: State Statistics Committee: 559.

Været, L., Kelbe, B.E., Haldorsen, S. et. al. 2009. *A modeling study of the effects of land management and climatic variations on groundwater inflow to Lake St. Lucia, South Africa.* Hydrogeology Journal. Berlin: Springer-Verlag, Vol 17, No 8: 1949 – 1967.

Yakovlev, V.V., Svirenko, L.P., Chebanov, O.Yu. & Spirin, O.I. 2002. *Rising groundwater levels in northern-eastern Ukraine: hazardous trends in urban areas.* In: Current Problems of Hydrogeology on Urban Areas. Urban Agglomerates and Industrial Centers. NATO Science Series: IV. Earth and Environmental Sciences, Vol. 8: 221 – 241.

Effect of obturation line on protective efficiency of dust half-masks

S. Cheberyachko, O. Yavorska & V. Tykhonenko
National Mining University, Dnipropetrovsk, Ukraine

ABSTRACT: Protection level of filtering half-masks is determined as the correlation of aerosol concentration in the surrounding air to its concentration under the masks of respiratory protective device (RPD). There are two ways for unfiltered air to enter human lungs: through filter and through outflows along obturation line or exhalation valve.

1 TOPICALITY

Filter respirators are widely used to protect respiratory organs against injurious additives in the air. Their efficiency is evaluated by certified laboratory tests according to two standards (DSTU EN 149-2003 and DSTU EN 143-2002) specifying range of specific procedures meant for determining their protective and ergonomic properties. Identifying penetration coefficient of half-masks, worn by people doing specific exercises, is the important element of this test. It is supposed that the results can be used to characterize quality of protective items in general. However, there are publications (Petryanov et al. 1984, Mironov 2002, Basmanov et al. 2002, Falshtynskyi et al. 2014) considering that protective properties of respirators obtained under manufacturing conditions are worse to compare with laboratory ones. There are many reasons for that. It can be inappropriate selection of respiratory protective device (RPD), lack of knowledge how to use them as well as the situation when working conditions are not taken into account. Nevertheless, we believe that the main reason is incomplete laboratory research.

First, the majority of procedures for determining aerosol penetration through the filter are carried out on mass production basis which cannot reflect real parameters in the context of pulsating respiratory mechanism. Second, while determining aerosol aspiration along obturation line in accordance with DSTU EN 140-2004 tested on volunteers, it is recommended to use filter simulator. Its resistance to airflow should correspond to the one of real filter. However, while testing, aerosol particles are accumulated on RPD increasing their pressure differential with the growth of aspirations at obturator leakiness. That is why it is important to evaluate protective efficiency of respirator taking into account simultaneous aerosol penetration through filter and obturation line.

2 STATEMENT OF THE PROBLEM

The aim of this paper is to evaluate the effect of aspiration along obturation line simultaneously with filter operation in actual practice cyclically on RPD penetration coefficient.

3 MATERIALS AND TESTING METHODS

Testing procedure included the following:
– determining respiratory modes of testees;
– determining penetration coefficient through respiratory filter on the attachment using sodium chloride aerosol (Figure 1);
– determining penetration coefficient of sodium chloride aerosol through respirator with the help of testees (Figure 2).

Identifying respiratory modes of testees is required to reproduce them while determining penetration coefficient of RPD filters. Air volumes being inhaled and exhaled during normal, deep respiratory modes, while walking at the speed of 6 h/km were recorded with the help of electronic manometer "Testo 512". Its technical characteristics allow determining speed of air exhaling, pressure differences, rarefaction, and differential pressure.

Penetration coefficient of respirators was determined using test-aerosol of sodium chloride. Plant scheme meets the requirements of DSTU EN 143-2002 (Figure 2). Operating principle of the plant is as follows. Aerosol mixture with the concentration of 8 mg/m^3 and particles distribution of 0.02 – 2 mcm is supplied inside testing chamber from aerosol generator. Respiratory device is used to simulate specified respiratory modes obtained during previous test. Air is sampled from under-mask space through sample fixed to face piece of respirator.

The results are automatically entered into computer program AAS-2009 with their following processing and resulting in the form of graphs.

Taking into account the fact that penetration coefficient of RPD filters located on head model was determined without aspirations according to respiratory modes reproduced by respiratory device according to specified patterns, aspiration coefficient is calculated using the formula:

$$K_a = K_p - K_{phm}, \qquad (2)$$

where K_a – aspiration coefficient due to leakiness beyond contour (obturation) line; K_p – penetration coefficient of respirator on head model; K_{phm} – penetration coefficient of half-mask on a human.

Disposable and nondisposable half-masks with different obturator structures were used. Thus, disposable half-masks have special-purpose gaskets to ensure better adjacency to face (Figure 4, a, b, c).

(a)   (b)   (c)

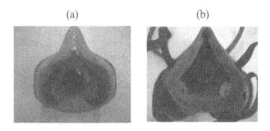

Figure 4. Obturator structures of disposable half-masks: (a) sample 1; (b) sample 2; (c) sample 3.

Nondisposable half-masks have standard obturator with similar perimeter width (Figure 5, a). New original obturators (Figure 5, b) have been developed owing to use of silicone, plastic as well as combination of different elastic materials (for example, plastesol, "croton" and others).

(a)   (b)

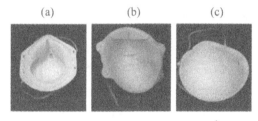

Figure 5. Modern nondisposable respirators: (a) sample 5; (b) sample 6.

4 THE RESEARCH RESULTS

Figures 6, 7 demonstrate graphs of determining penetration coefficient of respirators on volunteers cyclically. The obtained results show the available dependence of protective efficiency upon inhale and exhale phases. Disposable half-masks show almost similar oscillation of penetration coefficient when nondisposable half-masks show gradual increase of penetrating aerosol particles into under-mask space along with time. This fact can be explained by the available "dead" (non-ventilated) spot in heavy half-masks; as for light half-masks, they have minor "dead" spot. At exhale phase, part of aerosol entered under-mask space cannot come out through exhalation valve. It accumulates increasing concentration of aerosol entering with the next inhale.

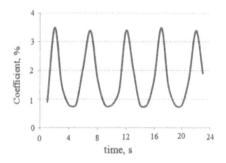

Figure 6. Cyclical dependence of penetration coefficient of light half-mask upon the time with 30 l/min loss.

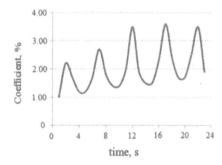

Figure 7. Cyclical dependence of penetration coefficient of nondisposable respirator upon the time with 30 l/min loss.

Minor "dead spot" of light respirators favors eliminating aerosol residues out of the mask at exhale phase or concentration of admixtures is so low that integral photometer cannot record them. However, there is another situation during deep breathing, conversations and body bending (Figure 8). In this case, exhale volume is not enough to eliminate penetrating particles from the space under respirator. It is also known that exhale volume is less to some extent than inhale volume as not all blood-saturating oxygen is replaced with the liberated carbon dioxide. Additional gasket on obturator decreases possible occurrence of leakiness during head and body movement or conversation.

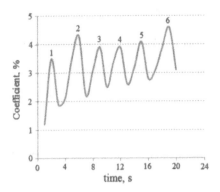

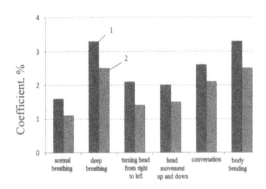

Figure 8. Cyclical dependence of penetration coefficient of light half-mask upon the time while doing different exercises: 1 – normal respiration; 2 – deep respiration; 3 – turning head from side to side; 4 – turning head up and down; 5 – conversation; 6 – body bending.

Respirators of the third degree of protection (sample 1) have practically no oscillations of penetration coefficient while doing different exercises (Figure 9, 10) contrary to half-masks of the first degree of protection (sample 3) where obturator is strengthen only with the strip of foam plastic within the area of nose bridge.

Evaluation of nondisposable respirators shows that standard half-mask has worse results comparing to the modern one. Sample 5 has obturator reproducing face features better to be confirmed by low aspiration coefficient. Area of nose bridge is a trouble spot of respirators and developers try to strengthen obturator in this very zone improving protective properties of RDP.

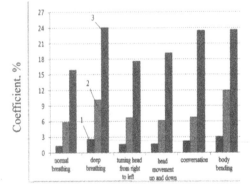

Figure 9. Dependence of penetration coefficient of light half-mask while doing various exercises: 1 – sample 1; 2 – sample 2; 3 – sample 3.

The results of determining aspiration coefficient of respirators calculated according to formula 2 are represented in Table 2.

Figure 10. Dependence of penetration coefficient of non-disposable half-mask while doing various exercises: 1 – sample 4; 2 – sample 5.

Table 2. Results of determining aspiration coefficient along obturation line.

Respirator sample	1	2	3	4	5
Protection class	FFP3	FFP2	FFP1	FFP3	FFP3
Average values of penetration coefficient on testees, C_p, %	1.4	7.5	22.3	2.3	1.8
Penetration coefficientof respirator on a model, C_r, %	0.8	4.1	12.6	0.05	0.05
Aspiration coefficient, C_a, %	0.6	3.4	9.7	2.25	1.75

Much more aerosol particles penetrate into non-disposable respirators through leaky obturation lines than through filtering components; it is contrary in disposable masks (Figure 11, 12). It is interesting that increase of air consumption in disposable half-masks results in decrease of aspiration value. Most of all it depends on half-masks sticking to faces due to increase of pressure differential. Similar effect has not been determined in case of nondisposable respirators as their half-masks are stiffer.

One of the reasons provoking the situation is the difference in standard requirements as for RPD quality. Thus, DSTU EN 149-2003 for half-masks of the third degree of protection specified that penetration coefficient through filter should be not more than 1% when penetration coefficient through the whole respirator per eight tests out of ten is not more than 2%.

Difference between them is the possible aspiration coefficient along obturation line, i.e. 4%. As for nondisposable respirators, DSTU 143-2002 specifies that penetration coefficient of protective third-degree filters is 0.05%.

322

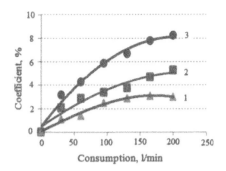

Figure 11. Dependence of penetration coefficient upon air consumption of disposable half-masks: 1 – aspiration value; 2 – filter; 3 – sample 2.

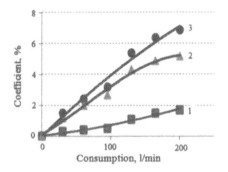

Figure 12. Dependence of penetration coefficient upon air consumption of nondisposable respirator: 1 – filter; 2 – aspiration value; 3 – sample 5.

Another standard (DSTU 140-2004) has the requirements as for aspiration value through half-masks of these respirators; it should be the same as disposable ones have, irrespective of protection class, i.e. not more than 2% per eight tests out of ten. At the same time, filters with low penetration coefficient are characterized by increased resistance to air flow. Consequently there is a possibility to increase aerosol aspiration through obturator of nondisposable RDP as air flow resistance in obturator airspaces between face and half-mask can be less than filter resistance. Thus, according to DSTU EN 140-2004 standard half-masks are not divided into different protection classes; as a result, there is the situation with increased aspiration through obturation.

The research shows that there is a possibility to improve protective efficiency of nondisposable respirators owing to more severe requirements as for aspiration coefficient according to protection class. These requirements can be met at the expense of introducing additional gaskets as well as by having equal distribution of pressure forces along obturation line, by reducing "dead spot" etc.

5 CONCLUSIONS

Determining aspiration coefficient of different half-masks has allowed establishing the following:
– each tested half-masks meets the requirements of standards within the determined research efforts;
– aerosol particles that are accumulated in under-mask space of nondisposable half-masks resulting in deterioration of protective properties;
– aspiration coefficient increases during deep breathing, conversation or body bending;
– value of unfiltered air aspiration in nondisposable half-masks along obturation line is larger than that of disposable ones;
– contrary to nondisposable half-masks, aspiration value of light ones through obturator begins to decrease at reaching air consumption of 150 l/min.

REFERENCES

Basmanov, P., Kaminsky, S., Korobeinikov, A. & Trubitsyna, M. 2002. *Respiratory protective device, reference manual.* GIPP "Iskusstvo Rossiyi": 399.
Falshtynskyi, V., Dychkovskyi, R., Lozynskyi, V & Saik, P. 2014. *Some aspects of technological processes control of in-situ gasifier at coal seam gasification.* Progressive technologies of coal, coalbed methane, and ores mining. The Netherlands: CRC Press/Balkema: 109 – 112.
Mironov, L. 2002. *Importance of contaminated air aspiration into under-mask space of filtering respirators and methods of its detection.* Rabochaya odezhda, No 3: 15 – 21.
Petryanov, I., Koshcheyev, V. & Basmanov, P. 1984. *Lepestok (light respirators).* Moscow: Nauka: 218.

Theoretical and Practical Solutions of Mineral Resources Mining – Pivnyak, Bondarenko & Kovalevska (eds)
© 2015 Taylor & Francis Group, London, ISBN: 978-1-138-02883-8

The use of improved dump trucks for substantiation parameters of the deep pits trench

S. Felonenko, K. Bas & V. Krivda
National Mining University, Dnipropetrovsk, Ukraine

ABSTRACT: Exploitation of the deep mine horizons faces the great problems due to reduced in-pit environment, minimal operational sites, limited accessed reserves of the deposit. Therefore, because of mineral deposit excavation the working flank flattens due to significant volumes of rock removal. In its turn, the angle of working flank depends on the cutting depth, operational sites and transport communications width, safety berms width. Hence, the parameters of the operational sites influence on the flank flattering and at the same time, tightly depend on in-pit environment at deep mine horizons.

The purpose of the paper is to calculate the minimal possible operational site parameters at deep mine horizons in constrained operational environment ant substantiation their rational parameters by the state-of-art improved dump cars with electromechanical transmission.

1 INTRODUCTION

At the design stage, while working drawings are in process, the calculation of operational site parameters is foreseen. However, in case when the maximum possible rock removal occurs on the upper level within the overburden bench in dependence on excavator pass width in order to assure maximal deposit's opening, but for operational site, where the deposit excavation occurs, the calculation conditions are another. The opencast flanks are situated, generally, on deep horizons, where at least minimal possible parameters for open-pit operation must be kept. Therefore, the paper purpose is calculation of the minimal possible parameters of operational site at the deep horizons in constrained conditions and their parameters substantiation while the usage of modernized dump car BelAZ.

2 PREVIOUS RESEARCH ANALYSIS

In the paper (Rzevskyi 1985) defines the interval of possible operational site's width as following:

$$B + 1.7R_S \geq W_W \geq B + B_1, \qquad (1)$$

where B – width, m; R_S – excavator cutting radius on the grade, m; B_1 – excavator width, m.

While designing work (unlike the classic meaning) the minimal operational site width of the pit with ring transportation scheme defines by the formula, where the core parameters are transport possibilities:

$$B + 1.7R_S \geq W_W = a + D_W + O_b + b, \qquad (2)$$

where a – the distance from the bottom flank to the road edge, $a = 2.5$ m; D_W – diameter of the turning site, m; O_b – width of the roadside verge, $O_b = 0.5$ m; b – width of the bulk protections under the embankment, m.

While calculation of the minimal width of the working area (road width) with dead-end reversal scheme (2) changes the basic design parameter by transport overall sizes:

$$W_W = a + W_{WS} + O_b + b, \qquad (3)$$

where W_{WS} – road width, m.

The minimum dimensions of the operation site width determines in dependence on technical parameters of the technological scheme and equipment (excavator – dump truck, excavator – train, excavator – conveyor).

3 THE RESEARCH RESULTS

Object of the research are drifts at different levels of iron ore excavation. Nowadays, for rock removal in pits of Kryvyi Rig dump cars are in exploitation. They are used both in mining operations and rock removal in constrained working environment in deep mine horizons. However, such working conditions lead to significant reduction of active deposits in open pit, slump of ready and prepared deposits, extremely closed operational space and lead to inappropriate operational sites as well.

The analysis of obtained data shows, that deposits excavation provides by parallel pushback in vertical and horizontal directions (step by step in all horizons' direction from top to bottom). Soft overburden rock removal provides on the top flanks (horizons 115 – 53 m), rock of raptured zones and overburden rock on horizons 89 – minus 280 m, deposit mining – on horizons 41 – minus 280 m.

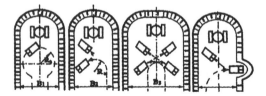

Figure 1. Variations while passing maneuvers dump trenches.

The trench work on the working flanks provides with excavator's mechanical shovel with bottom dumping into dump cars. The usage of dump cars while trench work significantly improves the drivage indexes. While dump cars usage the following transportation schemes could be applied: dead-end, double dead-end, dead-end with niches (Figure 1).

In case of ring arrival scheme of dump trucks, the width of a bottom of a trench is determined by a formula:

$$B_1 = 2 \cdot (R_a + 0.5a + c), \qquad (4)$$

where R_a – the minimal turning radius a dump truck, m; c – a gap between a dump truck and a trench board, $c = 1 - 3$ m; a – width of a dump truck.

At deadlock arrival scheme of dump trucks for:

$$B_2 = R_a + 0.5a + 0.5l + c. \qquad (5)$$

While application of two-deadlock arrival scheme:

$$B_3 = R_a + 0.5a + 0.5l + 2c. \qquad (6)$$

For mining narrow trenches and blocks cutting a scheme with dump cars loading and turning within niches can be applied. Distance between niches calculates taking into account necessary traffic capacity of a trench and usually accounts 50 – 60 m width.

The trench width in this case determines by a formula:

$$B_4 = R_a + 0.5a + 0.5l + 2c + m, \qquad (7)$$

where m – niche depth, m.

The ultimate trench excavation speed and blocks cutting reaches while deadlock dump trucks arriving scheme is applied. It is caused by the fact, that loading can be carried on with no stops or pauses. And the ultimate mining speed can be reached when single dead-

lock arrival scheme is used. Average trench excavation speed blocks cutting amounts 150 – 180 m/month while automobile transport using, that is 1.5 times more in comparison with railway transport. Excavators efficiency in this case increases up to 20 – 90% in comparison with railway transport.

The trenches excavation and cutting blocks for full section while bottom dump car loading has following advantages:

– rail-laying work absence;
– time consumption reduction while rolling stock change operations;
– possibility to cut off cycle time owing to excavator's turning angle decrease while loading.

At open-casts of Kryvbass the pit's majority are situated in constrained conditions and therefore there is necessary to be guided by technical possibilities of the excavator (digging radius, radius of body rotation) and motor transport (car dimensions, dynamical turning radius).

Efficiency of the LHD equipment considerably depends on the applied driving schemes and dump cars location for loading which should provide: operational safety, maximum excavator usage, minimum time waste for maneuvering, loading and car cars' change, maneuvering whenever possible of the empty cars, instead of the loaded, minimum width of a working platform.

Due to the huge cars applicability variety and high maneuverability there is a large number of vehicle supply circuit. Depending on the excavator maneuvers while operations, the LHDs can run either in the same direction or in counter directions. Depending on LHDs maneuvers the following supply circuits can be formed (Drizhenko 2011, Standard for the organization of Ukraine Ministry of Industrial Policy 2008):

– through, are used while one-way counter direction passing driving scheme on a ledge;
– ring scheme, can be used while oncoming and one-way passing traffic of LHDs and the excavator (Figure 1, a);
– deadlock (pendular), are applied in the constrained conditions (Figure 1, b);
– combined (Drizhenko 2011).

Scheme selection is determined by conditions of loading and movement of vehicles in his career, and the width of the working platform excavator stope, as well as the entrance way to vehicles for loading, which depends on the direction of motion-machines and excavator. Counter access safer and in most cases allows more efficient to set the machine for loading.

Analysis of the passports of mining in the quarry showed that the minimum width of the working area when the annular turn (calculated by (2)) depends only on the radius constructive career dump that is displayed in the Table 1.

At the dead end turn dump width of the working area (see Table 2) consists of the sum of two components: the width of the excavator and stope width of the carriageway. At the same time maneuvering of vehicles on the reversal is limited to the width of the excavator stope (see Figure 1, b).

Stope width of the excavator in all sources of literature (Rzhevskiy 1985) determined by the dependence of the radius of the excavator digging at the level of state. Re-calculated results stope width of the excavator according to the geological conditions of mining are summarized in Table 3.

Table 1. Summary data mining passports at the ring reversal pit dump trucks (Drizhenko 2011).

Type of excavator	Type of dump truck	Stope width of excavation, m	The diameter of the a reversal area, m	The width of the carriageway, m	Digging radius at state, m	Radius unloading, m	The minimum width of the working area, m
EKG – 5A	BelAZ – 7512	17	32	18	9	12.6	41.5
EKG – 6.3	BelAZ – 7512	25	32	18	13.5	16.3	41.5
EKG – 8	BelAZ – 7512	21.6	32	18	12	15.6	41.5
EKG – 10	BelAZ – 7512	22	32	18	12.6	15.4	41.5

Table 2. The minimum width of the working area at a dead end turn vehicles (Drizhenko 2011).

Type of excavator	Type of dump truck	Stope width of excavation, m	The diameter of the a reversal area, m	Width economic highway, m	The width of the carriageway, m	Digging radius at state, m	The minimum width of the working area, m
EKG – 5	BelAZ – 548	17	12	18	12	12.6	23
EKG – 8	BelAZ – 549	22	15	22	16	16.3	27
EKG – 10	BelAZ – 7512	22	15	25	18	15.6	30
EKG – 10	BelAZ – 7512	25	15	25	18	15.4	30

Table 3. The working area parameters (Drizhenko 2011).

Type of excavator	Digging radius at state, m	Digging radius at state, m		
		at the bottom dead end	normal stope	in the cramped conditions
The calculation formula		$2R_s$	$1.5 - 1.7R_s$	$0.5 - 1.0R_s$
EKG – 5	9.0	18.0	13.5 – 15.3	4.5 – 9.0
EKG – 5	11.2	22.4	16.8 – 19.04	5.6 – 11.2
EKG – 6.3	13.5	27.0	20.25 – 22.95	6.75 – 13.5
EKG – 8	12.0	24.0	18.0 – 20.4	6.0 – 12.0
EKG – 10	12.6	25.2	18.9 – 21.42	6.3 – 12.6
EKG – 12.5	14.8	29.6	22.2 – 25.16	7.4 – 14.8
EKG – 15	15.6	31.2	23.4 – 26.52	7.8 – 15.6
EKG – 20	16.0	32.0	24.0 – 27.2	8.0 – 16.0

The results obtained technological possibility of loading equipment graphically shows the histogram in Figure 2 from which seen the maximum and minimum width of stope for different types of excavator.

Taking into account the fact that dump truck maneuvering might be at a deadlock turn on the area of the excavator operation, we are providing the comparative analysis of the maneuvering scheme (Figure 1) with width of a working platform in various conditions. According to the technical characteristics of dump truck BELAZ – 7512 its working radius is 15 m, but constructive is 13 m. Therefore, in con-

strained conditions of the narrow drifts he problem of maneuvering still loud issue, because of smaller working radius of a dump truck.

The maneuvering problem of trucks in operational areas is offered to be decided by means of advanced design of dump trucks (BELAZ – 7512) that significantly improves its technical and operational indicators. Improvement of operational parameters of dump truck could be achieved by the usage of special mechanism of mass center and axle spacing variation (MMCV), which is depicted on the Figure 3 (Krivda 2013).

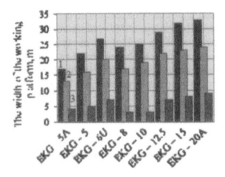

Figure 2. Maximum and minimum width of the stope for different types of excavator: 1 – working out at a dead-end; 2 – ordinary stope; 3 – narrow stope in the constrained conditions.

(a)

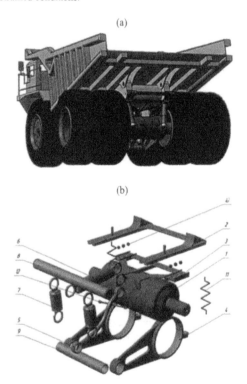

(b)

Figure 3. Mechanism of mass center and axle spacing variation: (a) MMCV on dump truck frame; (b) MMCV structure; 1 – the rear axle with in-built into wheel hub tractive motors; 2 – sledges; 3 – bar; 4 – trailing arm; 5 – bar; 6 – hydraulic cylinders; 7 – link; 8 – rod; 9 – bushing; 10 – sphere; 11 – elastic dissipative unit; 12 – track control arm.

MMCV consists the main element based on a frame – sledges and trailing arms. Sledges allows the longitudinal positioning of a wheel axle with the elastic elements fixed on it, and trailing arms facilitates the torque transmission.

While haulage the mechanical system synchronizes by 3 hydraulic cylinders which allow to change axle spacing of the dump truck.

While usage MMCV system in various BELAZ truck models (Krivda 2013) the turning radius can be reduced up to 10 m that significantly influences on the operational areas width. For the purpose of Kryvbass mines several research have been provided. The provided research concern the usage of improved dump trucks with MMCV system in the context of operational areas size definition while two maneuvering schemes (Figure 4).

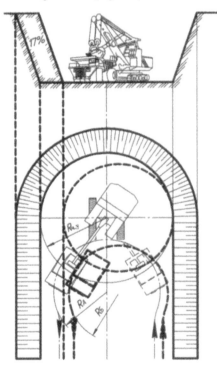

Figure 4. Mine operation according to design calculations and with application of the advanced of dump truck design while usual maneuvering scheme (——) and with MMCV system (- - - - -): R_s – digging radius; R_A – normal turning radius dump; R_B – turn radius of the dump with MMCV.

4 CONCLUSIONS

In the calculating the minimum width of the working area when the circular turn dump truck basic parameters is constructive turning radius, at a dead end turn – machine overall dimensions, and in the constrained conditions – turning radius dump truck. In the calculation of technological schemes of extraction and transportation of rocks by changing the distance between the centers (the base dump truck) decreases the width of jobsite during shunting ring configuration on 6 m, and a dead-end – 3 m, which is displayed on the width of the jobsite.

Thus, we can conclude that the passage of the trenches, split blocks and quarrying in the constrained conditions in the application of a dead-end or ring turn dump truck the minimum width of the working platform can be reduced through the use of advanced designs such as pit dump trucks BelAZ with MMCV.

Implementation of a complex research to rationalize career in space has shown that when the distance between the centers (the base dump truck) dump truck base turning radius is reduced by 23.1%, to reduce the width of the working platform in the constrained conditions at the dead end turn to 7% and 17% the circular.

REFERENCES

Drizhenko, A.Yu. 2011. *Career technology of mining and transportation system*. Abstract of D. Sc. Monograph, Open cast mining. State Higher Educational Establishment "National Mining University".

Krivda, V.V. 2013. *Justification operational and technological parameters of dump trucks*. System technology, Vol. 4: 56 – 62.

Rzhevskiy, V.V. 1985. "*Open cast mining. Part 2*". Production processes: Textbook for Universities, No 4th, ed. rev. and complementary. Russia: Nedra.

Standard for the organization of Ukraine Ministry of Industrial Policy. 2008. 73.020 – 078 – 2: Standards of technological design of mining company with open pit mining of deposits minerals, Part II, Vol. 1. Ukraine: Kyiv.

Determination of parameters effecting firedamp ignition while methane-saturated rock mass fragmentating

V. Golinko, A. Yavorskyi & Ya. Lebedev
National Mining University, Dnipropetrovsk, Ukraine

ABSTRACT: Results of research to improve mining safety in the process of methane-saturated rock mass fragmentation by end organs of winning equipment are demonstrated. Dependences to determine volume of heat accumulated and given-up by frictional thermal particle for time evaluation of the flammable mixture warmth providing consideration of methane-air medium induction properties are obtained.

1 THE PROBLEM DEFINITION

As a rule, rock fragmentation by cutter-loaders is connected with spark formation. In the context of gas-saturated rock mass it may exercise a significant influence on mining safety.

Due to accumulating heat in the process of rock mass fragmentation by winning equipment cutters, increase in contact temperature takes place; its value is determined by operating parameters of a cutter-loader end organ (cutting velocity V_p, cutting force Z etc.).

Increase in velocity V_p factors into contact temperature increase, and spark formation begins. A process of friction sparks formation both while impacting and rock fractioning depends on transformation of some part of kinetic energy of mechanical interaction into heat energy with subsequent exothermic oxidizing and particle heating.

Sparks are fragments of hot rock grains and metal particles which temperature is $500 - 700°C$. Following increase in V_p factors into spark temperature reaching $700 - 1400°C$; spark flow increases too (Shevtsov 2002).

Intensity of spark formation depends on properties of rocks being components of a mass under fragmentation.

2 TOPICALITY

Analysis of methane and coal dust ignitions and explosions happened in stopes and development workings (Shevtsov 2002) means that application of cutter-loaders and preparatory shearers is actually accompanied with danger of frictional sparkling as a result of rock mass fragmentation. Thus, on the 8th of February, 1985 an explosion took place in "Karagailinskaia" mine ("Kiselevskugol" production as-

sociation) when cutter-loader operated in mine workings. Methane explosion was a result of frictional sparking in the process of cutters friction on solid inclusions in coal seam. Methane ignition passed into coal-dust explosion. 630 m of mine workings were ruined.

One of the largest methane explosions happed on the 2nd of December, 1997 in "Zarianovskaia" mine (Kuzbass), where 67 people died. Methane ignition happened in a stope when cutter-loader operated, that is due to frictional sparking.

Similar accidents take place in Ukrainian mines as well. In 2009 a fire happened in 1021 longwall of "Zapadno-Donbasskaia" mine ("DTEK Pavlogradugol" Public Open Joint-stock Company). Expert commission concluded that the fire source was frictional sparking resulted from screw cutter friction on pyrite inclusion in a coal seam. In 2011 a fire happened in 158 longwall of "Stepnaia" mine ("DTEK Pavlogradugol" Public Open Joint-stock Company). The fire source is frictional sparking resulted from interaction of damaged cutter of cutter-loader end organ with pyrite inclusions in coal seam.

Frictional sparking ranks second among all possible ignition sources in coal mines (Shevtsov 2002). Two thirds of all ignitions in stopes of coal mines resulted from frictional sparking. Moreover, 55% of them were caused by friction of cutters on pyrite being inclusion in coal seams; 15% were sandstone cutting. Several ignitions resulted from steel instrument impact on hard shale rock. It is also known methane explosion due to frictional sparking of cutters on limestone; that is, origination of such a thermal source as frictional sparking depends heavily on rock mineralogy.

As for the coal fields of CIS, the number of coal seams with enclosing roof and floor capable of generating explosive sparking under the conditions of friction, is almost 19%; moreover, sandstones are 38% of all enclosing rocks (Shevtsov 2002). As a rule, coal seams have pyrite inclusions in the form of lenses and snake-stones of various dimensions as well as other solid inclusions which distribution in a seam is of random nature.

Up to now, mechanism of explosive mixture ignition by means of frictional spark formation stays to be poorly studied. However, it is known that both intensity and igniting ability of frictional sparks depend on friction mode and impact of two bodies; physical and mechanical, and physicochemical properties of contact surfaces, and a number of other factors. It means that mechanism of mixture ignition by such sparks depends on their nature.

3 RESEARCH ANALYSIS

Not all frictional sparks are able to ignite dust-methane-air mixture. At one time, any sparks resulting from friction or a cutter impact on quartziferous sandstone were considered as the key source of combustion initiation (Tkachuk et al. 2000). However, Blickenderfer, Bergess and Willer determined in laboratory environment that it was very difficult problem to ignite methane-air mixture with the help of such sparking; as a rule, single sparks can not have sufficient combination of duration, temperature, and surface area for ignition. Single sparks are quite enough for such gases as hydrogen, having low ignition temperature, to be ignited. But methane has a feature of ignition delay if thermal source is available; for this reason, dust-methane-air mixture may be ignited by frictional sparks under certain conditions only (Shevtsov 2002), when thermal energy and activity of sparks are sufficient to ignite explosive mixture of methane or dust with air (Falshtynskyi 2012). Such conditions are provided when cutters are frictioned on quartz, sandstone, granite or other coarse hard rocks ($f = 10 - 12$), when hot fine metal particles in the form of $8 - 10$ mm brilliant white trace which flames during 5 ms with $1200 - 1250°C$ stay on friction surface being a source of methane-air mixture ignition. Sparks resulting from friction of cutters on quartz and coarse sandstone (which hardness is $f = 10 - 12$ according to a scale by professor M.M. Protodiakonov) have the highest igniting capability. Friction of cutters on fine sandstone factors into initiation of much fine dust phlegmatizing explosive environment. As a result, no ignition takes place (research shows that if $f < 6$, then methane ignition is impossible, and if $f = 8 - 9$, then ignition probability is 0.16).

Besides, it is specified (Shevtsov 2002) that methane ignition happens in the process of new cutter friction; that is when button insert is worn, and steel holder is not in contact with rock yet. Furthermore, owing to durability, fractioning and impacting result in availability of brilliant white trace at the rock surface; the trace temperature is close to the alloy melting point ($1300 - 1350°C$). Hard alloy wear results in steel holder friction; then, high-temperature trace vanishes, and conditions to cool it down up to the temperature, which cannot ignite methane, are improved.

When cutters are frictioned on pyrite, a cloud of sulphide dust, ignited as a result of frictional sparking, is a source of dust-gas mixture ignition when a cutter trace is no less than 10 cm (Tkachuk et al. 2000, Lozynskyi et al. 2015). According to the data of high-speed photography, firing time for $15 - 20$ dm³ pyritic cloud is $200 - 300$ ms. Hence, in the context of relatively low value of minimum ignition point for pyrite aerosuspension ($350 - 400°C$), according to duration, heated surface, and firing temperature (over $1000°C$), pyrite cloud ignition may be considered as a secondary source with more powerful action than frictional sparking itself.

Dispersivity of frictional particles scattering within the environment with dust-gas mixture, their amount and energy parameters are determined with the help of rate of load application and its value as well as physical and mechanical properties of materials, reacting bodies, and surface coatings.

Thermal mass, accumulated by sparks, and time of the heat maintaining during induction period for methane are basic parameters, affecting the capability of frictional sparks to ignite methane-air mixture formed as a result of gas-saturated rock mass fragmentation.

4 THE MAIN PART

To determine igniting capability of frictional sparks for assessing their hazard within analyzable combustible medium, consider thermal processes, taking place when single heated frictional particle moves within methane-air environment.

According to Newton-Richmann law, thermal flux density q_f from the surface of heated frictional particle, moving toward environment, is proportional to temperature difference of frictional particle t_f and methane-air environment t_{mae}; that is (Reznikov & Reznikov 1990):

$$q_f = \alpha\left(t_f - t_{mae}\right), \text{W/m}^2, \tag{1}$$

where α – heat-transfer factor, W/(m²·°C); t_f – the frictional particle temperature, °C; t_{mae} – the methane-air environment temperature, °C.

Temperature of frictional particles t_f, cut by cutters of end organs of cutter-loaders, may reach $1400°C$ (Tkachuk et al. 2000).

Formula (1) makes it possible to identify quantity of heat q_f, conducting from surface unit to environment in unit time. According to Fouriers law (Reznikov & Reznikov 1990), following flux is conducted away from the frictional particle surface:

$$q_f = -\lambda_{mae} \, grad \, t_f = -\lambda_{mae} \frac{\partial t_f}{\partial n},$$ (2)

where n – a normal to isothermal surface.
Hence,

$$\alpha(t_f - t_{mae}) = -\lambda_{mae} \frac{\partial t_f}{\partial n} \quad \text{or}$$

$$\frac{\partial t_f}{\partial n} = -\frac{\alpha}{\lambda_{mae}}(t_f - t_{mae}).$$ (3)

Expression (3) is mathematical formulation of boundary conditions of type three.

To determine waste of the heat energy Q for moving particle during τ_f based on (1), we obtain:

$$Q = \alpha(t_{fs} - t_{mae}) S \tau_f, \, \text{J},$$ (4)

where t_{fs} – initial temperature of frictional particle, °C; S – surface area of heated particle, m²;

A process of accumulated heat transfer by frictional particle to methane-air environment in the context of forced convection is characterized by criterion equation (Kulinchenko 1990):

$$Nu = 2 + 0,03 \, Pr^{0,33} \, Re^{0,54} + 0,35 \, Pr^{0,36} \, Re^{0,58},$$ (5)

where N_u – Nusselt criterion; P_r – Prandtl number; R_e – Reynolds criterion.

Reynolds criterion R_e characterizes a motion of heated frictional particle in methane-air medium of a stope being determined by (Reznikov & Reznikov 1990):

$$Re = \frac{v_f d_{eq}}{v_{mae}},$$ (6)

where v_f – velocity of heated frictional particle as for methane-air medium, m/s; d_{eq} – typical size (equivalent diameter of the particle), m; v_{mae} – kinematic viscosity of methane-air medium, m²/s.

As for methane-air medium, velocity of heated frictional particle depends on cutting velocity being within 0.5 to 1.5 m/s (Khoreshok et al. 2012).

Frictional particles are hollow steel spheres which maximum diameter is about 0.8 mm, various-size chips, and sandstone cuttings which dimensions are several microns up to 2 – 3 mm. It is supposed in this context that mixture ignition depends on ignition within a point of impact and hot steel particles (Tkachuk et al. 2000).

Prandtl number characterizes capability of heat to emit within the environment. It is determined by the formula:

$$Pr = \frac{v_{mae}}{\chi},$$ (7)

where χ – temperature conductivity coefficient,

$$\chi = \frac{\lambda}{\rho \, c_p},$$ (8)

where λ, ρ, c_p – thermal conductivity, density, and specific heat of methane-air environment.

A value of thermal conductivity coefficient α related to a surface of heated frictional particle can be determined using dimensionless group called Nusselt criterion (Reznikov & Reznikov 1990, Isachenko et al. 1965):

$$Nu = \frac{\alpha d_{eq}}{\lambda_{mae}}.$$ (9)

According to equation (8), heat transfer coefficient is:

$$\alpha = \frac{Nu \lambda_{mae}}{d_{eq}}.$$ (10)

Substituting formula (10) for expression (4), and taking into account (5), (6) and (7), we obtain following expression:

$$Q = \left[2 + 0.03 \left(\frac{v_{mae}}{v_f} \right)^{0.33} \left(\frac{v_f d_{eq}}{v_{mae}} \right)^{0.54} + \right.$$

$$\left. + 0.35 \left(\frac{v_{mae}}{v_f} \right)^{0.36} \left(\frac{v_f d_{eq}}{v_{mae}} \right)^{0.58} \right] \frac{\lambda_{mae}}{d_{eq}} (t_{fs} - t_{mae}) S \tau_f.$$ (11)

After simple transformations, the expression will be:

$$Q = \left[2 + 0.03 d_{eq}^{0.33} \left(\frac{v_f d_{eq}}{v_{mae}} \right)^{0.21} + 0.35 d_{eq}^{0.36} \left(\frac{v_f d_{eq}}{v_{mae}} \right)^{0.22} \right] \times$$

$$\times \frac{\lambda_{mae}}{d_{eq}} (t_{fs} - t_{mae}) S \tau_f.$$ (12)

Without loss of accuracy, expression (12) may be assumed as:

$$Q = \left[2 + 0.38 d_{eq}^{0.345} \left(\frac{v_f d_{eq}}{v_{mae}} \right)^{0.215} \right] \times$$

$$\times \frac{\lambda_{mae}}{d_{eq}} (t_{fs} - t_{mae}) S \tau_f.$$ (13)

Average particle temperature in equation (13) is:

$$\frac{t_{fs} + t_{mae}}{2},$$

hence,

$$Q = \left[2 + 0.38 d_{eq}^{0.345} \left(\frac{v_f d_{eq}}{V_{mae}} \right)^{0.215} \right] \times$$

$$\times \frac{\lambda_{mae}}{d_{eq}} \left(\frac{t_{fs} + t_{mae}}{2} - t_{mae} \right) S\tau_f \text{ or}$$

$$Q = \left[2 + 0.38 d_{eq}^{0.345} \left(\frac{v_f d_{eq}}{V_{mae}} \right)^{0.215} \right] \times$$

$$\times \frac{\lambda_{mae}}{d_{eq}} \frac{(t_{fs} - t_{mae})}{2} S\tau_f. \tag{14}$$

Thermal energy, accumulated by moving frictional particle, may be shown as:

$$E = M c_{pf} t_f, \text{ kJ}, \tag{15}$$

where c_{pf} – specific heat of frictional particle, kJ/(kg·°C) (Kuzmichev 1989); $c_{pf} = 0.460$ kJ/(kg·°C) for steel; M – a frictional particle mass, kg.

$$M = V\rho_f = \frac{\pi d_{eq}^3}{6} \rho_f, \tag{16}$$

where V – a frictional particle volume, m³; ρ_f – a frictional particle density, kg/m³; $\rho_f = 7800$ kg/m³.
Substituting (16) for (15), we obtain:

$$E = \frac{\pi d_{eq}^3}{6} \rho_f c_{pf} t_f, \text{ kJ} \tag{17}$$

To determine τ_f time, required for frictional part to lose accumulated thermal power, equate expressions (14) and (17) to solve the equation related to time of particle τ_f cooling down up to t_{mae} temperature:

$$\tau_f = \frac{1.047 \cdot d_{eq}^4 \rho_f c_{pf} t_f}{S\lambda_{mae} (t_{fs} - t_{mae}) \left[2 + 0.38 d_{eq}^{0.345} \left(\frac{v_f d_{eq}}{V_{mae}} \right)^{0.215} \right]}. \tag{18}$$

Igniting capability of frictional sparks depends on amount of heat accumulated in them.

It is practically impossible to ignite methane-air mixture with the help of single frictional sparks as amount of heat given up to methane-air environment is not sufficient. Besides, methane has delayed ignition; that is, in addition to certain heat amount for methane-air mixture ignition it is required to control the heat during induction period. However, if there

are favourable conditions for a cloud of fine red-hot metal particles and rock formation, then, ignition of explosive methane-air mixture with standard oxygen content is quite real thing (Tkachuk et al. 2000). Methane-air mixture igniting with the help of a cloud of frictional sparks is possible if total amount of heat, given up to methane-air environment, meets the requirements of combustible mixture ignition.

$$Q_{f.cl.} = \sum_{i=1}^{n} \left[2 + 0.38 d_{eq}^{0.345} \left(\frac{v_f d_{eq}}{V_{mae}} \right)^{0.215} \right] \times$$

$$\times \frac{\lambda_{mae}}{d_{eq}} \frac{(t_{fs} - t_{mae})}{2} S\tau_f \geq Q_{ign}. \tag{19}$$

Expression (19) can be used to evaluate ignition hazard of methane-air environment. In this case, following ratio may be assumed as safety criterion K_{saf}:

$$K_{saf} = \frac{Q_{f.cl.}}{Q_{ign}}. \tag{20}$$

The less K_{saf} is, the less is the fair of methane-air environment ignition. If $K_{saf} = 1$, then methane-air environment is critical; if $K_{saf} > 1$, then it is risky; and if $K_{saf} < 1$, then it is non-hazardous.

5 CONCLUSIONS

The temperature of frictional particles formed while end organ of cutter-loader interacting with gas-saturated rock mass is much higher than the temperature of methane-air mixture ignition. In this context, amount heat given up by frictional particle to methane-air environment is not enough to ignite combustible mixture as it is required to control the heat during induction period.

Methane-air mixture ignition is possible when a cloud of frictional sparks is formed providing that total amount of heat, given up to methane-air environment, meets the requirements of combustible mixture igniting.

To evaluate effect of frictional sparking on methane-air mixture ignition in the process of gas-saturated rock mass fragmentation, a safety criterion of methane-air environment is proposed. The criterion makes it possible to identify a hazardous level while gas-saturated rock mass fragmentizing by means of a cutter-loader end organ cutter.

REFERENCES

Falshtynskyy, V., Dychkovskyy, R., Lozynskyy, V., Saik, P. 2012. *New method for justification the technological parameters of coal gasification in the test setting.* Geomechanical Processes During Underground Mining - Proceedings of the School of Underground Mining. The Netherlands: CRC Press/Balkema: 201 – 208.

Golinko, V., Yavorskyi, A., Lebedev, Ya. & Yavorskaya E. 2013. *Effect of frictional sparking on firedump inflammation while gas-saturated rock mass fragmentation.* Naukovyi Visnyk Natsionalnoho Hirnychoho Universytetu, No 6: 31 – 37.

Isachenko, V., Osipova, V. & Sukomel, A. 1965. *Heat transfer.* Moscow: Energia: 421.

Khoreshok, A., Mametiev, L., Tsekhin, A. & Borisov, A. 2012. *Mining machines and equipment for underground mining. Cutters of mining machines.* Study guide. Kemerovo: Kuzbass State Technical University: 288.

Kulinchenko, V. 1990. *Reference book on heat-exchange calculations.* Kiev: Tekhnika: 164.

Kuzmichev, V. 1989. *Laws and formulas of physics.* Kiev: Naukova dumka: 864.

Lozynskyi, V.G., Dychkovskyi, R.O., Falshtynskyi, V.S & Saik, P.B. 2015. *Revisiting possibility to cross the disjunctive geological faults by underground gasifier.* Naukovyi Visnyk Natsionalnoho Hirnychoho Universytetu, No 4:

Reznikov, A. & Reznikov, L. 1990. *Thermal processes in manufacturing systems.* Moscow: Mashinostroenie: 288.

Shevtsov, N. 2002. *Explosion protection of mine workings (course of lectures).* Study guide for institutions of higher education. Revised and enlarged edition 2. Donetsk: Donetsk national Technical University: 280.

Tkachuk, S., Kolosiuk, V. & Ikhno, S. 2000. *Fire and explosion safety of mining equipment.* Kyiv: Osnova: 694.

Theoretical and Practical Solutions of Mineral Resources Mining – Pivnyak, Bondarenko & Kovalevska (eds)
© 2015 Taylor & Francis Group, London, ISBN: 978-1-138-02883-8

The rise in efficiency of solid fuel thermal processing under nonstationary regimes of air delivery into fluidized bed

I. Dyakun
M.S. Poliakov Institute of Geotechnical Mechanics under the
National Academy of Sciences of Ukraine, Dnipropetrovsk, Ukraine

ABSTRACT: The problems of enhancement of fluidized bed technology by organization of pulsing air (gas) delivery into the bed are considered. The model and calculation procedure of rational parameters of pulsing flow as well as coal particle movement and combustion in fluidized bed are examined in the paper. The investigations of regularities of influence of amplitude-frequency characteristics of pulsing air discharge on completeness of combustion of coal particles in the bed are carried out. Rational parameters air flow pulsations are determined that enhance the combustion rate the solid fuel in a pulsating fluidized bed in comparison with the stationary to 50%. The configurations of techniques (pulsators) realizing delivery of pulsing air into fluidized bed and ensuring air flow oscillation shape and pulse ratio control are elaborated.

1 INTRODUCTION

The intensification of technological processes is one of important tasks of science and technology. The basis of increasing of the equipment productivity and reducing of the energy consumption for carrying out of processes is the creation and the implementation of effective technological devices with low energy intensity and materials capacity, high degree of influence on the substance that is processed.

Issues of the intensification are especially topical for the heat-mass-exchange processes, particularly for the thermal processing of the solid fuel. The aging of the mines in the country, the deterioration of the coal mining equipment, the use of imperfect production technologies, the complication of mining and geological production conditions with the transition to the deep horizons, all these factors have led to a significant worsening of the quality of the produced solid fuel. This situation worsens because of an inadequate technology excellence and an unpreparedness of power generating units of the main consumers of the solid fuel (power plants, industrial and mining boiler stations, coking plants) to work on the fuel which quality is much worse than it was calculated. In this regard using of "clean" coal technologies becomes more actual. One of such technologies is a method of the thermal processing of the solid fuel in the fluidized bed (Korchevoy & Maystrenko 2004). According to the solid fuel the method provides to process efficiently and environmentally friendly the fuel of different quality for example low-grade coal and wastes of coal preparation. The technology is also good because of the possibility of the creating of the compact fuel-burning equipment and the automation of the burning process. However despite of the significant advantages of these plants they are not universal and have a number of significant drawbacks. In particular during the processing serious difficulties arise generally because of the aerodynamic instability of the fluidized bed that is shown up in the formation of canals, the bedding of a part of the processed fuel on the grate etc. A complicating factor is often a considerable loss of small fuel particles from the bed.

The possibilities of the increasing of the efficiency of the thermal processing of the solid fuel in the fluidized bed are directly connected with the application of its variations particularly of the pulsating bed.

Using of the active non-stationary modes during the thermal processing of the solid fuel significantly intensifies the process (Bokun 2011) because it allows to increase the contact surface between the particles of the material and the fluidizing agent that influences the burning completeness of the fuel particles and also it allows to reduce unit costs of the thermal energy. In addition it was shown by a number of researchers (Gichev & Adamenko 2008, Nakorchevskiy 2000) that the imposition of the pulsating effects of the fluidizing agent on the processed material contributes to the stabilization of the aerodynamics of the bed, prevent the formation of large gas bubbles and loop-through canals in the bed and the formation of stagnant zones while maintaining the mobility of the particles in the bed and a considerable reduction of the carryover of small fractions of the material from the apparatus.

Thus the pulsating bed that has all the advantages of the fluidized bed allows to stabilize the aerodynamics, to improve the mixing of the material that is processing and to reduce its loss.

2 RESULTS OF RESEARCH

For calculating and designing of the technological devices it is important to know the laws of the impact of the carrier phase fluctuations to the intensity of the heat-mass-exchange processes. However a lot of issues both theoretical and practical that are connected with the pulsation process investigation in the literature are either insufficiently elucidated or characterized by a lack of scientific and methodological foundations.

In addition to the above the object of the study is to investigate the possibility of applying the pulsating bed technology for improving the process of the thermal processing of the solid fuel in the fluidized bed.

The issues of the burning speed and the burn-up time are of great theoretical and practical interest. Their solution provides discovering of the main features of the process and thus makes possible to manage it identifying ways of its intensification and creating the method of engineering calculation.

For using the pulsating modes that are the powerful intensifier for various thermal processes the interaction of the fuel particle with the pulsating gas flow should be an important factor.

An estimation of the trajectory and the speed of the particle motion is needed for the calculating of the fuel particles burn-out, the aerodynamic optimization of the combustion chamber for reducing the removal of unburned particles, for the calculation of the separation of fuel particles and their transportation and for the resolving of other practical tasks of the modern combustion technology. Another important task is to determine the relative velocity of the particles in the flow because it determines the intensity of the heat-mass exchange of particles in the flow and therefore the speed of the burn-out.

Let us look the mathematical model of the coal particle burning with the pulsating supply of the fluidizing agent (gas, air) in details. In this case for the description of the flow course it is used so-called "single-particle" approach. Dynamic characteristics that are the trajectories, the traverse speed of the particles are made from the analysis of the flow power status and are described by the system of equations (Dyakun 2014):

$$\frac{d\delta}{dt} = -\frac{1}{2 \cdot \tau_0 \cdot \delta} \cdot \left[1 + 0.276 \cdot \sqrt{\frac{d_0}{v_g} \cdot |V - u| \cdot \delta} \right], \quad (1)$$

$$\frac{du}{dt} = g \cdot \frac{\rho_g - \rho_p}{2 \cdot \rho_p + \rho_g} + \frac{36 \cdot \mu_g \cdot (V - u)}{d_0^2 \cdot \delta^2 \cdot (2 \cdot \rho_p + \rho_g)} \times$$

$$\times \left(1 + \frac{1}{6} \cdot \left(\frac{d_0 \cdot \delta}{v_g} \cdot |V - u| \right)^{2/3} \right), \quad (2)$$

$$\frac{dx}{dt} = u, \quad (3)$$

where $\delta = d/d_0$ – particle relative diameter; d_0 – particle starting diameter; u – particle traverse speed; x – particle coordinate; t – time; V – gas suspension variable speed, that describes the impulse form; τ_0 – particle burning time without slipping:

$$\tau_0 = \frac{\beta \cdot \rho_p \cdot d_0^2}{8 \cdot D \cdot (c - c_0)}, \quad (4)$$

where D – coefficient of oxygen and air diffusion to the surface of the burning coal particle; c_0, c – concentration of oxygen in the flow and on the surface of the particle accordingly; $\beta = 12/32$ – ratio of molecular weight of carbon and oxygen; ρ_g, ρ_p – density of gas and solid particles, accordingly; μ – viscosity of the gas suspension; v_g – kinematic viscosity of the gas.

The particle size is reduced during the burning process that leads to the reduction of the slip speed. However the power of the viscosity increases the slip speed. Therefore the effectiveness of the slip can be estimated by the calculating of a total time of the burning of the particle of the reducing size in the gas suspension pulsating flow that consists of the fluidizing agent (air) and the metric slug.

As a parameter that characterizes the influence of the pulsations to the particle burning speed with variable parameters of the pulsating flow relative to the constant flow should be used the relative decrease of the burning time:

$$\theta = \frac{\Delta t_g}{t_{g0}} = \frac{t_{g0} - t_g}{t_{g0}}, \quad (5)$$

where t_g – particle burn-up time in the pulsating flow; t_{g0} – particle burn-up time in the constant flow.

It is obvious that the bigger the value of θ is the more effective the appropriate parameters of the pulsating flow can be regarded.

338

The system of equations (1 – 3) can be solved by the Runge-Kutta method with the initial conditions $\delta = 1$, $u = 0$, $x = 0$ with $t = 0$.

The pulsating speed of the fluidizing agent is determined by the following dependence:

$$V = \begin{cases} V_{const} + V_a \cdot sin\left(\dfrac{\pi}{\psi} \cdot t\right) & \text{если} \quad t \leq T \cdot \psi \\ V_{const} & \text{если} \quad T \cdot \psi < t \leq T \end{cases} \quad (6)$$

where $V_{const} = V_m \cdot 20\%$ – the constant component of the well-balanced dependence, m/s; V_m – an average gas speed, m/s; $V_a = (V_m - V_{const}) \cdot \pi/2 \cdot \psi$ – the amplitude of the well-balanced dependence that is determined by the equality of the consumption of the constant and the pulsating flow, m/s; ψ, T –the off-duty ratio (the proportion of time the impulse of air flow in the period pulsation) and the flow period accordingly.

The object of the investigation was the determination of the influence of such main parameters of the pulsating flow as the off-duty ratio ψ and the period T on the reduction of the coal particle burn-up time in comparison with the constant gas suspension flow with the equal combustion during the period.

While calculating the values of the parameters following values were used: for the coal particle – starting diameter $d_0 = 2.5$ mm; density $\rho_p = 1.5 \cdot 10^3$ kg/m³; for the metric slug particles – diameter $d_i = 0.3$ mm; density $\rho_m = 2 \cdot 10^3$ kg/m³; for the fluidizing agent (gas) – density and viscosity are determined with the help of the table for dry air with the temperature $T_g = 900^\circ C$, $\rho_g = 0.301$ kg/m³; $\mu_g = 46.7 \cdot 10^{-6}$ P·s; an average consumed gas speed; $V_m = 5$ m/s; for the parameters of the pulsating flow – the period of pulsations $T = 0.5$ s; 1 s; 2 s; 4 s; the off-duty ratio $\psi = 0.1$; 0.25; 0.5; 1.0; the particle burn-up time in the flow of the constant speed ($\psi = 1$) is 5.435 s.

Figure 1 shows the results of the task.

As shown in Figure 1, at first the solid particle begins to rise gradually relative to the point of its entry into the flow but after some interval the particle mass is reduced so much that the viscous forces become much higher than the force of the gravity and the particle coordinate begins to increase sharply. The same happens with the amplitude of the fluctuation of the particle speed.

Furthermore it was investigated the influence of parameters of the gas suspension and the forms of the air pulsation speed on the increasing of the time when the coal particle is in the bed as compared with the constant flow of the gas suspension during the equal period: with the superposition of the poly-frequency component (Figure 2) and in the mode of the reverse-pulsating air supply (Figure 3).

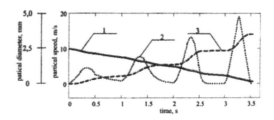

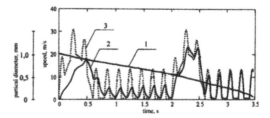

Figure 1. Parameters of the process of the coal particle burning with $T = 1$ s, $\psi = 0.5$: 1 – particle diameter; 2 – particle speed; 3 – particle track.

Figure 2. Parametres of the process in the poly-frequency pulsating flow: 1 – particle diameter; 2 – particle speed; 3 – fluidizing agent (gas, air) speed.

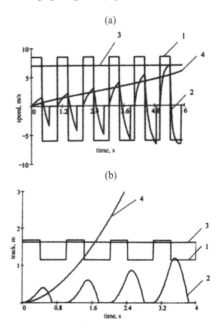

Figure 3. Parametres of the process: (a) particle speed; (b) particle track; 1, 2 – with the reverse-pulsating air supply into the bed; 3, 4 – with the constant air supply into the bed.

Analysis of the results of the research of the model with pulsating movement of the fluidized bed showed that the rational parameters of pulsations are the period $T = 2 - 10$ seconds, and the off-duty ratio $\psi = 0.2 - 0.4$.

339

The investigations allowed specialists of IGTM NAS of Ukraine to develop the pulsators constructions of the following types: the pulsator with the trilobite impeller (Bulat et al. 2003), the pulsator with additional impellers (Bulat et al. 2004), the trilobite pulsator with six identical holes on the body (Bulat et al. 2005) the pulsator with a slotted rotor (Bulat et al. 2005), the disc pulsator (Bulat et al. 2005) that implement the various impulse forms shown on Figure 4.

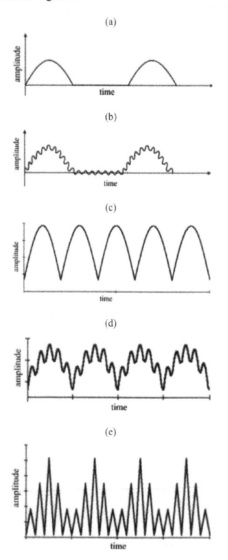

Figure 4. The pulsations the air flow output device (a) the pulsator with the trilobite impeller, patent UA 57662A; (b) the pulsator with additional impellers, patentUA 66646A; (c) the trilobite pulsator with six identical holes on the body, patent UA 4290; (d) the pulsator with a slotted rotor, patent UA 4855; (e) the disc pulsator, patentUA 7783.

Pulsators constructions that were developed provide the form of airflow fluctuations and the off-duty ratio of the flow, i.e. the relation of the duration of one pulse to the duration of one period of the pulsation within 0.2 – 0.5, and a possible air flow speed going through the pulsator that can be 10 – 12 m/s.

3 CONCLUSIONS

1. On the basis of the analysis of the results of the model of pulsating fluidized bed research it is shown that the rational parameters of pulsations are the period $T = 2 - 10$ seconds, and the pulse off-duty ratio $\psi = 0.2 - 0.4$. Also it is shown that in the whole range of investigated parameters of the gas suspension bed the pulsation influence intensifies the processes of thermal processing of the solid fuel in the fluidized bed till 50%.

2. Using of the pulsed mode with the rational amplitude and the frequency characteristics of pulsating air supply ensures the increasing of the efficiency of the thermal processing of the solid fuel in the furnace with the fluidized bed by the completeness of coal particles burning-out in the bed and reducing of the mechanical underburning.

3. For the first time the pulsators constructions of the following types were developed: the pulsator with the trilobite impeller, the pulsator with additional impellers, the trilobite pulsator with six identical holes on the body, the pulsator with a slotted rotor, the disc pulsator that providing the regulation of the form airflow fluctuations and the off-duty ratio of the flow within 0.2 – 0.5.

REFERENCES

Bokun, I.A. 2011. *Gasification of low-grade fuels pulsed layer*. Materials of the Republican Scientific and Practical Conference "Prospects for Energy Development in the XXI Century". Minsk, May 12 – 14.

Bulat, A.F., Chemeris, I.F. & Voziyanov, V.S. 2003. *Device for creating pulsating air flows in fluidized bed furnace*. Kiev: State Register of Patents of Ukraine, No 57662A.

Bulat, A.F., Chemeris, I.F., Voziyanov, V.S. & Slobodyannikova, I.L. 2004. *Device for creating pulsating air flows in fluidized bed furnace*. Kiev: State Register of Patents of Ukraine, No 66646A.

Bulat, A.F., Chemeris, I.F., Voziyanov, V.S. & Slobodyannikova, I.L. 2005. *Device for creating pulsating air flows in fluidized bed furnace*. Kiev: State Register of Patents of Ukraine, No 4290.

Bulat, A.F., Chemeris, I.F., Voziyanov, V.S. & Slobodyannikova, I.L. 2005. *Device for creating pulsating air flows in fluidized bed furnace*. Kiev: State Register of Patents of Ukraine, No 4855.

Bulat, A.F., Chemeris, I.F., Voziyanov, V.S. & Slobod-yannikova, I.L. 2005. *Device for creating pulsating air flows in fluidized bed furnace*. Kiev: State Register of Patents of Ukraine, No 7783.

Gichev, Yu.A. & Adamenko, D.S. 2008. *Analysis of the impact-acoustic pulsating combustion characteristics on the boiler and the development of technical solutions*". Metallurgical heating engineer. Dnepropetrovsk: The Scientific publication National Metallurgical Academy of Ukraine: 69 – 80.

Dyakun, I.L. 2014. *Increase of efficiency energy coal processing*. Kyiv: Naukova dumka: 126.

Korchevoy, Yu.P., Maystrenko, A.Yu. & Topal, A.I. 2004. *Ecologically clean coal energytechnologies*. Kyiv: Naukova dumka: 186.

Nakorchevskiy, A.I. 2000. *Dynamics pulsating monodisperse mixture*. Theoretical Foundations of Chemical Engineering, Vol. 34, No 1: 11 – 15.

Electrothermal stimulation of chemical reactions in mixture of calcite and silicon powders

V. Soboliev & N. Bilan
National Mining University, Dnipropetrovsk, Ukraine

ABSTRACT: The results of experimental investigation of phase transformation in ground to powder calcite with silicon additives during simultaneous heating and passing a low-amperage electrical current are given. Observed phenomenon of the abrupt increase of electrical conductivity in $CaCO_3 + Si$-specimens is conditioned by spontaneous forming of carbon phase – graphite, which has mainly electron type of conductivity. Treatment (heating and passing current through mechanically activated specimens of $CaCO_3 + Si$ and $CaCO_3 +$ quartz) decreases significantly the temperature of phase transformation, initiates formation of solid conducting phase of carbon. It is suggested that these processes are characteristic for rocks in the period of tectonic activity. Received data can be useful when interpreting the way of graphite forming in nature, physical and chemical formation conditions of its deposits and the formation of diamonds in graphite schists.

1 INTRODUCTION

Phase transformations in calcite specimens under the integrated action of temperature and the electric field are an important source of additional information about the role of mechanical effect in activation of physicochemical processes in rocks. Besides phase transformations are the object under the study of ways of formation of elemental carbon, necessary, for example, to form diamond, other carbon phases and their deposits. One of regularities of location of deposits is their association with the most deformed fault zones. It is estimated that during activation deep fault releases energy more than 10^5 J/year per kilometer of its length. The main fraction of this energy is transformed into the energy of rock deformation and movement of blocks along the faults, and another one goes into heat and electric energy. In such areas electric fields with the intensity of about 20 V/cm are arisen, herein the ionic conductivity was detected even when the field intensity is about 1 V/cm (Khairetdinov 1980). The electrical energy generated during deformations changes valence states of elements, helps break bonds in minerals (Nyussik & Komov 1981), activates chemical processes, stimulates the phase transformations and forming of new minerals, etc.

It is known that by ordinary heating in a closed system calcite (which can be represented as the derivative of sodium chloride) $Ca^{2+}[CO_3]^{2-}$ decomposes to CaO and CO_2, i.e. all oxygen of the anionic complex $[CO_3]^{2-}$ is used completely for oxidation of calcium and carbon. It is obvious that deviation from zero of oxygen balance to a negative value (lack of oxygen) should lead to the formation of free carbon. One of the minerals that can select a portion of oxygen during the decomposition of calcite may be silicon, which reacts easily with oxygen at a temperature above 800 K.

One method to induce phase transformation in siderite is simultaneous heating and the effect of weak electric current, wherein transitions occur at lower temperatures than by heating only (Sobolev et al. 1998). That work of reference presents the results of experimental studies of the phenomenon of the abrupt increase in electrical conductivity of siderite and its derivatives, due to the spontaneous formation of electrically conductive solids upon reaching a certain critical temperature – the temperature jump. It is assumed that graphite (phase with predominantly electronic conductivity) can be formed in the dissociation products of calcite.

The purpose of the research is to study the character changes in electrical resistance of mixture of powder $CaCO_3 + Si$ by heating and simultaneous passing of electric current, as well as to study the way of forming of new mineral phases.

2 MATERIALS AND METHODS

Crystals of pure Iceland spar and synthetic silicon were used for studies. These substances were ground separately in a mortar to fractions with an average size of about 20 microns. The obtained powders of $CaCO_3$ and Si were mixed in weight ratio (%) 55:45, 70:30 and 85:15. Specimens were shaped from the powder mixture directly in a ceramic container by pressing under a pressure of $4 \cdot 10^5$ Pa (Figure 1). The weight of specimens depending on the ratio of components was 1.41, 1.43 and 1.47 g respectively, a volume was 0.6 cm^3. The porosity of specimens did not exceed $10-12\%$. Glue "Avtostik" (silicate with kaolin filler) was applied on the ends of container and the hollow electrodes in order to hermetic sealing the specimen during thermal treatment. Surfaces coated by glue were joined and dried for an hour at 350 K until the action of high temperatures.

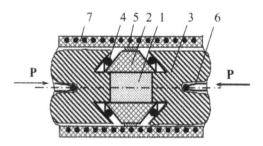

Figure 1. The lay-out diagram of the ground specimen $CaCO_3 + Si$ in the experimental device: 1 – a specimen; 2 – a ceramic container with an effective volume of 0.95 cm^3; 3 – a hollow steel electrode; 4 – a sealing ring; 5 – a steel ring; 6 – a thermocouple; 7 – an electrical furnace.

Treatment of specimens was carried out in a device with an electrical furnace, described in detail in (Sobolev et al. 1998). Heating of the specimens was operated up to 1000 K at an average rate of 14 K/min; electric field intensity was 250 V/cm, the maximum intensity of pass current – 0.3 A. The value of the potentials difference and the electric current was supplied by source of power B5-50. The values of electrical resistance in the process of specimens heating were recorded continuously using a universal voltmeter B7-46/1.

3 RESULTS AND DISCUSSIONS

Paramagnetic centers (PMC) are formed as a result of silicon destruction. They are reliably recorded by the EPR (electron paramagnetic resonance) method. Experience has shown that not all valence-unsaturated atoms are registered, but only part of them preserved after stabilization of the surface. Apparently, the valence-unsaturated atoms of silicon surface are centers of irreversible absorption of atmospheric oxygen. We recorded the value of surface concentration of centers – $2 \cdot 10^{17}$ m^{-2} (the number of atoms on the silicon surface is about 10^{18} m^{-2}), i.e., ~20% of the surface atoms. In

chemical reactions radical $\equiv Si^\bullet$ acts as a reducing center. Atoms with two broken bonds (silylene centers $= Si:$) react actively with different molecules.

The results of electron microscopic studies show that the diameter of the silicon active center formed at the end of dislocation averages about $3 \cdot 10^{-6}$ m, and the active surface area around the dislocation is ~$7 \cdot 10^{-12}$ m^{-2}. Given these parameters, the calculated value of the total active surface of silicon particles is 0.6, 0.4 and 0.2 m^2/cm^3 at the weight content of silicon in the specimen 45, 30 and 15% respectively. It is assumed that heating and simultaneous passing of electric current through the specimen of silicon with additives of calcite stimulates phase transitions leading to graphite formation.

The experimental dependence of the electrical resistance of $CaCO_3$ and mixtures of $CaCO_3 + Si$ on the heating temperature at a constant value of the electric field is shown in Figure 2. Parameters of the electrothermal treatment of the crystal of Iceland spar are electric field intensity of 250 V/cm and the maximum heating temperature of 1220 K (curve 1, Figure 2); the electrical resistance at the maximum temperature is decreased to $5 \cdot 10^4$ Ohms. The sharp drop in electrical resistance from $8 \cdot 10^5$ to $5.4 \cdot 10^5$ Ohms observed at temperatures 970…990 K. A monotonic increase in the electrical conductivity was recorded during ordinary heating and continuously measuring the electrical resistance (electric field intensity does not exceed 7 V/cm). Calcite and silicon was identified on the diffractograms of $CaCO_3 + Si$ specimens treated by the heat field only.

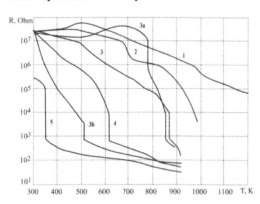

Figure 2. Dependence of the electrical resistance of $CaCO_3$ without impurities and $CaCO_3$ with additives of silicon by heating and simultaneous action of electric current and potentials difference: 1 – $CaCO_3$ crystal (Iceland spar); 2 – ground to powder Iceland spar; 3 – a mixture of powders of Iceland spar and silicon (85:15 wt%); 3a – a mixture of powders of Iceland spar and quartz (85:15 wt%); 3b – a mixture of ground jointly to powders Iceland spar and silicon (85:15 wt%); 4 – a mixture of powders of Iceland spar and silicon (70:30 wt%); 5 – the same mixture (55:45 wt%).

344

Ground to powder $CaCO_3$ (Figure 2, curve 2) behaves quite differently: the beginning of a sharp drop in the electrical resistance of 10^7 Ohms to $1.2 \cdot 10^6$ Ohm recorded at temperatures $670 - 720$ K, the minimum resistance $2 \cdot 10^3$ Ohm – at a temperature of 980 K. It should be noted that temperature of active decomposition while the ordinary heating of a similar specimen corresponds to 1075 K, i.e. 395 K above. Thus, additional effect of the electric field reduces greatly the temperature of the beginning of active chemical processes. The new phase of $Ca(OH)_2$ could occur during in the air abrasion of crystals.

In comparison to experiment 1 data of experiment 2 suggest that structural defects formed in the crystal lattice of $CaCO_3$ during grinding, considerably facilitate the subsequent chemical decomposition. It can be assumed that the abrupt drop in electrical resistance in the temperature range $970 - 990$ K is due to the forming of a new phase, which could not be fixed after the end of the experiment. Increased reactivity is caused by structural defects of calcite in the volume of crystal and by active states on its surface, which in nature may be formed during periodic transtension – compression and shearing strain.

There are jumps in conductivity in all cases (Figure 2, curves $3 - 5$) as a result of electrothermal treatment the mixture of powders $CaCO_3 + Si$ and $CaCO_3 + SiO_2$ activated during the pregrinding. There is a strong decrease in temperature jump from reducing the weight fraction of $CaCO_3$ in the specimen. Table 1 shows the parameters of some treatments and experimental data. It is seen that in the case of additional abrasion (mechanical activation) of a mixture of powders $CaCO_3 + Si$ (experiment 3b) jump of electric resistance is at lower temperature more than 300 K than in experiment 3. This effect may indicate that even at stage of mechanical impact plastic flow of mineral substance during a joint grinding of crystals of Iceland spar and silicon initiates a solid-state chemical reaction (decomposition of $CaCO_3$ and oxidation of silicon). In this case, formation of active surfaces in minerals results in the acceleration of chemical reactions between them.

Table 1. Parameter of treated powder mixtures $CaCO_3 + Si$.

Exp. No	Conditions of specimen preparation	The mineral composition in the specimen, wt%	Electrical resistance at 900 K, Ohms	Temperature of conductivity increase, K	The value of electrical resistance, Ohm	
					beginning of the drop	the end of the drop
1	Crystal $CaCO_3$	100	$5 \cdot 10^6$	$970 - 990$	$8 \cdot 10^5$	$5.4 \cdot 10^5$
2	Ground to powder $CaCO_3$	100	$1.2 \cdot 10^5$	$670 - 720$	10^7	$1.2 \cdot 10^6$
3	Separate grinding of $CaCO_3$ and Si in a mortar	85 and 15	204	jump 855	$5.6 \cdot 10^3$	$5 \cdot 10^2$
3a	Joint grinding of $CaCO_3$ and SiO_2 in a mortar	85 and 15	198	first jump – 770 second jump – 850	$1.2 \cdot 10^7$ $4 \cdot 10^3$	$6 \cdot 10^5$ $4 \cdot 10^2$
3b	Additional mechanical activation of a mixture of powders $CaCO_3 + Si$	85 and 15	45	jump 515	$2 \cdot 10^3$	$3 \cdot 10^2$
4	Separate grinding of $CaCO_3$ and Si in a mortar	70 and 30	47	jump 610	$8 \cdot 10^3$	$4.5 \cdot 10^2$
5	Separate grinding of $CaCO_3$ and Si in a mortar	55 and 45	32	350	$6 \cdot 10^3$	$4 \cdot 10^2$

Paramagnetic centers are located in a thin surface layer during the quartz destruction. The spectrum of the PMC corresponds to the radical type $\equiv Si^{\bullet}$ and $\equiv SiO^{\bullet}$ and their concentration is close to $0.5 \cdot 10^{17}$ m^{-2}. It is also known that there are active states of silicon atoms with two broken bonds $=Si:$ together with the radicals $\equiv Si^{\bullet}$ and $\equiv SiO^{\bullet}$. Deformed bonds, which are carriers of residual internal stresses, stimulate the chemical activity. According to their chemical properties deformed bonds are similar to the radical pairs $\equiv Si...OSi \equiv$. The concentration of active centers $\equiv Si^{\bullet}$ and $=Si:$ in quartz and silicon are closely spaced.

Reduction of calcite content in a specimen from 100 to 1% by ordinary heating leads to a decrease of decomposition temperature from 1275 to 1075 K. Abrupt temperature drop in electrical resistance in $CaCO_3 + Si$ at electrothermal treatment would be approaching this temperature range of calcite decomposition, however, in experiments with the specimens of $CaCO_3$ content 85% and 55%, the temperature jump is considerably reduced to 860 K (experiment 3) and to 350 K (experiment 5).

In experiment 3 the mixtures of $CaCO_3$ and SiO_2 powders were used. Here an abrupt drop in electrical resistance was observed twice: at temperatures of 770 and 850 K, and there is a sharp decrease in electrical resistance in the temperature range 810...850 K, which decreases abruptly at 850 K. The temperature of the second jump, marked on the curve 3a coincides with the beginning of the transition temperature α-quarts to β-quartz. Perhaps this is just a coincidence, but it is possible that the decrease in resistance due to the reaction of polymorphism ($\alpha \rightarrow \beta$ quartz). When ordinary heating the structure of unstrained quartz at $\alpha \rightarrow \beta$ transitions remains almost unchanged, there is only a change in the angle of chemical bond and a slight displacement of atoms. The effects of mechanochemical activation of quartz due to violation of long-range order in the arrangement of tetrahedrons can cause an abrupt decrease in the resistance in a mixture of $CaCO_3$ and SiO_2 powders. Virtually amorphization of quartz accelerates solid-state processes of its interaction with calcite, but the issue remains open.

Experimental data indicate that there is the oxidation of CO to CO_2 on the silicon surface at the time of mechanical activation in an atmosphere of carbon monoxide. Oxygen chemisorption is carried out on the short-lived active centers (lifetime 10^{-4} s); their concentration is close to the concentration of surface atoms. Using ground (activated) silicon as an additive to $CaCO_3$ in experiments with heating and passing an electric current makes impossible the oxidation reaction $2CO + O_2 \rightarrow 2CO_2$ on the quartz surface. This is probably due to the fact that several physical parameters affecting the half-closed system consisting of $CaCO_3$ and SiO_2, the main parameter that determines the direction of chemical reactions is the electric field. In this case, the formation at least one new phase of low electrical resistance is energetically favorable in the system. For example, graphite may be this phase.

Calcite decomposes to CaO and CO_2 at ordinary heating in the temperature range of 1075...1275 K. Additional effect of the electric field leads to the breaking of chemical bonds, what is more favorable energetically to the ions as product of breakdown:

$$CaCO_3 \rightarrow Ca^{2+} + [CO_3]^{2-}. \qquad (1)$$

Ion-carbonate $[CO_3]^{2-}$, which is formed at the initial stage of $CaCO_3$ decomposition, in the field of active center on the surface of silicon (or Ca^{2+} ions) is resolved to atomic carbon and oxygen. One part of the oxygen oxidizes the calcium ions (Ca^{2+}), and another is captured by active centers, resulting in the stabilization on the quartz surface, for example, $\equiv SiO_3$ radical. A similar way is considered by authors using physical and mathematical modeling of elementary chemical acts occurring on the surface of diamond particles. Upon reaching the critical concentration of atomic carbon in the intergranular space system due to fluctuations of the energy and density (and possibly resonance phenomena) new phase of condensed carbon is formed spontaneously, which is correspond to the kinetic theory (Folmer 1986).

The way of graphite nucleation directly related to the fluctuation phenomena occurring in the intergranular space, supersaturated by atomic carbon. Reactions "carbon dioxide $\rightarrow$ graphite" occur spontaneously when the supersaturation of carbon at certain temperature (here the temperature of conductivity jump). It is assumed in the statistical model that a collision of CO or CO_2 molecules to the surface with electric charges, the part of molecules comes into the field of an ion. If the energy of the molecule is large enough to approach for a critical distance to the ion, then its dissociation occurs, and in this process, there may be several ways:

$$I.\ CO \rightarrow C + O,$$
$$II.\ CO \rightarrow C^+ + O^-,$$
$$III.\ CO \rightarrow C^- + O^+,$$
$$IV.\ CO \rightarrow C^+ + O + e,$$
$$V.\ CO \rightarrow C + O^+ + e,\ etc.$$

The minimum energy that must energize the dissociation of CO molecule corresponds to the first reaction. Therefore, the calculation takes into account only the first reaction. As a further simplification of the real physical situation was considered that the surface density of ions is constant, and that atom of oxygen combines with calcium ion as a result of the dissociation of carbon oxide.

Calcium oxides are formed on the surface of calcite particles, and carbon oxides (mobile components) filled the intergranular space. Thus, the intergranular space between the surfaces is a zone of chemical reactions. With increasing temperature, the concentration of carbon oxides in the reaction zone increases, thus increasing the frequency of collisions and the probability of convergence of these molecules with point charges (e.g., calcium ions, broken bonds, etc.) on the distance at which the strong influence of the field ion leads to dissociation of CO and CO_2 molecules. Quantum-mechanical calculations suggests that at temperatures from 640 to

980 K dissociation of CO and CO_2 molecules in the field of the monovalent ion occurs at distances of $\sim(10...15) \cdot 10^{-10}$ m, and in the field of the bivalent ion $-(20...30) \cdot 10^{-10}$ m.

Virtually in all cases the temperature of the beginning of mass dissociation of carbon oxides in the zone of chemical reaction is in the temperature range of the beginning of calcite decomposition at ordinary heating. At some point the critical (maximum) concentration of atomic carbon occurs in the reaction zone, where is spontaneous nucleation of graphite (the new phase with the electronic conductivity). According to known Le Chatelier's principle, it can be assumed that in a case where the main influence on the system has an electric field, at least one of the phases, characterized by higher electrical conductivity as compared with any of the initial phases is formed. Formed graphite crystals give to the system a high electrical conductivity. At the steady-set maximum value of current flow the increase of temperature to the priority of its impact in comparison with other treatment parameters will lead to the forming of new phases. This forming is an endothermic reaction.

Characteristics of some mineral phase formed as a result of electrothermal treatment of mixture of calcite powder with silicon are presented on Table 2. According to the X-ray phase analysis the high electrical conductivity in the mixture is due to the formed hexagonal graphite; the highest content of graphite in studied specimens is less than 3%. The results of the high-voltage electron diffraction diagnostics of the treated specimens show that depending on the minimum size of the formed film (thickness exceeds 1.5 nm) graphite is characterized by a two-dimensional ordering. When the particle size is less than 1.5 nm, films are identified as amorphous. In some experiments, the graphite particles are presented by three-dimensional units.

The formation of $Ca(OH)_2$ is most likely to occur directly at the grinding step of Iceland spar crystals in air. Unidentified phases with interplanar distances 0.315 nm (I = 2...5); 0.236 nm (I = 5...60); 0.168 nm (I = 2...50); 0.162 nm (I = 2...10), and others remain in many specimens after the electrothermal treatment of ground $CaCO_3$.

Table 2. New mineral phases in the treated mixture of $CaCO_3$ + Si powders.

Exp. No	The mineral composition of the specimen, wt%	The maximum temperature of heating, K	New phases
1	$CaCO_3$ crystal (100)	1220	—
2	$CaCO_3$ powder (100)	980	$Ca(OH)_2$
3	Mixture of $CaCO_3$ (85) and Si (15) powders, separate grinding	920	Very weak lines of graphite; $Ca(OH)_2$
3a	Mixture of $CaCO_3$ (85) and SiO_2 (15) powders, joint grinding	920	Very weak lines of graphite; $Ca(OH)_2$
3b	Mixture of $CaCO_3$ (85) and Si (15) powders, additional joint grinding	920	Weak lines of α-SiO_2; $CaSiO_3$
4	Mixture of $CaCO_3$ (70) and Si (30) powders, separate grinding	920	Very weak lines of α-SiO_2; weak lines of graphite
5	Mixture of $CaCO_3$ (55) and Si (45) powders, separate grinding	920	weak lines of graphite; $Ca_3O(SiO_4)$

During the long-term passing of electric current with maximum intensity after an abrupt change in electric conductive, new phases decompose and form high-temperature secondary phases, probably due to the high local temperatures as a result of Joule heating. If after the jump the density of pass current was reduced by a factor of 10^2, the conductive phases in many specimens not only preserved, but also grew up.

4 CONCLUSIONS

It was established experimentally that in the ground to powder calcite when heating and passing an electric current the beginning of active chemical pro-

cesses corresponds to a lower temperature than upon the ordinary heating. Increased reactivity of calcite is conditioned on structural defects of various kinds in the crystal volume and the active states on its surface. After the preliminary grinding, jumps in conductivity in $CaCO_3$ + Si specimens are observed in all cases. There is a strong decrease in temperature jump as of an increase in the silicon content in the mixture.

Reducing the temperature of the beginning of chemical reactions may be due to a decrease in the energy barrier as the action of surface charges localized in areas of dislocations exit points.

The feature of the electrothermal treatment of calcite + silicon, calcite + quartz specimens is that increasing the silicon content decreases the temperature of the abrupt drop in the electrical resistance.

Structural defects formed in the crystal lattice upon grinding, greatly facilitate subsequent electrothermal decomposition. Additional mechanical activation of a mixture of powders $CaCO_3 + Si$ leads to the jump of electrical resistance at a temperature of more than 300 K lower than that in the experiments with a separate grinding. This indicates that the plastic flow of mineral substance at the joint grinding of crystals initiates solid-phase chemical reactions directly in the step of mechanical action. In this case, the formation of active surfaces in minerals and complementary stored energy accelerates reactions between minerals and reduces the energy barrier for origin of a new phase.

Thus, treatment – passing of current when heated the mechanically activated specimens of silicon and calcite, calcite and quartz – declines significantly the temperature of phase transformations, initiates formation of a new solid phase with electron conductivity. It can be assumed that the physical and chemical processes stimulated by mechanical processing with subsequent electrothermal influence take place in rocks in the periods of tectonic activity. Received data may be useful in studies of the carbon source and the ways of diamond and graphite formation in nature, physical and chemical formation conditions of their deposits and the factors of diamond formation in graphite schists.

REFERENCES

Folmer, M. 1986. *Kinetics of new phase formation.* Moscow: Nauka: 208.
Khairetdinov, I.A. 1980. *Introduction to electrogeochemistry.* Moscow: Nauka: 255.
Nyussik, Ya.M. & Komov, I.L. 1981. *Electrochemistry in geology.* Leningrad: Nauka: 240.
Sobolev, V.V., Orlinskaya, O.V. & Chernay, A.V. 1998. *The phenomenon of abrupt increase in the electrical conductivity of mineral carbonates class under the action of temperature.* Scientific Proceedings of National mining academy of Ukraine, No 2: 215 – 224.

Theoretical and Practical Solutions of Mineral Resources Mining – Pivnyak, Bondarenko & Kovalevska (eds)
© 2015 Taylor & Francis Group, London, ISBN: 978-1-138-02883-8

Substantiation of coal slurry thickening rate during dewatering on vibrating screen with multi-slope working surface

D. Polulyakh, O. Polulyakh & A. Tarnovskyi
National Mining University, Dnipropetrovsk, Ukraine

I. Eremejev
GP "Ukrniiugleobogashchenie", Dnipropetrovsk, Ukraine

ABSTRACT: Equation of motion including nonlinearity due to the dry friction force has been developed based on the analysis of the intraluminal stresses in material being subject to harmonic force. Harmonic linearization of the sgn function resulted in reduction of motion equation to the linear form. The solution of model equation allowed identification of evolutionary and oscillatory components of material vibration thickening process (without any change in its mass and at constant resistance). The obtained dependence of layer thickening rate on the material rheological characteristics includes the working surface vibration parameters in an explicit form. This study result could be useful at the screen operational conditions determination.

1 INTRODUCTION

During the coal slurry dewatering, solid phase volumetric concentration is increased due to water removal through the screen mesh. Thus coal slurry structural-mechanical properties are substantially changed. Effective slurry viscosity is subject to the most significant alteration due to its nonlinear dependence on volumetric concentration of solid particles.

Viscosity, plasticity and elasticity are the basic rheological characteristics of concentrated slurry.

2 MATHEMATICAL DEFINITION OF SLURRY PROPERTIES

Slurry elastic characteristics are resulted from the air bubbles presence. However, during dewatering on the screen sieve accompanied by the layer vibration thickening air bubbles are intensively released through the free layer surface and through its bottom boundary, i.e. through the screen. Thus, slurry layer on the screen may be assumed as a viscoplastic rheological body.

High-concentration slurry is characterized by the spatial structure resistive to stress not exceeding certain τ_c, value designated as shear stress or yield strength (Wilkinson 1964). In case that material stress exceeds the yield strength, its structure collapses, and shearing flow occurs, of which rate is proportional to excess shear velocity (Wilkinson 1964), i.e., material behaves as a Newtonian fluid at the shear stress $\tau - \tau_c$.

Deformation of viscoplastic material results in occurrence of stress:

$$\tau = \eta \dot{e} + \tau_c , \qquad (1)$$

where η – viscosity coefficient, and $\dot{e}$ – rate of deformation.

Let us consider a viscous material behavior on the harmonically oscillating horizontal sieve surface under the unseparated conditions.

Let us distinguish unit cross-section bar in the material layer with axis coinciding with the pressure load normal component on the screen side. The bar height is equal to the material layer thickness h and bar mass: $m = \rho h$, where ρ – the material density.

Material layer is subject to harmonic exciting force: $F \cdot \cos \omega t$, where F and ω – the exciting force amplitude and frequency, and t – the time. In addition, during the oscillation layer is subject to inertia forces due to the material density ρ, viscous friction forces and dry friction forces determining the material plastic deformation.

The screen surface vibration normal component results in layer deformation and facilitates its dewatering, while the tangent component provides layer vibratory displacement. Therefore, when taking account of vibration normal component only, we assume that layer inertia force will be: $m\ddot{y} = \rho h \ddot{y}$, where y – layer vertical displacement, $y = \varepsilon h$.

Viscous friction force in the material layer is equal to $\eta \dot{y}$, where $\dot{y} = h\dot{e}$ – the layer vertical displacement velocity.

Dry friction force $\vec{R}$ is constant in magnitude and is directed oppositely to displacement velocity

(Panovko 1976) $\vec{R} = -R\dot{y}/|\dot{y}|$, where R – constant depending on the friction coefficient and cohesive force. If force $\vec{R}$ is resulted from the stress exerted upon the lateral surface of the square bar, then $R = 4h\tau_c$. Assuming the permanent plastic deformation during the layer vibration thickening, we will represent dry friction force as $R(sgn\ \dot{y}+1)/2$ (Palmov 1976), where:

$$sgn\ \dot{y} = \begin{cases} 1\ during\ \dot{y} > 0 \\ -1\ during\ \dot{y} \leq 0 \end{cases}. \qquad (2)$$

Thus, layer plastic deformation occurs, if screen displacement velocity is directed upward, and the stress in layer exceeds ultimate shear stress τ_c. If screen velocity is directed downward, then dry friction force is equal to zero, and layer moves as an inertia body.

Figure 1 shows dynamic computational diagram of the inertia viscoplastic body being subject to harmonic exciting force.

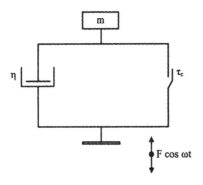

Figure 1. Dynamic computational diagram of the inertia viscoplastic material on the vibrating surface.

The equation for layer motion on the vibrating surface takes the following form based on the assumed dynamic design:

$$m\ddot{y} + \eta\dot{y} + 2h\tau_c(sgn\ \dot{y}+1) = F\ cos\ \varpi t. \qquad (3)$$

Layer height varies slowly at the final dewatering process stage, so that $dh/dt << 1$

The equation (3) includes nonlinearity due to the function $sgn\ \dot{y}$. Such type equations are solved with the use of methods of step-by-step integrating (Blekhman & Dzhanelidze 1964) or methods based on the motion equation linearization.

In order to reduce equation (3) to the linear form we will use the power balance method (Panovko 1971) whereby nonlinear dry friction force can be replaced with the energy equivalent linear force $b_o\dot{y}$ for which coefficient b_o is determined based on

the condition of equality of the works done by both forces over the oscillation period, i.e.:

$$\int_0^T \tau_c\ sgn\ \dot{y}\cdot\dot{y}\cdot dt = \int_0^T b_o\dot{y}^2 dt. \qquad (4)$$

We can assume for a first approximation that oscillatory process is harmonic in the steady conditions.

One can see from the equation (3) that nonlinear friction force presents at the positive velocity values only, and function sgn $\dot{y}$ takes on values 0 and 1. Therefore expected oscillatory law for the layer displacement velocity will be:

$$\dot{y} = -a\varpi\ sin\ \varpi t, \qquad (5)$$

where a – the layer oscillation amplitude.

Insertion of expression (5) into (4) results as follows:

$$\int_0^T b_o\dot{y}^2 dt = b_o a^2\varpi^2 \int_0^{2\pi} sin^2\ \psi d\psi = b_o\pi a^2\varpi, \qquad (6)$$

where $\psi = \omega t$.

Let us calculate an integral on the left part of (4) for the nonlinear resistance force:

$$\int_0^T \tau_c\dot{y}\ sgn\ \dot{y}\cdot dt = -\tau_c a \int_0^{2\pi} sin\psi\ sgn\dot{y}\cdot d\psi = 4\tau_c a. \qquad (7)$$

Equating results of calculations of (6) and (7) allows determination of equivalent friction coefficient:

$$b_o = \frac{4}{\pi a\varpi}\tau_c. \qquad (8)$$

Once the coefficient b_o – determined, the problem reduces to investigation of the equivalent linear dynamic system where the dry friction force is as follows:

$$2h\tau_c(sgn\ \dot{y}+1) \approx 2h\tau_c(b\dot{y}+1), \qquad (9)$$

where $b = 4/\pi a\omega$.

In case of the system with the non-elastic resistance, oscillatory motions lag behind the exciting force. Therefore, if oscillatory motions follow the low $y = a\cdot cos\cdot\omega\cdot t$ as a first approximation, then law of variation of exciting force can be on written as $F\cdot cos\cdot(\omega t +\varphi)$, where φ is phase angle. Then the motion of the linearized dynamic system will be described by equation:

$$m\ddot{y} + (\eta + 2bh\tau_c)\dot{y} + 2h\tau_c = F\ cos(\varpi t + \varphi). \qquad (10)$$

At the moments of maximum system departure from the equilibrium point when cos·ωt = 1:

$$F\ cos\ \varphi = 2h\tau_c - ma\varpi^2, \qquad (11)$$

and at the moment of equilibrium point passage cos ωt = 0 and

$$F\ sin\ \varphi = (\eta + 2bh\tau_c)a\varpi - 2h\tau_c. \qquad (12)$$

After the last equalities squaring and addition, we will obtain following expression associating the exciting force value and system oscillation amplitude:

$$F^2 = (2h\tau_c - ma\varpi^2)^2 + (Ba\varpi - 2h\tau_c)^2, \qquad (13)$$

where: $B = \eta + 2bh\tau_c$.

Thus oscillation phase is as follows:

$$\varphi = arccos\frac{2h\tau_c - ma\varpi^2}{F}. \qquad (14)$$

Equation (10) is linear with respect to $\dot{y}$, and it's solution is known (Bronstein & Semendyaev 1986):

$$\dot{y} = \ell^{-\frac{B}{m}t}.$$

$$\left\{ \int \left[\frac{F}{m}cos(\varpi t + \varphi) - \frac{2h\tau_c}{m} \right] \ell^{\frac{B}{m}t} dt + C_o \right\}, \qquad (15)$$

where C_o – the initial value.

Solving the integral (15) at the initial condition:

$$\dot{y} = 0 \ given \ t = 0, \qquad (16)$$

we obtain:

$$\dot{y} = \frac{2h\tau_c}{B}\left(\ell^{-\frac{B}{m}t} - 1 \right) + \frac{Fm}{B^2 + m^2\varpi^2} \times$$

$$\times \left[\frac{B}{m}cos(\varpi t + \phi) + \varpi \ sin(\varpi t + \phi) - \right.$$

$$\left. - \left(\frac{B}{m}cos\phi + \varpi \ sin\phi \right)\ell^{-\frac{B}{m}t} \right]. \qquad (17)$$

The solution (15) describes the layer velocity variation with account for the transient process in the initial time period. The expression (15) can be rewritten for the steady-state process $t \rightarrow \infty$ as follows:

$$\dot{y} = \frac{Fm}{B^2 + m^2\varpi^2} \times$$

$$\times \left[\frac{B}{m}cos(\varpi t + \varphi) + \varpi \ sin(\varpi t + \varphi) \right] - \frac{2h\tau_c}{B}. \qquad (18)$$

Integration of expression (17) at the initial condition $y = h_o$ at $t = 0$ results in determination of the vibrating layer displacement:

$$y = h_o - \frac{2h\tau_c}{B}\left[t - \frac{m}{B}\left(1 - \ell^{-\frac{B}{m}t} \right) \right] +$$

$$+ \frac{Fm}{B^2 + m^2\varpi^2} \times \left[\frac{B}{m\varpi}sin(\varpi t + \phi) \right] -$$

$$- cos(\varpi t + \phi) + \left(cos\phi + \frac{m\varpi}{B}sin\phi \right)\ell^{-\frac{B}{m}t} -$$

$$- \frac{B^2 + m^2\varpi^2}{m\varpi B}sin\phi. \qquad (19)$$

In steady-state process $(t \rightarrow \infty)$:

$$y = h_o - \frac{2h\tau_c}{B}t + \frac{Fm}{B^2 + m^2\varpi^2} \times \left[\frac{B}{m\varpi} \times \right.$$

$$\left. \times sin(\varpi t + \phi) - cos(\varpi t + \phi) - \frac{B^2 + m^2\varpi^2}{m\varpi B}sin\phi \right]. \qquad (20)$$

The solution (20) describes material layer surface displacement at the oscillatory shear flow and includes evolutionary and oscillatory components:

$$y_1 = h_o - \frac{2h\tau_c}{B}t - \frac{F}{\varpi B}sin\varphi, \qquad (21)$$

$$y_2 = \frac{Fm}{B^2 + m^2\varpi^2} \times$$

$$\times \left[\frac{B}{m\varpi}sin(\varpi t + \varphi) - cos(\varpi t + \varphi) \right]. \qquad (22)$$

The layer vibration thickening process takes place without its mass variation at the constant resistance. It is reasonable that the layer thickening is only possible when the stress in material exceeds shear stress τ_c.

Evolutionary component of thickening rate under the steady-state conditions $dy/dt = -2h\tau_c/B$. Here the layer height h is present as a parameter.

Concentrated slurry viscosity coefficient depends on the vibration parameters as follows (Rudenko 1971):

$$\eta = \eta_o + \frac{k}{a\varpi^3}, \qquad (23)$$

where k – the constant coefficient; η_o – the residual viscosity coefficient due to the oscillatory thixotropic destruction of dispersion medium.

The rate of layer thickening on the vibrating screen:

$$\frac{dy}{dt} \approx -\frac{2h\tau_c}{\eta_o + \frac{k}{a\varpi^3} + \frac{8h\tau_c}{\pi a\varpi}}. \qquad (24)$$

Parameters τ_c, η_o and k included in this formula are subject to the experimental determination.

Figure 2 shows plots of layer thickening rate against working surface vibration amplitude and frequency. Exemplary material parameters for the layer of $h = 0.1$ m are assumed as follows (Rudenko 1971, Fomenko & Kondratenko 1977): $\tau_c = 10$ N/m^2, $\eta_o = 10^3$ M·s/m^2, $k = 10^6$ N/m·s^2.

351

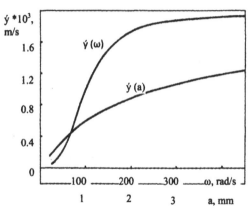

ẏ *10³, m/s

Figure 2. Viscoplastic material vibration thickening rate vs. working surface vibration amplitude and frequency: $\dot{y}(v)$ during a = 1 mm; $\dot{y}(a)$ during v = 75 rad/s.

3 CONCLUSIONS

These data enable length calculation for screen dewatering surface devices, where the first device is designed for the material dewatering with the prevailing use of hydraulic principles, while the second one is designed for dewatering due to the laws of inertial mixtures mechanics.

REFERENCES

Blekhman, I.I. & Dzhanelidze, G.Y. 1964. *Oscillatory displacement*. M.: Nauka.
Bronstein, I.N. & Semendyaev, K.A. 1986. *Reference book on mathematics for engineers and students of technical colleges*. M.: Nauka.
Fomenko, T.G. & Kondratenko, A.F. 1977. *Flotation tailings and properties thereof*. M: Nedra.
Palmov, V.A. 1976. *Elastic-plastic body oscillation*. M: Nauka.
Panovko, Y.G. 1971. *Introduction to the mechanical oscillation theory*. M: Nauka.
Panovko, Y.G. 1976. *Fundamentals of the applied oscillation and impact theory*. L: Mashinostroenie.
Rudenko, I.F. 1971. *Product forming with the use of vibroforming equipment*. M.: Stroyizdat.
Wilkinsonm, U.L. 1964. *Non-Newtonian fluids*. M: Mir.

Improving the assessment effectiveness methods of innovative industrial Leading Ukrainian Companies activity

D. Kabachenko

National Mining University, Dnipropetrovsk, Ukraine

ANNOTATION: Considered are the main assessment methods of effectiveness of innovative projects, highlighted are their strength and weaknesses. Separated are two groups of methods, which are mostly used in the process of assessment of economical effectiveness of projects: discount (static) and discounted (dynamic) methods. Considered are features, advantages and terms of using the method of real options for assessment of effectiveness of innovative projects, analyzed kinds of real options and methods of assessment of their value. Presented is an example of using of the method of real options for assessment of effectiveness of an innovative project, implemented on the acting enterprise, identified is the growth of the capital of the enterprise as a result of realization of the project in view of a real option.

1 INTRODUCTION

A problem of acceleration of economic development on the innovative basis is exclusively actual in Ukraine. In conditions of market relations, the innovative activity is the most progressive form of enterprise whose goal is – maximal satisfaction of need of society.

Taking into account strategic tasks, nationwide effect from creation and using of innovations, their effect on the common development of the country's economic, is important during choosing foreground directions of the innovational activity, which are supported by the state and are expressed by acting legal acts, which include Laws of Ukraine "About innovational activity", "About foreground directions of innovative activity in Ukraine", "About foreground directions of development of science and technic" etc. Assessment of effectiveness of innovational projects on the basis of science and technical, economic, social, ecological factors, is an integral background of their realization.

Today, innovative activity is a fundamental rod in strengthening of an enterprise on the market in conditions of rough competition, which must provide stable economic development, increasing of competitiveness of produced by an enterprise production and provided services.

The paradigm of innovational development is reflected in the theory of intellectual technology of F. Hayek, the theory of innovative economic and corporate society of P. Druker, technical and economic conception of J. Schumpeter, researches of M. Porter, B. Santo and other scientists.

2 MAIN PART

At the moment, not solved is a problem of creation of methods of management of innovative activity, which are adequate to the market conditions of economic management and optimization of mechanisms of complex assessment of effectiveness of innovative projects at Ukrainian enterprises.

A goal of this research is theoretical foundation and development of methodical recommendations concerning improvement of methods of assessment of innovative projects at an industrial enterprise.

It is necessary to highlight two groups of methods, which are mostly used in the process of assessment of economic effectiveness of projects, – these are discount (static) methods and discounted (dynamic) methods. (Volkov & Denisenko 2007).

We will consider the main methods of assessment of effectiveness of innovational projects and also will highlight their strength and weaknesses.

Discounted methods belong to the most famous and widely used methods of assessment of investments before the conception of discounting of cash flows was universally accepted. Thus, methods of assessment of effectiveness of a project, which are based on discounting assessments (without discounting) is a payback period of investments (PP) and the coefficient of effectiveness of investments (ARR).

The payback period of investments (PP) is a period from the beginning of realization until the recoupment of investments. The gist of the method is in calculation of the period, by which, the earnings are compared with the amount of initial investments:

$$PP = \frac{K_o}{CF_{cr}}, \tag{1}$$

where PP – term of recoupment of investments (years); K_o – initial investments; CF_{cr} – average annual value of net earnings from realization of an investment project.

This index allows to calculate, during which period will be received a net income, which, by value, will be equal to the value of advance capital. By comparing of projects, it is necessary to accept a project with the shortest term of recoupment. However, a simple term of recoupment of investment has a range of significant defects. It doesn't take into account value of money in time and doesn't accept discounting, and it means, that it doesn't take into account segregation of incomes by years, and thus, it will be used only for projects in a short-term period.

The method of calculation of the coefficient of effectiveness of investments (ARR) – provides discounting of cash flows and is equal to the ratio of the annual average estimated net income to the annual average volume of investments. Annual net income is calculated as a difference between the cash flow of this year and the sum of annual depreciation deductions, associated with this project.

Annual average net income is calculated as a part of segregation of the difference between incomes and costs, associated with this project for the foreseeable term of investments. The income in this case must be decreased to the amount of deductions to the budget.

If the amortization is charged linearly, the value of investments will decrease evenly with time flow. An average value of investments wherein will be equal to a half of the sum of initial investment costs, increased to a half of the liquidation value. If it is foreseen, that, after ending of the term of realization of the analyzed project, all capital costs will be written off, then the average value of investments will be in accordance with a half of the sum of initial investment costs.

In view of these conditions, a simple rate of profitability is modified into an index, called a coefficient of effectiveness of investments ARR:

$$ARR = \frac{PN}{\frac{1}{2}(IC - RV)}, \tag{2}$$

where PN – annual average net income; IC – initial investments; RV –liquidation (residual) value of a project.

Implemented is the project, whose discount profitability is higher. Wherein occurs its comparison with the market interest rate, in order to estimate – how better of worst result do these investments present in comparison with other investments of the capital.

A common defect of discount methods of assessment of effectiveness of innovation projects if that they don't take into account a range of factors, such as inflation, segregation of cash flows in time and risks. Furthermore, these methods are based on discount assessment of income, which are conditional, and depend on the opposite politics of a company. That's why these methods are not always applicable for assessment of effectiveness of innovation projects. However, they are important for rejection of obviously ineffective projects. They differ by simplicity in calculations. As it is stated in some sources, discount methods are used for assessment of projects of small enterprises, which don't implement large and long-term investment projects.

Dynamic methods of assessment of effectiveness of innovative projects ate based on the model of discounted cash flow (DCF). Discounting is adduction of value of a cash flow (CF) of a project into different time for a specific period of time. It is possible to do by means of a rate of discounting (RD). The meaning of DCF is that money lose their purchasing power, namely, money in the future period are cheaper, the in the real period.

Assessment by means of discount methods of assessment is more accurate, thus considered are different kind of inflation, changes of the interest rate, norms of profitability etc. These indexes include methods of the profitability index (PI), net discounted value (net present income NPV), the internal norm (rate) of profitability (IRR), the discounted payback period (DPP).

Net present value (NPV) is a sum of cash flows, connected with an innovational project, coerced by the factor of time to the moment of assessment and is calculated by the following formula:

$$NPV = -I_0 + \frac{CF_1}{1+RD} + \frac{CF_2}{(1+RD)^2} +$$
$$+ \frac{CF_j}{(1+RD)^j} + \frac{CF_n}{(1+RD)^n}, \tag{3}$$

where CF_j – cash flow, coerced to j-moment (interval) of time; I_0 – value of initial investments; RD – rate of discount; n – term of realization of a project.

As a rate of discount (RD) is used a necessary level of profitability, which is determined in view of a risk. If $NPV > 0$, this means that:
– effect of the project is a positive value;
– the project has a higher profitability, than a rate of discount (RD), which is needed on the market of capitals from investments with such level of risk.

If $NPV < 0$, this means, that the adjusted norm of profit is not provided and the project is unprofitable. If $NPV = 0$ the project just recoup the costs but doesn't provide income.

However, correct using of the NPV-method is possible only in case of compliance with a range of terms:

– a volume of cash flows in frames of an investment project must be estimated for the whole plan period and refer to specific time intervals;

– cash flows in frames of an investment project must be considered isolated from other industrial activity of an enterprise, namely, to characterize only payments and earnings, directly related to realization of the project;

– the principle of discounting, which is used at calculation of the net coerced income, from economic site of view, provides a possibility of unlimited implication and investment of financial means by the considered rate of discount (which doesn't't comply with reality);

– using of the method for comparing of effectiveness of several projects provides using of a single for all projects rate of discount and a single time interval (which is often determined as the largest term of realization from the available).

At calculation of NPV, usually, is used a permanent rate of discount, however, depending on circumstances (for instance, change of level of interest rates is expected) a rate of discount can differentiate by years. If during calculations are used different rates of discounting, the project which is acceptable at the permanent rate of discounting, can became unacceptable.

Being an absolute index, the most important property of NPV – is a property of additivity, namely, it is possible to summarize NPV of different projects. Other most important properties of this criteria include most realistic assumptions about a rate of reinvestment of entrants of means (in NPV method, it is not clearly provided that the means, which come from realization of a project, are reinvested by the adjusted norm of discount RD).

Using of NPV criteria is theoretically grounded and, in general, it is considered as the most correct measurer of effectiveness of investments. At the same time it has a range of defects:

– NPV is not an absolutely correct criteria at choosing between a project with bigger real net value and a continuous period of recoupment and a project with smaller real net value and a short period of recoupment. In other words, the NPV method doesn't allow to judge about a break-even point and a stock of financial strength of a project;

– the NPV calculation method takes into account only obvious cash flows and doesn't take into account effect of changes of the value of real estate, fixed assets and raw materials on the real net value of a project;

– using of NPV is complicated by difficulties of prediction of a rate of discount (a weighted value of a capital) and/or a rate of bank interest;

– a rate of discount is usually accepted unchangeable for the whole horizon of discount. However, in future, it can change in connection with changing of economic conditions.

A profitability index of a project (PI) displays what amount of units of the current volume of the cash flow refers to a unit of expected initial costs. The PI index can be calculated by the following formula:

$$PI = \sum_{n=1}^{n} \frac{CF_n}{(1+RD)^n} / I_o \; . \tag{4}$$

If the profitability index of a project exceeds one, the current cash flow is bigger than the initial investments and this, in its turn, provides a positive value of NPV, and thus, we accept the project.

If $PI < 1$, the project is refused.

In case if $PI = 1$, investment don't earn, – the project is neither profitable nor unprofitable.

So, the PI index shows, how effective are investments; namely this criteria is the best in case, when it is needed to put in order some independent innovational projects for forming an optimal portfolio.

This method is useful at choosing the best project from alternative variants.

One of the main defects of the profitable index is its sensibility to the project's scale. This index doesn't always provide a univocal assessment of effectiveness of financial investments and a project with the highest index may not comply with a project with a high value of NPV. In connection with the fact, that this index doesn't correctly estimate mutually exclusive innovational projects, it is used as an addition to the NPV method. The internal rate of return of investments (IRR) – is a rate of discounting (RD), at which $NPV = 0$.

Thus, IRR is determined by the formula:

$$IRR = RD_1 + \frac{NPV_1}{NPV_1 - NPV_2} \cdot (RD_2 - RD_1) \; , \tag{5}$$

The algorithm for solving is the next:

– we take two rates of discounting RD_1 and RD_2 (randomly selected), at that, $RD_1 < RD_2$;

– then, using value of the rates, we calculate NPV_1 and NPV_2.

The meaning of the calculation of this coefficient at assessment of effectiveness is that IRR shows the most permissible relative level of costs, which can be connected with an innovational project. For instance, if sources of the means of the project were loans of a commercial bank, then the value of IRR shows the upper border of the permissible level of bank interest rate.

If *IRR* < *R*, namely, profitability of a project is smaller than the rate of income on invested capital, needed by investors of the project, the project is refused.

Unlike NPV, which measures the absolute value of income, IRR shows an income for a unit of the invested capital.

Advantages of the method are:
– comparison of IRR with the "value" of a loan provides a possibility to determine advisability of attraction of loaned means;
– it allows to find a limit value of a rate of discount, which segregates investments to acceptable and unprofitable, and also, to rank projects by a level of profitability at identity of the main initial parameters;
– it serves as an indicator of a level of risk of a project: the more IRR exceeds the accepted level of recoupment, the bigger is a stock of strength of the project.

Defects of the IRR criteria:
– a possibility of existence of several meanings of IRR. In the general case, if we analyze a single or some independent projects with an ordinary cash flow, when after the initial costs come positive inflows of cash flows, using of the IRR criteria always leads to the same results as NPV. But, in case of alternation of the inflow of cash with outflows, for one project may exist several meaning of IRR;
– IRR is very sensible to the structure of the flow of payments and doesn't always allow to definitely estimate mutually exclusive projects;
– it is non-additive, it characterizes only a specific project, it is difficult to use it for assessment of an investment portfolio;
– IRR admits, that a rate of discount will be permanent during the whole term of realization of investments.

A range of economists at calculation of the term of the payback period of investment (PP) recommend to consider a temporal aspect. In this case, the cash flows are discounted into the calculation by the WACC index (the weighted value of the capital). So, determined is a moment, when discounted cash flows of incomes are compared with discounted cash flows of costs.

For the calculation of the discounted payback period of investments (DPP) is used the following formula:

$$DPP = min\ n,\ \text{wherein} \sum_{n=1}^{n} CF_n \frac{1}{(1+RD)^n} \geq IC\ .$$

The advantage of the method of discounting of payback period is that it takes into account a tempo-

rary value of money and a possibility of reinvestment of profits. A defect of such method is that it doesn't take into account the cash flows after the end of the payback period of the project.

3 RESULTS

Using of traditional approaches, based on discounting of the cash flows, doesn't take into account managerial flexibility at analysis of an innovational project. Identification of effectiveness of investments by them provides that after the project is started, its estimated parameters (value of the cash flow, size of costs etc.) are unchangeable, displaying these aftermath just due to variation of rates of discounting adequately to risks, which are estimated by the expert way. Wherein, their ability to perform as factors in conditions of flexibility of management, is ignored.

According to the author, it is possible to grade these defects and to estimate the value of flexibility of management at decision making in answer to unexpected market changes, with help of using of the method of real options.

Generally, by an option we mean a right to purchase or sell the base asset in the specific volume for a fixed price (a price of implementation) to a date of the end of an option or until this date (Ziyatlinov 2010). Depending on the situation, the owner of the option can either realize the option or not. In the real options, unlike the financial ones, the base asset is not a financial instrument (stock, bond), but real assets of a company (investments, resources, production capacity etc.).

The most famous works dedicated to the option approach include the works of F. Black and M. Sholes, A. Damodaran, R. Merton, D. Mooney, A. Dixit and R. Pindyck, N. Kulatilak, D. Ingersoll and S. Ross, L. Trigeorgis and a range of other foreign investigators. In Ukraine this issue is less learned, but its actuality is displayed in scientific works of I. Gavryschkov, G. Drahan, N. Petrov, O. Chechil and other.

At using of the method of discounted payback period of investments, the analyst tries to avoid the uncertainty at the moment of analysis of the investment project. As a result appears one ore several scenarios of the future march. However, the scenarios analysis doesn't solve the main problem – a static character, because in the result is taken the averaged variant, which shows, how will be solved the uncertainty in accordance with the laid prerequisites.

The method of real options provides a fundamentally other approach. The uncertainty stays and the management with flow of time makes optimal decisions in accordance with the situation, which changes. In other words, real options provides a possibility to change and to make optimal decisions in the future, in accordance with the new information, which comes. Wherein, the possibilities to change and to make decisions in the future are quantitatively estimated at the moment of analysis. This method considers a risk not only as a negative factor, but also as a possibility to receive an additional income and (or) to minimize loses in case of the adverse developments (Malyshev & Podoynichyn 2013).

Many investment projects include different kinds of options. For instance, a company considers a possibility of purchasing an option for development of a source of oil on the specific parcel. But, at the moment, the value of extraction of oil in this point will not be recouped by the incomes from its realization, that's why such project seems to be unprofitable. On the other side, taking into account the fact, that the prices for oil on the world market tend to serious oscillations, it is not difficult to assume, that in one or two years they will seriously grow and development of an oil source will provide serious profits. In such case, an option for development of oil provides a company with a right, but not an obligation, to realize a project, if conditions for this will be favorable (Bruslanova 2004).

Depending on time of implementation (realization of the right to purchase/sale), options are divided into the "American" and "European". The owner of the American option can use his right to purchase or sell the securities at any time before the end of the established term, and the owner of the European option can use the option only on one determined day.

Using of the method of real options is perspective at strategic investment projecting also because managers of the projects can more operatively take into account appearance of new information. For instance, a considered project at the moment is economically ineffective but, in case of favorable change of the market's conjuncture, it can became advisable. Then, making a decision not to invest into a project, which is based on the following information, is lost a possibility to receive income in future. Namely, in the method of real options, a risk is considered not as circumstance, which must be avoided, but, vice versa, as a potential source of increasing the value.

Assessment of real options provides important additional information and thus contributes validity of made decisions. Along with that, assessment of real options must be considered not as an alternative, but rather, as an addition to the traditional NPV (net current value) method of assessment of effec-

tiveness of project, specifying the marks, specified by discounting of the cash flows (Buzak 2010).

Wide injection of the method in management of projects will allow to the management to react more operatively to change of the market's conjuncture, to pay attention to different scenarios of development and not to loose new possibilities.

Using of the method of real options until the assessment of investment projects is advisable when the following conditions are satisfied:

– a result of a project tend to the high level of the uncertainty;

– company's management is able to make flexible managing decisions at appearance of new data of the project;

– a financial result of a project significantly depends on the decisions, made by managers.

There are two main types of options: CALL (a right to purchase for a fixed price) and PUT (a right to sell for a fixed price) (Gilbert 2004).

Depending on the fact, by which conditions an option gains a worth for a company, allocated are the following species of real options:

– an option for refusing from investments: this option is a right to sell or close a project. In case, when results from realization of a project don't comply with expectations, the manager can assess a liquidation value of a project.

Thus, the price of implementation of a project will be a liquidation value of the project. When the current value of assets will be lower that a liquidation value, then the option will be implemented:

– an option for extension of a project: this option is a CALL option for increasing of values of production and investments in case of favorable marsh. This situation can appear when the current demand on the production exceeds the estimated and the company's management decides to increase the issuance. The price of implementation of an option is equal to the following expenses for creation of additional production capacities;

– an option for reduction of investments – is a possibility of the staged reduction of business in case of a pessimistic scenario;

– an option for postponement of investments: this option is used in cases of uncertainty of demand on production. It is necessary to note, that in some cases, a postponement of investments can negatively affect the companies, which have some technological advantages, thus these investments will be postponed to the later term (Hull 2013).

For assessment of value of real options are used two main methods: Black-Scholes Option Pricing Model and binomial model.

The Black-Scholes Option Pricing Model is simple as in exposition and in using, however, it has a range of restrictions:

– the assessed asset must be liquid (availability of the market is required for the assessed asset);

– changeability of the price of an asset stays the same (namely, sharp price hikes don't occur);

– the option can't be realized until the term of its implementation (the European option).

Calculation of the value of the real option is made by the formula of Black-Scholes, developed for assessment of the financial options of the CALL type (McMillan 2003):

$$C = S \cdot N(d_1) - K \cdot e^{-rt} N(d_2), \qquad (7)$$

$$d_1 = \frac{\ln\left(\dfrac{S}{K}\right) + \left(r + \dfrac{\sigma^2}{2}\right) \cdot t}{\sigma\sqrt{t}}, \qquad (8)$$

$$d_2 = d_1 - \sigma\sqrt{t}, \qquad (9)$$

where: C – a price for an option to purchase; S – a current price of a base asset (a discounted value of cash flows from realization of the investment possibility, which is received by the company as a result of implementation of an investment project); K – a price of implementation of an option (capital investments); r – a risk-free interest rate, which complies with the life term of an option; σ – volatility (a mean-square deviation of the coerced cash flows of a base asset from the average meaning); $N(d_1)$ – a function of density of segregation of a standard normal random value; e – a number, which is a basis of a natural logarithm (roundish meaning 2.71828).

From the analysis of this formula it follows that the price of a real option growing if:

– a discounted value of cash flows increases;

– costs for implementation of the project decrease;

– time until the end of the term of realization of the project increases;

– a risk increases.

Wherein, the bigger effect on increasing of the value of the option makes a discounted value of estimated cash flows. So, for increasing the investment attractiveness of the project, it is advisable for the companies to pay attention to increasing of incomes, but not on decreasing of costs.

Technique of building of the binomial model is more complicated, that Black-Scholes method, but it allows to receive more precise results, when there are several sources of uncertainty or a big amount of variants of accepting a decision. Calculation of the value of an option is a movement through a "tree of

decisions", where, in each point, managers try to make the best decisions. As a result, the cash flows, which appear as an aftermath of the future decisions, are brought to the discounted value.

In practice. The main difficulties of using the binominal model, are related to definition of meanings of relative increasing and decreasing of value of business in each period, and also, probabilities of positive and negative variants of marsh.

Coefficients of increasing and decreasing are calculated by the following formulas (Shreve 2005):

$$u = e^{\sigma\sqrt{t}}, \qquad (10)$$

and

$$d = \frac{1}{u}. \qquad (11)$$

where u – a coefficient of increasing of cash flows; d – a coefficient of decreasing of cash flows; e – a number, which is a basis of a natural logarithm (roundish meaning 2.71828); σ – a mean-square deviation of cash flows of a base asset from the average meaning; t – a term of implementation of an option.

After that is founded a stage of a continuous probability (p), which is determined by a formula:

$$p = \frac{e^r - d}{u - d}, \qquad (12)$$

where r – a risk-free interest rate on the financial market (% of annual interests).

At using of the binominal model, at first is built a tree of the price of a base asset and then – a tree of the value of the option, which, eventually, fold. At building of the tree of the value of a base asset, it is possible to take into account risks of a project in scenarios meanings, but not in the discounting rate.

We will build a five-step grid of changing of the meaning of a base asset (Figure 1).

A grid of changing of the value of an asset shows allowed in view of the current uncertainty meanings of an asset during the next 5 years. A price of the CALL option is determined by the following formula:

$$C = \frac{p - d}{p} - \frac{u - E}{u - d}, \qquad (13)$$

$$E = \frac{K}{S}, \qquad (14)$$

where S – a price of a base asset; K – a price of implementation of an option; p – a stage of a risk-free probability.

We will consider using of the method of real options for assessment of effectiveness of an innovational project, implemented at PJS "Dniprotelekom".

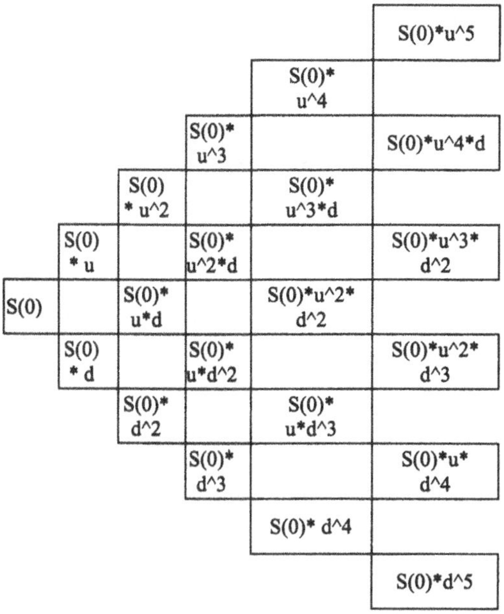

By making a decision to organize production of a heat-insulation material "ISOL" (a base project), PJS "Dniprotelekom" plans to enter the European market. For this, the enterprise must develop a new investment project.

This project starts after three years from the moment of start of a bas project. For implementation of the project, the enterprise plans to increase its productive capacities, the total amount of investments (K) is estimated in 1 million $.

The project is calculated for 5 years, the foreseen sales volumes by each year will make 70% from the similar meaning of the base project. In the new project, as in the base one, $WACC = 11.9\%$, however, in connection with different time frames in the new project will be changed meanings of coefficients of discounting, wherein, investments into the project must also be discounted.

For simplification of calculations, all other parameters of the new project can be compared with the base project. Besides that, a price for realization of 1 m³ of the heat-insulation material will be складе 140$.

For assessment of effectiveness of the project, we will build a table, which will display the cash flows by periods (Table 1).

Figure 1. A five-step grid of changing of the values of a base asset.

Grid values:
- $S(0)$
- $S(0)*u$, $S(0)*d$
- $S(0)*u^2$, $S(0)*u*d$, $S(0)*d^2$
- $S(0)*u^3$, $S(0)*u^2*d$, $S(0)*u*d^2$, $S(0)*d^3$
- $S(0)*u^4$, $S(0)*u^3*d$, $S(0)*u^2*d^2$, $S(0)*u*d^3$, $S(0)*d^4$
- $S(0)*u^5$, $S(0)*u^4*d$, $S(0)*u^3*d^2$, $S(0)*u^2*d^3$, $S(0)*u*d^4$, $S(0)*d^5$

Table 1. Dynamic of the cash flows.

Index	0	1	2	3	4	5
Investments, $	1000000					
Volume of realization of production, m³		8540	9074	9608	10141	10675
Revenue from realization, $ (page.2*150$)		1195600	1270325	1345050	1419775	1494500
Current costs, $		1010847	1049822	1088798	1127773	1166749
Net cash flow (CF), $ (page.3 – page.4)*(1–T)	–1000000	151498	180812	210127	239441	268756
Net cash flow + amortization (page.5+A1+A2)		255135	180812	210127	239441	268756
Coefficients of discounting at RD = 11,9%	0.71	0.64	0.57	0.51	0.46	0.41
Discounted net cash flow (DCF), $ (page.6*page.7)	–713614	207357	145439	147125	146806	144890

We will determine the net present value (NPV) by the formula (3): $NPV = -208382\$$.

As we see, the meaning of NPV of the project is negative, what witnesses about inexpedience of realization of this project. The main source of uncertainty in this project is a demand on the realized product.

The current analysis of the market witnesses that the demand on it is not high, but there is a potential for increasing of sales volumes in some time and, subsequently, the economic effect from the project will be positive.

Accordingly, in this situation the enterprise can use the CALL option for expansion with a view to determine a real economic effect from injection of the project. For calculation of value of an option we use Black-Scholes model.

According to this model., a price of an option is determined by the following formulas (7) – (9).

For assessment of the value of an option, we will provide the following base data: $S = 207357 + 145439 + 147125 + 146806 + 144890 = 791618\$$; $K = 1000000\$$; $r = 9.5\%$ (corresponds to profitability of the dollar bonds of the internal state loan); $t = 3$ years; $\sigma = 40\%$ (thus an enterprise did not implement its activity on the new market, accepted is the mean-square deviation of the discounted cash flows of the competing firms); e – a number, which is a basis of a natural logarithm (roundish meaning 2.71828).

$$d_1 = \frac{ln\left(\frac{791618}{1000000}\right) + \left(0.095 + \frac{0.4^2}{2}\right) \cdot 3}{0.4 \cdot \sqrt{3}} = 0.422$$

$$d_2 = 0.422 - 0.4 \cdot \sqrt{3} = -0.27$$

$$N(d2) = 0.3926$$

$$N(d1) = 0.663$$

$$C = 791618 \cdot 0.663 - 1000000 \times$$
$$\times e^{-0.095*3} \cdot 0.3926 = 229654.7$$

As some criteria of an index of economic effectiveness of a project we propose to use the known in the theory of real options index of the project's value (NPV$_{opc}$). This index is determined as a sum of the net discounted value of a project, calculated by the method of discounting of cash flows, and the value of options, which are purchased by an enterprise during implementation of the project:

$$NPV_{opc} = NPV + C_{opc}, \qquad (15)$$

where, NPV_{opc} – a value, generated by a project, growth of a capital of the enterprise as a result of realization of the project in view of a real option, \$; NPV – a net coerced value, which is equal to the discounted cash flow of the project, \$; C_{opc} – earnings from realization of a real option, \$.

$$NPV_{opc} = -208382 + 229654.7 = 21272.7 \text{ \$}$$

Thus $NPV_{opc} > 0$, the project is not unprofitable and at changing of the market's conjuncture it can be profitable, the meaning of this index, and also a value of the option and NPV value of the project will be displayed in the final Table 2.

Table 2. NPV value of the project.

Index	Meaning, $
NPV_{opc}	21272.7
C_{ocp}	229654.7
NPV	−208382.0

The difference between the indexes NPV$_{opc}$ and NPV is determined in the scientific literature as a stage of effectiveness of the real option. In many sources it is specified, that, according to the profitable approach of assessment of business, the market price of a company can increase for a sum of estimated net current values of investment projects, about which a company has specific advantages. (Kabachenko & Skrypka 2011). Accordingly, also a stage of effectiveness of a real option is at the same time a value of increasing of the value of an enterprise as a result of implementation of a real option.

4 CONCLUSION

Methods of assessment of effectiveness of a project, based on account assessments (without discounting), is a period of payback (PP) and an accounting rate of return (ARR). These methods differ by simplicity in calculations and are used for assessment of projects of small enterprises, which, as a rule, don't inject expensive and long-term innovational projects. The common defect of effectiveness of innovational projects is that they don't take into account a range of factors, such as inflation, segregation of cash flows in time and risks.

Dynamic methods of assessment of effectiveness of innovational projects are based on the model of a discounted cash flow. This group of methods also includes a method of the profitability index (PI), a net present value (NPV), an internal norm (rate) of return (IRR), a discounted payback period (DPP).

Assessment with help of dynamic methods of assessment is more precise in comparison with using of the account methods, because it takes into account different kinds of inflation, changes of an interest rate, norms of profitability. But, these methods have a range of defects:
– all these methods take into account only tangible, material factors and ignore non-material (future competitive advantages, potential possibilities and flexibility in management);
– thus the dynamic methods are focused exclusively on the future cash flows, they ignore external factors, which affect the value of the company, for instance, a ratio between a share's price and an income;
– at using of these methods is not taken into account the value of assets and other internal factors, able to decrease or to increase the value of the company;
– the methods provide a non-precise assessment of projects with a long term of realization, because at late stages of process of discounting, cash flows will almost completely depreciate, what conflicts with a real situation in business.

So, the described method of calculation of effectiveness of an innovational project allows to assess a value of flexibility of management at decision-making in response to unexpected market changes, and also, to take into account risks of the project and to make management solutions concerning choosing of a project and providing of its subsequent control.

REFERENCES

Bruslanova, N. 2004. *Methods of real options in assessment of investment projects*. Financial Director, No 7: 20 – 23.
Buzak, N. 2010. *Economic assessment of informational technologies*. Messenger of JDTU, No 3(53): 29 – 33.

Volkov, O. & Denisenko, M. 2007. *Economics and organ-ization of investment activity*. Kyiv: Textbook, 3rd edi-tion: Center of scientific literature: 662.

Ziyatlinov, A. 2010. *Method of real options for assess-ment of investment projects*. Economic sciences, No 3: 144 – 148.

Kabachenko, D & Skrypka, E. 2011. *Grounding of the conception of value of business as a criteria of effec-tiveness of an enterprise's activity*. Economic messen-ger of National Mining University, No 4: 64 – 71.

McMillan Lawrence, G. 2003. *Options as a strategic in-vestment*. Moscow, 3rd edition: Publishing house "Eu-ro": 1195.

Malyshev, E & Podoynichyn, R. 2013. *Method of assess-ment of investments on basis of real options*. Economics of the region, No 1: 198–204.

Hull John, C. 2013. *Options, futures and other derivatives*. Moscow, 3rd edition: Williams: 1056.

Gilbert, E. *An Introduction to Real Options*. 2004. Invest-ment Analysts Journal, No 60: 49 – 52.

Shreve Steven, E. 2005. *Stochastic Calculus for Finance I: The Binomial Asset Pricing Model*. Springer Verlag: 187.

Theoretical and Practical Solutions of Mineral Resources Mining – Pivnyak, Bondarenko & Kovalevska (eds)
© *2015 Taylor & Francis Group, London, ISBN: 978-1-138-02883-8*

Geological technology of grade mapping of non-metallic raw materials as the basis of its profitable mining

T. Volkova & K. Repina
National Technical University, Donetsk, Ukraine

ABSTRACT: Because of wide industry technological requirements to quality of nonmetallic raw materials it is necessary to develop of deposit in accordance with the map of industrial grades minerals distribution. Eventually, the distribution of useful components and harmful impurities are defined by the geological genetic processes which formed the deposit. We tested the industrial grade distribution by means of the spatial and statistical analysis on the Jelenovskyi field of carbonate raw materials. It is established that essential distinction of quality on every stratigraphic horizon is connected with intensity of secondary geological processes. The complex grade index is constructed on the basis of significant correlations between separate quality indicators. The development of the field is carried out according to the map of complex grade index for various consumers of raw materials. So the geological causes of spatial distribution of industrial grades is a key to improvement technology of field exploitation.

1 INTRODUCTION

Nonmetals have exclusive variety of material composition, a wide range of physical and chemical properties, which ensures their wide industry application. For certain types of raw materials industry requirements limit contents of the useful and harmful components in grades. The quality of minerals is studied at all exploration stages, especially, at feasibility stage. The result must provide the further selective mining, the optimal distribution of the high-quality productions to customers, also maximize the revenue of the mining company, the minimum volume of waste. The technology of gross extraction for non-metallic raw materials are less effective. So the main document of feasibility exploration is the authentic map of mineral grades distribution in the deposit. Some designers propose to reach required quality by modern technical devices and software during the development of deposit (Artan et al. 2011). In this case the quality of production is provided as result of mixing and testing of raw materials. Other researchers suggest to solve this problem by the mathematic methods which improve the accuracy of technological mapping (Chatterjee et al. 2005).

Making a geological map is a fundamental skill for any exploration or mine geologist. Any geological map can be considered as a model of the geological environment. The blunder of mapping is caused by inadequate model of geological object. Thus we need to create adequate model of geological processes interaction. Methodological basis of geological and technological mapping of mineral deposits directs to solution of this problem. The way of calculation of the grade block size was proved (Myagkov 1984). The further developing it is received in practical checking on different non-metallic raw materials (Volkova & Vershinin 1993, Volkova et al. 2006, Volkova & Rogachenko 2011). The geologic factors were different, but principles of modeling are similar.

2 MAIN SECTION

If the grade is presented by several quality indicators, the spatial grade position is defined by the superposition of several indicator maps. The result has a low accuracy due to the total error of the mapping of each indicator and their spatial discordance distribution. Possibility of grades delineation is defined by coherence of indicators variability. In this case complex geochemical index is usually used for increasing of map accuracy. The index formula is constructed on the basis of significant correlations between the grade indicators.

Authenticity of spatial distribution of complex index is determined by the aggregate of main and local genetic factors. The significant correlations between quality indicators really exist if their influence has the geological evidence. Thus, we need to prove the influence to quality of the main and local genetic factors within the deposit, calculate correlations and based on their receive the formula of complex grade index. The map of this index determines the spatial distribution of grades and can be used for selective mining.

Let's consider possibilities and practical realization of this way for carbonate raw materials of the known Jelenovskyi deposit. It is located in southwest part of articulation zone of Donbas with Pryazovskyi megablock of the Ukrainian shield. The geological structure of Jelenovskyi deposit is presented by three structural stages divided by unconformity surfaces. The lower stage is presented by the complex dislocated, deeply metamorphosed magmatic Pre-Cambrian formations. A middle stage includes the plicate complexes of Devonian and Carboniferous sediments. The upper stage consists the horizontally bedding sediments of Cenozoic. The productive strata of middle structural stage are presented by the sediments of Early Carboniferous. In structure plan it is gently pitching monocline that is the south wing of Kalmius-Torecky kettle. Monocline falls on a north-east under the angle of $8 - 12°$. She is broken on separate blocks by the transversal bends and numerous fractures. The Alexander anticline of the second order clearly appears on the general background of monocline. She has the north-eastern trend and asymmetric structure. On the southward the Alexander anticline is complicated by the system of the faults. They are oriented in longitudinal and diagonal directions in relation to an anticline. The geological structures of Jelenovskyi deposit are determined by three faults of the latitudinal trend. The North Volnovahskyi fault is traced in the north part of the deposit. Amplitude of displacement on him an about 50 m. Falling of fault is south-west under the angle of dip is $70 - 78°$. Along this fault the sediments of Early Carboniferous contact with the siliceous marls of $C_1^v e$ horizon and limestone of $C_1^v d$ horizon with sand-clay slates of $C_1^v d$ horizon. Volnovahsky and South Volnovahsky faults are traced through the Jelenovsky deposit more to south. The first fault falling is to south-west under the angle to 65° and displacement amplitude near 100 m. Along this direction early Tournasian rocks contacted with the limestone of $C_1^v f$ horizon. The South Volnovahskyi fault passes in parallel to Volnovahskyo fault more to south. The dip angle of the fault fluctuates from 35° to 70°. Amplitude of displacement reach an about 500 m. This fault is followed by the powerful zone of crushing and hydrothermal change of the rocks.

The limestone of Tournasian and Visean stages behave to the organogenic type. Tournasian stratigraphic stage (C_1^t) is located in low part of carbonate strata and divided on four stratum horizons (from a top to bottom) $C_1^t a$, $C_1^t b$, $C_1^t c$, $C_1^t d$. They

divide clearly enough on a fauna and differ in composition. Total thickness of Tournasian sediments changes from 100 m on east to 535 m on a west. Working thickness makes 72.4 m on the east to 90.3 m on the west of Jelenovskyi deposit. Tournasian stratigraphic stage differs by alternation of limestone, dolomitized limestone, dolomites with interbeds of clay, silicified and slaty limestone. On the $C_1^t d$ horizon the Visean carbonate strata concordant lies. The sediments are widespread in northern part of the Jelenovskyi field. The Visean strata are presented, mainly, the limestone beds and divided on seven stratum horizons. The $C_1^v a$ horizon is presented small – and medium-grained black bituminous limestone which intercalate with bituminous coaly-clay shale. Its thickness is low (5 – 6 m) but lithological composition is constant. Industrial interest is represented by the lower part of strata. It is the horizons of $C_1^v b$, $C_1^v c$ and, especially, $C_1^v d$. The composition of these horizons is similar, but the thicknesses are 10, 26 and 82 meters respectively. The sediments are presented by medium-grained grey and dark grey limestone. With rare interbeds of lime-clay slates and black flint nodules. Among the repetitive strata stand out strongly silicified thin layers and lenticular layers of shale power up to 5 cm. A particularly large amount of siliceous layers located at the top and bottom of the horizon. Technological requirements of industries to quality of carbonate raw materials are presented on table 1.

Table 1. Technological requirements of industries to quality of carbonate raw materials.

Industries	Qualitative indicator, %					
	CaO	MgO	SiO_2	$\frac{Al_2O_3+}{Fe_2O_3}$	S^* 10^{-2}	P^* 10^{-2}
Metallurgy	≥ 49	≤ 5	≤ 2	$\Sigma \leq 3$	≤ 0.06	≤ 6
Ferroalloy	$\Sigma \geq 51$					
Food	≥ 95	≤ 1.5	≤ 2.5	$\Sigma \leq 1.5$	≤ 0.02	
Cement	≥ 45	≤ 2.5	≤ 8	≤ 8	< 0.7	$2 - 4$
Paper	≥ 52	≤ 1.5	2	$\Sigma \leq 1$		
Chemical	≥ 90	≤ 1.4	≤ 3	$\Sigma \leq 1$	< 0.5	< 1

The useful components of a chemical composition are the content of CaO and MgO. If they have bigger value, the grade is higher and more expensive. The maximum contents of harmful impurity are limited (S, P, SiO_2, Al_2O_3, Fe_2O_3). The chemical composition, physical and other primary properties of limestone depend on facial conditions of their formation. But the secondary geochemical processes have a great influence on quality of limestone. They usually wide spread in carbonate rocks and significantly reducing the quality. The contents of quality indicators on every stratigraphic horizon of deposit are presented on Table 2.

364

Table 2. Means of qualitative indicators on the stratigraphic horizons of carbonate strata.

Qualitative indicators	Average on all thickness, %	Means of qualitative indicators on the stratigraphic horizons, %								
		$C_1'a_1$	$C_1'a_2$	$C_1'tb_1$	$C_1'b_2$	$C_1'c$	$C_1'd.$	$C_1^v a.$	$C_1^v b+c$	$C_1'd$
CaO	40.33	38.8	36.3	34.4	47.7	50.8	52.8	47.2	52.5	50
MgO	10.6	6.6	14.7	16.1	3	0.7	0.9	1.4	0.9	0.9
SiO_2	2.2	7.8	2.1	2.9	4.6	3.9	2.1	6.8	1.9	8.2
$Al_2O_3+Fe_2O_3$	0.67	0.1	0.05	0.04	0.8	0.5	0.8	2.1	0.5	0.6
$S \cdot 10^{-2}$	6	6	8	3	9	3	6	2	0.5	8
$P \cdot 10^{-2}$	3	3	3	3	3	3	3	3	3	7

The difference in means of qualitative indicators on the stratigraphic horizons and average weighted on the thickness proves the advantages of selective mining. The most high-quality limestone belongs to the horizon of $C_1'd$. It is the top part of Tournasian sediments. The most low-quality limestone belongs to the horizon of $C_1'a_1$. It is the lowest lay of the carbonate strata which is characterized by interbedding of clays, dolomite and the dolomitized limestone. There are the most common all secondary processes in deposit.

3 PRACTICAL RESULTS

One of the major processes is dolomitization. Content of magnesium increases in result of this process. This quality indicator is limited as harmful impurity for food and chemical industries. On the contrary, it is a positive factor for metallurgical industry because dolomites have bigger fireproof than limestone. Intensity of dolomitization is shown in different degree within the sediments of stages. The main part of the dolomitized limestone within Jelenovskyi deposit was formed by the process of contemporaneous dolomitization, less – epigenetic.

The next epigenetic process is formation of sulfates minerals (gypsum, anhydrite). They are formed in intergranular, microfractures, interporiferous and karstic space. The ferruginous minerals (marcasite, pyrite) occur in the same places. In carbonate rocks of the Jelenovskyi field the different form of contemporaneous flints and the frameworks of silica (radiolaria, sponges, etc.) find out. Silicon replaces calcium in fossils, reducing the content of main useful component. It leads to increase of sulfur and ferrum contents, decreasing calcium content, worsens physical properties and quality of carbonate raw materials.

The process of leaching is the most widespread and negative epigenetic processes in carbonate rocks. There are different karst forms – funnels, caves, cavities, cracks, channels in the fields. The less intensive process of leaching leads to loss of rock durability. There are friable dolomites, cavernous limestone formed in carbonate strata. Cavities are usually filled with friable ferruginous sandy-argillaceous materials, rarely remained empty. Karst is especially intensively shown in zones of tectonic dislocations. On the Jelenovskyi field the external and internal karst is observed. External karst reaches the greatest development in valleys of the rivers and on ravines. The internal karst is result of interaction between carbonate rocks and underground waters. There are many water-bearing horizons, suites, complexes of different age on the Jelenovskyi field. The Early Carboniferous water-bearing horizon and Grabovskyi water-bearing suite (C_1^2) of carbonate strata have the greatest influence on the intensity of karst process. The karst significantly complicates conducting mining operations as the material filling cavities and strongly destroyed carbonate rocks are transported in dumps. The part of karst materials mixes up with minerals and worsens quality of raw materials. This circumstance promotes decrease in mining and growth of operational losses. The operational exploration has the main target to delineate and calculate of karst zones volumes. Unfortunately regularities of karst spatial distribution aren't determined on the Jelenovskyi field. According to mining operations, the internal karst has irregular distribution both on the area of the field and on the separate stratigraphic horizons. Intensity of karst is estimated by means the linear coefficient of karsting. Its average value is 16.6% on the Jelenovskyi field. The carbonate raw materials aren't extracted if its value makes more than 50%. Therefore it is impossible to predict qualities of raw materials in mining work.

We carried out the spatial and statistical analysis of quality on every stratigraphic horizons (Table 2). It allowed to determine the main tendencies in change of quality and to prove their reason. The quality of carbonate raw materials deteriorates with increasing depth. The main trend in distribution of quality is determined by the geological structure of the field. The content of the main useful component (CaO) increases according to the submersion direction of layers from the south-west by the north-east. In general deterioration of quality is caused by the site of tectonic fault in the southern part of the field. There are raw materials with the lowest quality indicators especially in regional parts. High-quality limestone is located in northeast part of the field. Especially accurately these regularities are shown on the horizons of $C_1'c$ and $C_1'd$ where the content of the main useful component (CaO) increases from the northeast to the southwest. Analogically on the horizons of $C_1'a$, $C_1'b$, C_1^va and C_1^vb+c the calcium oxide content increases from the south-west to the north-east. Generally the distribution of qualitative indicators is more stable within one horizon than on several horizons together. It allows to recommend the size of scrap for the deposit working off. It is approximately equal to the thickness of the stratigraphic horizon.

For construction of complex geochemical indicators of the grade was calculated matrix of correlations between qualitative indicators. Data for research are submitted by chemical analyses of section tests on prospecting wells. Sampling of core on the horizons of $C_1'a_2$ and $C_1'b_1$ was made by two-meter sections, on other horizons – four-meter. It is given in Table 3 for data sample in 1328 arrays, the significant level equal 0.05.

The significant negative correlation between pair (CaO, MgO) is explained by replacement of $CaCO_3$ with $MgCO_3$ in the process of dolomitization. Positive correlation between $Al_2O_3+Fe_2O_3$ and SiO_2 is determined by their joint position in space lattice of kaolinite. The significant positive pair correlations was revealed between $(S, Al_2O_3+Fe_2O_3)$ and (S, SiO_2).

Table 3. Matrix of correlations of qualitative indicators.

Qualitative indicators	CaO	MgO	SiO₂	Al₂O₃+ Fe₂O₃	S	P
CaO	1	-0.82	-0.31	-0.42	0.01	-0.1
MgO	-0.82	1	-0.23	0.25	-0.1	-0.02
SiO_2	-0.31	-0.23	1	0.51	0.23	0.25
Al_2O_3+ Fe_2O_3	-0.42	0.25	0.51	1	0.53	0.03
S	0.01	-0.07	0.25	0.53	1	0.01
P	-0.1	-0.02	0.25	0.03	0.01	1

*significant correlations are shown in bold

The single significant positive correlation of phosphorus with silicon reflect the natural conditions of sedimentation. Contents of phosphorus are stable enough in strata. Paragenetic associations of the qualitative indicators are confirmed by the cluster analysis. Two associations are received. The first is presented by association of replacement calcium oxide with magnesium oxide. The second association including three indicators SiO_2, $Al_2O_3+Fe_2O_3$, S is caused by joint passing of processes of sulfatization, silicification and ferruginization in the caverns. We checked that these processes reach maximum intensity in the caverns. By means statistical analysis we determined the wells with the most amount intersections of leached zones and their correlations with contents of harmful impurities. The more power caverns have the higher the content of these indicators on the horizon. Their minerals are fined out with sandy-argillaceous material in the karst space. The horizon $C_1'b_2$ is the most karsting (Figure 1).

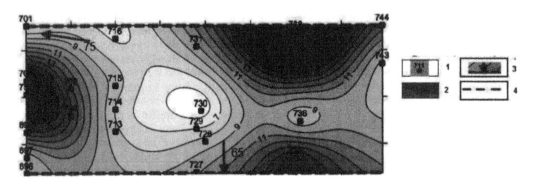

Figure 1. The power of karst on stratigraphic horizon $C_1'b_2$: 1 – number of well; 2 – isoline of indicator; 3 – fault; 4 – water-bearing horizon.

It was established that the increase of karst in the west connect with arrangement of the water-bearing horizon. Its depth is 60 to 80 meters on the fields. The tectonic faults on the north and south of the fields is the cause of widespread karst cavities in the carbonate rocks. Sediments of the C_1^vb+c и $C_1'd$ horizons are characterized by the smallest karsting spread therefore represent the carbonate raw materials the most high-quality. Thus, intensity of karst process is the action result of two genetic factors - tectonic and groundwater hydrology.

On the basis of the established associations created complex grade index of carbonate raw materials:

$$K_c = \frac{CaO}{MgO + (S + P) \cdot 10^{-2} + (Al_2O_3 + Fe_2O_3)}, \quad (1)$$

where K_c – complex grade index.

The boundary values of the complex grade index in accordance with technological requirements (Table 1) were calculated for different industries. The map of grades for ferroalloy industry are presented in Figure 2.

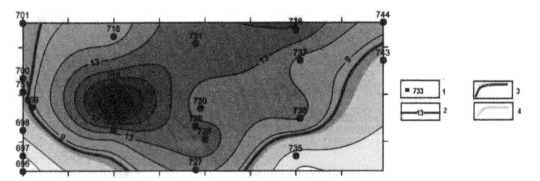

Figure 2. The grade map for steelmaking on horizon $C_1{}^tb_2$: 1 – number of well; 2 – isoline of grade index; 3 – the bound of grade g_1; 4 – the bound of grade g_2.

The maps of grade spread were built on each stratigraphic horizon for different industries. Similar maps are used for detailed planning of selective working off deposit. On the maps we try to allocate the maximum area for the selective mining. But we have averaged the grades if the tortuosity of the border is too great within horizons. In this case, selective extraction using the existing equipment would be difficult. Averages should be carried out taking into account the loss in value due to deterioration of quality. Reserves calculation and the profit from the sale of raw materials for grades showed that it is 1.5 times higher than the gross extraction.

4 CONCLUSIONS

In developing the deposits of carbonate resources on total thickness there is a significant decrease in quality. In this case, the raw material of Jelenovskyi field can only be used cement industry.

The variability of raw materials quality will be lower if the working off Jelenovskyi field would be based on the thickness of each horizon 3. Reduced quality of carbonate raw materials due to the intensity of the secondary geological processes - sulfation, silicification, ferrugination. The main negative process is karst as it essentially increases contents of all harmful impurities.

Spatial position of grade is determined by complex index which is constructed on the significant correlations of quality indicators. The maps of complex index provide the calculation of reserves according to grades. It is significantly expand the nomenclature of consumers and optimize distribution of production.

This technology provides effective use of nonmetallic resources, maximum the revenue of the mining company, the minimum volume of waste. Also it can be used for metallic raw materials which have essential difference in price of grades.

REFERENCES

Artan, U., Marshall, J. & Lavigne, N. 2011. *Robotic mapping of underground mine passageways*. Mining Technology. Research Article, Vol. 120, Issue 1: 18 – 24.

Chatterjee, S., Mishchenko, A., Brayton, R. et al. 2005. *Reducing structural bias in technology mapping*. Proc. ICCAD: 519 – 526.

Myagkov, V. 1984. *Geochemical method of paragenetic analysis of ores*. Moscow: Interior: 126.

Volkova, T. & Rogachenko, A. 2011. *Research of limestones quality for the purpose of exploitation optimization of the Rodnikovsky deposit*. Scientific works of Donetsk national technical university, Issue 13 (178): 74 – 77.

Volkova, T.P. & Vershinin, A.S. 1993. *The method of geologic technological mapping of the original kaolins deposits*. News of higher educational institutions. Mining journal (Moscow), No 4: 12 – 18.

Volkova, T., Buryak, G. & Otriyshko, E. 2006. *The geologic technological mapping of the refractory clay deposits*. Scientific works of Donetsk national technical university, Vol. 2, Issue 111: 136 – 144.

Theoretical and Practical Solutions of Mineral Resources Mining – Pivnyak, Bondarenko & Kovalevska (eds)
© 2015 Taylor & Francis Group, London, ISBN: 978-1-138-02883-8

Influence of hydrodynamic impact on degassing of steep coal seam at the top of the overhead longwall

V. Gavrilov & V. Vlasenko
M.S. Poliakov Institute of Geotechnical Mechanics under the
National Academy of Sciences of Ukraine, Dnipropetrovsk, Ukraine

O. Moskovskiy
SE "F.E. Dzerzhinsky Mine", Dzerzhynsk, Ukraine

ABSTRACT. On the basis of experimental studies of hydrodynamic impact process on the coal vein l_7 – "Pugachevka" 1146 m of adit level through production wells No 2 PK 47, No 3 PK 51 + 3, No 4 PK 54 + 6 at the top of longwall the dynamic pattern of methane mean daily concentration was determined in return ventilation air mixture before and after hydrodynamic impact on the virgin coal, and its size of working zone. Degassing index of coal seam in worked-out areas and the amount of extracted methane through underground wells were calculated, according to change of methane mean daily concentration with account of background concentration and amount of air, which is aerated to ventilation area. The efficiency of hydrodynamic impact was determined as a way to degassing intensification and reduction of gas-dynamic activity in the area of production wells influence.

1 INTRODUCTION

Intensification and concentration of mining operations, especially with mines deepening, are becoming more limited with hazardous properties of coal seams, such as gas emission, gas-dynamic phenomena, dusting, spontaneous combustion of coal in solid. (Puchkov & Slastunov 2006). Increasing load on breakage face, the pace of underground mining workings and mine safety, are particularly important in the mines of Central Donbass, where geological and mining conditions of coal mining are extremely unfavorable. In the work (Bondarenko et al. 2013) was developed the common scientifically-based points of forecasting technique for gas dynamics prediction in undermined massif during stoping area exploitation and after its isolation taking into account the mine rock movements appearing during the longwall development and desorption of gas from coal broken and crushed by shearer. The necessity of alternative energy sources research is dictated by the rising of natural gas, oil and coal costs (Bondarenko et al. 2010). Ukrainian proved coal reserves ranked on seventh place in the world according World Energy Council report (Falshtynskyi et al. 2014).

Reducing the coal demand in Ukrainian power engineering with extraction increasing associate with an imbalance of it usage in the energy and power sector, seasonally necessary, underdeveloped complex and industrial processing.

Due to the unstable coal extraction become the growth of the cost and complexity of the technical and economic conditions and environmental safety in mines (Falshtynskyy et al. 2012).

Among solving methods for these problems the main processes are processes of stimulation on coal seams which can change their condition and due to this they can prevent gas-dynamic phenomena during mining operations.

The efficiency of stimulation methods is determined by reduction level of essential hazardous characteristics of coal seam during coal extraction. A number of stimulation methods on seam make it possible to overcome the negative influence of main hazards and increase intensity of manufacturing process. However, the efficiency of these methods and stimulation tools are still insufficient, according to industrial disasters data. (Buhantsov & Porubai 1999, Briuhanov et al. 2003).

One of the most prospective lines of rock condition controlling in order to its degassing and decreasing gas-dynamic activity is hydrodynamic impact through wells from abandoned place. Forces of overburden stress and gas pressure can be used as active forces. As a result vibrational properties of the "coal seam – enclosing rocks" system are activated. (Gavrilov et al. 2011).

The purpose of this research was to study the processes of gas emission from coal rock mass and to determine the efficiency of the hydrodynamic method of intensification of degassing and to reduce gas-dynamic activity of gas-saturate coal formations at the top of overhead longwall.

2 MAIN BODY

The essence of the method of hydrodynamic impact consists on the application of the free surface of the coal seam alternating loads. They are summed at a time with the forces of the mining and gas pressure to overcome the tensile strength at break of coal, perform work on the destruction of the free surface and the formation of a wider system of cracks in the seam thereby contributing to an increase in the filtration volume, intensify the process of gassing and redistribution of loads, which leads to a decrease in the gas-dynamic activity of the seam (Agaiev et al. 2014).

In conducting mining and experimental works equipment was located in the circuit shown in Figure 1.

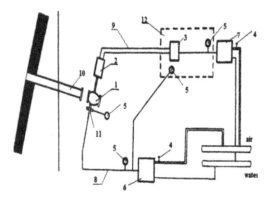

Figure 1. Location scheme of equipment, instruments and high-pressure armature: 1 – latch; 2 – cylinder with spring system; 3 – hydraulic control valve (ERA); 4 – valves; 5 – manometers; 6, 7 – pumping systems; 8 – high-pressure water hoses; 9 – High-pressure hoses for oil; 10 – casing; 11 – cross; 12 – remote control.

After checking on the well hermetic sealing possible produces a cycle of hydrodynamic impact.

Works on hydrodynamic impact on coal seam through the wells in the following sequence: closed latch device for the hydrodynamic pump unit and impact technology in the well created pressure 2 – 7 MPa, then produce it for the discharge of 0.1 – 1.0 seconds and has been issuing coal with water and gas. While hydrodynamic impact of water

flow measurements were conducted and the state of the gas situation in the mine working.

The subject of research on work on the top of the degassing overhead longwall of hydrodynamic impact is mountain overhead longwall which fulfills coal seam $l_7{}^f$ – Pugachevka-vostok: mine named after FE Dzerzhinsky, and ventilation development conducted to prepare it.

Seam thickness: geological – 0.86 – 1.10 m, useful – 0.82 – 0.98 m, excavated – 1.02 – 1.44 m, the dip angle of 58°.

Coal seam $l_7{}^f$ – "Pugachevka" throughout the excavation field is maintained by seam thickness and structure, one pack consists of coal, in which the layers are allocated different degrees of disturbance. The top layer thickness of 0.25 – 0.30 m is represented by coal semi-bright, layered textures, with the inclusion of sulfur pyrite lenses, fractured, cracked multidirectional fortress mean, I – II type of violation. Below – coal semi-bright to semi-mat consists of individual lenses with polished surface, no bedding, weak, prone to rashes, layer thickness 0.02 – 0.08 m, type of disturbance III. The lower layer thickness 0.55 – 0.60 m – coal semi-bright, layered textures, with the inclusion of sulfur pyrite lenses, fractured, medium strength, I – II type of violation.

The presence of carbonaceous shale at the top significantly reduces adhesion of seam with rock walls and can occur spontaneously collapse of overhanging coal massif. Hypsometry of seam is homogenous. Natural gas content 19 – 21 m³/ t.d.a.f.b., yield of volatile matter 29.4 – 33.4%. Fortress coal $f = 0.8 – 1.0$, volumetric weight 1.3 – 1.4 m³/t. The seam $l_7{}^v$ is dangerous to sudden outbursts of coal and gas and dangerous collapse of coal, dangerous explosiveness of coal dust, is not prone to spontaneous combustion, the mountain is not dangerous blows.

Overlying horizon worked out on the field haulage drift leaving interstorey pillar 6 – 7 meters. Plast $l_7{}^v$ worked out previously of spent seam m_3.

Mining works is complicated by the existence of zones of high rock pressure hazard category 3 from seam l_5 – Solenuj. The zone extends from 42 to 106 meters down dip, measured from the airway

For production of hydrodynamic from air fringe drift 65 – 1146 m four consecutive production wells were drilled on seam through mineral. Parameters of production wells for the production of hydro-dynamic impact are shown in Table 1.

As an example, Figure 1 illustrates arrangement of the well No 2 on the picket No 46.

Dead borehole ground No 2 with 3.5 m long was reamed to 150 mm diameter for well casing. Casing was made with metal pipes with 114 mm diameter. Total pipes length was 4.0 m.

Table 1. Parameters of production wells location at the top of the overhead longwall No 65 – 1146 m.

No of well	No of PK	Well length	Normally seam distance	Amount of inclination to upper area
		m	L_s, m	γ, degree
2	47	5.0	2.0	30
3	51+3	2.2	0.8	30
4	54+6	4.5	1.3	30

(a)

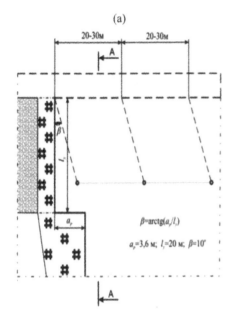

(b)

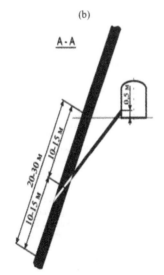

Figure 2. The arrangement of production wells in ventilating entry: (a) location scheme of boreholes; (b) cross-section location scheme of boreholes.

The device for hydrodynamic impact was installed on surface casing which length was 0.5 m. Water-proofing was made with cement-sand mixture in the ration of C:W:S-1:1:2 at a depth of 3.5 m. Total borehole length was 6 m.

For methane control on the upcast at remote-control station sensors with automatic gas protection (AGP) were installed.

The work on hydrodynamic impact on the coal seam was conducted as follows: valve device for hydrodynamic impact was closed and with the help of pumpset (PVI-pump and vacuum installation) in the production well the pressure was 1.5 – 2.0 MPa. After that depressurization time was about 0.1 – 1.0 with emission of gas, coal and water. Pump sets were plugged from remote-control station. Manometers which were installed on re-mote-control station were controlling water pressure in the gate valves operation system and in seam injection system.

Hydrodynamic impact through seam No 2 was made in 10 cycles. Gas was bleeding from seam: methane concentration in return ventilation air was up by 15% and made 0.617%. Yield of coal from seam No 2 was 2 ton.

Dead borehole ground No 3 with 2.4 m long was reamed to 150 mm diameter for well casing. Total pipes length of casing was 2.85 m.

Surface casing outbreak was 0.5 m. Water-proofing was made by cement-sand mixture at a depth of 2.4 m. Total borehole length was 3.4 m.

Hydrodynamic impact on coal seam was made in 15 cycles and 2.5 ton of coal was got. Methane concentration increased by 10% in return ventilation air. Total borehole length No 4 was 6.8 m. Borehole was drilled for standpipes full length. Casing was made with pipes about 100 mm in diameter.

Total length of pipes was 5.2 m. Water- proofing was made at a depth of 4.9 m.

8 cycles of formation stimulation through borehole No 4 got 0.5 ton of coal. Methane concentration in air current remained at the level of 0.689%.

The efficiency of hydrodynamic impact was determined by degassing index, which can be calculated according to formula:

$$K_d = \frac{V_f}{V_p},\qquad(1)$$

where V_f– volume of gas output from the worked-out area of coal seam, m^3; V_p– design gas volume in coal seam surface, m^3.

Volume of gas output from worked-out area was calculated according to formula:

$$V_f = 0.01 Q_u \left(C_C - C_f\right) N,\qquad(2)$$

371

where Q_u – quantity of air on the upcast zone, m³ per day; C_c – methane daily concentration in the upcast zone (ventilating entry), during performance on worked- out area, %; C_f – methane background concentration in return ventilation air in the zone (ventilating entry),before approach of face of longwall to worked-out area of seam and (or) after its approach, %; N – number of design days.

Design gas volume in the worked- out area of coal seam was calculated according to formula:

$$V_p = S_z m \gamma \chi ,$$ (3)

where S_z – area of worked-out zone, m³; m – net pay zone of coal seam, m; γ – specific weight of coal, ton/m³; χ – natural gas content of coal seam, m³/t.d.a.f.b.

Area of worked-out zone is calculated according to formula:

$$S_z = \pi R_0^2 ,$$ (4)

where R_0 – worked-out zone radius, m.

Worked-out zone radius of the top of longwall was calculated according to formula:

$$R_0 = \sqrt{\frac{V_f}{s_z pm \gamma \chi}} .$$ (5)

According to methane daily concentration the volume of total gas output was calculated to its concentration reduction in going headway till initial reading in Table 2.

Table 2. Volume of produced gas made.

No of PK	No of well	Volume gas, m³
47	2	28227
51+3	3	34481
54+6	4	58661

Total volume of produced gas made:

$V_f = 121369$ m³.

Volume of gas in effected zone amounted to:

$Vp = 9430 \cdot 1 \cdot 1.3 \cdot 22 = 269698$ m³.

Degassing index equals to:

$$K_d = \frac{121369}{269698} = 0.45 .$$

4 CONCLUSIONS

1. Effeciency of method was confirmed by degassing works and reduction of gas-dynamic

activity of coal seam l_7 – "Pugachevka" at the top of the overhead longwall No 65 – 1146 m Dzerzhinsky mine with hydrodynamic impact through production wells No 2, No 3 and No 4, which were drilled from fringe drift.

2. As a result of hydrodynamic impact on coal seam the total area of worked-out zones with boreholes No 2, No 3 and No 4 amounted to 9430 m², actual volume of produced gas amounted to 121369 m³ with calculated – 269698 m³, degassing index amounted to – 0.45.

3. Methane daily concentration monitoring in return ventilation air from mining area at working at longwall work in worked-out zone showed that, optimum borehole length is 15 – 20 m, and distance between boreholes along the length of ventilating entry – is at least two radius of treatment.

4. Shows of gas-dynamic phenomena during mining operations in worked-out areas were not observed.

REFERENCES

Agaiev, R., Vlasenko, V. & Kliuev, E. 2014 *Methane receiving from coal and technogenic deposits* Progressive Technologies of Coal, Coalbed Methane, and Ores Mining. The Netherlands: CRC Press/Balkema: 113 – 120.

Bondarenko, V., Tabachenko, M & Wachowicz, J. 2010. *Possibility of production complex of sufficient gasses in Ukraine.* New Techniques and Technologies in Mining. The Netherlands: CRC Press/Balkema: 113 – 119.

Bondarenko, V., Kharin, E. & Antoshchenko, M. 2013. *Basic scientific positions of forecast of the dynamics of methane release when mining the gas bearing coal seams.* Naukovyi Visnyk Natsionalnoho Hirnychoho Universytetu, No 5: 24 – 30.

Briuhanov, A., Mnuhin, A. & Busygin, K. 2003. *The analysis of methane explosion in mines and its preventive actions.* Coal of Ukraine, No 4: 37 – 40.

Buhantsov, A. & Porubai, V. 1999. *Increasing of mine safety on the basis of effective degassing.* Coal of Ukraine, No 10: 42 – 45.

Falshtynskyi, V., Dychkovskyi, R., Lozynskyi, V. & Saik, P. 2014. *Some aspects of technological processes control of in-situ gasifier at coal seam gasification.* Progressive Technologies of Coal, Coalbed Methane, and Ores Mining. The Netherlands: CRC Press/Balkema: 109 – 112.

Falshtynskyy, V., Dychkovskyy, R., Lozynskyy, V. & Saik, P. 2012. *New method for justification the technological parameters of coal gasification in the test setting.* Geomechanical Processes During Underground Mining The Netherlands: CRC Press/Balkema: 201 – 208.

Gavrilov, V., Vlasenko, V. & Pishchev, A. 2011. *Reduction of gas saturation and outburst hazard of coal seams in overhead longwalls with alternating impact.* Vibrations in technique and technologies: All-Ukrainian scientific and technical journal, No 4 (64): 94 – 96.

Puchkov, L. & Slastunov, S. 2006. *Effective solution of methanoic safety in Russian mines – today's task.* Coal, No 12: 24 – 28.

Theoretical and Practical Solutions of Mineral Resources Mining – Pivnyak, Bondarenko & Kovalevska (eds)
© 2015 Taylor & Francis Group, London, ISBN: 978-1-138-02883-8

Grounds of parameters of high concentrated pulps storage technologies

E. Semenenko & S. Kirichko
M.S. Poliakov Institute of Geotechnical Mechanics under the
National Academy of Sciences of Ukraine, Dnipropetrovsk, Ukraine

ABSTRACT: For the first time for the systems of the enrichment process wastes storage in a form of the high concentrated pulp it was suggested a method of pipeline diameter calculation based on universal dimensionless values that take into consideration the pulp, flow and pipeline specifications. The analytic dependence between the pipeline diameter, the upper yield point, the effective viscosity, and the volume flow rate of a suspension, coefficients of which do not depend on the solid phase properties, the diameter or material of the pipeline was grounded. It was obtained the estimation of dimensionless flow rate intervals where the high concentrated pulp viscosity or upper yield point input into the hydraulic gradient value are dominant.

1 INTRODUCTION

Mining Processing Plants (MPP) are the basis for industrial regions of Ukraine sustainable development and simultaneously the source of the greatest environmental threat. MPPs consume the considerable amount of the primary ecological resource – clean water and also require more land areas for enrichment process wastes storage (Semenenko 2011). However the amount of process water becomes less and there is no land allotment for new wastes storages.

In these circumstances the only way to save the functionality of MPPs and the sustainable development of regions where MPP are located is an implementation of the retrofit technologies which assume getting of the enrichment process wastes in a form of the high concentrated pulps (HCP) with more than 30% of the solid particles inclusion volume fraction.

Today the technologies are known that make it possible to thicken the enrichment process wastes up to such concentration but the resultant HCP has rheological properties and also plastic fluid properties appear. For those fluids the dependences of rheological properties and some of hydraulic characteristics while flowing over the pipeline are known. Modes and operating standards of hydrotransportation and storage technologies of such wastes are unknown that restrains those prospective technologies implementation at the homeland MPPs.

While calculating of the hydrotransportation parameters the goal of the calculation is the evaluation of the critical speed value, the hydraulic gradient or the pipeline critical diameter for the chosen materi-al, the specified pipeline diameter with the prescribed pulp charge or the stream of supply depending on the suspension concentration. However the definition of the critical speed for HCP does not correspond to the same definition for the low concentrated pulps (Semenenko 2011). The most part of HCP generated from the enrichment process wastes are plastic fluids and correspond the Bingham-Shvedov Rheological Law. Some researchers offer to calculate the critical speed for HCP according to the condition of the complete flow core destruction and structural flow regime finishing (Svitlyi & Krut' 2010, Krut' 2002). Other scientists suggest to use the evaluation of input into the hydraulic gradient of an item that contains the viscosity of the suspension (Semenenko 2011, Kirichko 2012).

This work has as its goal to ground the possible modes of operation for the hydrotransportational complexes that provide enrichment process wastes storage in a form of and also to develop the method of their designed calculation.

2 MODES OF OPERATION GROUNDS

For the grounding of the modes of HCP pumped flow regimes it is reasonable to overwrite the dependence for the hydraulic gradient using the dimensional quantities scales included into it (Svitlyi & Krut' 2010, Kirichko 2012):

$$i = \frac{\tilde{\tau}_0}{\lambda} + \frac{Q}{\mu}, \qquad (1)$$

$$\lambda = \frac{\rho_0 g D}{2},$$ (2)

$$\mu = \frac{\rho_0 g \pi D^4}{16 \tilde{\eta}},$$ (3)

$$\tilde{\tau}_0 = \alpha \tau_0,$$ (4)

$$\tilde{\eta} = \beta \eta,$$ (5)

where i – hydraulic gradient; λ – upper yield point module; μ – rate of discharge of hydrotransportational complex line (Semenenko 2011); ρ_0 – water density; g – gravity acceleration; τ_0 – upper yield point (UYP); η – pulp effective viscosity (EV); Q – HCP volume flow rate; $\tilde{\tau}_0$ – UYP effective value; $\tilde{\eta}$ – EV effective value; D – pipeline diameter; α, β – coefficients of approximation indices of the Buckingham's complete equation solution.

The expression (1) includes two items, each of them dominates in its flow regime. The first item dominates in the Shvedovskiy regime, the second item dominates in the pseudo-laminar regime. Thus the Bingham's regime is the flow regime where the influence of both of the items is comparable to each other (Krut' 2002).

It is not difficult to show that the Shvedovskiy regime is realized with the volume flow rates complying with the following inequation (Krut' 2002, Shvornikova 2010, Svitlyi & Krut' 2010, Kirichko 2012):

$$Q \leq Q_S,$$ (6)

$$Q_S = 0.1 \cdot Q_*,$$ (7)

$$Q_* = \frac{\pi D^3 \tilde{\tau}_0}{8 \tilde{\eta}},$$ (8)

the hydraulic gradient in such regime can be calculated by the formula:

$$i = 1.1 \cdot \frac{\tilde{\tau}_0}{\lambda},$$ (9)

where Q_* – line input equivalent value.

In much the same way for the pseudo-laminar regime the volume rate should be (Krut' 2002, Shvornikova 2010, Svitlyi & Krut' 2010, Kirichko 2012)

$$Q_L \leq Q,$$ (10)

$$Q_L = 10 Q_*,$$ (11)

the hydraulic gradient in this regime can be calculated by the formula:

$$i = 1.1 \cdot \frac{Q}{\mu},$$ (12)

after simple transformations it can be rewritten in this way:

$$i = \frac{\rho}{\rho_0} \frac{\lambda_* Q^2}{\pi^2 g D^5},$$ (13)

$$\lambda_* = 1.1 \cdot \beta \frac{64}{Re},$$ (14)

where λ_* – adjusted coefficient of hydraulic friction.

For the grounding of an adequacy of using the suggested formulas it is necessary to investigate their correlation with the limit values of Reynolds criterion:

$$Re = Kr \frac{Q}{Q_*},$$ (15)

$$Kr = \frac{\rho \beta D^2 \tilde{\tau}_0}{2 \tilde{\eta}^2}.$$ (16)

On the basis of the expression (15) and taking into consideration the inequations (6) and (10) the following limitations for Reynolds number in the Binghamovskiy flow regime were received:

$$0.1 Kr \leq Re \leq 10 Kr.$$ (17)

According to this the hydraulic gradient is calculated by the formula (13).

In HCP flow regimes for which the Reynolds number value exceeds the upper limit from the set of inequalities (17), the hydraulic grade can be calculated by the formulas (12) and (13). While the exceeding of the HCP consumption and the start of the turbulent conditions the hydraulic calculation for the hydrotransportational complex is calculated similar to the flow regime of the low concentrated hydromixture but the coefficient of hydraulic friction in the formula (13) is calculated by the following expression (Svitlyi & Krut' 2010):

$$\lambda_T = \frac{0.075}{\sqrt[8]{Re}},$$ (18)

where λ_T – coefficient of hydraulic friction for the turbulent flow regime.

Alternatively to the majority of all known recommendations the inequations (17) take into consideration not only the rheological parameters of HCP, the density of the solid phase particles but also the pipeline diameter. The adherence of Reynolds criterion limits (17) as it is not difficult to show make it possible to implement the core or structure flow regimes of HCP for the relative flow core radii over the range between $0.132 - 0.753$.

3 DESIGNED CALCULATION

The designed calculation is made with the aim to fix the pump applicability factors that are the pipe diameter and the pumps total pressure head and that are necessary for the scheduled parameters and operative modes for the hydrotransportational complex if to follow the conditions (17) and the given pulp intake (Krut' 2002, Svitlyi & Krut' 2010, Kirichko 2012).

With the given productivity of hydrotransportational complex for HCP:

$$Q_S < Q < Q_L , \qquad (19)$$

it is easy to introduce the limits in a form of the pipeline diameter limitations:

$$0.464 < \frac{D}{2R_*} < 2.154 , \qquad (20)$$

$$R_* = \sqrt[3]{\frac{\tilde{\eta}Q_P}{\pi\tilde{\tau}_0}} , \qquad (21)$$

and corresponding limitations for the hydraulic gradient:

$$\frac{0.511}{\rho_0 g}\sqrt[3]{\frac{\pi\tilde{\tau}_0^4}{\tilde{\eta}Q_P}} < i < \frac{23.729}{\rho_0 g}\sqrt[3]{\frac{\pi\tilde{\tau}_0^4}{\tilde{\eta}Q_P}} , \qquad (22)$$

where Q_P – scheduled suspension flow; R_* – pipeline typical radius.

Formulas (20 – 21) allow to choose the possible pipeline diameter variation range and the necessary pipe pressure due to the given consumption, the concentration and the rheological properties of the pulp. For that firstly the relative input of the installation is calculated:

$$q = \frac{Q_P}{Q_K} , \qquad (23)$$

$$Q_K = \frac{\tilde{\tau}_0^4}{\pi\tilde{\eta}(\rho_0 g)^3} , \qquad (24)$$

where q – relative input of the installation; Q_K – rheological input module.

Then due to the q value limits of the acceptable values for pipe diameters and pumps pressure are calculated:

$$0.43\sqrt[3]{q}\,\frac{\tilde{\tau}_0}{\rho_0 g} < D < 2.01\sqrt[3]{q}\,\frac{\tilde{\tau}_0}{\rho_0 g} , \qquad (25)$$

$$\frac{0.51 k_z L}{\sqrt[3]{q}} + \rho\Delta Z < \chi H(Q_P) < \frac{23.7 k_z L}{\sqrt[3]{q}} + \rho\Delta Z , \qquad (26)$$

where $H(Q_P)$ – flow pump pressure certified value with the input of Q_P; χ – coefficient of recalculation of consumption and pressure peculiarities (CPP) of the pump changing water to HCP; ΔZ – difference of line dead lifts; k_z – coefficient that takes into consideration local hydraulic resistances.

Formulas (23) – (26) make it possible to estimate the variation ranges of mentioned values. For calculating of the specific value it is necessary to choose the pump an operational field of which will obey all the parameters (26) and then to calculate the modified pressure linear head:

$$\tilde{i} = \frac{\chi H(Q_P) - \rho\Delta Z}{k_z L} , \qquad (27)$$

and to estimate the necessary pipeline diameter value with the solution of the equation (27):

$$\tilde{i} = \frac{2\tilde{\tau}_0}{\rho_0 g D} + \frac{16\tilde{\eta}Q_P}{\rho_0 g \pi D^4} , \qquad (28)$$

where $\tilde{i}$ – modified pressure linear head; ρ – relative density of the suspension.

The equation (28) according to D is not complete algebraic quartic equation that after simple substitutions can be reduced to:

$$y^4 + \theta y - 1 = 0 , \qquad (29)$$

$$y = \sqrt[4]{\frac{\tilde{\eta}Q_P}{\pi\rho_0 g \tilde{i}}}\,\frac{2}{D} , \qquad (30)$$

$$\theta = \frac{\tilde{\tau}_0}{\rho_0 g \tilde{i}}\sqrt[4]{\frac{\pi\rho_0 g \tilde{i}}{\tilde{\eta}Q_P}} . \qquad (31)$$

Real roots of the equation (29) are cross points of quartic parabola with turned up branch lines and parabola vertex in the coordinate origin and direct line that cross axis of ordinates and axis of abscises in points $(0; 1)$ and $(\theta^{-1}; 0)$ respectively. Function charts of such functions always intercross in the first and the second quadrants of the coordinate plane. In this case there can be only two cross points. On the ground of this it can be concluded that the equation (29) has two real and two complex roots thus it can be represented in the following form:

$$\left((y - e)^2 - f^2\right)\left((y - c)^2 + d^2\right) = 0 , \qquad (32)$$

where e, f, c, d – real numbers.

Removing the parentheses in the expressions (32), collecting similar and equating expressions for coefficients with argument coherent degrees to coefficients in the equation (29) the system of non-linear algebraic equations to figure out e, f, c, d is obtained. The solution of this set of equations resolves itself into finding the real roots of cubic equation with constantly positive discriminant. Thereby considering that the root of the equation (29) must be real and positive, analytic expression for calculation can be shown in the form of:

$$D = \frac{2^{\frac{7}{6}}\sqrt{\sigma(\theta)}}{\sqrt{\left(\frac{2}{\sigma(\theta)}\right)^{\frac{3}{2}} - 1 - 1}} \sqrt[3]{\frac{\tilde{\eta} Q_P}{\pi \tilde{\tau}_0}}, \qquad (33)$$

$$\sigma(\theta) = \sqrt[3]{\sqrt{\frac{256}{27\theta^4 + 1}} + 1} - \sqrt[3]{\sqrt{\frac{256}{27\theta^4 + 1}} - 1}. \qquad (34)$$

The solution that was obtained fixes considerably nonlinear connection between pipeline diameter that is prescribed by the consumption and rheological properties of HCP. It is possible to use this dependence only with the help of special tables or computing machinery. From the expression (30) it can be seen that pipeline diameter value that is inversely proportional to the root of the equation (29). The numerical analysis of the dependence (33) illustrates that value that is inversed to the equation solution can be approximated with engineering relevance as a linear function (Figure 1, 2), that makes it possible to use the following formula instead of the considered expression (Table 1):

$$\frac{D}{2} = \frac{\alpha' \tilde{\tau}_0}{\rho_0 g \tilde{i}} + \beta' \sqrt[4]{\frac{\tilde{\eta} Q_P}{\rho_0 g \pi \tilde{i}}} \qquad (35)$$

where α', β' – approximation indices.

Table 1. Coefficients of functions in the formula (35).

Value of variables	Values of coefficients		R^2
	α'	β'	
$1.2 < \theta$	0.9976	0.1048	1.0000
$\theta < 1.2$	0.3921	0.9721	0.9871

According to the value that is calculated using the formula (35) from the pipe range it is chosen the proximal larger diameter. Then the adherence to the specifications (25) is checked. In a case when it is satisfied it is calculated by the formulas (27) and (26) the pressure loss and the prescribed pipe pressure are adjusted.

The designed calculation is finished by the conditions checking (19) and the estimation of motors capacity of chosen pumps with the reserve in 30%.

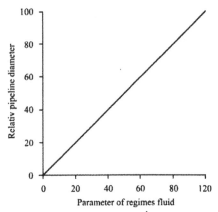

Figure 1. The dependence of value y^{-1} on the parameter θ for $1.2 < \theta$.

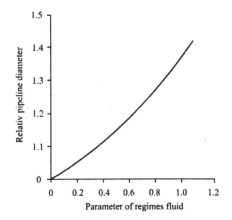

Figure 2. The dependence of value y^{-1} on the parameter θ for $\theta < 1.2$.

4 CONCLUSIONS

1. Thus for the first time for the systems of the enrichment process wastes storage in a form of HCP it was suggested a method of pipeline diameter calculation based on universal dimensionless values that take into consideration the high concentrated pulp, flow and pipeline specifications.

2. For the first time it was grounded the analytic dependence between the pipeline diameter, the upper yield point, the effective viscosity, and the volume flow rate of a suspension, coefficients of which do not depend on the solid phase properties, the diameter or material of the pipeline.

3. It was obtained the estimation of dimensionless flow rate intervals where the high concentrated pulp viscosity or upper yield point input into the hydraulic gradient value are dominant.

REFERENCES

Kirichko, S.N. 2012. *Calculation of hydrotransportation parameters of high concentrated hydromixtures in the conditions of Krivbass*. Geotechnical Mechanics: No 103: 101 – 106.

Krut', O.A. 2002. *Coal-water fuel*. Kyiv: Naukova Dumka: 172.

Semenenko, E.V. 2011. *Scientific basis of hydromechanization technologies for opencast mining of titanium-zircon placers*. Kyiv: Naukova Dumka: 231.

Shvornikova, G.M. 2010. *Reduction of energy costs for coal-water fuel transportation via optimization of processes and operating modes of hydrotransportation systems:* diss. Candidate of Science: 9.09.2010. Luhansk: East Ukrainian National University n.a. Volodymir Dal: 191.

Svitlyi, Yu.G. & Krut', O.A. 2010. *Hydraulic transportation of hard materials*. Donetz'k: Eastern Publishing House: 268.

Analysis to assess risk of occupational diseases at mining and preparation plants

I. Cheberiachko, Ya. Lebedev, Yu. Cheberiachko & D. Pustovoi
National Mining University, Dnipropetrovsk, Ukraine

ABSTRACT: Techniques to assess integral risk of diseases have been analyzed; it has been determined that any of the studied techniques cannot take into account use of means of individual protection of respiratory organs (MIPRO).

1 INTRODUCTION

Topicality. Coal dust is the basic factor of occupational diseases of respiratory organs for preparation plants staff. Large concentrations of coal dust can also decrease safety production level due to decrease in both natural and artificial light at working places increasing risk of high accident rate. Electrical fitters of crushers and conveyors operators are those who have the highest level of respiratory organs diseases.

It is impossible for the available dust control system to reduce dust concentration during basic operational processes down to the permissible level.

The problem statement. To decrease the disease level among mining and preparation plants staff analyze techniques to assess risk of occupational diseases origination and determine safe operation period in terms of dust contact (until pathological changes appear) for each job; calculate permissible dust burden being safe for human.

Analysis of recent published papers. Preparation plants were subject to chemical analysis of dust (Labour safety 2010); silicone dioxide, copper, iron, sulphur, calcium, aluminium, and cobalt have been found out. According to paper (Yanova et al. 2005) crystalline silicone dioxide and copper have fibrogenic effect on the workers. Air of crushing plant contains the highest level of silicone dioxide (23.4%). Disperse composition of dust at working places is of 2.1 – 5.0 mcm; thus it can penetrate deep into respiratory organs, besides if dust is present in pulmonary alveoli for a long time, then it results in occupational diseases (Golinko 2001). To determine safe dust contact working period (until pathological changes appear) foe each job connected with dust release, permissible dust burden is calculated to be further compared with a value of actual dust burden of a worker.

2 THEORETICAL STAGE

Several approaches to normalize dust burden are available today. Some of them are based on the roentgen diagnosing or on the principle of maximum changes in lungs (pathomorphological changes). Others are based on the determination of dust control mass depending upon permissible dust concentration and average working experience. In Russia they apply formula by Tkachov to determine dust burden (Labour safety 2010):

$$R = 8.6x_1 + 6.0x_2 + 19.4x_3k_1 + 6.4x_4k_2k_3, \qquad (1)$$

where R – integral coefficient of disease risk; x_1 – the age of a worker, years; x_2 – total working experience, years; x_3 – working experience involving dust contact, years; x_4 – dust burden, g; k_1 – coefficient involving SiO_2 content (being within 0.6 – 1.2); k_2 – coefficient involving mineral composition and dust concentration in the air (it is within 0.47 up to 2.2 depending upon the exceed of permissible dust concentration in terms of dust in working zone air being true for coal dust if it contains up to 5% of free silica); k_3 – coefficient involving labour intensity (being within 1.1 – 1.8). Disease risk of miners is determined depending upon integral index value (Table 1).

It is based on the results of medical examinations of workers' groups who had intensive contact with high dust concentrations in terms of ore mines. Moreover, disease risk R as well as critical values of dust burden can be calculated using formula by V.S. Beletsky (Yanova et al. 2005):

$$R = 3.88 \cdot 10^{-4} c^{3.02}. \qquad (2)$$

Disease risk can be assessed on the formula according to Kudinov technique (Golinko 2001):

$$R = 1 - exp(23.8 - 0.372PT) \cdot 10^{-3}. \qquad (3)$$

Table 1. Disease risk for the staff of mining enterprises depending upon integral index R.

Integral index	1000 1150	1151 1200	1201 1250	1251 1300	1301 1350	1351 1400	1401 1450	1451 1500	1501 1551	1551 1600	More than 1600
Disease risk, %	up to 2	5	10	20	30	40	50	60	70	80	90

Pathomorphological changes in lungs can help diagnose initial signs of pneumoconiosis originating long before roentgen diagnosing (according to V.V. Sukhanov). The researcher proposed following index to determine them (Testing technique for dust respirators 1996):

$$D = \left[exp\left(0.2473\left(\frac{0.1742 + 0.4 \cdot 10^{-5} \cdot}{\alpha(N - \beta)x_1 + 0.2345Qx_1 + 0.046CQ} \right) \right) - 1 \right] \times$$
$$\times \left[exp0.25\left(2.75 - \frac{5.8}{N+1} \right) \right] \frac{1}{K}, \tag{4}$$

where $a = exp\left(-0.3069 - \frac{1.1}{N} + \frac{2.226}{(V+2)^4} \right)$,

$b = -134.84 + 67.99\, exp \times$

$\times \left(0.1477V - 0.00252V^2 + 0.0113N - \right.$

$\left. -0.0000VN \right) + 134.03\, exp \frac{V}{N^2}$,

where x_1 – the age of a worker, years; V – the level of coal metamorphism, %; K – the coefficient involving dependence of coal metamorphism level upon pathogenic dust effect (being within 0.388 up to 1.26).

According to calculations by V.V. Sukhanov, safe level of changes when risk of pneumoconiosis is excluded is $D_P = 0.426$; the whole range in 0 to 6.

Using the value one can identify safe labour conditions and adequate control dust burdens.

To assess occupational disease risk A.I. Fomin proposed following formula having based upon numerous experiments (Sanitary regulations and standards 1996):

$$R = ae^{bx}, \tag{5}$$

where x – working experience; a and b are regression coefficients determined upon labour complexity with the help of certain tables (Sanitary regulations and standards 1996).

In foreign publications concerning the problem, works by M. Reisner and D.S. Muir should be noted; they proposed to calculate total disease risk according to the formula (Sololdilov 2006).

$$R = 1 - \left[exp - \left(\frac{X}{a} \right)^b \right], \tag{6}$$

where X – exposure dose of respirable dust, mg/m^3; α – scale parameter; β – a parameter of "Weibull" model form.

According to various experts values of α coefficient is within $31.5 - 92.8$ and as for β coefficient they are within $1.14 - 2.22$. Values of these coefficients being the most common ones are $\alpha = 54.0$; $\beta = 1.72$.

German scientists have determined the value of exposure dust dose for 35-year working experience. However, they stress that disease risk depends not only upon respirable dust concentration and working experience but also upon specific dust hazard. In Germany safety rules for coal enterprises separate working places not only according to respirable dust fraction concentration but also on quartz (SiO$_2$) content in the air. For example, if quartz content is less than 5%, then depending upon respirable fraction working places are categorized as follows: category 1 – up to 5 mg/m3; category 2 – 5 to 9 mg/m^3; category 3 – 9 to 14 mg/m^3. If quartz content is more than 5% then: category 1 – up to 0.25 mg/m^3; category 2 – 0.25 to 0.45 mg/m^3; category 3 – 0.45 to 0.6 mg/m^3. If dust content is more than permissible one of category 3, then operations are prohibited. It is determined that miners cannot work for more than 500 shifts at the working places of category 3 during 5-year period. Those who have worked for 500 shifts are transferred to the working places of category 1 or 2. They are under constant health control for the initial signs of pneumoconiosis to be diagnosed.

In our country safe operation period is calculated depending upon a value of CD control dust burden (Table 2) (Kozlova 1981) by:

$$T = \frac{R\Pi}{CNQ}. \tag{7}$$

3 RESULTS AND DISCUSSION

All analyzed techniques to prognosticate dust ethiology diseases are based upon the values of shift average dust concentration (some of them concerns only respirable one), working experience, and dust properties. However, each author has own mathematical form of disease risk dependence upon the given factors. Try to compare their calculation data to determine the most reliable technique to use it in future for substantiation of permissible working experience and assessment of disease risk level decrease if MIPRO are available. To do that various shift average dust concentrations with up to 2% free SiO content were taken; working experience of 15 and 39 years was considered. Lung ventilation value was 30 l/min; working shift was 6 hours; worker's age was 35 if working experience was 15 years and 50 if working experience was 30 years. Table 2 demonstrates calculation results concerning pneumoconiosis origination in terms of contact with dust having various shift average concentration; different formulas were involved.

Table 2. Calculation of pneumoconiosis origination according to data by Beletsky, Kudinov and Sukhanov.

Shift average dust concentration in the air, mg/m^3	Disease risk according to V.S. Beletsky, %	Disease risk according to Kudinov		Disease risk according to V.V. Sukhanov	
		15 years	30 years	15 years	30 years
5	0.050	—	—	0.26	0.35
10	0.406	—	—	0.32	0.43
30	11.213	—	—	0.39	0.52
50	52.447	—	—	0.44	0.57
62	100	—	—	0.47	0.60
142	—	—	0.20	0.63	0.78
150	—	—	0.69	0.65	0.80
155	—	—	0.83	0.66	0.81
200	—	—	1.00	0.75	0.91
285	—	0.25	—	0.93	1.10
300	—	0.69	—	0.96	—
350	—	0.99	—	—	—

Note that according to B.S. Beletsky it is impossible to calculate disease risk in terms of working experience as the formula expresses only dependence of disease risk upon dust content. Working experience has certain effect on the values of correction coefficients. It is difficult to understand how labour complexity according to dust concentration in the air is considered; it is also difficult to involve labour character of workers performing more difficult operations as more dust is inhaled. Moreover, mechanism of correction coefficients determining is not clear; it complicates calculation of working period both per shift and per the whole working period. Formulas by Kudinov and V.V. Sukhanov allow making more precisely prognosis of disease risk possibility comparing with the formula by the previous author. The formulas take into account working experience and the amount of dust accumulated in the lungs. However, according to Kudinov it is impossible to assess disease risk at low dust concentrations while according to V.V. Sukhanov if dust concentration is about 60 mg/m^3, then disease probability is about 50%. At the same time if there are high dust levels (more than 300 mg/m^3), then both techniques of disease risk assessment propose 100% of pneumoconiosis. It should be noted that while calculating probability of disease origination, technique by V.V. Sukhanov involves period of dust presence in worker's lungs, i.e. not only when a person was in direct contact with dust but also how much time passed after the contact.

Table 3 shows calculation results for integral index and pneumoconiosis disease risk according to formula by V.V. Tkachov. The obtained indices also illustrate that low concentrations of air dust within working area result in minor disease chance to compare with data by V.V. Sukhanov formula. However, it is impossible to say about more precise calculation technique for the disease risk as every author carried out research under specific conditions with certain disperse and mineral dust composition and its particles form, found out mathematical dependence which took into account labour organization, resting time rules, climate, health status and other factors typical for definite objects.

Analysis of data concerning the disease risk obtained according to A.I. Fomin formula (Table 4) helps conclude that the values of empirical coefficients have been developed for specific job in Kemerovo region (Russia). The data involved only average dust concentration which makes it impossible to use them to calculate the disease risk without extra specification of correlation dependences on labour conditions, dust mineral composition, and complexity of operations in terms of certain object. Table 5 illustrates calculation results using formula by Muir if $\alpha = 52$ and $\beta = 1.4$. Determination of the disease risk using the technique to compare with national ones is based upon respirable dust concentration in lungs. We can see that the obtained values coincide to some extent with the results obtained using a technique by Sukhanov.

Table 3. Calculation of pneumoconiosis origination risk according to V.V. Tkachov formula.

Shift average dust concentration in the air, mg/m^3	Disease risk according to Tkachov			
	Integral value of R disease risk if working experience is 15 years	Disease risk if working experience is 15 years, %	Integral value of R disease risk if working experience is 30 years	Disease risk if working experience is 15 years, %
40	988	up to 2	1162.6	up to 5
62	1089.4	up to 2	1263.9	10
80	1241.4	20	1416	50
90	1325.9	30	1500.5	60
100	1410.4	50	1585	70
110	1494.9	60	1669.5	90
120	1579.4	80	1753.9	100
140	1748.3	100	—	—

Table 4. Calculation of pneumoconiosis origination risk according to A.I. Fomin formula.

Job	Values of coefficients		Working experience, years	
	a	b	15	30
Shaftman	0.0009	0.1231	0.0057	0.03615
Shearer operator	0.0006	0.1382	0.00477	0.0379
Miner dealing with mine working support	0.0003	0.1571	0.0032	0.00334
Miner dealing with material delivery	0.00001	0.1871	0.000166	0.00274
Miner-electrical fitter	0.0003	0.0992	0.00133	0.00588
Overman	0.000009	0.2363	0.000312	0.01078
Mining adjuster	0.00001	0.1817	0.000153	0.00233

Table 5. Calculation of pneumoconiosis origination risk according to D.S. Muir formula.

Exposure dose of respirable dust, mg/m^3,	Disease risk if working experience is 15 years, %	Disease risk if working experience is 30 years, %
4	0.18	0.03
5	0.33	0.15
10	0.45	0.43
30	0.55	1.00
50	0.70	—
62	0.96	—
142	0.98	—
150	0.99	—
155	1.00	—

To assess labour conditions at mining enterprises an index of disablement professional risk taking into account factors of working environment and labour process, period of their effect on those who work, actual values of disability, number of workers with the recorded occupational diseases as well as average registered number of workers.

Index of $K_{pк}$ professional disability risk is determined as follows:

$$k_{pk} = \frac{\sum\limits_{1}^{n_{nki}} B\Pi_{ik}}{n_{ik}}, \qquad (8)$$

where $B\Pi_{ik}$ – calculated disability value of workers of kth job of ith subdivision; n_{nki} – the number of workers of kth job of ith subdivision having recorded occupational diseases; n_{ik} – average registered number of workers of kth job of ith subdivision.

Hazard of labor conditions in production workshops or within sites is determined by means of generalized index of professional disability risk:

$$k_{pч} = \frac{\sum\limits_{1}^{j} k_{pki}}{n_{ws}}, \qquad (9)$$

where K_{pki} – the index of professional disability risk for workers of kth job; n_{ws} – average registered number of workers of certain structural subdivision; j – the number of jobs in the subdivision.

To determine index of professional disability risk for the whole enterprise professional risk indices are calculated for each structural subdivision (K_{rs1}, K_{rs2}... K_{rs}) being identified as arithmetical average of indices calculated on structural subdivisions:

$$k_{pi.p.} = \frac{k_{rs1} + k_{rs2} + ... + k_{rsi}}{i_s}, \qquad (10)$$

where i_s – the number of the enterprise structural subdivisions.

The developed indices of professional risk make it possible to: give qualitative assessment of labour conditions effect on disability level; determine dynamics of changes in professional disability risk values depending upon working experience under hazardous conditions; develop technical solutions as for timely taking workers out of hazardous labour conditions as well as possibilities for rational staff use; have thorough study of labour conditions for jobs having highest levels of disability risk; assess social and economic efficiency of measured concerning labour conditions improvement to select the most important ones.

According the technique by W. Shewhart we propose implementation of criterion to react on professional risk index change in terms of mining enterprises. To do that average arithmetical values of risk indices have been calculated in the context of studied jobs groups as well as their σ mean-square deviations. Two control boundaries have been identified: both lower and upper warning limits have been established as:

$$K_{pk} - \sigma \leq K_{pk} \leq K_{pk} + \sigma, \qquad (11)$$

both lower and upper reaction limits have been established as:

$$K_{pk} - 2\sigma \leq K_{pk} \leq K_{pk} + 2\sigma. \qquad (12)$$

Table 6 gives calculation results of professional disability risk for different jobs at preparation plants as well as warning limits and criteria to react on changes in labour conditions.

Table 6. Calculation of professional risk index and reaction limits as for changes in labour conditions for basic jobs in field of mineral preparation.

Job	K_{pk}	σ	Improvement of labour conditions		Deterioration of labour conditions	
			$K_{pk} - \sigma$	$K_{pk} - 2\sigma$	$K_{pk} + \sigma$	$K_{pk} + 2\sigma$
Open-pit mining (crushing plant)						
Crusher operator	10.848	1.964	8.884	6.921	12.812	14.775
Conveyor operator	3.167	0.573	2.594	2.020	3.740	4.314
Electrical fitter	4.966	0.899	4.067	3.168	5.865	6.764
Control-panel operator	2.423	0.439	1.984	1.545	2.862	3.301
Crane operator	6.214	1.125	5.089	3.964	7.339	8.464

4 CONCLUSIONS

If values of risk index is close to $K_{pk} + \sigma$ limit, then σ labour conditions deteriorate and professional disease risk increases. If K_{pk} is close to $K_{pk} + 2\sigma$ limit, then level of occupational disease experiences rapid increase. To the contrary, if K_{pk} values approach $K_{pk} - \sigma$ mark, then labour conditions improve owing to implementation of efficient measures aimed at occupational disease prevention. If K_{pk} values are either equal to $K_{pk} - 2\sigma$ or exceed it, then labour conditions experience considerable improvement.

Thus, having analyzed calculation data according to techniques proposed by different authors we can conclude that unified assessment technique for disa-

bility risk taking into prompt account sharp changes in dust concentration, decrease in dust burden by means of MIPRO use, preventive medical measures is not available today. Some techniques are out of using a value of dust burden to predict disease origination.

REFERENCES

Golinko, V.I., Kolesnik, V.E., Ishchenko, A.S. & Cheberiachko, S.I. 2001. Development of facilities to test dust protection means and dust control. Scientific messenger of NMAU, No 3: 64 – 66.

Kozlova, A.V. 1981. Effect of fibrogenic dust on respiratory organs depending on concentrations and inhalation *period*. Dissertation abstract for the degree of Candidate of Technical Sciences. Moskow: 18.

Labour safety. Manual for mining students of higher educational institutions under the editorship of K.N. Tkachuk. Kyev: Osnova: 320.

Sanitary regulations and standards 2.2.3.570-96. 1998. Hygienic requirements for coal industry enterprises and organization of work. Calculation and regulation of personal doses of major hazard factors (dust, noise, vibration) as a compulsory measure to prevent diseases (time protection). Moskow: Ministry of Health of Russia: 52 – 57.

Sololdilov, A.I. 2006. *Pneumoconiosis disease of miners in Donetsk region coal mines*. Messenger of hygiene and epidemiology, Vol. 10, No 1: 37 – 40.

Testing technique for dust respirators applied at the enterprises of the Ministry of Coal Industry of Ukraine. 1996. Developed by "Nadezhda" JSC. Donetsk: 34.

Yanova, L.A., Pishchikova, E.V. & Sakhno, T.S. 2005. *Social protection and healthcare of workers at the enterprises of mining and metallurgical complex*. Collection of scientific papers "Quality of mineral raw material". Krivoi Rog: KTU: 514 – 517.

Theoretical and Practical Solutions of Mineral Resources Mining – Pivnyak, Bondarenko & Kovalevska (eds)
© 2015 Taylor & Francis Group, London, ISBN: 978-1-138-02883-8

Fundamental statements of selection mechanization structures in cyclic-flow technology

A. Glebov, G. Karmaev & V. Bersenyev
Federal state budget establishment the Institute of Mining Ural branch Russian academy of sciences, Russia

S. Kulniyaz
The Aktjubinsk regional state K. Jubanov university, Kazakhstan Republic

ABSTRACT: The urgency of the question of selection mechanization structures in cyclic-flow technology is grounded. In accordance with worked up fundamental statements the stages of selection basic equipment in cyclic-flow technology schemes are recommended. The tendency of varying selected estimate indices is determined depending on the annual transport volume and the height of rock mass hoisted by conveyors.

1 PREFACE

The employment of cyclic-flow technology (CFT) is one of the cardinal directions of technical re-equipment and intensification of mining operations in the open pits where deep-bedding deposits of mineral resources are developed. The improvement and rising the efficiency of this technology is grounded on progressive designing, technical and technological solutions.

Technological solutions should stipulate open pit space forming during the whole period of open pit development, providing the required movement of the SFT system's equipment without long-term failure its rhythmical pace of operation in the regime of technical potentialities maximum application. This is possible if fundamental solutions on open pit's development up to its final depth are taken, that is both the procedure and period of its construction are determined as well as the mode of mining operations, productive capacity, the duration and stages of development, open pit transport technological schemes, the moment of CFT introduction and others. At the stage of current mining operations performance it is highly necessary to foresee preparation of needed sites and workings for the crushing-conveying system's (CCS) equipment arrangement in case of its movement to a new place of disposition as well as transport lines for its installation and following maintenance.

These fundamental solutions, equally with geological and mining-technological conditions, present basic information for selection mining and transport equipment employed for development deep-bedding deposits of hard mineral resources.

Technical CFT re-equipment proceeds the path of making new equipment that meets the variety of geological and mining-technological conditions most completely as far as the open pit's depth increases. First and foremost, the development and making prospective CCS equipment should be mentioned, that is high-capacity mobile crushing and transferring sets (CTS) and steep-inclined belt conveyors well adapting in changeable open pit space. This direction is broadly developed in foreign countries. The efficiency of employment mentioned above equipment is approved by many researches (Martinenko & Elias 1996, Melnikov et al. 1995, Yakovlev et al. 1997, Yakovlev 2003, Vorriss et al. 2008 and others). The equipment of mining-loading and assembly transport CFT units is also being constantly modernized. New excavators and trucks' models are presented.

In the conditions of improving the technology of open pits development, permanent mining and transport equipment modernization the task of selection rational CFT mechanization structures and its methodological backing is always urgent.

2 THE RESULTS OF INVESTIGATIONS

A great lot of investigations are devoted to CFT equipment selection. In home practice the efficiency of employment in the structures of CFT mechanization is in most cases estimated basing on the known economic criteria (capital outlays, operating costs and reduced costs). For grounding the most efficient structure of CFT mechanization it is specified, what factors are to be considered and what criteria are to be followed when selecting basic equipment. But practically, however, the light is not thrown on the fact, by what procedure specific CFT features and some equipment operating indicators are considered when optimizing mechanization structures. In terms of methodological supporting the selection of truck-conveyor transport equipment the work (Vasiliev & Yakovlev 1972) should be mentioned. The economic-mathematical model of selection the optimal set of parameters of mining-loading, truck and crushing-conveyor equipment as well as fashions of considering various factors are cited in the work.

In the work (Simkin et al. 1985), devoted to integrated mechanization of CFT technological processes, principle requirements both to the equipment schemes and components and calculation the parameters of some kinds of equipment are adduced. Without stating scientifically grounded methodology of equipment selection, the system approach to estimate its operation is recommended with due regard for the parameters and adjoining technological processes interrelation. For this, the costs for final product manufacture are considered as a criterion.

The necessary condition of rational CFT systems forming is the coordination of equipment of adjoining interacting units of the system in their capacity and operation time. Keeping to this condition in practice is complicated. The difficulties consist in conversion the assembly transport discrete freight traffic into continuous freight traffic of certain value

in any moment of time planning the CCS operation. This is conditioned both by different reliability of transport facilities, rigidity of equipment link and the stochastic nature of adjoining units' interaction.

In the units of mining-loading equipment and assembly transport the rigid link among separate elements is absent, thus the failure in any elements in them doesn't bring to CFT system outage. It can operate with reduced capacity up to the replacement or re-conditioning serviceability of the equipment that has failed. The failure of CCS equipment with consecutive elements integration greatly influences on the whole CFT system operation, breaking down its rhythmic pace. This fact indicates that the CCS operation mode is defining for the whole CFT system and points to the necessity of considering the equipment reliability when determining the duration of efficient CCS operation.

The mode of CCS operation is described by both the time of its efficient operation and downtimes for different reasons. The analysis of CCS operation in the iron ore open pits has displayed that there exist downtimes in between shifts (casual, short-term, as a rule) and integral-shift downtimes, having the break duration in some unit of the equipment divisible to the time shift. The CFT system's downtimes in between shifts define the probable capacity decrease in separate units and in the whole system (Table 1).

Table 1. The equipment capacity decrease in adjoining technological units and in the whole SFT system.

Technological unit of the system	Average-shift capacity decrease by months, %											
	I	II	III	IV	V	VI	VII	VIII	IX	X	XI	XII
Excavators' unit	9.5	13.8	12.4	13.3	13.3	13.0	15.1	13.6	14.0	13.9	14.3	16.5
	2.1	2.7	2.3	1.8	2.0	1.9	2.9	1.9	3.0	2.7	2.3	2.1
Trucks' unit	0.85	2.25	2.05	1.2	1.2	0.85	0.65	0.65	0.95	0.65	0.7	0.2
	—	—	—	—	—	—	—	—	—	—	—	—
Crushing-conveyor systems	5.6	7.8	6.0	5.9	5.3	4.6	7.1	6.4	8.2	7.4	6.6	4.9
	5.4	7.0	5.4	5.6	4.0	4.0	5.6	5.6	7.3	6.1	5.4	4.0
Crushing mill	1.15	0.85	0.65	2.0	1.3	1.15	2.15	1.45	1.65	2.75	1.2	1.0
	—	—	—	—	—	—	—	—	—	—	—	—
The CFT system as a whole	17.1	24.7	21.1	22.4	21.1	19.6	25.0	22.1	24.8	24.7	22.8	22.6
	2.1	2.7	2.3	1.8	2.0	1.9	2.9	1.9	3.0	2.7	2.3	2.1

This indicates the urgency of considering the influence of adjoining technological units when both the CCS operation mode is determined and equipment is selected. The kind of downtimes in between shifts considered by calculation the efficient (net) time of CCS operation and their quantitative values are cited in Table 2 (Yakovlev et al. 2001).

Owing to stochastic nature of the CFT equipment interaction its downtimes in adjoining units are not possible to be completely matched. Thus, the task of cyclic units is to provide for the CCS hour-long

capacity that is necessary for execution the planned annual haulage volume during rated efficient operation time. It follows from this that not only the parameters of CCS equipment, but also the number of vehicles in cyclic units serving it are defined by necessary hour-long capacity. In this connection it is possible to divide the selection of basic CFT equipment into two blocks: the flow CCS rational equipment parameters are defined at first and then the ones of the cyclic system (excavator-truck system ETS). By this, the basic ETS equipment integrated parameters, that is the excavator's bucket capacity and truck's load-carrying capacity, are selected according to annual average haulage volume. These parameters rational combination is determined by mass module that presents itself the ratio of a truck's load-carrying capacity to load mass in an excavator's bucket (Kuleshov 1980). Operation modes of these units are defined in accordance with open pits development conditions.

Table 2. Downtimes in between shifts, considered when the crushing-conveyor system operation time is determined (expressed as a percentage from the annual time fund).

Kinds of down-times	The technological unit of CFT system	The reasons of downtimes	Downtimes duration
Operational-technological downtimes	Crushing-conveyor system	Ore absence Discharge site absence and crushing mill equipment downtime The absence or face stripping	0.7 – 2.8
	The excavators operation	The absence and accesses' arrangement, cross-overs driving, the face, level smoothing out	2.5 – 4.3
Organizational downtimes	Crushing-conveyor system	Shift change The absence of dump trucks	1.6 – 37
	The excavators unit	Shift change The excavator's change over or moving The absence of dump trucks Crew absence	3.4 – 6.1
Accidental downtimes	Crushing-conveyor system	The trouble mechanic and electrical conveyor equipment Monitoring equipment wear Drums pre-pressing, belts slipping and stuffing Filling in leaks Faults of crushers, feeders Fault of the excavator's mechanisms of hoisting, turning and pressure	Depending on the number of elements of the system (Kg of the system is considered)
	The excavators unit	Fault of the excavator's boom, bucket and dipper stick The fault of converter set The fault of hydro and pneumatic systems	3.4 – 6.0

Two approaches are possible for CFT systems designing:

1. The parameters of stock-produced conveyor equipment being preliminary specified (flow unit), the annual CCS transporting capacity is defined:

$$Q_t = Q_h \cdot T_{nt},\qquad(1)$$

where Q_t – passport hour-long capacity of conveyor equipment, t/h; T_{nt} – net operation time, h.

Then, proceeding from the hour-long CCS capacity, both the selection the necessary flow unit equipment and development the technology of mining operations that provides for mining and transportation the planned annual rock mass volume are performed.

2. Proceeding from mining-technological potentialities and necessary volumes of rock mass output from an open pit, the determination of expedient annual CFT system's capacity is carried out. According to the set volume of rock mass processing the required hour-long CCS capacity is calculated:

$$Q_{hc} = \frac{Q_{ac}}{T_r},\qquad(2)$$

where Q_{ac} – the annual CFT system capacity, t.

According to the Q_{hc} value the parameters of conveyors, crushers and the CFT system cyclic units' equipment should be selected.

The second approach is undoubtedly preferable when the variation of produced equipment standard size (parameters) is limited.

Besides, having an essential number of conveyors' standard sizes, it is not always possible to choose conveyors with parameters meeting specific conditions of CFT system operation. As a rule, individually produced equipment is required for them; this is approved by the practice of CFT systems designing in the open pits of CIS and far abroad countries.

The duration of efficient operation (net operation time) in the CFT systems with continuous working week is defined according to the expression:

$$T_r = (T_{tf} - T_{tm} - T_{kl} - T_{tech} - T_{org} - T_{de})K_{af},\qquad(3)$$

where T_{tf} – annual time fund, h; T_{tm} – the time for performing maintenance and CCS equipment repairs, h; T_{kl} – the time for equipment downtimes for climate reasons, h; T_{tech} and T_{org} – the duration of in-between shift incompatible downtimes of the system for technological and organizational reasons, h; T_{de} – the duration of in-between shift incompatible downtimes of the excavators unit, h; K_{af} – CCS equipment availability factor.

Estimating the efficiency of employment this or that kind of equipment in the structures of CFT systems mechanization the selection of comparing indices of efficiency function is of great importance.

The selection of CFT systems equipment refers to the tasks of comparison variants not very differing and more often the same: the production capacity and duration of construction. In this case the known indices, that are discounted profit, the profitability index, internal rate of return and payback period may not be used; the equipment selection can be limited by reduced capital outlays and operating costs for optimization period (7 – 10 years) that would be nearer to the CCS equipment service life, defined by depreciation rates. For the pointed out period the stock of assembly transport is also renewed.

If capital outlays and operating costs in comparable variants turn out to be slightly distinguished, then the additional criteria should be introduced: equipment metal hardware, required energy saturation, labor productivity per one working person and ecologic indices (pollution the atmosphere by dust and exhaust gases, etc.). It is expedient, for this, to use specific indices (per 1 ton of processed rock mass); possible changes in the equipment components for optimization period should also be considered.

The basic postulates of CCS equipment selection come up to the following:

1. Initial data preparation that allows to determine conveyor transport parameters, to choose the type and parameters of the adjoining units equipment, to calculate capital outlays and operating costs for processing (removing) 1 ton of rock mass and other indices according to selected criteria of SFT system efficiency.

2. The selection of conveyor lines equipment is put into practice with due regard for interaction of adjoining technological units of CFT systems. It is quite sufficient, for this, to consider the influence of excavator-truck and crushing-conveyor systems' influence on the efficiency of their mutual operation and on the system as a whole. Drilling-blasting operations are not first-hand studied as their influence is reflected indirectly: through the quality of preparation blasted rock mass, mining-loading equipment operation and is considered by way of changing the excavators' output and modes of operation. The adjoining technological units' influence is estimated to shift-average capacity loss value proceeding from the downtimes for accidental, technological and organizational reasons.

3. The crushing-conveyor systems operation mode should be considered as defining when the duration in the CFT systems equipment operation is calculated. The adjoining units' operation should provide for crushing-conveyor systems operation in the mode of their technical potentialities maximum realization by way of forming optimal equipment reserve to perform appropriate technological operations. The necessary hour-long capacity of CFT system is determined proceeding from specified annual rock mass haulage time and most probable time of crushing-conveyor system efficient operation.

4. According to hour-long capacity with due regard for conveyors loading irregularity in crushing-transfer points the selection of system's equipment is conducted. The equipment is selected with parameters that provide for annual haulage volume absolute realization.

5. In connection with various produce complexity and conveyor components longevity (drive station, idlers, linear flight bearing constructions, belts) their price is taken into account separately when capital outlays are calculated. Operating costs are determined with due regard for service life of quickly deteriorated elements as their longevity is considerably lower than conveyors depreciation period as a whole.

The selection of ETS mining-loading equipment is performed in the following order.

1. According to initial data on annual output of mineral resource or overburden, mined with CFT system employment, the excavator type is chosen in terms of analogues available with due regard for open pit production capacity as a whole.

It should be considered, by this, that close interrelation between open pit production capacity and basic excavator's parameter, that is bucket volume, doesn't practically exist, but the tendency of employing excavators of greater capacity with an open pit output increasing is not excluded.

2. The EAS mass module is calculated by the selected excavator type. It makes it possible to determine load-carrying capacity of a truck in its values rational interval.

3. The hour-long excavator output is defined, the number of workers per a shift as well as the inventory excavators' fleet are determined proceeding from the necessary hour-long CCS capacity.

4. Capital outlays, operating costs and other indices characterizing the efficiency of excavators' fleet operating as a part of ETS are calculated.

The best model is presented by the excavator of rational ETS type selected on the results of comparison estimate indices of several systems with equipment parameters suitable for the recommended mass module interval values.

The assembly truck transport equipment selection comes up to trucks' fleet forming providing for necessary hour-long and annual freight traffic of rock mass processed by CCS. This task is, as a matter of principle, solved by way of searching for trucks' fleet rational age and run structures that provide for the CFT system production program reliable realization with profitable economic indices. It follows from this that the assembly truck fleet structure should meet the requirements:

– obligatory performance of necessary rock mass haulage volume or transporting work;

– the unit cost of rock mass transportation or transporting work should ensure profitability of truck transport operation, that is its value should not exceed the admissible one that ensures break-even trucking operations.

The task of trucks fleet forming is solved in several steps:

1. The trucks' load-carrying capacity and model are selected for the CFT system in known open pit mining and technological conditions and set annual rock mass output and the necessary trucks quantity is defined according to necessary CCS hour-long capacity.

Trucks' load-carrying capacity is chosen by the rational ratio both of the excavator's bucket volume and truck's body. The ratio determines the degree of efficient CTS operation.

The best model among the trucks of one load-carrying capacity class of different firms-producers for concrete pit conditions is chosen in terms of estimation the level of consumer qualities and competitiveness (Glebov 2008).

2. The selection of the number of age groups per each truck model in the fleet.

The number of age groups in technological truck transport fleet is limited by trucks service life recommended by a plant-producer. Service life is defined by the ratio of 90% regulatory resource to annual-average operating time. The resource is interpreted as a truck operating time from the moment of putting into operation up to truck cancellation.

The structure of technological truck transport fleet possesses complicated dynamics of its parameters that change on a time basis under the influence of different factors. At any time moment there are trucks in the fleet that operate with both different load-carrying capacity, models and age groups, thus the number of age groups is defined analogously for all trucks' models available in the fleet and operating in different operating conditions. The number of age groups is not constant. It depends, to a certain extent, on trucks model, their operating conditions and can change as a result of trucks design improvement and increasing their operation reliability.

3. Setting up empiric dependences of basic technical-and-economic indices of trucks operation from their age.

During statistical data processing the plots of changing annual specific unit cost of 1 ton rock mass of 1 t·km transport work cost as well as the capacity for a definite time period are constructed. In terms of these plots the empiric dependence between specific unit cost of 1 t rock mass or 1 t·km transport work cost and years of a truck operation is set. It is described by the function $C_s = f(tt)$. The empiric dependence of the trucks capacity and years of operation is set analogously: $Q = f(tt)$.

4. The determination of weighted-average trucks age in the fleet that meets specified criteria.

Weighted-average trucks age twa is defined by solving the approximating functions equations S_s and Q relatively tt_a. Naturally, for each mining plant the dependences and approximating functions consequently, will be of different view. Thus, the equations of admissible tt_a by chosen criteria will be different too.

5. Technological transport truck fleet forming.

Trucks fleet forming should be put into practice on the principle of keeping weighted-average trucks age on the level that provides for required performance of annual rock mass haulage volume and transport plant's work profitability, that is when specific cost of rock mass transportation does not exceed the admissible value ensuring trucking operations profitability.

The fleet could be formed of new trucks and those, formerly employed. Thus, the set forth criteria permit to regulate the size of once-only investments as well as to distribute them evenly in time. Besides, it is expedient to predict the plan of trucks fleet forming for the period not more than 5 – 7 years.

General tendencies of changing estimate indices from basic influencing factors (annual trucking volume and rock mass hoisting height by conveyors) are set by the investigations with economic-mathematical model application. The model has been developed with due regard for stated above postulates on the CFT system equipment selection.

The analysis of calculation costs on the CFT systems displays that the costs share on assembly truck transport prevails in the equipment price. The shoulder of rock mass delivery from faces to CTP being rational, the delivery makes up 40 – 60% depending on both the annual haulage volume and hoisting height of the material transported by conveyor elevator (Figure 1).

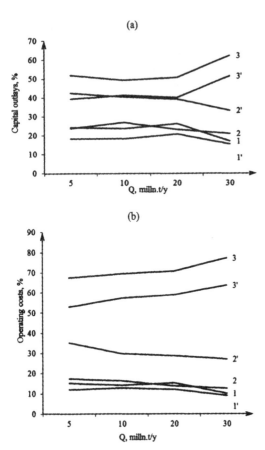

(a)

(b)

Figure 1. The change of capital outlays and operating costs on the CFT equipment depending on annual haulage volume: 1, 2, 3 – of excavator unit, CCS, assembly transport accordingly with hoisting height equaled to 100 m; 1', 2' 3' – the same with hoisting height being 600 m.

In operating costs the assembly truck transport share reaches 55 – 75% from aggregate costs on basic EFT equipment. The mining-loading equipment unit is less expensive, its share in capital outlays lies in the 15 – 24% interval and in operating costs – in 9 – 15% interval. Capital outlays share on the CCS equipment makes up 21 – 42%; operating costs – 13 – 35%. As the trucking volume increases, the costs share on equipment of accessory CFT units changes in non-single-valued nature. The annual CFT system's output increasing from 5 to 30 million tons, the share of capital outlays and operating costs on assembly transport equipment grows 1.2 – 1.3 and 1.15 – 1.2 times accordingly. At the same time capital outlays share on the CCS and mining-loading equipment decreases 1.1 – 1.3 and 1.2 – 1.4 times accordingly and operating costs share decreases 1.3 – 1.5 times.

Assembly transport, possessing dump trucks of minimum mass share, requires greater energy saturation that is determined by engines' capacity. When rock mass is hoisted at 100 meters height by conveyor elevator, the trucks' engines installed capacity makes up about 65% from the total required capacity of basic CFT systems' equipment. With rock mass hoisting height growth the redistribution of required energy saturation takes place that moves up its CCS share and moves down in its assembly transport unit share up to 50 – 55%. Thus, the excavation process is less energy intensive. The installed capacity share of mining-loading equipment unit makes up 10 – 15% (Karmaev & Tjulkin 2004).

Such relationship of costs on the systems' equipment indicates the CFT processes labor intensity and the necessity of assembly transport facilities improvement in the direction of the most complete conformity with consumption quality required level. The consumption quality level is a generalized indicator of private dump truck quality level that characterizes technical perfection, service maintenance facilities in the process of operation; maximum truck's operation serviceability in various mining-technological conditions and rock mass transportation productive efficiency.

3 DEDUCTIONS

The stated above materials permit to formulate the following results of investigations:
– the urgency of the problem of selection basic equipment for CFT mechanization structures is grounded;
– it is set up that the CCS mode of operation is a defining factor when the whole CFT system's equipment is selected. Its efficient operation time is calculated with due regard for adjoining units influence that proceeds from both planned and in-between shift downtimes of CCS and ETS equipment for accidental, technological and organizational reasons;
– basic postulates are elaborated and the stages of selection both the CCS equipment and the one of mining-loading and assembly units are recommended. Selecting the system's equipment, the interrelation is carried out by way of considering both the efficient operation time and basic operating indicator, that is hour-long unit capacity characterized by the greatest rigidity of elements connection;
– the principle scheme of selection is elaborated. It provides for optimization the equipment parameters of conveyor lines as well as qualitative and quantitative relationship of the equipment parameters in adjoining units of CFT systems;

– the nature of variation capital outlays and operating costs on CCS equipment as well as on mining-loading and assembly transport is determined depending on annual output and rock mass hoisting height elevated by conveyors.

The greatest costs fall at assembly truck transport. Their share in total capital outlays and operating costs reaches 40 – 60% 35 – 75% accordingly. Mining-loading equipment unit is the least expensive; its costs share makes up 15 – 24% (capital outlays) and 9 – 15% (operating costs).

The stated methodological principles could be applied for optimization the CCS equipment parameters that include both steep-inclined and traditional conveyors, as well as for grounding rational structures of CFT systems mechanization in concrete mining-technological conditions.

REFERENCES

Glebov, A.V. 2008. *The methodology of estimation the geo-technique consumption qualities and competitiveness level (open pit dump trucks as an example)*. Mining equipment and electro-mechanics, No 5: 49 – 55.

Karmaev, G.D. & Tjulkin, A.P. 2004. *The features of selection equipment of cyclic-flow technology systems in deep open pits*. Geo-technical problems of integrated mineral resources mining: coll. scient. works. Yekaterinburg: IM UR RAS: 437 – 444.

Kuleshov, A.A. 1980. *High-capacity excavator-truck systems of open pits*. Moscow: Nedra: 317.

Martinenko, V. & Elias, K.H. 1996. *The first FRUPP firm crushing-conveyor system in Poltavsky integrated works*. Mining industry, No 3: 27 – 29.

Melnikov, N.N., Usinin, V.I. & Reshetnyak, S.P. 1995. *Cyclic-flow technology with mobile crushing-transferring systems for deep open pits*. Apatites: the MI KSC RAS: 192.

Simkin, B.A., Dikhtyar, A.A. & Ziborov, A.P. and others. 1985. *Integrated mechanization of cyclic-flow technology processes in the open pits*. Moscow: Nedra: 195.

Vasiljev, M.V. & Yakovlev, V.L. 1972. *Scientific grounds of open pit transport designing*. Moscow: Nauka: 202.

Vorriss, P., Hustrulio, A. & Makeev, A. 2008. *Crushing and transportation systems in the open pits*. Mining industry, No 1: 24.

Yakovlev, V.L. 2003. *Prospect solutions in the field of CFT technology of deep open pits*. Mining magazine, No 4 – 5: 51 – 56.

Yakovlev, V.L., Karmaev, G.D. & Tjulkin, A.P. 2001. *Methodological principles of optimization conveyor equipment parameters in CFT systems*. Open pit transport: problems and solutions. Yekaterinburg: IM UR RAS: 142 – 152.

Yakovlev, V.L., Tjulkin, A.P. & Karmaev, G.D. 1997. *The progress of CFT systems with steep-inclined conveyor hoisting*. The problems of mining: coll. scient. works. Yekaterinburg: IM UB RAS: 194 – 205.

Theoretical and Practical Solutions of Mineral Resources Mining – Pivnyak, Bondarenko & Kovalevska (eds)
© 2015 Taylor & Francis Group, London, ISBN: 978-1-138-02883-8

Probability estimates for the operation modes of mining machinery and equipment overshooting the limits of their normal functioning

V. Slesarev & A. Malienko
National Mining University, Dnipropetrovsk, Ukraine

ABSTRACT: The variety and variability of mining and geological conditions, in which mining machinery and equipment (cutter-loaders, heading machines, conveyor transport, auxiliary equipment) operate, lead to the overshoot of their planned or optimal operation modes. In this connection, when developing the effective control systems for mining machinery and equipment, there arises the problem of determining the real probability of overshooting the predetermined limits for the mentioned operation modes, within the framework of which, from the technical or economic point of view, functioning of mining machinery is possibly still acceptable. The mathematical solution of this problem is reduced to determining the probability characteristics of the intersection of random variables of a given level i.e., it is reduced to solving "the problem of overshoots".

1 INTRODUCTION

In general, obtaining the probability characteristics of overshoots of a random function (probabilities of the predetermined number of overshoots in a given period of time, the law of distribution of the stay time of the random function above a predetermined level) encounters considerable mathematical difficulties. However, it is relatively easy to calculate such numerical characteristics as the average number of overshoots per unit of time, the average stay time above a given level, etc.

2 MAIN PART

General formulas for their determination were established by Stephen O. Rice (Sveshnikov 1968) in 1944 – 1945 and they are used for all (differentiable) random processes, although the numerical result can be obtained, for example, for the normal processes of determining the average number of overshoots and the average duration of an overshoot of a random function of a given level.

Let $X(t)$ be the differentiable random process and a – the value of the ordinate of the function $X(t)$ describing the level for the "overshoots" of interest (Figure 1). First, we will consider the determination of the probability that in the infinitesimal time interval dt, immediately following the time moment t, an overshoot will occur. In order that the overshoot can actually happen under the stated conditions, two events should take place: first, at the time moment t the ordinate of the random function should be less

than a i.e., $X(t) < a$, and secondly, at the time moment $t + dt$ the ordinate of the random function should be greater than a i.e., $X(t + dt) > a$.

Consequently, the probability of overshoots in the time interval dt can be written as follows:

$$P[X(t) < a; X(t + dt) > a],\qquad(1)$$

In the conditions of differentiability of the random function $X(t)$ (Fomin 1980), each of its coordinates in the small interval dt can be expressed via the previous value by using the approximate (up to the second-order infinitesimal) expression:

$$X(t + dt) = X(t) + V(t)dt,\qquad(2)$$

where $V(t) = \dot{X}(t)$.

Consequently, the inequality $X(t + dt) > a$ is equivalent to the inequality $a - V(t)dt < X(t)$ so instead of the two inequalities that condition (1) the availability of overshoots in the time interval dt, one can write one double inequality $a - V(t)dt < X(t) < a$, provided that:

$$(V(t) > 0).\qquad(3)$$

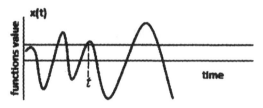

Figure 1. Random process.

To calculate the realization probability of this inequality in (Tikhonov & Samarskii 1999) the two-dimensional distribution law of an ordinate of a random function and its derivative at one and the same moment of time has been considered:

$$f(x, v|t),\qquad (4)$$

then for the desired overshoot probability we will have:

$$P[a - V(t)dt < X(t) < a] = \int_0^\infty \int_{a-vdt}^a f(x, v|t)\,dxdv,\qquad (5)$$

where the integration limits cover all the values $X(t)$ and $V(t)$, satisfying the inequality (3). The inner integral in (5) can be calculated immediately as its integration limits differ by the infinitesimal value V_{dt}, hence, by using the mean value theorem, we obtain $\int_{a-vdt}^a f(x, v|t)\,dx = dt \cdot vf(a, v|t)$.

Substitution of the equation (5) into (4) gives:

$$P[a - V(t)dt < X(t) < a] = dt \int_0^\infty f(a, v|t)v\,dv.\qquad (6)$$

The resulting formula shows that the probability of an overshoot in the infinitesimal time interval dt is proportional to the interval value. Therefore, it is expedient to introduce the notion of time-density for the overshoot probability (Emelanov 2004), denoting $p(a\,t)$ as the probability of an overshoot per level a, at the time moment t, calculated per unit of time i.e.:

$$P[a - V(t)dt < X(t) < a] = p(a)dt.\qquad (7)$$

Comparison of (7) with (6) provides the final equation for the probability density $p(a|t)$:

$$p(a) = \int_0^\infty f(a, v|t)v\,dv.\qquad (8)$$

Similarly, it is possible to calculate the time probability density $p'(a|t)$ for the intersection of a random variable of the level a top down. Summerizing the arguments given above we obtain:

$$p'(a) = -\int_{-\infty}^0 f(a, v|t)v\,dv.\qquad (9)$$

Adding and subtracting (8) and (9), we have:

$$p(a|t) + p'(a|t) = \int_{-\infty}^\infty f(a, v|t)|v|\,dv.\qquad (10)$$

$$p(a|t) - p'(a|t) = \int_{-\infty}^\infty f(a, v|t)v\,dv.\qquad (11)$$

On the other hand, since $f(a, v|t) = f(v|a, t)f(a|t)$, the right parts of the last two equations can be expressed via the conventional mathematical expectations of the rate v of change of the process and its absolute value i.e., it is possible to rewrite these equations in this way:

$$p(a|t) + p'(a|t) = f(a|t)M\big[|V(t)|X(t)|\big|X(t) = a\big],$$

$$p(a|t) - p'(a|t) = f(a|t)M\big[V(t)|X(t) = a\big].\qquad (12)$$

Using the expression (8) allows us to calculate the average stay time of the random function above a given level for any period of time T. Divide the period of time T into equal small time intervals dt_j, each of which is located near the time moment $t_j(j = 1,2...n)$. There is the probability that the random function ordinate is above a given level:

$$P|X(t_j)| > a| = \int_a^\infty f(x|t_j)\,dx.\qquad (13)$$

Assume the values of the time intervals dt_j so small that when calculating the total stay time of the random function above a given level, we could neglect the cases when the function within the time interval $[X(t)-a]$ changes its sign. Consider the system of random variables Δj, each of which is equal to the corresponding interval dt_j or 0 depending on whether the random function in this time interval is greater or less than a. Then it is obvious that the total stay time T of the random function above a given level a is equal to the sum Δj i.e.:

$$T_a = \sum_{j=1}^n \Delta_j.\qquad (14)$$

To determine the average stay time of the random function above a given level for the period of time T, we will find the mathematical expectation of both parts of the equation (14). Applying the theorem of the mathematical expectation of the sum, we find:

$$t_a = \sum_{j=1}^n M|\Delta_j|.\qquad (15)$$

By definition, the random variable Δj can take only two values (dt_j and 0), hence its mathematical expectation is equal to the product dt_j and the probability (13) i.e.:

$$M|\Delta_j| = dt_j \int_a^\infty f(x|t_j)\,dx.\qquad (16)$$

Substituting (16) into (15) and passing to the limit of $n \to \infty$ instead of $\sum_{j=1}^n$ we obtain the integral, and

394

for the average stay time of the random function above the level a, calculated for the period of time T, we have:

$$\bar{t}_a = \int_0^T \int_a^\infty f(x|t)dxdt . \tag{17}$$

Typically, there is an interest to the average stay time of the random function above a given level per overshoot. To determine this average stay time τ it is necessary to divide the time $\bar{t}_a$ into the average number of overshoots occurring during the time T.

To determine the value $\bar{n}_a$, we will again divide the period of time T into n equal time intervals dt_j and introduce the auxiliary random variables N_j each of which is equal to one if an overshoot occurred during the corresponding interval of time (due to small duration of the time intervals dt_j the possibility of having more than one choice can not be ignored), and zero in the opposite case. Then the total number of overshoots N_a in the time interval T will be equal to the sum of the values N_j:

$$N_a = \sum_{j=1}^n N_j . \tag{18}$$

By finding the mathematical expectation of both parts of the equation (18) and taking into account that the mathematical expectation of each of the variables N_j is numerically equal to the overshoot probability in the j-th time interval i.e., $p(a|t_j)dt_j$ we have:

$$n_a = \sum_{j=1}^n p(a|t_j)dt_j. \tag{19}$$

Increasing the number of time intervals dt_j and substituting the expression (8) instead of $p(a|t)$ we obtain:

$$n_a = \int_0^T \int_0^\infty vf(a,v|t)dvdt . \tag{20}$$

Finally, the division of (17) into (20) gives the desired average duration of the overshoot:

$$\tau = \frac{t_a}{n_a} = \frac{\int_0^T \int_a^\infty f(x|t)dvdt}{\int_0^T \int_0^\infty vf(a,v|t)dvd} \tag{21}$$

3 CONCLUSIONS

The obtained formulas are of great interest for stationary processes, since only for steady in time processes the average overshoot duration has a direct visual effect.

REFERENCES

Emelanov, V.E. *E-compatibility of electronic equipment.* Part 1. Features of the analysis. M: MSTUCA: 108.

Fomin, I.A. 1980. *Emission theory of stochastic processes.* M: Communications: 216.

Sveshnikov, A.A. 1968. *Applied methods of the theory of random functions, 2nd edition, revised and enlarged.* Home Edition Phys-Math. Literature. M: 457.

Tikhonov, A.N. & Samarskii, A.A. 1999. *Equations of Mathematical Physics.* Publishing house. M.: 798.

Theoretical and Practical Solutions of Mineral Resources Mining – Pivnyak, Bondarenko & Kovalevska (eds)
© 2015 Taylor & Francis Group, London, ISBN: 978-1-138-02883-8

The features of calculations of hydrotransport plants of geotechnological systems

E. Semenenko
*M.S. Poliakov Institute of Geotechnical Mechanics under the
National Academy of Sciences of Ukraine, Dnipropetrovsk, Ukraine*

N. Nykyforova
National Metallurgical Academy of Ukraine, Dnipropetrovsk, Ukraine

L. Tatarko
Ukrainian State University of Chemical Technology, Dnipropetrovsk, Ukraine

ABSTRACT: Modernization of hydrotransport plants of modern geotechnological systems with the purpose of decrease in power inputs and specific water consumption for hydrotransportation process is reasonable by using of polyethylene pipes and drag reducing agents. The method of economically sound diameter of pipeline of hydrotransport plant is elaborated for the case of using of the polyethylene pipes or drag reducing agents. The dependence of capital expenditure on flow hydraulic parameters for polyethylene pipelines is revealed. It is shown that identical approach to accounting of changing of hydrotransport parameters is feasible when using polyethylene pipes and drag reducing agents. This approach consists in using of relevant values of proportionality coefficient and of exponent in the dependence of drag coefficient on Reynolds number and of proportionality coefficients of summands in the dependence of critical velocity on slurry concentration. It is shown that average values of exponent in the dependence of drag coefficient on Reynolds number may be used in calculations of economically sound diameter of the polyethylene pipeline with engineering accuracy.

1 INTRODUCTION

The successful integration of all industrial enterprises of Ukraine into European and world markets nowadays is possible providing the cutting of production costs and competitive recovery. This is especially topical for ore-dressing and processing enterprises, because their production accounts the considerable part of Ukrainian export. Domestic ore-dressing and processing enterprises have to stand the competition with high-technology foreign mining companies and also to comply with international product quality standards and with environmental safety specification. The most effective method of implementation of marketing strategy of cutting of production costs and competitive recovery for ore-dressing and processing enterprises is a cut in expenditure for transporting of mined, processed and stackable bulk material inside of proper geotechnological systems (Alexandrov 2000). The pressurized hydrotransport is used first of all in the modern geotechnological systems for transporting of feed stock, intermediate products, concentrates and cleaning rejects. So in this case decrease in power inputs and specific water consumption is most reasonable (Alexandrov 2000, Semenenko et al. 2013).

The hydrotransport plant became one of the key fragments of geotechnological system and it determines stability of technological process and prime cost of finished concentrates in many respects (Semenenko et al. 2013). Using of water as a separation agent is typical for all ore-dressing and processing enterprises of Ukraine. The main technological units namely crushers, mills, screens, hydrocyclones and separators are meant for elutriation in water medium and for separation of slurries. These units are interconnected solely by hydrotransport and the situation at the foreign mining companies is technologically similar.

The problem of scientific substantiation of parameters and operating modes of the systems of cleaning rejects removal is especially topical and essential against a background of increased requirements of environmental safety, a tendency to implementation of resource and energy saving technologies and new economic actuals. This forces to change an attitude to hydrotransport plants of geotechnological systems, to install control and accounting equipment, to carry out the monitoring of parameters and operating modes, to automatize the regulation of operating modes of pumps.

So substantiation of parameters and operating modes of hydrotransport plants of geotechnological systems is very important for development of ore-dressing and processing enterprises of Ukraine.

It is necessary to take into account some aspects when solving formulated problem (Alexandrov 2000, Semenenko et al. 2013).

Firstly the existing field experience of hydrotransport complexes, which transport high concentration slurries, is negligible. This is caused by economic conditions, under which all hydrotransport plants and waste storages were projected at the ore-dressing and processing enterprises of Ukraine. These conditions determined low cost of electric power and indifference to natural resources of region where enterprise was located. As a result the hydrotransport systems transported and stored slurries with mass concentration of solid material not more than 5% that is nearly water.

Secondly the greater part of known experimental research of dependence of hydrotransport parameters on slurry concentration dates back to the 1970s and doesn't take into account the changes in properties of transporting materials in this space of time. These investigations were carried out at sectoral research institutes, which were liquidated for different reasons or were restructured for another function at the turn of the 1980s.

Thirdly the data about hydrotransport parameters and properties of cleaning rejects lain in storages from the end of 1970s are absent. Nobody assumed when projecting waste storages that cleaning rejects will be iteratively used or processed in future. Furthermore in the beginning there were research institutes handled hydrotransport but there were no cleaning rejects lain for a long time. When such cleaning rejects have appeared the research institutes have disappeared.

Fourthly the known methods of estimation of economic efficiency of hydrotransport plants and of substantiation of their rational parameters don't take into account existing economic actuals. All these methods were elaborated before 1990s and were oriented to the conditions of socialist economics. Furthermore cost minimization consisted in the choice of minimal hydraulic gradient when projecting hydrotransport systems at that time. Only a part of operating costs was minimized in so doing, but capital and current expenditure were disregarded.

Fifthly in the beginning of 2000s the new methods of drag reduction have appeared, which are effective for pipeline transport of water and gas and are promising for hydrotransport plants of geotechnological systems (Table 1) (Shvabauer et al. 2006, Market Report Company 2006). Using of polyethylene pipes in place of steel pipes and doping of drag reducing agents to slurries must be applied to such methods (Semenenko et al. 2013, Stupin 2000).

Table 1. Distribution of Ukrainian market size of polyethylene pipes according to their intended purpose.

Purpose	Market share of polyethylene pipes, %			
	2004	2005	2006	2007
Gas	59	44	41	33
Water	38	53	55	62
Other	3	3	4	5

USA and European countries have stored sufficient practices in using of polyethylene pipes for transport of water, gas and waste water. This experience shows that polyethylene pipes have less drag and running weight, greater thermal capacity and wear resistance in comparison with steel pipes. The cost of polyethylene pipes strongly depends on working pressure. Low pressure pipes are cheaper, but high pressure pipes are more expensive than steel pipes. The analogous experience is known also in Ukraine (Semenenko et al. 2013) and the trend of changing of production quantity of domestic polyethylene pipes allows extension of their range of application. Furthermore in Ukraine first in Europe polyethylene pipes were used for hydrotransport of ore sands at the Vilnogorsk mining and smelting enterprise and this instantiates a possibility of such pipes using in geotechnological systems.

Drag reduction at the expense of doping of drag reducing agents to water flow in pipelines is used in refrigeration plants, in fire-control units. The projects of use of drag reducing agents in large water supply systems were considered in the USSR (Stupin 2000). Doping of drag reducing agents to pressure flow results in changing of wall turbulence and this ensures friction reduction. Furthermore drag reducing agents are widely used as flocculants during water purification (Stupin 2000). The perspective of using of drag reducing agents for pressure flows of slurries consists both in drag reduction and in pulp thickening during its transporting. But this requires the substantiation of hydrotransport parameters and selection of drag reducing agent type depending on properties of transporting material and of carrying agent.

So it is necessary to take into account new features of working of hydrotransport plants of geotechnological systems during substantiation of rational parameters and operating modes.

2 MAIN PART

The purpose of the paper is elaboration of calculation method of economically sound diameter of pipeline of hydrotransport plant when using the polyethylene pipes and drag reducing agents with regard to features of geotechnological systems.

The known procedures of cost effectiveness evaluation are based on using of reduced annual expenses (Semenenko et al. 2013):

$$\Pi = C + (p + E)K , \tag{1}$$

$$C = (Z_N + (Z_E + Z_T)T)N + Z_0(1 - S)QT + C_E , \tag{2}$$

$$K = g\pi D_0 \delta_0 \rho_\Pi \left(1 + \frac{1 - \omega}{\omega}\frac{L_\bullet}{L}\right)LK_1' +$$

$$+ (DK_2 + K_3)L + 1.1K_4 N + K_0 , \tag{3}$$

$$N = (1 + ArS)\frac{\rho_0 gQ}{\eta}H' , \tag{4}$$

$$H' = k_Z LY\frac{Q^{2-n}}{D^{5-n}} + (1 + ArS)\Delta Z , \tag{5}$$

$$Y = \left(1 + (q - 1)\frac{L_\bullet}{L}\right)\left(1 + \frac{ArS(1 - S)}{1 + ArS}S\right)\times$$

$$\times \wp\frac{m2^{3-2n}v^n}{\pi^{2-n}g} , \tag{6}$$

$$q = \left(\frac{D_0}{D}\right)^5 \cdot \frac{0.30864\,Re^n}{m\,lg^2(0.1\,Re)} , \tag{7}$$

$$Ar = \frac{\rho_s - \rho_0}{\rho_0} , \tag{8}$$

$$\omega = \frac{D_0 \delta_0 \rho_\Pi}{D\delta\rho_T}\cdot\frac{K_1'}{K_1} , \tag{9}$$

where Π – reduced annual expenses; C – annual working costs; K – capital investment; p – annual assignment for current repair and depreciation of installations; E – comparative efficiency coefficient; Z_N – expenses for materials reckoned towards unit of operating power; Z_E – electricity charges for industrial enterprises; Z_T – pay with charge of staff of pumping station; T – annual average time of operation of the plant, hours; N – installed power of the pumps of the hydrotransport complex; Z_0 – cost of water for industrial enterprises; S – discharge concentration of pulp; C_E – additional charge not depending on pipeline parameters; D – pipeline diameter; δ – wall thickness of steel pipeline; ρ_T – density of pipeline material; δ_0 – wall thickness of polyethylene pipeline; ρ_Π – density of polyethylene pipe-

line; K_1 – cost of pipes, welding and transporting reckoned towards unit of steel pipeline weight; K_1' – cost of pipes, welding and transporting reckoned towards unit of polyethylene pipeline weight; K_2 – cost of earthwork reckoned towards unit of pipeline diameter; K_3 – cost of survey work, of construction of roads, of communication lines etcetera; K_4 – cost of pump units, buildings for them etcetera, reckoned towards unit of installed power; K_0 – expenses, which don't depend on pipeline parameters; g – gravity acceleration; H' – pressure loss in polyethylene main pipeline; D – internal diameter of steel pipes; D_0 – internal diameter of polyethylene pipes; m, n – constants of flow friction law for polyethylene pipes, $m = 0.000543$ and $n = 0.25$ (Dobromislov 2004, Shvabauer et al. 2006); k_Z – coefficient, which takes into account local drags; q – coefficient, which takes into account influence of pipe material on critical velocity (Figure 1, Table 2), (Semenenko et al. 2013); $L_\bullet$ – length of steel pipes; L – length of main pipeline; ΔZ – difference of geodesic heights of pipeline beginning and end; Re – Reynolds number; v – kinematic coefficient of slurry viscosity; S – bulk concentration of fine fraction particles; ρ_0 – water density; ρ_S – density of solid particles; $\wp$ – coefficient, which takes into account hydraulic gradient rise at the expense of particles of medium and coarse fractions; η – pump unit coefficient of efficiency.

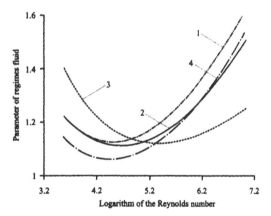

Figure 1. Dependence of q value on Reynolds number $lgRe$ for different calculation procedures (Semenenko et al. 2013).

Using of polyethylene pipes or drag reducing agents allows ensuring of required pulp discharge with lesser drags. This will result in decrease of installed power and will be taken into account by choosing of values m, n and q (Figure 1, Table 2).

Table 2. The values of coefficients used for hydraulic calculation of polymeric pipes in different calculation procedures.

Calculation procedure	$K \cdot 10^4$	$m \cdot 10$	$n \cdot 10$	q_{min}	q_{max}
SNiP 2.04.02-85	6.46	2.26	2.71	1.127	1.619
ISO TR 10501	5.37	2.40	2.73	1.110	1.508
ISO TR 10501	5.79	2.00	1.71	1.122	1.400
IGTM	5.43	2.50	3.16	1.065	1.530

Furthermore the capital investment in the case of using of polyethylene pipes will be dependent on flow hydraulic parameters (Semenenko et al. 2013):

$$K_1 = P_0 + P_1 \frac{\mu \varphi}{n_N} H', \qquad (10)$$

where P_0, P_1 – approximation parameters; μ – coefficient, which takes into account pressure boost at the expense of placement of concatenated pumps at a short distance; for pipes fabricated by LLC «UKRPLASTPEREROBKA» $P_0 = 186.88$; $P_1 = 4.47$; φ – safety factor, which takes into account pressure boost at the expense of hydraulic hammers; n_N – number of using pumps.

Thus with the regard for mentioned dependences and for flow features in polyethylene pipes or in steel pipes using drag reducing agents (Stupin 2000) the calculation of reduced annual expenses may be realized in that way:

$$\Pi = \frac{e_0}{D^{5-n}} + \frac{e_1}{D^{4-n}} + e_2 D + e_3, \qquad (11)$$

$$e_0 = [Z_N + (Z_E + Z_T)T + (p + E)1.1K_4] \times$$
$$\times (1 + ArS) \frac{P_0 g Q^{3-n}}{\eta} k_z LY, \qquad (12)$$

$$e_1 = (p + E) g \pi D_0 \delta_0 \rho_\Pi \left(1 + \frac{1 - \omega}{\omega} \frac{L_*}{L}\right) \times$$
$$\times P_1 \frac{\mu \varphi}{n_N} k_z L^2 Y Q^{2-n}, \qquad (13)$$

$$e_2 = (p + E) L \left[\begin{array}{c} K_2 + \left(P_0 + P_1 \frac{\mu \varphi}{n} \bar{\rho} \Delta Z\right) \times \\ \times g \pi D_0 \delta_0 \rho_\Pi \left(1 + \frac{1 - \omega}{\omega} \frac{L_*}{L}\right) \end{array} \right], \qquad (14)$$

$$C_E + (p + E)[K_3 L + K_0] + Z_0(1 - S)QT +$$
$$+ [Z_N + (Z_E + Z_T)T + 1.1(p + E)K_4] \times$$
$$\times \frac{P_0 g Q}{\eta} (1 + ArS)^2 \Delta Z, \qquad (15)$$

Equating to zero first-order derivative of expression (11) with respect to diameter, we will obtain equation for determination of pipeline optimal diameter:

$$y^{6-n} - \theta y - 1 = 0, \qquad (16)$$

$$y = 6 - n \sqrt{\frac{1}{5 - n} \frac{e_2}{e_0}} D, \qquad (17)$$

$$\theta = 6 - n \sqrt{\frac{e_1^{6-n}}{e_2 e_0^{5-n}} \frac{4 - n}{(5 - n)^{\frac{5-n}{6-n}}}}. \qquad (18)$$

The analysis show that the magnitudes in the equation (16) weakly depend on n value (Table 3) and so their average values may be used for calculations with engineering accuracy.

It is easily to show that single real positive root of equation (16) belongs to the interval:

$$\left(\sqrt[5-n]{\theta}; \sqrt[5-n]{\theta + \frac{1}{\sqrt[5-n]{\theta}}} \right). \qquad (19)$$

Table 3. Dependence of the magnitudes in the equation (16) on n value.

Value of n	$6 - n$	$5 - n$	$\dfrac{4-n}{5-n}$	$\dfrac{4-n}{(5-n)^{\frac{5-n}{6-n}}}$
0.18	5.82	4.82	0.793	0.610
0.20	5.80	4.80	0.792	0.610
0.22	5.78	4.78	0.791	0.610
0.24	5.76	4.76	0.790	0.609
0.26	5.74	4.74	0.789	0.609
0.28	5.72	4.72	0.788	0.609
0.30	5.70	4.70	0.787	0.608
0.32	5.68	4.68	0.786	0.608
Average value	5.75	4.75	0.789	0.609
Standard deviation	0.049	0.049	0.002	0.001
Constant of variation, %	0.85	1.03	0.28	0.1

Numerical analysis of solution of the equation (16) and of the plot at the Figure 2 shows that this solution may be approximated to such function:

$$y = 1.001 \sqrt[5-n]{\theta}. \qquad (20)$$

Taking into account the accepted symbols and dimensionless variables and jointly examining the formulas (16) – (20) we can obtain following expression for determination of rational pipeline diameter:

$$D = 5 - \sqrt[n]{(4-n)\frac{e_1}{e_2}} \; . \qquad (21)$$

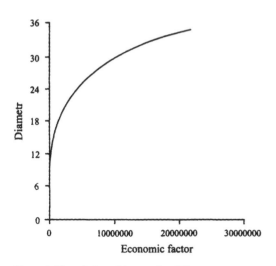

Figure 2. The solution of the equation (16).

The obtained solution is increasing function in all interval of argument changing. So when selecting pipeline rational diameter it is necessary to test whether this solution ensures overcritical flow regimes (Semenenko et al. 2013):

$$D \leq D_{kp} \; . \qquad (22)$$

Suggested dependences are partly approximate and require improvement with the purpose of accounting of a whole number of factors. However they allow estimation of potential of the systems under consideration and of complexity of assigned problem. They also let to outline possible ways of problem solution.

3 CONCLUSIONS

1. It is reasonable to realize modernization of hydrotransport plants of modern geotechnological systems with the purpose of decrease in power inputs and specific water consumption for hydrotransportation process by using of polyethylene pipes and drag reducing agents.

2. The presence of sections of steel and polyethylene pipes with different diameter in main pipelines as well as changing of hydraulic gradient and of critical velocity as a result of drag reducing agent action must be taken into account during hydraulic calculations of parameters and operating modes hydrotransport plants of modern geotechnological systems.

3. It is an identical approach to accounting of changing of hydrotransport parameters when using polyethylene pipes and drag reducing agents. This approach consists in using of relevant values of proportionality coefficient and of exponent in the dependence of hydraulic friction coefficient on Reynolds number and of proportionality coefficients of summands in the dependence of critical velocity on slurry concentration.

4. The distinctive feature of the calculation of economically sound diameter of the polyethylene pipeline as opposed to steel pipes is the dependence of capital expenditure on flow hydraulic parameters.

5. It is possible to use average values of exponent in the dependence of hydraulic friction coefficient on Reynolds number in calculations of economically sound diameter of the polyethylene pipeline with engineering accuracy.

REFERENCES

Alexandrov, V.I. 2000. *The methods of decrease in power inputs during hydraulic transportation of high concentration slurries*. St. Petersburg: SPSMI (TU): 117.
Dobromislov, A.Ya. 2004. *Tables for hydraulic calculations of pipelines of polymeric materials*. Moscow: VNIIMP: 209.
Market Report Company. 2006. *Ukrainian market of pipe polyethylene and of polyethylene pipes*. Polymeric pipes of Ukraine, No 1 (1): 10 – 14.
Semenenko, Ye.V., Nykyforova, N.A. & Tatarko, L.G. 2013. *Calculation of operating modes of hydrotransport complexes with main pipelines composed of polyethylene and steel pipes sections*. Geotechnical mechanics, No 111: 140 – 152.
Shvabauer, V., Gozdev, I. & Gorilovskiy, M. 2006. *Calculation of pressure loss in the pipeline of plastic*. Polymeric pipes of Ukraine. No 1: 46 – 50.
Stupin, A.B., Simonenko, A.P., Aslanov, P.V. & Bikovskaya, N.V. 2000. *Drag reducing polymeric compositions in firefighting*. Donetsk: DonNU: 198.

Theoretical and Practical Solutions of Mineral Resources Mining – Pivnyak, Bondarenko & Kovalevska (eds)
© 2015 Taylor & Francis Group, London, ISBN: 978-1-138-02883-8

Rosk mass assessment for man-made disaster risk management

O. Sarybayev, M. Nurpeisova & G. Kyrgizbayeva
Kazakh National Research Technical University named after K.I. Satpayev, Almaty, Kazakhstan

B. Toleyov
SAT & Company, Almaty, Kazakhstan

ABSTRACT: Results of long-term research by Kazakh National Research Technical University (KazNI-TU) employees on geomechanics processes study have been considered. It is shown that the problem of geomechanics processes control can be solved on the basis of methodology of rock mass condition geomonitoring viewed in this article ensuring complex accounting and analysis of all natural and man-caused factors.

1 INTRODUCTION

One of the pertinent problems in case with conduction of large-scale mining especially when it comes to roach mass is the technogenic seismicity causing not only catastrophic technical and economic consequences (technogenic earthquakes, rock bumps, landslides, etc.), but sometimes leading to death of people. All this is a direct consequence of change in geodynamic mode of geological environment under the influence of large-scale mining work which was convincingly confirmed by the results of scientific research on example of "Zhezkazganskayazona" of Kazakhmys Corporation" natural and technical system (NTS).

2 MAIN CONTENT

Favorable prices conjuncture at the global metal market creates new possibilities for mining sector of Kazakhstan. Leading industry player – Kazakhmys Group – intends to utilize this chance in full. Along with development of new large and medium-sized projects, Kazakhmys launches large-scale modernization programs for its enterprises, introduces advanced exploration and production technologies, and enhances the human resources management and environmental safety compliance policies. As of today, Kazakhmyscomprises: "Itauyz" and "Akchiy-Spaskiy" quarries, underground mines, 10 dressing plants, zinc plant, 2 copper smelting complexes, power plants, etc.

After having analyzed global trends, Kazakhmys has developed and launched the implementation of large-scale "reboot" plan. First of all, the company has returned to projects on development of Bozshakol and Aktogay gigantic deposits. Capital expenses on those will make $3.8 bln, and the expected production level will reach 60% of the current production volume. Besides, Kazakhmys develops small and medium-sized projects. In particular, it plans to invest $800 mn in development of Akbastau and Kosmurun copper mines (in due time these deposits were discovered by our mentor, academician A.Zh. Mashanov) where, by 2015, the construction of plant with production capacity of 2 mntn will be completed. Second stage of the project is the development of Kosmurunmine with a total volume of capital investment of $240 mn whichwill be completed by 2018. And all this comprises a single fold system of "Zhezkazganskayazona" which is a powerful subject of anthropogenic influence on the environment and representing huge possibilities for research of wide range of technogenic disasters and decreasing their risk (John 2004, Owen 2005, Nurpeissova et al. 2009, 2010).

Zhezkazgan deposit is dated by Zhezkazgan series of sedimentary rocks consisting of interbedded layers of grayback and rotten rocks and silt stones with conglomerates and chertinterburden and lenses. Total thickness of this complex is assessed as 650 m. Zhezkazgan mass of sedimentary rocks is divided into two suites: lower-Taskuduk and upper-Zhezkazgan. Lower-Taskuduk suite's thickness is 257 m and it consists of 16 layers of rotten and grayback silt stones united into 3 ore-bearing horizons. Development of tectonic fracturing in ore-bearing mass rocks is closely connected with the formation of folder structure of the deposit. In positioning orientation, two main cavity types can be marked out in the layers:

– according to stratum or gentle-intersecting stratums;

– transversal and large-intersecting fractures and crushed zones.

First type fractures often develop in earlier inter-stratum fractures, most probably of diagenetic origin. Main role in ore body field formation is played by faults represented by systems of high-angle faults of sublatitudinal bearing.

In order to take adequate measures on prevention of harmless influence from deposits development it is necessary to have knowledge of that particular technogenic component of total value of vertical earth settlement (VES). Untimely taking of measures on prevention of VES may lead to significant unjustified expenses and, probably, to emergency situation.

Geomechanics complex development of subsoil riches is solving a series of tasks with long (target period) sustainability of underground objects and control over in situ stress and distorted state of hosting rocks, identification of influence of mining work on surrounding natural environment and engineering constructions, both during construction period, operation periodand during their reconstruction and, in particular, liquidation. Its main purpose is prevention of emergency situations during development of deposits, increasing safety and efficiency of mining work, ensuring preservation and normal operation of buildings and structures falling into influence zone and protection of natural environment (Trubetskoy et al. 2012, Miletenko & Nurpeissova 2014).

For solid minerals deposits geomechanical ensuring of mining safety is usually based on engineering approaches adapted to given mining and geological development conditions. In this approach, local peculiarities of underworked series, variability of physical and mechanical properties of rocks and geomechanical features of geological environment are not taken into account. Whereas, all this influences the credibility of geomechanical assessments of real mining technical situation. Possibility of accounting of the abovementioned physical and geological factors in geomechanical calculations appear due to maintenance and accounting of results of geological, geophysical, seismological and especially geodesic and surveying research.

Analysis of state of surveying studies methodology and interpretation of obtained data first of all connected with absence of efficient ways for determination of values of earth settlement which leads to necessity to improve the methods of surveying and geodesic observations over deform-ation of rocks with use of advanced electronic devices in order to increase the credibility, speed during determination of earth settlement parameters for safe development of subsoil riches and protection of developed objects (Nurpeissova 2012).

In general, surveying with use of new generation devices allow revealing mass deformation which is significant for assessment of geomechanical situation in deposit development area. But they do not allow obtaining the full picture of deformation processes in time. This can be done only by using complex methodology of studying natural and technical system (NTS) the structure of which is presented in Figure 1.

On the basis of analysis of region's geology and tectonics, applied development system, control of roof and superincumbent rock, and experimental assessments of ICPstress condition, "energy-saturated" areas are being identified which define the boundaries of geomechanical monitoring. Then, a monitoring of troublesome zone is arranged which mainly includes control of deformation and mass structural fracture level.

Later on, all information on regularity of system strike shift and its critical state parameters are entered into the expert system where, on the basis of databases and knowledge of integration, the assessment of NTS state is performed and corresponding decisions on protection of subsoil and earth surface are being substantiated (Nurpeissova 2014).

Mining enterprises experience shows that conventional design parameters of pillar extraction drilling (pillars, dimensions, chamber distances) for given mining and geological conditions in general ensure long-term stability of mine goafs.

However, it must be taken into account that strength properties and pillars dimensions of the project are calculated based on average readings of mining and geological conditions. However, rock mass is not homogenous and, due to its natural structural variability, it is irregularly disrupted. The main cause of disruption in physical integrity of the mass is its cleavage or fracturing which is of probabilistic nature. Besides, blasting workings conducted during preparation of ICPserve as a source of additional, so-called technological cleavage, also decreasing the bearing capacity and stability of pillars and roof. Therefore, experience has proven that there are cases when separate (single) ICPor groups of ICPare destructed; cleavage and outburst of roof occur, and sometimes, depending on severity of exposure to various factors and intensity of occurring irreversible geome-chanical processes, a collapse of the whole superincumbent rock happens with outbreak.

This creates emergency situations, decreases safety and efficiency of underground mining, disturbs regular operation of mines and pits, which, eventually, negatively affects the main technical and economic indicators of ore production. Situation is even more complicated during mining of superimposed overlapping reserves as ore extraction is made in tiers which leads to emergence of new structural load-bearing element such as interchamber cap pillar which requires additional studying.

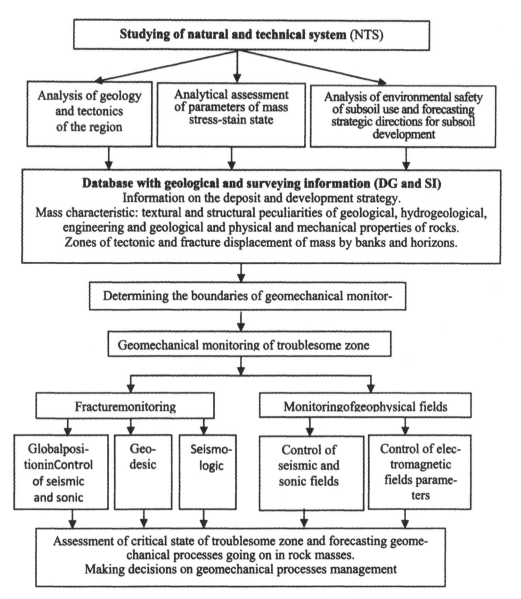

Figure 1. Scheme of methodology of studying the forecast of trouble events in NTS.

Improving stability of roof pillars in these conditions may be achieved by development and application of technology aimed on their hardening based on knowledge in specifics of destruction of components of pillar extraction drilling as a uniform geomechanicalmultileg structure (GMS) – "pillars – roof – superincumbent rock" and principle of processes of hardening of fractured roach, and improving the blasting workings technology ensuring maximum unequivocal use of the explosion energy during rock mass breakage which is an important and pertinent scientific and practical target solving of which will allow improving safety and efficiency of massive ore deposits mining.

Due to this, the assessment of stability of structural load-bearing elements of pillar extraction drilling means identifying the nature and mechanism of destruction of support columned interchamber pillars and "pillar-roof" system in order to reveal their prelimit, limit and out-of-limit (critical) state testifying to necessity of timely hardening of these elements (Nurlybayev 2014).

As it was shown in the results of field studies by several authors, at the initial stage of destruction, the bearing capacity of ICPis decreased either because of pillar shape change with preservation of its structural integrity or the weakening of ICPby various fracture and interbed systems occurs with preservation of their external shape. The most typical and widespread form of pillars destruction is destruction with formation of shear fractures and ruptures.

Cleavage in ICPmay be expressed in the form of separate large diagonal cracks of tectonic origin or in the form of angled and horizontal cracks and soft rock interbeds.

Top cover collapse happens after the rock breakdown point is exceeded both for stress and for compressing due to which the mass is broken into blocks by through cracks system. Mechanism of formation and development of technogenic cracks in the mass is shown in Figure 2.

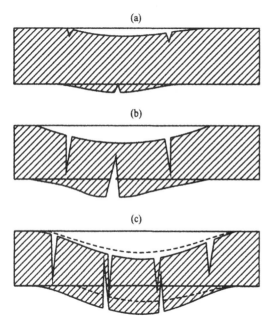

Figure 2. Scheme of cracks formation in the roof during overcut: (a) first emergence of cracks in the curve layer; (b) opening of crack depth; (c) destruction of layer during alternating strains.

As a result of layer depression it begins to have tension stresses and with a certain dimension of bay they reach the limit of rock tensile strength and on the upper and lower layer surfaces the transverse cracks begin to appear. Further increase of bay leads to the corresponding growth of tensile stresses and cracks development.

One of the main reasons for ICPdestruction is their unbalanced stress due to change in pillars rigidity caused by difference in their geometrical sizes and fracturing. Stress-stain state (SSS) of column interchamber pillar and the mechanics of its destruction depending on the cleavage were in detail studied in the paper (Aitaliyev & Khalmanov 1989) where various stress stages of pillar are considered characterizing its corresponding state. By studies (Nurlybayev 2014, Borsch-Komponiets & Makarov 1990, Kuznetsov & Miletenko 1989) it was revealed that the stress of interchamber pillars with a large variety of geometrical parameters is unevenly distributed.

In case of uneven distribution of stresses in the pillar its destruction is activated when the value of existing tension in the typical area along the section of pillar exceeds some admissible value. The faster stress increases in the pillar and the closer the values of existing tensions to the level of bearing capacity, the faster this pillar will destruct. In turn, activation of such process will lead to redistribution of stress to adjacent pillars and may lead to destruction of pillars and roof collapse as it happed in panel 29 lev. 220 m, pit 55 of West-Zhezkazgan mine which is only one example out of many.

Research performed on volumetric and flat modeling allowed obtaining experimental material on values and nature of stress distribution in pillars of different deposit types. For successful solving of the open-pit sides stabilityproblem means and methods were developed ensuring obtaining prompt and credible information on stressed state of near edge mass. For example, methods of automated identification of mass disturbance were developed and implemented (Nurpeissova et al. 2008, 2009).

Methods include placement of thermometers and ultrasonic acoustic transmitters in operated pits connected with information processing units. In the first caseasimultaneous measurement of temperature in mass areas by all thermometers is performed during cooling of mass and determination of interconnection of heat transmission by rock mass is done.

Second method is based on measurement of amplitude value of ultrasound impulse passing the rock between deposits and defining the availability and opening of cracks in the mass based on smaspodic change of amplitude of the ultrasound impulse.

Methods allow more precisely reveal disturbances and follow the dynamics of near edge mass. Novelty of developed means and devices is in revealing the potentially unstable areas of benches and edges of quarry for reasonable design of profile lines of observation station and prompt control over slope stability.

Besides, in order to supervise the shifting of roof rock during cleaning work a method of remote determination of roof pillar shifting and ICP (Nurpeissova et al. 2015).

In case of destruction of ICPits integrity is disturbed, rigidity is decreased and, accordingly, the stress on it is decreased with simultaneous increase of deformation which leads to rock lamination and formation of roof arch above the pillar. Scheme of formation of area of rock destruction in stope ore (chamber) roof is shown in Figure 3.

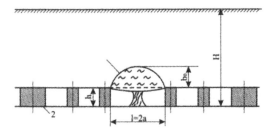

Figure 3. Scheme of formation of area of rock destruction in stope ore (chamber) roof in case of ICP destruction: 1 – destructing ICP with decreased stability; 2 – panel pillar; 3 – contour of roof arch (natural equilibrium); 4 – roof sag (deformation); b_0 – height of roof arch; l – length of roof arch bottom; h – height of mined space; H – depth of mining.

Practice of development of flat-dipping ore deposits has shown that collapse of top cover happens only when the bay of cropping in mine exceeds limit width rather certain for these mining and geological conditions and depth of bedding. If cropping bay of roof is less than limit, the collapse of overlying rock is of limited nature and happens within the contours of dome of natural equilibrium preserving its shape and stability within a long term.

Practice of development of flat-dipping ore deposits has shown that collapse of top cover happens only when the bay of cropping in mine exceeds limit width rather certain for these mining and geological conditions and depth of bedding. If cropping bay of roof is less than limit, the collapse of overlying rock is of limited nature and happens within the contours of dome of natural equilibrium preserving it's shape and stability within a long term.

Usually, stopes during cleaning work are developed with application of drilling-blasting work; in this case, rock in stope under influence of gravity, explosions and due to redistribution and concentration of stress are collapsing until the stope gets some stable shape of roof arch.

Therefore, one must distinguish among "dome of natural equilibrium" representing a theoretically cam contour in rock mass and "roof arch", that is, a cupola-shaped roof which may be observed visually after rock fall from the area limited by contour of dome of natural equilibrium.

Determination of dome of natural equilibrium height relates to solving of inverse problems of elasticity theory when one must find a shape of AB arch which is a part of $ABCO$ contour (Figure 4). In work

by G.P. Cherepanov (Cherepanov 1966) it is determined that when $ABCB$ square is close to lower half of equal in strength ellipse, and the mine is located far from earth ground, the sought AB arch in its shape shall not have a large difference in shape from the upper half of the same equal in strength ellipse; in this case, the height of arch the contour of which is approximated by ellipse shall be calculated by formula which is most appropriate for these conditions:

$$b_p = \frac{va}{1-v}, \tag{1}$$

where b_p – height of dome of natural equilibrium; v – poison ratio; a – width of stope.

Based on this and based on practical data, we can determine the height of roof arch for Zhezkazgan conditions with the formula (Cherepanov 1966):

$$b_o = \frac{2va}{1-v} \tag{2}$$

where b_o – roof arch height.

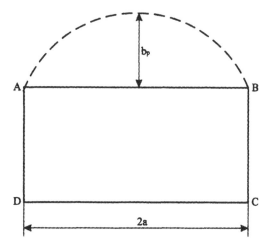

Figure 4. Scheme for definining the value of dome of natural equilibrium according to G.P.Cherepanov: b_p – height of dome of natural equilibrium; $2a$ – width of development (shrink stoping).

Shape of the dome of natural equilibrium ensures the most even distribution of compressing stresses along the arch contour which is a condition for the maximum ore strength and stability in the roof development area.

Currently, at industrial and technical Zhezkazgan area geomechanical monitoring is performed on a regular basis. One of the components of geomonitoring is systematic instrumental observations. Among modern methods and means of earth deformation research, GPS proved itself as rather efficient one. GPS satellite systems are necessary for creation of a basis of deformation calculation and geo-informational systems allowing forecasting the parameters of geomechanical processes (Nurpeissova et al. 2012, 2014). Along with GPS technologies, systematic observations are being conducted with electronic tachometers and digital leveling devices in areas with intensive extraction of solid minerals.

In Figure 5, a the result of monitoring is shown which testifies to:

– continuous nature of earth surface deformation, and deformation occurs rather unevenly in time and space (depressions have positive and negative values);

– maximum deformation velocities that were evidenced in 2008 – 2010 times to fault boundaries (Figure 5, b);

– tectonic nature of earth surface depressions (by positive values of shifts of benchmarks in height).

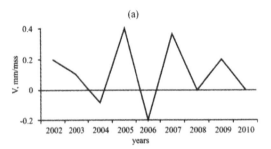

(a)

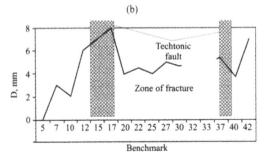

(b)

Figure 5. Schedules of benchmark shifts in observation station: (a) vertical shift; (b) benchmark shift speeds.

When observing deformations in underground developments, Leica Geosystems robotic electronic tachometers were used as sensors. Later on, all information on principle of system shift process and it's critical state parameters go to expert system where in the basis of databases and knowledge integration the assessment of mining enterprise is done and corresponding managerial decisions are substantiated.

Analysis of underworked mass state was made from two positions. First, the potential threat of

vertical cracks system development in the mass was estimated which may act as channels for penetration of underground waters into the underworked pit space. Secondly, the possibility of formation of weakened zones on the upper part of cut was considered which may be of real threat for surface facilities and engineering structures (Nurpeissova et al. 2013).

3 CONCLUSIONS

Complex assessment of rock mass condition allows taking into account the peculiarities of mass rock within the scope of geology, geodynamics, tectonics and, by this, improve the quality of geomechanical maintenance of mining. In turn, the results of geo-mechanical forecasts allow defining the most dangerous areas requiring monitoring of geophysical and surveying observations in order to localize technical faults areas.

Researches continue and the results will be published by means of:

– developing innovative methods of geomonitoring on the basis of satellite and laser technologies for creation of geospatial database on stress-stain state of rock mass allowing improving the accuracy and saving time on measurement work;

– submitting application with "Methods of acoustic forecasting of rock mass condition" and "Methods of geomechanical process management" subjects for issue of innovative patents of the Republic of Kazakhstan for invention;

– scientific publications with high impact factor jointly with foreign scientists.

REFERENCES

Aitaliyev, Sh.M. & Khalmanov, Kh.Zh. *Mechanics of cracks at pillar load management.* 1989. FTPRPI, No 2: 9 – 14.
Borsch-Komponiets, V.I. & Makarov, A.B. 1990. *Ground pressure at development of massive flat oredeposits.* M: Nedra: 271.
Cherepanov, G.P. 1966. *Equally efficient development in rock mass.* Materials from 1-All Union scientific conference on mechanics of rock. Almaty: 440 – 447.
Kuznetsov, S.V. & Miletenko, I.V. 1989. *Stress state andforecast of mass behavior at pillar extraction drilling.* Almaty: Nauka: 187.
Miletenko, N.A. & Nurpeissova, M.B. 2014. *Geomechanical approach to forecast of dangerous hydrogeological process during combined method of reserves development.* M: Surveying Bulletin, No 4: 55 – 58.
Nurlybayev, R.O. *Assessment of stability of structural elements of pillar extraction system.* 2014. Almaty: Mining journal of Kazakhstan, No 10: 19 – 23.
Nurpeisova, M.B., Aitkasinovam, Sh.K., & Kirgisbae-

va, G.M. 2014. *Geomechanical monitoring of the massif of rocks at the combined way of development of fields.* Geodesy & Mine Surveying 17 – 26 June: 279 – 192.

Nurpeissova, M.B. 2012. *Geomechanics of ore reserves of Kazakhstan.* Almaty: KazNTU: 324.

Nurpeissova, M.B. 2014. *Development of ideas by A.Zh. Mashanov, Academician in geomechanics area.* M: Surveying and subsoil use, No 4: 21 – 24.

Nurpeissova, M.B., Kasymkanova, Kh.M. & Kyrgizbayeva, G.M. 2009. *Change of geodynamic mode of geological environment during subsoil development.* Geology and protection of subsoil. Geologiya I ohrananedr, No 4: 80 – 85.

Nurpeissova, M.B., Kasymkanova, Kh.M. & Kyrgizbayeva, G.M. 2010. *Assessment of geomechanical state of rock mass during open-underground extraction of deposit.* Papers by international scientific environmental conference "Science and education" – leading factor of "Kazakhstan-2030" Strategy-Karaganda: 90 – 94.

Nurpeissova, M.B., Kyrgizbayeva, G.M. & Sarybayev, O.A.

2012. *Perspectives of use of modern devices for geomechnical monitoring of natural and technical systems.* Papers of international scientific environmental conference. "Problems of subsoil development in XXI from the view of young specialists". M: IPKON RAN: 126 – 130.

Nurpeissova, M.B., Kyrgizbayeva, G.M. & Sarybayev, O.A. 2013. *Forecast of technogenic threat of the territory based on surveying and satellite measurements data.* Karaganda: Surveying and geodesics, No 1: 21 – 24.

Owen, M.L. 2005. *Calibrating a semi-guantitative seismic risk model using rockburst case studies from underground metallifrous mines.* Controlling seismic risk. Australia: Australion Centre for Geomechanics: 191 – 204.

Simmons, J.V. 2004. *Geotechnical risk management in open pit coal mines.* Australian Center for Geomechanics Newsletter, No 22: 1 – 4.

Trubetskoy, K.N., Miletenko, I.V., Miletenko, N.A., Odintsev, V.N. 2012. *Analytical assessment of technogenic fault of high pillar.* M: IPKON RAN.

Theoretical and Practical Solutions of Mineral Resources Mining – Pivnyak, Bondarenko & Kovalevska (eds)
© *2015 Taylor & Francis Group, London, ISBN: 978-1-138-02883-8*

The temporary criterion account in the metal beam-rod structure design

A. Ivanova & L. Feskova
National Mining University, Dnipropetrovsk, Ukraine

ABSTRACT: In this work the inverse problem when the sizes of the beam for a predetermined period of its operation determine in an aggressive environment is solved. The optimization of the beam-rod structures is reduced to minimize the area of the metal structure in contact with this medium with the given parameters. Software upgrade Microsoft Excel Solver ("Finding Solutions") is used in this study to determine the optimal sizes of rod structure elements without using the complex software systems. The formulas for determining the height of the beam expressed in terms of a single parameter are obtained. The optimal ratio between the height and width of the beam is obtained. Thus, the saving material depending on the type of load is applied: the effect of a distributed load – 10.65%, by the action of a concentrated force – 6.25%.

1 INTRODUCTION

Creating a rational design schemes in the most advantageous values of the geometric parameters and the size of individual elements is the main condition of the metal structure engineering.

These issues can be resolved by using a variant design and optimization. As a rule, optimum metal structure as a whole cannot be obtained on the basis of the optimal partial solutions of nodes as the linkage in complex individual nodes may lose their optimum values of the parameters. The solution to this problem may be the selection of more efficient and technologically advanced materials of construction elements. With regard to metal to the fore the issue of reducing metal consumption while maintaining their strength, local stability and rigidity, since the mass of the structure largely determines its value (70% and above)

Elements of many engineering structures in operation under-exposed to not only subjected to loads and temperatures, but also a variety of corrosive media. And often these factors work together in the most unfavorable combinations, which significantly reduce the load-bearing capacity and shorten the life of the structure, and can also cause damage and even failure of the construction projects. Failures of various designs leading to financial losses, negative impact on the process, pose a threat to people's lives.

A metal structure for one of the main reasons for the refusal is material degradation due to corrosive wear with subsequent loss of strength. The user must assess the actual state of corrosion of metal structures and to predict its change to take timely measures to prevent failures in the work and determination of reserve service life. The mathematical model can perform its functions if it adequately reflects the real picture of the phenomenon, so it should describe the properties of an object to be available for research and work. In this case, various researchers often offer different models for the description of the same process. In this regard, the selection of one of a plurality of possible mathematical models of the corrosion process is caused by a specific situation the structure.

2 CALCULATION METHOD

In solving problems of stress-strain state (SSS) and durability corrosive designs traditionally used approach based on the combined use of a numerical method for calculating stress-strain state and numerical method for solving the same Cauchy problem for the complex differential equations (CDE) describing the process of damage accumulation (Ovchinnikov et al. 2012, Shashenko et al. 2013), and may also be involved in computer-aided design (CAD), realizing the finite element method for the determination of the SSS of the structures. One of the modern CAD tools is a software module such as SolidWorks, Ansys, Auto Desk Robot and others.

Thus, the optimization problem of beam-rod structures operating in aggressive environments is reduced to minimize the area of contact with this medium with the given parameters. As the objective function accept:

$$V = h_0 \cdot b_0 \cdot l, \tag{1}$$

where V – the volume of metal beam m³; h_0 – the initial height of the beam section, m; b_0 – the initial width of the beam section, m; l – the beam length, m;

The stress influence on the corrosion rate in the design calculations in the first time was considered by V.M. Dolinskiy (Dolinskiy 1967). The exponential dependence of the corrosion rate of the magnitude of stresses in structures was proposed by E.M. Gutman and R.S. Zaynulin (Gutman & Zaynulin 1984). The results of the optimization of the initial form of no prismatic compressed struts of rectangular and circular crosssections are performed in (Pochtman & Fridman 1997, Fridman 2001).

Problems of optimization of metal structures are covered in the papers by Y.M. Lihtarnikova (Lihtarnikov 1979), I.S. Kholopova (Pholopov 1992, Kholopov & Popov 1999, Alpatov & Kholopov 2009) and other authors (Mandelbrot 2002, Trofimovitch 1981). In general, the solution of such problems is very complex which is caused by the nonlinearity of their performances and certain mathematical description difficulties.

Specificity of optimization problems is that at the initial stage of design some initial parameters are given, for example: an applied load, the corrosive wear rate, the size of the beam section or its dependence, material properties, etc. In the future, the problem reduces to finding the time at which this design does not lose its strength properties.

The purpose of our research is to obtain expressions for determining the necessary size of the original metal beams for a predetermined period of operation. Based on the features of calculation and design, as well as taking into account practical issues, we can formulate the following problem of optimal design: to determine the optimal geometric beam construction parameters for a predetermined period of operation. Using the dimensionless parameter helps to unify the calculation of metal beam-rod structures which in turn reduces the time required.

Rod metal structures often are subjected to a corrosion in the initial period of operation (the first 10 – 15 years), further corrosion is slowed down but the process continues (Kovalenko 2012). In this regard, operation of the metal rod structure is taken during 10 years.

The loads acting on the structure are in relation to it by external forces. These forces are applied to particular construction elements according to some areas of the surface or distributed throughout its volume. On the beam can act either the uniformly distributed load or the concentrated force.

Determination of the optimal size of the beam is derived from the ratio of the height of the rod beam section to its width, that is h/b.

This paper analyzes the following most used ratio (Ivanova & Feskova 2014): $h = 3b$, $h = 2.5b$, $h = 2b$, $h = 1.5b$.

Consider a metal beam of, rectangular cross-section undergoing bending deformation from the distributed load q (Figure 1) and from the concentrated force, which can be expressed as $P = q \cdot l$ (Figure 2).

Figure 1. The action on the beam of the distributed load.

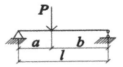

Figure 2. The action on the beam of the concentrated force.

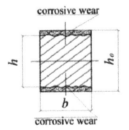

Figure 3. The cross-section of the beam.

For the calculations the safety factor is taken $n = 1.1$ (Ivanova & Feskova 2014). Inputs: the distributed load $q = 10$ kN/m, the length of the beam $l = 1$ m, the yield stress St.3 steel $\sigma = 240$ MPa corrosion rate $v = 0.25$ mm/year (DBN V.2.6-163.2010 2011) within $T = 10$ years.

Introduce the following assumptions:
– the corrosion is uniform;
– the corrosion lateral faces are neglected, since the side faces corroded significantly less than the top and bottom on Figure 3.

Corrosive destruction of metals has the following characteristics:
– destruction always starts at the surface and extends into the bulk metal;
– the destruction is often accompanied by changes in the surface of the metal. Damaged areas are seen on the metal in the form of asymmetric depressions, points, "pitting", etc.

During the period of operation of the original height of the beam − h_0, due to the corrosion process is reduced to a value − h. Based on this, write:

$$h_0 = h + 2 \cdot v \cdot T,\qquad(2)$$

where h_0 − the original height of the beam, m; h − the final height of the beam (at the end of operation), m; T − the period of operation of the beam, years; v − corrosion rate, mm/year; 2 − coefficient taking into account the corrosive wear of top and bottom edges of the beam (Figure 3).

A condition of bending strength is given by (DBN V.2.6-163.2010 2011):

$$\sigma = \frac{M}{W} = \frac{q \cdot l^2 \cdot 6}{8 \cdot b \cdot h^2} \le [\sigma],\qquad(3)$$

where M − the bending moment, kN·m; W − the moment of resistance, m^3.

From the condition of the strength (3) we obtain the height of the beam:

$$h = \sqrt[3]{\frac{q \cdot l^2 \cdot 6}{8 \cdot b \cdot \sigma}}.\qquad(4)$$

An important role is played by the height of the beam after the corrosive wear because after operation in the construction should not cause strain.

3 ANALYSIS RESULT

Substituting the height of the beam section h in (3) with the condition that $h_0 = 3b_0$ and replacing in (4) b to h_0, obtain:

$$h_0 = \sqrt[3]{\frac{n \cdot 9 \cdot q \cdot l^2}{4 \cdot [\sigma]}}.\qquad(5)$$

Similarly for: $h_0 = 2.5b_0$; $h_0 = 2b_0$; $h_0 = 1.5b_0$.

The results of calculation of the sizes ratio with the distributed load are shown in Figure 4, with the concentrated force – Figure 5.

Depending on the applied load the cross-sectional dimensions of the beam is changed, too (Figure 6). The action of the concentrated load on a beam the cross section height greater is 25% than under the action of a distributed load.

The program MS Excel Solver ("Finding Solutions"), the application of which is presented in (Ivanova & Feskova 2014), allows determining a rational ratio h/b, depending on the applied load (distributed and concentrated), of its sizes and length of the beam. In the research for optimal solutions many calls to the program (iterations) are performed.

Depending on that the program enumerates the possible values and produces some results.

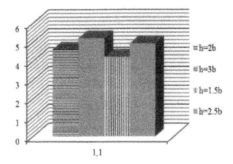

Figure 4. Graph of ratio of the beam section sizes with $T = 10$ years for the distribute load.

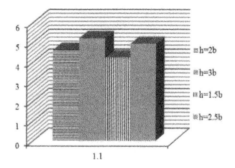

Figure 5. Graph of ratio of the beam section sizes with $T = 10$ years for the concentrated force.

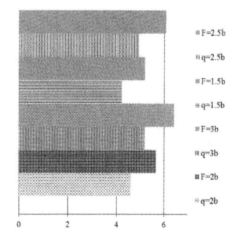

Figure 6. Dependence of the original height of the beam cross-section of the type of the applied load (the distributed and the concentrated).

The most optimal dependence: a distributed load − $h = 1.675b$, for concentrated load − $h = 3.201b$. At this the material savings is achieved.

Based on the obtained dependences, using the formula (3), we obtain the required beam dimensions, expressed in terms of one parameter, wherein the safety factor was $n = 1.1$ (Ivanova & Feskova 2014).

Formulas for determining metal beams sizes for the distributed load:

− with a certain beam section width b, the length will be:

$$l = \sqrt{\frac{4.1 \cdot b^3 \cdot \sigma}{q}} , \qquad (6)$$

− with a certain beam section height h, the length will be:

$$l = \sqrt{\frac{0.9 \cdot h^3 \cdot \sigma}{q}} , \qquad (7)$$

− with a certain beam length l and dependencies $h = 1.675b$, sectional height and width can be defined as:

$$b = \sqrt[3]{\frac{1.88 \cdot q \cdot l^2}{\sigma}} , \qquad (8)$$

$$h = \sqrt[3]{\frac{1.14 \cdot q \cdot l^2}{\sigma}} . \qquad (9)$$

Formulas for determining metal beams sizes for the concentrated force:

− with a certain beam section width b, the length will be:

$$l = \sqrt{\frac{7.5 \cdot b^3 \cdot \sigma}{q}} , \qquad (10)$$

− with a certain beam section height h, the length will be:

$$l = \sqrt{\frac{0.23 \cdot h^3 \cdot \sigma}{q}} , \qquad (11)$$

− with a certain beam length l and dependencies $h = 3.201b$, sectional height and width can be defined as:

$$b = \sqrt[3]{\frac{14 \cdot q \cdot l^2}{\sigma}} , \qquad (12)$$

$$h = \sqrt[3]{\frac{4.4 \cdot q \cdot l^2}{\sigma}} . \qquad (13)$$

4 CONCLUSIONS

The formulas are obtained for determining the size of the beam with the specified term of operation in an aggressive medium expressed in terms of a single parameter. It is obtained the optimal ratio between the height and width of the beam, thus saving mate-

rial, depending on the type of load, is applied: the effect of the distributed load − 10.65%, by the action of the concentrated force − 6.25%.

Under the action of the concentrated load on a beam the cross section height greater is 25% than its higher under the action of a distributed load.

The application of software add-ins Microsoft Excel Solver ("Finding Solutions") allows determining the optimal sizes of rod structures elements without using complex software systems.

REFERENCES

Alpatov, V.Yu. & Holopov, J.S. 2009. *Optimization geometric shape spatio-rod designs*. Metal structures, No 1: 47 − 57.

Dolinskiy, V.M. 1967. *Calculation of loaded tubes corroded*. Chemical and petroleum engineering, No 10: 21 − 30.

Fridman, M.M. & Zyczkowski, M. 2001. *Structural optimization of elastic columns under stress corrosion conditions*. Structural Optimization, Vol. 21 (3), No 11: 218 − 228.

Gutman, E.M. & Zaynulin, R.S. 1984. *The kinetics of mechanochemical destruction and durability of structural elements with stretched elastic-plastic deformations*. FHMM, No 4: 14 − 17.

Ivanova, A.P. & Feskova, L.V. 2014. *Optimal design of metal structures with rational use of safety factor in aggressive medium*. The collection of materials of regional scientific-practical conference "Problems of mining technology". Krasnoarmijsk, No 4: 203 − 206.

Ivanova, A.P., Shashenko, A.N. & Shtelmakh, A.S. 2012. *Modeling the processes of accumulation of geometric damage due to corrosion in the bars under axial tension*. "Metallurgical and mining industry", No 4: 75 − 78.

Kovalenko, V.V. 2012. *Protection against corrosion of the metal lining with shotcrete*. Dnipropetrovsk: National Mining University: 107.

Likhtarnikov, Yu.M. 1979. *Variant design and optimization of steel structures*. Moscow: Construction publishing: 319.

Mandelbrot, B. 2002. *The fractal geometry of nature*. Moscow: Institute of Computer Science: 856.

Metal constructions. Standards of design, manufacture and installation. DBN V.2.6-163: 2010 2011. Kyiv: Ministry of Regional Development of Ukraine: 201.

Ovchinnikov, I.I., Ovchinnikov, I.G., Zanin, A.A., Zelencov, D.G. & Short, L.I. 2012. *Problem of optimum design of the loaded constructions which are exposed to influence of aggressive environments (review)*. Online journal "Science of Science": 21.

Pochtman, J.M. & Friedman, M.M. 1997. *Methods for calculating the reliability and optimal design of structures operating in extreme conditions*. Dnipropetrovsk: "Science and Education": 134.

Slaves, I. & Popov, A. 1999. *Multi-criteria optimization of the elements of metal structures in a CAD system*. Modern constructions of metal and wood, No 9: 226 − 234.

Slaves, J.S. 1992. *Optimization rod systems as applied to CAD*. Thesis for academic Dr. Sci. Tech: Moscow: 05.23.17: 39.

Trofimovitch, V.V. & Permiakov V.A. 1981. *Optimal design of steel structures*. Kyiv: Builder: 136.

Theoretical and Practical Solutions of Mineral Resources Mining – Pivnyak, Bondarenko & Kovalevska (eds)
© 2015 Taylor & Francis Group, London, ISBN: 978-1-138-02883-8

Faults and linear crusts of weathering as a gold-controlling factor

K. Zmiyevskaya
M.S. Poliakov Institute of Geotechnical Mechanics under the
National Academy of Sciences of Ukraine, Dnipropetrovsk, Ukraine

ABSTRACT: This paper studies the results of faults separations and genetically connected linear crusts of weathering with them. The results were made by using the natural impulse electromagnetic field of Earth (NIEMFE) method on the Solonianske ore field fragments. Though, extracted structures were considered as gold-controlling on the Sergijivske deposit and Soniachne mineralizations. Their spatial orientation on the studied areas was analyzed, as well as their connection with faults which complicated the bedding of rocks on the Srednepridneprovskii megablock. Comparison of received trend azimuths of faults are demonstrated not only according to NIEMFE but to geomagnetics, geoelectics. It allowed to conclude that, deformation processes which formed observed net of faults within the Srednepridneprovskii megablock, provided appearance of faults on the Solonianske ore field.

1 INTRODUCTION

Faults and linear crusts of weathering on the East European and Siberian platforms are connected with deposits of iron ore, rare-earth minerals, nickel, manganese, gold and other minerals. Gold ore deposits on the Ukrainian shield which are linked with faults and linear crust of weathering: Balka Zolota, Balka Shyroka, Sergijivske, Klintsevske, Surozhske, Jurijivske and others were explored. Faults and linear crusts of weathering should be considered as one of the most important ore-controlling factor. Hence, it is a crucial task to study, map and analyze their spatial orientation.

2 MAIN PART

In studies of deep faults of the Ukrainian shield, which were defined with geophysical method by K. Tyapkin (Tyapkin & Dovbnich 2009), main six pairs mutually orthogonal system of deep faults were defined. They divide blocks, which are set with polytypic rocks complexes and are characterized by continuity of bearing of trend, wide width, depth and distances between them. Deposits of main minerals such as iron ore, manganese, nickel, gold and others are connected with their development.

Within blocks a wide net of faults are present. They are notable for smaller length, width and perhaps have smaller depth.

On the one of the most studied blocks of Ukrainian shield the Srednepridneprovskii megablock (SPMB) the main systems of intra-block faults are: 0° and 270°, 17° and 287°, 35° and 305°, 45° and 315°, 62° and 332°, 77° and 347°. Furthermore, another two pairs of directions with azimuths 25°, 56° and orthogonal to them 295°, 326° can be seen.

Perhaps, intra-block faults (higher- order faults) are appearing in the late phase of block formation and corresponding folded structure. These faults are presented with:

– tectonic contacts of rocks with different litho-logic-and-petrographic structure;

– different phases of metamorphism and/or with different structural plans of intrusive formations, mainly basic or ultrabasic structure in the form of dykes and elongated bodies, which are filling fracture openings;

– crushing zones, mylonitization, silicification and demonstrations of other secondary processes.

In most cases, mentioned aspects are characters of linear crusts of weathering.

Thus, gold ore deposits which are localized in-granitoid-gneiss greenstone terrain (GGT) zones are connected with faults and linear crusts of weathering. The most perspective and studied one in Serednioprydniprovskyi GGT is Surska greenstone structure (GSS) (Figure 1).

Today, within its territory group of related structural and compositional plan of deposits and ore occurrence is known. On its territory such gold ore deposits and mineralizations were discovered: Sergijivske, Balka Zolota, Balka Shyroka, Soniachne, Nove, Apolonivske, Shidno-Apolonivske, Raschetnoe, Pivdenno-Petrivske, Dorozhne, Tanino and others.

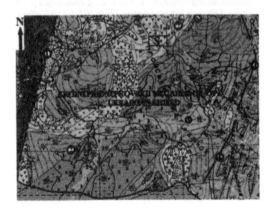

Figure 1. Scheme of tectonic structure of Serednioprydniprovskyi megablock of Ukrainian shield Sc 1:1000000 (Gurskiy et al. 2004), ppLegend: ✪ – circled figures indicate following faults: 1 – Kryvorizko-Kremenchutskyi; 2 – Derezovatskyi; 3 – Malokaterynivskyi; 4 – Gorikhovo-Pavlogradskyi; – – – – higher-order breaking failures; ——— Second-order breaking failures; ⬛S – study area (Solonianskyi ore field) within Sergijivske deposit and Soniachne ore occurrence.

V. Kravchenko and M. Ruzina in studies of faulted tectonic (Sc 1:1000000) and its connection with ore-bearing (Kravchenko & Ruzina 1997), discovered that by number of appearances of ore manifestations, confinedness to different system faults, which were detailed within the Surska structure, five of six systems are ore-bearing. The most productive, on their opinion, are the following: 45° and 315°, 35° and 305°, 17° and 287°. Less ore-bearing are – 62° and 322°, 77° and 347°. To prospect less system authors refer 0° and 270°. At the same time, intersection nodes of these systems are considered as the main structural ore-controlling factor.

In detailed studies of faulted tectonic of Sergijivske deposit (Sc 1:1000) and Solnechnoe ore manifestations (Sc 1:2000), according to current observations of natural impulse electromagnetic field of Earth (NIEMFE), author determined the gold-controlling role some of the following systems of dedicated damages, which is presented in Table 1.

For comparison, trend azimuths of faults and linear crusts of weathering of SPMB and deposit areas of Sergijivske, Soniachne ore manifestations are also presented in table 1 according to tectonic map of Ukraine (Gurskiy et al. 2004).

Sublatitudinal trend is the main development direction of faults and linear crusts of weathering on the Sergijivske deposit. It is determined by formation and by following activation stages of sub latitudinal Devladovskyi fault, in close vicinity of which the Sergijivske deposit area is located.

Table 1. Trend azimuths of faulted damages and linear crusts of weathering of SPMB and deposit areas of Sergijivske, ore manifestations of Soniachne, dedicated according to tectonic map of Ukraine (Gurskiy et al. 2004) and by observation results NIEMFE.

Serednio-prydniprovskyi megablock (Gurskiy et al. 2004)	Sergijivske deposit (Zmiyevskaya 2014)	Soniachne ore manifestation (Zmiyevskaya 2014)
0° & 360°		—
—	—	—
12°	—	12°
17°	—	—
25°	—	25°
30°	27°–30°	30°
35°	—	—
40°–45°	45°	40°–45°
50°	50°	—
—	65°	—
—	—	70°–75°
85°	—	—
90° & 270°		
—	—	—
290°	290°	—
300°–305°	300°–305°	—
310°	310°	—
315°–320°	315°–320°	315°
330°–335°	—	—
340°	—	340°–345°
350°	—	—

Furthermore, diagonal systems of faults 45° – 50° and 315° – 320°, 27° – 30° and 300° can be seen.

The gold ore mineralization is connected with dedicated structures. Gold ore mineralization gravitate towards systems of faults 70° – 75° and 340° – 345°, 45° – 315° and 12° on the Soniachne ore manifestation area.

By accomplishing a comparison of the main trend azimuths of previously dedicated structures for complex geophysical studies as: magnetic survey, geoelectric survey, made in Sc 1:2000 (Malinovskiy & Berzenin 2001) on the Sergijivske deposit and Solnechnoe ore manifestation, similar strikes of discontinuous structure were received (Table 2).

General trend azimuths of faults which were dedicated by different geophysical methods are: 0° and 90°, 12° – 17° and 287°, 25° – 30° and 300° – 305°, 50° and 320° – 325°.

3 CONCLUSIONS

Faults on the Sergijivske deposit area have following trend azimuths: 0° and 90°, 45° – 50° and 315° – 320°, 27° – 30° and 300°.

Table 2. Trend azimuths of the main tectonic structures and linear crusts of weathering, dedicated according to geophysical studies (Malinovskiy & Berzenin 2001).

Trend azimuths, dedicated on residual anomaly map $\rho_{cresid.}$ (AB/2) = 250 m	Trend azimuths, dedicated on residual anomaly map $\Delta g_{resid.}$	Trend azimuths, dedicated on isoresistivity map $\rho_{c.}$ (AB/2) = 250 m	Strikes dedicated on the map of anomaly geomagnetic field $\Delta T\alpha$
	0° & 360°		
12°	12°	—	10° – 12°
17°	17°	17°	—
35°	—	35°	30° – 35°
—	—	45°	—
—	—	—	—
—	—	—	60° – 62°
—	—	77°	—
—	—	—	—
	90° & 270°		
—	—	—	275°
—	—	287°– 290°	287°– 290°
305°	—	—	300° – 305°
—	310°	—	—
—	315°	—	315°
—	—	325°	327°
332°	332°	332°– 335°	—
—	340°	—	—
347°	347°	347°	—
—	355°	—	—

Faults on the Solnechnoe ore manifestation area have following trend azimuths: 70° – 75° and 340° – 345°, 45° – 315° and 12°.

Comparison of received trend azimuths values of faults which were dedicated by different geophysical methods, allowed to conclude that deformation processes, which formed an observed net of faults within SPMB on the whole caused an appearance of faults on the Solonianske ore field.

REFERENCES

Gurskiy, D. & Kruglov, S. 2004. *Tectonic map of Ukraine.* Scale 1:1000000. Kiev: The state geological service.

Kravchenko, V. & Ruzina, M. 1997. *The role of regional faults of Middle Dnieper in disposition of precious metals occurrence among greenstone structure.* Dnipropetrovsk: Register Mining Academy of Science in Ukraine, No 4: 104 – 105.

Malinovskiy, A. & Berzenin, B. 2001. *Forecast of a gold mineralization in Verkhovtsevsky and Sursky gold mining structures.* Naukovyi visnyk NGAU, No 5: 57 – 58.

Tyapkin, K. & Dovbnich, M. 2009. *New rotary hypothesis of structure and its geological and mathematical justification.* Dnipropetrovsk – Donetsk: 342.

Zmiyevskaya, K. 2014. *Comparative analysis of spatial regularities of the development of high-order breaking failures within the areal fragments of Srednepridneprovskii and Ingulskii megablocks of Ukrainian shield.* Collection of scientific articles of National Mining University, No 44: 17 – 21.

Theoretical and Practical Solutions of Mineral Resources Mining – Pivnyak, Bondarenko & Kovalevska (eds)
© 2015 Taylor & Francis Group, London, ISBN: 978-1-138-02883-8

On some mathematical models of facility location problems of mining and concentration industry

S. Us & O. Stanina
National Mining University, Dnipropetrovsk, Ukraine

ABSTRACT: The optimal location problem for mining and concentration enterprises in a given region has been studied. The discrete and continuous location models were analyzed. New mathematical models have been formed. They combined discrete and continuous task feature and took into consideration multi-stage character of production process.

1 INTRODUCTION

The production and use of energy resources are the most important prerequisite for translational and sustainable economic growth in Ukraine. The mining industry has a special role in this process. This industry includes the production, processing and concentration of mineral resources (energy, ore, mining and chemical, building materials).It is distinguished by a high degree of concentration and large scale of production, which allows to reduce the cost of production, and to apply modern technology.

Analysis of factors affecting the strategic planning of development of coal enterprises was held in (Melnikov & Pavlenko 2012), the economic assessment of the costs of raw materials in the processing of underground and surface ore deposits was given in work (Prokopenko & Tymoshenko 2004) and features of the development of these enterprises was emphasized. The need to develop new enterprises (for example development of new fields) and modernization or reorganization of the existing enterprises is a characteristic feature of development of this industry, and therefore there is the problem of optimal location of enterprises in the given area.

To solve this problem it is necessary to take into account various factors (economic, social, environmental) and features of the industry, but in modern conditions this is impossible without the use of mathematics and information technology. That's why, the creation of adequate mathematical models problem of locating production is relevant.

The purpose of study is scientific justification of mining and concentration industry location.

The task of investigation is to propose the system of mathematical models for concentration industry location.

2 FEATURES OF LOCATION PROBLEM FOR ORE PROCESSING ENTERPRISES AND EXISTING MODELS ANALYSIS

Facility location problem has been interesting for many researchers for a long time. They occur in a variety of practice areas, to solving a wide range of issues. Continual interest to this problem is confirmed by a great amount of papers dedicated to the creating mathematical models, efficient methods and algorithms of solving various facility location tasks (Mikhalevich et al. 1986, Montlevich 2000). A survey of facility location tasks and methods of this solving are given in (Drezner & Hamacher 2001, Farahani & Hekmatfar 2009, Brimberg et al. 2008).

Initially, researchers in the field of economics (Launhardt, Weber, Tyunen etc.) (Blaug 1994) have isolated groups of factors that must be considered in solving the problems of accommodation, namely natural-geographic, socio-demographic, techno-economic, socio-economic, environmental, transport, since as only consideration of the totality of unequal spatial conditions, resources and their properties and proper use will provide the best results when placing production facilities and development areas.

The most important principles (Shulgin & Elagin 2004) which govern by the issues of territorial planning are:

– approximation of production to sources of raw materials, fuel and consumers of products;

– environmental protection and rational use of natural resources;

– an excessive production concentration in cities has to restrict;

– alignment of economic development levels of regions and districts.

– strengthening the country's defense;

– consideration of the interests of economic integration to the European world market.

In common case location-allocation problem (LAP) can be formulated as follow: It is necessary to determine an optimal number and location of set facilities and assign customers to them such a way that the total transportation cost is minimized and customers demand is satisfy.

Classification and survey LA problem and solution methods are represented by (Drezner & Hamacher 2001).

A lot of papers formulate LAP such as: let the set $I = \{1, ..., m\}$ describe possible locations for the enterprises, which making one of some kind of products. Enterprises could be opened in any points of set I, and value $c_i \geq 0$ defines the associated costs. The opened enterprises can make unlimited quantities of product for customers. The list of customer is given by set $J = \{1, ..., J\}$, value $g_{ij} \geq 0$ is cost of production and transportation of the product to the customers for any couple ij.

The task is to determine the set $S \subseteq J$, $S \neq \emptyset$ of the installed new enterprises, and partitioning the set of clients J, so that all clients would be connected to enterprises, needs of all clients have been met and the cost of transportation would be minimized.

Using these notations the problem of optimization formulation can be written as follows:

$$F(S) = \sum_{i \in S} c_i + \sum_{j \in J} \min_{i \in S} g_{ij} \to min. \tag{1}$$

It should be noted that the placement of enterprises has significant features depending on the type of production.

Therefore, we can get a variety of mathematical models, depending on these factors.

One of the simple and natural ways to location is the approach, proposed by (Pupasov-Maksimov et al. 2013). It takes into consideration a number of enterprises location features on example HPP.

The authors consider such factors as the location, proximity to infrastructure, the distance to the consumer, demand and the overall cost.

The proposed mathematical model is formulated in such way:

$$rank = nw \left(wkP \cdot kP + wdf \cdot df - wRd \frac{Rd}{Rd\,max} - \right.$$

$$\left. - wLd \frac{Ld}{Ld\,max} - wFa \frac{Fa}{Fa\,max} - wPc \frac{Pc}{Pc\,max} \right), \tag{2}$$

where:

$$nw = \left(wkP^2 + wdf^2 + wRd^2 + wLd^2 + \right.$$

$$\left. + wFa^2 + wPc^2 \right)^{\frac{1}{2}}, \tag{3}$$

$w(i)$ – the weight of appropriate factor, kP – capacity factor, df – demand, Rd – the distance to the consumer, Ld – the length of the diversion canal, Fa – area of flooded land, Pc – the cost of the project.

This approach implies calculation of the rank for the proposed projects and the optimal project is chosen according to these parameters.

A clear advantage of this approach is simplicity and clarity of calculation, however this way can be applied only if some variants of placement are known and there is a person (or group of peoples), who can define real value of coefficients weight.

Other spread approach to facilities location is based on calculating total sum of spending and shown in (Rusyak & Nefedov 2012).

Generally it can be written down in such way:

$$F = \sum_i \sum_j g_{ij} z_{ij} + C_1 + C_2 + C_3, \tag{4}$$

where g_{ij} – are the cost of transporting raw materials (products) units from point i to point j, z_{ij} – are the amount of raw materials (products), which transported from point i to point j, C_1, C_2 – are annual cost of raw materials processing and production accountant, C_3 – are the cost of operation and service.

But, this approach can't be always used in practice, because it doesn't consider special features of industrial enterprises placement.

There are some models, which enable to solve the same tasks.

In the most general form the location problem of production can be formulated as follows (Velikanova & Ladoshkin 2013): let $I = \{1, ..., m\}$ is the set of regions, where enterprises can be located, $J = \{1, ...,n\}$ – the set of product consumer, $A_i \geq 0$ – the costs of organizing production in region i, $c_{ij} \geq 0$ transportation and product cost for demand satisfaction of client j by enterprise i, $p > 0$ the maximum number of enterprises, which can be stated.

It is required to find subset $\Omega \in S$, allowing satisfying of all consumers demand with minimum total cost.

Let:

$$x_i = \begin{cases} 1, & \text{if the plant is placed in position } i, \\ 0, & \text{otherwise} \end{cases}, \tag{5}$$

$$x_{ij} = \begin{cases} 1, & \text{if the plant } i \text{ services customer } j, \\ 0, & \text{otherwise} \end{cases}. \tag{6}$$

The following mathematical model corresponds to a set task:

$$F = \sum_i \sum_j c_{ij} x_{ij} + \sum_i A_i x_i \to min, \tag{7}$$

under constraints:

$$\sum_i x_{ij} = 1, \tag{8}$$

$$x_{ik} \geq x_{ij}, i \in I, j \in J, \tag{9}$$

$$\sum_{i \in I} x_i \leq p, \tag{10}$$

$$x_{ij}, x_i \in \{0,1\}, i \in J, k \in I. \tag{11}$$

Peculiarity of this model is that it doesn't suggest the raw materials delivery, that's why it can be used only for such enterprises, which don't demand its calling from outside. But in practice, this condition is often unfeasible; therefore models with consideration of transportation costs are more applicable.

In this case, signify:

$$x_i = \begin{cases} 1, & \text{if the plant is placed in position } i, \\ 0, & \text{otherwise} \end{cases}, \tag{12}$$

$$z_{ij} = \begin{cases} 1, & \text{if the plant } i \text{ services customer } j, \\ 0, & \text{otherwise} \end{cases}, \tag{13}$$

$$y_{ik} = \begin{cases} 1, & \text{if the plant is gets a row materials} \\ & \text{from the suppliern } k, \\ 0, & \text{otherwise} \end{cases} \tag{14}$$

Then mathematical model can be formulated as follows:

$$F = \sum_i \sum_j c_{ij} z_{ij} + \sum_i A_i x_i + \sum_i c_{ik} y_{ik} \rightarrow min, \tag{15}$$

under constraints:

$$\sum_i z_{ij} = 1, \tag{16}$$

$$x_{ik} \geq z_{ij}, i \in I, j \in J, \tag{17}$$

$$\sum_{i \in I} x_i \leq p, \tag{18}$$

$$\sum_{k \in I} y_k \leq k, \tag{19}$$

$$z_{ij}, x_i \in \{0,1\}, i \in J, k \in I. \tag{20}$$

These models correspond to case when the number of possible location places is finite and their positions have defined previously.

However, in practice task, we often face the situation, when enterprises can be located in any points of region. Such tasks, known as continual set partitioning problem (CSPP) were studied by (Kiseleva & Shor 2005).

Let us formulate the task of optimal partitioning of continuous set in the following way.

Supposing, there is a set of consumers of some homogeneous product allocated in the area Ω The finite number N of producers placed in the isolated points τ_i, $i = \overline{1, N}$ of the area Ω forms the system of points $\tau_1, \tau_2, ..., \tau_N$; besides, the coordinates of some points or all of them can be unknown early. The demand $\rho(x)$ for product at each point x of the area Ω and product delivery cost $c_i(x, \tau_i)$, $i = \overline{1, N}$ from a producer τ_i to a client x are known. Assume that producer's profit is only dependable on transportation costs. The capacity of a producer i is defined by a total demand of the service clients and should not be higher than specified volumes b_i, $i = \overline{1, N}$. The area Ω is required to be subdivided into service zones by each of the producers, i.e. into sets Ω_1, Ω_2, ..., Ω_N, to minimize the total costs spent on product delivery.

Mathematical model of the stated problem can be presented as follows (Kiseleva & Shor 2005): let Ω is a bounded Lebesgue measurable set in an n-dimensional Euclidean space E^n. It is required that the set is divided into N Lebesgue measurable subsets Ω_1, Ω_2, ..., Ω_N, and subset centers $\tau_1, \tau_2, ..., \tau_N$ are located in the area Ω to let the functional:

$$F(\Omega_1, \Omega_2, ..., \Omega_N, \tau_1, \tau_2, ..., \tau_N) =$$
$$= \sum_{i=1}^N \int_{\Omega_i} c_i(x, \tau_i) \rho(x) dx, \tag{21}$$

reach the minimum value under constraints:

$$\int_{\Omega_i} \rho(x) dx \leq b_i, \ i = \overline{1, N} \tag{22}$$

$$mes(\Omega_i \cap \Omega_j) = 0, \ i \neq j, \ i, j = \overline{1, N}, \tag{23}$$

$$\bigcup_{i=1}^N \Omega_i = \Omega. \tag{24}$$

It should be noted the existence of multi-stage LA problems. This class is a generalization of multistate transportation and production problems, which is actively studied for the last time.

Multistage facility location problem on a substantive level, is as follows. The set of businesses and consumers who need their products are specified. For production companies are united in technological chains. Thus products undergo several stages of processing. The set of feasible production chains are known. The cost of opening enterprises is given for each company. Production and transportation costs for each processing chains are known. For each user defined need to find a set of enterprises that with minimal total cost would allow to meet the demand of all consumers.

This task can be formulated as the linear integer programming task with Boolean variables.

Formalize the statement of the problem.

Let $N = \{1,...,3\}$ is set of point of the final product demand; M_1 is set of possible location of the first stage enterprises, M_2 is set of possible location of the second stage enterprises; g_i^r – are the cost of enterprises placement for 1 and 2 stage correspondly, $g_i^r \geq 0$; c_{ij} – are the transportation cost from point i to point j for unit of product, $c_{ij} \geq 0$, $i, j \in N$; b_j – demand volume in point j, $b_j > 0$, $j \in N$.

It is need to select subset of locations for each level (stage) and make settings chosen productions at points of demand so as to minimize the total cost of all the selected accommodation facilities and the transportation of the product.

To construct a mathematical model, we introduce the following notation: $x_i = 1$ ($y_k = 1$), if enterprises 1 (2) level is located in point $i \in M_1$ $(k \in M_2)$ and $x_i = 0$ ($y_k = 0$) otherwise; $x_{kij} = 1$ if the consumption point j is serviced by the 2-nd level point k through the i-th point of the 1-st level, and $x_{kij} = 0$ otherwise.

And the mathematical model is:

Minimize:

$$\sum_{i \in M_1} g_i^1 x_i + \sum_{k \in M_2} g_k^2 y_k +$$

$$+ \sum_{j \in N} b_j \sum_{k \in M_2} \sum_{i \in M_1} \left(c_{ki} + c_{ij} \right) x_{kij} , \qquad (25)$$

$$\sum_{k \in M_2} \sum_{i \in M_1} x_{kij} = 1, j \in N , \qquad (26)$$

$$\sum_{k \in M_2} x_{kij} \leq x_i, j \in N, i \in M_1 , \qquad (27)$$

$$\sum_{i \in M_1} x_{kij} \leq y_k, j \in N, k \in M_2 , \qquad (28)$$

$$x_i, y_k, x_{kij} \in \{0,1\} . \qquad (29)$$

This task and methods are considered in papers (Gimadi 1995).

Now there are many papers devoted to solving methods for multi stage problem, for example (Trubin & Sharifov 1992, Bischoff et al. 2009, Ageev et al. 2009). But the continuous multi stage problems remain almost unstudied, because their realization is more difficult. Nevertheless there are many arias, where such kind of task exists. Particularly, extraction and processing of natural resources (oil, ore) are the two-level process.

3 BUILDING MATHEMATICAL MODELS

To build the mathematical model for location problem of mining and concentration enterprises, we have to take into account such features of this industry:

1. There are complex enterprises and different stages of its production processes are territorially divided. For example, mining and processing plant consists of such basic production subdivisions:

– the mining division (quarries, mines);
– the transport unit that is designed for the mined ore delivery to the processing plant. Ore delivery to the processing plant can be carried out using a variety of transport systems and types of transport: road, rail, pipeline, cable cars, ore passes and other;
– the division for processing of mined minerals, which is usually represented by the processing plant;
– general manufacturing divisions: electric industry, repair and mechanical workshop and other necessary subdivisions.

2. Enterprises are located in points, which are not optimum in terms of availability of labor, material, energy and other resources of geographic areas, and tend to mineral deposits (Melnykov & Pavlenko 2012).

3. The production capacity size of the coal enterprise depends on the mineral resources quantities, their mining-geological characteristics, extraction technology possibilities are projected, used technique and technical capabilities of the several stage of the production process.

Thus, modeling of placement such enterprises we have to consider:

1. Resource allocation in given area is not concentrated in some points but it is placed in area continuously.

2. The production is multi stage.

Analysis of trends in the development of open cast mining in Ukraine and abroad shows that more and more importance in the formation production costs gets transportation of raw materials. The cost of the rock mass transporting reaches now 40 – 60% of the total cost of ore production, and when the depth of quarries nearly 500 – 1000 m it increase to 70% (Anistratov et al. 2007). Accordingly the transport efficiency affect significantly to the cost of the final product, and thus the main criterion for the building of mathematical model of a location problems is minimum of the total cost of transportation and the enterprise accommodation (although others are possible).

A characteristic feature of this production process is the presence of two phases implemented at enterprises of various types.

The organizational structure of production is shown schematically in Figure 1.

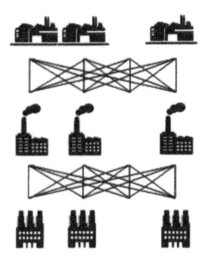

Figure 1. Scheme of organizational structure of production.

Ore is mined at a mine and it is then sent for concentration via hydraulic transport. The concentration plant directs the resulting commercial concentrate to the consumption point (e.g. warehouses, or other businesses). Thus, the manufacturing process involves several stages requiring a variety of production resources. Furthermore, when locating companies of the first stage it is necessary to consider the distribution of raw materials in the given area based on geological exploration.

Location criteria may be different: one of the most common is to minimize the transportation costs for the delivery of raw materials and finished products.

Informal statement of location problem for a multi-stage processing industry can be summarized as follows: it is necessary to place concentration production, which includes mines and processing plants in the region Ω, so that the cost of shipping raw materials and products was minimal. It is assumed that mines can be placed anywhere in the area Ω, possible location of factories have known early and the location of a finite number of users is known in advance.

To construct a mathematical model, we introduce the following notation: Ω – given domain; Ω_i – fixed area of i-th mine; A – the cost of enterprises placement (mine or plant); N – the number of mines; M – the number of plants to be placed in the area; K – finite number of consumers; λ_j – point of j-th plant location; τ^r_i – coordinates of i-th mine; $\tau^r_i = (\tau^r_{1i}, \tau^r_{2i})$; τ^p_j – coordinates of j-th plant; $\tau^p_j = (\tau^p_{1j}, \tau^p_{2j})$; τ^u_k – coordinates of the consumer, $\tau^u_k = (\tau^u_{1k}, \tau^u_{2k})$; $c^r_i = c^r_i(x, \tau^r_i)$ – cost of shipping a unit of raw material to i-th mine; $c^p_{ij} = c^p(\tau_i, \tau_j)$ – cost of shipping a unit of raw material from i-th mine to j-th plant; $c^u_{jk} = c^u(\tau_j, \tau_k)$ – the cost of delivery of products from j-th factory to k-th consumers; $\rho(x)$ – thickness

of a seam at the x point of Ω; v^p_{ij} – the volume of products delivered from i-th of the mine to j-th plant; v^u_{jk} – the volume of products delivered from j-th factory to k-th consumers; b^r_i – the yield of i-th mine; b^p_j – maximum production capacity of the plant; b^u_k – the demand of k-th consumer.

The total cost of transportation and the enterprise accommodation can be writing as follow:

$$\sum_{i=1}^{N} A^r_i + \sum_{j=1}^{M} A^p_j \lambda_j + \sum_{i=1}^{N} \int_{\Omega_i} c^r_i(x, \tau^r_i) \rho(x) dx +$$

$$+ \sum_{i=1}^{N} \sum_{j=1}^{M} c^p_{ij} v^p_{ij} \lambda_j + \sum_{k=1}^{K} \sum_{j=1}^{M} c^u_{jk} v^u_{jk} \lambda_j . \tag{30}$$

Constrains of the task can be formulate in such way.

The total reserves of the resource in the service area of the i-th company I stage are not less than the production capacity of the company:

$$\int_{\Omega_i} \rho(x) dx \geq b^r_i , \quad i = 1, 2, ..., N , \tag{31}$$

corporate demand of plants of the second stage must be satisfied:

$$\sum_{i=1}^{N} v^p_{ij} = b^p_j , \quad j = 1, 2, ..., M , \tag{32}$$

corporate demand of consumers must be satisfied:

$$\sum_{j=1}^{M} v^u_{jk} \lambda_j = b^u_k , \quad k = 1, 2, ..., K . \tag{33}$$

Then the above problem can be described by the following model:

Minimize:

$$\sum_{i=1}^{N} A^r_i + \sum_{j=1}^{M} A^p_j \lambda_j + \sum_{i=1}^{N} \int_{\Omega_i} c^r_i(x, \tau^r_i) \rho(x) dx +$$

$$+ \sum_{i=1}^{N} \sum_{j=1}^{M} c^p_{ij}(\tau^r_i, \tau^p_j) v^p_{ij} \lambda_j + \sum_{k=1}^{K} \sum_{j=1}^{M} c^u_{jk}(\tau^p_j, \tau^u_k) v^u_{jk} \lambda_j . \tag{34}$$

Under constrain:

$$\tau^r = (\tau^r_1, \tau^r_2, ..., \tau^r_N) \in \Omega^N , \tag{35}$$

$$\tau^p = (\tau^p_1, \tau^p_2, ..., \tau^p_M) \in \Omega^M , \tag{36}$$

$$\bigcup_{i=1}^{N} \Omega_i = \Omega , \tag{37}$$

$$\Omega_i \cap \Omega_j = 0, \ i \neq j, \ i, j = 1, n , \tag{38}$$

$$\int_{\Omega_i} \rho(x) dx \geq b^r_i , \quad i = 1, 2, ..., N , \tag{39}$$

$$\sum_{i=1}^{N} v^p_{ij} = b^p_j \quad j = 1, M , \tag{40}$$

423

$$\sum_{j=1}^{M} v_{jk}^u = b_k^u , \quad k = 1, K , \tag{41}$$

$$v_{ij}^p \geq 0 , \quad i = 1, 2, ..., N , \quad j = 1, 2, ..., M , \tag{42}$$

$$v_{jk}^u \geq 0 , \quad k = 1, 2, ..., K , \tag{43}$$

where the constraints (37 – 38) mean a partition of Ω into the service areas for mines, that is, to cover the whole area Ω (37), and to serve each point of the area by only one mine (38); (39) – ensures that limits for enterprises capacity are executed, (40 – 41) – mean that corporate demand of plants of the second stage and consumers must be satisfied.

This model can have various modifications, for example the restrictions on the plant capacity may be absent or enterprises any stage may be placed anywhere in the field. These models are shown in (Us & Stanina 2014).

A feature of the proposed model is a combination of discrete and continuous component and that's why it requires a special approach to a solution. Some of approaches to solving this type of problems are suggest by authors. They based on a sequential decision of continual optimal set partitioning problem and discrete multi-stage facility location problem.

4 CONCLUSIONS

In recent years, a growing number of researchers drew their attention to the solution of various problems in the field of economy of material and natural resources. In this work some of the mathematical models of locating production in the mining industry of were considered and analyzed taking into account various criteria. Specific characteristics of their use were considered.

Mixed discrete-continuous model of two-stage facility location problem was suggested for the problem of locating processing industry, which allows taking into account features of the industry. Namely, a continuous distribution of minerals in area and the production of multi-stage, which allowed reducing the cost of transportation of raw materials and final products, and eventually reducing the production cost.

REFERENCE

Ageev, A.A., Gimadi, A.H. & Kurochkin, A.A. 2009. *A polynomial algorithm for solving the problem of placing on the chain uniform capacities* (in Russian). Discrete Analysis and Operations Research. 16 (5): 3 – 18.

Anistratov, K.Y., Degrees, M.S, Stremilov, V.J. & Teterin M.V. 2007. *Economic-mathematical model of enter-*

prise technology career vehicles (in Russian). A mining industry. No 1.

Bischoff, M., Fleischmann, T. & Klamroth K. 2009. *The Multi-Facility Location-Allocation Problem with Polyhedral Barriers* (in Russian). Computers and Operations Research, Vol. 36, No 5p: 1376 – 1392.

Blaug, M. 1994. *Economic theory use of space and the classical theory of production location.* Economic thought in retrospect, http://gallery.economicus.ru/cgi-bin/frame_rightn.pl?type=school&links=./school/firm/lectures/firm_l2.txt&name=firm&img=lectures.jpg.

Brimberg, J., Hansen, P., Mladenović N. & Salhi. S. 2008. *A Survey of Solution Methods for the Continuous Location-Allocation Problem.* International Journal of Operations Research, Vol. 5, No 1: 1 – 12.

Drezner, Z. & Hamacher, H. 2001. *Facility Location: Application and Theory.* Berlin: Springer.

Farahani, R. Z. & Hekmatfar, M. 2009. *Facility Location. Concept, Model, Algirithms and Case Studies.* Springer Dordrecht Heidelberg London New York.

Gimadi, E.H. 1995. *Efficient algorithms for solving multistep problem in a string.* Discrete Analysis and Operations Research, October – December, Vol. 2, No 4: 13 – 3.

Kiseleva, E. & Shor, N. 2005. *Continuous optimal set partitioning: theory, algorithms, applications*: Monograph (in Russian). Kiev: Naukova dumka.

Melnykov, A.M. & Pavlenko, O.V. 2012. *On the issue of taking into account the specific features of coal enterprises in the strategic planning of their development.* Naukovyi Visnyk Natsionalnoho Hirnychoho Universytetu, No 6: 140 – 146.

Mikhalevich, V.S., Trubin, V.A. & Shore N.Z. 1986. *Optimization problems of production and transport planning: models, methods, algorithms.* M: Nauka: 264.

Montlevich, V.M. 2000. *The problem of locating enterprises with standard capacities and indivisible consumers.* Computational Mathematics and Mathematical Physics: 1491 – 1507.

Prokopenko, V.I. & Tymoshenko, L.V. 2004. *Development factor of direct material costs for stowing operations.* Journal of CTU, Kryvyi Rig, Vol 3: 173 – 175.

Pupasov-Maksimov, A.M., Orlov, A.V. & Fedoseev, A.V. 2013. *The problem of optimizing the location and structure of a small hydropower plant on the stage of investment rationale.* Naukovedenie: No 5.

Rusyak, I.G. & Nefedov, D.G. 2012. *Solving optimization problems location scheme of production of wood types of fuel by the criterion of the cost of thermal energy.* Computer studies and modeling, Vol. 4, No 3: 651 – 659.

Shulgin, E.P. & Elagin, V.S. 2004. *Location of productive forces.*

Trubin, V. A. & Sharifov, F. A. 1992. *Simple multistage location problem on a treelike network.* Cybernetics and Systems Analysis November – December, Vol. 28, Issue 6: 912 – 917.

Us, S.A. & Stanina, O.D. 2014. *On mathematical models of multi-stage problems of facility location* (in Russian). Issues of applied mathematics and mathematical modeling. Collected research works. Dnipropetrovsk: DNU: 258 – 267.

Velikanova, T.V. & Ladoshkin, A.I. 2013. *Using optimization methods in planning the location of production.* Bulletin of the Samara Municipal Institute of Management, No 2 (25): 67 – 74.

Theoretical and Practical Solutions of Mineral Resources Mining – Pivnyak, Bondarenko & Kovalevska (eds)
© 2015 Taylor & Francis Group, London, ISBN: 978-1-138-02883-8

Influence of dynamic processes in mine winding plants on operating safety of shafts with broken geometry

S. Iljin
*M.S. Poliakov Institute of Geotechnical Mechanics under the
National Academy of Sciences of Ukraine, Dnipropetrovsk, Ukraine*

V. Samusya, I. Iljina & S. Iljina
National Mining University, Dnipropetrovsk, Ukraine

ABSTRACT: In the article the analysis of real exploitation conditions of the systems "vessel – reinforcement" in vertical shafts is executed, recommendations providing the necessary increase of operating strength security of hoisting plants for the shafts with the broken geometry are represented. The researches are done on the base of application of modern hardware and software measuring complexes, calculation of the correcting impacts localized to address and controlling parameters. These measures are applied to prevent the emergency situations and provide transition of shaft equipment in the state of the equal spreading of set level safety indicator on the depth of shaft, at influence of external factors of difficult technical and mountain-geological conditions with minimum technological economical expenses.

1 INTRODUCTION

In the operative mine shafts in the area of mountain breeds movement guides' deviation from vertical line are the main source of excitation of strike-cyclic interaction.

Mine shafts are the continuously changing systems, in a different period of time found in different types of the technical state – from normal to emergency dangerous. With the increase of exploitation term speed of possible transition from one state to another increases because of the contemporary accumulation of the concomitant degradation occurrences in the all elements of shafts' equipment. It is the main source of failures in mine shafts with the heavy material consequences and threat for the people life.

The mine winding plant (MWP) is an extensive, multiple-link, multi-oscillatory system with multiple degrees of freedom where links consist of a combination of assumed hard limits of operating parameters and stochastic variables with high variability. During their operation exists strong correlations between mutual influenced dynamic processes where non-localized heterogeneous links influence each other.

The operational state of each link is sufficiently characterized by a set of diagnostic parameters with normative values that are both deterministic and stochastic in nature. The functioning deterministic state of the MWP links must work in the projects dynamic operation dampening the presence of the undue fluctuations influencing the dynamics of other system links.

2 DYNAMIC APPARATUS CONTROL OF STATE OF SYSTEM "VESSEL – REINFORCEMENT"

In the presence of local defects, the dynamic isolation of processes is disrupted by separate links distributing energy such that stochastic processes appears in temporal concentration in motion in links or within pairs of contiguous links. Such energy pumping corresponds to unforeseen resonance activation in the system of different types: external, internal including parametrical, "beating", auto-oscillations, etc.

The most dangerous behaviors are resonances which are revealed upon actual operation of the mine hoist and attempts to compensate are taken into account in an operating project documents. An array of other resonance effects is camouflaged from attention and only revealed indirectly via disturbances in the chain of dynamic links and operations of individual links. These defects appear as unexpected energy due to accumulation of system effects from the total influence of insignificant defects in the functioning of links. Such effects are not studied and not reflected in a normative documentation on servicing, adjusting, and diagnosing tests on the MWP equipment. This is especially true in the MWP shaft equipment where conditions of:

– intensive corrosive effects and mechanical wear are present;

– reinforcement with minimal permissible errors on amplitudes of oscillations of lifting vessels (20 – 30 mm) as compared to their overall sizes (10 – 20 m) exists;

– small deviations of guides from the vertical alignment (10 mm on contiguous tiers);

– large contact loading in a pair "vessel – guide" exist.

– transmissible on bunton and shaft's lining are present.

The main MWP working element and concentrator are subject to influences from different links within the plant while lifting the vessel where the links attempt to provide a safe and stable hoisting motion.

Objective information measuring the technical parameters of MWP equipment operational states experiencing long-term exploitation, consist of special inspections of apparatuses during dynamic tests of hoisting during operating and test conditions with subsequent system analysis. In Ukraine, such measurements are conducted in accordance with positions of normative documents and involve complex constructions described in articles (Cherednichenko et al. 2008). In a number of countries, dynamic control of the systems "vessel – reinforcement" is obligatory along with traditional surveyor observation of guides' shape.

Project tiers of reinforcement are designed to work with identical terms of loading and, at one level, resist loading from the side of lifting vessels and rock massif. During the first decade of exploitation, because of influence of aggressive environment, dynamic loading, moving of rock, and repair works there appears sharply heterogeneous changes on the depth of the shaft picture dispensing levels of bearing strength and reinforcement elements – guides & buntons.

As a result of data analysis, greater than one hundred and fifty (150) inspections and dynamic tests of systems "vessel – reinforcement" must be conducted on mine shafts to determine their technical state and operating security level. These tests are conducted by Diagnostic Laboratory of the mining hoist of IGTM in the name of N.S. Polyakov NAS of Ukraine, that states during the life cycle of shaft's functioning values of technical state parameters of elements of reinforcement get substantially heterogeneous character on its depth. They can be unlike on different areas of shaft in 5 – 10 times.

The level diagrams for remaining sections of guides and buntons, was derived on the basis of instrumental measuring during mine shaft inspection and can serve as illustration of this statement (Figure 1). An upper line shows a project level for new reinforcement (100 %), and continuous curves show

its value varies on the numbers of tiers. The dotted line represents the level of loss of section 20% where below it an object must be exposed to special inspections for estimation the operational technical state and security level of exploitation. As we see on the beams of the central buntom, there are anomalous areas of section loss up to 90% in the top part of shaft at tiers No 60 – 50 (with the proper level of loss of load-carrying ability), while on average the section shaft loss consists of 50 – 60%.

(a)

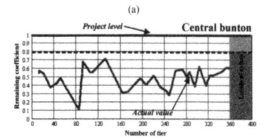

(b)

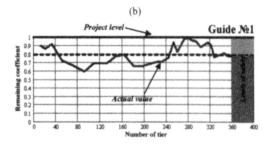

(c)

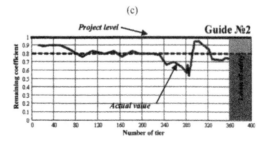

Figure 1. Diagram of levels of remaining section of reinforcement elements.

The picture is different for guides, where the first guide lost up to 40% section in top part of shaft, and the second has an obvious local anomaly on the area of tiers No 270 – 280 with a loss 50% of section at a general level of section's saving to 80% in the shaft.

A similar process of homogeneity loss concerning properties takes place in the geometrical parameters of guide shapes which the lifting vessels move, thus geometrical heterogeneity concerning reinforcement on the depth of shaft appears (Iljin et al. 2013).

Our research indicates that along points in the shaft, there exist sharp changes in the speed of vessel lifting especially with action covering the preventive brake. There are horizontal impacts of vessels on guides that are four – five times higher than during motion with constant speed, this is due to the excitation of vertical speed oscillations along a resilient rope. Repeating from cycle to cycle, they are instrumental in accumulation of tireless damages in guides and buntons, formation of cracks on the welds, weakening of attachment points of guides, slacking of buntons in the places of lining attachment (Iljina 2010).

All these heterogeneous anomalies which are simultaneously, permanently and slowly changing - affect the process of dynamic cooperation of vessel with reinforcement which it takes place on each tier area. Original dynamic heterogeneity of loading reinforcement on the depth of shaft with the middle level and areas of dynamic anomalies appears as a result of such imposition (Figure 2).

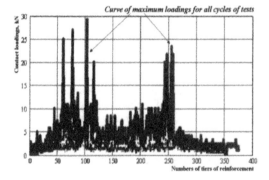

Figure 2. Diagrams of the dynamic loadings on guides during motion of lifting vessel.

The primary factor determining the technical state of shaft's reinforcement is remaining assurance coefficients concerning elements (guides and buntons) which are defined on every tier. Destruction of any even along one tier will inevitably result in a failure with drastic consequences.

Dynamic cooperation along vessels with reinforcement has an impact and cyclic causal character and this has been taken into consideration while estimating the technical guides and buntons states under effect during the actual operating loading cycles. It is necessary to adopt the minimal assurance coefficient (2.0) on the criterion to account for the accumulation of fatigue damages in metalware reinforcement. A combination of loading and remaining ruggedness for guides and buntons, provides an assurance coefficient (on the limited area of shaft) such that its technical state is considered

safe. As assurance coefficient declines from 2.0 to 1.0 (on a local area or even on separate tiers) the state becomes potentially dangerous, requiring special supervision after appearance of fatigue cracks and another defect. Tiers (with an assurance coefficient less than 1.0) are considered on the limit of stretching strain, thus it is necessary to considered them abnormal, dangerous and required immediate acceptance of measures or decline of the contact loading (due to the decline of motion speed on an area) or on the renewal of ruggedness of the element of reinforcement .

There are three independent and basic processes affecting the remaining assurance coefficients concerning bearing elements of reinforcement:
– wear of guides and buntons (corrosive and mechanical);
– curvature of spatial shape of guides (under influencing of moving of rocks and violations at permanent repair of reinforcement, formation of ledges on the units of guides);
– sharp change of altitude rate of lifting vessel.

Reinforcement wear causes a decline in the remaining load-carrying capability of its elements where the decline is due to the possible contact loading at the intended values of assurance coefficients. Curvature of guides' shape causes growth of the dynamic loading on the local areas of the shaft - on condition of absence of the resonance effect. Sharp changes in vessel altitude rater (emergency braking (EB), in particular) cause certain areas of instability along the shaft with growth in the horizontal dynamic loading on guides (Iljina 2010).

Indicated heterogeneities are the sources of potential danger during the long exploitation of lifting complexes.

The "Diagrams of safety" are the graphic form of this system and presented with measurments and calculations (Figure 3), which are built for each element of reinforcement. They include the results of mathematical data processing for instrumental measuring of guides' and buntons' wear, apparatus measuring of the contact loading, stress-strain analyses of reinforcement (Iljin et al. 2002).

Curves of maximum possible contact loading from the side of lifting vessel for the assurance coefficient of $n = 1$, $n = 1.5$, $n = 2$ are built on diagrams for each guide or bunton using the numbers of tiers of reinforcement. They take account of the actual remaining thickness of given section of the given element and curves of the actual maximal contact loading got for all test passways of vessel in a shaft according to the program of dynamic tests.

Diagrams depict:
– areas of assurance values / coefficients for each certain element of reinforcement on every tier in the

moment once inspection is exploited (emergency dangerous, potential dangerous, safe);

– because of technical reasons an element enters a negative area of safety according to the assurance coefficients. Appearance of areas of failures on the curves of possible loading is caused by the increased wear of the given element of reinforcement. Therefore, even at the moderate dynamic loading an element can be exploited in an under abnormal condition dangerous area. Appearance of splashes of the increased values of the actual operating loading is caused by the presence of large local disturbing of straightness of guides' shapes or ledges on their units.

From the analysis of diagrams we observe what technical measures are necessary in order to choose primary metrics and plan for translation of element exploitation in areas with high assurance coefficient values. The decline in the level of the contact loading on the set area of shaft (above all things) can be provided with decline of altitude rate of vessel on an area, or with correction of parameters of guides' shapes or ledges in units. The increase of level of the possible loading on element without a correction of shape and altitude rate of vessel is achieved by replacing or measures on strengthening of load-carrying capability construction.

(a)

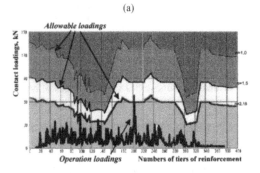

(b)

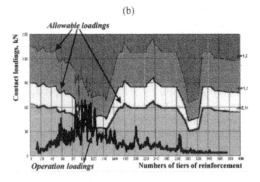

Figure 3. The diagrams of safety for the lateral boxlike guide of strong reinforcement in skip compartments of mine shaft in lateral (a) and frontal (b) loadings.

Doing this work on the mines reveals that even in the hard technical conditions, the effective management by exploitation safety of shaft equipment and its maintenance is possible.

3 METHODS OF DYNAMIC LOADINGS DECREASE IN SYSTEM "VESSEL – REINFORCEMENT"

Parameter "remaining strength margin" of reinforcement elements under action of the working loading is basic, determining the degree of danger emergency of shaft operation. Minimal possible value of strength margin is 1.0, which corresponds to equality of actual tensions in the most loaded section of reinforcement to the limit of stretching strain of its material. Therefore all measures of safe exploitation providing are directed on its increasing to the nominal value 2.15, which corresponds to actual tensions less than endurance limit. With such strength margin there is no accumulation of tireless damages in the metal under impact of cyclic alternative dynamic loadings from the vessels. Such level can be achieved in several ways.

The most widespread method of strength margin increasing is decline of hoisting speed on fixed area of shaft for reduction of the dynamic loadings to the safe level with same type of guides and remaining carrying capacity of reinforcement. For hoists equipped by computer operating systems, realization of the smoothed diagrams of drum rotation speed with minimal jump is the effective method of loadings decrease (Iljin et al. 2014).

The main characteristic of work of any mine hoisting plant is that control center of a mechanical motion of all its links is the drum (pulley) of a hoisting machine. During a descent and ascent cycle mechanical influences in the form of the rotating moment of the electric drive, the working or emergency brake moments with different intensity can be put to the drum. Thus horizontal interaction of lifting vessels with stationary metalwork of a mine shaft (guides and buntoms of reinforcement) takes place at a great distance from the drum.

From the drive or a brake of mine hoisting plant it is possible to carry out the operating impact on lifting vessels only with the drum by means of head ropes flexible and pliable in the longitudinal and cross directions. Such scheme of power transmission system of hoisting plant (especially for depths over 1000 m) is the main technical reason of complexity in ensuring synchronism between the movement of a drum and vessels (Samusya & Komissarov 2000).

The researches of the braking modes of MWP, and also experience of their exploitation, showed that at the sudden application of working brake the decrease of hoist speed can come with decelerations. The value of such deceleration sometimes exceeds the accepted one by Rules of safety and Rules of technical exploitation. That is why recommendations on adjustment of requirements to the mode of working braking are strongly need to be developed. The analysis of the results in experimental research proves the necessity of the existent brake systems modernization for providing of exploitation safety in hoisting plants in the modes of working and emergency braking.

More difficult method is making of the local profiling and smoothing of guides on the areas with major lack of perpendicularity. Sometimes the decrease of guides' lack of perpendicularity on contiguous tiers to the minimal value on technical terms (4 – 5 mm) does not provide no-impact interworking of skips with reinforcement because of the curvature of main line of shaft. Then to improve the dynamic situation it is possible to use roller directors with increased dissipative properties of shock dampers (Iljina 2010).

As a result of transition of the "vessel – reinforcement" systems on higher operational safety level on working modes it becomes possible to increase the maximum hoisting speed. Despite of some contact loadings escalation with growth of vertical speed, the balanced application of complex measures at the systematic apparatus control allows to retain the actual strength margins of steel structures at the level, providing the absence of unrespectable destructions and emergency situations.

In that case, when avoiding a failure in a shaft was not succeeded, using of independent safe mobile hoisting settings is the only way out. It allows solve the problem of wrecking of the miners from underground horizons, as well as cages at their hanging up in the barrels of mines. Because of the various conditions in shaft inset the application of such mobile settings in every case also requires the special analysis for determination of safe parameters for the operating systems of working stroke and emergency braking mode (Bondarenko et al. 2004).

The developed methods of diagnostics and renewal of the operational safety level of «vessel – reinforcement» systems have been successfully used in many mines in Ukraine. Technical solutions, described in the article, allow to increase the hoisting safe exploitation level for shafts with broken geometry in difficult technical and mountain-geological terms, to prevent great failure in operating mines or minimize damages after.

4 CONCLUSIONS

The exploitation experience and analysis of reasons of serious failures on operating hoists show that in a number of cases using of mechanical brake, being the executive branch of their safety system, results to transition of the MWP to emergency operation, which is impermissible for the safe exploitation. The necessity of continuing of modernization works of the mine hoists brake system is also determined by multiple requirements to the braking modes and complication of methods of their realization for different terms of modern MWP work.

The researches revealed that a perspective way to the decline of contact dynamic loadings on guides in the shafts with broken geometry lies in two directions: control of the guide profile and control of the elastic-dissipative parameters of the roller directors. The loadings decline due to the decline of vessel speed on the curve area is the last preventive technological measure. It can be used only in that case, when the previous solutions exhausted the possibilities because of technical term, or in the case, when it is necessary to avoid the emergency-dangerous of "vessel – reinforcement" system to provide next technical measures. Because such speed decline lowers the hoisting productivity of mine and causes serious material losses.

REFERENCES

Bondarenko, V.I., Samusya, V.I. & Smolanov, S.N. 2004. *Application of mobile hoisting settings for a wrecking in mine shafts.* Donetsk: Collection of scientific works of SRIMM by name of M.M. Federova, No 98: 28 – 32.

Cherednichenko, O.L., Iljin, S.R. & Radchenko, V.K. 2008. *Monitoring of safety of operation of mine shafts.* Dnipropetrovsk: Technopolis, No 12: 30 – 31.

Iljin, S.R., Dvornikov, V.I. & Krcelin, E.R. 2002. *Program complex "Reinforcement of a mine shaft".* Dnipropetrovsk: Collection of scientific works of national mining academy of Ukraine, Vol. 3, No 13: 40 – 43.

Iljin, S.R., Iljina, S.S. & Samusya, V.I. 2014. *Mechanics of mine hoist.* Dnipropetrovsk: NMU: 247.

Iljin, S.R., Posled, B.S., Adorskaja, L.G., Radchenko, V.K., Iljina, I.S. & Iljina, S.S. 2013. *The Experience Of Dynamic Apparatus Control And Estimation Of Exploitation System Safety.* Transport szybowy. Monografia. Instytut Techniki Gorniczej: 163 – 175.

Iljina, I.S. 2010. *Method of diagnostics of reinforcement work in the extreme modes.* Perm: Scientific researches and innovations, Vol. 3, No 4: 45 – 49.

Iljina, S.S. 2010. *Experimental researches of shock loadings dampening of mine shafts reinforcement with elastic roller damper.* Perm: Scientific researches and innovations, Vol 4, No 2: 59 – 63.

Samusya, V.I. & Komissarov, U.A. 2000. *Transition processes in a pneumatic system of working brake control of mine hoisting machines.* Moscow: Mining informational-analytical bulletin, No 2: 178 – 179.

Theoretical and Practical Solutions of Mineral Resources Mining – Pivnyak, Bondarenko & Kovalevska (eds)
© 2015 Taylor & Francis Group, London, ISBN: 978-1-138-02883-8

"Green" economy in mining

S. Bekbassarov
Almaty University of Power Engineering & Telecommunications, Almaty, Kazakhstan

S. Soltabaeva, A. Daurenbekova & A. Ormanbekova
Kazakh National Research Technical University named after K.I. Satpayev, Almaty, Kazakhstan

ABSTRACT: In this article the main tendencies of green economy development in the world are considered and the Concept of switching to this economy by our state. It is shown that green economy contributes to country's steady development. Preliminary results of work by Kazakh National Technical University on resource economy and development of new technologies of technogenic waste recycling with salable production are summarized. Waste disposal at mining enterprises allows decreasing the technogenic load to environment and ensure the efficient use of secondary raw materials.

1 INTRODUCTION

30.05.2013 Nursultan Nazarbayev, President of the country, has signed the Decree on approval of Concept on our state's transfer to "green economy. This Concept raises issues of efficient use of natural resources and of improvement of welfare of Kazakhstani citizens through economy diversification, creation of new jobs and improvement of life standards for our citizens. In the President's message to people of Kazakhstan dated by 17.01.2014 it is said that the transfer to green economy will be implemented according to the adopted Concept.

Therefore, in the Concept it is suggested to firstly utilize and store industrial waste in full volume. By 2050, a so-called non-waste economy shall be formed in Kazakhstan.

For development and implementation of effective environment-oriented measures on processing of waste credible information on influence on natural systems must be obtained: on surface, underground waters, atmosphere and disturbance of soils in industrial scale at growing production volumes. Generalized materials will allow obtaining the unbiased information on state of ecosystems in the studied region and plan priority environment-oriented measures the implementation of which will contribute to decrease of harmful influence on the environment.

According to national cadastre, on the territory of republic, in culm banks, tailing dumps and storage facilities of mining enterprises there is about 30 million ton of industrial waste including: 72% – debris of uncovering and not good ore, 20% – final tailings of concentration, 8% – other waste. With annual output of industrial waste in 1 million ton useful is not more than 100 million ton. The remaining part is gradually polluting the environment and is accumulated in it (Bekbassarov 2008).

2 MAIN PART

The main reason of professing waste accumulation in the country is a raw-material orientation of our economy. Currently, at 450 ranges in the country there is more than 22 million ton of solid waste (Figure 1).

Solid waste is characterized by variety of composition and properties and by wide range of application; therefore, it would be rational to create database on available and newly produced waste which will allow identifying the cost of secondary raw material, storage and secondary recycling requirements for optimal solving of issues of rational natural resources management.

A common problem for the whole waste management industry is the absence of system of regular collection and analysis of information. For today, we can only make an approximate assessment of mining volumes and accumulation of waste in the country. It is supposed that inventory of industrial and household waste performed for each industry sector and each region, may provide even more depressing figures.

It would be rational to divide the production waste as environment pollution source into two groups (modern and historical) and analyze situations relating to their management separately.

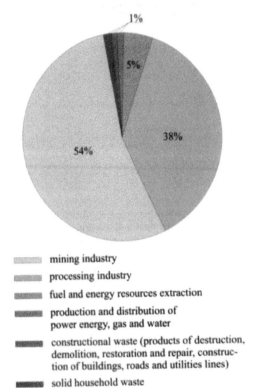

- mining industry
- processing industry
- fuel and energy resources extraction
- production and distribution of power energy, gas and water
- constructional waste (products of destruction, demolition, restoration and repair, construction of buildings, roads and utilities lines)
- solid household waste

Figure 1. Main figures on solid waste accumulated in the Republic by 2010.

Historical accumulations include production waste from oil and gas, mining enterprises, hear engineering, chemical industry, and waste formed at abandoned oil and gas well sites, mines and pits, tailing dumps and waste water dumps, military ranges from Soviet times (Nurpeissova & Bekbassarov 2006)

Many mineral deposits are abandoned or decommissioned with no compliance to environmental requirements and represent a threat to environment. Recultivation at closing enterprises is of rare matter.

According to data of State Control and Supervision over natural resources, the share of used waste in the republic is 18 – 20%. For example, in 2007 the percentage of waste utilization has made 16%, in 2008 – 18.98%, in 2009 – 20% (Figure 2). However, this figure in the recent past of former Soviet industry was making 29%. It remains really low also when compared to global practice. In Western Europe (France, Germany, Italy, England) this indicator makes up to 58%, in North America (USA, Canada) – up to 63%, in Japan – up to 87%, in China – up to 37%.

The largest accumulations of old industrial waste are in Karaganda and Pavlodar Regions, East Kazakhstan, Kostanai Region, Aktyubinsk Region where more than 11 million ton is accumulated at ranges and storage facilities. Thus, in Karaganda Region there is more than 4.5 million ton, in Pavlodar Region – about 1 million ton, in East Kazakhstan – more than 1.5 million, in Kostanai Region – 1.5 million and about 1 million in Aktyubinsk Region (Nurpeissova & Bekbassarov 2006).

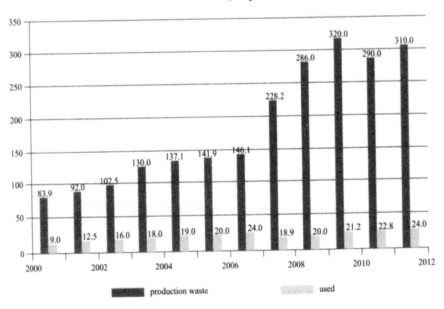

Figure 2. Dynamics of formation and use of production waste in the Republic of Kazakhstan.

The accumulated waste brings significant economic and ecological harm to environment and people numbered in dozens and hundreds of millions of dollars per year. Second place after mining industry is taken by volumes of metallurgical and power sectors. Accumulated waste is, on one hand, main pollution factor of the environment, and one the other hand, is a valuable product potentially fitting for recycling and secondary use with salable product as an output with high added value.

Complex use of raw and industrial waste of metallurgical, mining and construction enterprises is an important problem of not only Kazakhstan, but any economically developed state. As practice has proven, waste from particularly these industry sectors is produced in small amounts and represent a serious economic hazard. In these conditions the problem of environmentally rational use of production waste as secondary raw and development of scientific principles in creation of new technological regulations of production of salable product from industrial household waste becomes extremely pertinent. Thus, the implementation of environmentally clean technology on production of construction materials on the basis of waste of mining enterprise is very urgent for Kazakhstan.

For development and implementation of efficient environment-oriented measures on processing of waste credible information on influence on natural systems must be obtained: on surface, underground waters, atmosphere and disturbance of soils in industrial scale at growing production volumes. Generalized materials will allow obtaining the unbiased information on state of ecosystems in the studied region and plan priority environment-oriented measures the implementation of which will contribute to decrease of harmful influence on the environment.

Particularly from this point of view the objective was set, the idea was substantiated and targets of our research work were established.

3 CONCLUSIONS

On the basis of research done the following perspective schemes and technologies for industrial use of waste in construction materials (CM) production are suggested:

1. Production of small piece CM on the basis of granulated phosphorous slag (walkway slab, wall blocks, bricks, etc.). Distinctive feature of new suggested in the Project waste utilization technology from the previous one is that for the first time it will use new additives types significantly increasing the physical and mechanical binding properties (Yestemessov & Bekbassarov 2009).

2. Production of dry mortars with use of glass cullet and polymer additives in the form of bond plaster; dry mixed glue; dry heat insulating mixtures. Dry mixtures represent a mixture consisting of binding (cement, gypsum, lime) filler and various chemical complex additives (Patent No 19861).

In the republic, production of dry mortars in 2008 reached 500 thousand ton with the capacity of Kazakhstani market of 2 million. Taking into account the dynamics of construction market and the forecasted growth of the economy of Kazakhstan up to 2030, the demanded volume of annual use of dry mortars may reach 5 – 7 million. This will be one of the most rapidly developing directions in the construction materials industry.

Industrial production of dry mortars allows rapidly increase the use of technogenic waste. Most countries use from 35 to 70% of industrial waste, in Kazakhstan it is only a little bit more than 5%.

Expected results of the suggested "Development of environmentally clean technology of preparation of effective construction materials with use of mining and metallurgical companies waste" Project are the following:

1. Disposal by unique technology of extremely dangerous gas-emitting granulated phosphorous slag storage facility by producing small piece construction materials significantly beats in strength and durability known analogues on the basis of artificial cement which contributes to rational use of country's mineral resources, solving of environmental problems and introduction of innovative policy into the construction sector.

2. Uniqueness of small piece technology on the bass of granulated phosphorous slag is that the expected increased strength and durability of concretes on the basis of neutralized granulated phosphorous slag binder is conditioned by absence of free lime in them, hydroaluminates and high-basic calcium hydrosilicates, and – formation of exclusively low-basic calcium hydrosilicates of submicrocrystalline- and fiber structure with increased strength, density and durability.

3. Simplicity of technology, mobility of technological line, neutralization of hazardous gases of granulated phosphorous slag in the closed cycle opens big perspective for decrease of environmental pressure from the storage facility of phosphorous slag to environment of Zhambyl and South Kazakhstan regions and obtaining effective small piece materials with the designed durability, frost-resistance and durability with economic effect of 10 million USD per year if the production capacity of technological line makes 100 thousand m^3 of concrete mix per year.

Scientific results are executed in the form of scientific publications, applications for inventions. Scientific research are provided with methodical and normative documentation.

In the suggested project, a brand new, not polluting the environment technology of dump recycling and production of small piece construction materials will be developed for the first time which will allow decreasing the technogenic load to environment and ensure rational use of secondary raw.

We think that "Green economy" starts from each of us. It does not need catchwords and slogans, assurances preferring concrete and efficient actions coming from us, people based on internal beliefs.

REFERENCES

Bekbassarov, Sh.Sh. 2008. *Production waste and problems of its disposal.* Almaty, KazNTU Bulletin, No 3: 31 – 34.

Nurpeissova, M.B. & Bekbassarov, Sh.Sh. 2005. *Disposal of mining eterprises waste.* Materials from international conference "Innovative technology of industry development, Atyrau: 178 – 181.

Nurpeissova, M.B. & Bekbassarov, Sh.Sh. 2006. *Rational use of technogenic resources.* Almaty: KazNTU: 28.

Patent No 19861, Republic of Kazakhstan, *Compound for fixing of dust collecting surfaces in tailing dumps and other objects.* Nurpeissova, M.B., Kyrgyzbayeva, G.M. & Bek, A. Pub. 25.05.2008.

Yestemessov, Z.A. & Bekbassarov, Sh.Sh. 2009. *Solid waste and their use in production of construction materials.* Almaty: TcelSIM: 190.

Theoretical and Practical Solutions of Mineral Resources Mining – Pivnyak, Bondarenko & Kovalevska (eds)
© 2015 Taylor & Francis Group, London, ISBN: 978-1-138-02883-8

Dynamics of dry grinding in two-compartment separator mills

M. Adamaev & A. Kuttybaev
Kazakh National Research Technical University named after K.I. Satpayev, Almaty, Kazakhstan

A. Auezova
Almaty University of Power Engineering & Telecommunications, Almaty, Kazakhstan

ABSTRACT: The results of research the dynamic characteristics of the main and cross-technological control channels two-compartment separator ball mill dry grinding of raw materials. Analysis of these characteristics prescribes new structure develops highly effective automatic control system. It should provide independent regulation of load levels cameras mill (dual circuit system), to compensate for adverse interference of these circuits at each other by Inland technological cross-connection channels (invariant system of interconnected control circuits with independent properties), have varying parameters setting dual-core system with regulators and pounce load shedding (system with variable settings main regulators).

1 INTRODUCTION

Crushing and grinding are subjected to hundreds of millions of tons of ore and non-metallic minerals. For such large volumes of production even a small percentage increase of the efficiency will have a significant economic effect. Dry milling of raw materials is carried out mainly in two-compartment ball mill of dry grinding with a combined separation of the products of grinding from both chambers which are widely used, as basic milling-aggregates in the mining, chemical and cement industries, in the preparation of coal dust CHP and others.

It was found that in order to have efficient ball mills, it is necessary to maintain an optimal internal load (for grinding to go well), which is achieved by using systems of automatic control. Analysis of the results of the review of technical literature and patent research process automation systems for grinding showed no known developments for the applicability of the object due to the presence in it of a number of specific features of both technological and constructive novelties. Synthesis of new effective systems of automatic control is impossible without knowledge of dynamic control. The results of the investigations in this direction are given below.

2 MAIN PART

Two-chamber dry grinding ball mill ("Polysius-22" type) has a diameter of 2.2 m, comprising of two working chambers and a discharge chamber located between them. All three chambers disposed in a common drum mill. Camera of coarse grinding *3* has a length of 2.8 m and the camera of fine grinding *4* – 2.2 m (Figure 1).

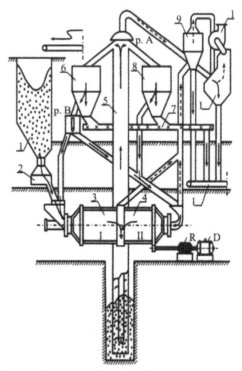

Figure 1. Flow sheet of ore grinding mills in the two-chamber dry grinding with a combined separation.

The discharge chamber is separated from the operating ones with unloading bars and has a length of 0.5 m. These mills are produced by the manufacturer, with the productivity of 10 t/h. Currently, they were reconstructed to increase their productivity. The camera of pre-drying mill was liquidated. Drying of ore is produced in special dryers. Ore, after three stages of crushing and drying is fed into a hopper of grinding units *1*, located directly in the grinding section. After, ore feeder poppet *2* is fed into the chamber of coarse grinding. The ground ore from both cameras through the discharge chamber of the mill falls on a tray in elevator section. Bucket elevator transports the milled product to two conical air separators, where it's classified according to the particle size with separation of ready product and the intermediate product – "grit".

In *point A* (Figure 1) milled product is distributed evenly for uniform loading and efficient operation of air separators.

The final product after separation *7* is transported on a conveyor *12* and then goes to a final product silo. "Grit" – intermediate product after the second separator – *8* goes to the second chamber, and the "grit" after the first separator *6* is regulated between the two chambers in *p. B*. Grinding cycle repeats (Adambaev 2004a). The described technology of crushing of minerals in the two-chamber ball-mills is classified as grinding circuit with a partially closed cycle in both stages of grinding, determined by dividing the total redistribution-circulating load between the stages (Adambaev 2014).

Considered bicameral ball mill grinding with a combined separation of product in conjunction with air separation set is applicable only for dry grinding and have a number of features that greatly complicate the development of management systems of this process (Adambaev 2010).

Drum of the mill includes two cameras – the camera of coarse and fine grinding, which differ in the geometric sizes, ball-mill loads (including weight of the balls, and size), physical and mechanical characteristics of the ore fed into these cameras. Summation of output streams of the first and second chambers in a mill itself (in the discharge chamber) excludes direct control of cameral productivity by known methods.

Return part of the stream "grit" in the camera of coarse grinding is an important sign in terms of the principle of building of the system with automatic control. In the case when the entire stream is sent to the camera of fine grinding, the only influence of the automatic load control of the chamber of course and fine grinding is reflected in control of the feed. Such technological-circuit of the camera of coarse grinding in the open cycle does not have back-flux

and has a dramatically reduced productivity of the grinding machine. It is well-known that the performance of the mills operating in a closed cycle is significantly higher than that of the mills operating in the open cycle (Adambaev 2004a). In addition, the presence of only one control lowers the quality of the regulatory process, because of the value of the transport delay between the control action and the controlled unit (original feed of the system – camera of course grinding – elevator – separator – load level of the camera of fine grinding), which significantly increases and can lead to unwanted buildup of the system. Under normal operating conditions, when there are two control actions (the change in the power source and change in redistribution of "grits" between the first and second chambers), there is a possibility of independent control of both chambers (Adambaev 2004a).

For objects with the return of the unconditional class to the secondary processing, it is usual to change the coefficient of object transmission and the time constant of the transition of the processes that lead to a reduction in speed and reduction in resistance (Adambaev 2010). The duration of the transition process of such objects is long. Thus, for test mills it reaches up to 3600 seconds. Changing the properties of the original product is a little longer than these time intervals (Adambaev 2004a). This leads to the fact that a change in grit characteristics takes a significant part of the time of the ore milling process in transient conditions, followed by quantitative fluctuations in the quality and quantity of final product. One of the tasks of automatic control system is elimination of these oscillations.

In developing high-quality systems of control of the operation of the investigated grinding technology, we should consider the presence of internal cross-links between technological cameras, which give rise to mutually influence on both input and output, coordinates of automatable mill. The solution of this question may be the development of invariant system (quasi) to cross-substitute coordinates of the system in the process of controlling. It is necessary to develop a system of autonomous control of this process, with the main unit – two-chamber ball mill with a combined separation-of the product.

Using the intensity of the noise spectrum generated by each camera for collateral control of these cameras (for example, output signals of microphone sensors, mounted in each camera), in the majority of cases is not true. This fact is explained by the fact that the signal, removable microphone, is the sum of the main signal (useful component) and signal interference. Solution of the problem is the development of automatic control system, which allows to compensate the interference (Adambaev 2012).

The grinding unit consists of the following technological devices, separated by their functions.

Plate feeder mounted at the top of the grinding process for feeding of raw material to the mill. Its input starts with the change of the drive speed n, and the output – quantity of power source Q_n.

Ball mill consists of two operating cameras. Coarse camera has two input signals Q_n and K_1S, which regulates the exposure to this chamber, and KS_1 that initiates the exposure. This camera has one option – to fill intracameral space M_1, controlled by the intermediate sound-signal Z_1M_1 regulated by the value S_1 and O_{r1} (S_1 – ield to the newly formed circulating load, a Q_{r1} – output of the newly formed final product). Consequently, the course camera has three output coordinates – Z_1, S_1, Q_{r1}. Note that the values S_1 and Q_{r1} cannot be controlled, so value: $S_1 = f(Q_n)$; $Q_r = f(Q_n)$; $S_1 = f(M_1)$, $Q_r = f(M_{11})$ are accepted in accordance with the results of statistical studies obtained previously (Adambaev 2010).

Input signal of the fine grind camera is determined on the other part of the circulating load $(1 - K_1)S = K_2S$. This camera also has three outputs Z_{11}, S_2, Q_{r2} similar to output signals of course grind camera.

Bucket elevator is designed to transport the crushed ore to the separator. Input and output are equal between them and shifted in time to value of net transport delay.

Air separator classifies crushed ore into two products: final product and circulatory load. This element has one input M and two outputs S, Q_r.

Unit of distribution of circulating loads S, which is the input to two streams K_1S and K_2S.

Pipelines designed to transport circulating load to the operating cameras, which are links of net transport delay.

Plate feeder is approximated by accelerating dynamic element without inertia (Figure 2, a) with transfer function $W(p) = K_n = 1$. A slight delay between the feeder and the mill is related to the delay of the camera of coarse grind.

Bucket elevator is a link for net transport delay with the transfer function $W_e(p) = e^{-p\tau_e}$ (Figure 2, b). From this point, due to the large values of the time constants T_i and delays τ_i values are given in their minutes. Separator approximated with uninertial link on both channels (Figure 2, c).

$W_c^s(p) = K_s$ – transfer function of the channel input-output circulating load.

$W_c^s(p) = K_r$ – transfer function for an input-output channel of the final product, and $K_r = 1 - K_s$.

A slight delay in the separator attributed to the delay of bucket elevators.

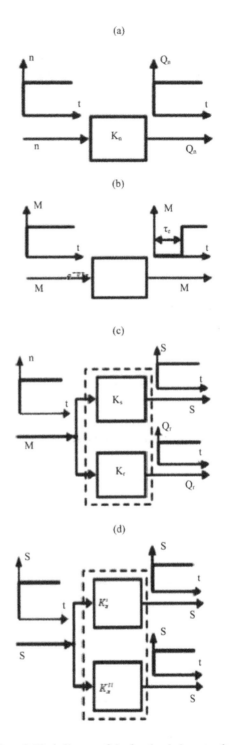

Figure 2. Block diagram of the functional elements of the grinding unit.

Unit of distribution of sand load between the chambers has one input and two outputs, and a separating element with transfer functions (Figure 2, d): $W_s^I(p) = K_s^I$ – transfer function of the gate on the channel input-output chamber of coarse grinding. $W_s^{II}(p) = K_s^{II}$ – transfer function of the gate on the channel input-output in the fine grinding chamber, and $K_s^{II} = 1 - K_s^I$.

Complete block diagram of a two-chamber walking mill is shown in Figure 2, where: $W_{11}^r(p)$ – transfer function of the camera of coarse grinding in the main channel input-output of the newly formed final product; $W_{11}^s(p)$ – the same channel output circulating load; $W_{22}^r(p), W_{22}^s(p)$ – the same as for fine grinding chamber.

$W_{21}(p)$ – transfer function of the first cross-communication channel (the influence of large-camera mode of operation of grinding of chamber of fine grinding).

$$W_{21}(p) = W_e(p) \cdot W_c^s(p) \cdot W_s^{II}(p) = K_s \cdot K_s^{II} \cdot e^{-p\tau_e}.$$

$W_{12}(p)$ – transfer function of the second crossing-communication channel (the influence of the camera mode of fine grinding of camera of coarse grinding):

$$W_{12}(p) = W_e(p) \cdot W_c^s(p) \cdot W_s^I(p) = K_s \cdot K_s^I \cdot e^{-p\tau_e}. \ W_{12}(p),$$

$W_{II}^z(p)$ – transfer functions of the mill chambers through channels of input – output sound-intermediate signals.

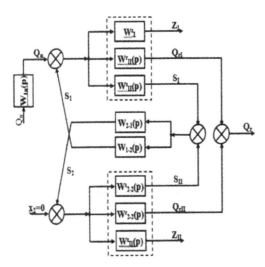

Figure 3. Block diagram of two-stage process of dry grinding with technological cross-links.

On the other hand, studies of many authors (Kosharsky et al. 1977) have established that the ball

mill is adequately described by the transfer function of the form $W_M(p) = \dfrac{K_M}{1 + T_M \cdot p} \cdot e^{-p\tau_M}$, i.e. approximated inertial units of the first-order with series of links of net transport delay. In this approach, a complete block diagram of this technology takes the form shown in Figure 3.

The parameters of the transfer functions of elements of the system are defined by transition characteristics with step-disturbance on their input. Disturbance (step) applied using two factor changes in the amount of the original supply and redistribution of circulating load between the chambers some dynamic characteristics obtained on the object are shown in Figure 4.

(a)

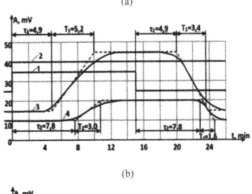

(b)

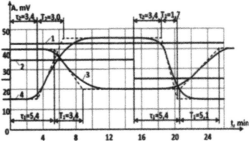

Figure 4. Mill acceleration capabilities.

3 CONCLUSIONS

Preliminary analysis of the experimental curves of acceleration at step disturbance leads to the following conclusions on the dynamics of a two-compartment ball mill of dry grinding:

1. Forms of transmission cameras of mills confirm the legitimacy of approximation of their dynamics with consistent inclusion of inertial links and link of net transport delay with sufficient for practical goals accuracy.

2. The time constants T_i for different polar disturbances of the same amplitude take different values.

During load disturbance (positive) T_i takes bigger values than those at unload of disturbance (negative).

3. The studied mill with its main regulated values and regulatory exposures is an object with interdependent values. Disturbances of the initial feed define mainly the load level of the first camera, but at the same time via the channel the output of the first chamber – elevator-separator-input of the second chamber is the disturbance that affects the level of load of the chamber of fine grinding. Similarly disturbance with redistribution of circulating load between cameras affect the utilization rates of both cameras (Figure 4).

4. During disturbances applied to the newly formed circulating load object returns to the camera inputs through process feedback channels and plays the role of additional disturbance, which significantly changes the shape of the transition process. During positive disturbances accumulation of circulating load increases the time constants and vice versa for negative disturbances, there is a sharp decrease in the circulating load, and reduction in the time constants of the object.

REFERENCES

Adambaev, M.D. 2004a. *Process control of dry ore dressing (monograph)*. Almaty: Complex.
Adambaev, M.D. 2004b. *Automatic control theory. Methods of identification of industrial facilities management*. Almaty: KazNTU.
Adambaev, M.D. 2010. *Determination of the dynamic structure and the parameters of industrial facilities management (monograph)*. Almaty: TST-Company.
Adambaev, M.D. 2012. *Sound-system of automatic control on the cell basis and the total load of dual-chamber of dry grinding ball mill*. Novosibirsk: Collective Monograph.
Adambaev, M.D. 2014. *Improving the efficiency of dry grinding. Identification and automation (monograph)*. Berlin: LAPLAMBERT Academic Publishing.
Balakirev, B.C., Dudnikov, E.G. & Tsirlin A.M. 1967. *Experimental determination of the dynamic characteristics of industrial facilities management*. Moskow: Energy.
Kosharsky, B.D., Sitkovskiy, A., & Krasnomovets, A.V. 2011. *Automation of processing plants*. Moskow: Nedra.

Theoretical and Practical Solutions of Mineral Resources Mining – Pivnyak, Bondarenko & Kovalevska (eds)
© 2015 Taylor & Francis Group, London, ISBN: 978-1-138-02883-8

Modeling of vibro screening at fine classification of metallic basalt

V. Naduty
M.S. Poliakov Institute of Geotechnical Mechanics under the
National Academy of Sciences of Ukraine, Dnipropetrovsk, Ukraine

Z. Malanchuk, E. Malanchuk & V. Korniyenko
National University of Water Management and Natural Resources Use, Rivne, Ukraine

ABSTRACT: Currently, intensive research on comprehensive development of basalt deposits in Volhynia with the use of non-waste technology is being conducted. This is due to the rich mineral composition of basalt raw material, which besides basalt contains a significant portion of zeolite-selective tuff and lava cluster breccia. All the three components of the fields contain a high percentage of iron, titanium and native copper. These components are of industrial interest, but their extraction requires special technology of pretreatment. In the Institute of Geotechnical Mechanics named after N.S. Polyakov of the NAS of Ukraine and the National University of Water Management and Natural Resources Use the research on the development of comprehensive technology for the processing of basalt raw material is being conducted. In addition, the dependence of technological parameters of each unit of the necessary equipment on the key forces of their impact has been investigated.

1 INTRODUCTION

The purpose of these studies is to determine the dependence of efficiency of basalt rock mass fine vibrating screening on the main dominant adjustable parameters.

Considering that the processes of fine classification vibration method are not fully studied, and existing vibroscreens demand adaptability of their characteristics to the requirements of technology for performing these purposes, the research has been conducted on fine classification screening with dynamically active working surface specially designed by IGTM NAS of Ukraine. Its peculiarity consists in the fact that the screening in the separation process is carried out in vibro-impact mode due to the impulses (strokes) of the supporting dynamically active rubber resonant band-string sieve at the top classifier of rigid metal or nylon mesh.

2 THE PROCESS OF VIBRO SCREENING

Due to the fact that physical-mechanical characteristics of the rock mass make a significant impact in the classification performance and the terms previously established according to a specific classification require clarification and correction, the present research has experimentally determined the efficiency of basalt raw material fine screening in their dependence on a number of variable parameters, in particular: γ – density of the rock mass ($\gamma_1 = 2,6$ –basalt, $\gamma_2 = 2.2$ – lavabreccia, $\gamma_3 = 1.4$ – tuff), β – angle of the perturbing force inclination, α – angle of inclination of the vibroscreen working mechanism, $\varDelta$ – screen opening size, q – screen unit loading, ω – perturbation frequency of screen drive and its length L. Variation of these parameters has a significant impact on the efficiency of screening. To analyze the effect of these factors on the system it is necessary to work out multivariate regression dependence. (Malanchuk & Malanchuk 2014a, Malanchuk & Malanchuk 2014b, Malanchuk & Korniyenko 2014).

The calculation of the linear regression dependencies has been carried out by the method of the least squares. This nonlinear dependence on the factor attributes were given to the change of variables by linear method. In the studied case, it is the dependence of the efficiency on the size of the sieve opening $E = f(\varDelta)$, and for the others listed above, parabolic dependence appeared to be more adequate. On the basis of the correlation coefficients R for each pair of attributes the correlation matrix has been designed, and the multiple coefficient of determination has been calculated, showing what portion of variation of the resultant attribute is explained by the factor attributes variation of the model. In its turn, the adequacy of the model has been evaluated using Fisher statistics F, which was compared with its critical value F_{kr} according to the adequacy level of $\alpha = 0.05$ or $\alpha = 0.01$.

This work has established the dependence of the efficiency of fine screening on the most influential factor characteristics in the form of:

$$E = f(\beta, \alpha, \omega, q, \Delta, L, \gamma). \qquad (1)$$

Stage by stage, there has been studied the model of effectiveness dependence of screening on individual factors at fixed sieve opening size Δ and variations of the type of rock characterized by the density γ. Then, there has been obtained a generalized model with density factor γ taking into account the influence of the angle of the perturbing force inclination. The calculation was performed with $\Delta = 2$ mm, 3 mm, 5 mm, the results of the calculations are given in Table 1.

Table 1. Calculation results of the regression parameters.

Δ	a_0	a_1	a_2	a_3	a_4	R^2	F
2	-25.12	-11.67	5.9	3.94	-0.038	0.979	153.2
3	-47.08	29.58	-4.17	3.35	-0.032	0.979	154.4
5	-35.82	3.33	2.43	4.07	-0.039	0.099	341.6

3 RESULTS OF CALCULATIONS

The regression model has been specified as:

$$\tilde{E}(\gamma, \beta) = a_0 + a_1\gamma + a_2\gamma + a_3\beta + a_4\beta^2. \qquad (2)$$

The generalized model with the inclusion of the sieve opening size Δ in the number of variable factors, other factors (α, ω, q, L) have been recorded:

$$\tilde{E}(\Delta, y, \beta) = a_0 + a_1\Delta + a_2y + a_3\beta + a_4\beta^2. \qquad (3)$$

The results of calculations of the parameters of this regression are given in Table 2.

Table 2. Calculation results of the regression parameters.

a_0	a_1	a_2	a_3	a_4	a_5	R^2	F
-45.45	2.83	7.07	1.39	3.78	-0.036	0.972	331.3

For each of the sieve opening size Δ the regression model was specified as variable parameters γ and a:

$$\tilde{E}(y, \beta) = a_0 + a_1\Delta + a_2y + a_3\alpha + a_4\alpha^2. \qquad (4)$$

The results of calculations of the parameters of this regression are given in Table 3.

The generalized model with the inclusion of the inclination angle of screening and the number of variable factors has been determined as:

$$\tilde{E}(\Delta, y, \alpha) = a_0 + a_1\Delta + a_2y + a_3y^2 + a_4\alpha^2 + a_5\alpha^2. (5)$$

Table 3. Calculation results of the regression parameters.

Δ	a_0	a_1	a_2	a_3	a_4	R^2	F
2	61.93	-8.0	5.21	-0.337	-0.036	0.991	271.7
3	68.12	-6.5	4.58	-0.6	-0.031	0.986	176.7
5	36.79	34.5	-5.83	0.439	-0.081	0.963	64.9

The factors not included in the model, are fixed at a constant level. The calculation results are shown in Table 4.

Table 4. Calculation results of the regression parameters.

a_0	a_1	a_2	a_3	a_4	a_5	R^2	F
39.4	4.71	6.67	1.32	-0.166	-0.49	0.973	282.5

The dependence of the screening efficiency on the perturbation frequency of the vibrator ω, the specific load on the screen q and the length of the sieve on the screen, similarly to the previous case, has been determined in stages.

The model with the influence factor ω has the following form:

$$\tilde{E}(y, \omega) = a_0 + a_1y + a_2y^2 + a_3\omega + a_4\omega^2. \qquad (6)$$

The calculation results are shown in Table 5.

Table 5. Calculation results of the regression parameters.

Δ	a_0	a_1	a_2	a_3	a_4	R^2	F
2	-50.88	10.42	0.0	0.156	-0.000058	0.866	93.2
3	-62.65	38.33	-6.42	0.146	-0.000057	0.949	60.2
5	-43.07	15.83	-0.694	0.159	-0.000065	0.934	46.1

Generalized model in this case is obtained in the form:

$$A^0 = -62,84 + 3,19\Delta + 21,53y -$$
$$- 2,37y^2 + 0,154\omega - 0,00006\omega^2. \qquad (7)$$

The model including the specific load on the screen q has the form:

$$E(y, q) = a_0 + a_1y + a_2y^2 + a_3q + a_4q^2. \qquad (8)$$

The results of the calculation of the coefficients in the model are given in Table 6.

Table 6. Calculation results of the regression parameters.

Δ	a_0	a_1	a_2	a_3	a_4	R^2	F
2	66.09	2.86	2.53	-6.06	0.302	0.986	279.4
3	86.25	-15.0	6.84	-3.34	-0.079	0.985	269.4
5	71.64	7.5	1.19	-3.46	0.012	0.977	170.9

In its final form the generalized regression model has taken the form:

$$\overset{o}{A} = 59{,}82 + 4{,}45\varDelta - 1{,}55y +$$

$$+ 3{,}52y^2 - 4{,}29q + 0{,}078q^2 . \qquad (9)$$

The model including the length of the sieve L is given in the form:

$$E = a_0 + a_1x + a_2y^2 + a_3L + a_4L . \qquad (10)$$

The calculation results are shown in Table 7.

Table 7. Calculation results of the regression parameters.

$\varDelta$	a_0	a_1	a_2	a_3	a_4	R^2	F
2	0.896	-3.214	2.98	24.59	-2.13	0.991	458.0
3	13.92	-11.43	5.36	24.49	-2.17	0.983	235.0
5	3.54	5.0	1.34	23.49	-1.97	0.993	541.0

The calculations found that the generalized model has the form:

$$E = -5{,}23 + 3{,}14\varDelta - 3{,}21y +$$

$$+ 3{,}22y^2 + 24{,}19L - 2{,}09L^2 . \qquad (11)$$

Presented models allow us to solve individual problems of the screen parameters choice and determination of efficiency. The ultimate goal of our research was the modeling of the dependence of the efficiency of screening on the complex of factor attributes in the form of:

$$E = f(\varDelta, y, \beta, \alpha, \varpi, q, L) . \qquad (12)$$

Estimates of the model coefficients have been carried out according to methods described above, in this case for factor $\varDelta$ there was adopted linear dependence (based on experimental data), and for other factors – parabolic. The total sample size was $n = 279$, the number of overlay links $m = 14$ and the number of freedom degrees $v = n - m = 265$.

Resulting from calculation there has been obtained the generalized model of screening efficiency dependence of vibrating screen based on the key factor attributes:

$$A^0 = -162{,}42 + 3{,}7\varDelta + 5{,}54y + 1{,}54y^2 + 1{,}67\beta -$$

$$-0{,}012\beta^2 + 2{,}12\alpha - 0{,}153\alpha^2 + 0{,}162\varpi -$$

$$-0{,}0006 5\varpi^2 + 4{,}18q - 0{,}863q + 17{,}28L - 1{,}25L^2 . \quad (13)$$

With the coefficient of determination $R^2 = 0.88$, and Fisher's statistics $F = 150.3$, the adequacy of the resulting model has been proved.

4 CONCLUSIONS

The results of the research allow to analyze the process of fine vibratory screening of basalt raw material when in separate complex processing, to predict rational or optimal settings of screen for the required efficiency of screening. This stage of ore pretreatment is important because fine size classes contain a significant amount of native copper, titanomagnetite and their clusters. Therefore, the efficient separation of fine fractions allows the use of extensive scheme of extraction of useful components. (Franchuk et al. 2011), (Malanchuk & Malanchuk 2014).

REFERENCES

Franchuk, V.P., Naduty, V.P. & Egurnov, A.I. 2011. *Mathematical modeling of vibrating screening with regard to sensitive, constructive and technological parameters of thunder*. Scientific-technical collection National Mining University, Vol. 45(86): 48 – 52.
Malanchuk, Z.R. & Korniyenko, V.Y. 2014. *Modern condition and problems of extraction of amber in Ukraine*. Canadian Journal of Science, Education and Culture. Toronto Press, No 2(6): 372 – 376.
Malanchuk, Z.R. & Malanchuk, E.Z. 2014. *Experience in the use of products processing of basalt raw material in Ukraine*. American Journal of Scientific and Educational Research. New York: Columbia Press, No 2(5): 642 – 648.
Malanchuk, Z.R. & Malanchuk, E.Z. 2014a. *The results of studies of the distribution of native copper in rock mass Volhynia (Ukraine)*. Materials ot the 1st International Academic Congress «Fundamental and Applied Studies in the Pacific and Atlantic Oceans Countries». Japan, Tokyo, 25 October 2014. Tokyo University Press: 322 – 325.
Malanchuk, Z.R. & Malanchuk, E.Z. 2014b. *The analysis of the state of the question on the content of nonferrous metals in basalt raw materials Volyn region of Ukraine*. Canadian Journal of Science, Education and Culture. Toronto Press, No 2(6): 361 – 365.

Theoretical and Practical Solutions of Mineral Resources Mining – Pivnyak, Bondarenko & Kovalevska (eds)
© 2015 Taylor & Francis Group, London, ISBN: 978-1-138-02883-8

On the division of leased land parcels

V. Riabchii
National Mining University, Dnipropetrovsk, Ukraine

V. Riabchii
National University "Lviv Polytechnic", Lviv, Ukraine

ABSTRACT: Current situation in land law is analyzed on partial lease of land with real property. Proposals to improve the procedures and legal acts of Ukraine on this issue are offered in order to significantly accelerate the conclusion and registration of new leases.

1 INTRODUCTION

According to the Lands Codes of Ukraine (1990, 2001), Civil Code of Ukraine (2003) and the Law of Ukraine "On Land Lease" (1998) a large number of businesses and individuals received land parcels for rent for the construction of different objects as well as for operation of existing property. With time, circumstances change and land parcel tenants – owners of real estate, sell part of their property. Sometime the whole area of rented land parcel cannot be rationally used or no new building can be constructed to expand the activities. In this case, the tenants want to get rid of the part of leased land parcel.

New property owners want to formalize documents on rent of the relevant part of the main land parcel. At first glance, all this can be easily solved. But procedural steps are rather complicated and it does not take less than during the first time privatization of the land parcel.

Article 50 of the Law of Ukraine "On Land Management" (2003) states that "land management projects on land allocation are developed in the case of change of use of land or the development of new land parcels". That is, if the boundaries of land parcel have changed, it is necessary to draft land management. In this case, the new land parcels will be within the main land and their total area is equal to the area of the main of land parcel.

Quite often the main land parcel tenant does not want to renew the lease because surveying and land management work should be paid, which results in violation of the regulations of part 4 – 6 articles 120 of the Land Code of Ukraine (2001) by the owner of some part of a building as they cannot arrange to rent the appropriate part of the land: "In case of acquisition of ownership of a residential building or structure by more persons, the right to land is determined in proportion to the share of persons in ownership of an apartment house, building or structure. In the case of acquisition of ownership of a residential building or structure by any persons who are unable to own land parcels, they transfer the right to use the land on which the residential building or structure are situated on a rental basis. The contract that provides for the acquisition of ownership of a residential building or structure involving the transfer of a part of land may be signed after detachment of this part as a separate plot of land and giving it a separate cadastral number".

Part 2 of Article 377 of the Civil Code of Ukraine (2003), the "Size and cadastre number of land parcel, title transfer due to the transfer of ownership of a house, building or structure are essential terms of the contract that provides for the acquisition of rights ownership of these items (except for apartment buildings)". This article is also violated if the seller hesitates or does not want to document a part of the land, especially when they have the state act for perpetual use of land.

Thus a buyer can not register only part of the land parcel in the state land cadastre and registry because there is lease contract or the act of the permanent use of the whole land parcel. And the person cannot have two different rights on one land parcel. The new owner of the building can apply to court, but it is also a complex and long procedure.

Thus, improvements in the procedure for signing and registering of new lease contracts is important for businesses and individuals, as well as for land reform in Ukraine as a whole. Such propositions will improve the social conditions of our society.

2 REVIEW OF RECENT RESEARCH AND PUBLICATIONS

Division of land parcels is the subject of many legal acts of Ukraine (Lands Codes of Ukraine (1990, 2001), Law of Ukraine "On Land Lease" (1998), Civil Code of Ukraine (2003), Law of Ukraine "On Land Management" (2003) etc.) and some scientific and academic papers by famous scientists of our country, but the detailed study of this issue is not found.

In article 31 of the Law of Ukraine "On Land Lease" (1998) are defined the cases of termination of the lease, however there are no cases of transfer of ownership of the property.

3 THE PURPOSE OF ARTICLE

Development of proposals for improving the procedure for signing and registering new lease contracts in cases when the landholder does not sell whole real estate, but only part of it.

4 THE MAIN MATERIAL

First we introduce the following terminology. The main land parcel is a land parcel on which the real property of the main tenant (landholder) is situated. The main tenant is a seller of the property. The buyer of the property is the purchaser of the right to lease (future tenant) on the part of the main land parcel.

In most cases, the seller at the time of the transaction or after its conclusion gives the buyer a notarized consent of withdrawal of the land parcel with specified area of land parcel. Everything seems to be fine, but at best the boundary during division of land parcel is firstly established visually. So after establishing a geodetic boundary there must be discrepancies in the area specified in the agreement and it is good if there is space to shift this boundary.

A thoughtful reader can fairly argue here: "What is the division of land parcel? Land parcel is not in ownership but in use! If it were his property, then it would be a different matter, divide as you want".

There is no doubt that this is correct. Indeed, according to article 56 of the Law of Ukraine "On Land Management" (2003) there must be the consent of the owner of land on the division. Moreover, these land parcels have been registered as state or municipal property since the beginning of 2013. But in reality the land parcels have always been divided and it used to be called – determining the order of use of land. In addition, it may be that land parcel, depending on the location of buildings in general is indivisible, however neither the seller nor the buyer

knows about it. Therefore, the following steps are proposed: the seller must order the draft of land-management division of a land parcel from the land surveying company, so that the work can be continued with the use of it.

After drafting the division of land parcel the seller and the buyer can address the notary to develop the agreement and authorize the project. Project division of land parcel is certified by a notary, because it indicates the boundary of the division of main land parcel and areas of parts obtained as a result of its division. After finalizing the transaction at the notary, both the seller and the buyer order technical documents from the land surveying company on land management for the division of land for the seller and the technical documentation of land use to install (restore) the boundaries of land parcel for the buyer. However, it is necessary to obtain the consent of an appropriate council for the division of land parcel and the land management documentation. It is also necessary to set the terms on which lease contracts will be signed between the seller and buyer.

To solve the issues, consider the following. In fact, part of the property from the seller is alienated without anyone's permission (consent), but only by mutual agreement. In accordance with part 2 of article 377 of the Civil Code of Ukraine (2003) the right to use land transfers with the sale. Obtaining permission or consent to perform these operations on land management may take long (a few months or longer). And then there is a conflict between the seller and the buyer: "Who has to pay the rent?" The tenant does not want to pay rent for the entire (main) land parcel, a future tenant has no reason to pay the rent. Ironically, both the buyer and the seller are right.

Thus arises the need to establish such a procedure that would meet regulatory legal acts of Ukraine and, at the same time, significantly reduce the period during which the seller and the buyer are able to establish the right to lease the relevant parts of the main land parcel.

Consider the experience of individual customers' considerations regarding authorization or consent for the division of land parcels. Which state legal acts of Ukraine and who can prohibit the rightful owner of property renting a land parcel from selling or donating part of it, rather than the whole? The answer is: "Nobody!". And then the seller and the buyer raise unnecessary questions: "Why then is there a necessity of obtaining a permit if land parcel can be divided or partially divided?" No comments.

In addition, there are strong arguments for not obtaining this permit, namely the boundaries of land parcel are approved and adopted, and there is signed and valid lease contract on the main land parcel. Two new land parcels do not go beyond the outer boundaries of the main land parcel. Purpose of all buildings of main tenant is not changed. And the terms of new lease contracts of parts of the main land parcel shall not exceed the term of the lease remaining for the main land parcel.

But there is a rule and it must be followed. Therefore, the main tenant and the purchaser of the right to lease must file a petition to the Chairman of the relevant Council to provide guidance to the appropriate services to perform the required operation for the conclusion of the lease. The council needs to know about the changes on the land parcel granted on lease, which have already occurred and are only planned.

The main thing is the development of technical documents on land management for the division of land parcel for the seller and the technical documentation of land management to establish the boundaries of land parcel for the buyer. Developed technical documentation of the project on land management division must be sent to departments of State Land Agency in cities to check and determine cadastral numbers.

After obtaining the extracts from the State Land Cadastre, relevant documentation is forwarded to the City Council to draft a decision. The decision must state that the lease on the main land parcel is terminated by agreement of the parties and new land parcels become the property of the respective council, and they must be registered by this Council and tenants, and the new lease contracts must be signed.

Consequently, it is necessary to make amendments to point "e" of Article 56 of the Law of Ukraine "On Land Management" (2003), namely, the expression in brackets as follows: "except for the cases of the division of land parcel concerning the acquisition of ownership of a residential house or part of it, and non-residential buildings for various purposes situated on it".

Now consider what documents are necessary for the lease. Under Article 15 of the Law of Ukraine "On Land Lease" (1998) "... an integral part of the land lease agreement is:

– plan or scheme of land parcel that is leased;

– cadastral plan of the land parcel with the mapping restrictions (encumbrances) in its use and prescribed land servitudes;

– act defining the boundaries of land parcel;

– act of acceptance and transfer of leased facility;

– project of land allotment if it is developed under the law".

If this list is compared to the list set out in article 56 of the Law of Ukraine "On Land Management" (2003), we can see that even in case of presence of all documents, the lease cannot be concluded, as no act of defining the boundaries of land parcel; and there should be a project of land management on

land allotment. In addition, the technical documentation of land management for the division of land parcel is prepared by order of the main tenant of land parcel. It is enough to get act of acceptance and transfer of boundaries marks. And for the purchaser of the right to lease it is necessary to have the act of acceptance and transfer of boundaries marks on storage for all boundaries of the newly formed land parcel, not just on the boundary of division.

Let us find out what documentation should be on land management and what it must contain for the purchaser of the right to lease. Obviously, this documentation on land management for the division of land parcel should be supplemented by the act of establishment (updating) of the boundaries of the land parcel. This act is signed only by the purchaser and lease rights tenant. There should be developed technical documentation on land management according to applicable laws and regulations of Ukraine, except for the coordination of boundaries of the land with the owners or users of adjacent land parcels, and as already noted, without obtaining permission. Once the main land parcel boundaries were agreed with the main tenant and they have not changed and the purpose of buildings has not changed as well. In addition the project of land parcel division should be added to this technical documentation of land management.

Consider the possible cases (options) the division of the main land parcels:

1. The best option is when buildings stand apart. There are two separate driveways and walkways to both owners' buildings, no common parts of buildings and structures. In this case, no comment, and the basic land parcel can be called "divisible".

2. In site conditions of placement of buildings on the land there are no passages, and they cannot be built. In this case, it is offered to divide the basic land parcel into three parts. Two separate land parcels are allocated for the seller and buyer, while the third one is common – in their joint lease for passage, walk and more. In this case, this basic land parcel might be called "partly divisible".

3. By mutual arrangement of buildings, structures, one cannot perform the division of land parcel. For example, the first floor belongs to the seller, and the second – to the buyer. In this case, the only possibility is a joint lease. The rental payment for the land in common use is proportional to the parts (areas) of real property owned by the seller and the buyer. This basic land parcel title – "indivisible".

Now we will define the list of main documents in the division project of land parcel and their functions.

1. A copy of the lease contract to the main land parcel will testify that the lease is valid (validity has not expired) and customer ordering the division is

really a lessee (the person who has the right to use this land). This allows the performer to sign a contract to implement the project division of land parcel.

2. Actual topographical plan scaled 1:500 with the boundaries of land parcel and signatures of the purchaser and the tenant of the right to lease. This plan is required for review, illustrative purposes and vision of relief throughout the land to be divided. This plan is a graphical application to the act of establishment of the boundary of the main land parcel.

3. The act of establishment of the boundary of main land parcel signed by the tenant and the purchaser of the right to lease with a graphical application. The signed act will attest that the latter agreed on the boundary of division of the main land parcel.

4. Cadastral plans for each part of the land on which it is divided, with the lengths of sides between the points of boundaries marks, the perimeter and area of the land parcels as well signed by the tenant and purchaser of lease rights. These plans will testify that these people know exactly the size of each project land parcel and their areas.

There is another important aspect to be done in the project and the appropriate division of land management documentation, namely, to determine restrictions and encumbrances that have been established; determine the boundaries and areas of land parcel parts which are subject of restrictions; development of appropriate plans. These plans must also be signed by the tenant and the purchaser of the right to lease.

Note that the project of division is necessary to timely determine the possibility of separation of the main land parcel; and the seller and the buyer will address the notary and write a petition to the local authority, which brings the clarity and precision of the possibility of separation and area of land parcel. In addition, the project will be the basis for the development of appropriate documentation of land division.

5 CONCLUSIONS AND SUGGESTIONS

Summing up, the proposals developed for the division of land parcel for lease significantly reduce procedure and time spent on issuing new lease contracts on part of the land. Also, these proposals can be applied to the land parcels which, according to the previous Land Code of Ukraine (1990), were granted for permanent use to private and notutility legal entities having state acts on the right of permanent use. But according to the Land Code of Ukraine (2001) newly created land parcels of the main land user and purchaser of land use rights may be granted only for rent for the maximum period.

The prospect for further research is to detail the procedures and the list of documents required in cases when the main land parcel is indivisible and to improve procedures for combining the land parcels.

REFERENCES

Civil Code of Ukraine from 16.01.2003 No 435-IV, as amended by the Law of Ukraine on 27.03.2014 No 1170-VII.

Land Code of Ukraine from 18.12.1990 No 561-XII, as amended by the Law of Ukraine on 08.06.2000 No 1805-III (lost its validity on 01.01.2002 under the Land Code of Ukraine of 25.10.2001 No 2768-III).

Land Code of Ukraine from 25.10.2001 No 2768-III, as amended by the Law of Ukraine on 24.10.2013 No 661-VII.

Law of Ukraine "On Land Lease" of 06.10.1998 No 161-XIV, as amended by the Law of Ukraine on 04.07.2013 No 406-VII.

Law of Ukraine "On Land Management" of 22.05.2003 No 858-IV, as amended by the Law of Ukraine on 02.07.2013 No 367-VII.

Theoretical and Practical Solutions of Mineral Resources Mining – Pivnyak, Bondarenko & Kovalevska (eds)
© *2015 Taylor & Francis Group, London, ISBN: 978-1-138-02883-8*

Substantiation of land management methods of industrial cities

M. Tregub & Y. Trehub
National Mining University, Dnipropetrovsk, Ukraine

ABSTRACT: The results of analysis and substantiation of management methods are given. Suggestions on the use of economic, planning, administrative, social and environmental land management methods in industrial cities are made.

1 INTRODUCTION

The achievement of rational and efficient use of land resources is essential for the development of any country. Land resources, like other natural resources, declining over time, lose their natural properties that often cannot be reproduced, changing the quality of the soil, thus become unattractive from an economic point of view. Unfortunately, the efficient and effective level of use of land resources in Ukraine is lower than in Europe. The main reason is irresponsible attitude to the distribution of land and its use, imperfect technologies of rational land use, significant indicators of anthropogenic pressure on land resources, outdated environmental regulations that establish values of maximum permissible exposure etc.

Land Management is an important component of the economic management of the state, which with proper implementation will lead to the desired results. In our country land resources are a major national wealth (The Constitution of Ukraine 1996), so they require special methods of management. Formation of Land Management is one of the important tasks of our time. As a result, it is necessary to identify, classify and approve the land management techniques for specific categories of land at the state level that are not yet regulated by any legal act of Ukraine.

The agricultural orientation of Ukraine is generally accepted, but it should be noted that in our country major industrial agglomerations are combined, which form the industrial zone. Donetsk-Makiivska, Gorlivka-Yenakiivska, Lugansk industrial agglomerations are included in the Donetsk industrial area; Dnipropetrovsk-Dniprodzerzhinsk, Zaporizhzhya, Kryvyi Rig – in the Prydniprovsk industrial area; Lviv agglomeration – in the Prykarpatskyi industrial area, Novovolynsk and Chervonograd – in the Prybuzka industrial area. Outside the industrial areas are Kyiv, Kharkiv, Odesa industrial agglomeration. All these elements of the territorial organization of industry are united by various relations forming the industrial complex of Ukraine. Above are the main areas of mining and heavy industry in Ukraine. It should also be noted that industrial sites in Ukraine are often located directly in the city. As industrial lands are inextricably related to the ecosystem and cannot be isolated from the use along with other categories of land, there is a need for a specific mechanism of their use and protection, as well as management methods.

2 ANALYSIS OF RECENT RESEARCH AND PUBLICATIONS RELATED TO THE SOLUTION OF THIS PROBLEM

The concept, objectives and methods of land management are not defined and are not regulated by any legal act of Ukraine, in contrast to the Land Use. However, there are a sufficiently large number of textbooks and scientific publications in professional journals of Ukraine concerning the theoretical foundations of land management and practices. Among the most important professionals dealing with the issues of land management in Ukraine the following authors can be pointed out: V. Gorlachuk, J. Hutsuliak, D. Dorosh, J. Karpinski, M. Lyhohrud, A. Liashchenko, A. Martin, L. Nowakowski, L. Perovych, O. Petrakovska, M. Stupen, A. Tretiak and others.

The article by O. Petrakovska (Petrakovska 2005) deals with the generalization of all international and domestic experience dividing methods of land management into three main groups: planning, economic and institutional ones. According to the author (Petrakovska 2005), planning methods help to work out long-term and short-term programs of socio-economic development of areas, urban planning and land management documentation that regulates development and land use at different planning levels.

Economic methods of land resources management are aimed at creation of economic conditions that trigger the rational use and protection of land without administrative measures. Institutional methods are based on taking legal action by the authorities, making various kinds of institutional arrangements and logistical operations and combining administrative and judicial activities.

A. Tretiak (Tretiak 2008) defines the following methods of land management: social, economic, legal, land-use, organizational and administrative.

In the book "Land Management" (Tretiak & Dorosh 2006) the methods of land management are divided into: methods for the study of management objects, methods of development of management solutions, methods for implementing management decisions.

The authors of the book "Land Management" (Horlachuk et al. 2006) distinguish between land management techniques, which are divided into two groups – the methods of direct exposure (active) and indirect effects (passive). The first group includes administrative methods, the second – economic, social and psychological ones.

A. Ohryi in his Ph.D. thesis (Okhryi 2006) suggests the need for marketing and logistics management of land resources, which, in our opinion, will provide an opportunity to significantly increase the effectiveness of their use, for these methods will make it possible to make optimal decisions.

The authors in their paper (Tretiak et al. 2013) grounded the necessity of introducing multifunctional complex model of land management in Ukraine, however did not explain what management practices they consider necessary as the base for the model.

Each of the authors considers the content of the problem to some extent; however each of them focuses their scientific study of scientific management of land resources on agriculture and sustainable development of large cities. None of the authors considers the problem of land management of industrial cities.

3 THE PURPOSE OF THE RESEARCH

The purpose of the research is to investigate and justify the land management methods for industrial cities of Ukraine.

4 THE MAIN MATERIAL

Management method is a set of techniques and methods of influence on the managed object to achieve desired goals or the method of influence on members of the management process. Classifications of management methods are mainly related to economics and management and there are a large number of them. Still there is no unique, definite and approved classification of management at any level, especially management of land resources. On the one hand, it appears that land management is an integral part of the economic system created by and subject to the same economic laws that are faced by any process control, on the other hand, it is a unique and specific legal group, which is mainly characterized by category and purpose of facility management.

We believe that special attention should be focused on the study of methods of land resource management in the context of the legal characteristics of different categories of land use. This should identify which management practices from classic management can be used, which should be finalized and justified, and which could not be used at all.

At the core of management techniques are identified the basic functions of management (Meskon et al. 2000):

– organization;
– planning;
– motivation;
– control;
– regulation.

According to them, in the management the following groups are defined:

– economic;
– administrative;
– social and psychological.

The main task of management is to focus on achieving the necessary goals, but with the obligatory account of rational use of certain resources.

Economic management is set of tools and instruments that specifically affect the creation of conditions for the functioning and development of the business (Meskon et al. 2000). These methods are based on the factors of market economy and take a leading place in the management.

Administrative management is a system of methods and techniques of organizational and administrative actions that are used for the organization and coordination of management facilities to carry out their mission. These methods are based on the current legal regulations.

The essence of *social and psychological management* is reduced to the methods of impact on an individual and groups to change their attitudes to work and creative activity, as well as social and psychological interests of companies and their staff (Meskon et al. 2000). The purpose of using these management practices is the study and application of the laws of human mental activity and putting it in real conditions for the functioning of the organization.

The above basic management techniques are often coupled with legal, ideological, technological, illustrative, research and other management techniques, depending on the specific object of management.

In terms of market relations the basic principles of the use of management practices are the following: economic independence, ensuring profitability, self-sufficiency, financial interest, moral satisfaction, competitive products.

Due to the specific nature of industrial Land Management, the general list may be added by the principle of maintaining ecological stability of ecosystems. The first and the most important method of land management of industrial cities is an economic method.

The economy is a fundamental component of the existence of the country and it contains all prospects for the state development. Economic methods of land resources management of industrial cities must provide optimal use of land in accordance with the specific capacity of the existing natural resources of a territory. It works under the following conditions:

– financial interest of entities of land management of parcels located in urban areas and that can be used for industrial purposes;

– pricing of land, loans, fines, tariffs for services on land that would directly depend on the category of land and its purpose;

– formation of payment for land;

– compensation for the use of land inappropriately and for violation of land caused by the effects of natural and anthropogenic factors;

– prediction of the long-term economic situation in the country;

– combination of financial interests of the state, owners and users of specific management facilities.

Particular attention should be paid to land lease payment, fines, etc., because these components form the source of material revenues to the state budget. For industrial land special tax status should be defined, especially in case of a new venture or equipment upgrade at the old one, which can significantly improve the environmental situation, and more. We believe that economic method for industrial lands should be used as a method of stimulation and only in extreme cases as enforcement and punishment.

One of the important methods of land management of industrial cities is the *planning method*. Planning for urban areas plays a significant role in land management. Informed and meaningful planning must be the basis of any management. As a method of land management in industrial cities, this method should include the following:

– public discussion of the industry location in existing urban conditions for certain urban areas;

– substantiation of urban planning and projects in accordance with applicable legal acts of Ukraine, building codes, regulations and standards;

– account of local development rules in accordance with specific area;

– substantiation of prospects for the use of each territory;

– planning of engineering and transport infrastructure around industrial facilities.

In general, exploration, development and zoning should be done in close collaboration of architects, planners and experts on land relations. Adherence to this principle will implement a systematic approach to this issue.

Administrative management based on the law relating to Ukraine and on land relations occupy equally important place in the management of land resources of industrial cities. Administrative methods of land management of industrial cities can be divided into three main components: regulation, administration and disciplinary ones.

The first component refers to all legal acts, regulations, standards, guidelines, instructions, statutes of companies on issues of land resources which do not contradict the current legislation and have a strong impact on objects and subjects of control.

The second component includes all administrative documents of local governments aiming to address arising specific situations.

The last component is a basis for discipline in all areas of life, which governs all existing types of liability of Land Management (personal, collective, financial and other liabilities).

Social methods of land management of industrial cities are also an important part of the control system. This social component is not affected directly by the management facility, but indirectly by the state and/or individual owners and users. Components of social methods include different types of motivation, raising awareness to solve a particular problem, government support for critical needs and more. Social management techniques must be supported by adoption of rules and regulation for activities, as well as by permissions and prohibitions.

The last method of land management, which is one of the major methods for industrial cities, is *environmental* one. This topic is barely covered in scientific publications. Although the application of this method is the key to the cities with advanced and powerful industry. Environmental problems in Ukraine are increasing every day, the main reason is the anthropogenic pressures and their solutions depend directly on funds invested in prevention of the damage to human health and the environment.

Taking into account the land legislation of Ukraine there is a category of land as industrial land, transport, communications, energy, defense and other purposes that form a separate category. In terms of environmental law, this category is distinguished as land under the objects of the significant sources of negative impact on the environment.

Important components that make the basis of ecological method of land management are the following:

– location of land for industrial purposes must be justified according to urban standards, rules and regulations;

– required definition of sanitary protection and security zones;

– the introduction of advanced equipment and technology that enables to reduce negative impact on the environment;

– design and construction of industrial facilities should be subject to mandatory environmental assessment, which should provide an assessment of environmental safety of the area.

Ecological method of land management in developed countries is a priority and we believe that this approach should be adopted in Ukraine.

5 CONCLUSIONS

In recent years, there has been a rapid change of legal acts of Ukraine, which affects the development of land and state land cadaster of the country. Unfortunately, until now there is no general information on land management and land use as part of the state land cadasters. In Ukraine there is no single body of land management as a macro system. At the same time the responsibility for the efficient use of land is divided between the executive authorities and local governments subordinated to various ministries and departments.

Modern fast development of science and technology complicates human pressure in industrial cities through diversity of ultra-modern equipment and advanced production technology. But taking into account costs of new technologies in production and modern equipment, the majority of industrial enterprises cannot afford such changes. Because the industry is inextricably linked to the land where respective company is located, and with the environment in general, effective management will allow indirectly influence the development of the industry.

We propose to use economic, planning, administrative, social and ecological methods for land management of industrial cities. Although most of the management methods are taken from classical management, they have been adapted specifically for use in the relevant field of science and industry.

Characteristic feature of management of land resources in industrial cities is that the consequences of their actions are observed in all spheres of human activity.

Further research has to shift from macro-to micro-level of land management, depending on the purpose within the category.

REFERENCES

Horlachuk, V., Viun, V., Peschanska, I.M., Sokhnych, A.Ya. & other. 2006. *Land Management*: Textbook. Ed. prof. Horlachuk V., 2nd ed., corr. and remade. Lviv: Mahnoliia: 443.

Meskon, M.Kh., Alber, M., Khedoury, F. & Evenko, L. 2000. *Principles of Management*. 3rd ed. Moskow: Delo: 704.

Okhryi, O. 2006. *Integrated land use strategy administrative-territorial formation*. Synopsis of the diss. c.s. PhD Public Administration under the President of Ukraine: 29.

Petrakovska, O. 2005. *Methods of Land Management. Town planning and spatial planning*. Ed. 20: 261 – 267.

Tretiak, A. 2008. *Land Management*: Textbook. Kyiv: 528.

Tretiak, A., Kuryltsiv, R. & Tretiak, N. 2013. *Conceptual Framework for Ukraine in modern multi-functional land administration systems*. Zemlevporiadnyi visnyk, No 9: 25 – 28.

Tretiak, A. & Dorosh, O. 2006. *Land Management*: Textbook. Vinnytsia: Nova knyha: 360.

The Constitution of Ukraine 28.06.1996 № 254k/96-VR.

Theoretical and Practical Solutions of Mineral Resources Mining – Pivnyak, Bondarenko & Kovalevska (eds)
© 2015 Taylor & Francis Group, London, ISBN: 978-1-138-02883-8

Choice of supplier as a basic constituent of the logistics system

M. Ivanova

National Mining University, Dnipropetrovsk, Ukraine

ABSTRACT: The paper examines current approaches to the definition of the logistics system in order to use its basic principles and approaches in the practical operation of industrial enterprises. It has been substantiated that the choice of supplier is one of the main problems in the management of procurement of raw materials and components. This includes the search for sources of supply and an assessment of possibilities of timely delivery. Common and special scientific methods were used in this study, e.g. the nature of the "logistics system" category was studied through the abstraction method; structural and genetic analysis and synthesis were applied to determine the inter-related components of the logistics system, and a pairwise comparison was used in the selection of raw material suppliers. It has been revealed that the purpose of the logistics system is to deliver goods and products the best prepared for production or personal consumption and at the lowest cost. A matrix of pairwise comparisons is recommended for the selection of raw material suppliers. This technique allows not only a selection of suppliers, but ranking them by importance. The proposed method of selecting suppliers can be used by industrial enterprises in the selection of suppliers as the main function of the logistics system.

1 INTRODUCTION

Formation of a logistics system allows combining logistics functions and logistics operations at all stages of the production process. In this case, the correct choice of a supplier is an important constituent of the enterprise's success, which can result in reducing costs, increasing the efficiency of procurement logistics and providing the industrial enterprise with competitive advantages and stability in the target market.

Analysis of the current publications on logistics has shown that today there is no single definition of the logistics system. A.N. Rodnikova's definition can be considered the most general: "Logistics system is an adaptive feedback system that performs logistics functions and logistics operations, consists of several subsystems, and has developed links with the environment" (Rodnikov 1995). According to A.G. Kalchenko, "logistics system is an organizational and economic mechanism of management of material and information flows (Kalchenko 1999). Although this is a widely spread definition, one cannot insist on its universality, as the author has not included financial flows in the logistics system concept, though the nature of these is radically different both from the material and information flows.

The purpose of the logistics system is to deliver to destination the required quantity and range of best prepared for industrial or personal consumption goods and products at the lowest cost. In addition to functional subsystems, the logistics system includes supporting units such as information, legal, and human resources subsystems (Kalchenko 1999).

Y.M. Nerush defines the logistics system as a system that provides uninterrupted distribution and exchange of products in such a way as to ensure the best supply – demand situation. It optimizes the operation of constituent subsystems: some departments determine the amount of product necessary for continuous operation of the enterprise (procurement), others are engaged in the distribution of products (sales), some departments deliver products from suppliers to consumers, and some collect information about suppliers, the market, and consumers (Nerush 2000). In his definition, Y.M. Nerush underestimates the functional purpose of the logistics system, as he only focuses on "the distribution and exchange of products"; he ignores both the procurement of raw materials and release of finished products, which significantly narrows the object of study.

A.M. Oklander defines the logistics system of an enterprise as an organizational-administrative mechanism of coordination of the different services that manage material flows. According to the authors, the logistics system is an organizational and managerial mechanism of achieving the desired level of integration of logistics functions; this can be done through organizational changes in the management structure and implementation of specially designed management procedures, the most important of these being planning of the procurement and chaining the production and physical distribution in a single material flow (Oklander 2000). This interpretation of the "logistics system" concept is applicable to the subject of management that influences the object, vis. the material flow. In this case, setting aside the information and financial flows will also affect the development of the logistics system.

Russian scientists V.I. Sergeev, A.A. Kizim and P.A. Elyashevich have tried to solve this problem. They describe the logistics system as a complex but organizationally completed (structured) economic system, which consists of material as well as related flows interconnected in a single control process; the set of the units, their boundaries and function objectives being united by internal and external business objectives (Sergeev et al. 2001). Thus, the authors include the information and financial flows in the "related flows".

D.D. Kostoglodov, I.I. Savvidi, and V.N. Stakhanov suggest defining the logistics system as a set of interrelated and interacting participants of economic flows, which are united by similar purposes and economic interests (Kostoglodov et al. 2000). However, the authors do not disclose the participants of the economic flows, which makes the definition vague and imprecise.

2 METHODS

In our opinion, the logistics system is an adaptive, self-organizing feedback system which consists of subsystems of procurement, production and sales; it performs basic logistics functions such as planning, organization, control and motivation and has well-developed internal and external relations.

The main constituent of the logistics system is a logistic operation (elementary logistics activity), which is an isolated set of actions aimed at transforming the logistics flows (warehousing, transportation, etc.). A logistic function (an integrated logistics activity) is an enlarged group of logistics operations aimed at implementing the objectives of the logistics system (procurement, production, and sales) (Krikavsky 2004).

In their research, Ukrainian authors have studied the formation of logistic processes and the general aspects of logistics management, while foreign authors pay more attention to the formation and optimization of process control systems in general.

However, the specificity of logistics management in industrial enterprises needs a more detailed study. Moreover, the studies have left unattended the following issues: the particulars of logistics quality management in industry; problems of modelling, development, making and implementation of management decisions concerning the quality of logistics processes in industrial enterprises, the development of organizational and economic mechanism of logistic processes in industrial enterprises in Ukraine, the criteria for evaluating the efficiency of logistics management and the development of a supporting information model.

Logistics provides technical, technological, economic and methodological integration of the individual parts of the logistics chain into a single system, an effective management of cross-cutting material flows, as well as the rationalization of economic activities through their optimization (Baujersoks 2001). A.G. Kalchenko believes that logistics balances the process of moving material and information flows in space and time from their primary source to the end user (Kalchenko 1999). I.e., logistics facilitates the efficient management of material and related information and financial flows with the economic cost of all the resources to fully meet customer requirements.

For this, it should cover and harmoniously chain up such diverse activities as production, communication, transportation, procurement and inventory management, warehousing, materials handling, packaging, and other (Mirotin & Tashbaev 2002).

Improved cooperation with suppliers can minimize the cost of inventories, and a better coordination of specifications can minimize the need for field changes of the products procured; better coordination of pricing can make the company and its suppliers more competitive and profitable. Savings can be achieved at the other end of the logistics chain, vis. in the distribution process. Improved efficiency of distribution may affect the overall price advantage of the company. High cost of storekeeping and low cost of decentralized transport allow many companies to reduce distribution costs by centralizing the stocks.

Savings due to the reduction of capital as a result of the decentralization of inventories offset the high cost of fast delivery by truck or by air (Gordon & Karnaukhov 1998). The high proportion of logistics costs in the final price of goods reveals the potential of improving the economic performance of business entities due to better material flow management as a result of improving individual logistics operations. This tool can be rather effective in competing for a stable market.

The operation of domestic industrial enterprises shows that all aspects of logistics operations should be directly linked to the strategic plan of the enterprise. This is a basic condition for achieving the goals and, in particular, maximizing the profit from the use of logistics. Logistics systems management provides for synchronizing operations and procedures, linking the different business units into a single managed system aimed at efficient satisfaction of the end user (Mirotin & Tashbaev 2002). Thus, special attention should be paid to the reduction of those activities, operations, or procedures in logistics processes that do not serve the creation of value added so as to reduce costs in general.

3 RESULTS AND DISCUSSION

The issues of commodity-material supply being of great significance, we have studied the problem of a supplier selection by a pairwise comparison matrix that is one of the main constituents of the logistics system. The matrix has been tested at PAO "Spectrum", which specializes in the manufacture of alkyd enamels, anti-corrosion primers, oil paints, water-dispersion paints (fungicide primers for outdoor and indoor applications; deep penetration primers; outdoor, indoor and acrylic paints).

The choice of a supplier is one of the major problems in the management of procurement of raw materials and components. This includes search for sources of supply and assessment of guarantees of a timely delivery.

At this stage, the choice of a supplier is of importance not because of a great number of potential suppliers operating on the market; it is imperative that the supplier was a reliable partner in the formation of the enterprise logistics system.

It is the process of selecting a supplier that is one of the basics of procurement. According to the tasks, the process is divided into several steps shown in general in Figure 1 (Anikin 2003).

In industrial plants, the suppliers are often chosen without appropriate recommendations, standards or documentation. This leads to negative consequences such as duplication of functions and reduced responsibility. The results concerning the necessary items are achieved as a compromise in the course of negotiations and they depend on the supplier's and the buyer's positions in the market.

PAO "Spectrum" regularly buys the following types of raw materials: bentonite (0.6 t/mo.); butanol (0.6 t/mo.); xylol (1.2 t/mo.); pentaerythritol-modified phthalic resin (0.5 t/mo.); alkyd resin (0.5 t/mo.); kaolin P-1 (2 t/mo.); toluene (0.6 t/mo.); titanium dioxide (1 t/mo.).

Table 1 shows the main suppliers of raw materials.

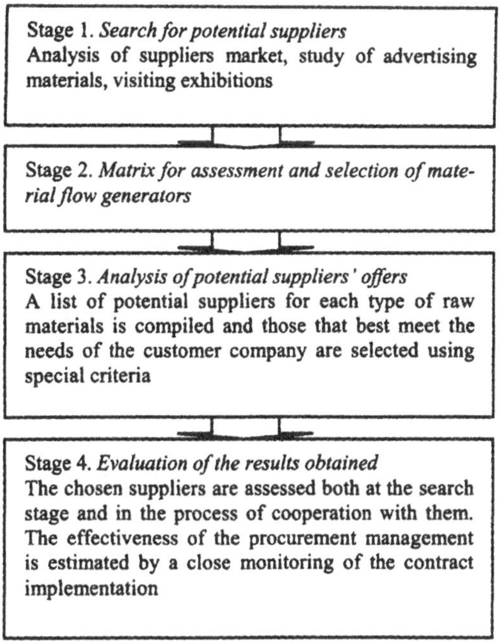

Stage 1. *Search for potential suppliers*
Analysis of suppliers market, study of advertising materials, visiting exhibitions

Stage 2. *Matrix for assessment and selection of material flow generators*

Stage 3. *Analysis of potential suppliers' offers*
A list of potential suppliers for each type of raw materials is compiled and those that best meet the needs of the customer company are selected using special criteria

Stage 4. *Evaluation of the results obtained*
The chosen suppliers are assessed both at the search stage and in the process of cooperation with them. The effectiveness of the procurement management is estimated by a close monitoring of the contract implementation

Figure 1. Steps in the process of a supplier selection.

Table 1. Volumes of supplies of key raw materials at PAO "Spectrum".

Raw Materials, tons / month	PE "Bilushenko A.V." Cherkasy	PAO Korostenskyj zavod Yantar ("Amber"), Korosten, Zhytomyr region	PE "Systema Optimum" Lviv	PE "Progressor" Bila Tserkva	OOO "Firma "Kaapri" Kyiv
			Suppliers		
Bentonite	0.5	—	—	—	—
Butanol	0.6	—	—	—	—
Xylol	—	—	1.2	—	—
Pentaerythritol-modified phthalic resin	—	0.5	—	—	—
Alkyd resin	—	0.5	—	—	—
P-1 kaolin	—	—	—	2	—
Toluene	—	—	0.6	—	—
Titanium dioxide	—	—	—	—	1

Table 2 shows the most important criteria of selecting suppliers for PAO "Spectrum".

Table 2. Matrix for assessment and selection of material flow generators.

Description	Selection criteria and scores		
	3	2	1
Distance from consumer to supplier	up to 100 km	up to 500 km	over 500 km
Order execution term	within 3 days	1 week	2 weeks
Intervals of supplies	once a week	once a month	every 3 months
Payment terms	1 month postponement	payment at the time of contract conclusion	prepayment
Possible discounts	15%	10%	5%
Supplier's share in covering transportation costs	transportation paid by supplier	transportation paid in equal parts by supplier and customer	transportation paid by customer
Completeness of assortment	complete	to order occasionally	to order
Terms of delivery	delivery	carriage mostly paid by supplier	shipping
Terms of risk sharing	at the time of goods transfer to the buyer	at the time of goods transfer to the carrier	at the time of goods transfer in the supplier's warehouse
Supplier's reputation	no complaints	occasional breaches	numerous complaints
Financial position of supplier	stable	unstable	crisis conditions
Price	below market price	market price	above market price

The study of potential suppliers results in a list of the most attractive suppliers who then become contracting parties. The list of suppliers is generally produced for each specific type of resources supplied. Figure 2 shows a comparison of the main suppliers for each type of raw materials.

Pairwise comparisons require uniformity of characteristics being assessed, the development of logically sound criteria and standards as well as clearly defined procedures for operating the criteria. The advantage of this method is the low labor input of a working group in the processing of the data. The method of pairwise comparisons can greatly simplify the work of an expert and reduce it to performing elementary repetitive operations of selection.

Drawbacks of the method can be related with the level of arrangement and execution of the examinations, the problems of selecting evaluation criteria, or experts' qualifications; they can be eliminated in the practical implementation of the method. The validity and accuracy of the results can be substantially increased by calling competent experts, discussion on the interim and final results.

The proposed method of pairwise comparisons is operational, which makes it possible to automate getting information from experts.

Suppliers' significance is calculated using the formula:

$$a_i = \frac{\sum Q_i}{\sum\limits_{i=1}^{m} Q_i} \qquad (1)$$

where $\sum Q_i$ – a sum of ranks (scores) of i-th supplier, $\sum_{i=1}^{m} Q_i$ – a sum of ranks (scores) of all suppliers.

Table 3 shows the recommended list of suppliers for PAO "Spectrum", based on the calculation of the suppliers' significance coefficients (Table 4).

The proposed method of pairwise comparisons reveals which of the suppliers are better, and how much better they are. This technique allows not only choosing suppliers, but ranking them by significance, that is, if the first-rank supplier leaves the market and is no longer able to supply the raw materials, the method permits to choose the next-in-the-rank supplier. This ensures an uninterrupted procurement within the logistics system.

Table 3. Recommended list of suppliers for PAO "Spectrum".

Raw material	Supplier	t/month
Titanium dioxide	OOO "Firma "Kaapri", Kyiv	1
Bentonite and butanol	PE "Bilushenko A.V.", Cherkasy	1.1
Xylol	OOO "Khimstatus", Kharkiv	1.2
Pentaerythritol-modified phthalic and alkyd resins	PAO "Elaks", Odesa	1
Kaolin P-1	PE "Progressor", Bila Tserkva	2

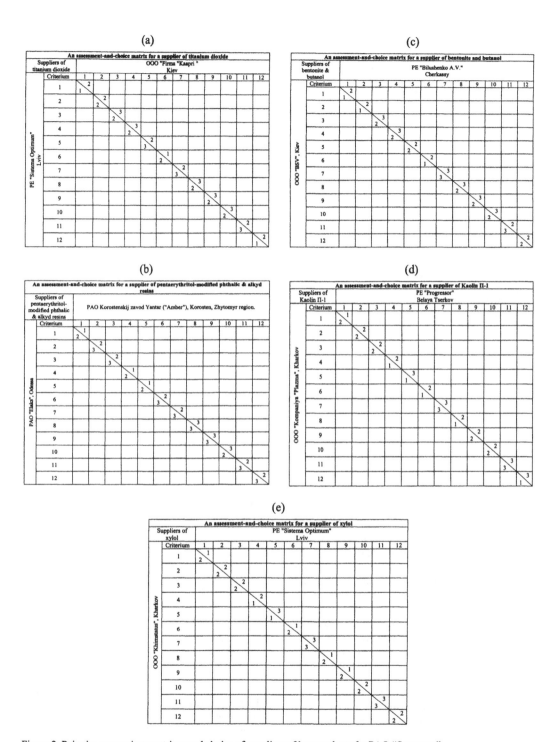

Figure 2. Pairwise comparison matrices and choice of suppliers of key products for PAO "Spectrum"

459

Table 4. Calculation of supplier's significance coefficient.

Raw material	Sum of ranks, $\sum Q_i$		Significance coefficient, a_i	
	First supplier	Second supplier	First supplier	Second supplier
Titanium dioxide	OOO "Firma "Kaapri", Kyiv	PE "Systema Optimum", Lviv	OOO "Firma "Kaapri", Kyiv	PE "Systema Optimum", Lviv
	[2+2+3+3+2+1+2+3+ +3+3+2+2] = 28	[1+2+2+2+3+2+3+2+2+2+ +3+1] = 25	28/53 = 0.53	25/53 = 0.47
Bentonite and butanol	PE "Bilushenko A.V.", Cherkasy	OOO "BSV", Kyiv	PE "Bilushenko A.V.", Cherkasy	OOO "BSV", Kyiv
	[2+2+3+3+2+2+3+3+ +3+3+2+2] = 30	[1+1+2+2+2+1+3+2+ +2+2++3+2] = 23	30/53 = 0.57	23/53 = 0.43
Xylol	OOO "Khimstatus", Kharkiv	PE "Systema Optimum", Lviv	OOO "Khimstatus", Kharkiv	PE "Systema Optimum", Lviv
	[2+2+2+3+1+2+3+2+ +2+2+3+2] = 26	[1+2+2+2+3+1+3+ +1+1+2++3+2] = 23	26/49 = 0.53	23/49 = 0.47
Pentaerythri- tol-modified phthalic & alkyd resins	PAO "Elaks", Odesa	PAO Korostenskyi zavod Yantar ("Amber"), Korosten, Zhytomyr region	PAO "Elaks", Odesa	PAO Korosten- skyi zavod Yantar ("Am- ber"), Korosten, Zhytomyr region
	[2+3+3+2+2+3+3+3+ +3+2+3+3] = 32	[1+2+2+1+1+2+2+3+3+ +3+2+2]=24	32/56 = 0.57	24/56 = 0.43
P-1kaolin	PE "Progressor", Bila Tserkva	OOO "Kompaniya "Plazma", Kharkiv	PE "Progressor", Bila Tserkva	OOO "Kom- paniya "Plaz- ma", Kharkiv
	[1+2+2+2+3+2+3+ +2+2+2+3+3] = 27	[2+2+2+1+1+1+3+1+ +2+2++3+1] = 21	27/48 = 0.56	21/48 = 0.44

4 CONCLUSIONS

1. One of the key concepts related to logistics is the logistics system, which is defined by the author as follows: logistics system is an adaptive, self-organizing feedback system which consists of sub-systems of procurement, production and sales; it performs basic logistics functions such as planning, organization, control and motivation, and has well-developed internal and external relations.

2. Logistics management allows synchronizing operations and procedures by linking disparate business units into a coherent managed system aimed at efficient satisfaction of the end user.

3. Commodity-material supply being of great importance, the authors have investigated the process of supplier selection using a pairwise comparison matrix as one of the main components of the logistics system. The technique was tested at PAO "Spectrum". The method offered allows ranking suppliers by significance, which ensures uninterrupted procurement in the frames of the logistics system.

REFERENCES

Baujersoks, Donald Dzh. 2001. *Logistics: the integrated chain of deliveries*. Olimp-Biznes: Moscow: Russia.

Gordon, M.P. & Karnaukhov, S.B. 1998. *Logistics of motion of commodity*. Tsentr ekonomiki i marketinga. Moscow: Russia.

Kalchenko, A.G. 1999. *Bases of logistics*. Kyiv: Znannya.

Kostoglodov, D.D., Savvidi, I.I. & Stakhanov, V.N. 2000. *Marketing and logistics of a firm*.: Moscow: PRIOR.

Krikavsky, Ye.V. 2004. *Logistics. Bases of theory*. Natsionalny universitet "Lvivska politehnika": Lviv: Intelekt-Zahid.

Anikin, B.A. 2003. *Logistics*. Moscow: Infra-M.

Mirotin, L.B. & Tashbaev, I.Ye. 2002. *Logistics for a businessman: basic concepts, positions, procedures*. Moscow: Infra-M.

Nerush, Yu.M. 2000. *Logistics*, 2nd ed. Moscow: YuNITI: DANA.

Mirotin, L.B. & Sergeev, V.I. 2002. *Bases of logistics*. Moscow: Infra-M.

Oklander, M.A. 2000. *Contours of economic logistics*. Kyiv: Naukova dumka.

Rodnikov, A.N. 1995. *The logistics: Terminological dictionary*. Moscow: Ekonomika.

Sergeev, V.I. Kizim, A.A. & Elyashevich, P.A. 2001. *Global logistics systems*. Saint Petersburg: Biznes-pressa.

Theoretical and Practical Solutions of Mineral Resources Mining – Pivnyak, Bondarenko & Kovalevska (eds)
© 2015 Taylor & Francis Group, London, ISBN: 978-1-138-02883-8

Differential system of the rope bolts loading during extraction drift support

V. Lapko, V. Fomychov & V. Pochepov
National Mining University, Dnipropetrovsk, Ukraine

ABSTRACT: The main purpose of computational experiment was to evaluate the effect of rope bolt pre-tensioning on the rocks of roof stability of mine working reuse for different geological conditions. Computations have allowed receiving dependence of rope bolt stress state in time, before and after the working face passage through the plane of the modeling bolts. The obtained results have allowed developing optimal differential conditions of rope bolt loading, depending on the strength characteristics of the rock layer that forms the rocks of roof above the mine working.

1 INTRODUCTION

Conduct the research to improve the stability of excavation workings in difficult geological and mine conditions aimed to identify the optimal operating conditions of steel support design. So it becomes possible to maintain their performance conditions during the maximum time term (Kovalevs'ka et al. 2011, Sally et al. 2014). Support of workings requires constructive solutions that provide the ability to quickly withdrawal or low material-output ratio of timber construction. Currently, the mines of Western Donbass widely used frame-roof bolting (Sotskov & Saleev 2013). This steel support provides a suitable level of stability of excavation drift, but only during prime cost of drift support. Increase the effectiveness of the frame-roof bolting is possible through the use of new schemes for roof-bolt setting that will reduce the convergence of rocks in the drift cavity and reduce metal consumption of frames (Bondarenko et al. 2013). In the mines of Western Donbass as new elements of rock bolt have been used rope bolts that can be installed by pairs in a one or two-level schemes (Lapko et al. 2013). Study of optimal characteristics of two-level schemes installation of rope bolts allow to revealed a number of dependencies between their geometric parameters and support capability change of rope bolts (Kovalevs'ka et al. 2011, Lapko et al. 2013). During these studies it was found that the effect of the pre-tensioning rope bolt extends to all geotechnical system consisting of rock bolt and rock layers that form directly generate the roof.

The aim of the work is to determine the optimal characteristics of pre-tensioning bolts that works in a unified extraction drift support system, change of their mutual influence during rock massif stratification.

2 STATEMENT OF THE PROBLEM

As part of the computational experiment perform a series of calculations of the rock bolt interaction with rock layer that forms the rocks of roof above excavation drift, for different combinations of strength characteristics of rocks and the strain is applied along the bolt axis.

Simulations of physical characteristics of the rocks perform in accordance with their rheological properties at a selected time interval. The interval was selected on the basis of data on the development of stress-strain state of the system "rock mass – rock bolt" obtained in previous studies (Lapko et al. 2013).

3 COMPUTATIONAL EXPERIMENT

During the simulation of state changes of geotechnical system was used three-dimensional model the size of which in vertical direction of 60 m in the horizontal plane – 180 m by 180 m. Through the geometric center of the model output is passed to the circular cross section of the arch, the base width of 4.5 m and height 3.5 m.

At a height of 0.6 m from the footwall extends horizontally coal seam with the thickness – 1 m. In all calculations under coal seam located bedrock – siltstone model, based on a rigid base. The side faces of the model are limited to rock mass conditions of symmetry. The top of the applied load is equal to 10 MPa.

Rock layer located in the top of the extraction drift is modeled using the mechanical characteristics of siltstone, sandstone and mudstone. The composition of the model includes features of elasticity, compressive strength and rheological curves description.

In the middle part of the extraction drift the roof bolting was simulated and was installed in increments 1 m. Interval of six meters along the drift. This lining consists of three elements with a cross section SVP 22 and two rope bolts, installed symmetrically in the top of the excavation drift at a distance 1.6 m from each other with inclination 74° to horizon. The cross section of the rock bolt was simulated with circle. The parameters of circle were 24 mm in diameter and 4.5 m in the length.

During the simulation calculations performed wallface advance at a speed of 4 meters per day. Thus was carried out 46 cycles of calculations, the first of which was carried out in elastoplastic formulation. With each subsequent cycle on one side of extraction drift, the name for which we choose the "right" was extracted the fragment of

coal seam with depth of 4 m and was made recalculation of the stress-strain state of geotechnical systems. "Left" side of the model during calculations did not change.

4 FINDINGS

At the first stage of the computational experiment two calculations was performed, wherein in rocks of roof simulated siltstone.

Impact assessments of pre-tensioning rock bolts on distribution efforts in the load-bearing system are carried out on the basis of the graphs are presented on Figure 1. These graphs show the temporal variation value P'. This indicator represents the ratio of the reduced internal efforts o rock bolt and tensile strength of rock of roof above drift excavation.

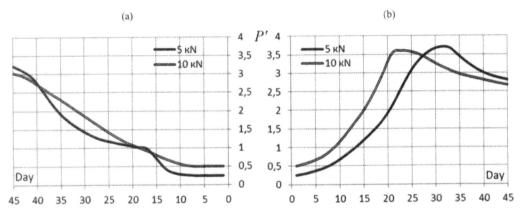

Figure 1. Changes of the maximum value of P' from the time to bolt set in siltstonerock: (a) the "Left"; (b) "Right".

The graphs show the change of internal rock bolt forces, pretension of which, in each case during calculation was the same. The graph shows that for different values of the pre-tensioning rock bolt the quantitative and qualitative indicators of bolt loading vary considerably.

In the graph (Figure 1), it is seen that the lower value of the rock bolt tension change of its state has the character of the offset. The presence of this offset occurs due to the passage of the wallface through the plane of the rock bolt installation that through the time in a 15th day. Diagram for the rock bolt with a greater magnitude pretension shows a steady uniform growth with less stress finite value. Thus, the development of deformations in the rock mass around the bolt does not lead to the formation of foci of stress concentrations that trigger the rapid growth of the main crack.

On Figure 1,b graphics have minor quantitative differences, but qualitatively repeat each other. Quantitative abnormalities associated with the difference in the pre-tensioning of bolt. The increase in pre-tensioning of bolt leads to advance start-bolt in the process of arch support. That is, in terms of computational experiment without differentiation initial efforts on roof bolt the resistance increases with volume. At the same time unloading of "right" bolt is observed in both cases, indicating that the activation of the rocks of roof subsidence process and its transition into the out-of-limit state.

We shall consider in detail the graph in Figure 1 by pairs in accordance with the magnitude pretension. With high tension the roof bolting perceives greater in magnitude. This load varies smoothly and uniformly with time.

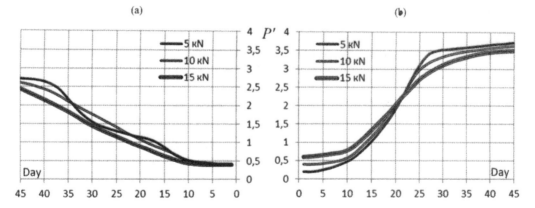

Figure 2. Changes in the maximum value of P' from the time to bolt set in sandstone: (a) the "Left"; (b) – "Right".

With less pre-tension the quantitative difference of internal forces for the "Left" and "Right" bolts increases, indicating that the activation of the destruction process in the central part of excavation drift. Thus, for consideration of computational model the large amount of pre-tensioning bolt increases the excavation drift stability due to the reducing of rock massif deformation in the rocks of roof.

In the next stage of the research was implemented differentiated system of pre-tensioning bolting in which the "Left" bolt was pulled with one effort and the "Right in three different efforts. In Figure 2,a and 3,a the "Left" bolt is represented by three charts for different tensions of the "Right" bolt. Accordingly, in Figure 2,b and Figure 3,b are shown the graphs of changes in the state of the "Right" bolt at different pretension values.

Calculations was carried out for sandstone and mudstone, demonstrate qualitative differences when changing the state of the bolt and thus changing the state of rocks of roof above drift excavation. Sandstone is sandstone is characterized by the preservation of a stable equilibrium of the rocks of roof from the "right" bolt even after a few days after the wall-face passage. But at the same time with a minimum of tension "right" bolt change the internal forces in the "left" bolt and has a stepped appearance. Therefore, during the relative stability of the entire computational of stress zone in direct rocks of roof above excavation oriented in the horizontal plane may lead to stratification of a rock massif.

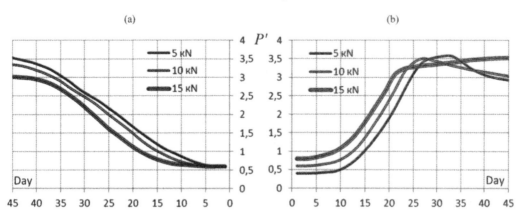

Figure 3. Changes in the maximum value of P' from the time to bolt set in mudstone: a) the "Left"; b) – "Right".

During mudstone calculation it is seen another character of the strains and stresses distribution in the top of the drift excavation. The values of the internal forces of the "left" and "right" bolt differ minimally (less than 6%). In two ways of calculating a transition of the roof generate a transcendental state, but at the maximum tension of the "right" bolt the rocks of roof keeps a steady state. That is, there is a gradual rock of roof subsidence while the value of this subsidence decreases with increasing pretension of the "right" bolt with a differentiated system loading.

Thus, the condition of the optimal differential loading of bolt is to analyze the difference of the efforts of the "left" and "right" bolts on the set of all available points for calculation and has following expression:

$$1,87 \frac{\sum_{i=1}^{n} |P_l' - P_r'|_i}{n} < 1, \tag{1}$$

where n – the number of days in the calculation; P_l' and P_r' – are the value of the internal forces in one cycle calculation for the "left" and "right".

5 RESEARCH CONCLUSIONS AND RECOMMENDATIONS FOR FURTHER RESEARCH IN THIS AREA

Pre-tensioning force of rope bolt determines the qualitative and quantitative development of deformations in the marginal rock massif. Selection of the optimal value of the bolt tension depends from rheological properties of rocks of roof above excavation drift. Differentiated system of rope bolt loading can improve the stable stability of the excavation drift by optimizing the distribution of non-linear deformation occurs in the rocks forming roof of a drift.

The results of computational experiments require verification in natural conditions. Necessary to carry out measurements of displacements at the roof of the drift fixed at the experimental site of the rope bolt with different pre-tensioning forces. Thus, our research will finally confirm the effectiveness of the system of differentiated load of bolt support combined into a single load-carrying system.

REFERENCES

Kovalevs'ka, I., Vivcharenko, O. & Fomychov, V. 2011. *Optimizationof frame-bolt support in the development workings, using computer modeling method.* XXII World mining congress & Expo. Istanbul, September 11 – 16, Vol. I: 267 – 278.

Sally, S., Pochepov, V. & Mamaykin, O. 2014. *Theoretical aspects of the potential technological schemes evaluation and their susceptibility to innovations.* Progressive Technologies of Coal, Coalbed Methane, and Ores Mining. The Netherlands: CRC Press/Balkema: 479 – 483.

Sotskov, V. & Saleev, I. 2013. *Investigation of the rock massif stress strain state in conditions of the drainage drift overworking.* Mining of Mineral Deposits. The Netherlands: CRC Press/Balkema: 197 – 201.

Bondarenko, V., Kovalevs'ka, I., Svystun, R. & Cherednichenko, Yu. 2013. *Optimal parameters of wall bolts computation in the united bearingsystem of extraction workings frame-bolt support.* Mining of Mineral Deposits. The Netherlands: CRC Press/Balkema: 5 – 9.

Lapko, V., Fomychov, V. & Pochepov, V. 2013. *Bolt support application peculiarities during support of development workings in weakly metamorphosed rocks.* Mining of Mineral Deposits. The Netherlands: CRC Press/Balkema: 211 – 215.

Theoretical and Practical Solutions of Mineral Resources Mining – Pivnyak, Bondarenko & Kovalevska (eds)
© 2015 Taylor & Francis Group, London, ISBN: 978-1-138-02883-8

Automated stabilization of loading capacity of coal shearer screw with controlled cutting drive

V. Tkachov & A. Bublikov
National Mining University, Dnipropetrovsk, Ukraine

G. Gruhler
Reutlingen University, Germany

ABSTRACT: A solution of topical scientific problem of coal shearer output increase providing minimum specific power supply for coal cutting, transportation, and loading in terms of thin seams has been proposed. The solution is based on the use of earlier proposed criterion of screw gumming for optimum cutting velocity-coal shearer feed rate ratio in the context of increased screw rotation owing to phase voltage frequency increase. Simulation results of automated control system for coal shearer operations with frequency-controlled cutting drive within thin seams have confirmed the efficiency of the system using proposed algorithm of smart analysis of coal shearer power signal.

1 INTRODUCTION

Today theoretical backgrounds of several approaches of automation for coal shearer operation modes with alternative cutting velocity are available. For example, one of them is an approach of parametrical optimization. It means constant electric motor power of P cutting drive at the specified level by controlling V_f feed velocity and keeping up optimum value of h_{opt} chip thickness by controlling V_c cutting velocity:

$$\begin{cases} V_f \rightarrow var \\ V_c \rightarrow var \\ P(V_f) \rightarrow const \\ h_{opt}(V_f, V_c) \rightarrow const \end{cases} \tag{1}$$

It has been proved that each distance between neighbouring cutters within a screw has its own optimum h_{onm} chip thickness value when coal breaking with cutters takes minimum specific power consumption (Pozin et al. 1984).

2 PROBLEM FORMULATION

Such an approach is true for coal shearers having increased loading capacity of end organs. When seams are thin, then end organs of coal shearers are of less loading capacity; if there is no balance between coal volume within working space and coal volume being unloaded from it, screw gumming happens which results in considerable in-

crease in the coal shearer specific power consumption. Thus, to implement an approach of parametrical optimization, operation within thin seams should have optimum value of chip thickness being adequate to the screw operation at the gumming boundary with its maximum loading efficiency. To do that it is proposed to apply the developed numerical criterion of screw gumming (Tkachov et al. 2013).

The aim of the paper is to modify criterion of coal shearer screw gumming to apply it at various rotation velocities of end organs as well as using it to develop algorithm of smart analysis of coal shearer power signal to stabilize loading efficiency of a screw.

3 RESEARCH TOOL

The research applies "face – screw – electromotor of cutting drive" simulation model basing upon known calculation methods for power characteristics of coal shearer (Pozin et al. 1984), mathematical description of power transformation in electric drive (Starikov et al. 1981), results of numerous research of coal shearer static dynamics to simulate loads within end organ (Dokukin et al. 1978) as well as method to calculate power characteristics of coal transportation and loading with the help of a screw (Boiko 2002). The simulation model is described in (Tkachov et al. 2013) work.

4 THE RESEARCH RESULTS

Analyze time behaviour of increased power value of cutting drive electromotor at the moment of coal power loading relative to quarters of screw rotation period in terms of low and high velocities of end organ rotation. To do that we will specify only the component of resistance moment within end organ connected with the process of coal transportation and loading with the help of a screw.

In Figures 1 and 2 "0" value of impulse signal (dot line) corresponds to the second and fourth quarters of a screw rotation period; as for "70" value it corresponds to the first and third quarters (Tkachov et al. 2013).

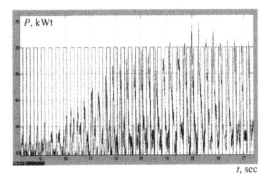

Figure 1. Time variations in cutting drive electromotor power P in terms of coal loading with the help of a screw when rotation velocity is 105.5 rot/min.

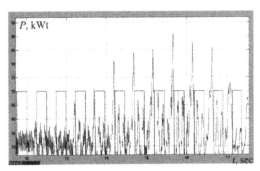

Figure 2. Time variations in cutting drive electromotor power P in terms of coal loading with the help of a screw when rotation velocity is 61 rot/min.

Figures 1 and 2 show that despite screw rotational velocities (both high and low) regularity of increased power value of cutting drive electromotor is available only at the end of the second and fourth quarters of a screw rotation period as the initial stage of gumming process. The fact is the condition for efficient use of a screw gumming criterion for those cutting velocities which differ from nominal

one; however, if so, numerical criterion of a screw gumming will be determined as follows:

$$K_i = T^{-1} \cdot \sum_{x=i-T}^{i} \begin{cases} \dfrac{P_{x-0,25\frac{60}{N}}}{P_x}, & if \left[\dfrac{x}{0,25\cdot\dfrac{60}{N}}\right] - (2\cdot k + 1); \\[4ex] \dfrac{P_x}{P_{x-0,25\frac{60}{N}}}, & if \left[\dfrac{x}{0,25\cdot\dfrac{60}{N}}\right] - (2\cdot k), \end{cases} \tag{2}$$

where T – the number of instantaneous power consumed by cutting drive electric motor to be used for moving averaging; N – a screw rotation velocity, rot/min; k – certain whole number to specify both pair and no-pair numbers; i – number of current instantaneous value of K_i criterion; P_x – instantaneous value of cutting drive electromotor power in terms of averaging, kW; $P_{x-(0,25\cdot60/N)}$ – instantaneous power being time-shift relative to P_x power value per a quarter of a screw rotation period, kW.

Earlier paper by (Tkachov et al. 2013) considered coal shearer control when loading capacity of a screw experiencing maximum stabilization at the expense of feed velocity control in terms of permanent cutting velocity. However, when coal shearer feed velocity is controlled it is not correct to tell about one critical value of feed velocity corresponding to a coal shearer operation at the boundary of a screw gumming. A condition of a screw location at the boundary of gumming in regards to controlled velocities of cutting and feed should be considered either as critical ratio of the given velocities or as critical value of a chip maximum thickness; if one of them is exceeded then balance between coal volume within a screw working space and coal volume being removed from it is disrupted. Thus, when cutting velocity is alternative then the task of coal shearer control on the criterion of maximum use of a screw loading capacity without its gumming is to determine and then maintain gumming critical maximum chip thickness.

To specify certain values of a chip maximum thickness it is quite enough to control the only velocity (either feed velocity or cutting one); the second velocity may be controlled using another criterion. It is better to use feed velocity as controlled value for extra control criterion as it determines both coal shearer efficiency and workload of cutting drive electromotor. Thus, if cutting velocity of coal shearers operating within thin seams is controlled, then optimum solution is to develop a system of automated control for cutting velocity using criterion of maximum use of a screw loading capacity without its gumming where feed velocity is the input parameter. In this context maximum chip thickness is controlled value and cutting velocity is controlling one. If so, then feed velocity is considered as disturbing action varying maximum chip thickness. It means that the problem of control can be now described as both determination and maintenance of gumming critical maximum chip thickness for various feed velocities of coal shearer at the expense of cutting velocity regulation.

Using "face – screw – electromotor of cutting drive" simulation model, identify critical cutting velocity values as for N_{cr} screw gumming for various values of V feed velocity within 1.6 to 5.1 m/min range (Figure 3).

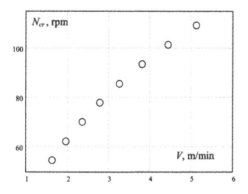

Figure 3. Dependence of coal shearer cutting velocity being critical on a screw gumming N_{cr} upon its feed velocity V.

The Figure 3 demonstrates that dependence between cutting velocity being critical on gumming and feed velocity is non-linear as the volume of unloaded coal corresponding to maximum screw load capacity decreases at reducing velocity of its rotation and visa versa. It means that maximum chip thickness being critical on chip thickness is a constant varying proportionally to a screw rotation velocity according to non-linear law (Figure 4).

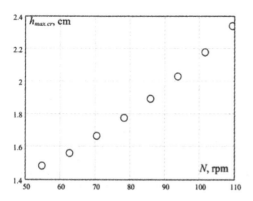

Figure 4. Dependence of maximum chip thickness being critical on gumming $h_{max.cr}$ upon screw rotation velocity N.

However, direct and practically linear dependence is also possible as for maximum chip thickness being critical on gumming (Figure 5).

Thus, procedure concerning determination of maximum chip thickness being critical on gumming is quite enough for two values of feed velocities with following approximation of the dependence using straight line equation.

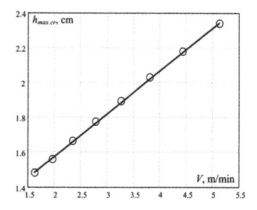

Figure 5. Dependence of maximum chip thickness being critical on gumming $h_{max.cr}$ upon coal shearer feed velocity V.

Further you can find the approximation result of maximum chip thickness being critical on gumming upon feed velocity according to two feed velocity values – 1.6 and 5.1 m/min (Figure 6). The obtained straight line equation is as follows:

$$h_{max.cr}(V) = 1.086 + 0.245 \cdot V .\qquad(3)$$

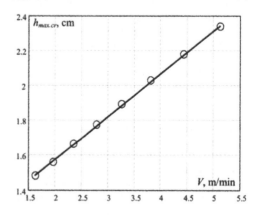

Figure 6. Result of dependence approximation of maximum chip thickness being critical on gumming $h_{max.cr}$ upon feed velocity V: graph of straight line equation.

As Figure 7 shows relative σ approximation error is not more than 0.6%.

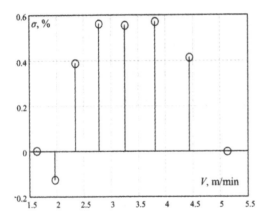

Figure 7. Result of dependence approximation of maximum chip thickness being critical on gumming $h_{max.cr}$ upon feed velocity V: relative approximation error.

As paper (Tkachov et al. 2013) demonstrates, to fix gumming process under the conditions of calculation experiment with probability equal to a unit, numerical criterion of screw gumming should be out of the variation range of $0.9\bar{K} - 1.1\bar{K}$ values where $\bar{K}$ is the value of numerical criterion averaged per long time period. To meet the condition, dependence between numerical criterion relative to variation value of its average value and increment of circulating coal volume per screw rotation has been determined; search of algorithm to determine screw feed velocity being critical on gumming has been proposed.

It should be noted that in (1) equation parameters of moving averaging of power ratios have been calculated basing upon spectral density of high-frequency component of cutting drive electromotor power (Pozin et al. 1984). That is why making specified confidence interval permanent in which gumming criterion will be with the probability of "1" under the conditions of calculation experiment in terms of screw gumming non-availability one cannot change the pitch of power measurement as well as time of moving averaging.

If control is performed on the basis of chip maximum thickness then to pass to a screw gumming mode to identify it providing permissible increment of circulating coal volume per screw rotation one should know dependence between maximum relative deviation of gumming criterion and increment of chip maximum thickness. Moreover, taking into account constant time of moving averaging the dependence should be obtained for various segments of cutting velocity as at high cutting velocities screw performs more rotations having similar increment of circulating coal per the same

period of moving averaging; circulating coal takes less time to be accumulated. According to (Tkachov et al. 2013) paper, screw gumming criterion is sensitive to accumulation velocity of circulating coal volume within working space of a screw; when permissible velocity value is exceeded it cannot identify screw gumming due to extremely short zone of redistribution of power and non-power coal load intervals.

To obtain dependence between maximum relative deviation of gumming criterion and chip maximum thickness increment as well as increment of circulating coal volume per second use "face – screw – electromotor of cutting drive" simulation model.

Note that maximum permissible increment of circulating coal volume per second is $0.72 \cdot 10^{-3}$ m^3 as paper (Tkachov et al. 2013) has identified. The Figure 8 shows that maximum permissible increment of circulating coal volume per second being $0.72 \cdot 10^{-3}$ m^3 is possible when maximum chip thickness exceeds its critical on gumming value by 0.155 cm (Figure 9).

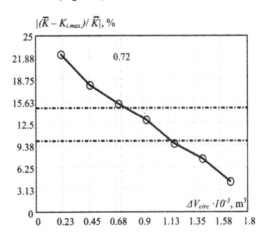

Figure 8. Dependence of maximum relative deviation of gumming criterion on increment of circulating coal volume per second ΔV_{circ} when feed velocity is 2.77 m/min and screw rotation velocity is 78 rot/min.

For velocities of feed and cutting being critical on gumming they are 2.77 m/min and 78 rot/min respectively to be adequate to cutting velocity decrease by 3.2 rot/min value.

The Figure 10 illustrates that in terms of reduced screw rotation velocity transfer maximum permissible increment of circulating coal volume per second being $0.72 \cdot 10^{-3}$ m^3 is observed at practically similar value of maximum relative deviation of gumming criterion from 14.7% average value.

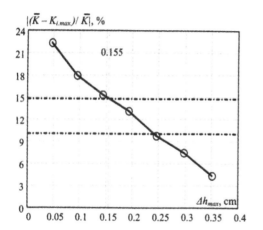

Figure 9. Dependence of maximum relative deviation of gumming criterion on increment of chip maximum thickness Δh_{max} when feed velocity is 2.77 m/min and screw rotation velocity is 78 rot/min.

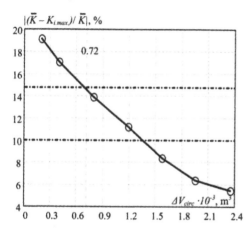

Figure 10. Dependence of maximum relative deviation of gumming criterion on increment of circulating coal volume per second ΔV_{circ} when feed velocity is 1.62 m/min and screw rotation velocity is 54.6 rot/min.

It means that time redistribution behaviour of power and non-power loading of coal using screw remains practically unchangeable; the only variation is in alternation frequency of the intervals and their duration.

However, as Figure 11 shows increment of circulating coal volume per second being $0.72 \cdot 10^{-3}$ m³ now provides deviation of chip maximum thickness from its critical on gumming value by 0.2 cm. That is due to 1.43 times decrease in screw rotation frequency to maintain the same velocity of circulating coal volume accumulation within working space of a screw, deviation of maximum chip thickness from its value being critical on gumming should be increased by 1.29 times.

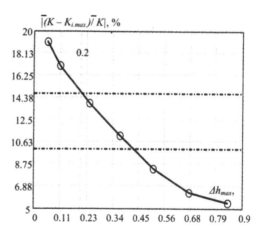

Figure 11. Dependence of maximum relative deviation of gumming criterion on increment of chip maximum thickness Δh_{max} when feed velocity is 1.62 m/min and screw rotation velocity is 54.6 rot/min.

Disproportion between changes in screw rotation frequency and value of maximum chip thickness deviation takes place due to decrease in screw loading capacity. For velocities of feed and cutting being critical on gumming (1.62 m/min and 54.6 rot/min respectively) 3.4 rot/min decrease in cutting velocity corresponds to deviation of maximum chip thickness by 0.2 cm.

Identify permissible deviation of maximum chip thickness from its value being critical on gumming for increased velocity of screw rotation (Figure 12).

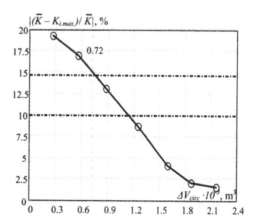

Figure 12. Dependence of maximum relative deviation of gumming criterion on increment of circulating coal volume per second ΔV_{circ} when feed velocity is 5.11 m/min and screw rotation velocity is 109.2 rot/min.

The Figure 12 shows that transfer to increased velocity of screw rotation cannot vary maximum deviation of gumming criterion in terms of maximum per-

469

missible increment of circulating coal volume per second ($0.72 \cdot 10^{-3}$ m³). However, as Figure 13 demonstrates, such an increment of circulating coal volume gives deviation of maximum chip thickness from its value being critical on gumming by 0.117 cm.

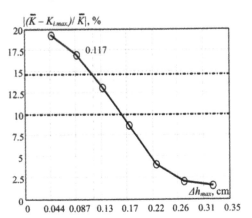

Figure 13. Dependence of maximum relative deviation of gumming criterion on increment of chip maximum thickness Δh_{max} when feed velocity is 5.11 m/min and screw rotation velocity is 109.2 rot/min.

That is due to 1.4 times increase in screw rotation velocity (to compare with 78 rot/min nominal value) maintenance of the same velocity of circulating coal volume accumulation in working space of a screw deviation of maximum chip thickness from its value being critical on gumming should be decreased by 1.32 times. Disproportion between changes in screw rotation frequency and value of maximum chip thickness deviation takes place due to increase in screw loading capacity. For velocities of feed and cutting being critical on gumming (5.11 m/min and 109.2 rot/min respectively) 2.6 rot/min decrease in cutting velocity corresponds to deviation of maximum chip thickness by 0.117 cm.

Identify permissible deviations of maximum chip thickness on its value being critical on gumming for other feed velocities.

The Figure 14 helps conclude that dependence between permissible deviations of maximum chip thickness on its value being critical on gumming and feed velocity is inverse and nonlinear due to dissimilar loading capacity of a screw in terms of various velocities of its rotation. However, to avoid development of the dependence under real conditions of coal shearer operation it is possible to assume boundary deviation of maximum chip thickness its value being critical on gumming when velocity of circulating coal volume accumulation will not be more than maximum permissible value with-

in the whole range of feed velocity variations. In our case, the boundary value of maximum chip thickness is 0.117 cm (dash-and-dot line in Figure 14).

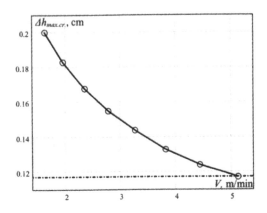

Figure 14. Dependence between permissible deviation of maximum chip thickness on its value being critical on gumming $\Delta h_{max.cr}$ and feed velocity V.

It is possible to single out two operation modes of automated control for coal shearer cutting velocity: a mode to search maximum chip thickness on its value being critical on gumming for various feed velocities and a mode of its tracing in a function of feed velocity variations. Consider both operation modes on the basis of the scheme of cutting velocity control algorithm represented in Figure 15.

A system of cutting velocity automated control always begins its procedure with operation mode one after V_{giv} feedrate-control setting was entered in unit 2; it requires determination of maximum chip thickness $h_{max.cr}$ being critical on gumming.

Moreover, cutting velocity setting N_{giv} is also calculated in unit 2 basing upon feedrate-control setting V_{giv} and specified T average value of maximum chip thickness $h_{max.giv}$

$$N_{giv} = \frac{V_{giv}}{h_{max.giv}} \cdot \frac{100}{n}, \qquad (4)$$

where n is the number of cutters within cutting line.

Initial average value of maximum chip thickness involves loading capacity of a screw mounted on a coal shearer; besides it should correspond to non-power mode of coal loading.

Research using "face – screw – electromotor of cutting drive" simulation mode has shown that non-power coal loading is typical for double-inlet screws which diameter is 0.7 m and above at 1m/min feed velocity and 78 rot/min screw rotation velocity (conveyor edge height is 0.228 m). Maximum chip thickness of 1.3 m is adequate to the data of feed and cutting velocities.

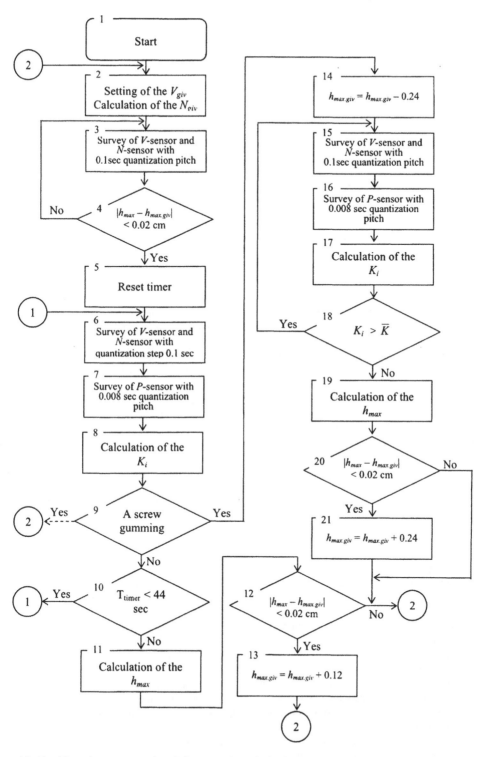

Figure 15. Algorithm scheme to control coal shearer cutting velocity in terms of parametrical stabilization of maximum chip thickness being critical on gumming.

Feed and cutting velocity scanning takes place in unit *3*; pitch is 100 milliseconds to check if the value of achieved maximum chip thickness corresponds to the specified one in unit *4*.

It should be noted that transition to a new value of maximum chip thickness depends on the direction of changes in feedrate-control setting. If feedrate-control setting decreases then coal shearer feed rate decreases first being followed by cutting velocity decrease. If feedrate-control setting increases, then cutting velocity also increases being followed by feed velocity increase. Thus, transition to new maximum chip thickness displays significant margin of a screw loading capacity; circulating coal is totally removed from working space of operating member.

As working conditions of coal shearer are constant, margin on circulating coal volume increase per screw rotation is kept invariable (Tkachov et al. 2013) due to its uncertainty at the initial stage of gumming and due to unbalance of real maximum chip thickness and the given one – 0.9% of screw working volume ($0.22 \cdot 10^{-3}$ m^3).

To involve actual average chip thickness unbalance to the specified value, circulating coal volume increase per screw rotation (0.3% of total screw capacity) is margined. Calculation experiment has determined that if feed velocity is 2.77 m/min and cutting velocity is 78 rot/min, the given increase in circulating coal volume per screw rotation corresponds to 0.02 cm deviation in chip thickness from the value being critical on gumming. Consequently, difference between actual average value of chip thickness and the given one should not be more than 0.02 cm. Otherwise, the beginning of screw gumming process will not be recorded.

When maximum chip thickness reaches the given value, timer zeroing and its start take place in unit *5*. Feed and cutting velocity scanning takes place in unit *6* with the pitch of 100 milliseconds and instant values recording takes place to check if the value of achieved maximum chip thickness corresponds to the specified one again. Cutting drive electromotor power sensor is scanned in unit *7* with the time pitch of 8 milliseconds.

Then criterion of screw gumming is calculated according to formula (2); a condition for a screw gumming reason is checked in unit *9*. If screw gumming is not available then timer records are checked in unit *10*. If 44 seconds have not passed since the moment of timer reset, then sensors of feed and cutting velocities are scanned in unit *6* and the cycle is repeated. If the condition in unit *10* is not met, then average value of actual maximum chip per time of gumming criterion analysis is calculated in unit *11*.

Conformance of average actual maximum chip thickness with the given value is checked in unit *12*. If deviation of average actual maximum chip thickness from the given one according to module is not more than 0.02 cm, then the given value of maximum chip thickness increases by 0.12 cm in unit *13*. To provide changes in maximum chip thickness by the given value following formula to calculate new cutting velocity setting is used:

$$N_{giv} = \frac{V_{giv} \cdot 100}{\dfrac{V_{giv} \cdot 100}{N'_{giv}} \pm 012}, \qquad (5)$$

where N'_{giv} – previous value of cutting velocity setting.

Then timer is reset in unit *5*; during following 44 seconds the above cyclic operation performance will take place. If the condition in unit *12* is not met, then the order of operations to be performed will be the same as if the condition in unit *12* is met. However, the given value of maximum chip thickness will remain the same. This approach will help to avoid cases when a system of automated control cannot record screw gumming due to error while stabilizing velocities of feed and cutting as a result of external disturbance effect.

The described closed operation cycle of automated system to control cutting velocity will take place until screw gumming happens while determining values of maximum chip thickness. After that automated control system will "remember" values of maximum chip thickness being critical on gumming for current feed velocity; transition to unit *2* will take place and feed velocity setting will be changed. Thus, as a result we obtain set of values of maximum chip thickness being critical on gumming for various feed velocities. However, as it has been shown before the determination of procedure of maximum chip thickness being critical on gumming is quite enough for two feed velocities only with following approximation of experimental dependence by straight line equation.

After obtaining dependences of maximum chip thickness being critical on gumming upon feed velocity, automated system to control cutting velocity transits into the mode of tracing maximum chip thickness being critical on gumming in the function of feed velocity change.

As in the previous mode, first feed and cutting velocity settings are specified in unit *2*; however, this time V_{giv} feed velocity setting is specified according to the taken control law according to random criterion and N_{giv} cutting velocity setting is calculated on the basis of obtained dependence of maximum chip thickness being critical on gumming upon feed velocity:

$$N_{giv} = \frac{V_{giv}}{1,086 + 0,245 \cdot V_{giv}} \cdot \frac{100}{n}. \quad (6)$$

Further operations in units $3 - 13$ are performed as it is described before; however, in this case screw gumming can be recorded within the second cycle.

If screw is not gummed in $11 - 20$ seconds from the moment of last timer reset, then the condition of non-available screw gumming in unit 9 will stop to be met. In this context 0.24 cm decrease in the given maximum chip thickness will take place in unit 14. Decrease in the given value of maximum chip thickness per double iteration pitch can be explained by the fact that in terms of previous given value being adequate to screw operation within gumming boundary there is no margin on screw loading capacity.

In unit 15 sensors of feed and cutting velocities are scanned to form velocity samplings of their instant values to check correspondence of actual maximum chip thickness to the given value. In unit 16 sensor of cutting drive electromotor power with 8 mc time pitch is scanned to calculate gumming criterion. Then criterion of screw gumming is calculated in unit 17. Condition of average gumming value exceed to compare with the current value is checked in unit 18. In such a way the quantity of coal circulating within working space of screw is controlled (Tkachov et al. 2013). Meeting the condition in unit 18 means that the level of coal circulating in screw working space is high; if so, then the system of automated control does not take any actions controlling only gumming criterion and gathering information on actual feed and cutting velocities. When circulating coal volume is decreased down to $10 - 14\%$ of screw working volume, the condition in unit 18 will cease to be met. Then, in unit 19 actual average value of maximum chip thickness is calculated and deviation value of the given one from the setting is checked in unit 20. If the difference between actual average maximum chip thickness and the given one is less than 0.02 cm, in unit 21 the given value of maximum chip thickness increases by double iteration pitch. Then, analysis of gumming screw criterion for maximum chip thickness is performed; screw gumming has been earlier recorded in terms of it. If error in terms of keeping the given value of maximum chip thickness is too large due to external disturbances (unmet

conditions in unit 20), then to prevent prolonged gumming decreased value of maximum chip thickness remains constant for the period of one more analysis of screw gumming criterion.

Thus, from time to time system of automated control for cutting velocity transfers coal shearer into the mode of gumming screw operation and check the set maximum chip thickness being critical on gumming to provide coal shearer operation with maximum screw loading capacity.

It should be noted that considered operation mode of cutting velocity automated control provides changes in feed velocity setting during the period of gumming criterion analysis; in this context transition to unit 2 takes place despite current operation. However, when feed velocity setting has been changed, screw gumming should experience at least one recording; thus, during the period of first analysis of gumming criterion (up to 20 seconds) feed velocity should remain constant. Hence, proposed algorithm to control cutting velocity according to a criterion of maximum use of screw loading capacity restricts to some extent the law of feed velocity change according to random control criterion.

Check correctness of coal shearer automated control system according to the proposed algorithm with the help of its simulation model which structural scheme is shown in Figure 16.

Two levels (upper and lower) can be singled out in a system of coal shearer automated control. Settings fro PID feed and cutting velocity regulators on the basis of smart analysis of coal shearer power signal are formed within upper level in ULCU (upper level control unit).

Lower level contains two contours of independent control for V feed velocity and N cutting velocity. Contour controlling feed velocity covers simulation models of coal shearer feed mechanism (FMCL), reducer and electromotor feed drive (EMFD) being the control object. Besides, simulation models of PID regulator, frequency converter (FC) as working component, and feed velocity sensor (V-sensor) are included into control contour.

Contour of cutting velocity control includes simulation models of reducer and cutting drive electromotor (EMCD) being the control object. Besides, simulation models of PID regulator, frequency converter (FC) as working component, and cutting velocity sensor (N-sensor) are included into control contour.

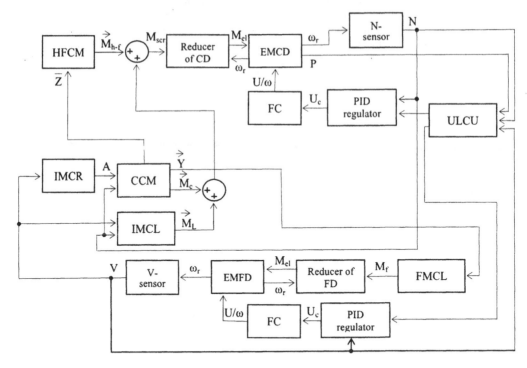

Figure 16. Structural scheme of simulation model for the system of coal shearer automated control.

To simulate load within shafts of electromotors of feed and cutting drives while coal cutting, simulation models of high frequency M_{h-f} (HFCM) and construction M_c (CCM) components of resistance moments within M_{scr} screw and M_f drive sprocket of feed mechanism are used. Low-frequency component of resistance moments within screw and drive sprocket are specified through simulation model of time change of A (IMCR) coal cutting resistance. Cutting force within single cut Z and feed force within screw Y are calculated on the basis of coal cutting resistance in CCM unit; they are required to determine resistance moments within shafts of feed and cutting drive motors.

To simulate load within shaft of cutting drive electromotor, simulation model of resistance moment component within end organ during the period of coal transportation and loading using M_L (IMCL) screw is also used.

Transients in reducers and electromotors of coal shearer feed and cutting drives are considered by means of simulation models of two-mass calculation circuit and mechanical diagrams and electro-mechanical energy transformation in asynchronous electric motor. Elastic moment in M_{el} reducer, angle velocity of ω_r rotor spinning and power con-

sumed by P electric motor are the initial parameters of the models.

Feed and cutting velocity sensors are specified by amplifying circuits due to fast character of transients taking place within these system components to compare with transients within other system components. Model of frequency transformation neglects transients as well as complex algorithm of powerful electric signal with alternative amplitude and frequency. Instead signals in a unit of frequency transformation are generated as follows:

$$u_\alpha^s = U_{max} \cdot cos(\omega_0 \cdot t), \, V, \qquad (7)$$

$$u_\beta^s = U_{max} \cdot sin(\omega_0 \cdot t), \, V, \qquad (8)$$

where u_α^s, u_β^s – real and imaginary components of phase voltage within U stator windings respectively, V; t – simulation time, sec; ω_0 – angle frequency of stator energizing, rad/sec. Moreover, U_{max} amplitude and ω_0 frequency of u_α^s, u_β^s voltages vary proportionally to U_c controlling signal coming to FC unit from PID regulators according to the law:

474

$$\frac{U_{max}}{\omega_0} = \frac{k_U \cdot U_c}{k_\omega \cdot U_c} = const , \qquad (9)$$

where k_U and k_ω – coefficients of proportionality between amplitude and frequency of u_α^s, u_β^s voltages, and U_c controlling signal.

The Figure 17 illustrates results of simulation of automated control system for cutting velocity in the mode of searching for maximum chip thickness being critical on gumming for 5.11 m/min feed velocity. The Figure 17, a, confirms that in the process of searching for maximum chip thickness being critical on gumming, feed velocity varies within

rather narrow range (5.1 to 2.12 m/min) with average value of 5.11 m/min.

Operation of cutting velocity automated control system is considered from the moment of analyzing criterion of screw gumming for 2.22 cm maximum chip thickness (4.5 to 24.5 sec) being adequate to 5.11 m/min feed velocity and 112 rot/min cutting velocity (Figure 17, b). According to Figure 17, e during the analysis, screw gumming criterion is within $0.9 \cdot \overline{K} - 1.1 \cdot \overline{K}$ range shown in dotted lines in Figure 17, e. It is the condition of screw gumming absence being proved by Figure 17, c which shows that within $4.5 - 24.5$ time interval circulating coal volume within screw working space is equal to zero.

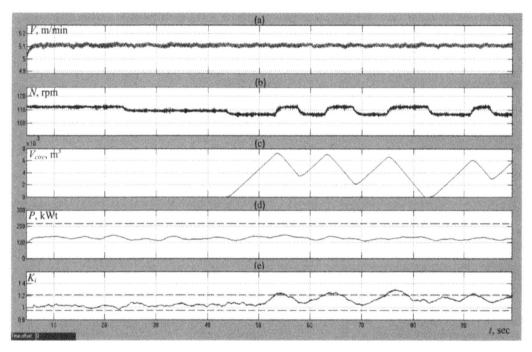

Figure 17. Time changes of: (a) – coal shearer feed velocity V; (b) – screw rotation velocity N (cutting velocity); (c) – coal circulating volume in screw working space V_{circ}; (d) – cutting drive electromotor power P; (e) – screw gumming criterion K_i

If value is 24.5 sec then maximum chip thickness increases up to 2.34 cm owing to decrease in screw rotation velocity from 112 down to 109 rot/min (Figure 17, b). According to Figures 17, c and 17, e the analysis of screw gumming criterion for the given chip thickness also demonstrated absence of screw gumming ($24.5 - 44.5$ sec). After that maximum chip thickness increases repeatedly taking into account limitations on permissible chip thickness increment up to 2.46 cm owing to decrease in screw rotation velocity from 109 down to 106 rot/min (Figure 17, b).

The Figure 17, c shows that transition to 106 rot/min screw rotation velocity originated screw gumming. Circulating coal which volume started increasing during the time appeared within working space of screw (Figure 17, c). Screw gumming criterion goes beyond the range of values $0.9 \cdot \overline{K} - 1.1 \cdot \overline{K}$ changes by 53.2 sec, when circulating coal volume reaches $7.2 \cdot 10^{-3}$ m^3 (28% of screw working volume). After screw gumming has been recorded, the system of circulating coal removal increases screw rotation velocity from 106 rot/min up to 112 rot/min, that is by double iteration pitch of maximum chip thickness (Figure 17, b).

Thus, system of automated control makes conclusions as for the fact that maximum chip thickness being critical on gumming for 5.11 m/min feed rate is equal to 2.34 cm; further it is traced by means of periodic decrease of screw rotation velocity for its gumming beginning. The Figure 17, c shows that in this context circulating coal volume is not more than $8 \cdot 10^{-3}$ m^3 (31% of screw working volume) and its average value is equal to $4.5 \cdot 10^{-3}$ m^3 (17.6% of screw working volume).

Note that time behaviour of power consumed by cutting drive electromotor after circulating coal has entered screw working space remains practically unchangeable (Figure 17, d) due to insignificant gumming process: automated control system records early screw gumming. Both before circulating coal entering (up to 44.5 sec) and after the process cutting drive electromotor power varies within 108 – 146 kW; average value is 124 kW (Figure 17, d).

The Figure 18 illustrates simulation results of automated control system for cutting velocity in a mode of searching for maximum chip thickness being critical on gumming in the function of feed velocity change.

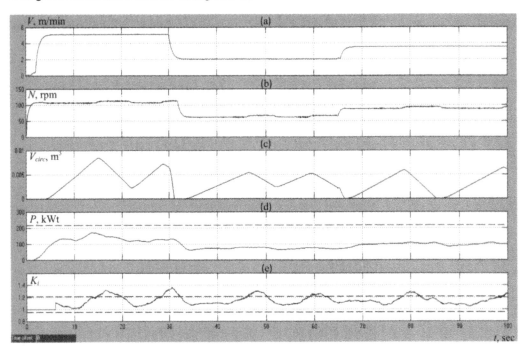

Figure 18. Time changes of: (a) – coal shearer feed velocity V; (b) – screw rotation velocity N (cutting velocity); (c) – coal circulating volume in screw working space V_{circ}; (d) – cutting drive electromotor power P; (e) – screw gumming criterion K_i.

Random rule of feed velocity time change is specified during the simulation process. Up to 30 sec feed velocity is stabilized at the level of 5.1 m/min; from 30 to 65 sec it stabilized at the level of 2 m/min and after 65 sec its level is 3.5 m/min (Figure 18, a). The Figure 18, b shows that cutting velocity changes practically synchronously along with feed velocity having minor retard when feed velocity decreases and advance when it increases. The Figure 18, c demonstrates that each transition to new feed velocity results in coal shearer operation with minor screw gumming and its following record by automated control system. That confirms correctness of cutting velocities setting calculation for each of three feed velocities from the viewpoint of maximum loading capacity for screw operation without prolonged gumming. Besides Figures 18, c and 18, e can help conclude that tracing of maximum chip thickness being critical on gumming by automated control system is correct. For each of three feed velocities screw gumming was twice recorded by a system of automated control; gumming criterion went beyond the range of such values as $0.9 \cdot \overline{K} - 1.1 \cdot \overline{K}$ change (Figure 18, e). The Figure 18, b illustrates that each time when screw gumming was recorded, automated control system accelerated cutting velocity to remove excess of circulating coal from screw working space with following cutting velocity decrease to maintain the mode of minor gumming. In this context circulating coal volume varies from 0 (moments of transition to new feed velocity when a screw has significant loading capacity margin) up to $8.4 \cdot 10^{-3}$ m³ when average value is $3.7 \cdot 10^{-3}$ m³ (14.5% of screw working space).

It should be noted that feed velocity change results in significant shift of the range of power consumed by cutting drive electromotor (Figure 18, d). However, it depends on coal cutting process with its cutters rather that the process of coal transportation and loading using screw. When feed velocity increases (Figure 6) maximum chip thickness being critical on gumming increases as automated control system maintains it. That results in cutting force increase on cutters. Besides when cutting velocity increases, path of cutters increases as well during similar time period which also factors into cutting forces increase. That very time when feed velocity remains unchangeable, range of power changes also remains constant (Figure 18, d). It means that screw gumming process is insignificant which cannot result in cutting drive electromotor power increase.

5 CONCLUSIONS

Under conditions of thin seams and alternative cutting velocity of coal shearers technique of parametric optimization is promising. The technique means that cutting drive electromotor power is maintained at specified level by means of feed velocity control; as for optimum chip thickness value corresponding to screw operation at the boundary of gumming with its maximum loading capacity – by means of cutting velocity control.

Both low and high velocities of screw rotation preserve regularity of cutting drive electromotor power increased value only at the end of second and fourth quarters of screw rotation at the initial stage of gumming process. It is the condition for efficient application of proposed screw gumming criterion for coal shearer cutting velocities differing from nominal one. However, time shift between measured powers of cutting drive electromotor corresponding to different angle intervals of screw rotation should be now alternative value varying inversely proportionally to cutting velocity.

Maximum chip thickness being critical on gumming is not a constant value varying proportionally to the screw rotation velocity according to non-linear law as its rotation velocity determines screw loading capacity. However, as for feed velocity of maximum chip thickness being critical on gumming direct and practically linear dependence is observed.

Dependence between permissible deviation of maximum chip thickness on its value being critical on gumming and feed velocity is inverse and non-linear due to different screw loading capacities in terms of its various rotational velocities. However to avoid the dependence building under real conditions of coal shearer operation one may take boundary deviation of maximum chip thickness on its value being critical on gumming when circulating coal volume accumulation is not exceed maximum permissible value within the whole considered range of feed velocity changes. This boundary value of maximum chip thickness deviation corresponds to maximum velocity of screw rotation.

Simulation modeling of cutting velocity automated control system both in a mode of searching maximum chip thickness being critical on gumming for various feed velocities and in a mode of its tracing in a function of feed velocity change confirms efficiency of the system operation using proposed algorithm of coal shearer power signal smart analysis.

REFERENCES

Boiko, N.G. 2002. *Coal loading by shearers*. Donetsk: DonNTU: 157.
Bublikov A., Tkachov V., Isakova M. 2013. *Control automation of shearers in term of auger gumming criterion*. Energy efficiency improvement of geotechnical systems – Dnipropetrovsk : Taylor & Francis Group: 137-145.
Dokukin, A.V., Krasnikov, Ju.D. & Hurgin, Z.Ja. 1978. *Statistical dynamics of mining machines*. Moscow: Mashinostroenie: 239.
Pozin, E.Z., Melamed, V.Z. & Ton, V.V. 1984. *Destruction of coal by stoping machines*. Moscow: Nedra: 288.
Starikov, B.Ya., Azarkh, V.L. & Rabinovich, Z.M. 1981. *Asynchronous drive of shearers*. Moscow: Nedra: 288.

Theoretical and Practical Solutions of Mineral Resources Mining – Pivnyak, Bondarenko & Kovalevska (eds)
© 2015 Taylor & Francis Group, London, ISBN: 978-1-138-02883-8

Optimization of the process of natural gas production stimulation from low permeable reservoirs

O. Kondrat & N. Hedzyk
Ivano-Frankivsk National Technical University of Oil and Gas, Ivano-Frankivsk, Ukraine

ABSTRACT: This paper presents an analysis of literary publications about improving the technology of hydraulic fracturing in unconventional, low permeable natural gas fields. Directions for further research to improve the technical and technological efficiency of hydraulic fracturing were analyzed. Using 3D hydrodynamic computer simulation the optimal parameters of horizontal wells and hydraulic fractures in low permeable reservoirs based on their technical and economic efficiency were estimated. Net present value (NPV) of cash income, depending on the parameters of the well, hydraulic fractures and reservoir permeability was evaluated. Recommendations for horizontal wells drilling and hydraulic fracturing in low permeable reservoirs were grounded. The results of the research were compared with international experience and the correct conclusions were drawn.

1 INTRODUCTION

In order to find new deposits of natural hydrocarbons it is necessary to increase the depth of oil and gas wells drilling. With the drilling depths increase the percentage of low permeable reservoirs will increase and it is not excepted that all productive layers will be low permeable. According to the field experience gas production from low permeable reservoirs using vertical wells is economically unprofitable (Arogundade et al. 2006). The wells are characterized by low initial flow rates which decrease rapidly during the field development. New, efficient processes design will enable to recover natural gas from the fields with low permeable reservoirs which will give rise to fuller development of hydrocarbon resources in deposits of Ukraine and ensuring its energy independence.

Field data analysis shows that natural gas production from low permeable reservoirs with economically profitable production rate can be achieved only by horizontal wells infill drilling with further gas flow stimulation (Cipolla 2009). Currently, the most effective method of enhanced gas recovery is multistage hydraulic fracturing (HF) of the formation. The main parameters that affect on the well performance using this technology is the length of the horizontal wellbore, the number and size of perforation intervals, the number of hydraulic fractures, their length, density and permeability.

2 LITERATURE REVIEW

Mayerhofer et al. (2010) suggested considering the volume of the reservoir being limited by transverse fractures as simulated (drained) reservoir volume (SRV). The authors identified the main factors that influence on the size of SRV: formation thickness, the length of the horizontal section of the wellbore, the distribution of stresses in the layer, the presence of natural fractures. It was determined that in order to increase the stimulated reservoir volume, which actually is the wells draining volume, it is worth considering the possibility of increase fractures density, wells perforating parameters, wellbore horizontal section orientation, open hole well completion and others.

Pope et al. (2010) investigated the effect of hydraulic fractures density, length and permeability on operating parameters of the wells based on the results of 3D computer modeling and field data from 56 wells in Haynesville shale gas field. Productive deposits are so-called black organic reach shale, with clay fractions containing less than 40%. Total organic content ranges from 3 to 5%, and the coefficient of thermal maturity $1.3 - 2.4$, which corresponds to the condition of "dry" gas formation. About 80% of the initial gas reserves are in the free state in the pore space, and the rest (20%) is adsorbed on the rock surface. The thickness of the formation $50 - 120$ m, reservoir permeability $5 - 800$ mD, open porosity coefficient $6 - 12\%$, initial water saturation $25 - 35\%$. For a detailed analysis of production data was chosed the well which was drilled from the beginning of the field development, with horizontal section length of 900 m, with 10 multistage hydraulic fracturing, the distance between them is 90 m. Using hydrodynamic simulation in the model of the part of Haynesville deposit with length of 1000 m and a width of 320 m, the authors investigates production history matching of the area in order to determine the real parameters of hydraulic fractures. For history matching two variants were considered. For the first variant for each interval of perforation two hydraulic fractures were formed (20 fractures), under the second - one fracture per each interval (10 fractures). Fracture length for each option varied from 500 to 0 m. Hydrodynamic simulation results showed that both variants were correlated with field data.

Cipolla (2009) considered the optimization of hydraulic fracturing parameters in horizontal wells by the example of the Barnett shale deposits. This paper analyzes the influence of fracture conductivity and their density on the dynamics of cumulative gas production. According to the simulation results it was estimated that with the increase of fractures conductivity the drainage area of the wells increases and cumulative gas production rises. With the decrease of the distance between fractures in the conditions of the uniform fractures conductivity the cumulative gas production from the deposit increases. In this connection, special attention should be paid to microseismic monitoring during hydraulic fracturing. These microseismic monitoring data will enable fuller consideration of the complex hydraulic fractures system and to predict the field development more accurately.

Work of Momeni at al. (2011) concerned with the features of practically dry gas extraction (C5 + fractions content of about 2%) from low permeable reservoirs and its main task was to study the influence of hydraulic fractures and the Darcy law violations for calculations of production well characteristics. The gas reservoir was simulated in the software CMG-GEM by two ways: the first one – the presence of natural fractures in the reservoir using a dual porosity model, the second one – without them using a single porosity model. Both variants were calculated with the presence of hydraulic fractures and without them. The porosity of the model of 2 – 13%, horizontal permeability – 0.12 – 1 mD, vertical permeability – 0.06 – 0.5 mD. Hydraulic fractures permeability is equal to 300 mD. The total volume of natural fractures in the reservoir was about 1% of the total pore volume. The simulation results showed that the hydraulic fractures have virtually no effect on wells productivity factor in the case of dual porosity

model (0.01%), but their presence significantly increases the well productivity (15%) in case of the calculation using single porosity model. The Darcy law violation has a significant impact on wells productivity factor, especially in the case of dual porosity model.

Bo Song at al. (2011) examines the economic efficiency of the horizontal wells drilling with the transverse hydraulic fractures. In this paper, the authors single out stimulated reservoir volume, as the area of the reservoir, which is not only limited by horizontal wellbore and hydraulic fractures, and protrudes at ½ of the distance between the fractures at the edges of the wellbore and at ¼ of horizontal distance between fractures at the ends of the fractures. In the initial period of time the gas flows independently for each linear fracture (fracture storage) then appears pseudolinear and pseudo pseudosteady gas inflow to the well, which is characterized by the gas inflow from drainage area of each fracture. With the further wells operation compound linear and radial inflow to the whole stimulated reservoir volume appears. To summarize, it is worth noting that in the design of well drilling and hydraulic fracturing in unconventional deposits not only operational parameters of the well, but also economic efficiency should be taken into account. This paper proposes a method for evaluating the effectiveness of horizontal wells with transverse fractures, according to which the profit from sales from each individual fracture should at least cover the costs for its formation and the drilling of necessary length of the borehole. If this condition is satisfied for a single fracture, it can also be implemented for its greater amount.

Clarkson et al. (2009), Freeman et al. (2009) and Al-Kobashi et al. (2006) also presented an analysis of the gas flow regimes to horizontal wells with multistage transverse hydraulic fractures for CBM production.

Using field data from Martsellus and Haynesvil shale deposits a study was conducted in order to determine the optimal wells spacing and their number for different matrix permeability ($5 \cdot 10^{-9}$, $500 \cdot 10^{-9}$ and $50 \cdot 10^{-9}$ D), fracture half-length (75, 150 and 275 m) and the distance between the fractures (12, 18, 25, 36 and 48 m) (Vivec Sahai at al. 2012). In the studies the following average parameters of Marcellus and Haynesville deposits were used: the depth of the reservoir – 3627 and 2095.5 m; formation thickness – 61 and 71.6 m; porosity – 8, 4.8%; initial reservoir pressure – 69 and 28 MPa; gas gravity – 0.593 and 0.57. The length of the horizontal sections of wells is the same for all versions – 1170 m. The experimental results show that when increasing the length of the hydraulic fracture the optimal number of wells in the area decreases, and with a decrease in the reservoir and fracture permeability – the number of wells increases. Also, the analysis of the research data show that the optimum distance between the fractures is 24 – 30 m at the fracture half-length of 150 m.

Khan & Al-Nakhli (2012) describe the specific features of gas production from tight sands. In particular, they include:

1. Identification of the most promising areas (sweet spot) in the productive layers which are areas with high porosity, permeability, increased reservoir pressure compared to the rest of the reservoir and the presence of natural fractures. In deposits developing gas flows from remote areas to the most promising zones where wells have been drilled. If there are no such zones, commercial gas production is not possible without hydraulic fracturing.

2. The increase of stimulated reservoir volume.

3. Low-permeable formation pollution and blocking the channels for gas inflow by drilling fluids, HF liquids etc.

3 PROBLEM FORMULATIONS

In Ukrainian and international publications widely considered is the issue of enhancing the technology of hydraulic fracturing by improving the technological properties of fracture fluids, proper selection of propant, openhole hydraulic fracturing, gas (energized) fracture fluids usage to provide a more completed removal of fluids to the surface etc. However, the issues of well parameters optimization and HF almost were not considered. There have been some attempts to establish such parameters for the specific fields, but a generalized problem solution of this nature, the results of which could be used for a number of deposits was not found. Therefore, taking into account the current state of the industry, the research of determining the optimal parameters of hydraulic fractures and wells is extremely important. The results of these studies will enable us to reduce the costs for carrying out hydraulic fracturing and to enhance current gas production and gas recovery factor from natural gas deposits with low permeable reservoirs.

4 THE BASIC MATERIALS OF THE RESEARCH

As a tool to assess the optimal parameters of horizontal wells with transverse hydraulic fractures was used hydrodynamic simulator ECLIPSE 300 in combination with geological simulator PETREL

which was donated and licensed to IFNTUOG by Schlumberger. The study was conducted in three stages. In the first stage horizontal well length was estimated, the second – the impact of fracture length in horizontal wells on technical and economic parameters were determined, and the third stage included the determination of the optimal fractures density.

These studies were conducted on the hydrodynamic model of the reservoir with a length of 3000 m and a width of 1000 m, which was divided into 21000 cells in the grid with fine grid in the near wellbore area. The depth of the formation is 2800 m, reservoir thickness – 100 m, reservoir temperature – 70°C, the initial reservoir pressure equals 300 bar. The reservoir permeability coefficient ranges from 0.01 to 10 mD, which allowed to assess the reservoir parameters impact on the efficiency of the horizontal wells drilling. As a base case in the studies the vertical well, which open 75% of the reservoir, was used. Other variants included well drilling with the length of its horizontal section from 50 to 2400 m. Calculations for all variants carried out for 20 years in case of the wells operation at a constant bottomhole pressure (200 bar). The technological efficiency was evaluated based on the analysis of cumulative gas production and recovery factor for different variants. The economic analysis was based on the determination of net present (discounted) value (NPV), which was conducted on the basis of national legal regulations acts and international publications (Hefley et al. 2011, Joshi 2011, Resolution 56 of the National Commission for State Energy and Public Utilities Regulation).

During the economic efficiency research some assumptions were adopted. In particular, the cost for vertical borehole drilling for all variants was considered the same. The cost of drilling of one meter of horizontal wellbore was adopted at 20000 UAH. Gas production costs were selected after the analysis of domestic and foreign publications (Hefley 2011, Joshi 2011), and in these studies they were 4100 UAH/thousand m^3. Sales price per gas volume unit was adopted in accordance with National Commission for State Energy and Public Utilities Regulation at 4874 UAH/thousand m^3. Also, economic analysis did not count the cost of wellhead equipment and vertical part of the well, because for all calculation variants it will be approximately equal and not affect the relative ratios. The cost of the horizontal section of casing and cementing was included in the drilling costs.

The results of hydrodynamic modeling for determining the optimal horizontal wellbore length is shown in Figures 1 and 2.

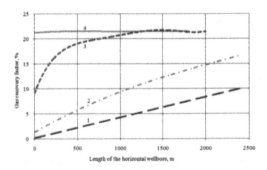

Figure 1. Graphic dependences of gas recovery factor on the horizontal well length for different values of reservoir permeability: 1 – 0.01 mD; 2 – 0.1 mD; 3 – 1 mD; 4 – 10 mD.

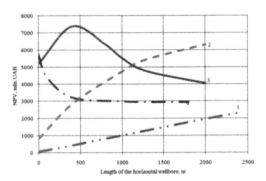

Figure 2. Graphic dependences of NPV on the horizontal well length for different values of reservoir permeability: 1 – 0.01 mD; 2 – 0.1 mD; 3 – 1 mD; 4 – 10 mD.

After obtained results analyzing it should be noted that it is not profitable to drill horizontal wells in layers with relatively high permeability (10 mD), because a significant gas recovery factor increases is not achieved. In case of horizontal wells drilling in these layers NPV will be lower than in case of vertical wells drilling. For layers with permeability of about 1 mD the optimum of the dependence of gas recovery factor on the length of the horizontal section is observed on the range at 800 – 1000 m. A further horizontal length increases does not provide a significant growth of recovery factor. NPV curve analysis for this variant testifies that the maximum net present value is achieved for the horizontal section length of 400 – 500 m. For low permeable layers (0.1 mD) the maximum gas recovery factor is observed at horizontal length of the well 3500 m. The maximum NPV is achieved at the length of horizontal section of 2400 m. For reservoirs with permeability of 0.01 mD the dependence of gas recovery factor on the length of the horizontal wellbore is almost straight-line and to determine its optimal value based only on simulation data is almost impossible. These results have also good correlation with the dependence for NPV.

The next stage of the study was the determining of the optimal hydraulic fractures length. The researches were conducted for vertical and horizontal wells with horizontal section length of 400 m in case of a fracture placement in the middle of perforated interval.

Simulated hydraulic fracture parameters are the following: fracture permeability 1 D, fracture fixed height 25 m, length of the fracture – from 0 (base case) to 600 m. It was made the assumption that fracture was formed perpendicular to the wellbore.

Figures 3 show the results of simulations for vertical well with hydraulic fractures.

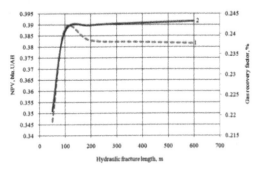

Figure 3. Graphical dependences of the NPV and gas recovery factor on the fracture length for vertical well for reservoir permeability of 0.01 mD: 1 – NPV; 2 – gas recovery factor.

In particular Figure 3 clearly shows the peak of economic efficiency, which corresponds to the length of the fracture of 120 m. Further fracture length increase will result in NPV curve decrease at almost constant gas recovery factor.

Figures 4 and 5 shows the results of simulations for horizontal well.

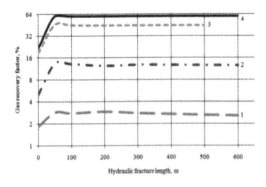

Figure 4. Graphical dependences of the gas recovery factor on the fracture length in horizontal well for the reservoir with different permeability: 1 – 0.01 mD; 2 – 0.1 mD; 3 – 1 mD; 4 – 10 mD.

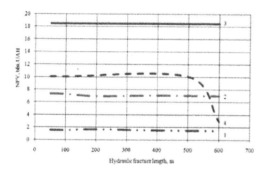

Figure 5. Graphical dependences of NPV on the fracture length in horizontal well for different reservoir permeability: 1 – 0.01 mD; 2 – 0.1 mD; 3 – 1 mD; 4 – 10 mD.

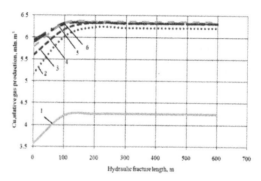

Figure 6. Graphical dependences of cumulative gas production on the fracture length for different distances between them: 1 – 15 m; 2 – 25 m; 3 – 45 m; 4 – 75 m; 5 – 120 m; 6 – 220 m.

Obtained results analysis showed that for the reservoir permeability of 0.01 mD the optimal fracture length is 150 – 200 m, for the reservoirs with the permeability of 0.1 mD – 70 m, for the permeability of 1 mD – 50 m. For the reservoirs with the permeability of 10 mD, sharp recovery factor increase is observed in the presence of hydraulic fracture length up to 50 m. Further fracture length increase does not affect the recovery factor. Such results correlate with the economic analysis, according to which the optimum fracture length for highly permeable reservoirs is 50 m. The further fracture length increase leads to NPV reduction.

In order to determine the optimum fracture density the hydrodynamic modeling was conducted with the following assumptions. In particular, for each variant the same number of fractures was adopted (6 fractures). The distance between fractures varied from 15 m to 220 m. The length of the horizontal wellbore and perforated interval also varied depending on the distance between fractures. For example, the maximum horizontal wellbore length was at a distance of 220 meters between fractures and equaled 1400 m, and the lowest one – in the absence of cracks (base case) and was equaled 200 m. The length of the hydraulic fractures also varied from 0 to 600 m.

In order to analyze the experimental results the graphical dependencies on Figure 6 – 8 were built. Best interdependence between different variants can be seen in Figure 8. This graph shows the absolute dependence of chain growth of cumulative gas production on the fracture length for different distances between them.

Figure 8 indicates graphical dependences for estimating the economic effectiveness of HF. In particular, the graph clearly shows that the highest economic effect is achieved when the distance between hydraulic fractures equals 25 m.

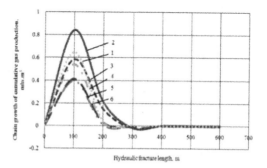

Figure 7. Absolute dependences of chain growth of cumulative gas production on the fracture length for different distances between them: 1 – 15 m; 2 – 25 m; 3 – 45 m; 4 – 75 m; 5 – 120 m; 6 – 220 m.

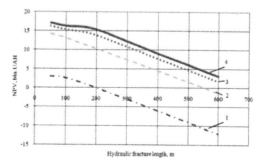

Figure 8. Graphical dependences of NPV on the fracture length for different distances between them: 1 – 15 m; 2 – 25 m; 3 – 45 m; 4 – 75 m.

Analyzing graphical dependences of the Figures 7 – 8 it can be concluded that with fracture length increasing increases the optimal distance between them. These results correlated with the publication of famous authors in this issue.

5 CONCLUSIONS

As a result of 3D computer modeling it was found that the optimal length of the horizontal wellbore is quite dependent on the reservoir permeability. For low permeable reservoirs the optimal horizontal wellbore length can be determined being exclusively based on the features of the specific conditions of the field and technical and economic indices of the company. For reservoirs with relatively high permeability (about 1 mD) the optimal horizontal wellbore length is about 1000 m, and its further increase does not give significant technical and economic effect. For reservoirs with lower permeability the dependence of recovery factor and NPV on the horizontal wellbore length is almost straight-line and horizontal wellbore increase lead to the growth of the recovery factor. These results are confirmed by the field experience of horizontal wells drilling in the unconventional natural gas fields of the United States, where oil and gas companies choose the length of the horizontal well not being based on technology reasonable parameters, but being based on the financial resources of company and mutual spatial arrangement of the wells and field boundaries.

In assessing the optimal transverse hydraulic fracture length it should be noted that it practically does not depend on the reservoir permeability and ranges within 100 – 200 m, and on average is 100 – 150 m. As an exceptions can be only high permeable reservoirs (~10 mD) for which transverse fracture up to 50 m is reasonable.

Further researches showed that the optimum distance between hydraulic fractures is 25 m. At this value of distance between fractures maximum NPV and cumulative gas production are observed. In lesser or greater distance between the fractures the technical and economic indecis are lower compared to the previous one.

Evaluation of technical and economic efficiency of the proposed solutions was made based on the assumptions that bottomhole formation zone is not polluted, and the reservoir is isotropic (porosity and permeability in all directions are the same).

To sum up it should be noted that for reservoirs with permeability of 1 – 0.01 mD it is recommended to drill horizontal section of the wells up to 1000 m long, and then treating it by multistage hydraulic fracturing with fracture length up to 150 m at a distance between them of 25 m.

The results do not contradict international experience of horizontal wells drilling and hydraulic fracturing, and in some cases confirm the results of known researches. However, in order to obtain the best results in the real Ukrainian fields specification of these parameters should be held on the basis of specific geological, technological and economic conditions.

REFERENCES

Al-Kobaisi, M., Ozkan, E., Kasogi, H. & Ramirez, B. 2006. *Pressure-transient analysis of horizontal wells with transverse, finite-conductivity fractures.* Alberta, Canada, June 13 – 15.

Arogundade, O., Sohrabi, M. & Holditch, S. 2006. *A review of recent developments and challenges in shale gas recovery.* Paper SPE 160869, JPT, Distinguished author series, June 2006: 86 – 94.

Bo Song, Michael, J. Economides, Christine Ehlig-Economides. 2011. *Design of multiple transverse fracture horizontal wells in shale gas reservoirs.* The Woodlands, Texas, USA, January 24 – 26.

Cipolla, C.L. 2009. *Modeling production and evaluating fracture performance in unconventional gas reservoirs.* Paper SPE 118536, JPT, Distinguished author series, September 2009: 84 – 90.

Clarkson, C.R., Jordan, C.L., Ilk, D. & Blasingame, T.A. 2009. *Production data analysis of fractured and horizontal CMB wells.* Charleston, West Virginia, September 23 – 25.

Freeman, C.M., Moridis, G., Ilk, D. & Blasingame, T.A. 2009. *A numerical study of performance for tight gas and shale gas reservoir systems.* Paper SPE 124961 presented at the 2009 SPE Annual Technical Conference and Exhibition. New Orleans, Louisiana, October 4 – 7.

Hefley, W.E. et al. 2011. *The economic impact of the value chain of a Marcellus shale well.* Working Paper. Katz Graduate School of Business. University of Pittsburgh.

Joshi, S. D. 2003. *Cost/benefits of horizontal wells.* SPE Western Regional/AAPG Pacific Section Joint Meeting held in Long Beach. California, U.S.A., May 19 – 24.

Mayerhofer, M.J., Lolon, E.P., Warpinski, N.R., Cipolla, C.L. & Walser, D. 2008. *What is stimulated reservoir volume (SRV)?* Paper SPE 119890 presented at the 2008 SPE Shale Gas Production Conference. Fort Worth, TX, November 16 – 18.

Momeni, A., Zargar, G. & Sabzi, A. 2011. *Simulating the effect of non-darcy flow and hydraulic fracturing on well productivity in a naturally fractured lean gas condensate reservoir.* Brazilian journal of petroleum and gas. Vol. 5, No 4: 189 – 196.

Pope, C.D., Palisch, T.T., Lolon, E.P. Dzubin, B.A. & Chapman, M.A. 2010. *Improving stimulation effectiveness-field results in the Haynesville shale.* Florence, Italy, September 12 – 22.

Postanova vid 30.09.2014 No 56 "Pro vstanovlennya hranychnoho rivnya ciny na pryrodnyj haz dlya promyslovyh spozhyvachiv ta inshyh sub'yektiv hospodaryuvannya".

Rashid Khan & Ayman R. 2012. Al-Nakhli. *An overview of emerging technologies and innovations for tight gas reservoir development.* SPE International Production and Operations Conference and Exhibition held in Doha Qatar, May 14 – 16.

Vivek Sahai, Greg Jackson, Rakesh Roshan Rai, Larry Coble. 2012. *Optimal well spacing configuration for unconventional gas reservoirs.* SPE 155751 presented at the SPE Americas Unconventional Resources Conference. Pittsburgh, Pennsylvania, USA, June 5 – 7.

Theoretical and Practical Solutions of Mineral Resources Mining – Pivnyak, Bondarenko & Kovalevska (eds)
© 2015 Taylor & Francis Group, London, ISBN: 978-1-138-02883-8

Resource-saving technology of selective mining with gob backfilling

V. Byzylo, O. Koshka, S. Poymanov & D. Malashkevych
National Mining University, Dnipropetrovsk, Ukraine

ABSTRACT: In the article, the question of the cost-effective use of resources while mining thin and very thin coal seams of Western Donbas is considered. The method of selective mining with gob backfilling is proposed. The estimation of operation condition the mine workings under gob pack protection is carried out. The investigation results of shearer power consumption that cutting tools form different longwall face shapes are given.

1 INTRODUCTION

The coal industry of Ukraine plays an important role in energy balance. Coal has 95% in the structure of Ukraine`s energy resources. According to estimation (Baker 2013), our country takes 7th place in the world on coal reserves – 33.9 billion tones or 4% of all world coal reserves that provide reliable function of power facilities and other consumers that use coal as the energy producing material.

In the balance of coal reserves only 20.4% comes on seams with thickness more than 1.2 m, 74% are thin coal seams (up to 1.2 m) including 33.3% of very thin coal seams (less than 0.8 m) (Topolov et al. 2004). Most of the proved reserves with thickness ≥ 0.8 m occurred in underworking and overworking of a coal, that make difficult or impossible exploration with traditional mining methods (Falshtynskyi et al. 2014, Lozynskyi 2015). Geo-technological conditions of mining thin and very thin coal seams are very difficult. While mining of these seams the level of labor intensity, amount of energy consumption, human and material recourses are constantly increasing (Byzilo et al. 2013). Therefore, underground mining of thin and very thin seams is actual and has to be based on resource-saving technology.

2 SUBSTATIATION OF THE PROBLEM

In many coal-mining countries the seams with thickness up to 1.2 m do not develop. The mechanized complexes, including a new technological level in most cases do not fit in the thickness of extracted seam. In a result, mines induce to extract the coal with wall rock undercutting. It leads to increase the level of ash content. On certain coal-mining enterprises, the ash content reaches 60%.

The main source of rock intake in coal are longwall and development faces. In total volume of hoisting rock from the mine, 70% comes from longwall and 22.5% development faces. The cause of coal clogging in longwall face are presence of false, unstable roofs and undercut of surrounding rocks (Byzilo et al. 2013).

Growth of ash content negatively influences on the main indicators of mine work, primarily on profit and profitability. Simultaneously, enrichment products are worsening and their output is reducing. Electricity consumption, labor and material cost due to increased volume of waste rock transportation and enrichment of high-ash coal are increasing (Bondarenko 2010).

Especially sharply, the problem of thin and very thin coal seams mining stands on Western Donbass mines. The distribution of coal reserves by thickness on PJSC "DTEK Pavlogradvugillia" mines is shown on Figure 1.

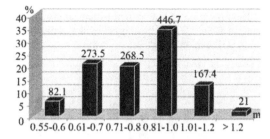

Figure 1. Distribution of coal reserves by thickness on PJSC "DTEK Pavlogradvugillia" mines.

The given data show that about 50% of coal reserves buried in seams with thickness less than 0.8 m that under current techniques and technologies do not possible to extract in Western Donbas conditions without wall rock undercutting.

In present time longwall faces are equipped basically by domestic mechanized complexes KD-80, KD-90, KD-99, DM types with shears KA-80, KA-200, UKD-200-250, UKD200/400. The minimum mining thickness in longwall equipped by these mechanized complexes is 1.0 – 1.05 m. On certain enterprises are introduced and successfully operated mechanized complexes of Czech production consisting from roof support Ostroj and shearer MB-410E

TMachinery company where minimum extraction thickness is 1.15 – 1.25 m.

The diagram of the average dynamic pattern of geological and mining seam thickness in considered region by years is shown on Figure 2. Coal seams of Western Donbas generally have simple structure. Therefore, the difference between mining and geological thickness is the size of wall rock undercutting in longwall faces.

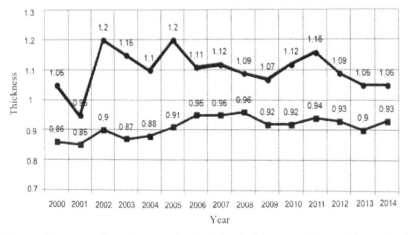

Figure 2. Diagram of the average dynamic pattern of geological and mining seam thickness in longwalls of PJSC "DTEK Pavlogradvugillia" mines: ■ geological; ● mining.

According the data presented in Figure 2, it is means the average size of a wall rock undercutting changes in the range from 0.1 to 0.29 m. However, on certain enterprises wall rock undercutting in stopping faces reaches 0.4 – 0.5 m (Bondarenko et al. 2013).

Thereby, introduction of new equipment the problem of improvement the quality of extracted coal did not solve, changes are aggravate. The bulk mining implementation on mines is the main cause of ash content increasing in produced coal.

The histogram of ash content pattern in produced coal on PJSC "DTEK Pavlogradvugillia" mines by years is shown on Figure 3.

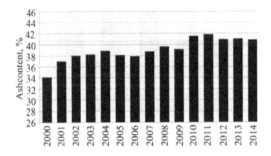

Figure 3. Histogram of ash content pattern in produced coal on PJSC "DTEK Pavlogradvugillia" mines.

Identifying feature of Western Donbass is the present in floor and roof of mining seams the soft clayish and siltstone-clayish rocks ($f = 1.5 – 2.5$ on M. M. Protodyakonov's scale) which under moisturizing get swell. Thus, 50 – 80% of rock strength properties are losting. In a result, the intensive shifts of mine opening contours appear, as well as deformations and support destructions, rock failures and sets on "rigid base" of mechanized complexes are happened.

Furthermore, the presence of hard and viscous coal with hardness coefficient on M.M. Protodyakonov's scale 3 – 5 having high-resistivity to cut 250 – 520 kH/m requires additional energy consumption.

While passing of mining operations to a great depth the loads on support are increasing. The displacements of mine opening contours increase approximately in three times. Conditions of mine workings remain unsatisfied, more over 40% are repaired before beginning of exploitation, and 52% of operational mine workings are deformed. The costs on repair and maintenance of section mine workings have been growing. They compound on different sources is 15 to 35% of total costs while coal mining.

The researches (Martovitskiy 2012), which were carried out on mines of considered region are shown that main part of costs put in works on repairing of section mine workings.

Thus, 77% from all repairs come on bottom ripping, 19% – alignment and 4% – support replacement. The actual volumes of repair operations in section mine workings of PJSC "DTEK Pavlogradvugillia" mines by types of works are shown on Figure 4.

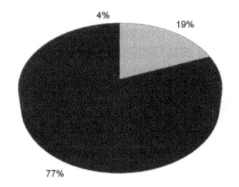

Figure 4. Actual volumes of repair operations in section mine workings of PJSC "DTEK Pavlogradvugillia" mines by types of works: ▪ aligment; ▪ bottom ripping; ▪ support replacement.

As a result, for ensuring operational condition of section mine workings it is necessary to conduct repair operations that include full remounting of particular sites or all mine working, application temporary strengthening support, increase of frame installation density, ripping of expanded rocks etc.

Having regard to the above, for coal seam mining in Western Donbas conditions the actual task is finding solutions that would allow to use for coal extraction the existing equipment providing coal separation from undercutting rock and leave the last one in the mine. Furthermore, simultaneously mined rocks to use in the case of material for protection construction building in mined-out area, reducing the negative impact of mine pressure and creating the conditions for maintenance-free supporting of mine workings.

3 THE WAYS OF PROBLEM SOLUTIONS

The retrospective analysis of existing experience of thin and very thin coal seams mining in Western Donbas shows that there are a number of scientific developments, especially the National Mining University, that can be implemented in the life and effectively use in production.

Thus, considerable contribution in mining science and techniques development, in our opinion, is development of mechanized complex "Western Donbas". This complex is based on the idea of using border part of hard and viscous coal seams in the case of base for installation of mechanized support

elements with advance cut. Such technical solution allows to mine the coal seams with unstable roof rocks (Kiyashko 1979).

4 TECHNOLOGY OF SELECTIVE MINING WITH GOB BACKFILLING

Today this idea can be used while selective mining thin and very thin coal seams with undercutting wall rock backfilling of mined-out area. The principle scheme of this technology is shown on Figure 5.

The complex includes modernized shearer KA-200 which has double separate drums – lower and upper. Mechanized roof support in front side has prop device that install on leaving after longwall pass the rock or coal bench. Scraper conveyor transports coal and rock to haulage drift.

Coal extraction is carried out by one longwall pass. Furthermore, the advance lower drum cuts coal seam and loads it on conveyor, lagging upper drum is put forward on cutting web and extracts roof rocks then broken rocks are placed on the bottom of seam between stopping face and scraper conveyor. The roof support is moved (frontal or consistently) on new way after longwall pass and coal transportation by scraper conveyor. Shearer colter squeezes all broken rocks onto conveyor for their transportation to haulage entry.

Using a method (Bondarenko et al. 2014) it is offered to leave the waste rocks from longwall face in mined-out area by vibration pan line installing under backside of mechanized roof support. As the filling work advances, the last pan disconnects and rolls over in vertical position. The vibratory pan line returns in horizontal position while support advancing on new way. The analysis of its methods influence on the surface subsidence is considered in work (Koshka et al. 2014)

5 EVALUATION OF EFFICIENCY AND REASONABILITY THE NEW TECHNOLOGY USE

Implementation of the new technology will allow to reduce the ash content of produced coal, cut down the cost on development, repair and protection of mine workings, decrease energy consumption on coal extraction, transportation and utilization of waste rocks.

The questions of the new technology efficiency are considered for conditions of 1068 longwall C_{10}^{u} seam of "Geriv Kosmosa" mine PJSC "DTEK Pavlogradvugillia". Mining-geological and mining-technological characteristics of development conditions are shown in Table 1.

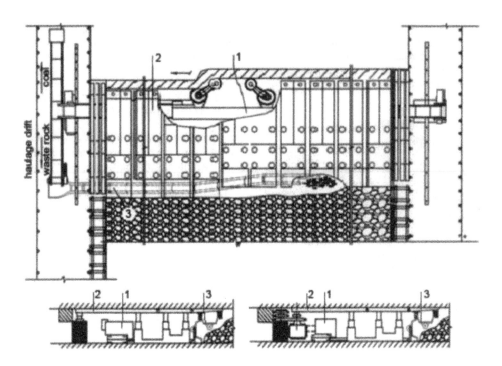

Figure 5. The principle scheme of selective mining of thin and very thin coal seam with waste rock backfilling: 1 – shearer with advance cut; 2 – mechanized roof support with reverse console; 3 – vibration pane line.

Table 1. Mining-geological and mining-technological characteristic 1068 longwall seam and 1068 haulage drift seam.

Mining and geological characteristics		Mining and technological characteristics	
Extracted seam thickness, m	1.05	Mining system	Advance longwall
Geological seam thickness, m	0.82	Roof control	Full collapse
Angle of dip, deg	0 – 3	Longwall length, m	270
Mining depth, m	370	Panel length, m	1748
Category of immediate roof by stability	B_{1-2}	Shearer	KA-200
Category of mine rocks by cavability	A_2	Scraper conveyor	SP-251
Maximum compressive resistance of coal seam, MPa	30	Mechanized roof support	KD-90
Average maximum compressive resistance of mine rocks buried at distance 20 m away from heading cross-section in roof, MPa	15.8	Form of cross-section of mine working	arch
Average maximum compressive resistance of mine rocks buried at distance 5 m away from heading cross-section in floor, MPa	15.2	Type of support	SAYL-11.7
Water inflow, m³/h	1.0 – 1.5	Support installation pitch, m	0.8

6 REDUCING ASH CONTENT

It follows from table 1 that the size of wall rock undercutting while geological thickness 0.82 m and extracted 1.05 is equal 0.23 m. Therefore, the ash content of produced coal from considered longwall face composes 45.8%. To reduce the ash content of coal to mother (A_C = 18%) will allow the technology of selective mining with gob backfilling. According to geometrical parameters of longwall face and shearer technological parameters, this volume of waste rocks will be enough for filling the mined-out area the width at least 65 m under longwall length 270 m or 30% of gob space. Besides, reducing the ash content of produced coal allows to decrease the amount of waste rocks hoisting from the mine on the surface. As a result, it is save energy consumption, simplify underground transport and preserve the land areas taken for waste rock pilling.

7 MINE WORKING PROTECTION

The haulage drift of 1068 longwall is supported during all term of panel exploitation and reused. For its protection are installed the wood chocks, organ timbers and yielding props in front and behind longwall. However, in spite of installation in mine opening with necessary calculated resistance and protected constructions the actual displacements of roof and floor are significantly exceed allowable sizes by technological requirements. Mine working is repeatedly driven and bottom ripping is carried out. Using the technique ARIMS (Instruction... 1986) we will carry out the exploitation condition of section mine working under backfilling of mined-out area and wood chock protection for 1068 haulage drift $C_{10}{}^{u}$ seam of "Geroyev Kosmosa" mine. Basic data for calculation of expected rock displacements in reusing mine working are shown in Table 2.

Table 2. Basic data for calculation of expected rock displacements.

Parameters	Value
Coefficient taking into account the method of working driving (combine method) C_{mw}	0.8
Coefficient taking into account roof type k_{rt}	0.8
Coefficient taking into account influence the cross-section of mine opening C_S	1.1
Coefficient characterizing the part of roof rock displacements in total rock shifts in mine workings, C_{pr}	0.4
Coefficient taking into account influence of protective construction C_{con}	
– gob pack	0.8
– wood chock	0.6
Rock displacements under influence mine working driving U_{dr}	
– roof	490
– floor	460
Average rock displacement velocity out of zone longwall operations V_0	
– roof	16
– floor	12
Time of mine working supporting out of zone of longwall operation influence t_0	10 mo.
Rock displacement in zone of temporary bearing pressure of longwall face U_1	
– roof	280
– floor	210
Settle down rock displacement velocity V_1	
– roof	76
– floor	71
Time of mine working supporting in zone of residual bearing pressure t_1	1 – 15 mo.

In our case mine working is fasten by SAYL support – 11.7 for which the acceptable size of working reduction U_{ac} is 890 mm. The calculated rock displacements of roof U_r and floor U_f of 1068 haulage drift $U_{10}{}^{u}$ seam subject to support time and method of their protection are shown in diagram on Figure 6.

According to given data it was found that under gob pack use the operational condition of mine working is provided while all term of panel mining. However, bottom ripping will be periodically need. The total sizes of roof and floor rock convergences under wood chock protection exceed accepted value U_{ac}. Therefore, the retimbering works of bottom ripping after first year of exploitation will need to carry out.

Thereby, while using in case of protection construction of gob pack the roof rock displacement for term of panel mining will be 236 mm. Under wood chocks, this value will be 277 mm. The roof rock convergence under gob pack protection is lower on 15% than wood chock. For floor rocks, the displacement will be 729 mm under gob pack protection and 826 mm under wood chock protection.

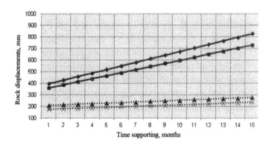

Figure 6. Diagram of roof and bottom calculated displacements of 1068 haulage drift $C_{10}{}^{u}$ seam under different protection during support time: ▬●▬ floor rock displacement under gob pack protection; ▬■▬ roof rock displacement under wood chock protection; ••▲•• roof rock displacement under wood chock protection; ••✕•• roof rock displacement under gob pack protection.

In this case, bottom heaving can be decreased on 15% using gob packing for mine workings protection by waste rocks from longwall face. Therefore, the technology of selective mining with gob backfilling allows not only solve the problem of produced coal quality but by the product of its worsening to create the condition for repair-free maintenance of mine workings.

8 REDUCING ENERGY CONSUMPTION

Another positive side of considered technology is energy consumption reduction on coal extraction due to change the form of longwall face from rectangular to heading-and-bench. Shearer with vertical axis rotation of drums is used for coal extraction. Lower advanced drum cuts coal and backward upper drum cuts rocks in roof. Under such scheme of drums` arrangement the heading-and-bench face is formed that used in case of floor for front roof support hydraulic prop installation. Direct influence of mining pressure on support elements leads to cracks formation in created after longwall pass the bench of coal. As a result, it is reduced the energy consumption on coal extraction.

These statements are confirmed by the results of underground investigations (Byzilo 2007). For example, on "Ternovska" mine PJSC "DTEK Pavlogradvugillia" was carried out investigation of energy consumption while coal extraction in longwalls with rectangular and head-and-bench forms of stopping faces. The researches were conducted while mining C_5 seam with hard and viscous coal. Applied shearers were the same type, operating tools of machines were drums with vertical axis of rotation but differences were only in the shapes of cutting tools. Extracted seam thickness C_5 in both faces was 1.05 m. Mining and geological characteristic did not change.

Measurements of coal extraction energy consumption on both shearers were carried out by self-registering kilowatt meter H379 under different regimes of electric motor loadings. At the same time the shearer advance speed was varied while constant cutting web and cutting web while constant speed. The diagram of energy consumption and power consumption of shearer electric motors depending on advance speed is shown on Figure 7.

The analysis of given dependences show that while increasing the shearer advance speed the specific energy consumption of coal extraction is decreasing. However, while shearer operation in heading-and-bench face this decreasing appears most intensively. In the range of advance speed of shearer KA-200 (with advance cut) from 0 to 2.76 m/min the energy consumption is changing from 42 to 98 kW that is much less than nameplate capacity of shearer (R = 200kW).

These investigations show that increasing cutting wed from 0 to 0.8 m leads to increase the shearer consume power from 35 to 155 kWt and reduce energy consumption of coal extraction up to 0.77 kWh/t.

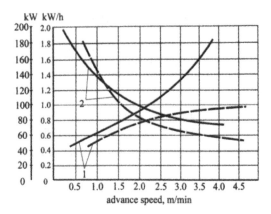

Figure 7. Diagram of energy consumption while coal extraction (1) and shearer power consumption (2) depending on advance speed: —— KA-200; – – – KA-200 with advance cut.

The specific energy consumption dependency while coal extraction and shearer consume power depending on cutting web while average advance speeds 2 m/min is shown on Figure 8.

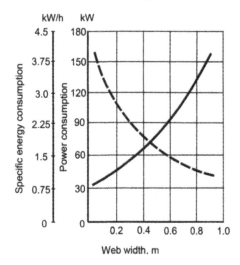

Figure 8. Characteristic of energy consumption: — power consumption; – – specific energy consumption per 1 t of coal.

It is also found, that while shearer advance speed 2 m/min and the same web width 0.5 m the specific energy consumption was 1.7 kWh/t, and experimental shearer was 0.78 kWh/t. That is less on 62% than while mining by serial shearer.

Therefore, changing the rectangular form of face on heading-and-bench and creating the advance cut on seam thickness lead to shearer output rise at expense of increasing advance speed and significantly reduction of energy consumption on hard and viscous coal extraction.

490

9 CONCLUSIONS

The rational use of available resources has especial importance. Thus, based on the developments of NMU it is possible to propose the technical solutions for mining thin and very thin coal seams directed on implementation in production non-waste and resource-saving technologies. The offered technological scheme with waste rocks leaving in the mine will allow to solve the problem of coal quality, reduce the cost on extraction and transportation of waste rocks on the surface, improve the mine working conditions, decrease the cost of their exploitation and reusing as well as decrease environmental impact in coal mining regions.

REFERENCES

Bondarenko, V., Kovalevs'ka, I. & Cherednichenko, Yu. 2010. *Substantiation of design and installation technology of tubular rock bolts by explosive method.* New Techniques and Technologies in Mining. The Netherlands: CRC Press/Balkema: 9 – 14.

Bondarenko, V., Kovalevs'ka, I., Svystun, R. & Cherednichenko, Yu. 2013. *Optimal parameters of wall bolts computation in the united bearingsystem of extraction workings frame-bolt support.* Mining of Mineral Deposits. The Netherlands: CRC Press/Balkema: 5 – 9.

Patent No 105458, UA. 2014. *Method of selective mining of coal deposits with gob backfilling.* Bondarenko, V.I., Koshka, O.G. & Malashkevych, D.S.

Buzilo, V.I., Sulaev, V.I & Koshka, A.G. 2013. *Technology of thin seam mining with gob backfilling.* Monograph: National Mining University: 124.

Buzilo, V.I., Poymanov, S.N., & Rastriga, V.P. 2013. *Analysis of influence of technology and system elements of mining on energy consumption in coal mines.* Mining of Mineral Deposits. The Netherlands: CRC Press/Balkema: 115 – 120.

Byzilo, V.I., Serduyk, V.P. & Koshka, O.G. 2007. *Influence of stope face shape on energy consumption and grate of mined production.* Materials of International scientific-practical conference "Scholl of underground mining". Yalta, September 03 – 07: 307 – 309.

Falshtynskyi, V., Dychkovskyi, R., Lozynskyi, V. & Saik, P. 2014. *Some aspects of technological processes control of in-situ gasifier at coal seam gasification.* Progressive Technologies of Coal, Coalbed Methane, and Ores Mining. The Netherlands: CRC Press/Balkema: 109 – 112.

Instruction on rational place, protection and support of mine working on coal mines of USSR. 1986. L: ARIMS: 220.

Kiyashko, I.A. 1979. *To develop technology, to create and implement the complex of stopping machines with advance cutting tools for thin flat seam mining with thickness 0.7 – 1.2 m with unstable wall rocks*: R&D report: Dnipropetrovsk: 180.

Koshka, O.G., Yavors'kyy, A.V. & Malashkevych, D.S. 2014. *Surface subsidence during mining thin seams with waste rock storage.* Progressive Technologies of Coal, Coalbed Methane, and Ores Mining. The Netherlands: CRC Press/Balkema: 229 – 234.

Lozynskyi, V.G., Dychkovskyi, R.O., Falshtynskyi, V.S., Saik, P.B. 2015. *Revisiting possibility to cross the disjunctive geological faults by underground gasifier.* Naukovyi Visnyk Natsionalnoho Hirnychoho Universytetu, No 4.

Martovitskiy, A.V. 2012. *Justification of complex of effective measures on increasing the stability of mine workings on PJSC "DTEK Pavlogradvugillia" mines.* Naukovyi Visnyk Natsionalnoho Hirnychoho Universytetu, No 3: 45 – 52.

Topolov, V.S., Topolov, V.S. & Bortnicov, A.A. 2004. *Problem of power-energy resources in the world and Ukraine.* Coal of Ukraine, No 5: 3 – 11.

Theoretical and Practical Solutions of Mineral Resources Mining – Pivnyak, Bondarenko & Kovalevska (eds)
© 2015 Taylor & Francis Group, London, ISBN: 978-1-138-02883-8

Technologies of overburden rock storing in depleted or operated iron-ore open pits of Ukraine

A. Dryzhenko
National Mining University, Dnipropetrovsk, Ukraine

N. Nikiforova
Dnipropetrovsk National University of Railway Transport named after academician V.Lazatyan

ABSTRACT: The geological and industrial characteristics and conditions of mining on iron-ore pits of Ukraine are presented. The prospects of improvement and parameters of overburden storing technology in accordance with the development of mining operations in outside dumps and gob areas are examined. The experience of gob area filling in iron-ore pits with increasing the depth of balance reserves mining is described. The economic and environmental efficiency of new technologies implementation are shown.

1 INTRODUCTION

Kryvorizhskyi iron-ore basin (Kryvbas) is one of the most studied and industrial development areas for the iron ore extraction within the province of the Ukrainian Craton. He is almost located within the Dnipropetrovsk region. Explored reserves of iron ore basin are 21.8 million tons. Predicted resources estimated at more than 19 million tons (Braun 1970, Afronin & Nesterov 1972). In metallogenic aspect Kryvbas included consistently in Kirovohrad metallogenic subprovince. This pay streak of different metamorphic characteristics of Kryvyi Rih series run for 85 km with a width of 0.5 – 0.7 km. In structural plane it is Saksahanska synclinal, complex folded structure. On its wings developed a short, often overturned on the east synclinal and anticlinal folds of the second and higher orders. Synclinal wings generally sink to the west, and the west wing sheared with fault. The structure of the synclinal is complicated by disjunctive faults and displacements. Kryvorizhska series with a total thickness up to 4700 m is divided into series of strata: Novokryvorizhska with a length – 1300 m; Skelevatska with a length – 300 m; Saksahanska with a length up to 1500 m and Hdantsivska – up to 1600 m. Previously attributed to this series, the uppermost, also Gleievatsk series of strata with estimated thickness from 850 to 3500 m.

Geographically, the most common productive is Saksahanska (average) series of strata, which a large industrial deposits of ferruginous quartzite and rich iron ore. It consists of seven series in the sequence of alternating horizons of ferruginous quartzite and shale. Their total thickness reaches 1500 meters. Ferruginous quartzite and rich ores are mostly tab

bedding place diposits, subconcordant, less columnar, nesting, and other forms. Their thickness varies from a few meters to 100 meters. Mostly horizons of ferruginous quartzite merge with each other or pinch out. In particular, in the core of Pervomaysk-Saksahanska synclinal (in the same ore field) merges at a depth into a single ore deposits with total thickness up to 260 m, which gently sink to the north and generally confined to the nucleus Saksahanska synclinale. They stretch the length of 110 km and developed to a depth of 2250 m. The depth of iron ore in Kryvbas basin together with immersion of Saksahan synclinale make 3.5 km.

The main types of iron ore in Kryvbas which develop open-pit-mining, are ferruginous quartzites (hornfels, jaspilites), non-oxidized, mostly magnetite with hematite and others. This is raw ore with an average content of $Fe = 33.3 – 34.2$ % to a depth of $500 – 800$ m. Reserves category $A + B + C_1 + C_2$ make more than 13 million tones. Increase in reserves of ferruginous quartzites is possible due to additional exploration flanks and deep horizons of the exploited, and new backup reconnoitered deposits. Unoxidized quartzite deposits are localized mainly in the folded structures of horizontal strata of productive thickness $400 – 1000$ m (Inguletsk, Skelevatsk Magnetit, Novokryvorizhsk deposit et al.) in the zones of transverse strains thick widths of up to 1.600 m (Pervomaysk et al.), in the wings of folded structures, where the power of ferruginous quartzite varies from $30 – 400$ m (Valyavkinsk, Annovsk et al.).

The chemical composition of iron ore is exposed on existing variations that determined by the content in the ore minerals and their specific proportions. Industrial ore types have the content (%) of the main chemical components: $SiO_2 = 32 - 50$; $TiO_2 = 0.03 - 0.1$; $Al_2O_3 = 0.3 - 17$; $Fe_2O_3 = 3.9 - 55$; $FeO = 0.5 - 30$; $Fe_{total} = 22 - 41$; $Fe_{mag} = 0.8 - 36$; $MnO = 0.04 - 0.6$; $CaO = 0.1 - 12$; $MgO = 0.1 - 6$; $Na_2O = 0.2 - 1.2$; $K_2O = 0.1 - 1.2$; $K_2O + Na_2O = 0.1 - 2.4$; $P_2O_5 = 0.09 - 0.4$; $S = 0.01 - 0.3$; $CO_2 = 0.1 - 16$.

2 MAIN PART

Iron ore of Kryvyi Rih-Kremenchug zones are characterized with relatively small content of trace elements. An exception is the germanium of Poltava ore mining and dressing plant. In some areas are high concentrations of Sc, V, U, Zr, TR, and the concentration of Au especially in rock refuse.

From south to north in Kryvyi Rih iron ore basin located exploited and unexploited (UE) field (areas of the mine): Inguletske (1), Szymanowske (UE), Skelevatske Magnetitove (2) Skelevatske, Valyavkinske, Novokryvorozhske (3) mines: Severna, Valyavko (4) Hihant Hlyboka (5), Zhovtneva, Kirova (6) Saksagan (UE), Rodina (7) Velyka Hlievatka (8), Zhovtneva (9), Bilshovyk (UE), mines: Frunze (10), Juvileyna (11), Lenina (12), Hvareyska (13), Ordzhonikidze, Pervomaiskyi Annovskyi pits (14) (Figure 1).

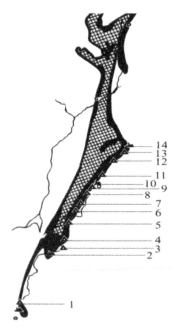

— 14
— 13
— 12

— 11
—10
— 9
— 8

— 7
— 6
— 5

— 4
— 3
— 2

— 1

Figure 1. Scheme of ore enterprises location in Kryvyi Rih basin. 1 – 9 – operating ore enterprises.

Along with the aforementioned deposits in the right-bank district of Kryvbas in the direction from south to north contains the following iron objects (deposits fields): Petrovske, Zhovtorechenske, Zahidne (UE), Artemivske, Popelnastivske (UE) and others that are located in the remnants thick of Artemivsk series of Ingulov-Inguletsk strata. In particular, the Petrovske deposit contains new for Kryvbas type of ore – rich metasomatic magnetite ore and contain large reserves of iron ore.

In Kryvyi Rih basin ferruginous quartzites (unoxidized and partially oxidized) mined by the open pit at 11 fields and processed in fiveore mining and dressing plant: Ingulets, Juzhnyi, Novokryvorizhsk, Centralyi and Severnyi. Five fields are mined by underground methods, along with the rich ores. Reserves life of ferruginous quartzites of these enterprises ranges from 32 to 85 years. A significant part of the reserves, proven even in the years 1955 – 1975, is located in the protective pillars, under the river Ingulets, urban development, mines and dumps of ore mining and dressing plant and located at depths down between 500 – 800 m, where their production is difficult because of the complex geological conditions (Maleckyi et al. 1986, Technical and economic…2002).

During ore mining and dressing plant operation wastes – overburden rocks and tailings – cannot be completely disposed of due to the large volume of their output. In this regard, such storage should be considered as man-made deposits suitable in future for re-development. At the same time the process of storage should provide measures for re-development, to prevent adverse effects on the environment, in particular, the exclusion of surfaces dusting and minimal surface disturbance.

It should be noted that in the initial period of deep pits development almost the entire volume of overburden rock are directed to external capping plateau-shaped rock-disposal dump and at least a gob release their part can be used for filling it.

The main parameters of dumps are: the buttom area S_0 (han), the amount of storing waste (mln m³), the side surface area S_s (ha) and the specific landg S_0' (ha/mln m³), depending on the ratio of the length L_0 (m) and short B_0 (m) sides dump field on surface are determined by the expression:

$$S_o = 1.57 \cdot 10^{-4} K_f (B_o + 2H_o ctg \varphi)^2, \tag{1}$$

$$W_o = 1.57 \cdot 10^{-6} K_f K_{f.i} + \\ + (B_o + H_o L_o ctg \varphi + 2H_o^2 ctg^2 \varphi), \tag{2}$$

$$S_s = \frac{10^{-4} H_o (5.14 B_o + 2L_o + 7.14 H_o ctg \varphi)}{sin \varphi}, \tag{3}$$

$$S_o' = \frac{2 \cdot 10^{-2} K_f (B_o^2 H_o L_o ctg \varphi + 4H_o^2 ctg)}{H_o (B_o^2 + H_o L_o ctg \varphi + 2H_o^2 ctg^2 \varphi)}, \tag{4}$$

where H_0 – the high of the dump, m; $H_0 = h_l n_l$; n_l – number of dump layers; h_l – the high of dump layer; L_0, B_0 – long and short side of dump on top pit, m; φ – stable slope of dump, degree:

$$\varphi = arcctg\frac{n_l h_l}{n_l h_l ctg\varphi_{o.l} + (n_l - 1)w_l}, \qquad (5)$$

where $\varphi_{o.l}$ – stable slope of dump later, degree; w_l – width between the layer, m; $K_{f.i}$ – fragmentation index; for hard rock $K_{f.i} = 1.12 - 1.2$; for combined rock and heavy clay $K_{f.i} = 1.05 - 1.12$; for fragmentation and clay rock $K_{f.i} = 1.05 - 1.07$.

Increasing the height of rock dump in general improve the angle slope, capacity of assigned area and hence to reduce the total specific landgrab. Analysis of the external dumps parameters of Kryvyi Rih ore mining and dressing plant shows that the greatest landgrab are inherent to soft rock dumps at a height $35 - 40$ m that make $6.4 - 7.15$ ha/mln m³. An increase H_0 up to $100 - 120$ m the magnitude S_o' is reduced to $1.26 - 1.93$ m/mln m³. It is established (Figure 2) that increase in area of the rock dump base reduce a specific landgrab during rock storage.

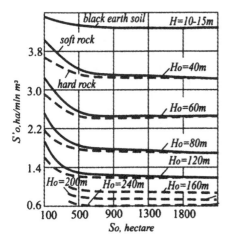

Figure 2. Diagram of specific landgrab S' depending on the height H_0 and rock dump base area S_0'.

Figure 2 show that during S_0 increases from 100 to 500 ha S_0' reduced in $15 - 17\%$. Specific landgrab of soft rock dumps is high compared to hard rock in $11 - 19\%$ at the height of 40 m and in $5 - 6\%$ at $H_0 - 80$ m. With a further increase of S_0 $1800 - 2500$ hectares the specific landgrab changes slightly. However, it should be noted that according to the stability conditions the height of soft rock dump usually does not exceed $40 -$

60 meters, while the hard rock reaches $160 - 240$ m. In this regard the specific landgrab during hard rock storing in comparison with soft rock in $3 - 3.5$ times.

During joint stored of different types of rock the dump base increases accordingly. However, because of the appearing is an opportunity to further storing of the rock (m³) in the recess between the adjacent slopes of different types of dumps:

$$W_{f.s} = H_o ctg\varphi(L_o - 2H_o ctg\varphi). \qquad (6)$$

By increasing the overall height of the dump during the joint storage the dump capacity also increases:

$$W_o = H_o K_{f.o}(L_o - H_o ctg\varphi)(B_o - H_o ctg). \qquad (7)$$

Subsequent development of stored different types of rocks of technogenic deposits in order to use them as raw material is not significantly different from the separate development during open pit mining. A contacts adjacent rock at the joint storage has a regular geometric shape. They may be worked in any direction. As a consequence, a rock loss after redevelopment is minimal.

During separate storing of different types of rocks in separate rock dumps the landgrab area increased in $30 - 50\%$. However, if the separate storing is carried out on the existing rock dumps together there are no significant increase of the landgrab area. Moreover, during storing different rock types in dump the overall height of the rock dump can be increased in $20 - 30$ m. Consequently, the specific landgrab of such dumps is reduced. The cost of storage of overburden Z_0 (UAH) is generally determined from the expression:

$$Z_0 = A_0((C_t + K_{c.e}E_n)l_{a.d} + S_0 K_l), \qquad (8)$$

where C_t – the cost of transportation of 1 ton of overburden with pit rail transport, UAH; $K_{c.e}$ – capital expenditures for the purchase of rolling stock and auxiliary equipment for operation of rail transport UAH; E_n – normative coefficient of effectiveness; $l_{a.d}$ – the average distance of transportation, km; K_l – the cost of landgrab of 1 hectare, UAH.

The determining factor in choosing the construction site of rock dumps are the distance of waste production transportation and the cost of compensation for land allotment. Expenses for waste rock transportation surpass by far for the land area disturbance. Consequently, in order to reduce the cost of waste production storing it is necessary to locate rock dumps and tailings dump as close to the open pit and ore mining and dressing plant. Also it is necessary to maximize their capacity.

Commercial by-product should be stored separately in the temporarily abandoned horizons in the mined out space of open pit forming technogenic deposits (Figure 3).

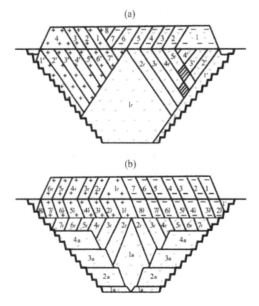

Figure 3. Scheme of technogenic deposits formation in the mined-out space of deep open pits (a) and during development (b): 1r – 3r – the sequence of dumping during railway transport; 1a – 4a – the sequence of dumping during automobile transport; 1' – 8' – stope dump below the earth's surface; 1" – 7" – stope dump above the earth's surface; " ", + +, v-v – type of rock.

Thus, according to technology of mining development, the bottom side of mined out space carry out waste rock filling, in the upper part – nonmetallic minerals. Depending on the volume of nonmetallic mineral raw materials, start fill from the end of face, is stored on the side of foot wall and handing wall, waste rock – in the middle.

The sequence of stope dumping it is necessary to made formation of linear contacts between heterogeneous types of rocks, in other word on previously formed pit slopes. Otherwise, the border between them will be mixed (Figure 3 – hatched). Formation of the line of contact between the heterogeneous rocks at an angle close to the natural slope, allows during re-development reach minimal dilution.

In order to reduce the land for overburden storing in mining enterprises the height of rock dumps make 300 – 500 m or even more. The best results for rock dump filling and reduce the distance for overburden rocks transporting achieved through the organization of their formation one floor from the top working platform. However, to ensure sufficient stability of slopes and prevented the large-scale landslides in the body of high dumps stored distinguished varieties of hard overburden rock. In the world practice there is enough experience in the construction of high dumps in mountainous terrain. When filling deep depleted iron-ore pits or individual sections the positive results of the overburden rock storing in one tier achieved in Kryvbas.

As we know the experience of deep Kryvbas pits characterized by great depth, high advancing intensity of front bench and dump placement at the top of heavy mining equipment for providing safety work and stability of internal dump in the bottom should be created additional efforts impeding shift. This condition is ensured by focusing of the lower edge of the inner dump in the opposite side, or in a special prism as is shown in technological scheme (Figure 4).

Depending on the angle of pit slopes schemes of its filling during different work intensity and width of the bottom are divided into one- and two-sided, transversal and longitudinal (Table 1).

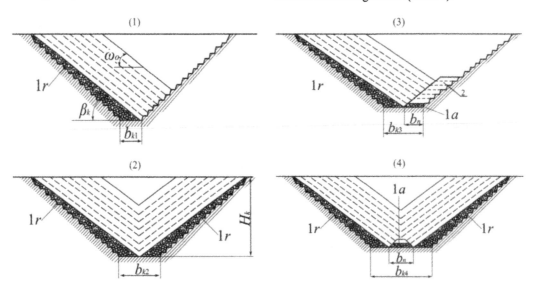

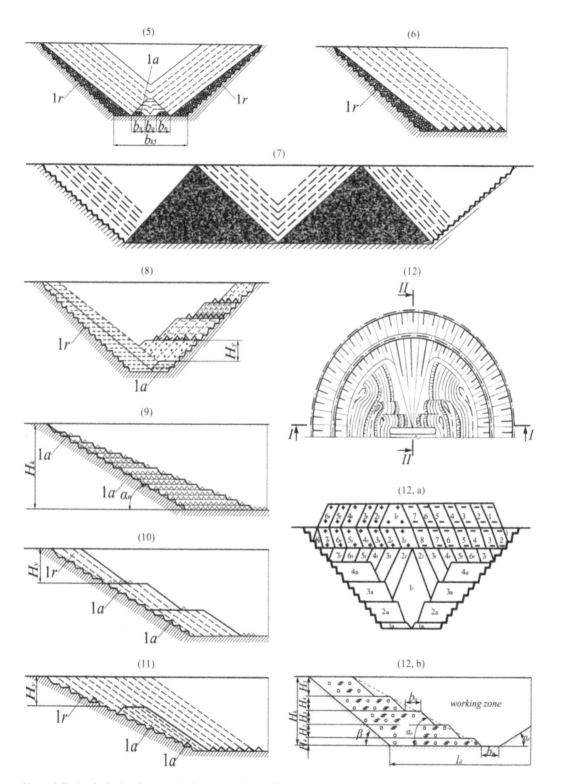

Figure 4. Technological schemes of deep pits back filling 1 – 12: 1_r – the first stope during rail transport; 1_a – the first stope during automobile transport; 12 (a) cross section I – I; 12 (b) cross section II – II.

Table 1. Pit slopes filling schemes during different work intensity and width of the bottom.

No	The method of filling	Direction of filling	Sustaining of internal dump	Gob parameters
1	gross	one-sided, transversal	opposite side of pit	l_{K1} – unlimited, during $\beta \geq w_0$ $$b_{K1} = H_K\left(ctg\ w_0 - ctg\ \beta\right) + b_E$$
2	gross	two-sided, transversal	opposite dumping bench	l_{K2} – unlimited, during $\beta \geq w_0$ $$b_{K2} = 2\left(H_K\left(ctg\ w_0 - ctg\ \beta\right) + b_{\ni}\right)$$
3	gross	one-sided, transversal	cutoff part of thrust prism	l_{K3} – unlimited, during $\beta \geq w_0$ $$b_{K3} = H_K\left(ctg\ w_0 - ctg\ \beta\right) + b_E + b_P$$
4	gross	two-sided, transversal	central continuous thrust on the bottom of the pit prism	l_{K4} – unlimited, during $\beta \geq w_0$ $$b_{K4} = 2\left(H_K\left(ctg\ w_0 - ctg\ \beta\right) + b_E\right) + b_P$$
5	gross	two-sided, transversal	central dispersed resistant prism at the bottom of the pit	l_{K5} – unlimited, during $\beta \geq w_0$ $$b_{K5} = 2\left(H_K\left(ctg\ w_0 - ctg\ \beta\right) + b_E\right) + 2b_P + b_0$$
6, 7	gross, selective	combined	parallel thrust prism filled from the surface	l_{K6} – unlimited, during $\beta \geq w_0$ $$b_{K6} = \left(2H_K ctg\ \alpha - b_E\right)\!n_{UP},$$
8	gross	one-sided, longitudinal	individual thrust prism for each layer of dump bench	l_K and b_K – unlimited, during $\delta \geq w_0$
9	selective	two-sided, longitudinal	Opposing dumping bench	l_K and b_K – unlimited, during $\sigma \geq w_0$
10	gross, selective	one-sided, longitudinal	individual thrust prism for each layer of dump bench	l_K and b_K – unlimited, during $\alpha_H \geq w_0$
11	gross, selective	one-sided, longitudinal	advance oval bench and individual thrust prism of each layer	l_K and b_K – unlimited, during $\alpha_H \geq w_0$
12	gross, selective	two-sided, longitudinal	Opposing dumping bench	$\left.\begin{array}{cc} l_K\ u\ b_K \\ \alpha_H\ u\ w_0 \end{array}\right\}$ – unlimited

In operated iron-ore pit in worked out or temporarily mothballed parts may be selective overburden rock storing followed by the development and removal on the surface (schemes 7 – 10). The height of the internal dump layer h_l (m) is regulated with minimum distance transportation of overburden with pit transport and can be varying from 15 to 60 m and more.

Thrust prisms at the base of the inner pit slope or dump tier are formed with bulldozed using waste rock delivered by road. Their width b_n (m) shall be equal to the width of the stope b_s (m). Depending on the required volume resistant prism is formed of one or three or more layers of tightly adjacent to each other cone rock unloaded from dumping trucks.

On the front of the construction of the prism should be hard to outpace the development of depleted open-pit bench. On the front of the slope construction the prism should outpace the development of depleted open-pit bench.

In Kryvyi Rih basin during filling the depleted open pit or some pit sections the overburden to the top-floor internal dump often is delivered by rail transport. Rock storing in the dump can be performed with bulldozer or shovel. To protect the surface of the dump his body should be laid only hard rocks without clay inclusions. Application of dragline for backfilling pits preferable as they slept stope width is in 3 – 5 times more than in excavators. Greater step refolding railways and removing them from the top edge of the dump allow make the internal dump with dragline more economically and safely. As an example, the creation of internal dump during the development of steep deposits revealing experience backfill career in ore mining and dressing plant No 1, where according to the recommendations of Dnipropetrovsk mining institute (Authors' license No 825771) since February 1979 the backfill of southeastern region carried by excavators EKG-6.3us, working in conjunction with the rail transport on the horizon +66 m (Figure 5).

(a)

(b)

Figure 5. The gob backfilling in ore mining and dress plant No 1 using hard rocks of overburden: (a) – 1979; (b) – 1983.

In the initial period the height of pit was 174 m and at the end of 1983 after filling of lower bench already reached 100 – 110 m. The stability of the excavator dump layer provides with device at its bottom where overburden delivered from the northern end of the pit No 1 with automobile transport. Then, delivery of rocks carried with trucks from the pit No 2-bis, and laying it in the depleted layer – using excavator with an extended working equipment EKG-4u.

Before entering of excavator layer dump in operation at the bottom of the waste land of pit using dump trucks BelAZ-548 and bulldozers D-385a for 2.6 months was formed the inner dump with the height of 23 m and the volume of 217 thousand m^3 of overburden rock.

Then, excavating layer advance along the gob makes 0.16 m/day. Placement of overburden rock at the bottom of the pit has reduced the distance transportation in the internal dump rock on 0.9 km as compared to their delivery to the transshipment point, and when transportation in the outside dump with rail transport on 12 km. In addition, the overload of overburden has been eliminated.

The observations show that during hard rock using excavator the height of dump layer decreases and hence reduces the width of possible landfall-prism, which allows the dump to increase the width of the stope. As a result of the described technology in gob are of the pit No 1 till 1996 was laid about 20 million m^3 of hard overburden rocks. It allowed reducing the amount of disturbed land area (44 hectares) and with the end of inside dump storing to return urban affairs about 17 hectares of remediated site. During mining lowering of pit in ore mining and dress plant to maintain its production capacity required to systematically advance of northern edge. However, placing within it spent Skelevatskoho pit hindered the normal development of the above-lying horizons and laying them on the train tracks road haulage. In this regard by employees were asked to place the railroad workers' horizons rebuilt on the northern side of the berm Skelevatskoho pit with the formation of the transport platforms from the overburden delivered to his temporary filling (Authors' license No 825771, Authors' license No 1121434). At the same time simplified the development of bridges between the pits (Figure 6).

Figure 6. Temporary storage of oxidized quartzite from a pit of ore mining and dressing plant in mined-out space of Skelevatskoho pit (1981).

The depth of gob varied from 60 to 110 m. The length of the dump front makes 250 – 300 m. The bottom width of the Skelevatskoho pit was small, and when filling the slope of the internal dump rested on the opposite side, which increases the resistance of storing rocks. On some areas the overburden rocks was stored on an inclined or horizontal base without stop. In order to increase their stability should be made additional surcharge. For this purpose storing of overburden rocks was carried out with the front counter of the opposite side of the pit, which contributed to the stability of the inner dump.

The rock was delivered by rail transport and fit into dump using excavator ECG-8y. The width of the stope make 16 m. Annually in the dump was laid 800 thousand m³ of rock. By reducing the operating costs for the transportation of railway transport on 11.8 km and reduce track packing work it has been a significant economic effect. The practice of open pit mining shows that extraction of steep deposits of long length most intensively conducted in areas with a maximum thickness of seams minerals, which are characterized by minimal stripping ratio, the considerable size of the working area and the project depth of the main pit. Areas of pit deposits with less seams thickness of minerals have higher stripping ratio.

In order to reduce the total volume of overburden excavation is seeking to increase the speed of mining operations in the horizontal direction along the strike of the deposit at a low intensity of push-back in pit.

The great length of pit fields allows to apply for rock mass transportation highly economical rail transport. Locate of train tracks at both front sides of the pit it is possible to ensure the supply of trains at relatively greater depth in the deepening area. Opening bench of first and second stages produce from the deepening area (Authors' license No 1330313, Patent No 1836561, Patent No 8679). Temporary internal dump is formed on the base of bench average queue on distance no less than $l_{n.1}$ (m) from the project boundary which is determined by the formula:

$$l_{n.1} = h_2\left(\frac{w_w}{h_b} + ctg\alpha_b\right) + 2\sqrt{h_o w_w ctg\varphi_d} + w_s, \quad (9)$$

where h_2 – the height of the bench of the second stage, m; w_w – the width of the work area, m; h_b – the bench height, m; α_{sb}, φ_s – angle of bench slopes and internal dump; h_o – the height of the primary internal dump, m; w_s – the width of safety area between the lower curve of the primary bench and the sole medium of the pit field, m.

For example, in Poltavskyi ore mining and dressing found that the average height of the line is 170 – 180 m. Mining were achieved without increasing the depth at the back for directions recommend measures to limit border at the end of pit fields (Figure 7). Then, mining of the middle part was planned to a depth of 320 m.

In the temporary inner dump was laid 60 million m^3 of overburden. The annual volume of inside dump was 4 – 5 million m^3. The use of internal storing allowed to reduce transportation distance of overburden with automobile transport on 1.2 – 2.5 km, rail transport – 5 km. Saved from external disturbances dumps 6 hectares of arable land.

Figure 7. The system of pit development in Poltavskyi ore mining and dressing plant with temporary storing of overburden rock.

Filling of depleted pits or areas is a major event, not only promotes the rational use of land resources during mining operations, but also significantly reduce the cost of mining. Placement of overburden rock in the gob area of deep pits minimizes the distance of transportation and protects the land area of the external disturbances dumps. Therefore, during deposits or pit fields operating the significant size should provide the first stage of a pit, work to limit of their contours, and then lead to the subsequent development with storing of overburden rocks in gob area.

So, overburden rocks from the pit No 2-bis of Ltd. "ArselorMittal Kryvyi Rih" are delivered by trucks with carrying capacity of 120 t and stored in a gob area in pit No 1 using EKG-4U. The depth of pit is 300 m. The amount of gob makes 74.4 million m^3. Its northern border is adjacent to the previously paved and remediated Burschytskomu dump. The size of pieces about 400 mm makes 40%.

According to the technical solution (Patent No 56822A) in the final stage of each dump stope dumping excavating state from one place along an arc with a maximum radius of discharge at the top thereof warehoused lump rock, which by its own weight slides down. From the collapsed rock mass is formed resistant prism at the bottom dump stope slope with an angle of less natural. To improve the stability of stope

dump moving depleted front works are around the board spent a pit or a portion thereof with the transition of each of the current stope to the previous.

Dumping of slope sites were with the length of 200 – 300 m. During the formation of dangerous displacement of rock dumping operation on the job site was discontinued and transferred to the adjacent. After stabilization of overburden within 2 – 3 months the work on the idle site was resumed. Common front of dumping works should be divided into 3 – 5 sections. Monitoring of subsidence of overburden surveying measurements carried out continuously during the entire period of work.

It should be noted that the dumps in the complex of "AMKR" will soon be filled, and even now are seek opportunities to compensate for their disposal containers. Therefore, for intensification of filling in pit No 1 in 2006 commissioned dragline excavators ESH-6/45 with the annual output of 1.17 million m^3 (Patent No 28816) (Figure 8, 9) provides accommodation for the total volume of overburden rock in pit No 2-bis to its full mining. At the same time after dump advance in the pit front No 1 on 60 – 80 m from the original position, there is the possibility of organizing its square further to 2 – 3 dump lines and increase their level from top previously slept Burschytskym dump.

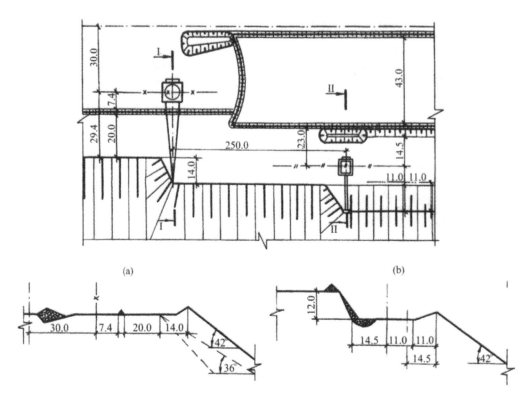

(a)   (b)

Figure 8. General view of the joint work of excavators ESH-6/45 and EKG-4U during formation of the basic layer of internal waste dump in career No 1: (a) cross section I-I; (b) cross section II-II.

Figure 9. Storing of overburden in the depleted open pit No 1 with using EKG-4U and ES-6/60.

For safe operation of the excavator EKG-4U at main layer dumping using dragline ES-6/45 is formed stable stope with the width of 14 m using crushed overburden rock from central terminal of pit No 2-bis. They are delivered by railway trains. ES-6/45 is located on the distance – 30 meters from the axis of the rail track. The direction of its advance guided enclosing the shaft, which is built up by moving the excavator in the direction dumping and removed from the opposite side. Axis motion dragline is 29.4 m from the boundary of subsidence and collapse of dumped stope.

3 CONCLUSIONS

At a distance not less than 250 m from the dragline excavator moves EKG-4U and backfilled stope width of dumping of 10 – 12 m. It is positioning on the consolidated massif of dumping excavator stope EL 6/45. Possible border of rock subsidence and collapse in stope removed from the axis of ECG-4U motion 8 – 9 m. Dumping stope formed the bulk of 5 – 6 m above the working platform ECG-4U. This rocks is used subsequently to repair and seal cracks and subsidence in the bulk rock at its consolidation.

Hard rock for EKG-4U is delivered by trucks from the pit No 2-bis. The site of unloading is on the level of the working platform ESH-6/45, with safety shafts up to 2 m from the gob and the railway. When depleted front advance shafts are increased by dragline ESH-6/45. The width of the discharge area is 43 m, height up to 12 m. Rock faces into a hopper for EKG-4U bulldozer. The total width of the dump stope that form together excavators ESH-6/45 and EKG-4U, is 25 m. The length of the front of their work is 1.75 km.

Start-up operation of dump for large-scale storage of overburden and oxidized quartzite in the deplated pit No 1 and the adjacent surface to the top of the closure with the previously paved Burschytskym damp allows to solve the problem in the long term stacking when working pit No 2-bis and No 3 for a considerable period. This primarily relates to the use of land allotment for storage overburden and oxidized quartzite instead of new areas of alienation from land users.

REFERENCES

Afronin, V.H. & Nesterov, P.H. 1972. *Mining of Ukraine in 1971 – 1975 years*. K: Technika: 148.

Authors' license No 1040152 USSR, IPC E21 C41/00. *A method of pit recultivation by inner level*. Dryzhenko, A., Shebko, V., Kuc, V. et al. Pub. 7.09.83, Bul. No 33.

Authors' license No 1121434 USSR. IPC E21 C41/00. *The method of steep deposits development*. Dryzhenko, A., Shebko, V., Kuc, V. et al. Pub. 30.10.84, Bul. No 30.

Authors' license No 1330313 USSR. IPC E21 C41/02. *The method of dip pit recultivation*. Dryzhenko, A., Charpko, N., Hardash, N. et al. Pub. 28.10.87, Bul. No 30.

Authors' license No 825771 USSR. IPC E21 C41/00. *A method of pit recultivation from the inner rampart-ledge, including its dumping*. Dryzhenko, A., Prosandeev, N., Martynenko, V. et al. Pub. 30.04.81, Bul. No 16.

Braun, H.A. 1970. *Iron-steel Base*. M: Nedra: 312.

Maleckyi, N.A., Kabanov, A.V., Baryshpolec, V.T. 1986. *Integrated use of mineral resources for the enrichment of ferrous metals*. M: Nedra: 192.

Patent No 1836561 RU. IPC E21 C41/26. *The method of steep deposits development*. Dryzhenko, A., Martynenko, V., Badahov, V. et al. Pub. 23.08.93, Bul. No 31.

Patent No 28816, UA, IPC E21 C41/32. *The method of recultivation of depleted dip pit*. Dryzhenko, A., Drobin, H., Hrycyna, O. et al. Pub. 25.12.07, Bul. No 21.

Patent No 56822A, UA. IPC EC21 C41/32. *The method of recultivation of depleted dip pit*. Dryzhenko, A., Symonenko, A., Sokurenko et al. Pub. 25.12.07, Bul. No 2.

Patent No 8679, UA. IPC E21 C41/00. *The method of open pit development*. Dryzhenko, A., Trubitsyn, M., Shebeko, V., et al. Pub. 30.09.96, Bul. No 3.

Technical and economic indicators of mining enterprises for 1990 – 2001. 2002. Ekaterenburg: 379.

Theoretical and Practical Solutions of Mineral Resources Mining – Pivnyak, Bondarenko & Kovalevska (eds)
© 2015 Taylor & Francis Group, London, ISBN: 978-1-138-02883-8

On the problem of operation schedule reliability improvement in mines

O. Mamaykin
National Mining University, Dnipropetrovsk, Ukraine

ABSTRACT: This article represents the idea of determining technological reliability of a mine. It include economic criterion to evaluate reliability of operation schedules of mines, which depend on management methods of reliability of certain mine subsystems. Principle one means use of improved that is more capital intensive mining equipment, transport systems etc. Principle two is component redundancy.

1 THE PROBLEM DEFINITION

Increase of coal share in the structure of energy resources is the world tendency. Currently coal satisfies as much as 25% of the world's need for energy resources. Before 2030 coal share in power production will increase up to 45%. Our country is no exception in terms of coal production. Even after the process of all non-profitable and ruined mines closure due to military operations, in the near future coal will stay the key energy resource in Ukraine. The main reason is instability of gas deliveries from the Russian Federation which dictates the necessity to oust gas from each segment of power industry and household. Moreover, instability in Eastern regions affects adversely the plans for shale gas production relegating to the background opportunity to produce great quantity of the fuel. That is why use of national coal as fuel for power stations is the most rational idea to cut demands for gas from Russian Federation in the Ukrainian market (Official site of Ministry…).

2 FORMULATION OF THE PAPER OBJECTIVE

The objective of the paper is to generalize and develop scientific and methodological background, to develop tools and practical recommendations as for regulation of streams of governmental support for unprofitable enterprises; it aims at improving potential of specific coal-mining region.

3 STATEMENT OF BASIC MATERIAL

Today control of operational reliability of mines from the viewpoint of investment addressness in terms of certain segments restraining productivity enhancement is one of the basic problems of coal industry. Generally, to determine a level of technological reliability it is required to consider capacity of basic segments of a mine: mining operations, underground transport, hoisting, ventilation, and technological complex of surface. Each production unit of a mine is evaluated in terms of potential of equipment or structure of construction. First of all, it concerns mine working or interaction of the two components (for example, main ventilation fans power and section of mine workings as well as their aerodynamic resistance). At first thought it seems expedient to assume ratio between actual mining (that is actual loading level of specific production segment or process) and its potential or capacity. However, such a factor will characterize actual use of capacities rather than them. Strictly speaking it is of subjective nature. Moreover, a coefficient determined in such a manner will give distorted information from the viewpoint of a mine future (Pivnyak et al. 2004).

One example can illustrate that. Suppose that a mine having 2000 t/day capacity of underground transport mines 1000 tons in fact. In the context of the approach, coefficient of its technological reliability is 0.5. Such a value means that it is required to modernize underground transport to improve coefficient of reliability. However, improvement of the mine performance (that is use of subjective factors) is required rather than its modernization.

Potential of a mine is maximum capacity of basic subsystems of operation schedule. For example, preliminary work as a component of the segment is not limiting factor as in terms of actually perspective (even high) load on stopes it is always possible to provide required paces to construct mine workings being adjacent to a longwall. It also concerns all other mine workings. Ventilation is general mine limiting factor despite it is not a component of direct production process. Main ventilation device is its mobile component which can be reshaped with relative ease. However, mine workings are much more inert. Increase in a mine service life results in their extension; thus, potential capacity in terms of ventilation factor reduces.

Abovementioned make it possible to propose following idea to determine technological reliability of a mine. Sequence of such basic production processes is considered: mining; transport from a longwall to main mine working or block one; transport within the mine working (inclined drift, slope, mother entry); within mine workings which transport coal to main haulage roads (drifts); transport within main haulage roads (common drifts, main drifts, and panel ones), vertical or inclined haulage, surface coal technological chain including loading facilities. Each component's capacity should be determined. Moreover, capacity of a mine as for ventilation factor is also identified. In this context, capacity is determined in such a way to meet the demands of current standards or those reserves taken depending upon other thoughts. Maximum capacity within considered sequence and ventilation is taken as a unit. Ratio between capacity of any other process and maximum value taken as described is a coefficient of technological reliability of the production process.

Processes aimed at the system maintaining also effect technological reliability of a mine and a level of the whole operation schedule passport. However, the effect is not manifested as directly as in comparison with basic production processes. In addition, one more important reason is available: processes to maintain the system are less inertial to compare with basic production processes; they need not so much expenses and time.

Effect of segments (subsystems) on technological reliability may be multidirectional. Some of them (for example, degassing, air cooling, mine working maintenance, support transport etc.) may effect extent of production if they are of poor capacity (to be more exact, if they are not maintained properly). If possibilities of the processes are higher to compare with possibilities of basic processes (for example, when substantial decrease of mining takes place) the reserves will hardly be used: nobody will increase capacity of conveyor transport if discharge pumps are underutilized. State of certain processes (maintenance of mine workings, repair of equipment and electrical facilities) depends on organizational factors rather than technological ones.

There are several cases when factors of system maintenance may exercise important effect on technological availability of a mine. As an example indicate a case of extremely unstable rocks including severe rock heaving which may involve specific expensive and labour-intensive timbering as well as intensive activities connected with maintenance of mine workings. In this respect, settle upon a process of development workings construction. Rather often they are the factor limiting mining in operative mines. However, they are not generally supported by objective reasons. In practice such a situation results from bad mine management. In the context of Donbas very unstable rocks are few exceptions. Different from basic processes, effect of processes of system maintenance on technological reliability of a mine involves specific techniques to evaluate the effect.

Level of specific operation schedule passport (Pivnyak et al. 2004) compared with a standard may become economic criterion to evaluate reliability of operation schedules of mines. Study of operation schedule target function extremum is possible when adequate management of reliability of certain mine subsystems is available. Two basic of them have been used in the study. Principle one: modernization means use of improved that is more capital intensive mining equipment, transport systems etc. Principle two is component redundancy. Their application needs more single expenditures at different stages of a mine life cycle. That very time they are connected with increase (to compare with base variation) in initial investment. If redundancy takes place within operating level, expenditures connected with planned operational measures aimed at maintenance of basic segments of operation schedules experience their increase. Thus, in the process of mine operation schedule control in terms of certain level working period, "investment-reliability" dependence $S(P)$ is increasing function. That is:

$$\left[\frac{dS_F(P)}{d(P)}\right] > 0. \tag{1}$$

Specific value of one of $dS_F(P)$ components – semi-variable costs – does not depend on accepted technique to control reliability, and specific value of semi-fixed operating expenditures component decreases with reliability growth. In other words, both absolute value of 3rd component (losses for deviation of actual reliability level of operation schedule from ideal one) as well as its specific value is decreasing function.

Thus, efficiency criterion of operation schedule of a mine is additive function of specified values; that is it has its extremum for minimization. As a result of reliability control in terms of certain components of operation schedule of a mine, extra investment aimed to improve basic reliability level are distributed among components and segments of operation schedule. As a consequence of this equipment modernization or redundancy improvement increase passport level of operation schedule of a mine and preferences are determined as for the enterprise capacity financing depending upon demand for end coal products. Many studies have proved (Salli et al. 1994, Bondarenko et al. 2012, Lapko et al. 2014, Bondarenko et al. 2012, Sally et al. 2014) that low capacity of technological segments is the key problem of the majority of mines with long-life performance. That is also true for limited possibilities in the context of "mining operations" factor. Moreover, it generally depends on decentralization of operations and availability of several stopes with relatively low (less than 700 tons) daily output within mine field. Such a situation is absolutely rejectable as support for capacity of promising mines depends on substantial investment of each technological segment just to reconstruct mining line. In the context of limitations in the context of the state and industry that will result in further worsening of economic conditions of a group of promising mines. Thus, longwalls and adjacent segments need primary investment with similar scheduling for mining under the terms and conditions of maximum concentration of mining operations.

Reliability which means efficiency of working area depends on technological segments connected with it as their downtime or low capacity limits efficiency of the working area. When seam mining with enclosing roof and floor undercut is applied operation schedule of a mine as complex system is characterized by a number of interacting components and multistage structure to be the main reason of deconcentration of mining operations and rock mass quality degradation. As a rule, reliability analysis of such schedules involves application of reliability logistic scheme. In this context operation schedule of a mine is a system of elements in series-in parallel in terms of reliability. It depends on event probability – trouble-free performance of each component of the system. Such an approach makes it possible to identify optimum reliability values of components and segments as well as the whole operation schedule of a mine. The values ensure single expenditures involving possible loss resulting from deviation of actual reliability level from ideal one when certain site of a mine field (panel, block) is under operation.

Algorithm of stepped development of reasonable analysis scheme is shown on Figure 1 (Wagner 1973, Sotskov & Gusev 2014). Qualitative reliability in the form of failure-free operation probability in terms of specified period (shift, day) is determined for each ith component of operation schedule.

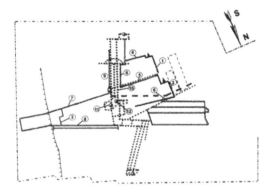

Figure 1. Development plan for "Fominski" h_8 seam of "Progress" mine: 1 – 3 – longwalls; 4 – 8 – site mine workings; 9, 10 – permanent mine workings; 11, 12 – shafts.

Optimum values of reliability factor ensuring minimum expenditures connected with mining involving possible loss due to reliability level deviation in terms of seam mining according to actual data on reliability of components and "investment-reliability" dependences use above procedural principles. Money resources to support capacity of mines are distributed among components of operation procedure in such a way to make its general reliability maximum. Notion "resource" means increase in expenditures connected with innovative activity that is facilities for stopes, construction and maintenance of mine workings as a result of reliability improvement, process-oriented differentiation, and new equipment instead of old as well as transition to higher organizational and technical production level.

Each operation schedule of a mine (Figure 1) consists of m levels and classes within which r of operational components are available. For each ith component of opening and development plan, reliability qualitative characteristics (P_{ij}) as well as cost "investment-reliability" functions in terms of all processes and activity types have been preset. Moreover, tolerance zone for functional and dependence of a mine operation schedule functioning reliability on reliability qualitative characteristics of components of technological elements and segments are known. The key idea of the technique is as follows. Extra investment to improve basic, actual reliability level of operational schedule for opening and development are divided into M in the form of discrete set.

Following step is to identify that segment of operation schedule which investing in the form of extra capital costs ΔK will result in maximum (in terms of the step) reliability improvement of the whole performance. In this context, components of the segment experience their increase; reliability characteristics of technological components of other classes of the structure are unchanged. Reasonable distribution of limited extra capital investment among components and segments of operation schedule of a mine is performed with the help of discrete programming methods and ideas of relaxation search (Wagner 1973).

Thus, currently national budget still regulates potential of mines. That is why they underachieve due to lack of drastic changes in a system of target investment in certain segments. Therefore, a technique to control reliability of operation schedules of a mine is proposed. Together with adequate index of resource potential application it will favour loss ratio lowering from the viewpoint of investment support.

4 CONCLUSIONS

1. A mine has five basic technological segments. One of them is of maximum capacity, and another one is of minimum. Ratio between minimum capacity and maximum one is technological reliability coefficient. At every instant the coefficient has universal deterministic value which depends upon technical state of a mine and characterizes the state: minimum value determines actual possibilities of the mine; maximum value determines it's potential. The fewer coefficient of technological reliability is the greater is gap between chief segments. Thus, implementation of potential of a mine becomes more difficult.

2. Labour capability of a mine depends on reliability of components involving value of reserves and their amount. Moreover, the fewer reserves are the tougher should be policy in the context of keeping achieved extent of mining at the expense of concentration of production. In this statement, optimality criterion is a value which extremum reflects boundary (in terms of specified limitations) possibility to mine reserves with cost minimization up to border breakeven result.

3. The greatest reserves for productivity enhancement of mines (together with their resource potential management) is in focused control of operational reliability of certain technological segments to increase extent of production through decrease in breakdown of machines and mechanisms as well as implementation of target investment system for components basing upon "Investment-reliability" approach.

REFERENCES

Bondarenko, V., Kovalevs'ka, I. & Fomychov, V. 2012. *Features of carrying out experiment using finite-element method at multivariate calculation of "mine massif – combined support" system.* Geomechanical Processes During Underground Mining. The Netherlands: CRC Press/Balkema: 7 – 13.

Bondarenko, V., Symanovych, G. & Koval, O. 2012. *The mechanism of over-coal thin-layered massif deformation of weak rocks in a longwall.* Geomechanical Processes During Underground Mining. The Netherlands: CRC Press/Balkema: 41 – 44.

Salli, S., Bondarenko, Y. & Tereshchenko, M. 2009. *Management of technical-and-economic parameters of coal mines.* Dnipropetrovsk: Gerda: 150.

Lapko, V., Fomychov, V. & Fomychova, L. 2014. *Modern technologies of bolting in weakly metamorphosed rocks: experience and perspectives.* Progressive Technologies of Coal, Coalbed Methane, and Ores Mining. The Netherlands: CRC Press/Balkema: 347 – 350.

Official site of Ministry of Energy and Coal Industry of Ukraine: [E-resource]: http://mpe.kmu.gov.ua/.

Pivnyak, G., Amosha, A., Yashchenko, Y. and other. 2004. *Reproduction of mine fund and investment processes in coal industry of Ukraine.* Kyiv: Naukova Dumka: 331.

Salli, V., Malov, V. & Bychkov, V. *Maintenance of coal mine capacity in the context of new construction limited possibilities.* Moskva: Nedra, 1994. 272.

Sally, S., Pochepov, V. & Mamaykin, O. 2014. *Theoretical aspects of the potential technological schemes evaluation and their susceptibility to innovations.* Progressive Technologies of Coal, Coalbed Methane, and Ores Mining. The Netherlands: CRC Press/Balkema: 479 – 483.

Sotskov, V. & Gusev, O. 2014. *Features of using numerical experiment to analyze the stability of development workings.* Progressive Technologies of Coal, Coalbed Methane, and Ores Mining. The Netherlands: CRC Press/Balkema: 401 – 404.

Wagner, G. 1973. *Foundations of operational research.* Moskva: Mir: 502.

Theoretical and Practical Solutions of Mineral Resources Mining – Pivnyak, Bondarenko & Kovalevska (eds)
© 2015 Taylor & Francis Group, London, ISBN: 978-1-138-02883-8

Elements of the technology of storage gases in the gas hydrate form

L. Pedchenko & M. Pedchenko
Poltava National Technical University named after Yu. Kondratuk, Poltava, Ukraine

ABSTRACT: The technology of production of the gas hydrates, suitable for the non-equilibrium conditions for storage, is developed and tested. Large blocks are formed from pre-cooled mixture of crushed and the granulated mass of gas hydrate. Technology preservation of gas hydrate blocks by the layer of ice at atmospheric pressure is designed to improve their stability (two repeats application of water and its crystallization basically at the expense of the energy of the gas hydrates). Formed in such a way the blocks of gas hydrate are suitable for long-term storage and transportation without additional cooling in converted vehicles (tankers, barges, cars). Gas hydrate can be stored in improved the terrestrial inflatable structures.

1 INTRODUCTION

The reserves of natural gas is rapidly exhausted. Nearly 80% of deposits opened into category small and medium-sized distant. However, much of it is not being developed, as traditional transportation technologies are often ineffective. At present gas is transported mainly with pipelines or LNG-tankers and storage – in underground storage. However, in recent years, technology based on the ability of gas molecules and water form gas hydrates, is actively developed. In the composition of gas hydrate considerable volumes of gas can long be stored at atmospheric pressure and a slight negative temperature. This technology has undeniable prospects for implementation in the near future, but needs improvement and testing of its elements.

At present several concepts the transportation of gas in hydrate form are considered. The technology for the transportation of non-equilibrium conditions (small negative temperature and atmospheric pressure) is the most attractive. It needs the production of gas hydrates in the most stable form under these conditions. At present granules of the gas hydrate is proposed transport today (Gudmundsson 1996).

However, over time freezing granular hydrates, complicating unloading (Dawe et al. 2003). Also granulated hydrates only 78% fills the volume vehicles or storage (Gudmundsson & Graff 2003). Besides, much of the total surface area of the granules and system of open channels between channels granules stimulates the process of volumetric dissociation of the gas hydrate mass. Keeping them stable at atmospheric pressure needs additional costs for cooling to temperatures below a 258 K. Monolithic blocks of large size are a good option. However, today the industrial technology of production does not exist (Yakushev et al. 2008).

2 MAIN PART OF THE ARTICLE

Due to thermal properties and features process of transportation and storage, gas hydrates is proposed to produce in the form of gas hydrate of blocks maximum cooled large, preserved with the layer of ice and refrigerated in the production process to the desired level.

This technology provides intensive production synthetic gas hydrates. Moreover it must have a maximum gas content (up to 160 m^3/m^3). According to the research the formation of hydrates proposed to perform in the contact devices on the basis of jet devices (ejectors with elongated mixing chamber or jet devices with free stream). Their application allows increasing efficiency and simplifies the process of technological design of technology (Pedchenko & Pedchenko 2014). Possibility of organization of continuous production of gas hydrates when using the jet devices as contact devices also substantiated and confirmed experimentally (Figure 1).

In addition, the problem of section of water in the process of separation from the produced hydrates of free water (captured between the crystals and the film) was recorded during the research. Therefore, after the separation the gas hydrates requires draining by transferring of the balance of free water into the composition of gas hydrates (with stirring the mixture is blown cooled gas). Blocks will be formed from the previously cooled mixture of crushed and granulated hydrates of minimum porosity in the respective proportion. This solution allows to obtain blocks of uniform density and waive the need them cooled (Figure 2).

The manifestation of creep deformation is observed at the study of process of formation of the gas hydrate mass (Figure 3). Its consideration will allow reduce on the order the pressure of formation, and hence the energy consumption for this operation. For example, with the continuation of the time of exposure of efforts to 8 minutes, the pressure required to formation of the hydrates to established porosity (0.08 – 0.1), reduced from 57 MPa (forming with blow) to 2.7 MPa.

(a)

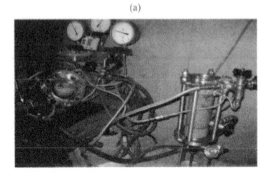

(b)

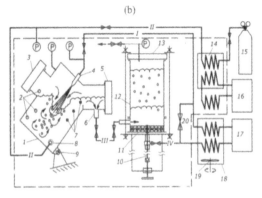

Figure 1. The laboratory unit for continuous production of gas hydrates: (a) photos; (b) scheme; 1 – reactor; 2 – temperature sensors; 3, 5, 13 – observation windows; 4 – inkjet apparatus; 6 – branch pipe; 7 – LEDs; 8 – bubbler; 9 – hinge; 10 – rod; 11 – plunger with filter; 12 – separator; 14, 18 – exchanger; 15 – gas bottle; 16 – refrigerator; 17 – pump; 19 – mixer; 20 –valve; flows: I, IV – water; II – gas; III – mixture of hydrate and water.

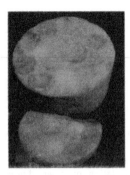

Figure 2. Transverse section of the gas hydrate block after the formation.

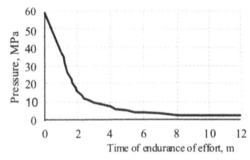

Figure 3. The dependence of the pressure necessary for the formation of gas hydrate mass on exposure time of the effort.

The next mechanism of the creep of deformation of gas hydrate mass offered: the compressing load – the destruction of the crystal lattice of gas hydrates and ice – conversion of mechanical energy into heat – partial dissociation of gas hydrates (with the release of gas and free water including supercooled and absorption of energy at dissociation) – formation by the gas and water of the gas hydrate masses in other pores and heat release. The processes take place at the same time and preferably in the areas of contact structural units. The processes of dissociation and destruction of crystals dominate at a temperature higher than point of water of the crystallization and atmospheric pressure. The processes recrystallization dominate at low temperatures: reformation of the gas hydrates with the dissociation products and crystallization of supercooled water.

In non-equilibrium conditions manifestation of the effect preservation slows the dissociation of gas hydrates (Yakushev & Istomin 1990). The balance of power in the surface layer of hydrate at the preservation is determined by its porosity. If for dissociation of layer of gas hydrate (with heat exchange) consumed more energy than on the crystallization of water that evolved as a result (on the surface or in the pores of the surface layer of the sample), self-preservation with the layer of ice occurs.

To establish the conditions of its manifestation the expression is proposed:

$$\frac{L\rho_{ice}}{\rho_w} \leq \left(t_{freez} - t_{gh}\right)\frac{c_{gh}(1-m)}{m} +$$

$$+ \frac{Hm\left(t_{273} - t_{gh}\right)}{0.87\alpha_{air}\left(t_{gh} - t_{air}\right)}\sqrt{\frac{\lambda_{gh}c_{gh}\rho_{gh}(1-m)}{\pi\tau}}m, \qquad (1)$$

where t_{gh}, t_{air}, t_{273} and t_{freez} – temperature of the sample, of air, of the crystallization of water and gas hydrate after crystallization of water, respectively, K; τ – time of heating of the sample, s; ρ_{gh}, ρ_{ice} and ρ_w – density of the gas hydrate, ice and water, respectively, kg/m³; m – porosity; c_{gh} – specific heat of gas hydrate, J/(kg·K); L – specific heat of crystallization water, J/kg; λ_{gh} – thermal conductivity of the gas hydrate, W/(m·K); α_{air} – heat irradiation, W/(m²·K); H – enthalpy of formation of hydrates, J/kg.

To increase stability and the mechanical strength of hydrate blocks necessary force to preserve with the layer of ice. For this purpose, on their surface freeze crust 1 – 2 mm is sufficient (ice crust, formed as a result self-preservation is thinner) (Gudmundsson & Parlaktuna 1991). However, dissociation is negligible if the energy of the wet layer enough for the crystallization of water (Figure 4). For example, for quality preservation is necessary that the initial temperature of the sample of gas hydrate porosity within 0.07 – 0.15 should be between 248 – 259 K.

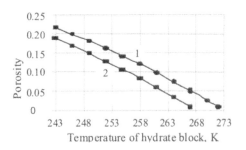

Figure 4. The dependence of the porosity hydrate on the initial temperature for the implementation of its forced conservation: curves: 1 – at a pressure of 0.8 – 0.9 MPa; 2 – at atmospheric pressure; (temperature of the hydrate after crystallization of water: at a pressure of 0.8 – 0.9 MPa – 273 K, at atmospheric pressure – 268 K).

In the study of the force preservation of the gas hydrate blocks the need re-application of water from endurance to crystallization of the previous layer was established. In the first application main part of the pores was blocked and «cementing» of the surface of the sample to a depth of water penetration. In the second – pores are blocked completely, surface defects are smoothed and an ice layer are frozen (Figure 5).

To the experimental establish of the dependence of time of crystallization of moistened layer of the gas hydrates on its thickness was applied at water samples with an initial temperature of 259 K to 1.5×10^{-5}, 3.0×10^{-5} and 4.5×10^{-5} m³ for 6 seconds, and to establish of the dependence of time of crystallization moistened layer of the gas hydrates on its initial temperature of – 1.5×10^{-5} m³ (Figure 6 – 9, Table 1 and 2). Crystallization was considered completed when the growth of temperature of the surface layer stopped.

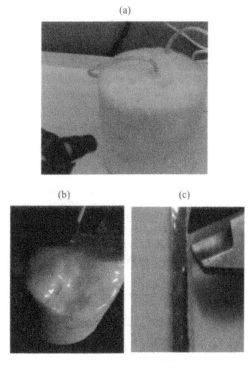

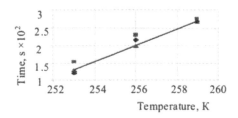

Figure 5. Forced conservation of sample of gas hydrates with the ice layer: (a) application of water; (b) "shine" of a layer of ice on a sample; (c) ice layer on a section of a sample.

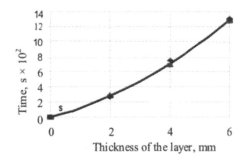

Figure 6. Dependence of time freezing of a layer of ice from reference temperature of a sample.

Figure 7. Dependence of time freezing of a layer of ice from its thickness.

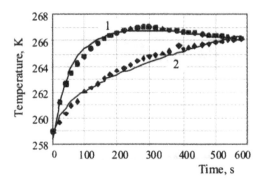

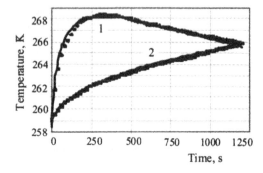

Figure 8. Dynamics of change of temperature of a surface (curve 1) and centre of a sample (curve 2) during cementation of its surface (the first drawing of water).

Figure 9. Dynamics of change of temperature of a surface (curve 1) and centre of a sample (curve 2) in process freezing of a layer of ice (repeated drawing of water).

Table 1. Approximation of experimental data of dependence of time of crystallization of water in the moistened layer from its thickness and from the initial temperature of gas hydrate.

Factor	Time of crystallization, s	r	v	Number of the formula
Initial temperature, K	$\tau = 23.2 \cdot t - 5745.8$	0.97	0.093	(2)
Thickness of the layer, mm	$\tau = 18.4 \cdot 1^2 + 103.4 \cdot l$	0.997	0.049	(3)

Table 2. Approximation of experimental data change of temperature of a superficial layer and at the centre of a sample in process crystallization of water.

Submission of water	Temperature, K	The equation of dependence	r	Criterion F	Number of the formula
First	superficial	$y = -1.99 \cdot (ln(\tau /330 + 0.128))^2 - 6.24$	0.990	1418.41	(4)
	centre	$y = 0.60 \cdot (\tau + 2.0)^{0.41} - 15$	0.989	833.51	(5)
Second	superficial	$y = -1.53 \cdot (ln (\tau / 330 + 0.08))^2 - 4.77$	0.992	2061.34	(6)
	centre	$y = 0.42 \cdot (\tau + 6.20)^{0.41} - 15$	0.992	2587.66	(7)

During research of the kinetics of dissociation of samples ($d = 0.08$ m, $h = 0.11$ m) made according to the proposed technology for isobaric and isochoric conditions, we found that they a long time (390 min.) were in the non-equilibrium conditions (atmospheric pressure, $T = 276$ K). The process takes place without damaging the integrity of the sample. We found that the forced preservation gas hydrate blocks can increase the maximum storage temperature at atmospheric pressure to $270 - 273$ K.

The mechanism of dissociation of samples is described as follows.

In the case of isochoric of heating process surface of the sample to the melting point of ice and the corresponding value of its inner part occurs first. Further heating will go inside, and the surface temperature remains constant (273 K). More heat goes inside of it, and the surface temperature remains constant (273 K). This layer of ice melt, reaching a certain value but its thickness is stabilized. With further energy intake surface ice layer melts, melt water flows down, and on the other side of it is the crystallization of supercooled water formed due to hydrate dissociation surface. Gradually contact «gas – water – ice – supercooled water – gas hydrates» moves toward the center of the sample. However, under the pressure of gas released is temporary chaotic violation of the integrity peel. The next moment supercooled water crystallizes and its integrity is restored. In the isobaric conditions at achieving equilibrium pressure in volume process terminated.

Consequently, the existence of the sample in equilibrium conditions (for example, as a result of application of water) thickness of the ice crust is determined by the balance of energy and amount of water in its pores and on the surface. Thus, forced preservation allows you to create for of gas hydrate under a layer of ice conditions its stability while the sample itself may be in equilibrium conditions. The method of production of gas hydrate in the form of blocks with internal energy source, preserved layer of ice is proposed, considering the properties of gas hydrate and experimental results (Figure 10 and 11).

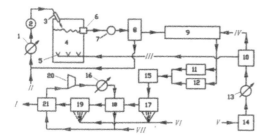

Figure 10. Method of production of gas hydrate blocks:
1 – cooling of water; 2 – water pumping; 3 – feed water stream into the reactor; 4 – formation of gas hydrates; 5 – bubbling gas; 6 – selection of gas hydrates; 7 – vibration treatment; 8 – separation; 9 – concentration and cooling; 10 – crushing of hydrate; 11 – pelleting of hydrate; 12 – separation of gas; 13 – cooling of gas; 14 – preparation of gas; 15 – formation of blocks; 16 – cooling of air; 17 – the first water supply to the blocks; 19 – the second water supply on the blocks; 18 – cementing surface; 21 – freezing of layer of the ice; 20 – air circulation; stream I: – blocks; II, VI – water; III – methane; IV – condensate; V – gas; VII – air.

It includes: gas hydrate production with a significant water content, its separation, enrichment of the gas hydrate mass by transferring of the residue water (be-tween the crystalline and captured) to the hydrate and increasing filling with the gas molecules of the crystal lattice, single the cooling of gas hydrates (T ≤ 258 K) granulating one part and crushing of the gas hydrates another, the formation of blocks with the mixtures forced their conservation with the layer of ice.

For realization of this method project pilot plant with a capacity of 20×10^3 m^3 gas per a day (140 tons per day hydrate) is developed (Figure 12). Pilot plant is designed for production of gas hydrate blocks weight to 250 kg. Gas consumption for technological needs of the pilot plant during the summer installation is 2400 m^3 per a day (or 12% of its capacity). At low temperatures the gas consumption is reduced to 568 m^3 per a day (or 2.8%).

To increase the effectiveness of the proposed technology hydrate blocks must be stored under a layer of polyurethane (thickness of 0.5 – 0.7 m) in terrestrial Inflatable structures covered with double-layer soft shell with nonflammable gas locking layer (the nature of changes in the composition of this gas will indicate breach of integrity external or internal shell). These structures are closed constructions that "lying" on the gas cushion pressure which exceeds atmospheric only for compensation the weight of the shell (Figure 13).

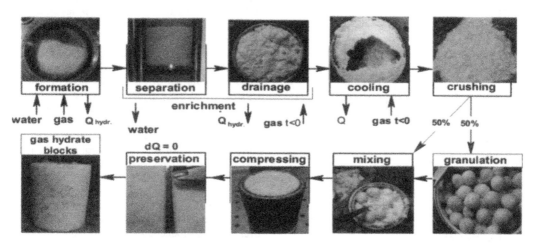

Figure 11. Scheme of method of production of gas hydrate blocks.

To place such storage we propose near of gas consumers. This will allow give gas to distribution networks of low pressure. Therefore hydrate dissociates at a pressure of 0.3 – 0.4 MPa, and therefore at a much lower temperature (coolant temperature to melt the hydrate will not exceed 283 K). Calculation of parameters of storage of the gas hydrates in the terrestrial Inflatable structures is given in the Table 3 and Figure 14 – the dynamics of change of the temperature in surface of the hydrate. Energy savings for cooling of the inflatable structures (Table 4) during storage of 5.2×10^6 m^3 of gas in the form preserved with the layer of ice of gas hydrate blocks is 49.3% (6.95×10^3 m^3 in terms of gas), including by: the temperature difference between storage of blocks, covered with layer ice and without (ΔT = 12 K) – 17% (2.69×10^3 m^3 of gas); improve the efficiency of the cooling system by an amount proportional to ΔT – 26.9%. Cost of storage during the year in terrestrial Inflatable structures of the gas hydrate blocks amount to in terms of gas 15.86×10^3 m^3, representing only 0.3% of the content terrestrial inflatable structures.

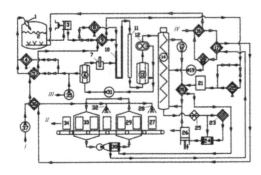

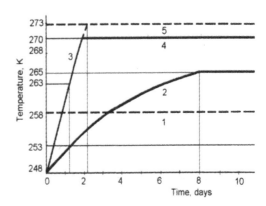

Figure 12. Principle scheme of the installation for the production of hydrocarbon gas hydrate in the form of blocks capacity of 140 t per a day (20000 m³ per a day of gas) 1 – inkjet apparatus; 2 – reactor; 3, 7, 19, 31, 35, 37 – pump; 4, 8, 22, 30 – device of air cooling; 5, 9, 16, 17, 20 – heat exchanger; 6, 13 18 – separator; 10 – coiled pipe; 11 – moderator; 12 – squeezing device; 14 – column of draining of gas hydrates; 15 – compressor; 21 – refrigerator; 23 – granulator; 24 – mixer; 25 – straw chopper; 26 – press for the formation of gas hydrate blocks (GHB); 27,34 – (GHB); 28, 32 – nozzle; 29, 33 – zone of blowing GHB; streams: I, III – water; II – GHB; IV – gas.

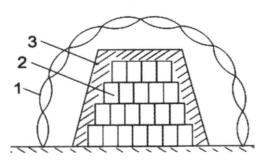

Figure 13. Scheme storage of gas hydrates: 1 – elastic the shell bilayer; 2 – hydrate blocks; 3 – layer of polyurethane.

Table 3. Calculation of parameters of storage of the gas hydrates in terrestrial Inflatable structures.

The thermodynamic parameters	January	July
Thermal resistance, $(m^2 \cdot K)/W$:		
– transition heat, R_{a};	0.23	0.09
– coating of the layer of PVC, R_{pvc};	0.025	0.025
– locking layer, R_{loc};	0.22	0.18
– covering (no layer of polyuret.), R_{cov};	0.34	0.28
– covering (with a layer of polyuret.), R_{pu}	17.0	16.6
The heat flow to the hydrates, W/m^2:		
– without insulation polyurethane, q;	20.58	104.1
– insulated with polyurethane, q_{pu}	0.64	92.4
The heat flux in the Inflatable structures to of gas hydrates, kW:		
– without insulation polyurethane Q_{exi};	80.77	409.0
– insulated with polyurethane, $Q_{ext\,pu}$	2.53	9.36
The flow of heat from the earth, Q_{ear}, kW	9.0	9.0
Energy consumption for cooling, Q_{cool}, kW	6.6	15.3

Figure 14. Dynamics of change of surface temperature of gas hydrates in ground storage: limit of hydrate stability 1 – without preservation; 2 – winter without additional cooling; 3 – summer without additional cooling; 4 – summer with additional cooling; 5 – forced preservation with the layer of ice.

Therefore, forced preservation of blocks with the layer of ice and separated in time of production operations and dissociation of gas hydrates in time allows on 42.9% (in terms of gas – at 466.5×10³ m³) to reduce energy consumption of the proposed technology and thus significantly improve its competitiveness.

To improve efficiency of the technologies, dissociation of the gas hydrate blocks in the summer must be carried out by solar energy, thus saving in terms of gas 594.26×10³ m³ (Table 5). Expenses for dissociation of blocks not covered with a layer of ice at the expense higher temperature of storage in terms of the gas will be higher at 33.16×10³ m³.

In the production of gas hydrates to 80% of the energy is consumed in recycling heat from the process. Before the gas consumption of the same amount of energy expended on melting hydrate. Use of gas hydrate blocks (large of the size, refrigerated and preserved with the layer of ice) and inexpensive terrestrial Inflatable structures allows to divided over time its production and dissociation. To produce of the gas hydrates useful in the cold season (T < 278 – 280 K), and melt in the warm (T > 280 K), respectively using natural energy from the sun and cold.

Accordingly, the proposed technological chain, which involves: production of gas hydrates in the form of gas hydrate blocks its transportation without additional cooling, storage vaults in the ground at a temperature of 270 K, hydrate melting by solar energy. (The gas structured than fresh water, cold, compressed gas energy receive.) This solution will improve the efficiency of gas transportation technologies to hydrate form to a competitive level.

Table 4. Comparison of costs for 5.2 million gas storage in the form of gas hydrate blocks force preserved with layer of ice and without it, in terrestrial Inflatable structures.

Forced preservation	Proceeds heat into the terrestrial Inflatable structures, MJ / day	Days of the storage	Energy consumption for cooling of storage		Consumption of gas for storage 5.2×10^6 m^3 in the composition of gas hydrate		Temperature of the storage, K	The savings in terms of gas from the temperature difference storage of blocks covered with ice and without ($\Delta T = 12$ K)				
			Day, MJ	Entire periode, MJ				Straight		Improve the efficiency of the cooling system		
					$\times 10^3$ m^3	% gas		$\times 10^3$ m^3	% gas	$\times 10^3$ m^3	%
					Storage for "warm period" (15.04 – 15.10)							
+	1285.6	180	1071.4	192845	6.43	0.12	270	1.50	9.5		
–	1586.3	180	1321.9	237946	7.93	0.15	258	–	–		
					Storage for the "cold period" (15.10 – 15.04)							
+	620.4	180	413.6	74442	2.48	0.05	270	1.19	7.5		
–	916.7	180	611.1	110005	3.67	0.07	258	–	–	4.26	26.9
					Storage during the year							
+		360	742.5	267287	8.91	0.17	270	2.69	17.0		
–		360	966.5	347951	11.6	0.22	258	–	–	4.26	26.9
		360	1321.9	475884	15.86	0.30	258	2.69	17.0	4.26	26.9

Table 5. Comparison of energy consumption for dissociation of 2870 tons (typical contents of terrestrial Inflatable structures) of the gas hydrate blocks force preserved with the layer of ice and without in different seasons.

Initiale temperature of the blocks. K	Forced preservation of the blocks	Solar radiation MJ/m^2 at the day, ($T_{air} \geq 285$K)	Energy consumption for dissociation Content terrestrial Inflatable structures (2870 tons hydrate) due to the energy				Save of gas due, $\times 10^3$ m^3 / year				
			Sun in terms of gas		Of the gas		Preservation	(higher temperature storage)	Energy of sun	Total	
			$\times 10^3$ m^3	%	$\times 10^3$ m^3	%					
258	–	21.7	594.26	100	0	0	–		594.26	594.26	
270	+	summer	561.10	100	0	0	33.16		561.10	594.26	
258	–	3.0	82.52	13.9	511.7	86.1	–		82.52	82.52	
270	+	spring – autumn	82.52	14.7	478.6	85.3	33.16		82.52	115.68	
258	–	–	0	0	594.3	100	–		–	–	
270	+	winter	0	0	561.1	100	33.16		–	33.16	

Table 6. Comparison of energy consumption for the different technology of marine transportation of natural gas (5.2×10^6 m^3) in the form of gas hydrate blocks: preserved layer of ice when separated in time manufacturing operations and dissociation of gas hydrates and without forced preservation of blocks in continuous production process.

Composite of technological chains	Energy consumption for the different technology of the transportation in the form of gas hydrates blocks				Energy savings in terms of gas due			
	Separated in time manufacturing operations and dissociation, preservation of the blocks with layer of ice		Continuous process, absent preservation of blocks		Force preservation	Natural cold and solar energy	Total	
	%	$\times 10^3$ m^3	%	$\times 10^3$ m^3	$\times 10^3$ m^3	$\times 10^3$ m^3	$\times 10^3$ m^3	%
Production	4.33	225.54	11.0	572.98	–	347.44	347.44	6.67
Transportation	4.86	253.32	4.90	255.24	1.92	–	1.92	0.04
Storage	0.06	3.13	0.30	15.63	2.69	9.81	12.5	0.24
Dissotiation	2.69	140.11	4.7	244.82	33.16	71.55	104.71	2.01
Total	11.94	624.02	20.9	1088.67	37.77	428.8	466.57	8.96

Comparison of energy consumption of two variants of technologies of the marine transportation of 5.2×10^6 m^3 of natural gas in the form of gas hydrate blocks: 1) with forced preserved blocks in separated of production and dissociation of gas hydrates with time; 2) without forced conservation blocks in continuous production process is presented in Table 6.

The comparison of expenses of energy of a components of technological chain of sea transportation of natural gas in the form of the gas hydrate blocks and other known technologies (LNG-, CNG-, NGH-based on granular hydrate) is given in the Table 7 (Nogami et al. 2008), (Khamehchi et al. 2013).

Table 7. Compare energy consumption for the components of technological chains of maritime transportation of natural gas in the form of gas hydrate blocks and other known of technologies (LNG-, CNG-, NGH-based on granular hydrate (Nogami et al. 2008), (Khamehchi et al. 2013), %.

Components of technological chains	LNG - technology	CNG- technology	NGH- technology (granules)	NGH-technology (force preservation of the gas hydrate of blocks) Operating mode	
				continuous	seasonal
Production	11.0 – 24.0	1.4	11.0	6.63	4.33
Transportation	0.94	5.7	4.9	4.90	4.86
Storage	1.6	1.4	1.2	0.08	0.06
Dissociation	0.31	3.7	4.7	6.34	2.69
Total	13.85 – 26.85	12.2	21.8	17.95	11.94

3 CONCLUSIONS

Thus the gas hydrate blocks manufactured in accordance with the proposed technology can be regarded as "devices" for concentrate the gas with an internal energy source. They suitable for long-term storage and transportation at atmospheric pressure and a slight negative temperature.

To increase the stability of gas hydrates have formed a pre-chilled mixture of crushed and granulated hydrates in blocks and forced preservation at atmospheric pressure with the layer of ice.

Hydrate blocks is proposed to keep under a layer of polyurethane foam thickness of 0.5 – 0.7 m in terrestrial Inflatable structures covered with double-layer soft shell with a layer of nonflammable gas for the locking.

Production of refrigerated gas hydrate blocks allows them storage to carry out at atmospheric pressure with minimal energy and forced preservation with the layer of ice – increase the maximum temperature of storage on the 12 – 14 K.

Proposed gas hydrate technology creates important preconditions of development of small and medium-sized remote of gas fields (including gas-hydrate), creating a network of terrestrial Inflatable structures, improve the efficiency and competitiveness of marine technologies of transporting of natural gas in hydrate form.

In addition, the use of alternative energy sources (natural cold and solar energy) at the production and dissociation of gas hydrate in terrestrial Inflatable structures allows considerably reduce capital and energy costs of the technological chain of the transportation and storage of hydrocarbon gases in hydrate form.

REFERENCES

Dawe, R.A., Thomas, M.S. & Kromah, M. 2003. *Hydrate Technology for Transporting Natural Gas*. Engineering Journal of the University of Qatar, Vol. 16: 11 – 18.

Gudmundsson, J.S. & Parlaktuna, M. 1991 *Gas-in-ice: Concept evaluation*. Technical report. Department of Petroleum Engineering and Applied Geophysics: Norwegian University of Science and Technology.

Patent No 5536893, US, MPK C07C 7/20. *Method for production of gas hydrates for transportation and storage*. J.S. Gudmundsson. Pub. 16.07.1996.

Gudmundsson, J.S. & Graff, O.F. 2003. *Hydrate non-pipeline technology for transport of natural gas*. Mode of access: http://www.igu.org/html/wgc2003/WGCpdffiles/10056_1046347297_14 776_1.pdf.

Khamehchi, E., Hamidreza, S, Y. & Sanaei, A. 2013. *Selection of the Best Efficient Method for Natural Gas Storage at High Capacities Using TOPSIS Method*. Gas Processing Journal, Vol. 1: 9 – 18.

Nogami Tomonori, Oya Nobutaka, Ishida Hiroshige & Matsumoto Hitoshi. 2008. *Development of natural gas ocean transportation chain by means of natural gas hydrate (NGH)*. Proceedings of the 6th International Conference on Gas Hydrates (ICGH 2008), July 6 – 10.

Patent No 105208, UA. *Application of liquid-gas jet apparatus with an elongated a camera of mixing as a contact device for the formation of gas hydrates*. Pedchenko M.M. Publ. 25.04.2014, Bul. No 8.

Yakushev, V.S & Istomin, V.S. 1990. *Features of the existence of gas hydrates in the rocks of at low temperatures*. Geochemistry, No 6: 899 – 903.

Yakushev, V.S., Gerasimov, Y., Kwon, V.G. & Istomin, V.A. 2008. *The current state of the gas hydrate technology. Review*. M: 48 – 52

© 2015 Taylor & Francis Group, London, ISBN: 978-1-138-02883-8

Dynamic stability of balancing rope of skip hoists

V. Pochepov, L. Fomychova & V. Salli
National Mining University, Dnipropetrovsk, Ukraine

ABSTRACT: On the basis of the research of the system of differential equations were considered two possible cases of deflectionmotion of the balancing rope, that define different options of the loss of stability (turning loops and twisting branches).

1 INTRODUCTION

Required capacity of multiropeskip hoisting plants in the development of seams at great depths is achieved by increasing the diameter of the drive pulleys or the number of head ropes. Diameters of pulleys currently reach 5 m. Increase the number of head rope leads to additional structural and operational disadvantages. In addition, the use of balancing ropes with a large mass per unit length increases their bending rigidity, which negatively affects the operation of the skip hoists. Improving skip hoists requires the creation of reliable recommendations on the choice of balancing ropes to ensure trouble-free operation.

Analysis of performance of balancing ropes shows that, despite the sufficient high strength and no significant static loads, ropes relatively quickly fail. This can be explained only by the intensity of the transverse and torsional oscillations, which resulted that ropes hit with elements of reinforcement and lining of shaft.

Let's consider some cases of possible unstable oscillations of the balancing rope.

2 TRANSVERSE OSCILLATIONS OF A FLAT ROPE

Figure 1 line represents a balancing rope, ends of which move with constant velocities V in opposite directions. The length of the rope l is much greater than the distance a by horizontal between points of the rope suspension to skips A, C. Neglecting the bending rigidity of the rope, was foundthat in the static state the angle α', which is formed by axis of the rope and the vertical line, is described by the relation:

$$tg\alpha' = \frac{\beta_1}{\beta_2} - \xi , (\xi \in [0,1]),$$

(1)

where ξ – longation coordinate of an arbitrary point of the rope, measured from point A, (Figure 1), $\xi = \frac{s}{l}$; β_1, β_2 – constant of integration, defined as the roots of two transcendental equations:

$$\frac{a}{l} = -\beta_1 \ln \frac{\sqrt{\beta_1^2 + (\beta_2 - 1)^2} + (\beta_2 - 1)}{\sqrt{\beta_1 + \beta_2} + \beta_2} ,$$

(2)

$$\frac{H_0 - 2Vt}{t} = \frac{b}{l} = \sqrt{\beta_1^2 + \beta_2^2} - \sqrt{\beta_1^2 + (1 - \beta_2)^2} ,$$

(3)

where H_0 – lifting height with a uniform velocity V; t – time in lift from the beginning of uniform motion.

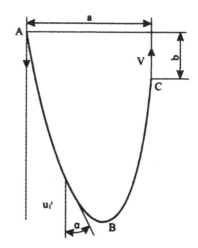

Figure 1. Account circuit of balancing rope.

Since $\frac{a}{l} \ll 1$, then $\beta_1 \ll 1$; besides that $\beta_1 \ll \frac{a}{l}$. Therefore, we as sum (is assumed) that $\dfrac{\partial \beta_1}{\partial t} \approx 0$ and in those cases, where it does not lead to violations of regularity, we will as sum (will be assumed) $\beta_1 = 0$. Then from equation (3) is obtained:

$$\beta_2 = \frac{1}{2}\left(1+\frac{b}{l}\right), \quad \frac{\partial \beta_2}{\partial t} = -\frac{V}{t}. \qquad (4)$$

Further, oscillations of the rope in the plane of its static balance are considered. Thus, in accordance with expression (1) the horizontal coordinate u_1' (Figure 1) of any point of the rope is given by ratio:

$$\frac{u_1'}{l} = -\beta_1 \ln\frac{\sqrt{\beta_1^2 + (\beta_2 - \xi)^2} + (\beta_2 - \xi)}{\sqrt{\beta_1^2 + \beta_2^2} + \beta_2}. \qquad (5)$$

Differentiating twice by t expression (5) considering (4), is obtained:

$$\rho\frac{\partial^2 u_1'}{\partial t^2} = \frac{\rho V^2}{l}\frac{\beta_1(\beta_2 - \xi)}{\left[\beta_1^2 + (\beta_2 - \xi)^2\right]^{\frac{3}{2}}}. \qquad (6)$$

Formally it is centrifugal force per unit of length of the rope and significantly different from zero only in the vicinity of the point $\xi = \beta_1$, that is near point B.

We write the equation of plane small oscillations relative to the position of static balance of the rope:

$$\frac{\partial}{\partial \xi}\left[\frac{\rho g l(\beta_2 - \xi)}{l^2 \cos \alpha'}\frac{\partial \tilde{u}_1}{\partial \xi}\right] = \rho\left(\frac{\partial^2 u_1'}{\partial t^2} + \frac{\partial^2 \tilde{u}_1}{\partial t^2}\right), \qquad (7)$$

where ρ – the linear density of the rope; g – free fall acceleration; u_1' – locally transverse deviation of rope points from the position of the balance.

Since rope branches are symmetric with respect to the point B, that is ensured by the condition $\beta_1 \ll 1$, centrifugal forces $\rho\dfrac{\partial^2 u_1'}{\partial t^2}$ cannot cause excitation of the vibrational rope.

Let's consider the solution of the homogeneous equation (7), which, with the replacement of the time coordinate t on τ using the relation $\tau = \dfrac{Vt}{l}$, takes the form:

$$\frac{\partial}{\partial \xi}\left[\sqrt{\beta_1^2 + (\beta_2 - \xi)^2}\frac{\partial \tilde{u}_1}{\partial \xi}\right] = \mu^2\frac{\partial^2 \tilde{u}_1}{\partial t^2}, \qquad (8)$$

where $\mu^2 = \dfrac{V^2}{l_D}$.

The solution of the equation (8) will be searched in the form of a first approximation, the Galerkin method, assuming:

$$\tilde{u}_1 = \omega(\tau)Z_0(\xi), \qquad (9)$$

where $Z_0(\xi)$ – some cylindrical function of zero order, satisfying the boundary conditions:

$$Z_0(0) = Z_0(1) = 0. \qquad (10)$$

Substituting (10) into equation (8), multiplying both sides of the equation by a function $Z_0(\xi)$ and integration by the entire length of the rope, considering the boundary conditions (10) is obtained:

$$\omega(\tau)\int_0^1 \sqrt{\beta_1^2 + (\beta_2 - \xi)^2}\,\frac{dZ_0}{d\xi}\,d\xi +$$
$$+ \mu^2\frac{d^2\omega(\tau)}{dt^2}\int_0^1 (Z_0)^2 d\xi = 0. \qquad (11)$$

The variable $x^2 = \sqrt{\beta_1^2 + (\beta_2 - \xi)^2}$ is introduced. In this case, with sufficient approximation, considering the infinitesimal β_1, we have:

$$d\xi = \begin{cases} -2xdx \text{ if } \sqrt[4]{\beta_1^2 + \beta_2^2} \leq x \leq \sqrt{\beta_1} \\ 2xdx \text{ if } \sqrt{\beta_1} < x \leq \sqrt[4]{\beta_1^2 + (1-\beta_2)^2} \end{cases}. \qquad (12)$$

Function $Z_0(\xi)$ can be represented as: $Z_0(\xi) = I(vx)Y_0(vx_0) - Y_0(vx)I_0(vx_0)$, where I_0, Y_0 – the first race Bessel function of zero order.

According to the boundary conditions (10), the parameter v is the smallest root of the equation: $I_0(vx_1)Y_0(vx_0) - Y_0(vx_1)I_0(vx_0) = 0$, where:

$$\begin{aligned} x_1 &= \sqrt[4]{\beta_1^2 + (1-\beta_2)^2} \approx \sqrt{1-\beta_2} \\ x_2 &= \sqrt[4]{\beta_1^2 + \beta_2^2} \approx \sqrt{\beta_2} \end{aligned}. \qquad (13)$$

Approximately, taking into account (13):

$$v = \frac{\pi}{-\sqrt[4]{\beta_1^2 + (1-\beta_2)^2} + \sqrt[4]{\beta_1^2 + \beta_2^2}} \approx$$
$$\approx \frac{\pi\left(\sqrt{\beta_2} + \sqrt{1-\beta_2}\right)}{\dfrac{H_0}{l} - 2\tau} \approx \frac{\pi\sqrt{2}}{\dfrac{H_0}{l} - 2\tau}. \qquad (14)$$

In (14) was taken into account, that the value $\sqrt{\beta_2} + \sqrt{1-\beta_2}$ is slightly different from the constant value under the condition of changing β_2 from 0.2 to 0.8, which corresponds to the actual limits of the area of the vessel movement with a steady rate.

Integrating the equation (11) with (12), (14), is obtained:

$$K_1\left(\frac{H_0}{l}-2\tau\right)^2\frac{d^2\omega}{d\tau^2}+\omega(\tau)=0,\qquad(15)$$

where $-\ K_1=\dfrac{\mu^2}{2\pi^2}\dfrac{\int\limits_{x_0}^{x_1}x\left(\dfrac{dZ_0}{dx}\right)^2dx}{\int\limits_{x_0}^{x_1}xZ_0^2dz}=const$.

Two particular solutions of the equation (15) have a common multiplier $\left|\dfrac{H_0}{l}-2\tau\right|^{0,5}$. From this it follows that fluctuation of the rope in the time interval $t\in\left(0,\dfrac{H_0}{2V}\right)$ occurs with decreasing amplitude over time, and in time interval $\left(\dfrac{H_0}{2V},\dfrac{l-H_0}{2V}\right)$ of the oscillation amplitude increases, that means the amplitude of the oscillation of the rope increases with turn-out moment of the vessels. This means that any perturbations, acting on the rope (air jet pressure, transverse oscillation of lifting vessels, etc.) lead to a buildup, increasing in the second period of the motion.

3 TURNING OF THE LOOP OF FLAT ROPE OR TAPE

Let's consider the differential equation of small torsional and transverse vibrations of a flat rope in the direction, perpendicular to the plane of the loop:

$$EI_1\frac{\partial^2}{\partial s^2}\left(\tilde\gamma\frac{\partial\alpha'}{\partial s}-\frac{\partial^2\tilde u_2}{\partial s^2}\right)+(EI_3-EI_2)\times$$

$$\times\frac{\partial}{\partial s}\left\{\frac{\partial\alpha'}{\partial s}\left(\frac{\partial\tilde\gamma}{\partial s}+\frac{\partial\tilde u_2}{\partial s}\frac{\partial\alpha'}{\partial s}\right)\right\}+$$

$$+\frac{\partial}{\partial s}\left(\tilde\gamma F_1'+\frac{\partial\tilde u_2}{\partial s}F_3'\right)=\rho\frac{\partial^2\tilde u_2}{\partial t^2},\qquad(16)$$

where $EI_3\dfrac{\partial}{\partial s}\left(\dfrac{\partial\tilde\gamma}{\partial s}+\dfrac{\partial\tilde u_2}{\partial s}\dfrac{\partial\alpha'}{\partial s}\right)+(EI_2-EI_1)\dfrac{\partial\alpha'}{\partial s}\times$

$$\times\left(\tilde\gamma\frac{\partial\alpha'}{\partial s}-\frac{\partial^2\tilde u_2}{\partial s^2}\right)=I_3\frac{\partial^2\tilde\gamma}{\partial t^2};\ \tilde\gamma\ \text{– the angular rota-}$$

tion of the rope section about the longitudinal axis; $\tilde u_2$ – small displacements of the rope in the perpendicular direction of the plane of static equilibrium of the rope; α' – the angle, which is formed by the axis of the rope at any point with the vertical; EI_1,

EI_2 – the smallest and largest bending rigidity of the rope; EI_3 – torsional rigidity of the rope; I_3 – the inertia moment of the rope segment of unit length relative to own central axis; F_1', F_2' – respectively, the shear and longitudinal forces, acting on sections of the rope in its static balance, $F_1'=[(\xi-\beta_2)\sin\alpha'-\beta_1\cos\alpha']\rho\,g\,l$; $F_3'=-[\beta_1\sin\alpha'+(\xi-\beta_2)\cos\alpha']\rho\,g\,l$.

Here β_1 , β_2 are defined by transcendental equations and the angle α' is determined from the nonlinear differential equation:

$$\varepsilon_2\frac{d^2\alpha'}{d\xi^2}+(\xi-\beta_2)\sin\alpha'-\beta_1\cos\alpha'=0,\qquad(17)$$

where $\varepsilon_2=\dfrac{EI_2}{(\rho g l)l_2}$.

The solution of the equations system (16) will be sought in the form of a first approximation the Galerkin method, assuming:

$$\tilde u_2=U(s)\frac{u(t)}{t};\ \gamma=G(s)\gamma(t).\qquad(18)$$

Substituting the expression (18) into the system (16), multiplying the first equation system by $U(s)$, and the second – by $G(s)$ and integrating by s from 0 to t , with subject to the boundary conditions:

$$U(0)=U(l),\ G(0)=G(l)=0,\qquad(19)$$

we have result (is obtained the result):

$$b_1\frac{d^2u}{d\tau^2}+a_{11}u+a_{12}\gamma=0,$$

$$\mu b_2\frac{d^2\gamma}{d\tau^2}+a_{21}u+a_{22}\gamma=0,\qquad(20)$$

where $\tau=t\sqrt{\dfrac{g}{l}}$; $\mu=\dfrac{I_3gl}{EI_3}$; $a_{11}=\int\limits_0^1\dfrac{F_3'}{\rho gl}\left(\dfrac{dU}{d\xi}\right)^2d\xi+$

$$+\varepsilon_1\int\limits_0^1\left(\frac{d^2U}{d\xi^2}\right)^2d\xi+(\varepsilon_3-\varepsilon_2)\int\limits_0^1\left(\frac{dU}{d\xi}\frac{d\alpha'}{d\xi}\right)^2d\xi;\ a_{12}=\int\limits_0^1\frac{F_1'}{\rho gl}\times$$

$$\times G\frac{dU}{d\xi}d\xi+\varepsilon_1\int\limits_0^1\frac{dU}{d\xi}\frac{d}{d\xi}\left(G\frac{d\alpha'}{d\xi}\right)d\xi+(\varepsilon_3-\varepsilon_2)\int\limits_0^1\frac{dU}{d\xi}\frac{d\alpha'}{d\xi}\frac{dG}{d\xi}d\xi;$$

$$a_{21}=\int\limits_0^1\frac{d\alpha'}{d\xi}\frac{dU}{d\xi}\frac{dG}{d\xi}d\xi-\frac{\varepsilon_1-\varepsilon_2}{\varepsilon_3}\int\limits_0^1G\frac{d\alpha'}{d\xi}\frac{d^2U}{d\xi^2}d\xi;$$

$$a_{22}=\int\limits_0^1\left(\frac{dG}{d\xi}\right)^2d\xi+\frac{\varepsilon_1-\varepsilon_2}{\varepsilon_3}\int\limits_0^1\left(G\frac{d\alpha'}{d\xi}\right)^2d\xi;\ b_1=\int\limits_0^1U^2d\xi;$$

$$b_2=\int\limits_0^1G^2d\xi\xi;\ \varepsilon_i=\frac{EI_i}{(\rho g l)l^2}.$$

519

To calculate the coefficients a_{ik}, b_i it is necessary to specify the form of transverse and torsional oscillations of the rope, which must satisfy the boundary conditions (19) and correspond to the actual motions of the rope, resulting in turning it in the bottom. By logical reasoning, is concluded that the turning of rope loops can occur when the loop is inclined from the static balance plane and maximum twisting of transverse sections of the rope occurs at the lowest point of the loop.

Forms of rope oscillations, corresponding the occasion of its turning, should have the form, shown in Figure 2. In order to establish more specific forms of the oscillation, let's consider a simplified calculation model of the balancing rope, representing two branches fastened at the bottom.

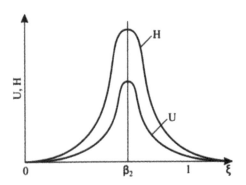

Figure 2. Forms of rope oscillations, corresponding the occasion of turning loop.

For the same length of branches, without considering bending stiffness the first form of oscillations of the rope like this has the form:

$$U(\xi) = I_0 \left(2\mu \sqrt{\left|\frac{1}{2}-\xi\right|} \right), \quad (\xi \in [0,1]), \text{ where } \mu$$

the smallest root of the equation $I_0\left(\mu\sqrt{2}\right) = 0$.

The function $I_0\left(2\mu\sqrt{\left|\frac{1}{2}-\xi\right|}\right)$ is approximated by the simpler, type:

$$U(\xi) = \frac{1}{4}\xi(1-\xi)e^{\left|\frac{1}{2}-\xi\right|}, \qquad (21)$$

which is taken as the desired shape of transverse oscillations of the rope.

To determine the shape of torsional oscillations, is considered a simplified equation of own forms of torsional oscillations derived from the second equation of the system (16) in the form:

$$EI_3\frac{d^2G}{d\xi^2} + l^2 I_3\omega^2 G = -\frac{d}{d\xi}\left(EI_3\frac{dU}{d\xi}\frac{d\alpha'}{d\xi}\right). \qquad (22)$$

Finding G as the solution of the inhomogeneous equation (22), is obtained precision up to the constant factor:

$$G(\xi) = \int_0^\xi \sin k(\xi - x)\frac{d}{dx}\left(\frac{dU}{dx}\frac{d\alpha'}{dx}\right)dx,$$

$$\left(k^2 = \frac{l^2\omega^2 I_3}{EI_3}\right). \qquad (23)$$

Since the expected frequency of the lateral-torsional vibration is negligible (at high rope pitches), the parameter k is small and $\sin k(\xi - x)$ – slowly varying function. At the same time the value $\frac{d}{dx}\left(\frac{dU}{dx}\frac{d\alpha'}{dx}\right)$ is practically non-zero only in the vicinity of the lowest point of the loop. Therefore, approximately $G(\xi) = \left|\frac{dU}{dx}\frac{d\alpha'}{dx}\right|$ and as $\frac{dU}{dx}$ is slightly different from the constant, the shape of torsional oscillations takes the form:

$$G(\xi) = \frac{d\alpha'}{dx}\frac{d\alpha'}{dx}. \qquad (24)$$

Taking as the basic shape of the lateral-torsional vibration of the balancing rope expression (21) and (24), can be calculated the required coefficients of equations (20).

The phenomenon of turning the rope is, obviously, directly related to the emergence of complex inherent numbers of equations (20), which are equivalent to the vanishing of the radicand in the relation:

$$\omega_{1,2}^2 = \frac{\mu b_2 a_{11} + b_1 a_{22}}{2\mu b_1 b_2} \pm$$

$$\pm \sqrt{\left(\frac{\mu b_2 a_{11} + b_1 a_{22}}{2\mu b_1 b_2}\right)^2 - \frac{a_{11}a_{22} - a_{12}a_{21}}{\mu b_1 b_2}},$$

where $\omega_{1,2}^2$ – the inherent numbers of the system (20), through which the solution of these equations can be written in the form $u = Ae^{i\omega\tau}$, $\gamma = Be^{i\omega\tau}$.

Let's consider the definition of parameters for two types of balancing organs structures. For rubber-rope tape bending and torsional rigidity with sufficient accuracy are determined by the relations:

$$EI_1 = \sum_{i=1}^{n} EI_{TPi} + EI_{RES}\,\frac{hb^3}{12},$$

$$EI_2 = \sum_{i=1}^{n} EI_{TPi} + EI_{RES}\,\frac{h^3 b}{12},$$

$$EI_3 = G_{RES}\,\frac{bh\!\left(h^2 + b^2\right)}{12}.$$

Flexural rigidity of cables is considered as the sum of the flexural rigidity E_{TPi} and rigidity of the rubber shell itself, having a modulus of elasticity E_{RES} (b, h — width and thickness of the tape). In the rigidity on the torsion tape the bulk is torsional rigidity of the rubber shell, having the shear modulus G_{RES}.

The inertia moment per unit of the length of the tape $I_3 = \rho\,\dfrac{\left(h^2 + b^2\right)}{12}$, where ρ — the linear mass of tape.

For the steel strip:

$$EI_1 = EI_{ST}\,\frac{hb^3}{12},$$

$$EI_2 = EI_{ST}\,\frac{h^3 b}{12},$$

$$EI_3 = G_{ST}\,\frac{bh\!\left(h^2 + b^2\right)}{12},$$

$$I_3 = \rho\,\frac{\left(h^2 + b^2\right)}{12}, \quad \text{where } E_{ST},\ G_{ST} - \text{the modulus}$$

of elasticity of steel.

4 CONCLUSION

As the result of calculations was found that turning of the rubber-rope loop tape is significantly dependent on the shape of the tape namely the flexural rigidity EI_2 and the ratio $\dfrac{a}{l}$ (Figure 3).

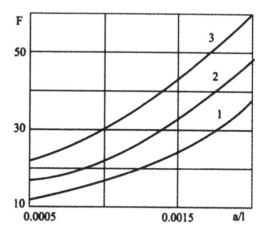

Figure 3. Dependence of values μ from $\dfrac{a}{f}$ if it is possible turning the loop: 1, 2, 3 – $\varepsilon = 9 \cdot 10^{-10}$; $5 \cdot 10^{-10}$; $2 \cdot 10^{-10}$.

If parameters of the rope are so, that $\mu = \dfrac{I_3 gl}{EI_3}$ reaches values that are shown in Figure 3, then the loop turning is possible, which is undesirable during the operation of the skip hoists. Steel tape is more stable. With existing relations b, h, steel tape is not undergone to the possibility of turning the loop.

Theoretical and Practical Solutions of Mineral Resources Mining – Pivnyak, Bondarenko & Kovalevska (eds)
© *2015 Taylor & Francis Group, London, ISBN: 978-1-138-02883-8*

Influence analysis of cross slip on friction coefficient of mine section locomotive on curvilinear rail track with intermediate medium

V. Lytvyn
National Mining University, Dnipropetrovsk, Ukraine

ABSTRACT: The problem of traction coefficient variation of mining sectional locomotive while motion on the curvilinear rail track is observed. The relation of force of hydrodynamic wheel unloading taking into account the intermediate layer existence is defined. Quasistatic model of mining locomotive negotiation in turnings is developed. Verification of the developed model is provided by the system of non-linear differential equations. Empirical friction coefficient relations subject to the rail track radius for velocities 3 and 5 m/s are defined.

1 INTRODUCTION

The interaction of a wheel and a rail is a physical basement of mining locomotive motion on the rail track. It generally defines the safety and the most important technical-exploitation parameters such as locomotive's mass, its motion velocity and exploitation indexes. The problem of wheel-rail grip still exists since the development and exploitation of the first rail vehicles (Isaev & Luznov 1985).

The core of the problem was definition of the relations between average friction coefficients of mining locomotive motion velocity in different exploitation environment. Also the relations of the friction coefficient in dependence on physical properties of contacting materials and frictional pair surfaces conditions, axle load, wheelband and wheel profile are obtained (Babichkov et al. 1971, Isaev 1973). The experimental study of frictional coefficient for electric and diesel locomotives of different types, which were used on the rail roads of former USSR, showed that frictional coefficient lies in the range of 0.1 – 0.4 according to different authors (Isaev 1970).

However, nowadays there is no existing methodic for frictional coefficient calculation while mining locomotive motion in the rail track of high curvature subject to lateral slip and rail surface conditions. In the paper, an attempt to solve the urgent problem is provided.

2 THE RESEARCH RESULTS

While mining locomotive motion on the rail track surfaces of rail and wheel are separated by the intermediate layer, which includes wear debris of friction bod-

ies, particles of hauling rock, condensed moisture and another components that form multicomponent environment. Existence of dispersive layer between friction surfaces facilitates to study the wheel-rail interaction process as liquid friction when friction surfaces and separated by the layer of liquid or solid lubricant.

On the described features of wheel-rail interaction, the model of a physical process is based. Between wheel rolling surface and rail surface nonstationary motion of viscous incompressible liquid arises. This leads to the necessity to take into account the offloading force F_a because of hydrodynamic component of intermediate layer. Therefore, friction coefficient will be defined by the following relation:

$$\Psi = \Psi_0 - \left(1 - \frac{F_a}{N}\right), \tag{1}$$

where ψ_0 – frictional coefficient of clean rolling surfaces of wheel and rail; N – real force, acting on the wheel, N.

To define the force F_a there is necessary to find out the law of intermediate layer flow, which induced by the lateral wheel skid V_y relative to the rail. In this purpose let study the wheel motion on the intermediate layer surface after being subjected to the pseudo skid forces, which result in the relative displacement of the environment's layer.

Let the wheel travels along the y axle direction (perpendicular to the locomotive motion axis) with the velocity V_y, and the layers of intermediate environment with velocity V_s (Figure 1). Due to the fact that wheel bandage has conicity, thus it is under the influence of the upward force (in this case the offload force Fa), that induced by the pressure differences ΔP:

$$p = \frac{6\mu V_s L}{h_{min}^2}\left[\frac{1}{\left[\xi-(\xi-1)\frac{x}{L}\right]} - \frac{\xi}{(\xi+1)\left[\xi-(\xi-1)\frac{x}{L}\right]^2} - \frac{1}{\xi+1}\right], \tag{2}$$

where μ – dynamical viscosity of the intermediate layer, Pa s; L – the length of wheel rolling area, m; ξ – the relation between maximal and minimal gaps between polluted rail and mining locomotive wheel; x – current distance in the range $h_{min}...h_{max}$, m.

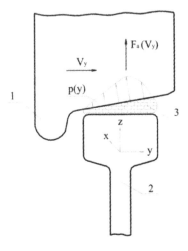

Figure 1. Calculation scheme of wheel rolling subject to intermediate layer: 1 – wheel, 2– rail, 3 – intermediate layer.

According to known pressure variation law (2) forces of hydrodynamic offload can be defined:

$$F_a = \int_A pdA \text{ or } F_a = \frac{\mu V_s L}{h_{min}^2}F_a, \tag{3}$$

where A – total wheel square, m²; F_a – loading coefficient, Pa.

From the formula (3) we can conclude, that offloading force F_a direct-proportionaly depends on lateral wheel slip V_y while locomotive motion in the curve rail track. To define the traction coefficient 1 we need to obtain the relation of lateral slip V_y in dependence on rail track bend radius.

The problem of rail vehicle negotiation on rail curve is being researched for a long time (Isaev 1970).The most detailed study of the problem is observed in in paper (Garg & Dukkipaty 1984), where the nonlinear nature of lateral and longitudinal frictional forces are considered and longitudinal force of the train. For this purpose that task is down to the system of transcendental equations, which solution can be obtained numerically.

Let wright a quasistatical simulation model of mining locomotive driving in curvilinear rail track in order to define lateral slip V_y and then to verify the results by the usage of differential equations, that have been obtained in (Garg & Dukkipaty 1984).

For this purpose lets assume, that the locomotive velocity is constant and bogie mass is distributed by the axles. Such assumptions are developed by the other scientists while lateral oscillation of rail vehicle study (Figure 2).

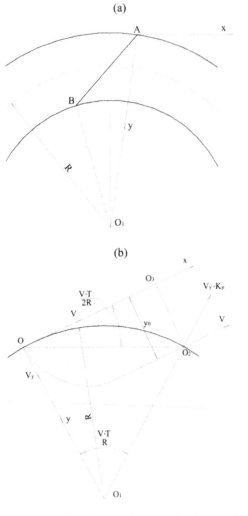

Figure 2. Equivalent (a) and calculation (b) scheme while lateral oscillations.

Let substitute the real scheme by equivalent, which represents two masses (wheels), that connected by the rigid bar (wheel pair). So, we have the following calculation scheme: some mass m, that drives with velocity V, meets the obstacle, and obtains some lateral velocity y_0, follows along the trajectory.

The obstacle has length and is bend with radius R. Let choose the coordination system, where the positive are of X-axis is motion direction. O – is a point of start. So, the motion direction is tangent to obstacle curve and Y-axis is oriented down along the radius. The value of lateral slip velocity V_y we will define according to assumption of periodicity- mass m at the moment of repeating shock might have lateral slip equals to $K_p V_u$, where k_p – the coefficient of elastic imperfections of the rail track, which value must be more then unity (for rail track made with rail P34 $k_p = 1.9$). The velocity $k_p V_y$ defines as a component of longitudinal velocity V directed to the radius, that guided through the shock point from the curvature center: $k_p V_y \approx \dfrac{V}{R} T$:

$$V_y \approx \frac{V}{k_p R} T , \qquad (4)$$

where T – the hunting period, that defines from the formula:

$$T = \frac{2R}{V} \sqrt{\frac{3\dfrac{2-k_p}{k_p}}{1+\dfrac{12\Psi_0 gR}{V^2 k_p^2}}} . \qquad (5)$$

Filling the (5) and (4) necessary to take into account that the rail P34 is the most wide spread across the coal mine of Ukraine. Thus:

$$V_y = 0{,}4 \sqrt{\frac{3{,}6V^2}{3{,}6V^2 + 12\Psi_0 gR}} . \qquad (6)$$

As a result, the quadrate of lateral slip (and offloading force F_a correspondingly), which induced by the influ ence of intermediate layer existence and wheel conicity is direct-proportional to square of locomotive speed and inversely proportional to the square of speed and curve radius.

On the Figure 3 the relations between velocity of lateral slip and motion velocity, curve radius which have been calculated according to formula (6) are depicted.

As above mentioned, the results adequacy is verified by the full dynamical model of spatial oscillations 8 that accounts lateral geometrical imperfections of the rail track.

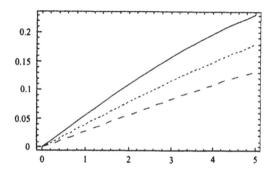

Figure 3. Relation of lateral wheel slip on velocity and track curvature: – R5, ---R10, - - - R20.

As the study showed, in curvilinear rail track, despite the insignificant oscillations, the value of lateral slip V_y can be assumed as stable. The deviation is 10, that is sufficient result for such models.

Thus, after the offload force F_a have been obtained, let define traction coefficient of mining locomotive with rail track, that is mudded finely-divided environment as a function of radius of constant curvature while different motion velocities according formula (1). For calculations let take $\psi = 0.3$. The results are depicted on the Figure 4.

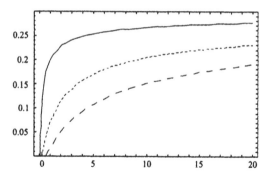

Figure 4. Relation of frictional coefficient in dependence on velocity slip on velocity and track curvature: — V = 1 m/s, --- V = 3 m/s, - - -V = 5 m/s.

3 CONCLUSIONS

The analysis of obtained relations shows that while mining locomotive motion through the rail track (R5...R20) with velocity less than 2 m/s there is no the significant friction coefficient variation. The relative difference between ψ (R6) and ψ (R20) is no more then 9%. However, while the velocity 3 m/s and 5 m/s this difference can be 26 and 45% correspondingly that results in significant tractive effort loss on the denoted rail track sections.

In order to facilitate the engineering calculations the author have defined empirical relations $\psi(R)$ for real curvatures subject to denoted intermediate layer properties, which are well-approximating by the simple logarithmic relations.

REFERENCES

Babichkov, A.M., Gusarskiy, P.A., Novikov, A.P. 1971. *Locomotive traction and tractive calculations*. M: Transport: 280.

Garg, V.K., Dukkipaty, R.V. 1984. *Dynamics of Railway Vehicle Systems*. Academic Press: New York.

Isaev, I.P. 1970. *Random factors and friction coefficient*. M.: Transport: 184.

Isaev, I.P. 1973. *On the problem of wheel and rail contact* MIIT. M: Transport: Vol. 445: 3 – 12.

Isaev, I.P., Luznov, Ju.M. 1985. *Problem of wheel-rail interaction*. M: Mashinstroenje: 240.

Shakhtar, P.S. 1982. *Mining locmotives*. M: Nedra: 296.

Vershinskiy, S.V., Danilov, V.N., Husidov, V.D. *Wagons' Dynamic*. M: Transport: 360.

Theoretical and Practical Solutions of Mineral Resources Mining – Pivnyak, Bondarenko & Kovalevska (eds)
© 2015 Taylor & Francis Group, London, ISBN: 978-1-138-02883-8

Influence of metasomatism on formation and criterion of relictness of comprehensive ore deposits confined to deep faults

M. Ruzina, N. Bilan, O. Tereshkova & N. Vavrysh
National Mining University, Dnipropetrovsk, Ukraine

ABSTRACT: Metallogenic forecasting of comprehensive deposits localized in deep fault zones has to substantiate the criterion of relictness of these deposits. It was established that metasomatism carries ambiguous action on qualitative characteristics and the criterion of relictness of formed deposits. Due to multistage activation of deep faults metasomatism can play a positive role in the forming of precious metals and, at the same time, be unfavorable factor for some occurrences of metallic and non-metallic minerals (chromite, chrysotile asbestos, etc.). Thus, a differentiated approach in evaluation the prospects for given raw materials is needed. The theoretical justification for the effect of metasomatism on the formation conditions and the criterion of relictness of comprehensive deposits within deep fault zones is given in paper.

1 INTRODUCTION

The relevance of studies is caused by the requirement to improve prediction technique for mineral resources in order to ensure sustainable development of the mineral raw material base of Ukraine. Metasomatically altered rocks are abundant in most ore provinces of the world, where associate with hydrothermal deposits of metallic and non-metallic mineral resources.

Occurrences of rare, precious and non-ferrous metals, as well as non-metallic minerals of hydrothermal origin were revealed in the Middle Prydniprovie megablock. Thereby, polychronous forming of these raw materials were defined; in addition it has been established that occurrences of metasomatic processes are spatially confined to deep faults and their intersection knots.

However, a differentiated approach is needed in forecasting of various types of metallic and non-metallic minerals, which are localized in deep fault zones and associated (spatially or genetically) with zones of metasomatism.

It was defined that occurrences of multi-stage metasomatic processes are some ways ambiguous in relation to qualitative characteristics and the criterion of relictness of the formed deposits.

Due to multi-phase activation of deep faults metasomatism can play a positive role in the formation of complex anomalies of precious metals and, at the same time be unfavorable factor for some occurrences of metallic and non-metallic minerals (chromite, chrysotile asbestos, etc.).

The purpose of the research was to justify theoretically the influence of metasomatic processes on formation conditions and the criterion of relictness of deposits.

2 MATERIALS AND DISCUSSION

The Middle Prydniprovie megablock is a granite-greenstone terrain located in the south-eastern part of the Ukrainian shield. As for tectonic structure the granite-greenstone region under consideration most closely matches the dome-depression type. Synformal, seldom monoclinic attitude of fragmented greenstone structures enclosed by granite domes and arches and limited by deep faults corresponds to this type. There are 13 greenstone structures in the Middle Prydniprovie megablock. According to (Tyapkin & Gontarenko 1990), there are six pairs of mutually orthogonal systems of deep faults with trend azimuths 0° and 270°, 17° and 287°, 35° and 305°, 45° and 315°, 62° and 332°, 77° and 347° within the megablock. Ore-bearing metasomatic formations are controlled by deep fault zones in the Middle Prydniprovie megablock and represented by 17 mineragenic types of pneumatolytic-hydrothermal, metamorphogenic-hydrothermal, plutonogenic- and volcanogenic-hydrothermal genesis.

Metallogenic forecasting is the science-based prediction of probable sites of mineralization founded on regional formation analysis used to determine the regularities of deposits distribution in relation to space and age. Putting to use the formation analysis allows:

– to establish the causality of mineralization with specific geological formations;

– to identify reliable factors that control the location of commercial mineralization in various geological structures;

– to develop criteria for prediction of mineral resources.

It is established beyond controversy that development of prospecting criteria for mineralization includes the analysis of terms of relictness of the formed deposits and ore types.

Recent metallogenic research of location of ore deposits established that about 84% of the total number of studied world's post-magmatic ore deposits is confined to faults or their intersections. The main features that determine the depth of the fault is now considered to be:

– the prevalence of basaltic volcanism at all stages of the development of fault zones;

– the differentiation of basaltic melts with the appearance of layered intrusions or contrasting basalt-rhyolite series;

– the presence of bodies of ultramafic rocks;

– the high concentration of elements, such as potassium, chloride, boron, fluorine, hydrogen, and a number of radioactive and rare earth elements;

– shows of mobilization of magmatic and ore material;

– the forming of extended metasomatic zones confined to deep faults.

Metasomatically altered rocks are spatially and genetically associated with mineral deposits. These rocks contain information about the structural-tectonic and lithological conditions of mineralization location. Most of them are enclosing rocks for ore bodies, but some of them are the separate types of mineral raw materials (talc, magnesite, chrysotile and amphibole asbestos).

Presence of metasomatites is also a kind of indicator of physical and mechanical properties of enclosing rocks, favorable or unfavorable for the localization of ore bodies.

Spatial position, shape and dimensions of the metasomatite bodies fixed the paths of ore-bearing flows and allow to judge the intensity of their influence on the enclosing rocks.

The most prospective for ore content should be considered the thick and extended metasomatic bodies characterized by complete metasomatic zones including interiors, as they indicate intense hydrothermal-metasomatic processes.

Age and genetic relation of metasomatic rocks and ores are ambiguous and underexplored. Sometimes ore formed almost simultaneously with the metasomatic rock, and in some cases they are separated by a considerable time interval during which there were repeated tectonic shifts with the intrusion of dykes of different material compositions. On the one part, various metasomatic formations are confined to a specific type of geological structures, on the other part, they have well-defined ore specialization.

Each metasomatic formation is characterized by a specific family of ore elements. However metasomatic processes in rocks rarely occur in a single stage, and are mostly multistage. In some cases, the multiple effect of metasomatic processes play an accumulating role, resulting in the formation of comprehensive ore deposits, in others – a substracting role, leading to the natural attenuation of formed mineral units, including ore ones.

Forming of chrysotile asbestos deposits in the ultrabasite mass of deep fault zones may be an example of a dual role of metasomatism in ore formation. Forming of monomineral chrysotile rocks in ultrabasite occurs after the effect of regional serpentinization, especially in the most permeable for ore-bearing fluid zones of crushing and mylonitization in ultrabasite. During the asbestos forming metasomatic processes leading to the formation of chrysotile rocks, have a devastating effect not only on the main rock-forming, but also on the accessory, including the ore minerals.

Metamorphism of chromium spinels in chromite ores is treptomorphism and isofacial metamorphism in relation to the alteration of their enclosing rocks. Due to this, under the serpentinization – leading to removal of FeO, MgO, Cr_2O_3 and to transition Fe^{2+} and Fe^{3+} – FeO, MgO, Cr_2O_3 are also subtracted from chromium spinel, which results in its attenuation, disappearance and subsequent substituting by magnetite. As a result, serpentinization, leading to the formation of asbestos occurrences, has a devastating impact on chromite occurrences, which were formed on earlier stages. This is probably why there are usually not large deposits of chromite ore in areas of intensive forming of asbestos. In turn, intensive forming of talc deposits and carbonatization in ultrabasite mass is a negative factor in prospecting this rock mass for asbestos.

Occurrences of chrysotile asbestos and talc-magnesite within the Middle Prydniprovie mega-block were exposed in several greenstone structures (GS); particularly in Bilozerska greenstone structure (BGS) and Surska greenstone structure (SGS).

Bilozerska greenstone structure is located at the intersection of four systems of deep faults. Occurrences of chromite, complex anomalies of Au, Ag, Pt, occurrences of talc, magnesite and chrysotile asbestos, as well as occurrences of Co, Cu, Ni and cinnabar were found in the southern part of the structure within the ultrabasite mass.

Almost all occurrences are confined to zones of tectonic dislocation and areas of intense metasomatism as several stages of listvenitization and serpentinization (lizardite, antigorite and chrysotile stages), talc forming and carbonatization. The origin of these occurrences is defined as dislocation-metamorphic.

In this case, the effect of several stages of metasomatic processes at the intersection of deep faults led to the development of favorable conditions for the forming of teleskoped ore formations, but the commercial value of these spatially combined different types of minerals are ambiguous. This ambiguousness is following:

– chromite occurrences in platinum-bearing chromite formation are considered as unpromising due to intensive replacement of chromite by serpentine and magnetite, which resulted in attenuation of already maggie ores and in subtraction of the platinoids;

– occurrences of cross-fibrose chrysotile asbestos were evaluated in general as perspective because of favorable structural conditions of localization (BGS is located within the intersection of four systems of deep faults), of greenschist facies metamorphism and because of pre-ore effect of continuous antigoritization. However, only small occurrences of labinsky and bredinsky types could be detected within the rock mass. Negative factor reducing the background for discovery occurrences of the largest commercial bazhenovsky type is intense multi-stage serpentinization, as a result of which the remnants of primary rock of bazhenovsky type are completely destroyed;

– occurrences of talc-magnesite associated with the latest metasomatic processes (listvenitization) are recognized as the most promising, and probable identification of medium-scale deposits of shabrovsky or medvedevsky morphogenetic types is substantiated. These types of deposits are characterized by association with fully serpentinized ultrabasite formations. Thus, intense multi-stage serpentinization reduces the prospects of platinum chromite ores, as well as the extent of occurrences of chrysotile asbestos materials, at the same time increases the prospects for talc-magnesite raw materials.

Pravdinsk deposit of talc-magnesite and carbonized serpentinite is located within Surska greenstone structure in the Middle Prydniprovie megablock. The ultrabasite mass is composed of chrysotile with the relicts of olivine, chrysotile-antigorite, antigorite talcose, carbonized serpentinites and talc-magnesite rocks (Ilvitsky 1989).

Unlike Bilozerska GS where antigoritization in serpentinites is pre-ore process in relation to the occurrences of chrysotile asbestos, antigoritization within Pravdinsk deposit is exposed in the form of several generations: earlier one predated the forming of chrysotile fibers and later one developed due to the growing substitution of chrysotile aggregates until their complete disappearance. Therefore, later antigoritization within the Pravdinsk rock mass is a negative factor for chrysotile asbestos occurrences, due to destruction of primary relict peridotites and aggregates of earlier chrysotile.

Talc-magnesite deposits are confined to zones of tectonic dislocation or peripheral areas of serpentinite bodies. Talc and magnesite are the latest units and replace both chrysotile veinlet and aggregates of antigorite.

Photomicrographs of carbonized and talcose chrysotile veinlet in chrysotile-antigorite serpentine are given in Figures 1, 2. There is distinct selective substitution of chrysotile veinlet by carbonate and talc in main rock-forming mass of antigorite composition. Eventually it led to the complete disappearance of chrysotile asbestos veinlet and formation of talc-magnesite aggregates. According to most researchers, talc-magnesite deposits in ultrabasites are hydrothermal metasomatic units.

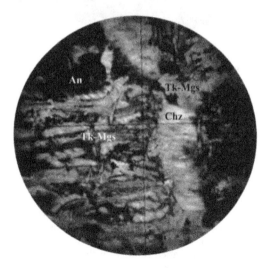

Figure 1. The initial stage of selective replacement of chrysotile asbestos veinlet (Chz, light grey) by talc-magnesite aggregates (Tk-Mgs, grey) within antigorite aggregates (An, black); relicts of asbestos mineralization are retained. Thin section, magnification 100x, nic. +.

There are following stages of forming and alteration of rock within the ultrabasite mass in Surska GS and Bilozerska GS of the Middle Prydniprovie megablock:

– igneous stage (dunite, peridotite);

– autometamorphic stage – serpentinites with the relicts of olivine;

– allometamorphic stage – recrystallized and carbonized antigorite serpentinites;

– hydrothermal-metasomatic stage – carbonatization, talc forming.

Talc-magnesite bodies are confined to areas of most intense effect of metasomatic processes, i.e. to zones of high permeability.

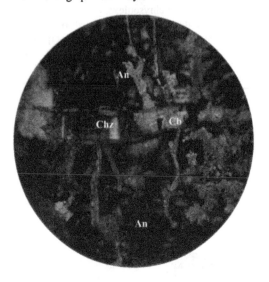

Figure 2. Almost complete replacement of chrysotile veinlet (Chz, white) by talc and carbonate (Cb, grey) within antigorite aggregates (An, black), ultimately leading to the destruction of chrysotile asbestos occurrences. Thin section, magnification 100x, nic. +.

When comparing the greenstone structures of Middle Prydniprovie with zones of deep faults systems it was established that location of Bilozerska GS is at the intersection of four fault systems, and Surska GS is at the intersection of six systems of faults, which indicates a higher degree of permeability of tectonic zones in Surska GS and, as a consequence, more intense and continuous hydrothermal-metasomatic alteration resulted in the formation of talc-magnesite deposits.

According to (Stul'chikov 1991), a number of occurrences of comprehensive mineralization were found within Verhovtsevska greenstone structure. Forming of this mineralization is spatially and genetically related to ore-bearing metasomatic rocks in the deep fault zone:

– copper-cobalt mineralization with gold, confined to the area of the eastern contact of Varvarovsky ultrabasite deposits; mineralization is associated with complex alteration in basic rocks (propylitization, carbonatization, silicification, tourmalinization), with alteration of ultramafic rocks (listvenitization, carbonatization, talc forming) and with beresitization of acid rocks;

– polymetallic mineralization (galena, sphalerite, arsenopyrite, berthierite) with gold in listvenitized ultrabasites spatially associated with the zone of Central-Verhovtsevsky deep fault;

– gold-arsenic mineralization established in beresites on keratophyres.

As a result of studies of the relationship of deep faults systems with occurrences of precious metals, it was established that high ore content in the greenstone structures of Middle Prydniprovie is caused by their alignment with the knots of intersection of deep faults (Surska, Chertomlykska, Bilozerska, Konkska greenstone structures) or overlapping of greenstone structures by the most productive subsystems of deep faults (Verhovtsevska GS).

These regularities are explained by the spatial alignment and repeated activation of deep faults of different systems. This activation is always associated with magmatic and metamorphic hydrothermal activity, as a result of which polychronous, comprehensive occurrences of rare and precious metals are formed. These occurrences are spatially associated with zones of teleskoped metasomatic formations.

Ore control of deep faults especially multiplies in the knots of intersection, where the so-called clusters are formed. According to (Tomson 1988) clusters are the areas (blocks) of intense tectonic processing, which are like punctured by large regional systems of dislocations and are specific blocks with a special mode of neotectonic movements.

The probability of deposits forming increases with the number of deep faults intersecting at a knot, which can become the ore cluster, concentrating deposits of different age and composition of precious metals and other minerals, including non-metallic minerals.

Thus, Bilozerska, Konkska, Chertomlykska and Surska greenstone structures can be related to the category of clusters, due to their location in the knots of the intersection of several systems of deep faults in the Middle Prydniprovie megablock. Derezovatska, Sophievska and Shirokovska greenstone structures are intersected only by deep faults of one direction.

Perhaps this is the reason why the most promising gold occurrences and deposits are located within the first four greenstone structures, but other greenstone structures are unpromising.

4 CONCLUSIONS

This dependence can be used not only as an ore control, but also a prospecting criterion for a number of mineral deposits, including gold occurrences and non-metallic minerals in the Middle Prydniprovie megablock. Thus, there are not only gold deposits, but also the large Pravdinsk deposit of serpentinite and talc-magnesite within Surska greenstone structure confined to the intersection of six systems of deep faults. Therefore, as a result of studies, the following conclusions are drawn.

1. Formations of teleskoped metasomatites with the comprehensive mineralization are spatially confined to the knots of intersection of deep faults in the Middle Prydniprovie megablock. The probability of forming of comprehensive deposits increases with the number of deep faults intersecting at a knot, which can become the ore cluster concentrating deposits of different age and composition of precious metals and other minerals, including non-metallic minerals.

2. The spatial alignment (telescoping) for various types of metasomatic processes in the knots of deep faults within the Middle Prydniprovie megablock has a dual significance for mineralization:

– on the one hand, it causes the forming of the valuable comprehensive mineralization and contributes to the forming of secondary concentrations of valuable components (mineralization of precious metals, copper-cobalt mineralization with gold, antimony-arsenic-zinc and gold-arsenic mineralization);

– on the other hand, it leads to attenuation of concentrations and to losses of quality of minerals associated with the earlier stages of metasomatic processes (substitution of chromite by magnetite during serpentinization of ultrabasites, the destruction of chrysotile asbestos during epigenetic talc forming).

REFERENCES

Ilvitsky, M.M. 1989. *Ultramafites of the Dnipropetrovsk granite-greenstone region*. Geology and metallogeny of Precambrian mafit-ultramafites in Prydniprovie. Dnipropetrovsk: DMU: 77 – 92.

Stul'chikov, V.A. 1991. *The regularities of metamorphism and metasomatism in the greenstone belts of the Ukrainian Shield (on example The Verhovtsevska synclinal)*. Kyiv: Naukova dumka: 171.

Tomson, I.N. 1988. *Metallogeny of ore districts*. Moscow: Nedra: 215.

Tyapkin, K.F. & Gontarenko, V.M. 1990. *Systems of deep faults of the Ukrainian Shield*. Kyiv: Naukova dumka: 184.

Theoretical and Practical Solutions of Mineral Resources Mining – Pivnyak, Bondarenko & Kovalevska (eds)
© 2015 Taylor & Francis Group, London, ISBN: 978-1-138-02883-8

Monitoring of mass blasting seismic impact on residencial buildings and constructions

O. Strilets, G. Pcholkin & V. Oliferuk
National Mining University, Dnipropetrovsk, Ukraine

ABSTRACT: The issue of seismic safety of mass blasting in nonmetalliferous construction raw materials quarries is considered. The influence of blasting operations parameters on the velocity of seismic vibrations of the ground is determined, and the main recommendations concerning their mitigation are developed.

1 INTRODUCTION

Topicality. In primary mining for rock in industrial non-metallic quarries, blasting is one of the main operating processes. Seismic vibrations of the soil and shock wave during blasting operations tend to damage residential buildings and constructions located in immediate vicinity of the quarry. Therefore, the quality and safety of conducting blasting operations, especially in non-metallic quarries, are of great importance.

The existing procedure of defining seismic-safe parameters of drilling-and-blasting operations and safe distance (National Standard 2008, National Standard 2009, Ukranian National Construction 2014) does not provide any objective picture. Complex structure of rock massifs on the way of seismic wave, as well as the structure of borehole charge and the ignition method can be pivotal for the intensity of seismic vibrations. The decrease in the volumes of mass blasting does not reduce the intensity of seismic vibrations, and in some cases it even increases them, which could sometimes lead to the issue of the mining enterprise closure.

Monitoring of seismic vibrations allows defining factual vibrations of the soil in the foundations of residential buildings and constructions. Basing on the monitoring results, it is possible to develop optimal parameters of drilling-and-blasting operations and anti-seismic measures to preserve buildings and constructions.

1.1 Analysis of research and papers

Sadovski M.O., Shemiakin Ye.I., Yefremov E.I., Baranov Ye.G., Kusheriavi F.I., Mosinets B.M., Khanukaiev O.N., Tseitlin Ya.I., Pergament V.K., Shvets V.Yu. and other researchers determined the nature and laws of seismic-wave propagation during blasting activities in different rocks. At the same time, a significant number of factors influencing propaga- tion of seismic waves do not allow developing a universal dependence that would be capable of pre- dicting considerably accurately the seismic hazard of industrial blasting.

Many ways of mitigating seismic effects of blasting are developed by different authors; however these methods are not taken into account in the procedure of defining seismic-safe parameters of blasting and the distances, which considerably decreases the prediction accuracy. In addition, the procedure (National Standard 2008) does not provide for the connection of blasting seismic effect and the quality of ore breaking by blasting. Moreover, long-term research demonstrates that the decrease in seismic effect results from the redistribution of blasting energy in effective capacity. One of the reasons of such redistribution is the increase in the time of explosive loading of rock massif with simultaneous reduction in initial impulse of explosion in a borehole, which conditions the critical stability state of rocks under smaller loadings.

1.2 Aim of the paper

The aims of the paper are: the analysis of methods and seismic effect measurement tools of mass blasting on residential buildings and constructions subject to conservation; determination of permissible charge mass limit and safe distance under which the vibration of the soil does not exceed the allowable standards stipulated by current legislation; develop- ment of drilling-and-blasting operations parameters aimed at increasing the quality of breaking rock and decreasing seismic vibrations of the soils in founda- tions of residential buildings and constructions.

To achieve the aims it is necessary to solve the following problems:

– based on the experimental research of seismic vibrations of soil in foundations of residential buildings and constructions, to determine real joint coefficient (*K*) depending on the conditions of blasting and propagation of shock wave;

– to study the intensity of seismic vibrations of soil in foundations of residential buildings and constructions during mass blasting operations in quarries;

– to study seismic and breaking effect of blasting borehole charges in redistribution of blasting energy by changing the direction of detonation, borehole diameter and borehole charge design.

2 MATERIALS AND RESULTS OF THE RESEARCHES

Because seismic effect of blasting in rock mass is multiple-factor process and it cannot be described analytically with high precision, the problems mentioned above were solved experimentally.

Experimental research was conducted according to national standards: DSTU 7116:2009 "Industrial blasting. Method of determining factual seismic stability of buildings and constructions", and DSTU 7117:2009 "Industrial blasting. Method of determining pressure at the front of air blast and boundaries of safety area". To register seismic vibrations and pressure at the front of air blast, the following equipment was used (Figure 1):

– ZET 048-E seismic station with BC 1313 three-component acceleration indicator and receiver for synchronization with GONASS/GPS and laptop with ZETLab Seismo software installed;

– BlastMate III digital seismograph with microphone and two three-axis geophones and laptop with BlastWare software installed.

Processing of the results was conducted using ZETLab Seismo and BlastWare software, with additional software for:

– monitoring and diagnostics of buildings and constructions according to GOST 53778-2010 national standard "Buildings and constructions. Rules of examination and monitoring of technical state";

– monitoring system for buildings and constructions, as well as environment to determine the period and logarithmic decrement of the fundamental tone of natural oscillations according to HOST P 54859-2011 national standard "Buildings and constructions. Determination of the parameters of fundamental tone of natural oscillations", and determination of parameters of seismic effect according to GOST P 53166-2008 national standard "Effect of natural external conditions on technical products. General characteristics. Earthquakes".

The results of monitoring seismic vibrations at the bottom of foundations of residential buildings and constructions showed the exceedance of permissible values 1.5 – 3 times, which was confirmed by objective complaints of the residents.

(a)

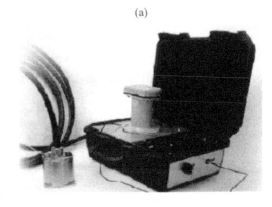

(b)

Figure 1. Equipment for registering seismic vibrations and pressure at the front of air blast: (a) ZET 048-E seismic station; (b) BlastMate III seismograph.

In all the above mentioned quarries a range of emulsion and granulated explosives were used: ukrainit-PP-2B, anemics, grammonite-79/21, igdanite, and recycled explosives in boreholes with diameter from 105 to 250 mm. Initiation of borehole charges was conducted by non-electrical initiating systems.

According to the results of soil seismic vibrations monitoring, at the bottom of foundations of residential buildings and constructions for each case, a real joint coefficient (K) is determined which depends on the conditions of blasting operations and propagation of seismic shock waves. This allowed establishing permissible mass of explosives per one stage of slowing down, developing the zoning map of quarries, as well as developing seismic measures for preserving residential buildings and constructions in the vicinity to the quarry.

As a result of introducing the developed anti-seismic measures at the quarries and complying with "Recommendations concerning seismic safety of mass blasting for industrial and residential buildings and constructions located in the vicinity of a quarry",

534

seismic vibrations of soil was reduced to the permissible level (National Standard 2008, National Standard 2009, Ukranian National Construction 2014) and the quality of blasting operations was improved.

3 CONCLUSION

Monitoring of seismic vibrations reflects the real effect of blasting operations on buildings and constructions. The results of monitoring allow establishing the attenuation coefficient of seismic vibrations and developing seismic safety parameters of drilling and blasting operations. The development of anti-seismic measures reduces the vibrations to the permissible norm, which allows retaining the general mass of explosives. The mass of explosives for the stage of slowing down is adjusted depending on the location of the object under protection, and parameters of seismic vibrations attenuation. This allows redistributing the intensity of seismic vibrations and mitigating their effect on residential buildings and constructions.

As the joint coefficient (K) changes under changing parameters of borehole pattern, design of borehole charge, properties of explosives and conditions of initiating borehole charges, as well as while developing new levels and at the vicinity of blasting operations to residential buildings, "Recommendations concerning seismic safety of mass blasting" are to be revised and adjusted on the results of seismic vibrations monitoring. For this, it is necessary to carry out control measurements of velocity and frequency of seismic vibrations during mass blasting with operational conclusions, recommendations and adjustments concerning further mass blasting operations depending on the realization conditions.

REFERENCES

National Standard DSTU 4704: 2008, 2009. *Carrying out industrial blasting. Seismic safety regulations:* Kyiv: 11.
National Standard DSTU 7116: 2009, 2010. *Industrial blasting. Method of determining factual seismic stability of buildings and constructions.* Kyiv: 6.
Ukrainian National Construction Regulation DBN B.1.1-12: 2014. *Construction in seismic areas of Ukraine:* Kyiv: Ministry of regional development of construction and communal service of Ukraine: 110

Theoretical and Practical Solutions of Mineral Resources Mining – Pivnyak, Bondarenko & Kovalevska (eds)
© 2015 Taylor & Francis Group, London, ISBN: 978-1-138-02883-8

Mathematical modeling of processes in transmissions of mining machines

V. Pavlysh, A. Grebyonkina & S. Grebyonkin
Donetsk National Technical University, Donetsk, Ukraine

ABSTRACT. The task of mathematical model construction of dynamic processes in the transmissions of mining machines as an object with the distributed parameters and its numerical realization of nets method is considered.

1 INTRODUCTION

The questions of design and calculation of drive of mining machines are closely related to the mathematical design of processes in the systems with the distributed parameters. In the article the mathematical model of transitional process is examined in the long transmissions of mining machines on the base of the system of equations in particular derivatives and its realization of nets method.

2 THEORETICAL PRE-CONDITIONS

Dynamic processes in a transmission are described in the system of equations, plugging in itself wave differential equations in particular derivatives, reflecting the turning vibrations of link with the distributed parameters, and equations of motion of engine and working instrumentas scope terms.

The purpose of work is development of theoretical base for research and calculation of parameters of transmissions of mining machines for further perfection of technology of the mechanized development of coal stratums.

3 THE MAIN CONTENTS

The system of two equations of the first order is in the basis of mathematical model (Grebyonkin et al. 2006, Grebyonkin et al. 2007):

$$-\frac{\partial \omega}{\partial x} = k_1 \cdot \frac{\partial M}{\partial t}, \quad -\frac{\partial M}{\partial x} = k_2 \cdot \frac{\partial \omega}{\partial t}, \quad 0 \le x \le L, \quad (1)$$

where ω – angular speed in examined section of transmission, s^{-1}; M – resultant twisting moment in a transmission on the billow of engine, N·m; k_1 – a coefficient of turning pliability of length unit of

billow, (N·m^2) $^{-1}$; k_2 – a coefficient of inertia of length unit of billow, N·s^2.

Scope terms:

$$x = 0 : I_\partial \cdot \frac{\partial \omega_H}{\partial t} = M_\partial - M_H, \quad (2)$$

$$\text{at } x = L : M_K = M_C(t); \quad \omega(t) = \omega_K, \quad (3)$$

where ω_H and ω_K – current values of angular speed accordingly in initial and eventual sections of transmission; M_H and M_K – current values of twisting moment in initial and eventual sections of transmission; I_∂ – moment of inertia; M_∂ – permanent loading moment; $M_C(t)$ – a variable loading moment.

Initial conditions look like:

$$\omega(x;0) = \omega_s, \quad M(x;0) = M_s, \quad (4)$$

The solution of the resulted system is executed on the basis of method of eventual differences, grounded A.A. Samarsky (Samarskiy & Sobol 1963). For approximation of particular derivatives it is necessary to apply a net model with the error of order O. For this purpose the system is taken to the kind:

$$\frac{\partial \omega^2}{\partial t^2} = \alpha \frac{\partial^2 \omega}{\partial x^2}, \quad \alpha = \frac{1}{k_1 k_2}, \quad \frac{\partial^2 M}{\partial t^2} = \alpha \frac{\partial^2 M}{\partial x^2},$$

$$x \in [0; L], \quad t > 0. \quad (5)$$

As a result of approximation net turns out:

$$x_i = i \cdot h_x, \quad i = 0.1, ..., n,$$

$$t_j = j \cdot h_t, \quad j = 0.1, ..., n,$$

$$t_j = j \cdot h_t, \quad j = 0.1, ..., n. \quad (6)$$

The ending-difference system looks like:

$$\frac{\omega_{i,j+1} - 2\omega_{i,j} + \omega_{i,j-1}}{h_t^2} = \alpha \cdot \frac{\omega_{i+1,j} - 2\omega_{i,j} + \omega_{i-1,j}}{h_x^2},$$

$$\frac{M_{i,j+1} - 2M_{i,j} + M_{i,j-1}}{h_t^2} = \alpha \cdot \frac{M_{i+1,j} - 2M_{i,j} + M_{i-1,j}}{h_x^2} . \quad (7)$$

Initial conditions:

$$\omega_{i,0} = \omega_s , \quad i = 0.1,...,n-1 ,$$

$$\omega_{n,0} = \omega_K ,$$

$$\omega_{i,1} = \omega_{i,0} , \quad i = 1.2,...,n-1 ,$$

$$M_{i,0} = M_s , \quad i = 0.1,...,n-1 ,$$

$$M_{i,1} = M_{i,0} , \quad i = 1.2,...,n-1 . \quad (8)$$

Scope terms:

$$\omega_{0,j+1} = \frac{h_t}{I_\partial}(M_\partial - M_H) , \quad j = 1.2,... ,$$

$$\omega_{n,j+1} = \frac{h_t}{I_\partial}(M_H - M_\partial)(\omega_{n,j} - \omega_{n,j-1}) ,$$

$$M_{0,j+1} = M_s , \quad M_{n,j+1} = M_C(t_j) .$$

A calculation chart, which is a result of four-point approximation:

$$\omega_{i,0} = \omega_s , \quad i = 0.1,...,n-1 ,$$

$$\omega_{n,0} = \omega_K ,$$

$$\omega_{i,1} = \omega_{i,0} , \quad i = 1.2,...,n-1 ,$$

$$\omega_{0,1} = \frac{h_t}{I_\partial}(M_\partial - M_H) ,$$

$$\omega_{n,1} = \frac{h_t}{I_\partial}(M_K - M_\partial) ,$$

$$\omega_{0,j+1} = \frac{h_t}{I_\partial}(M_\partial - M_H) , \quad j = 1.2,... ,$$

$$\omega_{i,j+1} = \frac{h_t^2 \alpha}{h_x^2}(\omega_{i+1,j} - 2\omega_{i,j} + \omega_{i-1,j}) + 2\omega_{i,j} - \omega_{i,j-1} ,$$

$$i = 1.2,...,n-1 ,$$

$$\omega_{i,j+1} = \frac{h_t^2 \alpha}{h_x^2}(\omega_{i+1,j} - 2\omega_{i,j} + \omega_{i-1,j}) + 2\omega_{i,j} - \omega_{i,j-1} ,$$

$$i = 1.2,...,n-1$$

$$\omega_{n,j+1} = \frac{h_t}{I_\partial}(M_K - M_\partial)(\omega_{i,j} - \omega_{i,j-1}) . \quad (10)$$

$$M_{i,0} = M_s , \quad i = 0.1,...,n-1 ,$$

$$M_{n,0} = M_K ,$$

$$M_{i,1} = M_{i,0} , \quad i = 1.2,...,n-1 ,$$

$$M_{0,j+1} = M_H ,$$

$$M_{i,j+1} = \frac{h_t^2 \alpha}{h_x^2}(M_{i+1,j} - 2M_{i,j} + M_{i-1,j}) + 2M_{i,j} - M_{i,j-1} ,$$

$$i = 1.2,...,n-1 ,$$

$$M_{n,j+1} = M_C(t_j) . \quad (11)$$

4 ANALYSIS OF RESULTS AND RECOMMENDATION ON THEIR USE

Realization of this model allows to get the array of information about the conduct of parameters of object in space-time and also grounds for conclusions about further perfection of construction.

As far as the receipt of new results there is a necessity of the further working out in detail, involving of new groups of parameters that complicate mathematical models. At the same time, modern facilities of computing engineering, due to the increase of fast-acting and volumes of memory, allow to successfully decide the put tasks, and it determines the prospect of mining technique perfection.

5 CONCLUSIONS

The mathematical model of transmissions functioning process of mining machines is based on the system of nonlinear hyperbolic equations of mathematical physics, which complemented the group of regional terms, reflecting the technological features of the chart.

Application of the offered model is provided by possibility of theoretical researches of processes with the preliminary calculation of parameters of technology, and also construction of computer-aided technological charts of drive of mining machines designs.

REFERENCES

Grebyonkin, S.S., Agafonof, A.V., Kosarev, V.V. et al. 2006. *Mining machines and complexes for underground coal mine.* Donetsk: 353.

Grebyonkin, S.S., Ryabichev, V.D., Pavlysh, V.N. et al. 2007. *Mathematical models and calculation methods of processes parameters of underground mining and mine equipment.* Donetsk: 385.

Samarskiy, A.A. & Sobol, I.M. 1963. *Examples of numerical calculation of temperature waves.* Journal of computational mathematics and mathematical physics, Vol. 3, No 4: 702 – 719.

Theoretical and Practical Solutions of Mineral Resources Mining – Pivnyak, Bondarenko & Kovalevska (eds)
© 2015 Taylor & Francis Group, London, ISBN: 978-1-138-02883-8

Case management of product quality at the mining enterprises

A. Melnikov & T. Herasymenko
National Mining University, Dnipropetrovsk, Ukraine

O. Kovalenko
Prydniprovska State Academy of Civil Engineering and Architecture, Dnipropetrovsk, Ukraine

ABSTRACT: Situational approaches to products quality management at the mining and metallurgical enterprises, which are based on the mechanism of situational modeling, where components are the mining and metallurgical combines are presented. The mechanism of cost management aims to achieve efficiency of the costs regulatory process, in which the function of the optimization of the cost level is a criterion for it's procured. In the proposed approach, the mechanism of managing of the costs associated with providing the required level of quality. The quality assurance process is put on the basis of such mechanism aimed at elimination of the mismatch of quality in the production and consumption of products of mining and metallurgical enterprises. Method that allows to take into account this fact, is the method of total economic effect, which covers that at the optimal mismatch and corresponding to this cost, savings in fixed conditions of preparation of crude ore and steel consumption are optimal.

1 INTRODUCTION

Iron and steel industry in Ukraine is one of the most powerful in the world. At present, the production of iron, steel and rolling metals it occupies a leading position in the global economy. In Ukraine, located 36% of the world's reserves of manganese ore, and it ranks the second place after South Africa. The provision of mining and metallurgical enterprises of manganese raw materials, based on their project capacity is more than 95 years. The level of production of manganese products will enable to provide not only the needs of the domestic market, but also allow to export abroad their significant amount. The quality of mined ore do not satisfy the requirements of consumers, in connection with which all raw manganese ores undergo enrichment and metallurgical conversion, which significantly increases the cost of preparing them to consume. One of the most important ways of increasing the effectiveness of the use of manganese ore resources in the steel industry is to establish the optimal structure of production and consumption on the basis of further development of economic methods of cost management in the mining and metallurgical industry.

Solving issues related to the management of quality of ore raw materials based on cost optimization and related adjacent tasks research of domestic and foreign scientists are devoted So, M.I. Agoshkov,

F.G. Grachev, Ya.Sh. Roizen, M.G. Novozhilov and other focused on solving the economic problems associated with the management of quality of ore raw materials; V.V. Rzhevskii, G.G. Lomonosov, V.F. Byzov, V.I. Ganitsky and other solved issues aimed at ensuring organizational quality management of ore raw materials; A.M. Erpert, L.P. Shupov, L.A. Barsky and other in their study focused on mathematical procuring of ore raw materials quality management; L. Kaas, J. Kelet, K. Lemke, J. Mosteran worked on quality management of crude ore treated from the perspective of market factors.

2 FORMULATING THE PROBLEM

The purpose of this publication is that situational product quality control at mining enterprises based on rational management decisions based on economic mechanism of cost management, situational conditions of production and processing accounted in the market mechanism of production and consumption of products of the mining enterprises.

With the transition to a market economy, a lot of attention is paid to the issues related to product quality and management. The increased interest in the problem of quality control is explained by the fact that there is a direct relationship between the level of product quality and the growth of national wealth. High product quality is fundamentally important in terms of long-term prospects for a steady increase in national wealth. The higher the quality of the products, their reliability and durability, the greater its use value and longer the values are in circulation, and this ensures the growth of national wealth.

The quality of products is to be understood as a set of properties causing its suitability to satisfy certain requirements in accordance with its value. Formation of quality occurs at all stages of product manufacture: quality is founded at projection stage, is provided at the production and is maintained at the operation stage.

Quality management is an important part of the overall industry management system and is carried out at the three hierarchical levels: national, sectoral, and enterprise. At the national (inter-branch) level to the functions of the system of quality management include: general business planning of product quality, cross-sectoral coordination of the largest events for improvement, balancing the necessary resources, standardization, organization of state quality control, state certification of products, etc. At the sector level is provided industry-quality planning, industry standardization and certification of products, activities of departmental quality control inspections, development of moral and material incentives, etc. Quality management system at the enterprise-level provides such management services as planning and technical departments, reliability standardization, metrology services and technical control departments. The process of projection, production, sale and consumption of products is included to the control system in this case. Each of the stages of product quality formation presents its requirements to the management. Very important problem of determining the optimal quality characteristics of future products and their planning at the projection stage is solved. The most important function of the quality management system is providing the necessary technical, organizational and operating conditions for the planned level of quality at the production stage. To the fore advances the task of maintaining the achieved level of quality at the sale and consumption stage is coupled.

The process of product quality control is reduced to the following issues (Mel'nykov et al. 2003):

– establishment of quality indicators and measuring the impact on them of various factors;

– establishing an optimum level of quality;

– control and regulation of the optimal level of quality in industrial environments;

– organization of feedback to determine the actual level of quality;

– decision making on the basis of the received information.

Quality of products of mining enterprises can be evaluated in many ways. These include useful content and related components, the characteristics of mechanical and physicochemical properties (strength, mostly), etc. Dominant qualitative characteristic should be recognized mineral content – concentrate and agglomerate in the final product of the enterprise.

In the future, we can assume that in the considered range of variation of manganese provides an acceptable level of all other quality characteristics, so that the main task of quality management is to establish acceptable limits and to determine the optimal value of the manganese content in the final product (Mel'nikov et al. 2007).

Let us consider the conditions which define the upper and lower boundaries of the mineral content in the concentrate and agglomerate with the requirements for the company's products from the metallurgical industry of the need to increase the mineral content in the concentrate and agglomerate. However, this will inevitably lead to higher costs. Improving the quality characteristics is due to increased costs. And eventually costs exceed market prices even with the supplements for excessive content of manganese in the product as compared with the base. The company's profit is equal to zero or negative, and the company becomes unprofitable. Thus, the upper limit of the content of the useful component in the product is determined by the conditions of increasing base prices for high quality products over the cost required to produce high quality products.

On the other hand, if you are not interested in quality problem and avoid the additional costs associated with its increase, such company will lose markets. This condition determines the lower limit of the content of the useful component in the final product of the mining enterprise.

The problem of quality management is closely related with presently very relevant requirement of orientation at final results. Quality is formed at all stages of the mining process and is become apparent in the end, not the interim results. Output characteristics (including qualitative) such a large system such as ore processing production, are not final for even greater system – mining and metallurgical complex. Requirements to be met by metallurgical production have an impact on the qualitative characteristics of the products of the mining enterprise, so that the level of quality that is acceptable when considering the operation of the mining enterprise alone may be insufficient in an integrated approach (Kovalenko 2008). Thus, the scientific solution to the problem of product quality and management is only possible when considering the mining complex as a whole.

The advantage of situational modeling over other methods is the possibility of considering a large number of alternatives, a more accurate reflection of the complex structure of the objects and more accurate prediction of the effects of various management decisions. The idea of situational modeling is simple and attractive. It is the ability of situational experiments to create a model in cases where such experiments on the real object are impossible or impractical. Virtually every model or representation of things is a form of imitation of the situation, so the concept of situational modeling is very broad, as well as its scope.

The main feature of situation models is to carry out these experiments in order to obtain information about the options of the real system.

Valuable quality of situational modeling is the ability to "act out" on the model different situations that can be used situational-simulation models as a tool for professional development specialists.

Functional dependencies describe the behavior of the variables and parameters within the component or between components effects. Usually, functional dependencies are mathematical equations that establish the relationship between endogenous and exogenous variables. The relationship between the variables can be deterministic or stochastic. In the deterministic case, the output of the system is uniquely determined by the information on the input. Stochastic relationships conform to the laws of probability theory. In this case, input information has indeterminate results on the output. Accounting of stochastic relations is difficult and requires prior prediction of random parameters.

There is a definite functional relationship between the improvement of product quality, production costs and production targets. If you concentrate on excessive improving of product quality, that disproportionate increase unit costs and it becomes almost impossible to plan output. At the other extreme is output of extremely poor quality (defect), as a result – loss of the market. Between these extremes is the best option and quality control, ensuring the highest economic efficiency. This option is achieved by compliance with such quality standards, which ensure the manufacture of products with high use value and the lowest cost and guarantees the fulfillment of the production tasks. The production process at the mining plant is a complex of closely related technological combinations. Therefore, the initiation of quality control in one area automatically entails the need to manage the process on the remaining parts of the production cycle. That is, quality management should be comprehensive and cover all production processes, tying them together to give a comprehensive economic evaluation of the proposed recommendations.

In order to take account of such contradictory and multilateral requirements for quality control problem of production, it is necessary firstly to determine the situation related to the distribution of the effect of reducing the cost of quality between manufacturers of quality raw materials and its customers. For this purpose, an economic mechanism was created that describes the situation for the effect of improving the quality of products between producers (mining company) and consumer products (metallurgical company).

3 RESULTS OF RESEARCH

At the economic mechanism of cost management to ensure the quality of crude ore, optimal economic value of the parameter is the cost, reflecting the process of quality assurance. Formation of economic-mathematical model of optimization in this case differs from the traditional models, which are based on the method of minimal costs associated with the production and exploitation of production (Mel'nikov et al. 2007).

In the proposed approach the management costs associated with the provision of (increasing) the quality of products, quality is the main argument in the present economic and mathematical model. This fact contradicts the statement of the problem, which explores the process of quality assurance and associated with the elimination of mismatch of quality in the production and consumption of products of the mining enterprise. Method that allows to take this into account, is the method of total economic effect, which comprises that at the optimal mismatch and corresponding to this cost, savings in fixed conditions of preparation of crude ore and steel consumption is optimal.

The method of maximum economic impact adequately reflects the quality assurance process ore raw materials, i.e., for the increment costs to improve product quality mining enterprise decreases mismatch and reduced costs of metallurgical enterprises. This method allows to consider the best use of the general possibility of achieving the adjusted level of quality assurance of crude ore, as the private criterion maximum savings in this case meets the requirement of the general criterion – cost savings of social labor as a result of the quality assurance of mining industry. According to the critical value of the error, indicators of quality can determine the maximum allowable costs increase and limit the error, which is important in terms of the overall effect. This method allows to determine not only the point, but also the area of cost-effective values of the error indicators of the quality of ore raw materials (Figure 1).

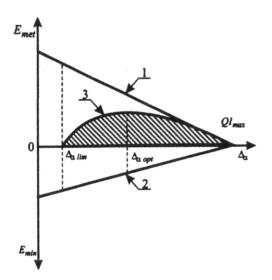

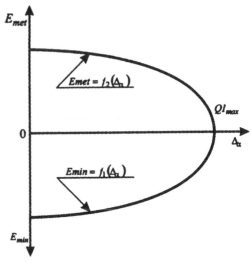

Figure 1. Graphical representation of the economic impact maximum from the preparation and consumption of raw ore (E_{min} – changing mining enterprise impact at improving the quality of ore raw materials; E_{met} – changing metallurgical enterprise impact at improving the quality of ore raw materials; QI_{max} – the maximum of mismatching of quality indicator; $\Delta\alpha_{lim}$ – the limit (minimum possible) of mismatching of quality indicator; $\Delta\alpha_{opt}$ – the optimal indicators of the quality mismatch; 1 – changes in costs of metallurgical enterprises; 2 – changes in costs of mining enterprises; 3 – curve for the total economic impact of metallurgical and mining enterprise.

Let us consider the relationship of the effect E, the resulting quality assurance ore raw materials and mismatch $\Delta\alpha$, actual and predetermined levels using the method of total economic effect.

Function $E = f(\Delta\alpha)$ is a dependence of the effect E on the error $\Delta\alpha$ of quality level in different variants of its approximation, which allows (Akof et al. 1971):

– to determine the optimal value of the error indicator of ore raw materials quality;

– to determine the area of the effective values of the error indicator of quality.

Quality management involves the reduction of crude ore mismatch $\Delta\alpha$, so its reduction leads to decrease cost and to increase losses, i.e. $E_{min} = f_1(\Delta\alpha)$ and $E_{met} = f_2(\Delta\alpha)$. Let us examine the situations, characterizing its limit indexes.

Situation 1. Dependences $E_{min} = f_1(\Delta\alpha)$ and $E_{met} = f_2(\Delta\alpha)$ are presented by symmetrical parabolas with axis Δ and apex QI_{max} (Figure 2).

The functions are symmetric, the optimum for a given situation is absent.

Situation 2. Dependence $E_{min} = f_2(\Delta\alpha)$ is presented by parabola which described by the equa-

tion $E = \sqrt{a_2 \Delta\alpha + b}$ with axis $\Delta\alpha$ and apex $\Delta\alpha_{max}$ (Figure 3).

Figure 2. Graphical representation of dependencies $E_{min} = f_1(\Delta\alpha)$ and $E_{met} = f_2(\Delta\alpha)$ as symmetrical parabola.

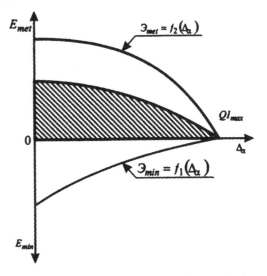

Figure 3. Graphical representation of dependencies $E_{min} = f_1(\Delta\alpha)$ and $E_{met} = f_2(\Delta\alpha)$ as asymmetrical parabolas.

Dependence $E_{min} = f_1(\Delta\alpha)$ also is presented by parabola which is described by the equation $E = \sqrt{a_1 \Delta\alpha + b}$ with axis $\Delta\alpha$ and apex $\Delta\alpha_{max}$. Expression of the general effect is equal to the algebraic sum E_{min} and E_{met}, i.e.

542

$E = \sqrt{a_1 \varDelta_a + b_1} - \sqrt{a_2 \varDelta_a + b_2}$.   At

$\varDelta_a = \sqrt{b_3} = \sqrt{b_1} - \sqrt{b_2}$ or $b_3 = b_1 + b_2 - 2\sqrt{b_1 b_2}$;

whereas $\varDelta_{a\,max} = \dfrac{b_3}{a_3}$, so $a_3 = -\dfrac{b_3}{\varDelta_{a\,max}}$. Let's calcu-

late the value of a_3 and b_3, we can estimate the overall effect. Substituting for the argument values from 0 to QI_{max}, it is possible to determine the total effect E_Σ that defines the area of the effective values of the error indicator of quality raw ore.

Situation 3. Dependence $E_{met} = f_2(\varDelta_a)$ is presented by exponential function, $E_{min} = f_1(\varDelta_a)$ – by parabolic function. For exponential function dependence is described by equation $E = a_1 e^{-b\varDelta_a}$, for parabolic function – $E = -\sqrt{a_2 \varDelta_a + b_2}$ (Figure 4).

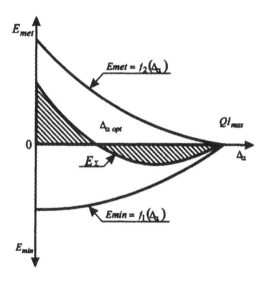

Figure 4. Graphical representation of dependencies $E_{met} = f_2(\varDelta_a)$ as function parameters and $E_{min} = f_1(\varDelta_a)$ as parabola.

The study data dependencies shows that when $\varDelta_a \leq \varDelta_{a\,opt} a_1^2 e^{-2b_1 \varDelta_a} \geq a_2 \varDelta_a + b_2$, the vertex of the

parabola corresponds to $QI_{max} = \dfrac{b_2}{a_2}$. At $\varDelta_a = QI_{max}$

functions are equal, then $a_1^2 e^{-b\varDelta_{a\,max}} = a_2 \varDelta_{a\,max} + b_2$. To find the coefficient a_2 and b_2 dependencies use the following conversion $(-E)^2 = -(a_2 \varDelta_a + b_2)$, at –

$E_{max} = 0$ и $\varDelta_a = 0$; $-b_2 = (-E^2)$;   at   E = 0;

$a_2 = -\dfrac{E^2}{\varDelta_{a\,max}}$ from the theory of Mathematical analy-

sis implies that $QI_{max} = 5b_1$ function is approximately 1% of the initial values, so for practical recommendations point $\varDelta_a$ sufficiently positioned at the location where its value is equal to 0.01. Therefore

$a_1 e^{-b_1 \varDelta_{a\,max}} = 0.01E = 0.01 a_1$. Graph $E = a_1 e^{-b_1 \varDelta_a}$

at $\varDelta_a = 0$ is characterized as $E_{max} = a_1$.

At $\varDelta_{a\,max} = \varDelta_a a_1 e^{-b_1 \varDelta_{a\,max}} = 0.01 E_{max} = 0.01 a_1$.
Following the appropriate steps and reducing by a_1 we receive $e^{-b_1 \varDelta_{a\,max}} = 0.01$. Doing the logarithm of the expression we obtain $b_1 \varDelta_{a\,max} = \ln 0.01$ $b_1 \varDelta_{a\,max} = \ln 0.01$, from where

$b_1 = \dfrac{\ln 0.01}{\varDelta_{a\,max}}$. At the transition point axis $\varDelta_a$ func-

tions are equal, therefore $a_1 e^{-2b_1 \varDelta_a} - a_2 \varDelta_a - b_2 = 0$.
Found from this equation determines the limiting value $\varDelta_a$ error value quality, limiting the scope from $\varDelta_a = 0$ to $\varDelta_{a\,lim}$, in which a positive effect is achieved.

Situation 4. Dependences $E_{met} = f_2(\varDelta_a)$ and $E_{min} = f_1(\varDelta_a)$ are approximated by exponential functions (Figure 5) having the form of the following equations $E = a_1 e^{-b_1 \varDelta_a}$ and $E = a_2 e^{-b_2 \varDelta_a}$.

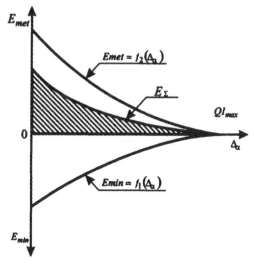

Figure 5. Graphical representation of dependencies $E_{met} = f_2(\varDelta_a)$ and $E_{min} = f_1(\varDelta_a)$ as the exponential functions.

Let us investigate the situation at $\varDelta_a = 0$, $a_1 = +E$, $a_2 = -E$. It is known from mathematical analysis, that at $QI_{max} = 5b_1$ function is about 1% of the initial values, so for practical purposes it is

543

sufficient to locate the point QI_{max}, where the value is 0.01. Therefore $0.01 a_1 = a_1 e^{-b_2 \Delta_{\alpha\, max}}$; reducing by a_1 and the logarithm, we obtain $\ln 0.01 = b_1 \ln e$. From

where $b_1 = \dfrac{\ln 0.01}{\Delta_{a\, max}}$. Converting the expression

$-E = a_2 e^{-b_2 \Delta_a}$, we obtain $b_2 = \dfrac{\ln 0.01}{\Delta_{a\, max}}$,

$E_\Sigma = a_1 e^{-b_1 \Delta_\alpha} - a_2 e^{-b_2 \Delta_\alpha} = e^{-b_1 \Delta_\alpha}(a_1 - a_2)$.

Situation 4. Dependences $E_{met} = f_2(\Delta_\alpha)$ and $E_{min} = f_1(\Delta_\alpha)$ are approximated by parabola functions $+E = \sqrt{a_1 \Delta_a + b_1}$ and an exponential curve $-E = a_2 e^{-b_2 \Delta_\alpha}$ (Figure 6).

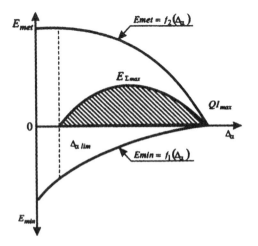

Figure 6. Graphical representation of dependencies $E_{met} = f_2(\Delta_\alpha)$ and $E_{min} = f_1(\Delta_\alpha)$ in the form of a parabola and an exponential curve.

The limiting value of the positive effect in reducing the dispersion reaches the final value at the approach of the error to zero. Theoretically, this final value is the cost of metallurgical enterprises, where the characteristics of batch preparation are determined by not ore components.

The limiting value of the negative effect in reducing the dispersion has finite value at the approach of the error to zero, in theory require certain costs from mining enterprise associated with the natural characteristics of crude ore. At $\Delta_a = 0$, $+E = \sqrt{b_1}$,

$b_1 = \sqrt{E}$. Vertex of the parabola is QI_{max},

$QI_{max} = \dfrac{b_1}{a_1}$, $a_1 = \dfrac{b_1}{QI_{max}}$. At $\Delta_a = 0$, $a_2 = -E$,

$a_2 e^{-b_2 \Delta_{a\, max}} = 0.01 E$. Taking the logarithm we obtain $\ln a_2 - b_2 \Delta_{a\, max} = \ln 0.01 + \ln E$ Wherefrom

$b_2 = \dfrac{\ln a_2 - \ln 0.01 - \ln E}{\Delta_{a\, max}}$. For the practical use of this formula point QI can be positioned where its value is 0.01. The total effect is determined by the expression: $E_\Sigma = (a_1 \Delta_\alpha + b_2)^{1/2} - a_2 e^{-b_2 \Delta_\alpha}$.

The choice between one or other dependencies E_{met} and E_{min} is determined by the degree of their compliance with the character of the economic effect and initial conditions, as well as specific to each case mismatch indicators of quality assurance of crude ore. Dependence $E_{met} = f_2(\Delta_\alpha)$ can be approximated by the function $E = \sqrt{a\Delta_a + b}$ and dependence $E_{min} = ae^{b\Delta_a}$ $E_{min} = f_1(\Delta_\alpha)$ by the curve of the exponential function $E = ae^{b\Delta_a}$. Whereas the curves E_{met} are positive and E_{min} are negative can be equally as parabolic curves, as well as an exponential function.

Research of functions $E = f(\Delta_\alpha)$ depending on the change of its parameters makes it possible to determine the area of the effective value of the quality index, corresponding to the cost of quality assurance, which will make it possible to determine the boundaries of the effective functioning of the organizational-economic mechanism of management quality ore processing.

To confirm the adequacy of the performed theoretical research and practical use of the conclusions in the quality management of the crude ore audited on empirical material. Initial data were used in relation to the quality of the concentrate produced at the Chkalov concentrator Ordzhonikidze Mining and Nikopol Ferroalloy plant producing silicomanganese. The analogue of indicator of the mismatch in the empirical model is based on the standard deviation indicator of manganese content σ_β in the concentrate Chkalov concentrator Ordzhonikidze Mining supplied for the production of sinter at Nikopol Ferroalloy plant. Decreasing the standard deviation for the mining enterprise associated with averaging useful component in the ore, which in turn holds additional costs associated with reexcavation of concentrate prior loading into rail cars. Dependence of cost on the error of standard deviation index of concentrate quality for the processing plant and ferroalloy plant expresses indicator by function:

$$C = -E = a_1 \Delta_\beta^{-b_1}, \tag{1}$$

544

where E – losses of processing plant associated with the stabilization of the quality of the concentrate, UAH/t; a_1 and b_1 – empirical coefficients; Δ_β – the value of the error of standard deviation of concentrate quality for the processing plant and ferroalloy plant, %.

Cost savings from ferroalloy plant occur as a consequence of the lack of need of averaging concentrate and of reducing gas consumption associated with the sinterability of sinter mix.

4 CONCLUSIONS

The results of studies aimed at deepening the situational approach to the formation of the economic mechanism of cost management to ensure the required quality of products of mining enterprises, which, unlike traditional approaches, based on the method of cost minimization in the production and consumption of products. Proved that the overall economic effect occurs when a mismatch of the proposed optimal quality and consumption and related costs that in which the savings in fixed conditions of preparation of crude ore and steel consumption is optimal. This approach makes it possible to determine the range of the effective values of quality index, as well as the corresponding quality costs, which make it possible to determine the boundaries of the effective functioning of the economic mechanism of management quality ore processing.

REFERENCES

Akof, R. & Sasieni, M. 1971. *Fundamentals of Operations Research*. Moscow: Mir.

Kovalenko, O.A. 2008. *Classification of organizational and economic problems of quality management of ore at mining enterprises. Ekonomichnyj prostir*, No 15: 59 – 69.

Mel'nikov, A.N. & Kovalenko, O.A. 2007. *Economic mechanism of cost management to ensure the quality of crude ore*. Materialy mezhdunarodnoj nauchno-prakticheskoj konferencii "Dni nauki – 2007": 5 – 12.

Mel'nikov, A.N., Kovalenko, O.A. 2007. The economy of quality manganese ore. *Informacionno-analiticheskij bjuleten' MGGU*, no.143, pp. 245 – 249.

Mel'nykov, A.M., Kovalenko, O.A. 2003. Economic projection of quantitative indicators of production and consumption of ore. *Ekonomika pidryjemstva: problemy teorii' ta praktyky*, no.45, pp. 127 – 135.

Theoretical and Practical Solutions of Mineral Resources Mining – Pivnyak, Bondarenko & Kovalevska (eds)
© *2015 Taylor & Francis Group, London, ISBN: 978-1-138-02883-8*

Pulse method of magnetite demagnetizing

O. Berezniak & O. Berezniak
National Mining University, Dnipropetrovsk, Ukraine

ABSTRACT: The article presents experimental results of classification of magnetized and demagnetized suspension of magnetite concentrate. It shows a high degree of demagnetization in a pulsed mode. The aim of this article is to evaluate effectiveness of magnetite suspension classification with implementation of pulse demagnetizing method, using experimental results.

1 INTRODUCTION

To provide disclosure of magnetite grains during the enrichment of magnetite quartzite there is a need to mill them to a particle size less than 50 μm. This process is carried out in ball mills with three-stage central discharge. Presence of the finished class in grinding zone reduces speed of the process and leads to undesirable overgrinding of magnetite grains. As a consequence, it increases the cost of magnetite quartzite enrichment. Discharge of the finished class is performed by external devices such as spiral classifiers at first stage of grinding and hydrocyclones at the second and third stages. The value of circulating load coefficient of the mill, which exceeds 200% in existing enrichment schemes, depends on the efficiency of the classifying devices. Efficiency of classification of magnetite particles with fineness of 50 microns in hydrocyclones of HC-500 series does not exceed 80%, if not to take into account their magnetic flocculation.

Magnetic separation is applied after each stage of grinding. As magnetite particles have residual magnetization, it leads to their magnetization. Magnetic flocculation, occurring in suspension of magnetized particles, increases their effective size. It further decreases the classification efficiency and increases the coefficient of circulating load.

Furthermore, magnetite particles less than 74 μm are widely used in coal preparation as a weighting agent for dense-media separation. Magnetic flocculation of the weighting particles leads to faster delamination of suspension. It increases lower size limit of enriched coal and enhances error of separation.

In general, it can be assumed that residual magnetization of magnetite plays a negative role in process of minerals' enrichment. Its reduction, or demagnetization of particles, is a burning task.

2 DEMAGNETIZATION METHODS

Ferromagnetic loses its magnetic properties when it is heated to a temperature above the Curie point. For magnetite it is equal to 580°C. Such a method of magnetite demagnetization is the most comprehensive, but it is unacceptable in mineral processing due to its high energy expenditure.

Another demagnetization method is to place a particle in an external alternating magnetic field, with a smooth induction decrease from a maximum value, which should be greater than the residual magnetization, to zero. In this case, demagnetization of fixed particles will occur according to the hysteresis curves shown in Figure 1 (Vikulov et al. 2012).

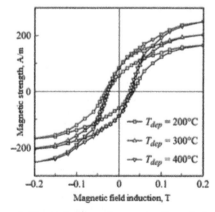

Figure 1. Hysteresis loops of magnetite films obtained at different substrate temperatures.

This mode of external field influence can be achieved in special devices, which are used for demagnetizing. Their concept is shown in Figure 2. Structurally, it is a solenoid – tube of non-magnetic material on which sections of specially designed electromagnet are situated. It provides a smooth reduction of alternating magnetic field induction during the magnetite particles moving through the tube (in the direction shown by arrow) while the solenoid is constantly connected to an AC power source. Degaussing occurs according to the hysteresis curves during the ejection of particles from the solenoid, wherein the ejection period should be at least an order of magnitude greater than the period of alternating current. Capacitor is connected in parallel to the solenoid, providing currents resonance at mains frequency to reduce power consumption. Quality factor (Q-factor) of the formed oscillation circuit must be over 10, so the coils are made of expensive copper. For example, weight of the coil with the pipeline diameter of 450 mm reaches up to 500 kg.

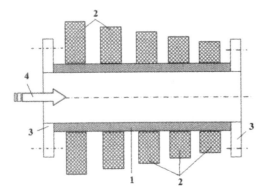

Figure 2. Demagnetizing apparatus: 1 – pipe of non-magnetic material; 2 – coils of electromagnet; 3 – flanges; 4 – direction of magnetite slurry flow.

On the other hand, damped current oscillations can be set up in the solenoid coil, and the number of oscillations should be greater than 5, which is ensured by the appropriate quality of the oscillatory circuit. In this case, all the particles located inside the solenoid, where maximum magnetic field induction is greater than their residual magnetization, will be subjected to demagnetization (Berezniak et al. 2012).

For this purpose we designed laboratory apparatus for degaussing of magnetite suspension. It follows the design of a typical solenoid, where damped oscillations are excited and through which a magnetite suspension passes. The outer and inner coil diameters are 20 mm and 12 mm respectively, and its length is 130 mm. The solenoid is a three layers winding of copper wire with a diameter of 1.2 mm and comprises 300 turns. Its inductance is 220 μH and active resistance is 0.5 Ω.

Damped oscillations in the solenoid are excited by a generator. See the change of the magnetic field induction in the center of the solenoid in time in Figure 3.

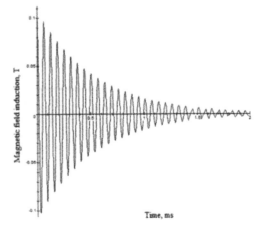

Figure 3. Dependence of magnetic field induction in the center of the coil on time.

3 EXPERIMENTAL RESULTS

We carried out a research and studied suspension demagnetization on magnetite concentrate after flotation from Ore Dressing Plant "Poltavskiy" (Poltava ODP). Suspension of magnetite with a solid content of 350 kg/m³, pre-magnetized in a constant magnetic field with induction of 0.35 T and stirred, was passing through a glass tube with inner diameter of 6 mm, which was located inside the solenoid.

Submission of suspension in the tube was implemented via funnel. Volumetric flow rate of slurry was $1.5 \cdot 10^{-5}$ m³/s, the speed of its movement in solenoid was 0.53 m/s. Thus, residence time of the particles of magnetite inside solenoid was 245 ms. Cycle time was 120 ms, so every particle of magnetite passing through solenoid was demagnetized at least twice.

Effects of alternating magnetic field on the slurry flow were evaluated by optical testing of pulp samples. See the magnetized slurry optical testing in Figure 4 and optical testing of slurry which has undergone demagnetization in Figure 5.

Analysis of the results shows that performed magnetic exposure leads to demagnetization of pulp particles and significantly reduces the size of their aggregates. At the same time, physical impact on the suspension does not affect the structure and size of aggregates.

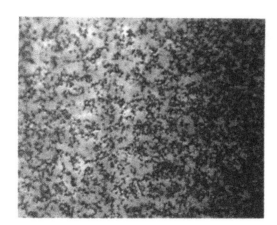

Figure 4. Magnetized slurry optical testing.

Figure 5. Demagnetized slurry optical testing.

Study suspension was stirred in glass cylinder with a diameter of 55 mm and a height of 23.5 cm, and then the time of movement of sediment and clarified zone interface was recorded to determine the settling velocity. Lack of experimental points on demagnetized magnetite settling curve is caused by formation of the transition zone, consisting of magnetite and quartz fine particles. Only quartz particles were remaining in clarified zone (Berezniak et al. 2012).

Results of research in the form of charts are shown in Figure 6.

As can be seen from graphs of magnetite settling, the average settling rate of demagnetized magnetite suspension is 3.4 times less than that of magnetized and 1.7 times less than that of initial suspension.

Furthermore, we performed the study of classification of magnetized and demagnetized magnetite suspension, prepared from Poltava ODP concentrate, on a laboratory model of hydrosizer. Content of magnetite in concentrate was 69%, and content of less than 50 microns particles was more than 94%.

Results of research are shown in Tables 1 and 2.

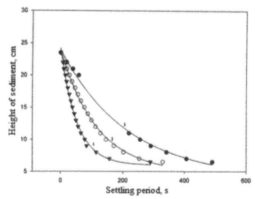

Figure 6. Dependence of magnetite sediment height on time. 1, 2 and 3 – magnetized, initial and demagnetized magnetite suspension respectively.

Table 1. Dependence of critical size of minerals on the flow rate in hydrosizer.

Mineral	Density, kg/m³	Flow rate, mm/s	Calculated boundary grain size, μm
Magnetite	5400	17.4	85.2
		6.63	52.6
Silica	2700	17.4	137.1
		6.63	84.6

Table 2. Distribution of magnetite suspensions during the separation in hydrosizer.

Calculated boundary grain size for magnetite, μm	Product yield, %			
	Magnetized		Demagnetized	
	over-flow	under-flow	over-flow	under-flow
85.2	51.2	48.8	93.1	6.9
52.6	23.1	76.9	79.7	20.3

Suspension was classified at two upstream rates. As can be seen from Table 1 at a rate of 17.4 mm/s virtually all magnetite and silica particles should fall into the overflow product. However, after magnetization of slurry the yield of the overflow product was 51.2%. With nearly all the silica particles were carried into the overflow (with the exception of a small amount of beans entrapped in magnetic floccules), we can conclude that more than a half of magnetite particles forms floccules larger than 85 microns. After demagnetizing the yield of overflow product was more than 93%. It indicates fairly complete demagnetization of magnetite particles.

At a flow rate of 6.63 mm/s more than a half of the magnetite and all silica particles should fall into the overflow. However, less than 20% of magnetite particles fell into the overflow product as a result of magnetic flocculation. As follows from Table 2, quantity of magnetite with particle size of less than 53 microns that fell into the overflow product after demagnetizing is 3.75 times greater than without demagnetizing. That also indicates a sufficiently complete demagnetization of magnetite particles.

4 CONCLUSIONS

To sum up, putting the demagnetization apparatus before the operation of classification of the ground product will significantly reduce the circulating load of the mill. Thus, the application of the pulse method of magnetite demagnetizing in the flow chart of magnetite quartzite enrichment will improve its overall efficiency and reduce energy costs.

REFERENCES

Berezniak, O., Berezniak, O. & Gumerov, M. 2012. *Calculation of required parameters of process of magnetite demagnetization*. Mineral Dressing: Scientific and Technical Collection, No 48 (89): 105 –109.

Berezniak, O., Berezniak, O., Gumerov, M. & Polyga, D. 2012. *Experimental results of magnetite demagnetization in a pulsed mode*. Mineral Dressing: Scientific and Technical Collection, No 50 (91): 111 – 114.

Vikulov, V., Balashev, V. & Pysarenko, T. 2012. *Influence of synthesis temperature on the structural and magnetic properties of Fe_3O_4 films on the surface of SiO_2/Si (001)*. Letters to the Journal of Technical Physics, No 38: 73 – 80.

Theoretical and Practical Solutions of Mineral Resources Mining – Pivnyak, Bondarenko & Kovalevska (eds)
© 2015 Taylor & Francis Group, London, ISBN: 978-1-138-02883-8

Functions and properties of a logistics system in sustainable enterprise development

L. Shvets
National Mining University, Dnipropetrovsk, Ukraine

ABSTRACT: The main objective of the article is to point out the importance of correlation between sustainable enterprise development and economic and logistics development of the enterprise. The case study of sustainable enterprise development in the tumultuous environment lead to the number of important conclusions – the restrictions factor and ignoring its negative effect should be taken into account in the enterprise management and efficient logistics system building. The author created conceptual framework of sustainable enterprise development and described elements of sustainable enterprise development as a foundation for logistics system of enterprise. This issue should be investigated further. Basic principles system should be formed for a logistics system in the context of sustainable enterprise development. As well as conceptual approaches should be formed to logistics system building based on sustainable development concept.

1 MAIN PART

In the modern management context most enterprises are managed logistically and strategically to minimize logistical channels costs and provide marketability. Enterprise development strategy may be used to build an efficient logistics system for consistent management of information and commodity flows between suppliers, enterprise and consumers. In the context of changing market and incomplete appropriate strategy standards, a logistics system is often built abruptly and inefficiently. Such logistics systems are built ignoring important economic, social and environmental factors.

To summarize basic modern theoretical and methodical achievements in sustainable enterprise development and to consider the proven experience of leading enterprises special mention should go to factors withstanding restrictions and being important in logistics system building, such as: expeditious science and technology advancement, organization and realization of environmental, eco-friendly and resource-saving measures, innovative activity increase in enterprise with new resource saving and other factors of socioeconomic, productive, ecological and institutional influence.

The case study of sustainable enterprise development in the tumultuous environment lead to the number of important conclusions – the restrictions factor and ignoring its negative effect should be taken into account in the enterprise management and efficient logistics system building. The efficient implementation of economic, social and environmental plans of sustainable enterprise development and the following logistics system building demands appropriate investment, organization and institution.

The enterprise activity may be determined as a repetitive process, thus logistics management may be distinguished as a closed managerial cycle, which is also repetitive. Logistics management is distinguished as the cyclic process by structural, processing and functional approaches, which are closely related. Logistics management is naturally related to the theory and practice of marketing and management, meaning connection of production and realization with the actual effective demand, the sales promotion through advertisement, the flexible pricing policy (margins and discounts), the search for new sources of income generation, etc. Dialectic engagement and convergence of logistics, marketing and management result in the economic impact (Khadzhynova 2013).

The main condition of sustainable enterprise development is sustainable financing of measures, arranged for harmonious enterprise development in social, environmental and economic aspects with due regard to concerned parties' expectations, as well as economic instruments enhancement. While noting the value of investment provision of the sustainable enterprise development, it's important to remark that the sustainable development foundations cannot develop without active governmental involvement in regulation of innovative investment support for enterprises that actively implement environmental and social programs, and have social and economic value for the civil society development (Prokopenko 2014). Thus the building of the enterprise logistics system should be founded on the conceptual framework of sustainable enterprise development (Figure 1).

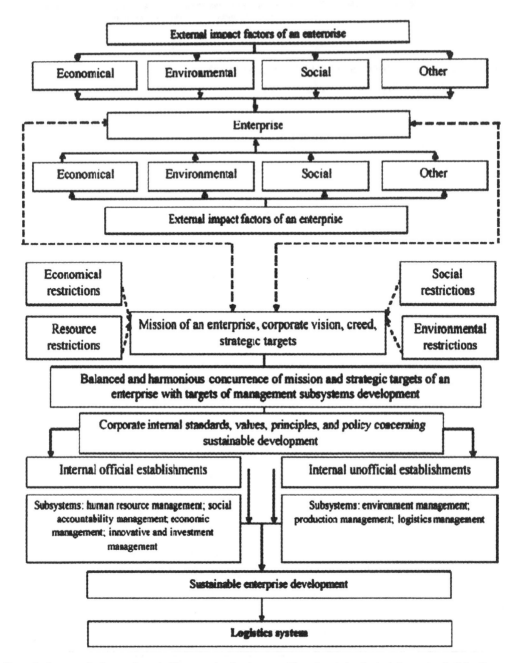

Figure 1. Conceptual scheme of sustainable enterprise development, like a foundation for logistics system building*.
*developed by the author

In the Modern economical dictionary sustainable enterprise development is defined as "financial status of an enterprise, where business activity ensures under the usual conditions fulfillment of all responsibilities towards employees, other organizations and government through sufficient income and according to income and expenditure" (Rayzberg 1999). So the sustainability of the enterprise level is usually determined by its financial status.

A.V. Sevastyanov defines sustainable development as sustainable excess of income over expenditure through its efficient management, which promotes production and realization of goods (Sevastyanov 2004).

According to O.V. Zetkina sustainable development is profitable production and trade activities of an enterprise, promoted by improving production resource efficiency and business management, sustainable financial status, promoted by improving capital structure, as well as sustainable development of business capacity and social development of personnel while keeping self-financing in conditions of external environment (Zetkina 2003).

Logistics system is business management controlling mechanism of product and information flows, which includes real assets enabling goods flow along logistic chain (warehouses, shipping and receiving mechanisms, transport), production supplies and control over all links in the chain. Logistics system is an adaptive, closed cycle system, which performs certain logistic functions and operations. Usually it consists of several subsystems and is widely connected with external environment (Koblyanska 2011).

According to L.V. Dubchak logistics system is a system with adaptive properties, which performs logistic operations and functions, is widely connected with external environment (Dmytrychenko et al. 2007).

T.N. Skorobohatova defines logistics system as assembly of functionally relative elements, which aim to fulfill the main objective of logistics – delivery of goods in the right quantity and of the right quality, in the right time and place with the minimum expenditures (Skorobohatova 2005).

According to T.V. Alesynska logistics system is a dynamic, open or stochastic, complex or big closed cycle system, which performs certain logistic functions (Alesynska 2005).

Thus logistics system building in conditions of sustainable enterprise development may be defined as an assembly of closely related elements, which are adapted to internal and external environment and perform logistics operations consistently.

Logistics system in conditions of sustainable enterprise development is defined as consistent management controlling mechanism, which consists of an assembly of elements (warehouses, shipping and receiving mechanisms, transport) and their relations, performed in order to promote sustainable economic (commercial sustainability, financial sustainability, organization sustainability, production and technical sustainability), social (personnel sustainability, income sustainability) and environmental (resource sustainability, ecological sustainability) enterprise development.

Internal economic factors include: production and delivery of goods, and enterprise's marketability.

Other internal factors are resource availability, resource saving technologies, etc. Internal factors also include rate of turnover, ratio of enterprise salary to industry average salary, personnel qualification and motivation, etc. (Yeletenko 2008).

Thus sustainable production enterprise development and logistics system building on its foundation need more than static enterprise sustainability and dynamic production enterprise sustainability (Trachuk 2014). Sustainable enterprise development also need constant increase of production, sales growth and improved environmental and social elements of sustainable development.

In the context of logistics system building, constant sales growth should lead to increase of income, which is one of the sources of company development, like wealth funds building, dividend growth, increase in net cash flow, rise in company.

However, logistics system, which helps to expand production, should also increase resource efficiency at enterprise (Honcharov & Kostyuk 2012).

Besides, environmental development is also important, as well as all its elements: resource saving, mitigation of environmental impact, and environmental remediation (Figure 2.).

Dynamic enterprise sustainability is formed as a result of various influence (as well as logistics influence) of following tools: personnel development; personnel training; improvement of social and economic level if personnel; ecological production; marketing tools for sales growth that accommodate economic growth; Innovations and R&D; investment activity (Trachuk 2014).

Thus logistics system should provide costs minimization and marketability, as well as external (market) and internal sustainability (internal enterprise development).

External enterprise sustainability (external enterprise development) is external environment factors, which have "positive" influence on enterprise performance. External sustainability includes global, regional and national sustainability.

Market enterprise sustainability (market enterprise development) forms market capacity, potential sales in conditions of competition, which influences on internal sustainability.

Internal enterprise sustainability (internal enterprise development) is defined as state of an enterprise, that is formed with managerial tools, which provide efficient work of an enterprise, and consists of economical (production, technological, investment, financial and organization sustainability), social and environmental sustainability.

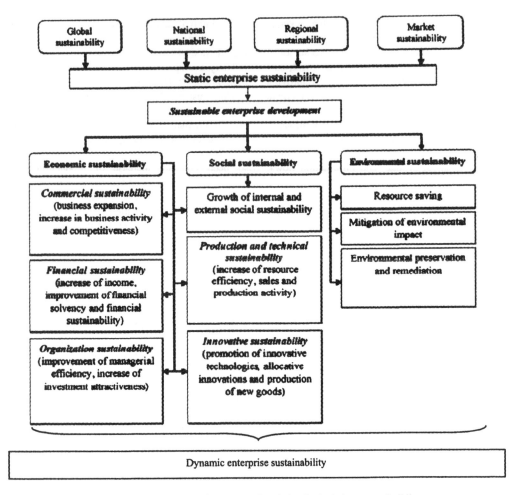

Figure 2. Elements of sustainable enterprise development, as foundation for logistics system building.
* developed by the author

2 CONCLUSIONS

Strategic management of logistics system in conditions of sustainable development is defined as highly professional management with its own logistic structural specialization, which aims at survival and development of logistics enterprise system in indefinite external environment, which includes corporate planning as strict task and specific subsystem.

Sustainable development as a foundation for logistics system should be comprised of three basic principles:

– provided balance between economic and environmental enterprise developments;

– provided balance between economic and social community spheres, which means maximal use of resources, provided by economic development;

– solution of development problems that influence on the present generation, as well as following generations, which have the same rights for resources.

REFERENCES

Alesynska, T.V. 2005. *Logistics basics: general issues of logistics management*. Taganrog: 121.

Dmytrychenko, M.F, Levkovets, P.R., Tkachenko, A.M., Ignatenko, O.S., Zayonchik, L.G. & Statnyk, I.M. 2007. *Transport technologies in logistics systems*. Kyiv: Informavtodor: 167 – 168.

Honcharov, Yu.V. & Kostyuk, H.V. 2012. *Methodological aspects of logistics building at light industry enterprise*. Access mode: www.problecon.com/pdf/2012/3_0/97_107.pdf.

Khadzhynova, O.V. 2013. *Sales management logistics strategy of diversified production enterprise*. Economics, enterprise organization and management. Donetsk: 23.

Koblyanska, I.I. 2011. *Methodological fundamentals of environmental logistics management of production enterprise.* SumDU: Economics of natural resource management and protection of the environment: 21.

Prokopenko, O.V. 2014. *Sustainable development of the enterprise, region, society: innovative approaches to its providing"* Poland: Drukarnia i Studio Graficzne Omnidium: 474

Rayzberg, B.A. 1999. *Modern economic dictionary.* Moscow: INFRA-M: 479.

Sevastyanov, A.V. 2004. *Methods and mechanisms of industrial integration processes management.* Moscow: Delo ltd: 243.

Skorobohatova, T.N. 2005. *Logistics: work book.* Simferopol, 2nd edition: 116.

Trachuk, A.R. 2014. *Energy potential as element of potential sustainable development of production.* Access mode: http://tef.kpi.ua/konf/tez_2014_1_179.pdf.

Yeletenko, O.V. 2008. *Controlling mechanism of enterprise logistics system.* Bulletin of the Lviv Polytechnic National University, No 628: 494 – 498.

Zetkina, O.V. 2003. *About management of sustainable enterprise development.* Moscow: Audit-UNITY: 134.

Theoretical and Practical Solutions of Mineral Resources Mining – Pivnyak, Bondarenko & Kovalevska (eds)
© 2015 Taylor & Francis Group, London, ISBN: 978-1-138-02883-8

Personal's industrial behavior management as a tool to improve company's effectiveness

O. Grosheleva & O. Usatenko
National Mining University, Dnipropetrovsk, Ukraine

ABSTRACT: An effective mechanism of managing the personal's industrial behavior to achieve the goals and best results of company's activity is developed. Proposed classification of potential's and motivation factors, as well as the model of personal's industrial behavior management allows establish the relationship between these variables and offer managerial tools applied, ensuring achievement of company goals.

1 INTRODUCTION

The key feature of modern economic condition is the saturation of markets in goods and services. It is a sound problem for any company that wants to gain a foothold and develop its market share, finding additional sources of long-term competitive advantages, activation of which leads to increased competitiveness.

Today, experts advise to look for sources of this kind of advantages in the areas of business, which are the least amenable to being copied by competitors. As priorities the intangible assets, including company's staff are considered. So, it's a topical question for managers to develop and implement such mechanism that can transform a passive performer of duties into an interested participant of the production process encouraging him to implement his potential to company's favor. Other words now it's very important to develop an effective system, which allows to detect and to form a true source to achieve the high level of company's competitiveness in a long term.

Since the 50th years of XX century academics and executive managers as a source of high level of company's performance consider organizational behavior (Latfulin & Gromova 2004, Spivak 2000, Ashirov 2006). An important part of the organizational behavior is an industrial behavior (Shekshnya 2002), which is formed under a number of factors. In particular it is important to develop the managerial tools which allow activating appropriate levers to learn and form the relevant to the strategic objectives type of the industrial behavior.

To achieve the company's objectives managers use all available recourses. It means that personal should perform certain operations, it is called industrial functions. It is more correct to use such term as "industrial behavior".

2 PRESENTATION OF THE MAIN RESEARCH

The main duty of human recourses management (HRM) is to ensure each staff's industrial behavior, which is necessary to achieve company's objectives. The effectiveness of HRM is determined by the level of implementation of company's objectives.

The efficiency of each individual employee and the team as a whole depends on the potential and motivation (Figure 1).

According to (Maslov 1999) employee's potential determines as labor capacity, resource capabilities in the labor, determined on the basis of age, physical abilities, existing knowledge and professional skills and qualification.

An important assignment for management is to form and to use employees' potential and to substantiate the optimal quantitative and qualitative characteristics corresponding to the chosen strategy.

One of the key tasks, the basics of the staff's industrial behavior management is to propose to the employees such sort of duties which fit the best to their skills, abilities, aptitudes, interests, individual socio-psychological characteristics.

However, only a strong combination of motivation and professional skills provide the necessary staff's industrial behavior, and through it – the achievement of the results (company's objectives).

Employees' motivation is one of the most important tools to increase the productivity. While there are quite different performers in the company (according to their age and qualification as well as to the occupied positions/work places and the stages of their career, staff's motivation should include several systems, due to which it is possible to encourage to the labor activity all employees of the company.

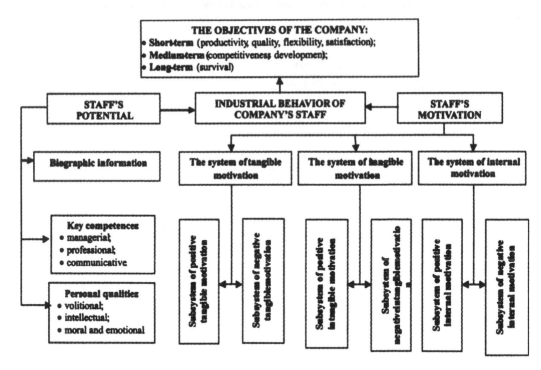

Figure 1. The model of industrial behavior management in the context of company's objectives achievement.

No doubt that there is a tight dependence between the staffs' performance and the level of their monetary interests. Eloquent in this context is the structure of incomes of Ukrainian citizens (Figure 2) (http://www.ukrstat.gov.ua).

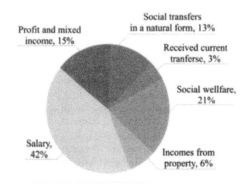

Figure 2. The structure of the Ukrainian citizens' incomes.

It is known that the salary has two main functions: reproductive and stimulating. Surely if the level of the salary is almost equal to the living wage (at the end of the 2014th it was 1218.00 UAH per able bodied person) the salary plays just the reproductive role. If its level is much higher it could play also a stimulating role.

The development level of the modern human suggests that the variety of his/her needs, which motivate him/her to the high level of performance, is much more complicated than just the monetary ones. But the real activation of other reasons is possible just in case if the amount of the received remuneration is not less than the industry's average level (Usatenko & Skripkina 2011).The statistics describing theaveragesalaryof different types of economic activitiesinUkrainein2014 is shown in Figure 3 (http://www.ukrstat.gov.ua).

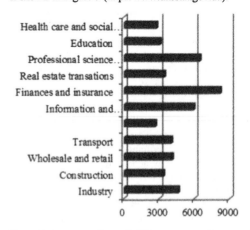

Figure 3. Average salary of different types of economic activities in Ukraine in 2014.

The practical experience shows that an important condition to increase the interest of employees to work is to use methods of intangible incentives.

There are a number of methods of intangible motivation:

– favorable socio-psychological climate in labor collective;

– favorable working conditions;

– employee's career promotion;

– additional leave;

– oral and/or written acknowledgment from the management for the effective operation;

– corporate events etc.

It is necessary to stress out that tangible and intangible motivation could make strong, but not a lasting impact on employee. Tangible and intangible motivation implies the existence of incentives provided by the third side that makes the employee to perform a set of actions, often not for inner satisfaction but to get the promised compensation in material or immaterial nature. The strongest kind of motivation which directs an employee to achieve the best result is the internal motivation. But there is one compulsory condition: it should be in coordination with the tangible and intangible motivations.

The system of internal motivation is connected with the content of the activity and with the employ's personal interest in the results of this activity. Also it is connected with the performer's understanding of the meaning of his/her activity, with the possibility to implement his/her knowledge, abilities, experience or to develop skills. First of all inner-motivated person strives to implementation of his/her potential. That's why such employ constantly improves his/her professionalism, enhances knowledge, and enriches experience.

Before the economic crises of 2008 companies and personal in Ukraine, Russia and other well developed countries accustomed to stimulate any activities by material incentives. Today it's a sound problem to restrict the budgets. So such interpersonal skill of managers as the ability to inspire their employees to work hard without using material incentives is becoming more and more important. The research of Kelly Services (http://www.kelly-services.com) – one of the biggest international companies providing decisions in the sphere of HRM – proves, that 44% of respondents considers in the conditions economic crises intangible motivation as the most important one. Only 40% of respondents think that the more important is the tangible one.

A number of research works are dedicated to learn the factors of tangible motivation. But much less attention is paid to learn the factors of intangible and internal motivation. To overcome this con-tradiction Joint Stock Company "Dnipropetrovsky Plant of Repair and Construction Passengers Railroad Cars" (JSC "Dniprovagonrembud") was researched. "Dniprovagonrembud" is one of the oldest industrial companies of Ukraine, the main industrial basics of Ukraine to meet the needs of railroad in capital repair and modernization of passenger railroad cars, wagons for narrow international traffic. The passengers railroad cars are converted into the wagon – saloons, technical and touristic wagons in the Plant. The company performs thorough – reconditioning passenger railroad cars of all types, operated on the railroads of CIS and international traffic. During the economic crises till 2009 the company was unprofitable. But since 2010 it starts to get profit. The level of the profitability during this period was from 4 to 7%. The workers and executive officers of the company were interviewed. On the whole 51 employees of the company were surveyed (6% of the average company headcount). It proves the validity of the sample. The main results are as the follows. At least two factors classified as the internal motivation motivate employees. The most universal factors are the follows:

– the interest to the work (64% of the respondents);

– granting the right to make decisions (50%);

– the possibility to improve the qualification (36%).

The meaning of the factors is almost independent of the position of the performer and of the content of his/her duties. On the other hand the significant impact on staff motivation causes relationships in the team and the manager's recognition of the subordinates' labor merits. The last two factors belong to intangible motivation. Taking into account all possible factors of motivation (accept for tangible motivation), plant's staff is more interested in those which supports the internal motivation (68%), than the intangible one (32%) (Figure 4).

Intangible motivation, 32%

Internal motivation, 68%

Figure 4. The ratio of the factors of internal and intangible motivation of the employ.

The integral index, which reflects the well-being/distress of the individual's position in the workplace, is job satisfaction, which includes an assessment of interest in the work, satisfaction with relationships with employees and management, the level of claims in professional activities, the satisfaction of conditions, labor and others (Fetiskin et al. 2002). This index includes factors of internal and intangible motivation. The average level of job satisfaction of workers and executive officers is correspondently 60% and 68.2%. The higher the index is, the better the professional feelings of the employees in the company are, and vice verse.

Taking into account that 86% of the employees consider the level of their job satisfaction as the "average" and the 7% consider it as the "low", so improving it to the level of "high" it is possible to impact the industrial behavior of the staff as well as the results of the staff's activity (Figure 5).

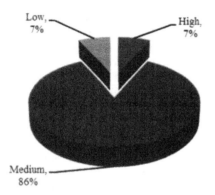

Figure 5. The level of the staff's job satisfaction.

During the research the correlation (direct) dependence of productivity of employees by job satisfaction was proved (Figure 6). The last one is calculated as follows:

$$K_{js} = 1 - \frac{Q_t}{Q_{an}},\qquad(1)$$

where Q_t – the number of employees who retired voluntarily or at the initiative of the company's administration entities; Q_{an} – the average number of employees, entities.

Any motivation system (tangible, intangible or internal) can be implemented in the context of a positive or a negative reinforcement. Most of all management concentrates on a positive reinforcement (wage increases plans, improving working conditions, the right to make decisions, etc.). However such situation could happen, when in order to solve specific production problems it is necessary to use the negative reinforcement (for example, staff's violation of working discipline is being punished by a fine, which is an element of negative reinforcement).

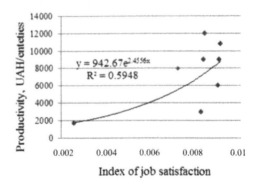

Figure 6. The dependence between the productivity and Index of job satisfaction.

The advantage of any research is the possibility of its results implementation into the practice of real business. The difficulty of constructing an adequate motivational system is the need for a balanced inclusion of elements of each of the three above mentioned systems of motivation. The overestimation of any of the subsystems that ignores the current situation of the company's activity, as well as the real incentives which reflects an exact industrial behavior of the employees, could led not only to decreasing of the motivational level, but also to deterioration of the efficiency indexes of the company. That's why it's dramatically important to managers to understand what exactly is the really meaning for the performers, what encourages them to the activity and forms a true desire to achieve the company's goals.

It is important to solve these problems at any stage of the staff's life cycle in the company, beginning with the moment when the position is just planning in the company and finishing the moment, when the employ is leaving the company. Timely resolution of the matter at the earliest stage of the life cycle reduces the costs associated with incomplete or non optimal utilization of labor resources caused by improperly constructed motivational system. The experts of the management recommend using the conception of HRM according to which the work steals up to the person with the most complete view of his abilities and desires. Incorrect assessment of the staff's abilities (its potential) makes it impossible perform its duties efficiently, even if the performer would desire it very much (the high level of motivation). Otherwise, even if the level of potential is very high, but the level of motivation is too low, it means that the company overpays foe the high quality recourse, which is not used efficiently.

Consequently, managers require specific applied tools that would allow on the stage of selection to identify those factors of motivation that will be important for the employee. This need is possible to meet if we consider a relationship between the level of potential and motivation factors (Figure 7).

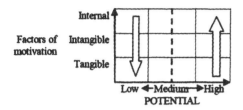

Figure 7. The correlation between the level of potential and the motivate factors in the process of staff's industrial behavior management

It was shown that it is much easier to assess the potential objectively, so it's important to point out the directions of its influence on the structure of the factors, which will encourage employees to the high level of the productivity (it means that it will help the company to form the staff's desirable industrial behavior).The carried out researches have revealed the presence of these dependencies:

– for the employees with the low level of potential equitable ratio is as follows:

| The factors of tangible motivation | > | The factors of intangible motivation | > | The factors of internal motivation |

– for the employees with the high level of potential equitable ratio is as follows:

| The factors of internal motivation | > | The factors of intangible motivation | > | The factors of tangible motivation |

– the staff with the medium level of potential takes the middle position: if the potential is increasing, it is necessary to focus on the second correlation. Otherwise, if the potential is decreasing it's more correct to focus on the correlation 1.

In the figure the direction of the arrow shows indicated correlations in the context of changes the significance of the motivate factors: from the least to the most significant factors. Obviously to improve the efficiency of the industrial behavior management the indicated correlation should be implemented in the developed human recourses policy.

3 CONCLUSIONS

To achieve a high level of competitiveness, which is a basic to the viability of the company in the long term in a market economy, it is important for management to focus on creating and maintaining long-term competitive advantages.

The staff's industrial behavior directed to the company's objectives achievement is to consider as the source of the long term competitive advantages.

The factors which determine the staff's industrial behavior are their abilities and desire to work hard on their positions. It is important for management to have an applied instrument to asses these factors as well as to manage theme to implement organization's strategy.

The developed classification of the factors of the motivation and of the potential allows setting the relationship between these variables and offering management the applied tools, ensuring the efficiency of industrial behavior of employees.

The further researches should be conducted in the direction of finding quantitative characteristics that measure the level of potential and motivation of employees.

REFERENCES

Ashirov, D. *Organizational behavior.* Moscow: Velbi, 360.
Fetiskin, N., Kozlov, V. & Manuilov G. 2002.*Integral satisfaction of labor.*Socio-psychological diagnosis of personality and small groups' development. Moscow: Publishing House of Institute for Psychotherapy: 470 – 473.
How to motivate staff in crisis [Electron resource] / Workforce Management Solutions KellyServices. Mode of access: www / URL: http://www.kellyservices.com/web/ global/services/en/pages/index.html.
Latfulin, G., Gromova, O. 2004. *Organizational behavior.* St. Petersburg: CJSC Publishing House "Piter": 273.
Maslov, E. 1999. *Human Recourses Management of the Company.* Moscow: Infra – M: 312.
Shekshnya, S. 2002. *Human Recourses Management of Modern Organization.* Moscow: CLSC "Business School", Intel-Syntez: 355.
Spivak, V. 2000. *Organizational behavior and Human Recourses Management.* St. Petersburg: CJSC Publishing House "Piter": 412.
The average monthly wage by types of industry in Ukraine [Electron. resource] – Access: http://www.ukrstat.gov.ua/
Usatenko, O. & Skripkina, D. 2011. *Salary: when the money kills motivation.* Journal of Social and Economic Research Odessa State Economic University. Scientific Papers, Issue 41, Part 2: 133 – 138.

Theoretical and Practical Solutions of Mineral Resources Mining – Pivnyak, Bondarenko & Kovalevska (eds)
© 2015 Taylor & Francis Group, London, ISBN: 978-1-138-02883-8

Formation of marketing strategy of nonprofit organizations in Ukraine

O. Varyanichenko & M. Ivanova
National Mining University, Dnipropetrovsk, Ukraine

ABSTRACT: The study examines the main approaches to the formation of marketing strategy of nonprofit organizations, sources of finance and evaluation of social efficiency in terms of Weisbrod's value.

1 INTRODUCTION

International experience requires the formation of marketing strategy in non-profit organizations (NPOs) in Ukraine. The marketing approach to non-commercial activity allows on the one hand, meeting important social needs of society at a level of quality, and on the other hand, the best use of the sources of NPO funding.

2 METHODS, RESULTS AND DISCUSSION

All kinds of marketing planning in the noncommercial sphere are closely linked with the formation of corporate social responsibility (CSR). NPOs need public relations with the media, visitors, contractors, and concerned parties to create their own unique image and strong reputation, which in turn allows them to attract the interest of private, corporate and public investors (Figure 1).

The development of a PR-campaign program, information materials, non-financial reports and training programs for potential donors (both public and private), sponsors and partners in Ukraine should be the first and vital step in NPO's marketing strategy of conquering the national market. The second step is working-out partner proposals to large companies that are leaders in their field, for joint CSR activities.

Analysts from the Corporate Social Responsibility Centre have analyzed the web sites of 100 largest companies in Ukraine, whose shares are being quoted on international stock exchanges. The authors have developed the so-called Index of transparency and accountability of companies in Ukraine. The companies' activities were estimated according to four criteria: accountability (availability of non-financial reports); transparency (disclosure on their key CSR activities); navigation (freedom of access to any information about the company's CSR activities); accessibility (ease of using the website from various view-points such as the language, the availability of contact information, or accessibility for people with disabilities).

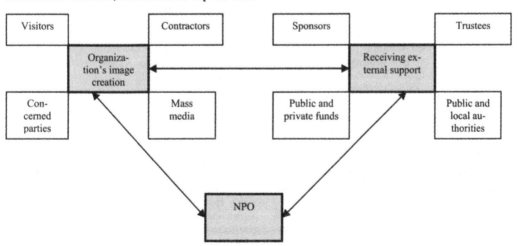

Figure 1. Directions of marketing planning in NPOs.

According to the investigation, eleven companies (10.3% of the total) have reached 50% of openness, which may be considered an "adequate" level of openness and transparency. In addition to the non-financial reporting available, the top ten companies on the Index have also met the other estimation criteria (description of social and environmental projects). The top ten companies that could be given the highest Index are shown in Table 1.

Table 1. The top ten companies according to the Index of transparency and accountability in Ukraine*.

Place	Company	Points (200 max)	Percentage
1	DTEK	160	80.0
2	Metinvest	146	73.0
3	Obolon	140	70.0
4	ArselorMittal Krivoy Rog	122	61.0
5	SCM	119	59.5
6	Kyivstar	117	58.5
7	Interpipe	106	53.0
8	Ernst & Young Ukraine	105	52.5
9	METRO Cash & Carry Ukraine	104	52.0
10	Galnaftogas	103	51.5

*Research Center "Development NPO" (2010)

In our opinion, partner relations may be effectively started by involving a company's employees into NPOs' volunteering activities with joint elaboration of an appropriate PR-campaign, both local and international.

The following activities are promising for cooperation: 1) Pro Bono / free services to NPOs and their beneficiaries; 2) humanitarian aid, discount programs for an NPO and its beneficiaries; 3) attracting resources from sponsors.

In addition, NPOs can offer services (including paid services) to their potential business partners, such as consulting and arrangement of charitable activities. These include:

– consultations on choosing fundraising ideas and projects that are individually suited to specific problems; development of projects and assessment of their efficiency; legal and accounting consultations;

– arrangement and planning of charitable activities for donors;

– preliminary assessment of projects for donors;

– information on the most pressing social problems of the region for donors;

– development and selection of individual fundraising ideas and projects that best suit the customer (the donor) on the basis of mini strategic planning;

– workshops for business (CSR themes, charitable giving);

– PR-services for business (charitable giving seasons, NPOs' and business fairs, sponsor of the year nominations, etc.);

– special measures to promote a business organization via NPOs (thematic weeks and days, local holidays, social contests, support for socially important initiatives of the business organization);

– information support through NPOs' / partners' information resources;

– fundraising services with a joint use and involvement of resources in social projects and services of both business organizations and NPOs.

Thus, as a public charity organization, an NPO can offer its business partner the following competitive advantages: improving the company's image at the local and national level; demonstration of social responsibility of the company; improvement of the intra-relationships; improved relations with investors; coverage by the media; access to certain markets; appeal to the target groups; advertising of goods and services; being associated with high-quality and prestigious events; tax benefits (when calculating an income tax, taxable income is reduced by the amount of contributions to charity - but not more than 3% of taxable income); attraction of new employees; arrangement of cultural activities for the company's employees, clients, customers and partners.

Diversification of funding sources is, in our opinion, a very promising part of the marketing strategy of sustainable development, which will make NPOs more self-sufficient and able to initiate and implement their own initiatives; through the diversification, NPOs will also become more attractive and reliable partners for foreign and Ukrainian stakeholders.

Non-commercial purposes of NPOs' economic management require both attracting external funding, and developing their own profitable activities. Sources of financing of non-profit organizations are divided into three groups (Shekova 2002, Shekova 2004), vis. attracted, public, and equity funds.

Attracted financing involves donations, sponsorships, grants from various funds, membership fees, etc. Public financing combines direct and indirect subsidies from state. Equity funds include income from primary and commercial activities (Figure 2).

The final stage of shaping NPOs' marketing strategy is an assessment of their social efficiency, which include Weisbrod's social index. The social index (publicness index) PI was introduced to determine the level of social effects production in non-profits to display the "relationship between the types of financial revenues of the enterprise and the nature of its services or output" (Steurer et al. 2005). In other words, the social index is calculated as the ratio of revenues from producing collective goods to the income from producing private goods. Income from collective goods in non-profit organizations is in the form of donations, grants, government subsidies and so on.

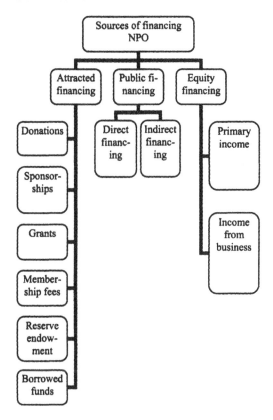

Figure 2. Sources of financing NPOs in developing the marketing strategy.

The production of private goods includes the sale of goods and services, as well as membership fees and sponsorships. In response to the sponsorships, NPOs offer advertising or PR-activities, while in exchange for membership fees they offer a range of their goods and services.

Absolute values of the index serve as the unit of its measurement. Social index can take values from 0 to infinity. The index is equal to zero in organizations that fully live by producing private goods, i.e. the fees of their members, or revenues from the sale of consumer goods and services. Infinitely large social index values are typical for non-profit organizations entirely involved in the creation of collective goods, i.e. whose activities are aimed at achieving social effects. Hence, these organizations are only funded by donations, sponsorships, grants and subsidies.

Social index may be used to assess the level of self-financing in organizations. If the index equals zero, the organization is in full self-financing. The higher the social index, the lower is the level of self-financing in the organization.

Most NPOs should direct their attention to planning the strategy of attracting sponsorships that are the revenues from the production of private goods, thus automatically increasing the level of the organization's self-financing. In contrast to the charitable giving, which is a form of voluntary unconditional support of nonprofit activities by individuals and legal entities, sponsorships are characterized by solely commercial interests.

The optimal form of possible sponsorships is corporate social investment, which is a deduction of money by commercial organizations for socially useful purposes. Through the sponsorship, NCOs receive material and financial resources necessary for their primary activities, along with the opportunities of using up-to-date means of communication, equipment, and of staff development. In turn, the sponsors may consider their corporate social investment as "components of PR-programs aimed at expanding external contacts and creating a favorable image of the company" and "inexpensive means of advertising and access to prestigious entertainment programs" (Hjejvud 1999). In response to the sponsorship, NPOs can include the name of the company in their publications; put the company's logo on the information brochures and entrance tickets for mass cultural activities; give the company the right to provide its brand and facilities for various events and participate in these, etc.

Thus, sponsors may obtain both direct and indirect benefits. Direct benefits include advertising, access to modern sources of information etc. Indirect benefits are improving the company's image in society, which would improve the business environment on the whole.

A reserve fund or endowment (reserve endowment) may become another progressive and innovative source of financing NPOs. This is a special form of fundraising by individuals and businesses in the non-profit organization. Unlike other types of funds, a reserve endowment is placed by NPOs in the accounts of banks and investment institutions to obtain stable interest. It is important that NPOs cannot withdraw the whole amount or any part of the endowment for their current needs, but accrued interest on the amount.

4 CONCLUSIONS

The marketing strategy of NPOs should be developed by the introduction of advanced management techniques, which involves a combination of marketing planning, selection of optimal funding and assessment of the effectiveness of the measures recommended.

REFERENCES

Hjejvud, R. 1999. *All of Public Relations: How to succeed in business using public relations.* Moscow: Laboratorija bazovyh znanij.

Index transparency and accountability of companies in Ukraine (Research Center "Development NPO"). 2010. Kyiv: Farbovanyj lyst.

Shekova, E.L. 2002. *On the question of capital structure optimization nonprofit organization.* Finansovyj menedzhment, Vol. 5: 25 – 27.

Shekova, E.L. 2004. *Economics and management of nonprofit organizations.* Saint Petersburg: Lan'.

Steurer, R. Langer, M. Konrad, A. & Martinuzzi A. 2005 *Corporations, Stakeholders and Sustainable Development: A Theoretical Exploration of Business-Society Relations.* Journal of Business Ethics, Vol. 61, No 3: 263 – 281.

Development and exploitation of storages of enrichment process wastes as anthropogenic deposits

O. Medvedeva
M.S. Poliakov Institute of Geotechnical Mechanics under the
National Academy of Sciences of Ukraine, Dnipropetrovsk, Ukraine

ABSTRACT: The paper shows the topicality of the development of storages of enrichment process wastes of mining processing plants (MPPs) in Kryvbas and that the only one solution that preserves the competitiveness of the production and the stable development of the region is their transformation into the anthropogenic deposits with the production beginning in the process of their exploitation and after the wastes storing finishing for the subsequent rehabilitation and conservation. The studies show that the implementation of the valuable component by-product extraction from the beach areas near the dike dam together with the technology of the combined tailings storing makes it possible to extract the part of the valuable component that remaining in the tailings storage at the stage of the storage filling.

1 INTRODUCTION

Mining processing plants (MPPs) of the iron-ore basin of Kryvyi Rig (of Kryvbas) were started to exploit from the middle of the last century and for the work term that is more than a half of the century they have accumulated in their storages significant amounts not only of the waste rock, of the process water but also of the valuable components that were not extracted from the feed mineral stock for a variety of reasons. Today the cost price of the mineral raw materials production in the quarries increased and it is equal to the cost price of these components production out of the wastes of the last century. And the wastes storages themselves reached the highest possible marks and their further exploitation becomes impossible. Herewith there is no land for the new wastes storages and the storages that are closed constitute a serious environmental danger to the whole region of Kryvyi Rig.

Thus the transformation of the wastes storages of mining processing plants (MPPs) of Kryvbas into the anthropogenic deposits with the production beginning in the process of their exploitation and after the wastes storing finishing for the subsequent rehabilitation and the conservation is the only one solution that preserves the competitiveness of the production and the stable development of the region.

However the existing world experience in the use of the recoverable resources refers to the industrial wastes of nonferrous, precious and rear-earth metals processing such as gold, platinum, copper, aluminum, plumbum and zinc that is caused by the high demand for these metals and the high cost of their concentrates (Bluss et al. 1999, Gumenick et al. 2001, Alexandrov 2000). In addition, in the world practice the anthropogenic alluvial deposits production from the wastes storages is carried out after the finishing of the storage exploitation and the stopping of the wastes storing in them. In the homeland conditions it is required a technology that enables the development of the anthropogenic alluvial deposits in the operating wastes storages without the interrupting of the ingress of new tailings. Thus, international experience cannot be transferred to the homeland conditions without adapting it to the peculiarities of the mining processing plants (MPPs) of Kryvbas and profitability of such technology should be evaluated taking into account the profitability of the completely mining processing plant (MPP).

2 THEORETICAL STUDIES

Studies of this issue show that for the further functioning of mining processing plants (MPP) of Kryvbas the modernization of the existing technologies of the wastes storing with the elements of the reorganization of the storage and the implementation of the by-product development of the anthropogenic alluvial deposits are needed (Baranov et al. 2006, Medvedeva 2014). In this case it is suggested to extract the concentrate that gets into the enrichment process wastes and deposits on the beach area near the floodwall (Figure 1) and to return it into the concentrating bound after the map alluviation and the draining of the area near the dike dam. The worked-out area can be refilled by the enrichment process wastes that were spissated to the paste concentration

that makes it possible to introduce the technologies of the inspissation and the storing of such pulps into the wastes storages that are exploited (Semenenko 2011, Levanov et al. 2005, Alexandrov 2000).

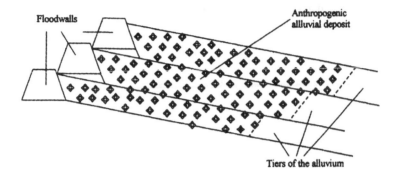

Figure 1. The structure of the anthropogenic alluvial deposits formed during the wastes storing.

By this decision the following algorithm of the transition to a new level of the pulp output is assumed (Medvedeva 2014): on the perimeter after finishing of the map alluviation to start mining work of excavating in the beach area near the dike dam and then to fill the area with the highly concentrated pulp consisted of the slim particles and only then to move to the next level of the wastes storing. As a result the slim particles that were previously deposited in the pond at this tier of the alluvium do not get into it that ensures the successful clarification of the recirculated water to the required standards. It is reasonable to implement this technology at the storages where the first or the second tier of the alluvium is sluiced because it makes the extraction of the anthropogenic alluvial deposits possible only at the current tier.

During the filling of the wastes storages that are exploited by the homeland mining processing plants

(MPPs) the technology of the storing is used that based on splitting of the perimeter of the current level of the retaining prism into the sections of the alluvium where the enrichment process wastes that were delivered by one hydrotransport installation are accumulated (Semenenko 2011). It is recommended to select the length of this section of the alluvium in a range from 1000 to 3000 m that is based on the experience of the exploitation of the previous wastes storages. It is recommended to divide the sections of the alluvium by the subsidiary dike dams that were built inside the wastes storages perpendicular to the floodwall minimum into 5 alluviation maps. The operating procedures at the alluviation maps within the same section of the alluvium are carried out in accordance with the process schemes that are based on the following sequence of the processing steps. (Table 1) (Semenenko 2011).

Table 1. The current process scheme of the maps of the section of the alluvium.

The stage number	The sequence number of a map in the section of the alluvium				
	I	II	III	IV	V
1	alluvium	reserve	installation	dike dam	drying
2	drying	alluvium	reserve	installation	dike dam
3	dike dam	drying	alluvium	reserve	installation
4	installation	dike dam	drying	alluvium	reserve
5	reserve	installation	dike dam	drying	alluvium

It is assumed that the duration of each of the processing steps is the same because while one alluviation map is dried up, diked, equipped with the pipelines and is in reserve, the other four maps are sluiced sequentially by the tailings. Thus when the first map goes out of the reserve the works on the wastes storing are transferred to a new tier of the alluvium.

With the implementation of the by-product extraction of the anthropogenic alluvial deposits the operating procedures with the alluviation maps within one section should be changed. In this case with the regard to the necessity of the extraction of the anthropogenic alluvial deposits and the renewal of the carrying capacity of the beach in the basis of the process scheme the following sequences of steps should be: enrichment process wastes alluvium; drying of the beaches; extraction of the anthropogenic alluvial deposit; filling of the worked-out area; drying of the newly formed beach or the shrinkage; installations of the floodwall; installation of the instrument desk; being in reserve.

While changing of the technological procedures with the new set of the processing steps it is rational to save the time of the map alluvium with the enrichment process wastes. The time of the map alluvium with the enrichment process wastes is determined by the capacity of the existing hydrotransport installations. It is possible to change this value only in the process of the modernization of the hydrotransport complex of the mining processing plants (MPPs) by changing of the pumping equipment. Not significant changing in the pump delivery can be achieved by setting of the impeller of the other diameter, by changing of the rotation frequency of the impeller or by air supply in the case of using the scheme "in one lift". However, none of these methods is widely used in the conditions of the homeland mining processing plants (MPPs) (Baranov et al. 2006, Semenenko 2011, Medvedeva 2014). Especially not appropriate is the productivity decrease of the hydrotransport installations that take away the tailings into the storage. Because their main task is to provide that the concentrating industry was not flooded. Thus the actual task is the task to keep constant the total pulp consumption that comes into the alluvial tier when the process scheme is changed.

If the pulp consumption that is given to the map remains permanent then the number of the maps at the section together with the length of the section increase. In this case it is need to use eight maps instead of five. And the extraction process is started in every map after its drying and then the worked-out area is filled by the secondary processing wastes, the area near the dike dam is dried, the floodwall is dumped, the distributing pipelines are installed and put into reserve. The working-off time of the tier of the alluvium increases on 40% but it is possible to decrease the number of sections because one section can be broken up and two others can be complemented by the maps that released.

For this approach it is important to coordinate the work modes of the hydrotransport installations that give the wastes into the maps, of the extraction equipment and installations that provide the storing of the secondary wastes. The quantity of the sections of the alluvium corresponds to the quantity of the hydrotransport installations that perform the enrichment process wastes withdrawal. In this regard it is possible to reduce the quantity of the sections of the alluvium but the increasing of their quantity requires a capital expenditure for the pumping stations equipment. In this case it is possible to increase the pulp concentration so that a smaller quantity of stations can provide the determined traffic and to correct slightly the parameters of the maps. But if it is necessary to save the quantity of sections of the alluvium then it is need to save the quantity of maps that is corresponded to them and that is possible only while performing simultaneous processing steps.

After the analysis of possible options the process scheme with the simultaneous processing steps at the five maps is possible to represent as follows (Table 2).

Table 2. The current process scheme of the maps of the section of the alluvium with the by-product extraction of the anthropogenic alluvial deposits and the simultaneous processing steps.

The stage number	The sequence number of a map in the section of the alluvium				
	I	II	III	IV	V
1	alluvium	reserve	installation shrinkage	dike dam filling	drying extraction
2	drying extraction	alluvium	reserve	installation shrinkage	dike dam filling
3	dike dam filling	drying extraction	alluvium	reserve	installation shrinkage
4	installation shrinkage	dike dam filling	drying extraction	alluvium	reserve
5	reserve	installation shrinkage	dike dam filling	drying extraction	alluvium

For implementing of such process scheme it is necessary to fulfill several conditions:

– the extraction processes should be carried out from the dike dam of the map without lowering of the extraction equipment to the beach or the extraction equipment that is used should not require the drying of the beach to get started;

– the distance from the inner slope of the dike dam from which the wastes are considered to be the anthropogenic alluvial deposits makes it possible to start dumping of the floodwall of a new tier without filling of the worked-out area by the secondary processing wastes.

An application of this technology at the third and higher tiers of the alluvium limits the capacity that was released for the storing of new particles. The length of the beach area near the dike dam where it is expected to make the extraction of the anthropogenic alluvial deposits is 20% of the length of the surface alluvium. In consideration of that the height of the dike dam does not exceed 3 m and the length of the beach may vary from 50 to 500 m the extraction from the workable section at the current tier of the alluvium can be carried out from several tiers that are situated below (Figure 2). In this case the extracted anthropogenic alluvial deposit is sent for the recycling, its wastes are stored in the newly created area (Figure 3). However in the case of the extraction from several tiers of the alluvium a part of the anthropogenic alluvial deposit remains in the body of the retaining prism under the floodwalls (Figure 3) and to extract this part of the alluvial deposit is possible with the technology that provides the reduction of the capacities of the storage (Figure 4).

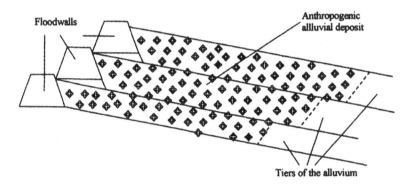

Figure 2. The release of the additional capacity for the wastes storing through the development of the anthropogenic alluvial deposits at several tiers of the alluvium.

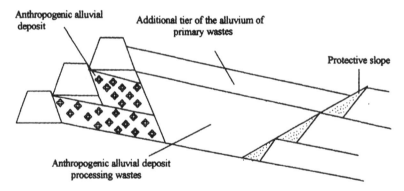

Figure 3. The storing of the additional amount of the wastes after the development of the anthropogenic alluvial deposits at several tiers of the alluvium.

For the extraction from several tiers of the alluvium (Figure 2 and 3) the slope of the worked-out area that is far from the dike dam is strengthened by the dumping that protects the material of the beach from dusting and getting wet because it contains clay and pulverescent particles. The anthropogenic alluvial deposits processing wastes should not be mixed with the primary deposits of the enrichment process wastes because the first do not contain thin, clay and pulverescent particles and can be stored on the beach. The remaining place allows to organize an additional tier of alluvium and to store the additional amount of the primary deposits wastes as a pulp with a low concentration or with the concentration of a paste (Figure 3). For the drainage of the material of the additional tier of the alluvium the protective dumping is broken by the drilling of a number of wells. However in the case of the extraction from several tiers of the alluvium a part of the anthropogenic alluvial deposit remains in the body of the retaining prism under the floodwalls (Figure 3).

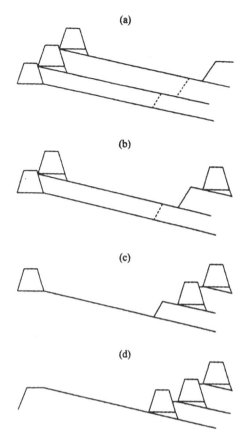

(a)

(b)

(c)

(d)

Figure 4. The technology of the anthropogenic alluvial deposits extraction from the wastes storages that were taken out of service.

In this connection for wastes storages of this type that were taken out of service the technology of the anthropogenic alluvial deposits extraction can be used that assumes the reduction of the capacity of the wastes storages (Figure 4). In the case of the anthropogenic alluvial deposits extraction from the wastes storages that were taken out of service the extraction processes are started with the excavation of a part of the beach near the dike dam of the upper tier of the alluvium (Figure 4, a). Then it is implemented the reexcavation of the floodwall to a new location (Figure 4, b). After that the corresponding processes are repeated in the subsequent tiers of the alluvium (Figure 4, c and 4, d). Recycling wastes of the anthropogenic alluvial deposits are used for the dumping of the remaining beaches and the sections of the underwater alluvium to prevent dusting and their complete conservation. This makes it possible to refuse from the process water intake for the maintaining of the pond level, to protect beaches from dusting and then completely to dehydrate the core of

the storage covering it from above by the wastes of the anthropogenic alluvial deposits processing after the extension of the floodwalls.

In all cases the extracted anthropogenic alluvial deposits can be transported to the concentrating factory by the vehicles or the hydrotransport using as a carrier fluid the clarified or not completely clarified recycled water. Taking into account the geodesic level difference typical for the wastes storages conditions (Table 3) the hydrotransport of the anthropogenic alluvial deposits to the concentrating factory can be carried out by the gravity (Baranov et al. 2006, Semenenko 2011, Gumenick et al. 2001). The extracted anthropogenic alluvial deposit can be added to the primary deposits at some point of the concentrating repartition or for its processing the section near the wastes storage can be equipped. In the second case the delivery costs for the anthropogenic alluvial deposit and the removal of wastes of its processing will be minimal.

Table 3. The geodesic level difference of wastes storages of mining processing plants (MPPs) in Kryvbas.

The title of the wastes storage	The dike dam height, m
"Vojkovo" SMMP	from 50 till 74
"Objedenjonoje" SMMP and ArselorMetall	from 40 till 59
"Mirolubovskoje" ArselorMetall	55
wastes storage of the InMPP	112
wastes storage of the CMPP	10
wastes storage of the SevMPP	76

To select the parameters of mining in all the technological solutions it is important to estimate the amount of material at the tier of the alluvium and the amount of the anthropogenic alluvial deposits at the tier that can be calculated using the following formulas:

$$W = BH^2\left(1-\frac{a}{H}\right)\frac{sin(\alpha-\beta)}{sin\,\alpha}\left(1-\frac{\sigma}{\eta}\right)\frac{\eta}{\mu}, \qquad (1)$$

$$W_0 = BH^2\left(1-\frac{a}{H}\right)\frac{sin(\alpha-\beta)}{\mu\,sin\,\alpha}(1-\sigma), \qquad (2)$$

$$\sigma = \left(1-\frac{a}{H}\right)\left(tg(\alpha-\beta)+tg\varphi\right)\frac{sin(\alpha-\beta)}{2\mu\,sin\,\alpha}, \qquad (3)$$

$$\mu = \frac{L}{H}, \qquad (4)$$

where W – the amount of the anthropogenic alluvial deposits in the tier of the alluvium; B – the length of the zone of the alluvium; H – the height of the

571

floodwall; a – excess of the floodwall crown over the alluvial beaches; α – the declivity angle of the external slope of the floodwall to the horizon; β – the declivity angle of the alluvial beach to the horizon; σ – the dimensionless thickness of tier of the alluvium; η – the part of the beach length where there is the anthropogenic alluvial deposit; μ – the horizontal equivalent of the tier of the alluvium; W_0 – the capacity of the tier of the alluvium; φ – the angle of the natural slope of the particles of the anthropogenic alluvial deposit; L – the beach length.

Taking into account that for the conditions of the wastes storages of the mining processing plants (MPPs) in Kryvbas ratios $\varphi \approx \alpha$ and $2\alpha - \beta \approx 2\alpha$ are correct it is possible to calculate with a high level of the accuracy that:

$$\sigma \approx \left(1 - \frac{a}{H}\right)\frac{sin(\alpha - \beta)}{2\mu}. \quad (5)$$

In this case instead of the formulas (2) and (7) the following relations can be used:

$$W = \frac{2(\eta - \sigma)\sigma}{sin\,\alpha}BH^2, \quad (6)$$

$$W_0 = \frac{2(1 - \sigma)\sigma}{sin\,\alpha}BH^2. \quad (7)$$

The effectiveness of the technologies can be estimated by the value of the relative volume of the anthropogenic alluvial deposit that is calculated as the ratio of the anthropogenic alluvial deposit amount to the capacity of the tier of the alluvium (Figure 5):

$$w = \frac{\eta - \sigma}{1 - \sigma}. \quad (12)$$

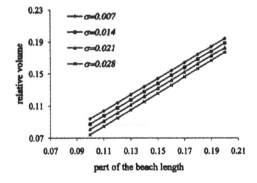

Figure 5. The dependence of the relative volume of the anthropogenic alluvial deposit on the part of the beach length where there is the anthropogenic alluvial deposit with different values of the dimensionless thickness of the tier of the alluvium.

3 CONCLUSIONS

As a result of the studies the technical solutions concerning the anthropogenic alluvial deposit development in the mining processing plants (MPPs) of Kryvbas were developed, they make it possible:
– to return to the concentrating repartition up to 20% of the concentrate that gets into the enrichment process wastes and deposits on the beach area near the floodwall after the map alluvium and the drainage of the zone near the dike dam;
– to refill the worked-out area with the enrichment process wastes that were thickened to a concentration of the pasta and to increase the capacity of the wastes storages;
– to use rationally the anthropogenic alluvial deposits processing wastes for the dumping of the beaches that left and areas of the underwater alluvium with the aim to prevent the dusting and their complete conservation during the anthropogenic alluvial deposits development in the wastes storages that were taken out of service;
– to abandon the process water intake for the maintaining of the pond level, to protect beaches from dusting and then completely to dehydrate the core of the storage covered it from above by the wastes of the anthropogenic alluvial deposits processing after the extension of the floodwalls.

For the first time it was proposed and grounded the technology of the valuable component by-product extraction from the beach areas near the dike dam based on the experimentally established fact that with the allocated alluvium because of the solid material fractionating the particles containing ferrous minerals deposit immediately after the outlet of the pipeline on the section that does not exceed 20% of the length of the formed beach. The idea of the technology is in recess this part of the beach after the map drying until the formation of new tier of the floodwall.

The studies show that the implementation of the valuable component by-product extraction from the beach areas near the dike dam together with the technology of combined wastes storing makes it possible: to extract a part of the valuable component remaining in the enrichment process wastes at the stage of storages filling; to abandon the exploitation of the storage as anthropogenic deposit after storing finishing; to avoid significant capacities of the reexcavation of the floodwalls of the upper levels; to improve the environmental safety and the resource-saving of the existing technologies.

REFERENCES

Aleksandrov, V.I. 2000. *Methods of energy consumption reducing in the hydraulic transportation of the mixtures of high concentration.* St. Petersburg: SPNMI (TU): 117.

Baranov, Y.D., Blyss, B.A., Semenenko, E.V. & Shurigin, V.D. 2006. *Grounds of parameters and system working modes of the hydrotransport of mining enterprises.* Dnipropetrovsk: New ideology: 416.

Blyuss, B.A., Sokil, A.M. & Goman, O.G. 1999. *Problems of the gravity concentration of titanium-zircon sands.* Dnipropetrovsk: Poligrafist: 190.

Gumenick, I.L., Sokil, A.M., Semenenko, E.V. & Shurigin, V.D. 2001. *Development problems of gravel.* Dnipropetrovsk: Sich: 224.

Levanov, N.I., Yaltanets, I.M., Melnikov, I.T. & Dyatlov, V.M. 2005. *Operating parameters of drag head devices of floating dredges and their design features.* Moscow: State Mining University: 236.

Medvedeva, O.A. 2014. *Technological solutions for the development of anthropogenic deposits of MPP in Kryvbas.* Materials of the mathematics international conference "Forum of miners", Dnipropetrovsk.

Semenenko, E.V. 2011. *Scientific bases of the hydromechanization of titan-zirconium alluvial deposits openworking technologies.* Kyiv: Naykova Dumka: 232.

Theoretical and Practical Solutions of Mineral Resources Mining – Pivnyak, Bondarenko & Kovalevska (eds)
© 2015 Taylor & Francis Group, London, ISBN: 978-1-138-02883-8

Intensification of coalbed methane recovery through the surface degassing borehole

R. Agaiev, D. Prytula & O. Katulskyi
M.S. Poliakov Institute of Geotechnical Mechanics under the
National Academy of Sciences of Ukraine, Dnipropetrovsk, Ukraine

ABSTRACT: The article presents the topicality in the development of coalbed methane fields mining. Conditions of the occurrence and potential reserves of alternative energy resources in Ukraine are presented. The main tasks of commercial methane extraction are shown, namely an increase of permeability in the mountain massif and raising durability of the surface degasification wells. Method a pneumatic hydrodynamic impact for degassing intensification of methane on gas-saturated coal rock massif through the surface degasification wells was described in the article, which was created in IGTM NAS.

1 INTRODUCTION

Fuel-energy supply of the economy of Ukraine is more related to the level of energy production, which is provided by through its own production around 35 – 40%. Natural gas is one of the main consumed power resources. It is satisfied by its own production of only 20 – 23%, and therefore the economy of natural gas or partial replacement of its other available source of energy, such as methane, is actual problem. The necessity of alternative energy sources research is dictated by the rising of natural gas, oil and coal costs (Bondarenko et al. 2010). Great attention is paid to off-balance energy sources. The most important is coalbed methane in modern conditions (Smorschok et al. 2001). In terms of conventional fuel reserves of coalbed methane (CBM), after oil, coal and natural gas ranks fourth among the world's fossil fuels. Ukrainian proved coal reserves ranked on seventh place in the world according World Energy Council report (Falshtynskyi et al. 2013). Resource CBM of Ukraine ranked fourth in the world after China, Russia and Canada. CBM reserves in Ukraine are from 4 to 12 trillion m^3, which technically possible to produce around 30%, namely from 1.2 to 3.6 trillion m^3 on different data (Krasnyk & Toropchin 2005).

The development of CBM deposits to a greater extent with coal mining in Ukraine. More than 60% of coal mine methane emitted from air flow and it is almost never used because of the low concentration of methane of 0.2 – 0.7%, which excludes the possibility of burning them automatically with traditional methods in the development of coal deposits of Ukraine.

Coal mines in Ukraine, especially Donbas, are characterized by very complex geological conditions. The average depth of the reservoir development is more than 700 meters. 33 mines are developing at a depth of 1000 – 1400 m. Of the 190 active mines 90% are dangerous content of methane gas, 60 – on coal-dust explosion, 45 – to sudden outbursts shocks and mining, 22% – by spontaneous combustion. Ukraine takes the 10th place among the leading countries of the world in coal production (Gur'yanov 2004). The methane content in coal seams is an average of 8 m^3 per 1 tone of dry ash-free rock mass, and in the host rock (sandstone) an average of 5 m^3/m^3(Antsiferov et al. 2004), in some cases up to 40 m^3 per 1 tone dry ash-free rock mass. Explored reserves of methane are about 1.2 trillion m^3 and with the separation of methane from coal seams on thicker which contains of coal – 25.4 trillion m^3 (Kasyanov 2000). From other sources of methane resources in the Donbas reached 11.86 trillion m^3, including 0.46 trillion m^3 in an aqueous state, 1.46 trillion m^3 coal bed capacity of more than 0.3 and 9.82 trillion m^3 in coal breed array, of which only 0.5 – 1.5% are free gas valued at depths from 500 to 1800 m. By results of testing for methane in geological prospecting shares in rocks and coal seams up 22.2 trillion m, and industrial – 11,6 trillion m^3, including suitable for extraction – 3 – 3.7 trillion m^3(Pudakov et al. 1996).

In the Ukrainian part of the Donets Basin in the state balance accounted for 286.4 billion m^3 of methane in the mines of the Donetsk region, 3.1 billion m^3 – in the mines of the Dnipropetrovsk region and 165.7 billion m^3 – in the mines of Luhansk region in 01.01.2011. This is ten times less than the forecast of geological methane resources within the boundaries of the mine fields. On the most reasonable optimal estimate of the volume of methane Donbas coal-bearing deposits that are suitable for its production are about 4 trillion m^3 (Zhykalyak et al. 2012).

Methane is found in three states: free, adsorbed and dissolved (in water) in the context of the development of coal deposits. Most of it (over 80%) – carbon and adsorbed rock mass, about 10% of the gas in the free state fills the natural voids and pores and only a small amount is dissolved in water (Bokiy 2006, Glatkay 2012, Glatkay et al. 2014).

Thus, the gas-bearing coal deposits are considered as CBM. Therefore, production and utilization of CBM have very important economic significance which is reflected in the following areas:

– large-scale mining CBM can add significantly to the production of natural gas in Ukraine and increase energy security of the state and to save financial resources that are spent on imports of natural gas (Bokiy 2011);

– production and utilization CBM makes it possible to significantly improve the safety of miners works by reducing the risk associated with explosions and ignition of methane in the mines. The same CBM production will help to reduce the cost of ventilation mines which will allow to increase the load on the gas factor on the working face and thus not only increase productivity but also improve the safety of operations miners (Efremov 2012). With the help of new technologies CBM can be used for the production of: electricity and heat; as fuel for boilers and refueling of automobiles; for biomass production (gaprin, paprin), etc. (Bulat 2014, Mjaken'kaja 1992). Generally all this allows to significantly increase the overall efficiency of the coal industry;

– creation of methane mining industry allows to reduce social tension in the mining regions in the conditions of (Zozulya 2011) restructuring of the coal industry. The jobs problem can be partially solved by refocusing on the mining and using of coal mine methane and CBM;

– methane is a greenhouse gas with a high global warming process. It takes one of the first places in the list of gas emissions into the atmosphere therefore this is followed to establishing firm control, because methane is 21 times more active than carbon dioxide. Attractiveness of mining projects and use of methane is associated with the ratification of the Kyoto Protocol on climate change and where under the reserves, based on the emissions of CO_2 in the Ukraine in 2008 – 2012 was $ 565 million m^3 (Lukin 2007, Dovbenko 2009, Makogon 2006, Krasnyk & Toropchin 2005).

– one of perspective directions is the use of coal bed methane in the chemical industry. Out of it can produce carbon black, hydrogen, ammonia, methanol, acetylene, nitric acid, various derivatives and formalin. As well as it is the basis for the production of plastics and synthetic fibers.

2 THE MAJOR PROBLEM

For commercial methane production a major problem is decision to increase the permeability of the rock mass and to increase the service life of surface degassing boreholes (SDBH). The reason for this is colmatation pore space near-boreholes area, due to the absorption of drilling fluid in the drilling process and the dynamics of the evolution of gas in the SDBH. The formation pressure is usually lower than hydrostatic pressure of the fluid and promotes deeper penetration of the drilling fluid in the productive horizons in conditions of Donbas.

The washing liquid always contains solids while drilling boreholes. Availability of solid phase in the washing liquid causes colmatation borehole clay crusts. Two types of filtering are shown at drilling. Static filtering occurs in the absence of circulation when during the drilling fluid does not prevent the growth of the filter cake. Dynamic filtering occurs in conditions where the drilling fluid is circulated and growth of drilling fluid filter cakes is limited due to the erosive action of drilling fluid flows. Thickness of the crust and filtration rate remained practically unchanged when the growth rate of filter cake drilling fluid becomes equal to the rate of its destruction.

After opening of the reservoir bed in borehole there remains a certain volume of the filtrate which is not forced into the borehole. After colmatation of the reservoir bed near-borehole and formation of filter cake the pressure of washing liquid is increasing and is followed by absorption of the filtrate liquid. Reduction in static pressure to the filtrate followed by displacement washing liquid and with it the formation fluids. During the drilling process of interaction between hole and reservoir bed the colmatation part of collector and the filter cake serves as a self-regulating dispenser absorption-displacement (Glatkay & Sergienko 2014).

3 SOLUTION OF THE PROBLEM

In IGTM NAS of Ukraine the method of pneumo-hydrodynamic impact (PHDI) for degassing intensification of methane on gas-saturated coal containing massif through the SDBH was established. The essence of the method is that: with the earth's surface drilled SDBH to the coal seam and stopped at a distance of five of its thickness. Over the entire length borehole is lined with metal pipes plugging of annular space at the site of productive horizon lowered pipe perforation. Borehole-head equipped device of pneumohydraulic impact (PHI).

Tubing is lowered in borehole. With the help of compressor compressed air is supplied with in the SDBH for some time under pressure (Figure 1).

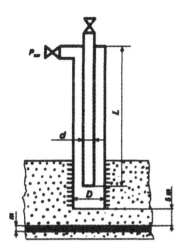

Figure 1. Surface degassing borehole: $P_{cas.pres.}$ – casing pressure, MPa; L – down hole length, m; D – diameter of perforated pipe on face; d – diameter of tubing string, m; m – seam thickness.

During filtration through the mudcake and col-matation of reservoir zone the injecting compressed air come to the reservoir bed and thereby increasing the reservoir fluid pressure. Once the value of down-hole pressure stopped increasing (for $t = 15$ minutes by an amount $\Delta P = 0.01$ MPa) is performed fast reset. Pressure in wells at the reservoir level is reduced almost instantly to the initial pressure. Formation pressure in the near-borehole part the collector is still higher than in the borehole. From this moment the collector starts to displace in the borehole fluid which is infiltrated into it. Displacement fluid partially destroys the filter cake, and its thickness is reduced. Formation pressure decreases to the near-borehole part the collector. Finally comes a moment when the pressure gradient and the initial gradient aligned and restores the initial equilibrium state of the system pneumodynamic borehole – collector in the end. The set and reset pressure present one cycle of impact.

Cycles of pneumohydrodynamic impact are repeated several times. The permeability coefficient is increasing with every subsequent impact. Discontinued of PHDI is subject: when the production rate is increasing to 500 m^3 day; if you follow the cycle of any condition $Q_n \geq 2Q_0$ (Q_0 – flow rate before the start of work, m^3/day; Q_n – production rate after the n-th cycle of PHDI, m^3/day).

This method allows to recover PHDI flow rate to the SDBH. The test method is carried out in an industrial environment in the field of mine named by A.F. Zasyadko in 8 SDBH. By the results of this work were produced 41.86 million m^3 of methane, which leads to the conclusion about the technical capabilities of commercial production of methane from CBM.

4 CONCLUSIONS

Coal deposits are CBM with a possible industrial gas extraction.

Established fundamental possibility of recovery with the use of PHDI on to restore the flow rate gas-saturated coal containing massif through SDBH.

Use of the PHDI method on gas-saturated coal containing massif enabled to extract about 42 million m^3 of methane out of 8 SDBH.

REFERENCES

Antsiferov, A.V., Tirkel, M.G., Khokhlov, M.T., Privlov, V.A, Golubev, AA, Maiboroda, A.A. & Antsiferov, V.A. 2004. *Bearing coal deposits of Donbas.* Kyiv: Naukova Dumka: 232.

Bokiy, B.V. 2006. *Recovery and use of coal mine methane.* Coal Ukraine, No 6: 3 – 7.

Bokiy, B.V., Gunya, D.P., & Efremov, E.V. 2011. *Coalbed methane extraction by applying pnevmogidrodinamicheskogo and pneumodynamic exposure via surface degasification wells for coal containing an array.* Geotechnical Mechanics, No 92.

Bondarenko, V., Tabachenko, M & Wachowicz, J. 2010. *Possibility of production complex of sufficient gasses in Ukraine.* New Techniques and Technologies in Mining. The Netherlands: CRC Press/Balkema: 113 – 119.

Bulat, A.F., Sofiiskyi, K.Km. & Bokiy, B.V. 2014. *The office of atmospheric methane mining methanotrophic bacteria.* Donetsk: LLC "Shidny Vidavnichiy Dim": 256.

Dovbenko, M. 2009. *The Kyoto Protocol as a promising source of additional investment Visn.* Kyiv: Scientific publication of The National Academy of Sciences of Ukraine, No 3: 26 – 34.

Efremov, I.A. 2012. *The concept of integrated degassing and the creation of heat and power systems with the use of coal mine methane in the mine to them "AF Zasyadko".* Visti Donetskogo girnichogo institutu, No 1: 287.

Falshtynskyi, V., Dychkovskyi, R., Lozynskyi, V. & Saik, P. 2014. *Some aspects of technological processes control of in-situ gasifier at coal seam gasification.* Progressive Technologies of Coal, Coalbed Methane, and Ores Mining. The Netherlands: CRC Press/Balkema: 109 – 112.

Gur'yanov, V.V. 2004. *State gazobezopasnost and direction of works to improve drainage and utilization of coal mine methane in Ukraine.* Mining informational and analytical bulletin, No 8: 184 – 187.

Glatkay, E.V., Sergienko, L.V., Pavlov, I.O. & Zhitlenok, D.M. 2012. *Peculiarities of zones of accumulation of methane in coal rock mass moonlighting with the interaction of natural and man-made fractures.* Geotechnical Mechanics, No 101: 56 – 62.

Glatkay, E.V. & Sergienko, L.V. 2014. *Development of a technique to determine the zones of accumulation of methane in the layers-satellites in a tectonic disturbance.* Materials of the international conference "Forum girnykiv", Dnipropetrovsk: NMU: 48 – 55.

Kasyanov, V.V. & Art, Lambert. 2000. *Prospects of development of methane industry in Ukraine.* Geotechnical Mechanics, No 17: 6 – 11.

Krasnyk, V.G. & Toropchin, O.S. 2005. *State of m production prospects for coal mine methane in Ukraine.* Ukrainian Coal, No 11: 16 – 18.

Kuzyara, S.V., Drozdnik, I.D., Kaftan, Y.S. & Dolzhanaskaya, Y.B. 2005. *Coal bed methane extraction and environmental protection (review).* Coal of Ukraine, No 6: 13 – 15.

Lukin, V.V. 2007. *Mining and geological conditions of extraction of methane from coal deposits of Ukraine.* Ekologiya i prirodokoristuvannya, No 10: 66 – 70.

Makogon, Y.U., Ryabchyn, A. & Role, M. 2006. *The Kyoto Protocol and modern eco-technologies in the energy saving policy of Ukraine.* Strategic Priorities, No 1: 135 – 143.

Mjaken'kaja, V.I., Kurdishm, I.K., Demchenko, V.B., & Petuch, A.P. 1992. *The efficiency of microbial oxidation of methane in coal mines goafs.* Microbiological magazine, No 1: 67 – 73.

Pudakov, V.V., Konarev, V.V. & Alekseev, A.D. 1996. *Research, development of technology and the industrial use of Methane from Coal mine Donbas.* Coal Ukraine, No October – November: 68 – 71.

Smorschok, V.S., Bokserman, Y.A. & Pyatnychko. 2006. *Software for the development of technological solutions and implementation of effective utilization of coal bed methane.* Oil and gas industry, No 6: 23 – 26.

Zhykalyak, M.V., Avdeev, A.M. & Kravchenko, N. 2012. *Gas potential of abandoned mines of Donbas.* Geotechnical Mechanics, No 102: 18 – 26.

Zozulya, A.D., Chakvetadze, F.A. & Maltsev, L.I. 2011. *Development of technology for supply to consumers of coal mine methane degasification wells coal mines.* GIABA, No 7: 181 – 187.

Theoretical and Practical Solutions of Mineral Resources Mining – Pivnyak, Bondarenko & Kovalevska (eds)
© 2015 Taylor & Francis Group, London, ISBN: 978-1-138-02883-8

The findings of visual-optical method of diagnosing of rock disposal process from mine cars

A. Shyrin
National Mining University, Dnipropetrovsk, Ukraine

ABSTRACT: Under coal resources mining which are located near the borders of the mine fields, the length of traffic arteries are up to 6.0 – 7.0 km. Possible reserves are mainly concentrated in the zones of influence of tectonic abnormality, heavy water cut and intense rock heaving. In such circumstances of mining, well-timed rock transportation from development headings by the means of rail transport is intractable problem, and mostly even impossible.

1 INTRODUCTION

Under active rock heaving at Western Donbas mines floor rope railway are used as a single hauling unit during preliminary development workings. However, even using technological transport schemes with the application of high-technology floor railways such as DKNP-1.6, self-dumping wagons VD-2.5 and accumulating rock pockets (Figure 1) there is a failure to comply with the planned rate of development workings because of tardy rock transportation from the pitfaces.

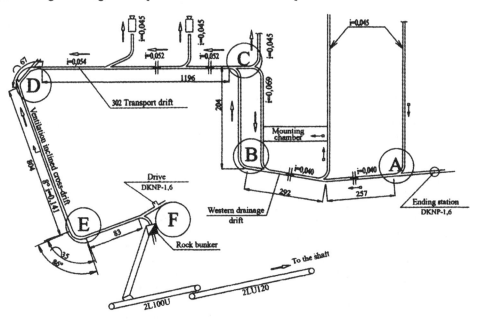

Figure 1. The technological scheme of rock transportation DKNP-1.6.

Analysis data of technological schemes current capacity of mine district transport showed that one of the reasons of decrease rate in roadway construction due to a transport fault is performance parameters defection of mine wagons because rock sticking to the walls and floor of the body. Changing the performance parameters of mine wagons such as volume (V) and loading capacity (G) is due to execution technology of zonal development workings.

2 MAIN PART

In accordance with the requirements of accident prevention the preliminary development process during combine mining entails dust suppression (moistening). As a result mined rock (waste, coal) are mined from the shortwalls in mining wagons with humidity rate about 50 – 80% and with stickability 75 g/cm^2 (Gavrilov et al. 2002). Studies showed that during mined rock transportation in extended curved roadways with alternating grading of track there are different progressive and rotating construction elements vibrations of mine wagons. The main progressive vibrations are jerking – along the x-axis; lateral motion – along the y-axis and bouncing – along the axis OZ. The rotational vibrations include: rolling motion – around the x-axis; pitching – around the y-axis and wagging – axis OZ. As a consequence of multiple spatial vibrations there is intense moist rock solidifying and sticking it on the walls and bottom of trolleys, as a result it leads to usage rate reduction of mine wagon body.

According to the analysis of works in the field of cargo management of mine rocks it was found that diagnosed problem remains insufficiently explored up to date. In such situations significantly increases the role of operative production diagnosis and transport process control of overburden recasting in underground workings during the performance of mining operations.

The purpose of operative production diagnosis and transport process control of overburden recasting is time and space coordination of searching and eliminating of imperfections during rock disposal from mine wagons to pocket (the final operation of transport process) and delivery of current information about carriage rolling stock level of readiness for work execution of next cycle – rock excavating and loading into wagons.

To justify the method of operative process control of rocks loading from mine cars battery of studies for setting out causes and nature of the rocks sticking to the trolley body, control methods and means of diagnosing this process were carried out.

Mine studies determined with advance speed of development heading $v \geq 2.4$ m/shift (three or more cycles), such events resulted in the end of the shift to a decrease in the useful volume of the trolley body at 25 – 30%.

Consequently, the schedules of current processes and operations in development headings were troubled, the temp of working decreased but the specific energy consumption increased incidental to with transportation of "dead weight", i.e. adhering rock mass.

Figure 2 shows the results of mine experiment to establish indicators and character of changes the useful volume of the body of mine cars VD-2.5 and VG-3.3, after three successive cycles of loading, unloading and transport operations.

It should be noted that in the practice of mechanical systems technical diagnostics the state of analyzing object is determined by the results of incumbent functions, i.e. using the methods of functional diagnosis (Kollakot 1998).

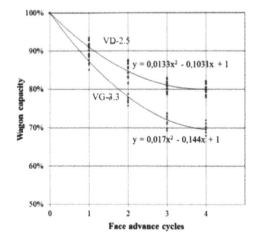

Figure 2. The reduction character of the useful volume of the mine cars body when rocks sticking in the back trolley.

Algorithm for functional diagnosis involves consideration of input actions and output responses, the sequence of operations execution, as well as external and indirect indicators.

The known methods of processes and operation diagnosing during loading and unloading operations in underground conditions are cost approach which requires the superinduction of power-intense stationary equipment in transport processing chain. For example, foreign mines of non-ferrous industry, extracting rare and precious metals, the volume of not unloaded rock mass is determined by the weighing of the loaded and empty trains. In works are described the control methods of mining mass winnings using modules APCS-R and stationary weighing devices of tubs AKRV-87 produced in Bulgaria (Sitnikov et al. 2006).

The mining technology in coal mines does not provide performing operations related to weighing wagons with waste rock (mining waste). For this reason, the enterprises have no experience in such procedures.

For the purpose of the little-explored scientific and practical tasks is recommended to use visual-optical method for monitoring the process of unloading rock from the mine cars. The nature of the proposed technical solution is to install a system of stereovision in the rock unloading point from ore train to a hopper in order to diagnose the dumping process of mine wagons.

The idea is to use the computer vision effect for rapid recognition of thickness layer of the rock mass sticking to the walls and the bottom of tub, and according to obtained visual analysis of 3D images make appropriate decisions in order to coordinate work schedule of transport-technological system.

The computer vision is theory and technology of creating artificial systems which provide detection, tracking and classification of objects by analysis and synthesis of 3D images and videos sequentially obtained from different cameras, sensors and devices. Video data can be represented by different forms such as video sequence of images from different cameras or 3D data, for example, from the sensors device of Kinect or stereo cameras.

The nature of stereovision is to imitate human binocular vision. Stereovision system allows the computer by two pictures taken with the two cameras to read out the information in 3D form about the environment and create a depth image map. In the case under consideration it is the thickness of the rock mass sticking.

Under the map it is entailed the depth image where each pixel does not represent a color but the distance from the object to the camera. The darker the object is on the map, the farther it is away from the camera.

The integrated study method provides with software package of a computer monitoring system and operation control DKNP-1.6 with the application stereovision should combine the three important functions:

– box-wagon movement registration with simultaneous search of tub which is installing for unloading;

– treatment of the received frame with the following search of the departure from the set standards;

– transfer of the processed information to the operator DKNP-1.6 to make a final decision.

In accordance with the listed functions the laboratory-based experiment program was developed which included the realization of specific studies on the working equipment of the rail transport laboratory of transport system and technologies department of National Mining University.

To build and install stereosystem it was necessary to determine the cameras positions – to calculate the length of stereobase, the camera installation height and direction. Under the calculations performance the illumination room indicators were used in the place of stereo installation but also technical characteristics of camera and mine car were used.

It should be noted that in practice the process of installing another tub for unloading rocks into the hopper is performed automatically, by pushing full tubs without uncoupling. Because of this for the qualitative detection of tubs movement in the both camera frames it is necessary to install the stereo system so that at one unit of time the tub would be simultaneously in the center of both camera frames (Figure 3).

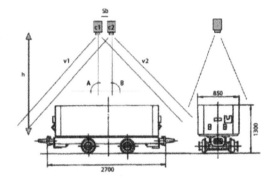

Figure 3. The scheme how to calculate the height of the camera: $c1$, $c2$ – stereosystem cameras; $v1$, $v2$ – vision area of each camera; h – the height of stereosystem relatively the rail track.

3 CALCULATIONAL EXPERIMENTS

The height between the cameras and tub depends on the length of the tub but because of various camera parameters, their matrices and resolution abilities, the calculations show only the approximate distance that is why system needs to be further checked and adjusted on the site.

The calculation of cameras height h cameras is made in accordance with the formula:

$$h = \frac{A_{tub} \cdot f}{m \cdot h} + H_{tub}, \qquad (1)$$

where A_{tub} – horizontal tub size, m; f – focal distance, mm; mh – horizontal matrix size, mm; H_{tub} – tub height, m.

When performing natural experiment the cameras with a focal distance of ~4.8 m and with horizontal dimension matrix size – 4 mm were used. In conformity with the considered conditions for the tub VG-2.5 minimum height of cameras installation, according to calculations is $h_{2.5} = 3.1$ m and for the tub VD-3.3 is $h_{3.3} = 4.3$ m.

The performance expectations of the frame capacity and the height of cameras installation with different focal lengths are given in Table 1.

In order to create initial database the calculation were performed to determine the rational height of stereocameras installation for the series-produced mine supplies cars.

Table 1. Frame capacity of cameras with $mh = 4.8$ mm.

Focal distance, mm	Distance from camera to the object, m		
	1.5	3.0	4.6
	Horizontal frame capacity, m		
1.78	7.6	15.1	22.7
2.45	5.3	10.5	15.8
3.00	3.5	7.5	11.3
3.60	2.3	4.6	6.9
4.00	2.0	4.0	6.0
8.00	0.9	1.8	2.7
16.00	0.4	0.9	1.3

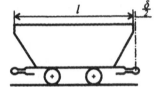

Figure 4. The scheme to determine the volume of utilization of coal mine wagon type VG-2.5 (with bottom-discharge-type).

According to the results of calculations it was found that in order one tub simultaneously was in the picture of each cameras, the cameras Logitech C110 used in the experiment are needed to be installed at the height of $h \geq 4.0$ m which do not respond to the mine working parameters. In order to lessen the height under the system will be able to function it is necessary to use the cameras a wider viewing angle and a correspondingly with a shorter focal distance.

To calculate the rational focal distance of the camera f, where it will be possible to get full-sized shots of mine car in underground mines the following expression was used:

$$f = \frac{v \cdot h \cdot S}{H}, \qquad (2)$$

where v – vertical matrix size, mm; h – horizontal matrix size, mm; S – distance to the object, m; H - horizontal object size, m.

The importance factor of natural experiment determined the nature of the utilization factor change of interval volume of the tub body at different stages of the mining industry. In this regard, the condition was set in order stereocameras qualitatively track events related with the rocks sticking processes inside the body of mine wagon, i.e. quickly determine the amount of rock mass remaining on the body at the bottom of the tub after each cycle of its unloading.

The results of experimental studies of the volume determination of stuck rocks and the utilization factor of the body tub volume had served as a basis for technique development of mine cars parameters coordination and stereo system in the conditions of the mine environment.

The definition of rational distance from the camera to the tub floor pan is performed due to the scheme given in Figure 4 by subtracting the size of the gap δ between the top of rail and floor pan of the total height of camera installation.

The need to perform such calculations is stipulated by the fact that the capacity of performance parameters volume (V) and capacity (G) is also expressed by the linear dimensions and utilization factor of coal mine wagon volume (Shyrin 2013):

$$V = m \cdot h \cdot l \cdot \mu , \text{m}^3, \qquad (3)$$

$$G = V \cdot \gamma = m \cdot h \cdot l \cdot \mu \cdot \gamma , \text{t}, \qquad (4)$$

where m – wagon width, m; h – wagon height from the top of rail to the top edge of body, m; l – length body on to of edge, m; γ – product density in stocking, t/m^3; μ – utilization factor of mine car volume depending on the geometry of the body.

Using design circuit shown in Figure 3 and 4, it was found the original distance from the camera to the upper edge of the tub which during the experiment was being corrected taking into consideration the design parameters of the wagon which is under the investigation. The reason is that for successful recognition of the rock mass volume remaining in the tub body after its discharge, it is necessary to made stereosystem oriented so that at one unit of time the tub was simultaneously in the frame center of the both cameras (Figure 3). On this basis the height of stereosystem installation which was calculated by the formula (1) should be coordinated with of design parameters of operated cars, cameras, their matrices and cameras resolution abilities:

$$h = \frac{f}{m \cdot h} \cdot L_{tub} + H_{tub}, \qquad (5)$$

where h – distance to the object, m; f – focal distance, mm; mh – horizontal matrix size; L_{tub} – tub length, m; H_{tub} – tub height, m.

The results of experimental studies have established that in order to have qualitative determination of the rock mass volume sticking in the back of the wagon of type VG-2.5 is it necessary to install the stereo-system at the height $h \geq 2.5$ m with a width of ~35 mm and a focal distance of ~17 mm.

At this distance the camera should be directed parallel to each other and at an angle 90° relatively to the rail track and tubs. Such installation of cameras direction is stipulated for the correlation of the basic parameters of carriage rolling stock and underground transportation workings.

In order to obtain the high-quality images of mine wagons content at a distance of 1.0 – 4.0 m it is enough to keep a distance between cameras within the standard stereo (50 – 80 mm). More accurate results are regulated by lighting and camera features.

The scheme of cameras location of in the conditions of laboratory experiment is shown in Figure 5.

Figure 5. Workplace programs with the active program window.

Applied to laboratory conditions stereo system consisting of cameras Logitech C110, was connected to a computer with a processor Intel Core i5 1.7 GHz and it was installed at a height of 3.0 m.

The fragments of a laboratory experiment are presented in Figure 6.

According to the results of the natural experiments the operation algorithm of diagnostic system was developed, according to which the preparation stage of stereosystem end with the camera direction installation and basic data record source into the computer, such as camera installation height; the distance from the bottom and the top edge of the body to the cameras and the main tub design parameters.

After storing the original data stereosystem algorithm is considered to be ready for work. The Diagnosis process of rock loading from wagons begins with commissioning of the cameras, sequential processing of each image from the video stream and motion detection.

Figure 6. The active window with depth map.

If the movement was detected in the picture, the system allocates movable areas and with the help of horizontal lines search divides the area into individual objects. If the size of one of the objects is comparable to the tub proportions the image is saved for processing and for the following construction of the depth map based on it.

The image processing and construction of depth map are carried out step-by-step. Firstly, in the recognized and found in the video stream tub, the received image area is marked out. The procedure of region selection is stipulated by the need to reduce computer utilization, because during diagnosis process we are interested in the whole image, but only a part of the tub body with sticking rock.

After receiving stereoimage of the tub execution algorithm of obtaining depth starts. As an example, Figure 7 shows the rocks depth map in half-uploaded tub.

Figure 7. The depth map of a half-filled tub.

After receiving and correcting of images gained from the two cameras the phase of their correlation begins, i.e. the function of two images into one single stereo-image by searching paired points on the input image. Subsequently, according to found points three-dimensional coordinates of the points and coordinates of their prototype image in three-dimensional space are being determined. Knowing the three-dimensional coordinates of the prototype image the depth image is being calculated.

4 CONCLUSION

The depth map is created on the basis of the received matrix depth. The matrix depth is the distance value of each pixel to the camera. The larger the value is, the closer pixel in to the camera. The matrix dimensions are depended on the size of the image. The results of the research are implemented in software products for process control of rocks withdrawal out of the wagon and the initial conditions on the use of stereo vision system for computer diagnostics of technical and operational parameters of transport and technological systems using the ropeways of new generation.

REFERENCES

Gavrilov, Y.I., Usanov, V.V., Stepanenko, V.F., Plehanov, Y.V. & Milohin, Y.P. 2002. *Complex of technical measures of weigh accounting of underground rail transport work*. Materials of meeting "Condition and prospects of processes automatization on open-pit and underground mines". Ordzhonikidze: 45 – 47.

Kollakot, R. 1998. *Diagnostics of damages*. Translated from English. Moscow: Mir: 512.

Shyrin, A.L. 2013. *Visual-optical method of rock unloading process control from mine cars*. Thesis of All-Ukrainian scientific-and-practical conference of post-graduates, young researches and students "Informational technologies in education, technics and industry". Ivano-Frankivsk.

Sitnikov, D.M., Potapenko, G.D. & Sautov, S.Y. 2006. *Self-cleaning mine cars with bottom unloading on mines of Podmoskovny basin*. Express-inform. Moscow: CNIEIugol: 15.

The stress-strain state of the belt in the operating changes of the burdening conveyor parameters

D. Kolosov & O. Dolgov
National Mining University, Dnipropetrovsk, Ukraine

O. Bilous
Dniprodzerzhynsk State Technical University, Dniprodzerzhynsk, Ukraine

A. Kolosov
Moscow State University of Technology and Management, Moscow, Russia

ABSTRACT: The analytical dependences for the stress-strain state of the steel cord rubber belt due to the changes in operating geometric parameters of the burdening conveyor into blast furnace are obtained. These results can be used in engineering practice to make decisions on elimination of wear defects of conveyor parts.

1 INTRODUCTION

Modern delivering systems of burden materials, including the delivery system of the burden materials on the largest blast furnace number 9, "ArcelorMittal Kryviy Rih", Ukraine (Bolshakov et al. 2005), are based on conveyor belts. Compared to skip charging, the conveyor system has a number of advantages. Specific capital and operating costs are less by 15 – 20%. The system performance is much greater and it is simpler. Such a system allows the reliable coordination of all the components of the preparation and of the charge delivery system. It can

be operated without human intervention, for example by means of automation.

The main blast furnace conveyor (Figure 1) has a length of several hundred meters. It is equipped with a discharge drum *1*, tension drum *2*, driving drums *3* and *4*. The drums are enveloped by the belt *5*, which rests on the rollers *6*. The charge material is fed to the belt from the tanks *7*. The conveyor, which has horizontal and inclined sections, operates continuously. The discharge drum *1* is displaced from the driver to the axis of the furnace. The drive drum is mounted on a special structure.

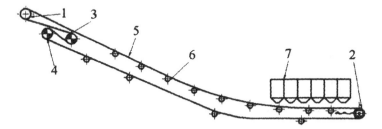

Figure 1. Diagram of the charge on the feeding conveyor of the blast furnace.

There are substantial tensile forces acting on the conveyor belt. To provide sufficient strength of the belt, it is made of the ropes, which are vulcanized into rubber band.

Charge materials are fed by conveyor in accordance with a predetermined program. Ore, coke, agglomerate are supplied in portions at predetermined intervals. It requires the use of increased belt width

(up to two meters), provided supplying in portions a significant amount of material.

A blast furnace is the unit of continuous action. Therefore a rhythm disturbance in delivery of raw materials to the blast furnace leads to disruption of the continuous iron smelting process and significant financial losses.

2 PROBLEM STATEMENT

In the operation some parts of the conveyor wear, their parts change the form. For example, the diameter of the cylindrical drum becomes variable along its length. A right angle changes between the axis of the belt and the drum. The described changes in the geometric parameters of the components of the conveyor lead to an uneven distribution of tensile forces over the width of the belt. The use of the belt with considerable width in the conveyor, feeding the blast furnace increases this unevenness. This uneven distribution of forces reduces the real traction capability of the belt and its service life. There appear conditions which lead to the destruction of ropes, further affecting the strength of the steel cord belt.

Thus, definition of the deviation influence in geometric parameters of the conveyor components from the design occurring during its operation is an actual scientific and technical problem of providing performance of the conveyor system feeding the burden to blast furnace top.

Deviation in geometric parameters of the conveyor design induces an additional deformation of the belt. Spatial deformation of conveyor belts is analyzed in (Belmas 1993). Effect of deviation from the cylinder-shaped of the conveyor drum on the stress state of the belt is studied in (Belous 2010). In the mentioned works the effect of the angle between the axes of the belt and the drum on the distribution of forces between the ropes in the belt was not studied. The influence of the shape of the conveyor drum on the distribution of forces within the belt was considered without taking into account the ruptures of rope in interacting with non-cylindrical drum.

3 PROBLEM SOLUTIONS

In this paper, the stress-strain state of the steel cord belts for previously unexplored changes in individual geometric parameters during the conveyor operation is studied. The problem is resolved in a general form. The number of ropes in the belt is taken equal to N. The lengths of the portions on which the belt does not interact with the drum, we assume infinite.

To determine the effect of disturbances of mutual location of the belt axis x and the axis of the drum, put the origin of the axis into the belt cross section, interacting with the drum. Due to the turning of the normal section of the semi-infinite belt by an angle φ, the k-th rope is moved by the value:

$$\delta_k = \left(\frac{N}{2} - k + 0.5\right) t \sin(\varphi),\qquad(1)$$

where t – a step of the ropes laying in the belt.

Expand the resulting dependence (1) into Fourier series of cosines:

$$f_i = D_0 + \sum_{m=1}^{\infty} D_m \cos(\mu_m(i-0.5)),\qquad(2)$$

where $D_0 = \sum_{k=1}^{N} \dfrac{\delta_k}{N}$; $D_m = \sum_{k=1}^{N} \dfrac{2\delta_k}{N}\cos(\mu_m(i-0.5))$.

In order to determine the effect of turning the belt only, the value of the tensile force of the conveyor belt will not be taken into account. External action (1) is performed in one section. According to the principle of Saint-Venant, the local disturbance (impact) leads to a local redistribution of the stresses and strains in solids. In this case, the deformation of sections of the belt and rope tensile force must be reduced to zero at an infinite increase in the distance from the cross section of local influences. With regard to the latter and in accordance with (Tantsura 2010), displacements of the rope sections u and internal forces P are determined from expressions:

$$u_i = \sum_{m=1}^{N} B_m e^{-\beta_m x} \cos(\mu_m(i-0.5)) + c,\qquad(3)$$

$$P_i = -EF \sum_{m=1}^{N} B_m e^{-\beta_m x} \beta_m \cos(\mu_m(i-0.5)),\qquad(4)$$

where B_m, c – constants of integration; E – the reduced modulus of elasticity of the material of ropes; F – cross-sectional area of the rope; $\mu_m = \dfrac{\pi n}{N}$;

$$\beta_m = \sqrt{2\frac{Gd}{hEF}(1 - \cos(\mu_m))}\; ;\; G - \text{the shear modulus}$$

of the rubber; d, h – the diameter of the ropes and the distance between them.

Considering the condition of the ropes deformation (1) in the section $x = 0$, from relation (3) we have:

$$B_m = \frac{2}{N}\sum_{k=1}^{N}\left(\frac{N}{2} - k + 0.5\right) t \sin(\varphi)\cos(\mu_m(k-0.5)),$$

$$c = 0.$$

To illustrate the results, the belt with parameters corresponding to the rubber-RTL-3150 is considered. The deviation of the axis of the belt from the right angle is taken equal to 1°. Calculations are represented in Figure 2 and 3.

Figure 2 and 3 show that significant changes in tensile forces and displacements of the ropes occur at the lengths up to 5 – 7 m. The latter conclusion permits to consider the length of portions of the belt up to seven meters as infinite. The lengths of portions of the conveyor belt located between the drums are no less than this distance. Thus, the above assumption of

586

infinite belt segments that interact with the drums of the conveyor can be considered acceptable. Changes of the forces along the belt width have skew-symmetric character. This character is the result of skew-symmetric deformation of the belt in the section which separates the non-interacting and interacting portion of the belt with the drum.

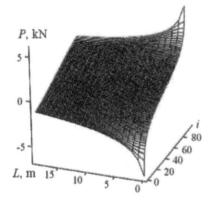

Figure 2. Distribution of the forces between the ropes due to a change in the angle between the belt and a drum.

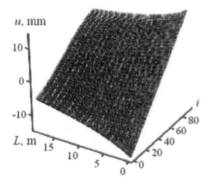

Figure 3. The surface describing the displacements of the ropes points, due to the drum turning.

The maximum deformations and efforts are realized in the last ropes. The maximum additional force in the last rope at small angles of deviation of the belt axis from the right angle to the axis of the drum is:

$$P_{max} = \frac{t\varphi\sqrt{8GEF}}{N} \sum_{m=1}^{N} \sqrt{1-\cos\frac{\pi m}{N}} \cos\left(\pi m\left(1-\frac{1}{2N}\right)\right) \times$$

$$\times \sum_{k=1}^{N}\left(\frac{N+1}{2}-k\right)\cos\left(\frac{\pi m}{N}(k-0.5)\right).$$

The obtained additional force that occurs in the last rope must be taken into account when developing the allowable standards for the accuracy of mounting and the changes of geometric parameters of the conveyor supplying charge materials into a blast furnace during its operation.

In the Figure 2 and 3 and the next it is shown that in the belt ropes may arise tensile and compressive forces. In resolving the problem we conventionally disregarded the tensile force, on which we design the conveyor belt. In a real conveyor, this force acts on the belt; it causes equal loads and deformations in all the ropes. Compressive forces can occur only when the loads caused by uneven deformation of the belt would exceed the average values of tensile forces in the ropes. The latter corresponds to one hundred percent excess of the maximum stresses in the belt compared with calculated.

The uneven distribution of the forces across the width of the belt causes uneven wear of the drum surface. One edge of the drum wears faster than the other. The cylindrical shape of the drum is changed to a conical. The ropes of the one belt edge are extended more than the ropes of the other edge. Their tensile forces are increasing more than in the other ropes; the likelihood of failure increases.

The nature of the belt deformation on the drum differs from its strain state in the area where the belt does not interact with the drum. In order to account the latter, divide the belt into the area of interaction with the drum and the area in which the belt does not interact with it. Consider conditionally the plane through the axis of the drum, bisecting the angle on which the belt encircles the drum. Accordingly, the drawn plane divides the belt into two symmetrical segments. Excluding the impact of the external load, the load of both sections will be symmetrical. This makes it possible to consider a single piece of the belt.

We use equations from (Tantsura 2010) and determine the stress-strain state of the belt with damaged (broken) ropes:

a) for the belt section interacting with the drum:

$$u_{[k]} = -\left[\sum_{m=1}^{\infty}\left(A_{[m]}e^{\beta_m x} + B_{[m]}e^{-\beta_m x}\right)\times\right.$$

$$\left.\times \cos(\mu_m(k-0.5))\right] + \frac{a_{[2]}x}{EF} + c_{[2]}, \tag{5}$$

$$P_{[k]} = \left\{\frac{f(k)+R}{R} - \left[\sum_{m=1}^{\infty}\beta_n\left(A_{[m]}e^{\beta_m x} - B_{[m]}e^{-\beta_m x}\right)\times\right.\right.$$

$$\left.\left.\times \cos(\mu_m(k-0.5))\right]\right\}EF + a_{[2]}; \tag{6}$$

b) for the length of the belt which does not interact with the drum:

$$u_k = \sum_{m=1}^{\infty}\left(A_m e^{\beta_n x} + B_m e^{-\beta_m x}\right)\cos(\mu_m(k-0.5)) +$$

$$+ \frac{a_2 x}{EF} + c_2, \tag{7}$$

587

$$P_k = \sum_{m=1}^{\infty} \beta_m \left(A_m e^{\beta_m x} - B_m e^{-\beta_m x} \right) \times$$

$$\times cos(\mu_m (k - 0.5)) EF + a_2, \tag{8}$$

where $f(k)$ – a function describing the dependence of the radius of the drum on the number of the ropes; R – average radius of the drum; A_m, a – constants of integration.

The quantity $\dfrac{f(k)+R}{R} EF$, according to (6), defines tensile forces of the ropes which deform on the drum provided the hypothesis of the plane cross sections of the belt (cross sections do not deform). The component

$$\left[\sum_{m=1}^{\infty} \beta_n \left(A_{[m]} e^{\beta_m x} - B_m e^{-\beta_m x} \right) cos(\mu_m (k - 0.5)) \right] EF$$

in the same equation describes the nature of the deformation of the belt cross sections, i.e. removes restrictions imposed by the use of the plane sections hypothesis. Along the length of the belt the ropes may be ruptured in arbitrary sections.

In the course of the conveyor, the ropes pass all its areas. The most dangerous is the moment when the cross section of the belt with a damaged rope runs the conveyor section with the most uneven loaded ropes. In our case, this cross-section corresponds to the section $x = 0$. Assuming that the k-th rope is damaged we have the following uniform condition for a cross-section $x = 0$:

a) $u_{[k](x=0)} = 0 \quad (k \neq \mathbf{k})$,

b) $P_{[k](x=0)} = 0$. \tag{9}

Write the boundary condition (9, a) in the following form:

$$u_{[k](x=0)} = K \begin{cases} 0 & (k \neq \mathbf{k}) \\ 1 & (k = \mathbf{k}) \end{cases}, \tag{10}$$

where K – unknown constant, which should be determined from the condition (9, b).

The length of the section where the belt interacts with a drum is denoted by l. In the section $x = l/2$ should be satisfied the strain compatibility condition:

$$u_{[k](x=l/2)} = u_{k(x=l/2)},$$

$$P_{[k](x=l/2)} = P_{k(x=l/2)}. \tag{11}$$

In accordance with the hypothesis of Saint-Venant, a difference of displacements and tensile forces of the ropes with increasing the distance from the local perturbation is reduced to zero. Thus, we have another boundary condition:

when $x \to \infty$ $\begin{cases} u_k - u_{k \pm 1} \to 0, \\ P_k \to 0. \end{cases}$ \tag{12}

The condition (10) will be fulfilled having taken:

$$u_{[k](x=0)} = K \left[\frac{2}{N} \sum_{m=1}^{N} cos(\mu_m (k - 0.5)) \times \right.$$

$$\left. \times cos(\mu_m (\mathbf{k} - 0.5)) + \frac{1}{N} \right]. \tag{13}$$

The function $\dfrac{f(k)+R}{R}$ is discrete and is performed for the values $1 \leq k \leq N$. We define it as the sum:

$$\frac{f(k)+R}{R} = \sum_{m=1}^{N} y_m \, cos(\mu_m (k - 0.5)). \tag{14}$$

Considering expressions (13) and (14), relations (5) – (8), from the boundary conditions (9) – (10) determine the unknown constants. Defined unknown constants allow determining the forces acting in ropes and their displacements. The results of calculations for the case of exceeding the maximum radius of the drum over a minimum by 0.2% are shown in Figure 4 and 5. The graphs are drawn in relative coordinates $x_0 = \dfrac{x}{l}$, $y_0 = \dfrac{y}{b}$.

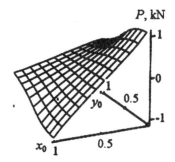

Figure 4. Distribution of forces between the ropes in the case of the interaction of the belt with the drum, the surface of which during operation has acquired a conical shape.

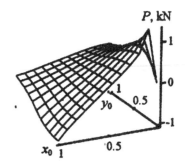

Figure 5. Distribution of the forces between the ropes in the case of the interaction of the belt with the drum, the surface of which in operation has acquired a conical shape when the last rope is ruptured.

Figure 4 and 5 shows that significant change of the tensile forces of the ropes also occurs at the lengths up to 5 – 7 meters. The change of the forces along the belt width has a skew-symmetric character due to skew-symmetric nature of the wear of the drum. The maximum strains and forces are realized in the last ropes. The rupture of the last rope leads to an increase in the maximum effort by 23%. The case of the rupture of the middle rope is not shown, as on the average rope the loads do not increase when the drum acquires a conical shape – they remain equal to zero. As a consequence, the rupture of the unloaded rope does not involve the changes in the loading of other cables.

The driving drum of the feed conveyor 3 of burden materials into the blast furnace (Figure 1) interacts with the belt surface, by which the raw materials are loaded. Abrasive dust, left on the belt surface, accelerates the wear of a drum surface. When loading, the material on the belt is located closer to the axis of the conveyor. Therefore, the wear of the drum in the middle part is more intense.

Equations (5) – (14) are written for the case when the shape of the drum generator is known. We assume that the generator of the drum as a result of the wear had the shape of a quadratic parabola. The greatest radius of the drum exceeds the lowest by 0.2%. Considering the drum shape in accordance with the function $f(k)$ was defined the stress-strain state of the conveyor belt during its interaction with the worn drum. The calculation results are shown in Figure 6 – 8.

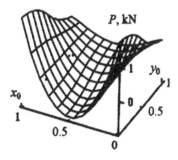

Figure 6. Distribution of the forces between the ropes for the case of interaction between the belt and a drum which has a parabolic shape of generator.

The maximum force due to rupture of the extreme rope, increased by 5%. At the instant of the rupture, the rope lost the ability to resist external loadings. In (Belmas 1993) it is shown that the damage to rope leads to the change of loads only in a few its adjacent ropes. Since the ropes, adjacent to the damaged extreme rope are less loaded, the additional load changes the greatest load of the belt rope insignifi-

cantly. This feature has led to the fact that damage to the rope, located closer to the middle of the belt, has changed a load of adjacent ropes by 9%.

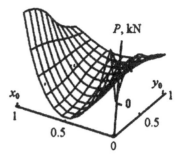

Figure 7. Distribution of the forces between the ropes for the case of interaction between the belt and a drum which has a parabolic shape of generator when the last rope is ruptured.

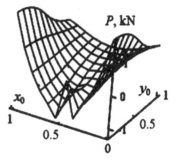

Figure 8. Distribution of the forces between the ropes for the case of interaction between the belt and a drum which has a parabolic shape of generator when the middle rope is ruptured.

In conveyors of a large belt width the bulging drums are used. This facilitates the centering of the belt relative to the conveyor axis and relative to the middle of the bulging drum. The use of such a technical design also leads to uneven distribution of the internal forces in the belt ropes. In the case of a convex drum at an invariance of other conditions, the expression (14) takes the form $\frac{-f(k)+R}{R} + \sum_{m=1}^{N} y_m \cos(\mu_m(k-0.5))$. The surfaces that describe the distribution of the forces, shown in Figure 6 – 8, change their sign.

We have considered only the impact of the deviation from the geometric parameters of the conveyor design. The linear formulation of the problem allows applying the principle of superposition. Therefore, to determine the effect of the belt tension it is sufficient to add evenly distributed forces to the loads discussed above. The influence of the ropes rupture should be taken into account through the stress concentration factor due to rope rupture. This ratio for the belt of the large width, loaded only by the force of stretching, equals 1.6 and 1.4 in cases of the rupture of the last and middle ropes (Belmas 1993).

4 CONCLUSIONS

Misalignment of the drum axis and the deviation from the cylindrical shape of the drum, results in a redistribution of the internal stretching forces in the ropes. The changes in the redistribution of the forces occur at a length of 5 – 7 meters. The assumption of infinity of the belt segments that interact with a drum of the conveyor, taken in this paper, may be considered acceptable and not affecting the accuracy of the results. The ruptures of the ropes alter the stress-strain state of the belt: the destruction of the edge and middle ropes (except the skew-symmetric deformation) leads to an increase in extreme values of tensile forces of the ropes.

The obtained mathematical relationships allow defining the stress-strain state of the belt due to changes in operating parameters of the feeding conveyor of the blast furnace. On the basis of the results one can reasonably make decisions on the elimination of defects against wear and tear of the conveyor components.

REFERENCES

Belmas, I.V. 1993. *Fundamentals of theory and calculus of spatial rubber-rope cables of conveyor.* Dnipropetrovsk: PhD thesis: 360.

Belous, E.I. 2010. *Flexible traction bodies in mechanical engineering. Analysis of deviation of the cylinder form working surface of the drum.* Dniprodzerzhynsk: Monograph: 123.

Bolshakov, V.I., Vishnjakov, V.I., Porubova T.P. & Lavrik, L.I. 2005. *Evaluation of without cone charging device wear.* Fundamental and applied problems of ferrous metallurgy: Collection of Scientific Papers of Institute of Ferrous Metallurgy of the National Academy of Sciences of Ukraine, No 10: 275 – 284.

Tantsura, A.I. 2010. *Flexible traction bodies in mechanical engineering. Butt-joints of conveyor belts.* Dniprodzerzhynsk: Monograph: 127.

Theoretical and Practical Solutions of Mineral Resources Mining – Pivnyak, Bondarenko & Kovalevska (eds)
© 2015 Taylor & Francis Group, London, ISBN: 978-1-138-02883-8

Study of the effects of drilling string eccentricity in the borehole on the quality of its cleaning

O. Vytyaz, I. Chudyk & V. Mykhailiuk
Ivano-Frankivsk National Technical University of Oil and Gas, Ivano-Frankivsk, Ukraine

ABSTRACT: The solution to the problem of determining the eccentric location of the drill string in the borehole using analytical studies and modern systems of mathematical computer modeling was analyzed. The practical implementation of technological problem solving was parameters calculation of the formation and changes in the area of stagnant zone in the cross section of the wellbore annulus in set out put characteristics of its washing. As a result of studies it has been found that at lower eccentric placement of both straight and deformed drill strings in the well drilling mud moves by crescent-shaped core flow. The maximum velocity of mud is inside the core, and the minimum one-in the contact zones of a flexible pipe with the borehole wall. Thus, the dependences in changes of hydraulic pressure loss in pumping by the mud pump and the core flow rate distribution along the axis of the drill string at its eccentric location in the wellbore were determined. The conditions of drilling mud stagnant zones formation in the annular space were defined and the solutions to eliminate them were proposed.

1 RELEVANCE OF THE RESEARCH PROBLEM

Energy consumption during drilling and the formation of cylindrical excavation in rock is divided into interrelated technological processes of rotation and axial movement of the drill string in the wellbore, the destruction of the downhole rock and excavation flushing. Thus, to overcome the frictional forces of the drill string with the borehole wall the rig consumes about 10 – 15% of its energy. On the bottom-hole rock destruction a drill bit consumes from 10 to 15% of this energy, and to perform tripping operations we need 5 – 10% relatively. The bulk of the consumed energy in the process of drilling (60 – 75%) is spent on flushing the well (cleaning the downhome from drilled cuttings, removal of drilled solids to the surface, overcoming hydraulic resistance of drill mud friction on the way of its movement, and other) (Makovei 1986, Chudyk 2008). Considering large energy losses of the well drilling process on its washing, there is a need to study the factors of their reduction, and their mechanisms of their control.

The tendency to increase drilling ultra-deep, directional and horizontal wells to a considerable degree determines the increase in the share of unproductive losses of energy in order to overcome the resistance of the medium and its reduction concerning rock disintegration. It leads to the need of improving the approaches to the design, selection and implementation of efficient drilling technologies, considering important mining and geological, technical and technological factors and problems. The latter includes, in particular, the problem of drilling mud stagnant zones formation in the annular space of horizontal wells due to specific (eccentric) location of the drill string (Makovei 1986, Chudyk 2008). This leads to the accumulation of sludge in the well annular space, the deterioration of axial displacement and rotation of the drill string during drilling, drill string drag and sticking, drilling speed decrease and reduction of power supply to the bit (Hulizade et al. 1967, Chudyk 2007).

Taking into account multifactorial process of washing wells, the great difficulty in establishing the degree of influence of the eccentric position of the string in the well and its impact on changing hydraulic resistances, this question remains poorly investigated. This article summarizes some of the result of earlier research in this area.

Thus, for the first time (Chudyk 2010) under consideration was the problem of Newtonian fluid flow investigation in a channel with full eccentric annular cross-section area. From the result of the research it was established that under laminar fluid flow mode, the same pressure drop and equal section of the channel the fluid flow rate in eccentric channel will always be higher than under the concentric tubing placement. The higher is the value of the ratio of inside and outside diameters of the pipe, the higher is the flow rate.

In studies (Chudyk 2010) it is stated that in existing wells the contact of internal and external surfaces of the circular channel for drilling mud circulation is not along the line, but along some surface, that leads to the formation of stagnant zones. This causes a decrease in cross-sectional area of drilling mud flow.

One of the practical ways of eliminating stagnant zones formation elimination in the borehole now is the way of increasing the mud pump performance, which has a number of significant drawbacks (Makovei 1986):

– differential pressure increase on the downhole and hydraulic losses growth in the annular space;

– enhanced erosion of the walls of the well which leads to the formation of cavities and channels;

– enhanced absorption of mud and wear out of mud pump elements, swivels, threaded connections of the drill string and bit nozzles;

– energy expenses increase on hydraulic flushing program implementation and well drilling in general.

Due to the negative impact of the above mentioned factors on the process of well drilling there is a need to find new and effective approaches to reduce the impact of drilling mud stagnant zone in the annular space on the drilling process, which is the urgent problem that needs solving.

2 THE PURPOSE OF THE STUDY AND FORMULATION OF THE PROBLEM

In the borehole of horizontal wells the axial movement of the drill string is provided with the longitudinal component of weight of the above located part of its design to overcome the friction force of the pipe with the wall. Under the influence of axial and transverse forces the elastic axis of the drill string takes the complex shape of bend (Figure 1) at the eccentric location in the borehole – e.

Figure 1. Diagram of the drill string placement and its deformation.

As a result of the drill string deformation in the annular space of the borehole during wells flushing there is a change in the cross-sectional shape of the flow core and its flow velocity distribution which is described by such schemes (Figure 2).

Taking it into consideration, the aim of this study is to examine the impact of the eccentric location of the drill string in the borehole of a horizontal well on the shape of drilling fluid flow core in the annular space.

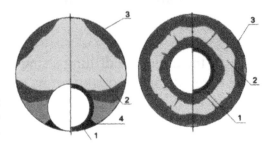

Figure 2. Scheme of drilling mud flow core orientation and distribution its velocity in the borehole annular space of horizontal wells: 1 – drill string; 2 – core of drilling fluid flow; 3 – hole wall; 4 – stagnant zone of drilling mud.

Also its flow velocity distribution due to the changes of the well and the drill string diameters, drilling fluid parameters, power characteristics of a mud pump and so on are examined.

The study relates to computer modeling using software environment SolidWorks Flow Simulation and mathematical modeling by means of calculation method. The drill pipe is examined in the borehole of horizontal wells under different boundary conditions according to Figure 1 and 2.

Input research parameters are the following: the diameter of the hole; the outer diameter and the length of the drill pipe; mud pump performance and pressure of pumping; roughness of the walls with which drilling mud contacts; its density and viscosity according to one of the rheological models.

3 NUMERICAL CALCULATIONS

To determine the effect of eccentricity of the drill string location in a horizontal section of the borehole by mathematical modeling (without simplifications and assumptions) a number of dependences were got relating to the drilling fluid velocity changes in the annular space sickle-shaped channel and pressure loss by using the following initial data were got (Table 1).

Table 1. Input parameters.

Parameter	Value
The diameter of the borehole, m	0.136
The outer diameter of flexible pipes, m	0.06
Mud pump performance, m³/s	0.0015
Pumping pressure of mud pump, MPa	20
The roughness of the walls, with which mud contacts, μ	300
The density of the drilling fluid (*model Herschel-Balkly*), kg/m³	1100
The length of the investigation area of the annular space, m	5

As a result of conducted research of drilling fluid movement in the annular space using computer modeling of the model "borehole – drill string" a number of features were established. At first, the eccentric location of the drill string in the borehole does not change the pumping pressure values in its cross-section area (Figure 3, a). Secondly, the eccentric location of the drill string in the borehole enables the reduction of the hydraulic resistance gradient. Therefore, for a concentric arrangement of pipes in the well it is about 1.36 kPa/m, and for eccentric – by 30% less, or 0.96 kPa/m (Figure 3, b). This difference has a significant effect on reducing both the hydraulic resistance and the energy consumption for wells flushing, but negatively affects the formation of drilling mud stagnant zones and sludge packs formation and other (Makovei 1986, Chudyk 2008).

drilling fluid in the annular space. In the extreme eccentric position of the drill string in the borehole in the center of the flow core we can observe the drilling mud flow velocity is $V_{df} \approx 0.8$ m/s (Figure 4, a) in the clearly visible stagnant zone. Most attention is caused by the intermediate eccentric position of the drill string, which in the flow core of drilling mud the flow velocity is $V_{df} \approx 0.86$ m/s with no signs of the stagnant zone. This justifies centering the drill string in the borehole of horizontal wells for maximum flow care velocity drilling mud in the annular space with the absence of the stagnant zone.

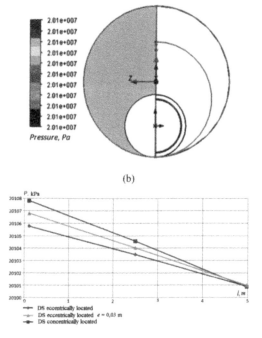

(a)

(b)

Figure 3. Distribution of mud pump pumping pressure under eccentric location of the drill string in the wellbore: (a) in the longitudinal section of the wellbore; (b) in the cross-section of the wellbore.

Ascending to the result of charting drilling fluid flow velocity changes in the annular space of the model "wellbore – the drill string" V_{df} it was found that stagnant zone is formed provided eccentricity e increase between the axes of the hole and the drill string (Figure 4). Under concentric arrangement of the drill string in the hole (for given input data) $V_{df} \approx 0.7$ m/s (Figure 4, b), without a stagnant zone

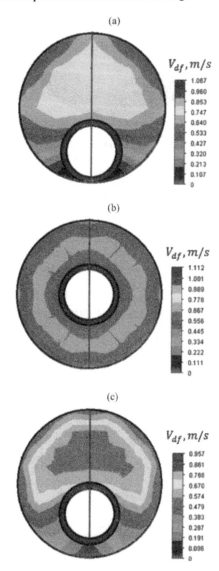

Figure 4. Average flow velocity of drilling fluid in the annular space of the borehole at different options of the drill string location in it.

To study the effect of the drill string eccentricity in the borehole of a horizontal well on the parameters of drilling mud pumping in the annular space the design scheme, shown in (Figure 5), is proposed.

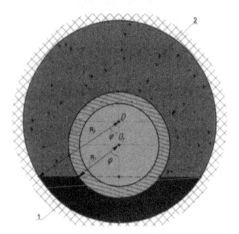

Figure 5. Calculation scheme of the drilling mud stagnant zone formation in the annular space with the eccentricity of the drilling string: 1 – drilling mud stagnant zone; 2 – drilling mud movable zone R_1 – the radius of the drill string pipe; R_2 – well radius; φ – the value of the semi angle; ψ – the value of the semi angle.

The ratio between the geometric dimensions of the drilling mud stagnant zone in the annular space at a pumping pressure of the mud pump is expressed:

$$P \cdot A_t + \tau_0 \cdot A_1 = \tau_0 \cdot A_2 , \qquad (1)$$

where A_t – the cross-sectional area of the drilling mud stagnant zone; P – pumping pressure of the mud pump; A_1 – the surface area of contact between the stagnant zone and moving drilling mud; A_2 – the surface area of drilling mud contact with the pipe and wellbore walls in the area of movement; τ_0 – the dynamic shear stress of drilling mud.

$$A_1 = 2\left(\begin{array}{c} R_2 \sin(\varphi) - \\ - \sin\left[arccos\left[\dfrac{R_2 \cos(\varphi) - e}{R_1} \right] \right] R_1 \end{array} \right) l , \qquad (2)$$

$$A_t = \frac{1}{2} R_2^2 (2\varphi - \sin(2\varphi)) - \\ - \frac{1}{2} R_1^2 \left(\begin{array}{c} 2\,arccos\left[\dfrac{R_2 \cos(\varphi) - e}{R_1} \right] - \\ - \sin\left(2\,arccos\left[\dfrac{R_2 \cos(\varphi) - e}{R_1} \right] \right) \end{array} \right) , \qquad (3)$$

$$A_2 = \frac{\pi}{180^0} \left[(360^0 - \varphi) R_2 + \right. \\ \left. + \left(360^0 - arccos\left[\dfrac{R_2 \cos((\varphi) - e}{R_1} \right] \right) R_1 \right] l , \qquad (4)$$

where l – the length of the drilling string.

To ensure the effective hydro transport of sludge by the drilling fluid flow through the annular space of the directional or horizontal well the necessary condition is to provide its required velocity and decrease the preconditions of stagnant zones formation.

For planning the required drilling mud flow velocity in the annular space V_{df} the mathematical model was built concerning sludge spherical particle movement between the borehole wall and the drilling string within the range of zenith angle forming and which is subjected to the influence of the following forces (Figure 6) (Chudyk 2010).

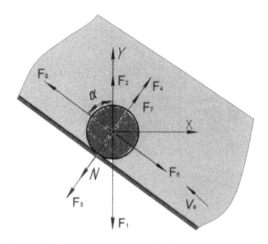

Figure 6. The scheme of forces distribution acting on the sludge particle in drilling mud inside annular space of the directional well: F_1 – gravity; F_2 – Archimedes; F_3 – centrifugal; F_4 – centripetal; F_5 – viscous medium resistance; F_6 – friction; F_7 – lifting.

These gravitational and Archimedes forces acting on a sludge particle under such conditions are defined as:

$$F_1 = \rho_r\, g\, V_p , \qquad (5)$$

where V_p – the maximum volume of sludge particle, which is removed by mud; ρ_r – rock density.

$$F_2 = \rho_{dm}\, g\, V_p , \qquad (6)$$

where ρ_{dm} – drilling mud density.

$$F_3 = \frac{2\rho_r\, V_{df}^2\, V_p k}{(2 + \kappa D_w)} , \qquad (7)$$

594

$$F_4 = \frac{2\rho_{dm} V_{df}^2 V_p k}{(2 + \kappa D_w)}, \qquad (8)$$

where k – curvature of the borehole; D_w – diameter of the well.

Talking into account the fact that the drilling mud stagnant zones formation leads to a decrease in the cross-sectional area of the annular space, which correspondingly causes an increase in the mud flow velocity, determined by:

$$V_{df} = \frac{4Q}{\left[\pi\left(D_w^2 - d_{dp}^2\right) - 4A_t\right]}, \qquad (9)$$

where Q – mud pump performance; d_{dp} – diameter of the drill pipe.

$$F_5 = \frac{c\,\rho_{dm}\,S_p\,V_p\,V_{df}^2}{2}, \qquad (10)$$

where S_p – the maximum cross-sectional area of the sludge particle being the largest one in the fractional composition formed on the bottom hole; c – empirical coefficient:

$$c = \left[\left(\frac{36\mu}{d_r^{1.5}\sqrt{3\,g\,\rho_{dm}\left(\rho_r - \rho_{dm}\right)}}\right) + 0.67\right]^2, \qquad (11)$$

where μ – the absolute viscosity of mud; d_r – conditional diameter of sludge particle.

Sludge particle friction with the wall of the well considering sticking it to the mud cake:

$$F_6 = N \cdot f, \qquad (12)$$

where f – coefficient of sludge particle friction with the drilling string or the borehole wall; N – normal component of the force pressing sludge particle to the wall of the borehole:

$$N = V_p\left(\rho_r - \rho_{dm}\right)\left[\left(\frac{2\kappa V_{df}^2}{2 + \kappa D_w}\right) + g\,sin(\alpha)\right] - F_7, \quad (13)$$

where α – the zenith angle of the borehole.

During the drilling string rotation at an angular velocity $\omega = const$ the following force acts on a sludge particle entering the zone of turbulent drilling mud:

$$F_7 = \frac{\rho_r V_p\, \omega^2 d_{dp}}{12}. \qquad (14)$$

According to Figure 6 we obtain the equation of equilibrium of forces acting on the sludge particle during its movement in drilling fluid:

– force projection on the axis OX:

$$\left(F_5 - F_6\right)cos(\alpha) + F_2 - F_1 +$$
$$+\left(F_7 + F_4 - N - F_3\right)sin(\alpha) = 0, \qquad (15)$$

– force projection on the axis OY:

$$\left(F_7 + F_4 - N - F_3\right)cos(\alpha) -$$
$$-\left(F_5 + F_6\right)sin(\alpha) = 0. \qquad (16)$$

Combining (1) – (16) in the system of equations and having done the appropriate calculations we obtain the values of quantity F_i and V_{df} at given technical and technological parameters of the process under the condition $A_t \to 0$.

4 CONCLUSIONS

1. Taking into consideration the eccentricity of the drill string in the borehole it leads to the introduction of significant changes in the quality of wellbore cleaning and to the specification of the existing methods of design and the selection of optimum mud pump performance which is of a considerable practical interest.

2. To determine the quantitative and qualitative impact of the drill string eccentricity in the wellbore the hydrodynamic system of flushing the well in the application module Flow Simulation program SolidWorks was proposed and appropriate studies were conducted. This uses nonlinear visco-plastic rheological model of Herschel-Balkli drilling mud considering mud pump performance and pressure, roughness of the borehole walls.

3. For designing the efficiency of the mud pump performance and to provide the sludge removal by the drilling mud flow through the annular space of the borehole the improved mathematical model was suggested. It takes into account a number of technical and technological parameters and factors that prevent from the drilling mud stagnant zones formation in the wellbore annular space of arbitrary curvature.

REFERENCES

Chudyk, I. 2007. *On hydraulic power costs for wells flushing.* IFNTUNG: Oil and gas energy, No 4 (5): 54 – 59.

Chudyk, I. 2008. *Effect of eccentric location of the drill string in the borehole on its flushing.* IFNTUNG: Exploration and development of oil and gas fields, No 1 (26): 44 – 48.

Chudyk, I. 2010. *Investigation of the pump supply value for directional wells washing.* IFNTUNG: Exploration and development of oil and gas fields, No 4 (37): 39 – 46.

Hulizade, M., Mihaliev, F. & Ilyasov, A. 1967. *Problem of determining hydraulic losses in the eccentric annular space under structural mode of visco-plastic fluid movement.* Moscow: Inform. bulletin of High Schools "Oil and Gas", No 11: 51 – 54.

Makovei, N. 1986. *Hydraulics of drilling.* Translation from Romanian. Moscow: Nedra: 537.

Influence of different factors and physical impacts on the power of flowing supersonic jet during slag spraying in the converter

P. Kharlashin, R. Kuzemko & V. Sinelnikov
Pryazovskyi State Technical University, Mariupol, Ukraine

ABSTRACT: To section nonisobaric supersonic jet jointly solved the equation of conservation energy and constancy of momentum, received distribution-average temperatures and power along the length of the jet flowing in the cavity of the converter.

1 INTRODUCTION

Introduction of the technology of the final blowing slag allows several times to increase the resistance of the lining of the oxygen converter. The technology of applying a protective lining on the scull oxygen converter based on the fact that a specially prepared slag is supplied and he passed a large enough energy to melt flowing nitrogen jets. The total capacity of the jets issuing from the nozzles of the lance $350t$ converter at the spraying reaches ~4 MW. Proper use of the power of energy makes it possible for application scull dramatically increase the resistance of the lining.

To properly calculate the energy flowing into the melt jets to be considered the expiration feature gas-jet at supersonic pressure drop. When applied to the lining of the slag scull $350t$ converter "AZOVSTAL IRON & STEEL WORKS" (Mariupol, Ukraine) nitrogen pressure at the inlet of the nozzle block of the lance is reduced by 1.6 MPa at the beginning of blowing up to 1.1 MPa at the end slagging. So how to maintain a constant pressure of nitrogen to the nozzle can not be unchanging form, the structure of blowing off-design supersonic jets always shock – wave and gas in them periodically accelerated and braked. On the direct and oblique shocks energy is dissipated compressed to high pressure nitrogen.

2 INTENSIFICATION OF THE PROCESS

2.1 Physical model

Figure 1 is a photograph of two shadow jet into the atmosphere at the same pressure P_o upstream of the nozzle lance same critical diameter D_{kr} and the same flow rate V_r, but using a different nozzle shapes –

cylindrical (left) and Laval (right). It can be seen that the gas-dynamic structure of these jets are completely different.

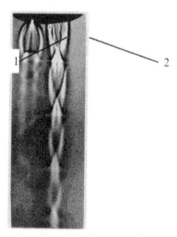

Figure. 1. Gas-dynamic picture of air jets issuing from the cylinder (1) and the expanding (2) jets double-nozzle lance (excessive pressure to the nozzle $p_o = 1.2$ MPa).

At supercritical pressure drop over the jet at considerable length accompanied by shock events and periodic change of thermodynamic parameters in her longitudinal and cross sections.

The static pressure at the nozzle exit p_1 is usually not equal to the pressure in the environment p_g. The degree of pressure ratio flow regime characterized by the ratio $n = p_1/p_g$, which is usually different from the unit due to fluctuations in the flow of nitrogen to the converter and the pressure on the network. In addition, the lance shortened divergent nozzle unplanned expiry provided solely to protect the trailing edge of the nozzle. Since $n < 1$ the flow within the nozzle exceed and can break away from the nozzle walls.

Supersonic jet consists of two parts: a supersonic core with a complex system of shock waves, gradually collapsing from friction and subsonic shell thickness is constantly increasing as a result of joining the environment and mass transfer with the supersonic core.

If design mode expiration supersonic core, formed by the interaction of weak waves discharge, similar in shape to a cone. The length and diameter of each successive "barrel" is smaller than the previous, straight races seal perpendicular to the velocity vector of the incoming flow is absent. The more direct the shock wave, the greater dissipation of mechanical energy in the shock wave, the shorter the stream, the less its long range on the same distance x.

A promising technology is when the skull using a special lance, which combined operation blowing slag and flare shotcrete (Kharlashin 2010, Chernyatevich 2010). In this case, the cooling water casing is not required, and the lance becomes efficient heat exchanger in which the heated nitrogen. As a result of an industrial experiments in (Chernyatevich 2010) that the wall temperature gas-cooled lance body after blowing slag and extracting it from the $160t$ converter reaches $\sim 500°C$.

Given the importance of the issue, to increase the energy flowing streams of nitrogen are possible heating scheme with an internal heat source.

2.2 The purpose of this work

Determine the effect of nitrogen flow rate V_H through a nozzle, temperature before nozzle t_o and in the cavity of the converter t_g, pressure ratio n to connected mass g, temperature distribution t_x and power of N_x in different sections of a supersonic jet introduced in the molten slag.

2.3 The computational model to determine the added mass g

In drawing up the model, the following assumptions. Due to the variability of the pressure p_o nitrogen lance nozzles of a given size expiring supersonic jets are variable degrees of pressure ratio n. The size added mass of slag is small ($g_{sh} \leq 1$), so the law of accession gas (N_2) is weakly dependent on the presence of slag in the gas. Such assumptions allow in this work to use the model calculation the added mass of gas provided in (Kharlashin 2010).

Added mass was calculated as:

$$g = m_g / m_1, \tag{1}$$

where the flow rate through the nozzle $m_1 = \rho_1 w_1 F_1$; m_g – the mass flow of the acquired gas jet from the cavity of the converter; ρ_1, w, F_1 – density, speed and area of exit section the nozzle.

Methods for calculating g nonisobaric supersonic jet includes a joint solution 12 of equations, as further described in (Kharlashin 2010).

This value is defined as:

$$g = \frac{2\bar{r}_{max}\bar{X}}{D_S} C_i \left(1 - C_i^2\right)^{\frac{1}{2}} \left(I_{1R} - I_{2R}\right). \tag{2}$$

In the formula (2) the number of Crocco is a value

equal to $C_i = \sqrt{1 - \left(1 + \frac{k-1}{2}M_i^2\right)^{-1}}$. The similarity

parameter, which characterizes the degree of turbulence of the jet, calculated by the formula $\sigma = 12 + 2.58M_i$, where M_i – the number Mach at the boundary of the jet and the surrounding gas. Coefficient D establishes a link between the Mach number M_1 at the nozzle exit and the degree of pressure ratio n flowing supersonic jet

$$D = \left(\frac{k-1}{2}\right)^{1/2} M_1 n^{\frac{k+1}{2k}}.$$ Relative to the maximum

radius of the first "barrel" nonisobaric jet with shock waves $\bar{r}_{max}$ calculated as in (Kharlashin

2010). In the formula (2) $\bar{X} = \frac{x}{r_1}$ – the distance

from the nozzle to the caliber to control section along the axis of the jet; r_1 – the radius of the nozzle in exit section. The numerical values of the integrals I_{1R} and I_{2R} defined as in (Kharlashin 2010).

$$I_{1R} = \int_{-\infty}^{\eta_R} \frac{\varphi d\eta}{\theta + \varphi(1-\theta) - \varphi^2 C_i^2},$$

$$I_{2R} = \int_{-\infty}^{\eta_R} \frac{\varphi^2 d\eta}{\theta + \varphi(1-\theta) - \varphi^2 C_i^2}, \tag{3}$$

where $\theta = T_G / T_0$ – relative temperature. The variable η_R – chosen so, to abide by the terms $1 - \varphi(\eta_R) < \varepsilon$, where ε – small quantity; R – index, relative to the inner boundary of the mixing zone.

2.4 Calculation model to determine the temperature t_x and power N_x jet

Consider the plot nonisobaric supersonic jet disposed between the trailing edge of the nozzle and the purge section xx located, for example, at the entrance of the molten slag. Neglecting speed w_g

ejected gases write for this section of the law of conservation of energy in the form of:

$$m_1 c_{p_1} T_0 + m_g c_{pg} T_g = (m_1 + m_g) c_{px} T_x +$$
$$+ \frac{(m_1 + m_g) w_x^2}{2 \cdot 10^3}, \tag{4}$$

where T – thermodynamic temperature; c_p – isobaric thermal capacity; w – averaged flow rate; indices parameters mean: o – isentropic flow deceleration; 1 – at the nozzle exit; g – gas cavity converter's; x – in the control section of the jet; n – under normal physical conditions.

From equation (4) can easily receive a weight average temperature jet in the distance $\bar{x} = \dfrac{x}{d_1}$ from the nozzle exit:

$$T_x = \frac{\left(c_{p0} T_o + g c_{pg} T_g - (1+g) w_x^2 / 2\right)}{(1+g) c_{px}}, \tag{5}$$

where 1 – relative cost (associated mass surrounding the gas jet); d_1 – the diameter of the nozzle the exit section.

It is assumed that the specific heat c_{pg} depends on the temperature T_g and T_o. In view of the apparent mass g average heat capacity of gas jet is:

$$c_{px} = \frac{c_{p1} + g c_{pg}}{1+g}. \tag{6}$$

The temperature of the of braking T_o, that along the nozzle remains constant, was calculated as $T_o = T_1 + w_1{}^2 / 2 c_{p1}$, where static temperature T_1 at the nozzle exit of the pre-determined value found gazodinamic consumption function $q(\lambda_1) = F_{\kappa p} / F_1$, and then $\tau(\lambda_1)$:

$$T_1 = \tau(\lambda_1) T_o. \tag{7}$$

Weight average speed of a jet w_x found by solving the equation of constancy of momentum recorded for the area between the jet nozzle and cut section xx:

$$\eta w_1^2 F_1 + F_1(p_1 - p_g) = m_x w_x + F_x(p_x - p_g). \tag{8}$$

Under the power N_x understand the kinetic energy of the jet, which is determined by the formula:

$$N_x = \frac{m_x w_x^2}{2}, \quad m_x = m_1 + m_g = m_1(1+g). \tag{9}$$

Power N_x can be expressed through mass adjoint g and the degree of pressure ratio n. So, knowing the diameter of the nozzle in critical d_{kr} and the exit section d_1 determined by gas-dynamic function flow $q(\lambda_1) = d^2{}_{kr} / d^2{}_1$, and in table by $q(\lambda_1)$ found gas-dynamic pressure function $\pi(\lambda_1) = p_1 / p_{op}$. Then the

design pressure before the nozzle (when $p_1 = p_g$) is $p_{op} = p_g / \pi (\lambda_1)$. As the actual pressure p_o' upstream of the nozzle is known, the degree nonisobaric will

equal $n = \dfrac{p_1}{p_G} = \dfrac{p_0'}{p_{0p}}$.

Starting from cross section of the jet xx where static pressure is equal to atmospheric p_x, the expression for power N_x can be written as:

$$N_x = \left[\frac{n-1}{n} \frac{k-1}{k} \frac{\tau(\lambda_1)}{1 - \left(\dfrac{n p_g}{p_0'}\right)^{\frac{k-1}{k}}} + 1 \right]^2 \frac{m_1 w_1^2}{2(1+g)}. \tag{10}$$

Power supersonic jet, calculated under the terms of the nozzle exit, determined by the formula

$$N_1 = \frac{m_1 w_1^2}{2}.$$ Speed w_1 at the nozzle exit from the condition $w_1 = \lambda_1 a_{\kappa p}$, and λ_1 we found in the tables of gasdynamic functions consumption function.

Thus, one of the main problems to be solved in this task – how to determine the added mass g taking into account a number of factors and physical influences – p_o, t_o, t_g, V_μ, n and many others.

3 BASELINE DATA

The calculations were made applied to the lance 350t converter "Azovstal" (Mariupol). The diameter of the nozzle is equal to a critical $d_{kr} = 43$ mm and the exit section $d_1 = 63$ mm. The stagnation temperature of nitrogen (i.e. at the nozzle inlet) varied in the range $t_o = 25 - 600°C$, and the temperature of the gases in the cavity of the converter – in the range $t_g = 300 - 1500°C$. When nonisobaric modes nitrogen flow through the same nozzle was changed within $V_H = 180 - 380$ m³/min (under normal physical conditions). The parameters were calculated at different distances from the nozzle exit $\bar{x}$.

4 THE RESULTS OF CALCULATION AND THEIR ANALYSIS

How to use the above technique is not quite a simple calculation will show the example of definition of power N_x. Let the nitrogen flow rate through the nozzle is $V_H = 220$ m³/min, actual pressure upstream of the nozzle $p_0' = 1.5$ MPa, temperature $t_o = 30°C$, the pressure in the cavity of the converter $p_g = 0.1$ MPa.

Determine the capacity N_x of the jet at a distance $\bar{x} = 20$ calibres, where adjoint mass, calculation defined by the formula (2) is $g = 0.32$.

For nitrogen adiabatic index $k = 1.4$. By

$$q(\lambda) = \frac{F_{kr}}{F_1} = \frac{(43)^2}{(63)^2} = 0.466 \text{ in tables gas-dynamic}$$

functions, we find that at the nozzle exit speed reduced $\lambda_1 = 1.748$, gas-dynamic function of pressure and temperature $\pi(\lambda_1) = 0.083$, $\tau(\lambda_1) = 0.49$. Then the design pressure before the nozzle is equal to

$$p_{op} = \frac{p_g}{\pi(\lambda_1)} = \frac{0.1}{0.083} = 1.2 \text{ MPa. Degree nonisobar-}$$

ic flowing the jet $n = \dfrac{p_1}{p_G} = \dfrac{p'_0}{p_{0_p}} = \dfrac{1.5}{1.2} = 1.25$

critical speed:

$$a_{kr} = \sqrt{\frac{2k}{k+1}} RT_0 = 1.08\sqrt{297 \cdot 303} = 324 \text{ m/s, (11)}$$

speed expiration of the nozzle $w_1 = \lambda_1 a_{kp} = 566$ m/s. Then the jet power at a distance $\bar{x} = 20$ calibres will be:

$$N_x = \left[\frac{1.25-1}{1.25} \frac{1.4-1}{1.4} \frac{0.49}{1 - \left(\dfrac{1.25 \cdot 0.1}{1.5}\right)^{\frac{1.4-1}{1.4}}} + 1 \right] \times$$

$$\times \frac{220 \cdot 1.25 \cdot (566)^2}{60 \cdot 2 \cdot 10^6 \cdot (1+0.32)} = 0.66 \text{ MW.} \quad (12)$$

The power of the jet at the nozzle exit

$$N_1 = \frac{m_1 w_1^2}{2000} = \frac{1.25 \cdot 220 \cdot 566^2}{60 \cdot 2 \cdot 10^6} = 0.734 \text{ MW.}$$

Using the above method of we will show how impact on the determinants of the associated mass g, power N_x and temperature t_x jet, flowing in the cavity of the converter.

4.1 Length jet

Represents scientific interest calculation results shown Figure 2, from which it follows, that by an amount the attached mass g, significantly affected the relative temperature $\theta = T_g / T_0$. Thus, when $\bar{x} = 20$ and the increase θ with 1 ($t_g = 300°C$) to $\theta = 7$ it reduces the mass the attached g with 0.58 to 0.18 (i.e. in 3.22 times). From Figure 2 also shows that the higher the value θ, the more w_x / w_1, i.e. more speed in the section xx. This is because when the jet is introduced into the heated, less dense space, its speed w_x, more, than when $\theta = 1$.

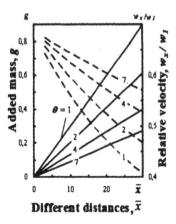

Different distances, $\bar{x}$

Figure 2. The dependence of a added mass g, and the relative velocity w_x / w_1 the relative temperature θ at different distances x.

4.2 The nitrogen flow rate through the nozzle

From Figure 3 that expense V_H through the nozzle thereby substantially affects the temperature change t_x and power N_x along the length of the jet $\bar{x}$. For example, by increasing the with increasing V_H with 180 to 300 m³/min on distance $\bar{x} = 20$ relative power N_x / N_1 increases with 0.49 to 0.76 , and the relative temperature T_x / T_1 decreases with 3.4 to 1.8.

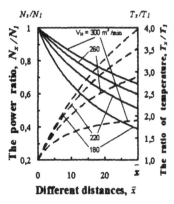

Different distances, $\bar{x}$

Figure 3. The dependence of power relations N_x / N_1 and temperatures T_x / T_1 of the flow rate of nitrogen through the nozzle V_H at different distances $\bar{x}$ from the nozzle exit.

We remark, that as a result of the deep extension within supersonic nozzle the static temperature T_1 in its output section is substantially reduced. In the example above, this temperature is $T_1 = T_0 \tau(\lambda_1) = 303 \odot 0.49 = 148$ K ($t_1 = -125°C$). But the higher the flow rate of cold gas V_H, the jet attaches less weight strongly heated gas from the cavity of the converter ($t_g = 1500°C$). Therefore, with increasing V_n relative temperature T_x / T_1 decreases.

600

4.3 The heating of the nitrogen before nozzles

As it is known, for one and the same nozzle discharge velocity w_1 does not depend on the initial pressure p_0. In the same time, heating the gas before the nozzle unit increases the discharge velocity w_1, and power N_1 flowing out the jet nozzle (N_1 proportional to stagnation temperature T_0). As for the power of the jet N_x, then as follows from the equation (10), it yet depends on the added mass g, which is greater, higher nitrogen before the heating nozzle. The calculations using this model, all these physical effects on the jet into account.

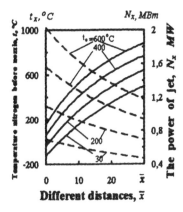

Figure 4. Temperature effect of nitrogen before the nozzle t_0 on the change of temperature t_x at different distances x.

Figure 4 shows, that the heated nitrogen before the nozzle range $t_0 = 30 - 600°C$ at a distance from the nozzle x, temperature t_x and speeds w_1 and w_x jets also increase. With heating of the nitrogen in the lance increases and power N_x in each arbitrary cross-section x jet. For example, in $t_0 = 30°C$, $V_H = 210$ m³/min at distance $\bar{x} = 20$ calibres, temperature $t_x = 330°C$, and power $N_x = 0.53$ MW. When heated nitrogen to $t_0 = 600°C$, temperature $t_x = 743°C$, and power $N_x = 1.38$ MW.

Thus, due to heating in a nitrogen lance, only increase in power ΔN_x jet (in this case 0.85 MW) maybe more, than the power N_x (0.53 MW) at expiration the cold gas jet.

We note one important fact – if a constant flow rate $V_H = $ const heated nitrogen gas cooled lance, the power of the jet N_x (and the impulse I_x) increase as a result of two effects – due to growth the speed of the jet w_x, as well as the inevitable (due to the heating of the gas) pressure rise p_0 before the lance nozzles.

4.4 The temperature in the cavity of the converter

Investigate the effect of temperature t_g in the cavity of the converter on-average temperatures t_x and power N_x jet. As follows from Figure 5, with an increase $\theta = T_g / T_0$, for example, with 1 to 6 on distance $\bar{x} = 20$, temperature t_x increases with -7 to 362°C, and the power N_x increases with 0.41 to 0.57 MW. The temperature rise t_x explained with increasing θ the jet eject heated gas cavity and the converter and very heats. In the same time, with increasing θ, i.e. with increasing t_g, added mass g less dense surrounding gas decreases and therefore, power N_x, it follows from formula (10), increases.

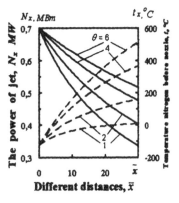

Figure 5. The dependence of power jet N_x and temperature t_x on the relative temperature θ at different distances x.

5 VERIFICATION MODEL ADEQUACY

Given the complexity of the physical experiment in the cavity of the converter, model adequacy us verify, using the test method verification solutions, considered particular cases. For example, it was confirmed that if:

– $x = 0$ (nozzle exit), that $g = 0$ for any value of the relative temperature θ, that is not in doubt (Figure 2);

– the control section of the jet is removed from the nozzle, that added mass g throughout different distances x increases at all temperatures, t_0, t_g, which is obviously (Figure 2);

– $\bar{x} \to 0$ (nozzle exit), where the surrounding gas connection $g \to 0$ the speed ratio $\dfrac{w_x}{w_1} \to 1$ for any θ, which naturally (Figure 2);

$-\bar{x}=0$, the relative power $\dfrac{N_x}{N_1}=1$ and tempera-

ture and relative $\dfrac{T_x}{T_1}=1$, which corresponds to the generally accepted physical concepts (Figure 3);

– heats the gas before the nozzle, the expiring jet has a higher temperature in any section $\bar{x}$ for any values θ (Figure 4);

– gas is warmed to the nozzles, in any cross-section $\bar{x}$ power N_x above, than expiration the nozzle of the cold gas (Figure 4);

– if $\bar{x}=0$, then $w_x=w_1=\lambda_1 a_{kr}$, $\dfrac{N_x}{N_1}=1$, and $t_x=t_1=-125\,°C$, which also follows from the formula $T_1=T_0\tau(\lambda)$ at $t_0=30\,°C$ (Figure 5);

– $x=0$, then performed a special case – power in an arbitrary section of the jet and the temperature is the parameters at the cut $N_x=N_1$, $T_x=T_1$ (Figure 5);

– value of the added mass ranges $g=0.2-1.5$ (Figure 2), which corresponds to a mass concentration, for example, CO to oxygen jet, equal

$$[CO]=\frac{g}{1+g}=0.17-0.6.$$

Thus, private decisions confirm the accuracy of the results of calculations using the above model.

6 CONCLUSIONS

1. The heating of the nitrogen attached to the lance increases added mass of the environment. Regeneration of heat from the cavity of the converter to heat the nitrogen supplied to the air blast nozzle – the real path to increase the added mass of gas from environment and power planes at their introduction into the melted slag.

2. Temperature increase in the cavity of the converter reduces ejecting ability nonisobaric supersonic jet.

3. The change in temperature before the nozzle of the lance and in cavity of the converter significantly affects the power introduced by the outflowing jet.

4. When heated nitrogen to 600°C, not reducing the power flowing into the melt jets, we can reduce the amount of nitrogen, for example, from 210 m³/min to 125 m³/min.

REFERENCES

Kharlashin, P., Kovura, A. & Kuzemko, R. 2010. *The model of calculation termo-gaz-dynamic jet parameters in the converter*. Metallurgical and Minind Industry, 2 (260): 97 – 100.

Chernyatevich, A., Sigarev, E. & Chubin, K. 2010. *Development lances devices and technologies gaspowder slagging lining 160t converter of "Arcelor Mittal Krivoy Rog"*. Metallurgical and Minind Industry, 2 (260): 134 – 137.

Theoretical and Practical Solutions of Mineral Resources Mining – Pivnyak, Bondarenko & Kovalevska (eds)
© 2015 Taylor & Francis Group, London, ISBN: 978-1-138-02883-8

Some technological specialties of rail steel melting in converter

P. Kharlashin, V. Baklanskii, S. Gerasin & D. Didenchuk
Pryazovskyi State Technical University, Mariupol, Ukraine

ABSTRACT: Some specialties of rail steel melting technology in oxygen converter are represented. Research and technology analysis of high-phosphorous iron reduction in converter rail steel in 350 tons oxygen converters with top blowing are given. Creation of conditions that conduct removing of phosphorous part by the beginning of intensive carbon acescence is given. Few variants of slag adjustment impact on steel phosphor removal are represented.

1 INTRODUCTION

It is well-known that the melted steel in oxygen converters has higher quality than martin steel by means of lower content of nonmetallic inclusions and gases, and the technology is energy efficient, i.e. it is more preferable to produce rail steel by a converter method. However, melting of such steel in converters is interfaced to certain technological difficulties. Basic provisions were considered for detection of some technological features of converter rail steel production from phosphorous iron (Voropayev 1982).

Features of rail converter steel melting technology from high-phosphorous iron are following:

– determination of grow iron quantity in furnace burden of the converter melting that allows to provide necessary temperature of steel in a ladle after cast;

– determination of slag skimming quantity that provides of a metal bath phosphor removal;

– determination of lime consumption and the mode of its additives that provides necessary basicity of slag during blowing;

– determination of methods of extra oven processing that allows to receive high-quality steel;

– determination of deoxidation methods, metal carbon penetration that provides receiving of necessary parameters of the melted steel.

On quality indicators, rail steel can belong to the first or second group of quality. The quiet steel deoxidated in a ladle by complex deoxidants without usage of aluminum or other deoxidants, which form inclusions, belongs to the first group.

Steel of the second group of quality is quiet steel, which is deoxidated by aluminum or manganese-aluminum alloy.

Melting technology and teeming of rail steel has to guarantee absence of nonmetallic inclusions and flakes, freckles and rippled surface aren't allowed (Bornatskyi et al. 1974).

The analysis of literary data allowed to choose some receptions and few the most perspective and effective technological solutions of melting processes and teeming of converter rail steel.

These include leaving of final slag part of the previous converter melting, which brings it into inert state additives of lime and carboniferous materials, and double skimming and adjustment of new slag. It is used for increase of degree of a phosphor removal of a bath during processing of iron in the converter with a mass fraction of phosphorus more than 1%.

As rail steel is highly-carbon, its melting happens in two ways:

– with interception of carbon at the set level;

– with carbon elimination of a bath of the converter with further carbon penetration to the set level.

2 INTENSIFICATION OF THE PROCESS

2.1 Process description

One of problems of converter rail steel melting is the raised gain of a phosphorus share which depends upon a mass fraction of phosphorus in iron (Izhmuhamedov 1987).

During converter rail steel melting due to the necessity of combination of a deep metal phosphorous removal, the analysis of sequence of iron processing technology with the high content of phosphorus in converter rail steel was executed.

2.2 Basic of analysis

Features of technology are developed in relation to steel melting conditions in 350 tons oxygen converters with the top blowing, at the content of phosphorus in iron of 1% and are as follows:

– leaving of final slag part from the previous melting;

– slag densification by lime;

– slag neutralization by materials with the content of carbon;

– blowing in three periods, with two intermediate skimming of slag, after scrap charge and iron filling;

– stopping of refosforation processes during steel oxidation in a ladle and during processing by its argon, by means of lime additive in the converter upon termination of blowing, a slag cut-off at production of metal from the converter and lime additive in a ladle after metal production (Izhmukhamedov 1987).

On the specified technology the number of the crowbar in furnace charge was determined by calculation of material and thermal balances of melting with taking into account temperature and composition of iron.

If the content of phosphorus in iron is 1.5 – 1.6%, and its temperature of 1300°C makes number of the crowbar by calculation will be equal to 100 tons with iron mass of 270 – 280 tons.

Blowing was conducted by five-nozzle injection lance, with oxygen 1300 – 1500 m³/min supply. Injection lance height over the level of a quiet bath at the beginning of blowing was equal to 3.9 – 4.1 m with the subsequent lowering it to level of 1.7 – 1.9 by the end of the first period (Voropayev 1982).

At first minutes of blowing, after melting ignition was added lime that allowed to direct slag with good fluidity. On the expiration of the first period slag was skimmed completely, with the subsequent targeting of the new one. For slag formation improvement in the second and third periods for fluxion of slag it is more preferable to use materials on the basis of CaF_2. These impacts of slag mode of melting on degree of a phosphorous removal of converter rail steel are given on Figure 1.

The general duration of blowing was 17 – 19 minutes, consumption of lime was in limits of 46 – 50 tons on melting; thus received slag with basicity of 3.0 – 3.5.

Usage of such technology allowed to conduct definitely high velocity of phosphorous deoxidation.

At accepted structure of metallic charge materials that are specified in Table 1, was obtained settlement data of expenses of the main metallic charge materials during rail steel melting which are given in Table 2.

The content of oxygen in metal first of all depends upon concentration of carbon in it. Oxidizers in the oxygen converter are oxygen of a gas phase in metal and iron oxide in slag.

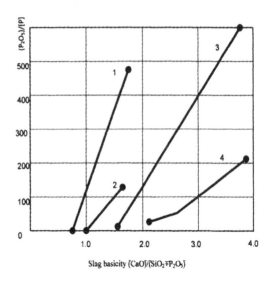

Figure 1. Impact of slag mode of melting on degree of phosphorous removal of converter rail steel, without usage o fluxing oil, with leaving final slag from previous melting: 1 – with usage of fluorspar equal to 5 – 10 tons; 2 – without final slag from previous melting; 3 – S_k = 5 – 20 tons; 4 – with usage of fluorspar and S_k, 15 – 20 tons.

Table 1. Composition of metal burden stocks that was chosen for calculation of its consumption during rail steel melting.

Metallic charge component	Mass fraction of elements, %					
	C	Mn	Si	S	P	As
Liquid iron	3.20	1.15	0.85	0.03	1.57	0.15
Scrap steel	0.20	0.65	0.25	0.03	0.03	—

Table 2. Consumption of the main metallic charge materials during melting of steel from high-phosphorous iron (calculation data).

Charge materials	Consumption of liquid steel that was melted according to technology			
	With carbon interception on the level of 0.6 – 0.7%		With slag leaving and full oxidation of carbon with further carbon penetration	
	without slag leaving	with slag leaving	iron	coke
Iron	0.894	1.045	0.965	0.960
Scrap	0.192	0.131	0.109	0.125
Lime	0.133	0.106	0.105	0.106
Coke	—	—	—	0.009

For instance, according to A. Samarin and N. Levents, phosphorus can be oxidized both in metal, and on limit of the section slag-metal. In metal oxi-

604

dation happens with formation of triphosphate of iron passing into slag on reaction:

$$2[P] + 8[O] + 3[Fe]=(3FeO \cdot P_2O_5). \qquad (1)$$

Deepening of phosphorous removal process during melting converter rail steel can be received owing to deeper decarbonanization of converter's bath.

The behavior of phosphorus in a bath of oxygen converter showed that composition of metal there are changes according to the content of phosphorus. For cancellation of this process in all skilled trunks add at the end of a purge 2 tons of lime for a slag thickening (Bornatskyi et al. 1974).

During release, particles of converter slag are got into a ladle. On argon installation where happened averaging and operational development of metal, also raised content of phosphorus is observed.

In final slags, which contain $15 - 22\%$ of FeO, concentration of P_2O_5 in the course of oxidation and the subsequent purge argon, content of phosphorus in ready steel increased on $0.007 - 0.018$. Data about received values is provided in Table 3.

Degree of a refosforation is influenced by amount of slag which gets to a ladle, P_2O_5 in slag, intensity and time of processing by argon. For its decrease it is necessary to resolve an issue of a cut-off of converter slag and additive of lime in a ladle.

Prevention of a refosforation of metal promotes not only to improvement of metal quality, but also increase of oxide content of phosphorus in slag (Voropayev 1982).

Depending on amount of phosphorus which contains in initial iron, in the course of working off parameters of technological process 2 options of melting of rail converter steel are developed:

– if a mass fraction of phosphorus in iron less than 1.2% a bath blowing by oxygen conducts before achievement in metal of a carbon mass fraction $0.6 - 0.8\%$;

– if a mass fraction of phosphorus in iron more than 1.2%, achievement of necessary degree of phosphorous removal is possible only at deeper carbonization of converter bath $(0.2 - 0.3\%)$.

During rail steel melting with blowing stopping at the set carbon level $(0.6 - 0.7\%)$, the number of scrap in furnace charge has to be no more than $10 - 12\%$ for ensuring temperature of metal before release of $1610 - 1630°C$.

After loading of the converter, blowing is carried out during three periods, with intermediate double skimming of slag on $6 - 7$ and $10 - 12$ minutes. Upon termination of every period, of metal and slag tests are selected and metal temperature is measured.

After blowing stopping, we made converter turndown during receiving necessary components of carbon composition and phosphorus in metal, also obtaining necessary temperature. Production of metal from converter is realized (Izhmukhamedov 1987).

During steel melting from iron with a mass share of phosphorus more than 1.2% on technology with oxidation of carbon in the course of blowing without "interception", steel carbonization at the level of $0.6 - 0.7\%$ is happened. Degasification of rail converter steel is done on installation of pumping out at discharge in a vacuum vessel not more than 200 Pa. Quantity of operation cycles not less than 30. After pumping out, temperature of metal was equal to $1520 - 1530°C$.

Table 3. Changing of phosphorous content in steel before oxidation and after processing by argon.

Melting*	Phosphorous content in metal, %		Imbedded with ferroalloy of phosphorous,%	Renewable phosphorous,%	Rephosphorization degree,%
	a	b			
1	0.009	0.021	0.002	0.010	111
2	0.010	0.022	0.002	0.010	100
3	0.009	0.019	0.002	0.008	89
4	0.015	0,035	0.002	0.018	120
5	0.011	0.023	0.002	0.010	91
6	0.009	0.027	0.002	0.016	178
7	0.009	0.029	0.002	0.018	200
8	0.007	0.017	0.002	0.008	114
9	0.009	0.029	0.002	0.018	200
10	0.007	0.019	0.002	0.010	143
C	(0.0095)	(0.0231)	(0.002)	(0.0126)	(132.5)

* in brackets – average value

3 CONCLUSIONS

1. Developed technology of converter rail steel with processing of high-phosphorous iron in 350 tons in converter provides receiving of carbon steel with lower content of phosphorus.

2. During development of technology features oxygen converter production of rail steel from high-phosphorous cast iron it was supposed realization with leaving of final slag from previous melting. With its double skimming at a mass fraction of phosphorus in iron more than 1%, with interception of carbon at the set level and with oxidation of all carbon of a converter bath with the subsequent carbon penetration of metal by solid carboniferous materials or conversion iron.

3. It was established that soft blowing could provide by the period of readiness of metal according to the content of carbon with combination of two such important parameters of process as temperature of metal and degree of its phosphorous removal.

4. Results of conducted melting showed that processing of high-phosphorous iron into converter rail steel in a steel-melting bathwith the top blowing on technology with usage only of lime and "interception" of carbon at the set level of 0.6 – 0.8% is possible at two-fold skimming of slag, as a fluxing oil, usage of fluorspar and final slag leaving from the previous melting, in a case when the mass fraction of phosphorus in cast iron doesn't exceed 0.9 – 1.2%. If content of phosphorus in iron is within higher limits, blowing should be conducted to the content of carbon in metal 0.2 – 0.3% with the subsequent carbon penetration of metal in a ladle by materials that containing carbon.

REFERENCES

Bornatskyi, I.I, Baptyzmanskyi, V.I. & Isayev, E.I. 1974. *Modern of oxygen-converter process*. Kyiv: Technics: 264.
Izhmuhamedov, N.K. 1987. *Oxygen-converter processing of phosphorous iron into steel*. Alma-Ata: Science: 71.
Voropayev, V.A. 1982. *Production of steel from phosphorous iron in steelmaking vessels*. Moscow: Metallurgy: 120.

Printed and bound by CPI Group (UK) Ltd, Croydon, CR0 4YY

24/10/2024

01778287-0005